Communications in Computer and Information Science 482

Fuling Bian Yichun Xie (Eds.)

Geo-Informatics in Resource Management and Sustainable Ecosystem

Second International Conference, GRMSE 2014
Ypsilanti, MI, USA, October 3-5, 2014
Proceedings

Volume Editors

Fuling Bian
Wuhan University, China
E-mail: flbian@whu.edu.cn

Yichun Xie
Eastern Michigan University
Ypsilanti, MI, USA
E-mail: yxie@emich.edu

ISSN 1865-0929 e-ISSN 1865-0937
ISBN 978-3-662-45736-8 e-ISBN 978-3-662-45737-5
DOI 10.1007/978-3-662-45737-5
Springer Heidelberg New York Dordrecht London

Library of Congress Control Number: 2015931253

Typesetting: Camera-ready by author, data conversion by Scientific Publishing Services, Chennai, India

Printed on acid-free paper

Springer is part of Springer Science+Business Media (www.springer.com)

Preface

The 2014 International Conference on Geo-Informatics in Resource Management and Sustainable Ecosystem (GRMSE 2014) was held in Ypsilanti, Michigan, USA, during October 3–5, 2014. GRMSE aims to bring together researchers, engineers, and students working in the areas of geo-informatics in resource management and sustainable ecosystem. GRMSE 2014 featured a unique mix of topics on smart city, spatial data acquisition, processing and management, modeling and analysis, and recent applications in the context of building a healthier ecology and better resource management.

We received 296 submissions from various parts of the world. The Technical Program Committee worked very hard to have all papers reviewed before the review deadline. The final technical program consisted of 73 papers. There were five keynote speeches and five invited sessions. All the keynote speakers are internationally recognized leading experts in their research fields, who have demonstrated outstanding proficiency and have achieved distinction in their profession. The proceedings are published as a volume in Springer's *Communications in Computer and Information Science* (CCIS) series. Some excellent papers were selected and recommended for publication in the *Sensors & Transducers Journal* and *International Journal of Embedded Systems*. We would like to mention that, due to the limitation of the conference venue capacity, we were not able to include many fine papers in the technical program. Our apologies to these authors.

We would like to express our sincere gratitude to all the members of Technical Program Committee and organizers for their enthusiasm, time, and expertise. Our deep thanks also go to many volunteers and staff for the long hours and hard work they generously gave to GRMSE 2014. We are very grateful to Eastern Michigan University and Wuhan University for their support in making GRMSE 2014 possible. The generous support from the Joint International Center for Resource, Environment Management and Digital Technologies (JIC-REDT), International School of Software, Wuhan University is greatly appreciated. Finally, we would like to thank all the authors, speakers, and participants of this conference for their contributions to GRMSE 2014.

November 2014

Fuling Bian
Yichun Xie

Organization

Second Annual International Conference on Geo-Informatics in Resource Management and Sustainable Ecosystem

GRMSE 2014

Eastern Michigan University

Ypsilanti, Michigan, USA

October 3-5, 2014

Organized by:

Eastern Michigan University, USA
Wuhan University, China

Published by:

Advisory Committee

Michael Batty	Centre for Advanced Spatial Analysis (CASA), University College London, UK
Chenghu Zhou	Chinese Academy of Sciences, Institute of Geographic Sciences and Natural Resources Research, Chinese Academy of Sciences, China
Jianya Gong	Chinese Academy of Sciences, State Key Laboratory for Information Engineering in Surveying, Mapping and Remote Sensing, Wuhan University, China

Organizing Committee

Honorary Chair, Tom Venner	College of Arts and Sciences, Eastern Michigan University, USA
Honorary Chair, Xiaohui Cui	International School of Software, Wuhan University, China
Honorary Chair, Fuling Bian	Wuhan University, China
Honorary Chair, Zhiguo Huang	Wuhan University, China
Co-chair, Elisabeth Morgan	Eastern Michigan University, USA
Co-chair, Richard Sambrook	Eastern Michigan University, USA
Co-chair, Shuang Li	Wuhan University, China
Co-chair, Xiaoliang Meng	Wuhan University, China
Kim Kozak	Eastern Michigan University, USA
Hugh Semple	Eastern Michigan University, USA
William Welsh	Eastern Michigan University, USA
Yichun Xie	Eastern Michigan University, USA
Edwin Joseph	Eastern Michigan University, USA
Wei Wei	Xi'an University of Technology, China
Secretary, Siyu Fan	Eastern Michigan University, USA
Secretary, Lin Sun	Wuhan University, China

Scientific Committee

Co-chair, Yichun Xie	Eastern Michigan University, USA
Co-chair, Fuling Bian	Wuhan University, China
Co-chair, William Welsh	Eastern Michigan University, USA
Wenzhong Shi	The Hong Kong Polytechnic University, SAR China
Guobin Zhu	Wuhan University, China
Zongyao Sha	Wuhan University, China

Editorial Committee

Table of Contents

Smart City in Resource Management & Sustainable Ecosystem

Spatial Data Acquisition through RS and GIS in Resource Management & Sustainable Ecosystem

Ecological and Environmental Data Processing and Management

Advanced Geospatial Model and Analysis for Understanding Ecological and Environmental Process

Applications of Geo-Informatics in Resource Management & Sustainable Ecosystem

Economies of Scale, Economies of Scope and Financial Holding Companies Appropriate/Moderate Scale Management

Li Zeng

South China University of Technology, School of Economics and Commerce
zengli23@qq.com

Abstract. Economies of scale and economies of scope are two big business advantages for financial holding companies. This paper is based on the operating data of 8 domestic financial holding companies from the years 2007-2012. Analyzing the effect economies of scale have on the returns and costs of financial holding companies, and analyzing the effect of economies of scope in terms of fixed effect and random effect. Research shows that both economies of scale and economies of scope exist within domestic holding companies; the effect of economies of scale is much more profound in moderate-sized financial holding companies; whereas in large financial holding companies the effect of economies of scope is comparatively weaker. Thus, this paper believes that the government should actively promote the development of financial holding companies, and that financial holding companies should maintain appropriate scale management.

Keywords: Financial holding company, economies of scale, economies of scope, appropriate scale.

1 Introduction

Financial holding companies are one of the primary financial institution reforms at the end of the 20th century. This reform came about as a result of the diversified development of social financial demands and profit seeking on the behalf of financial enterprises. Because financial holding companies enjoyed operating advantages such as those of economies of scale and economies of scope, mixed operation models best represented by the financial holding companies, gradually became the development trend in the global financial industry. In foreign theoretical and practical fields, a series of research was thus conducted specifically aimed at financial holding companies' economies of scale and economies of scope. In China, with economic and social development, increasingly fierce market competition, and the policy restrictions of separate operations, financial holding companies become the main method for financial enterprises to grow bigger and stronger, get around policy restrictions and expand into other business areas. In terms of future development trends, mixed operating is an inevitable development trend for the domestic financial

F. Bian and Y. Xie (Eds.): GRMSE 2014, CCIS 482, pp. 1–13, 2015.

industry, and financial holding companies will also become a kind of crucial carrier. At present, large domestic state-owned commercial banks and a portion of relatively large-scale joint stock banks have already established financial holding companies, while many others, including even some municipal commercial banks, are actively planning to do so in order to expand their operational scale and business scope. Set against this development boom of domestic financial holding companies, whether or not economies of scale and economies of scope exist within these companies, is a research topic that has theoretical and practical significance. Based on this, this paper will conduct comparative analysis on the economies of scale and scope of current domestic financial holding companies in order to provide theoretical guidance for the development of domestic financial holding companies.

2 Related Literature

In the 20th century, the 70's and 80's saw the relaxation of financial regulations, and the emergence of a large number of large scale and cross-border multi-operational financial institutions. Because of this, the foreign academic world began researching relevant theories on financial holding companies. Such research was primarily focused on two areas: the operational benefit and (operational) risks of financial holding companies. The primary focus of operational benefit was to argue the effects of economies of scale and scope on financial holding companies. The majority of studies showed that the financial industry had a distinct characteristic whereby returns increased with scale, the industry had, to some degree, a natural tendency towards monopolies, and that the average cost of financial product factors represented a relatively gentle U-shaped curve. According to these characteristics, financial holding companies could gain long-term reductions in cost by developing diversified operations and expanding their business scale; in other words, the effects of economies of scale. Numerous works of scholarly research backed up this idea. Benston (1972) believed that in a situation where conditions remained constant, if a financial institution expanded three times in size, then the operational average cost would fall by 5-8%. While Rodriguez (1993) was researching the Italian banking industry, he discovered that an interval effect existed within economies of scale, that medium-sized banks had obvious economies of scale, whereas in large-sized banks, it was much more common to find diseconomies of scale. Saunders and Walter (1994) conducted research on the largest 200 banks (groups) in the world, and discovered that banks with loan scales that exceeded USD 5 billion showed obvious economies of scale. Similar phenomena also existed in EU banking industry (Vennet, 1994). Clark (1998) conducted empirical research on the economies of scale of the early American finance industry, and discovered that when applied on a relatively low level there were no obvious economies of scale, but as restrictions were loosened and as information technology advanced, financial institutions' economies of scale operational return interval will continue to increase. Having applied translogarithmic cost function empirical surveys on European banks, Cavallo (2001) discovered that European banks showed economies of scale. Ivanov (2004) believed that financial holding companies with a scale of less than US $100 million would find it difficult to realize the effects of economies of scale, while companies with over US$ 100 billion worth of assets would also have diseconomies of scale, and those falling in between these two categories would have relatively evident effects of economies of scale. Chung (2009), and others, described the effects of economies of scale on financial holding companies from three dimensions: the internalization of product trading, the improvement of product development capability, and the reduction of costs for product sales.

The reduction in costs achieved by the financial institutions through expansion of business operation scope and improvement of their allocation efficiency of production input, that is, economy of scope. Some scholars maintain a positive view on the economies of scope of financial holding companies. In his research on Japanese, Israeli and European banking institutions, Murdur (1992) discovered that economies of scale evidently existed in the joint production of financial services. Rodriguez's (1993) research came to the conclusion that economies of scope were the same of those of scale: both had an interval effect, economies of scope were apparent in moderate-scale banks, whereas in large-scale banks the existence of economies of scope was not apparent. Allen and Rai's (1996) research showed that operating costs of mixed operation financial enterprises was lower than that of separate operation enterprises, and economies of scope existed clearly in financial carriers working in mixed operations. The results of Saunder and Walter's (1997) empirical research using American financial holding companies as an example showed that the economies of scope existed in the majority of institutions. Sanotomero (2001), and others, pointed out that financial institutions can offer clients one-stop service by diversifying their operations, at the same time as reducing their own operational costs, this also saved the customer information collecting time and transaction costs. Vennet (2002) conducted research based on 2375 financial institutions from 17 countries across Europe. By applying translogarithmic cost function the research demonstrated that an interval effect existed in the economies of scope of financial institutions, that economies of scope were clearly evident in moderately sized financial institutions, whereas the effect of economies of scope in operations that were too big or too small was not obvious, or even displayed diseconomies of scope. Schmid and Walter (2009) conducted research on American financial holding companies, and via empirical analysis discovered testimony for the existence of economies of scope.

In conclusion, because the object of research and method of research were different, there are discrepancies of views within the foreign academic world's views on the economies of scale and scope of financial holding companies, however the general trend supported the view that economies of scale and scope existed in financial institutions that had diversified operations, the degree of economies of scale and economies of scope also differed depending on the scale of operations. In other words, being of a larger scale did not signify that the effect of economies of scale and scope would be stronger, and that it was easier to realize economies of scale and scope in relatively modest-scale financial institutions.

An universal definition of financial holding companies has never been agreed on in the domestic academic world, and the majority of research on financial holding companies has been conducted from determining the nature, with the quantitative aspect generally focusing on inspecting the operational efficiency of commercial banks set against the background of financial holding companies (Xiong Peng, 2008,Fudan University – Shenzhen Development Bank Supply Chain Studio, 2011, etc.), and only a relative minority of research has been conducted from a quantitative perspective on the entire operational effect of financial holding companies. Xie Ping etc. (2004) elaborated a systematic description of the development situation of financial holding companies and correlating theories. By simulation of combining China Merchants Bank, Shenyin Wangguo Securities and China Life, and the economies of scope after the combination of China Construction Bank, Shenyin Wangguo Securities and China Life, Xu Wenbin and Wang Daqing's (2008) research reached an empirical result concluding that economies of scope did exist. Zhang Dixin and Deng Bin (2013) found solid evidence that not only did the effect of economies of scale and economies of scope exist within the operations of financial holding companies, but that they also helped to boost the ability to withstand risk.

Thus, the domestic academic world's appraisal of the economies of scale and scope of financial holding companies is far from perfect, this paper will investigate from an empirical point of view the economies of scale and scope of domestic financial holding companies, on the one hand trying to verify whether these two economic effects exist, and on the other hand trying to verify whether an interval effect exists within these two economic effects so as to complement the domestic academic world in this field of research.

3 Research Plan

3.1 Method of Research

Of the current domestic scholars' quantitative research on financial enterprises' operational economies of scale and scope, the majority are conducted from the perspective of cost reduction, and those companies involved are all independent financial enterprises (banks, insurance companies, securities companies). Attributable to the expansion of operational business, the intricacy of institutional setup, complexity of management and coordination processes and difficulty in quantifying many indicators, it is very difficult to measure the economies of scale and scope in terms of cost. At the same time, in order to estimate the accuracy requirements of cost function, it is necessary, on the one hand, for cost function specification to take into account to the greatest extent the operating cost, that the set form of cost function can become complex; on the other hand, cost function estimates also need the support of large samples. Thus, in order to solve the problems of cost function setting, this paper will use financial indicators to analyze the economies of scale of financial holding companies. As to economies of scope, borrowing from Walter's research (1994), this paper will approach it from the perspective of income, by constructing the income growth models of financial holding companies, estimate a set of parameters and analyze the economies of scope. The set forms of specific income growth models are as follows:

$$\ln AS_t = \alpha + \beta_1 NNII + \beta_2 LTD + \beta_3 LA + \beta_4 \ln AS_{t-1} + \varepsilon_t \quad (1)$$

$$\ln CA_t = \alpha + \beta_1 NNII + \beta_2 LTD + \beta_3 LA + \beta_4 \ln CA_{t-1} + \varepsilon_t \quad (2)$$

In which AS_t and AS_{t-1} represent the current and previous periods of asset scale respectively, CA_t and CA_{t-1} represent current and previous periods of capital scale respectively, NNII, LTD, LA represent the proportion of non-interest income of banking income, the proportion of loan funds derived from deposits, and the proportional total assets of loans. $\beta_1, \beta_2, \beta_3$ represent various factors influencing the logarithmic regression coefficient of a financial holding company's asset growth, if $\beta_1 + \beta_2 + \beta_3 > 0$, representing a positive influence of a financial holding company's business operations on its bank assets or capital growth, then there are economies of scope at work, otherwise this indicates the existence of diseconomies of scope.

3.2 Sample Selection and Data Sources

According to Wang Mei's (2009) classifications, China's financial holding companies can be divided into thc following five types: holding companies formed by bank financial institutions, holding companies formed by non-bank financial institutions, financial holding companies formed by industrial capital, foreign financial holding companies, national (governmental) financial holding companies. The banking industry is very important to China's financial economy. There are relatively large discrepancies between the financial enterprises and other holding companies established by other subjects in terms of operational management and business operations. Referencing Wang Mei (2009), and Zhang Dixin and Deng Bin's (2013) research, this paper selects financial holding companies, which consist of banks and non-bank financial institutions, as its research object. The specific samples include China CITIC Bank (CITIC), China Everbright Group (CEG), Ping An Group (PAG), China Merchant Bank (CMBC), Industrial and Commercial Bank of China (ICBC), Agricultural Bank of China (ABC), Bank of China (BOC), China Construction Bank (CCB), and the Bank of Communications (BOCM). The time period is from 2007-2012. Sample data comes from the annual reports of these companies.

3.3 Variables

In this paper, the measure of financial holding companies' economies of scale is based mainly on income index and cost index in financial indicators, including return on assets, return on equity, cost on assets and cost on income. The variables related to the measure of economies of scope are total assets, total capital, non-interest income ratio, loan-to-deposit ratio, and loan-to-assets ratio. For a statistical description of all these variables, see Table 1.

Table 1. Statistical Description of Related Variables

Variables	Description	Sample Size	Average	Maximum	Minimum	Variance
AS	Total assets (100 million)	48	53926	175422	6513	47440
CA	Total capital (100 million)	48	3274	11285	247	3921
NNII	Non-interest income ratio	48	0.2652	0.9773	0.0599	0.2646
LTD	Loan-to-deposit ratio	48	0.6866	0.9195	0.4945	0.0953
LA	Loan-to-assets ratio	48	0.4675	0.5987	0.0969	0.1275
ROA	Return on assets	48	0.0108	0.0239	0.0012	0.0031
ROE	Return on equity	48	0.1767	0.2625	0.0107	0.0031
COA	Cost on assets	48	0.0321	0.2274	0.0115	0.0484
COI	Cost on income	48	0.5729	1.0165	0.4274	0.1422

4 Empirical Results Analysis

4.1 Analysis of Scale Economies

In order to compare the operational effectiveness and operating cost of financial holding companies of different scales, this paper divided all sample financial holding companies into three kinds; the first being large-scale financial holding companies, including the Industrial and Commercial Bank of China (ICBC), China Construction Bank (CCB), and Bank of China (BOC); the second is medium-sized financial holding companies, including the Bank of Communications (BOCM), China Merchants Bank (CMBC), China CITIC Bank (CITIC); the third is small financial holding companies, including China Everbright Group (CEG) and Ping An Group (PAG). For operation effectiveness index and operating costs of different sized financial holding companies, see Table 2 and Table 3.

Table 2. Operation Effectiveness Index of Financial Holding Companies

	ROA					
	2007	2008	2009	2010	2011	2012
ICBC	0.94%	1.14%	1.10%	1.23%	1.35%	1.36%
CCB	1.05%	1.23%	1.11%	1.25%	1.38%	1.39%
BOC	1.03%	0.73%	0.73%	0.92%	1.04%	1.10%
Average	1.01%	1.03%	0.98%	1.13%	1.26%	1.28%
BOCM	0.98%	1.06%	0.91%	0.99%	1.10%	1.11%
CMBC	1.16%	1.33%	0.88%	1.07%	1.29%	1.33%
CITIC	0.82%	1.12%	0.82%	1.05%	1.12%	1.06%
Average	0.99%	1.17%	0.87%	1.04%	1.17%	1.17%

Table 2. (*Continued*)

CEG	0.68%	0.86%	0.64%	0.86%	1.04%	1.04%
PAG	1.09%	0.12%	1.55%	1.53%	0.99%	0.94%
Average	0.89%	0.49%	1.09%	1.20%	1.02%	0.99%
ROE	**ROE**					
	2007	2008	2009	2010	2011	2012
ICBC	15.06%	18.31%	19.05%	20.21%	21.76%	21.15%
CCB	16.37%	19.81%	19.11%	19.27%	20.75%	20.39%
BOC	13.63%	17.71%	18.96%	17.50%	18.77%	19.32%
Average	15.02%	18.61%	19.04%	18.99%	20.43%	20.29%
BOCM	16.03%	19.58%	18.37%	17.51%	18.63%	15.33%
CMBC	22.42%	26.25%	19.65%	19.23%	21.89%	22.58%
CITIC	9.85%	13.97%	13.61%	17.49%	17.25%	15.45%
Average	16.10%	19.94%	17.21%	18.08%	19.26%	17.79%
CEG	20.41%	22.02%	15.88%	15.70%	18.81%	20.66%
PAG	14.27%	1.07%	15.79%	15.35%	13.18%	12.76%
Average	17.34%	11.54%	15.83%	15.53%	15.99%	16.71%

According to Table 2, in 2008 the return on assets (ROA) of large-scale financial holding companies averaged at 1.03%, smaller than the figure of 1.17% of medium-sized financial holding companies. For the other years in the table, the average ROA of large financial holding companies was larger than that of medium-sized financial holding companies; In 2009 and 2010, the average ROA of medium-sized financial holding companies was 0.87% and 1.04% respectively, slightly smaller than the figure of 1.09% and 1.20% of small-sized financial holding companies. For the rest of the years in the table, the average ROA of medium-sized financial holding companies was larger than that of small-sized financial holding companies. Therefore, based on the results of ROA comparison, it can generally be concluded that the bigger the scale of financial holding companies, the stronger its profitability is. Further comparison of the data shows that in large financial holding companies, CCB's profitability is stronger than the rest of the two banks. CMBC's profitability is stronger than BOCM's. Both CCB and CMBC's profitability are stronger than BOC's. The profitability of all medium-sized financial holding companies is stronger than that of the small-sized financial holding companies. This shows that to a certain extent, in terms of a single company the larger the scale of a financial holding company, the stronger its profitability is, but when the scale expands to a certain degree, the profitability will decline, namely, there exists a size limit.

In concern with return on equity (ROE), in 2007 and 2008 the ROC of medium-sized financial holding companies averaged at 16.10% and 19.94% respectively, larger than the figure of 15.02% and 18.61% of large financial firms. For the rest of the years in the table, the average ROE of large financial holding companies was bigger. The average ROE of small financial holding companies was smaller than that of medium-sized financial companies in all the years except 2007. The distribution of a single company's ROE is not obvious.

Table 3. Operating Cost Index of Financial Holding Companies

	COA					
	2007	2008	2009	2010	2011	2012
ICBC	1.62%	1.70%	1.22%	1.24%	1.32%	1.31%
CCB	1.80%	1.97%	1.35%	1.39%	1.46%	1.51%
BOC	1.54%	2.27%	1.67%	1.65%	1.89%	1.77%
Average	1.66%	1.98%	1.41%	1.43%	1.56%	1.53%
BOCM	1.50%	1.53%	1.29%	1.38%	1.35%	1.38%
CMBC	1.53%	1.84%	1.43%	1.60%	1.77%	1.59%
CITIC	1.46%	1.90%	1.22%	1.34%	1.28%	1.62%
Average	1.49%	1.75%	1.31%	1.44%	1.47%	1.53%
CEG	1.57%	1.95%	1.15%	1.25%	1.26%	1.25%
PAG	22.7%	20.1%	13.7%	14.3%	9.6%	9.4%
mean	12.2%	11.0%	7.43%	7.76%	5.42%	5.32%
	COI					
	2007	2008	2009	2010	2011	2012
ICBC	55.47%	53.66%	46.36%	43.68%	42.97%	42.74%
CCB	54.19%	55.66%	48.50%	46.30%	45.18%	45.68%
BOC	51.18%	75.34%	66.82%	58.88%	58.51%	55.69%
Average	53.61%	61.56%	53.89%	49.62%	48.89%	48.04%
BOCM	50.50%	53.32%	52.78%	52.32%	48.88%	49.52%
CMBC	48.94%	52.25%	57.41%	53.82%	51.52%	47.86%
CITIC	52.93%	56.13%	53.13%	49.84%	46.16%	53.59%
Average	50.79%	53.90%	54.44%	51.99%	48.86%	50.32%
CEG	57.77%	67.09%	56.62%	52.21%	47.59%	47.41%
PAG	89.6%	101.7%	86.8%	88.2%	80.1%	89.2%
mean	73.69%	84.37%	71.69%	70.21%	67.79%	68.30%

According to Chart 3, in 2010 and 2012, the average COA of large-scale financial holding companies was 1.43% and 1.53% respectively; the number for medium-sized ones had not much difference, 1.44% and 1.53% respectively. However, the average COA of medium sized financial holding companies was lower than that of the large-scale ones in any other years in the chart. By contrast, the average COA of small-sized financial holding companies was far higher than those of large- and medium-scale ones. According to Table 3, in 2007, 2008, and 2011, the COI of large-scale financial holding companies was 53.6%, 61.56%, and 48.89% respectively, which was higher than that of medium-sized ones, at 50.79%, 53.90%, and 48.86%. In the other three years, the figure of medium-sized financial companies was higher than that of the large-scale ones. Generally speaking, the COI difference between these two kinds of companies was slight. In contrast, the average COI of small-sized financial companies was higher than the other two kinds of companies. There was a tendency of decrease in operating cost with the expansion of business scope, which shows the existence of economies of scale. But when the scale expands to a certain degree, there was a tendency of increase in cost. It indicates that larger scales do not necessarily mean bigger economies of scale.

To sum up, in concern with income and cost, the domestic financial holding companies all have obvious economies of scale, yet the economies of scale also has interval effect. Medium-sized financial companies enjoy the most obvious economies of scale, especially in terms of cost. This conclusion is similar to the ones of other scholars, such as Vennet (2002) and Ivanov (2004). Behind this phenomenon, there are mainly three reasons. First, the financial industry itself has obvious characteristics of economies of scale. As the business scale expands, and outlets increase, the operating costs of companies will have significant decline. Compared with other kinds of enterprises, the long-term cost curve of financial enterprises is smoother, and they have more potential in economies of scale. Second, through business diversification, financial holding companies can make financial product deals within their single group, and consequently reduce transaction expenses caused by uncertainties and the transaction costs of diversified investors, so as to achieve economies of scale more easily. Meanwhile, it can organize and control massive sales of products of its financial companies, achieve central selling, and reduce the unit advertising costs and marketing costs (Chung et al., 2009). Third, the U-shaped cost curve of financial enterprise shows the existence of a size limit. When the scale of operation reaches a certain degree, further expansion of the scale would not bring scale economies. To be specific, with the expansion of enterprise scale, the internal organization levels will increase, and disadvantages usually will appear in internal institutional arrangements, as a result, the goals to minimize costs and maximize profit will be difficult to achieve, and it will further decrease internal resource allocation efficiency.

4.2 Analysis of Economies of Scope

As this paper does not select many samples, in order to raise the accuracy of the estimates, small and medium-sized financial holding companies are put into one group and can be compared as a whole with large financial holding companies. This paper applies the fixed effect method to estimate, the specific estimates are as follows in Table 4.

According to Table 4, if asset scale (AS) is set as the dependent variable, the R^2 of three models is 0.9509, 0.9908, and 0.9471 respectively. The figures are rather high, which shows that the explanation variables can well explain the dependent variable and the model setup is relatively reasonable. The sum of coefficient of all samples is 1.8738, which indicates that the business of domestic financial holding companies have effect of scope economy. Large financial holding companies' coefficient is 0.5879 in total, which is less than the total coefficient of 2.6390 of small and medium-sized financial holding companies. It shows that small and medium-sized financial holding companies' scope economic effect is more obvious. If take capital scale (CA)

Table 4. Fixed Effect Estimates

	AS			CA		
	All	Large	Medium & Small	All	Large	Medium & Small
CONS	2.8463*** (3.10)	4.3267*** (12.14)	2.8135* (2.47)	1.5877* (2.07)	1.3687 (0.91)	2.1071 (0.69)
NNII	2.2615* (2.08)	1.3428 (2.38)	3.0229 (2.00)	4.3911*** (3.17)	3.2392* (2.45)	2.5587 (2.13)
LTD	-1.8454* (-1.90)	1.0301* (2.96)	-2.0732 (-1.65)	-2.4137*** (-3.68)	-2.2007 (-2.09)	-0.2498 (-0.06)
LA	1.4577* (3.24)	-1.7850 (-0.95)	1.6893* (2.22)	2.1449*** (4.33)	1.7498** (3.97)	7.2720 (0.82)
$LnAS_{t-1}$	0.8206*** (13.87)	0.7404*** (9.26)	0.8050*** (11.04)			
$LnCA_{t-1}$				0.8396*** (14.55)	0.8821*** (9.64)	0.5388** (4.20)
$\beta_1+\beta_2+\beta_3$	1.8738	0.5879	2.6390	4.1223	2.7883	9.5809
R^2	0.9509	0.9908	0.9471	0.9017	0.9248	0.8889

Note: Figures in brackets are of t-value. Symbol ***, **, and * stand for 1%, 5%, and 10% level of significance respectively.

as the dependent variable, all the R^2 values are comparatively big. The estimated coefficient of all the samples is 4.1223 in sum. The large financial holding companies' coefficient is 2.7883 in total, and small and medium-sized financial holding companies' is 9.5809, which proves the aforementioned conclusion.

From aspects of both assets and capital, it is confirmed that the business of financial holding companies exists effect of scope economies; furthermore, large financial holding companies' effect of scope economies is weaker than that of small and medium-sized financial holding companies. This is similar to research conclusions of scholars such as Vennet (2002), Schmid and Walter (2009). Financial holding companies can achieve economies of scope for two reasons. On the one hand, the input production factors of financial enterprise have obvious "public goods" nature, namely the exclusiveness of these factors are in a lower level; the use of production factors in one product does not exclude it from being used for other purposes. Hence, input production factors can be transformed in different areas of business, achieving low cost. Therefore, companies can better allocate resources to meet the need of their business, and improve resource utilization. To be specific, as for human resources, the knowledge structure of financial industries, such as banking and securities, is overlapped in a quite high degree. As a result, financial companies' internal personnel, especially high-level talents, can be shared, thus reducing the manpower cost. In the aspect of fixed assets investment, financial industry's fixed assets, which consist of electronic network platform and physical outlets, can be shared between different business areas, so as to achieve apportioned fixed costs in a wide range of business scope. While the utilization rate of fixed assets is improved, at the same time it also reduces the fixed assets input per unit output. As to product sales, a company's sales channels can be shared by different internal business areas reducing cost. Besides, when one financial product of a group company comes to market and achieve successful sales, the credibility spillover effect lowers the selling cost of other products of that group company, expanding the other products' market demand. On the other hand, the company can design "package" products, to provide one-stop service at point of sales, so as to improve the "stickiness" of clients to the company's service, to expand product sales, and to reduce marketing costs.

However, large-scale financial holding companies' scope economies effect is weaker, this is because with the expansion of operation scale, business overlap increases, professionals and management personnel increasingly work in the service of their non-professional, unskilled areas, leading to the growth of coordination costs between business areas, thus, operation and management efficiency decreases, further weakening the scope economies effect. This means that financial holding companies should not pursue business scale blindly. A moderate scale can achieve better operation effect.

To test the accuracy of the above estimates, this paper further uses the random effect method to have a robustness check on model (1) and (2). The results are as shown in Table 5.

Table 5. Random Effect Estimates

	AS			CA		
	All	Large	Medium & Small	All	Large	Medium & Small
CONS	1.4821*** (2.99)	3.7947*** (6.19)	1.4517** (2.23)	0.5420 (0.69)	1.3620 (0.98)	3.0949*** (3.69)
NNII	0.5356* (1.69)	1.4378*** (4.47)	0.5402 (1.49)	0.7189** (1.97)	0.4094* (1.66)	5.0582* (1.80)
LTD	-0.8027 (-1.54)	0.7548* (1.75)	-0.7901 (-1.13)	-0.6274 (-1.26)	-0.9933 (-0.90)	-1.6079 (-0.36)
LA	0.8521 (1.37)	-2.0469** (-2.46)	0.8664 (1.19)	1.6075** (2.22)	0.8967** (1.93)	0.7342 (0.16)
$LnAS_{t-1}$	0.9143*** (27.54)	0.7990*** (25.96)	0.9152*** (20.87)			
$LnCA_{t-1}$				0.9275*** (20.10)	0.9171*** (12.01)	0.7470*** (15.51)
$\beta_1+\beta_2+\beta_3$	0.5850	0.1457	0.6165	1.699	0.3128	4.1845
R^2	0.9397	0.9904	0.9312	0.8553	0.9039	0.8586

Note: Figures in brackets are of Z-value. Symbol ***, **, and * stand for 1%, 5%, and 10% level of significance respectively.

According to Table 5, whether the dependent variable is assets or capital, domestic financial holding companies all have obvious effect of scope economies, but large financial companies' scope economy effect is weaker. Therefore, it can be concluded that the aforementioned model estimates is sound.

5 Conclusions and Suggestions

Based on a sample of 8 financial holding companies between 2007 and 2012, and by the method of financial index, this paper analyzes the economies of scale of domestic financial holding companies from two angles: benefits and costs. Based on the fixed effect model and the random effect model, it analyzes the economies of scope of domestic financial holding companies from two aspects: assets and capital. Through comparative analysis and empirical study, it concluded that domestic financial holding companies have economies of scale and economies of scope. Medium-sized financial companies enjoy the most obvious economies of scale, especially in terms of cost. Large-sized financial holding companies have weaker effects of scope economies.

Based on this conclusion, this paper argues that the government should vigorously promote the development of financial holding companies, gradually liberalize financial regulations, promote business diversification in financial institutions, and improve supervision coordination mechanism of financial holding companies. Financial holding companies should actively carry out diversified business, get rid of the management philosophy of one-sided pursuit of scale expansion in the process of development, and keep an appropriate scale of operation according to the actual operating conditions.

References

1. Supply Chain Financial Studio of Shenzhen Development Bank, Fudan University, The Analysis of Listed Banks' Performance and Its Synergistic Effect under the Control of Financial Holding Companies, Shanghai Finance, No. 12, pp. 44-50 (2011)
2. Peng, X.: A Research on Business Performance of Banks with Backgrounds of Financial Holding Companies. Collected Essays on Finance and Economics 1, 51–58 (2008)
3. Wenbin, X., Daqin, W.: An Empirical Analysis on the Economies of Scope of Universal Banks. Commercial Research 12, 143–147 (2008)
4. Ping, X.: The Development and Regulation of Financial Holding Companies. CITIC Press, Beijing (2004)
5. Mei, W.: A Study on the Evolution, Risk Control and Development of Financial Holding Companies, PhD thesis of Southwestern University of Finance and Economics (2009)
6. Dixin, Z., Bin, D.: The Business Performance of Chinese Financial Holding Companies at the Time of Financial Crisis. Journal of Management Science 7, 66–79 (2013)
7. Allen, J., Rai, A.: Operational Efficiency in Banking: An International Comparison. Journal of Banking and Finance 20(4), 665–672 (1996)
8. Benston: Economics of Scale of Financial institutions. Journal of Money, Credit and Banking 4(2), 312–341 (1972)
9. Cavallo, L.: Scale and scope economies in the European banking systems. Journal of Multinational Financial Management 11(4), 515–531 (2001)
10. Clark, J., Paul, J.S.: Economies of Scale and Scope in Banking: evidence from A Generalized Translog Cost Function. Quarterly Journal of Business and Economics 33(2), 3–25 (1998)
11. Ivanov, A.I.: Bank Deregulation Revisited: Relationship Banking and the Bank Merger Wave in the U. S. Working Paper, Reed College Economics Department (2004)
12. Vennet, R.V.: Economies of Scale and Scope in EC Credit Institutions. Brussels Economic Review 144, 507–548 (1994)
13. Vennet, R.V.: Cost and Profit Efficiency of Financial Conglomerates and Universal Banks in European. Journal of Money, Credit and Banking 34(1), 254–282 (2002)

14. Saunders, A., Walter, I.: Universal Banking in the United States. Oxford University Press, New York (1994)
15. Saunders, A., Walter, I.: Universal Banking Financial System Design Reconsidered. Irwin Professional Pub., Chicago (1996)
16. Schmid, M.M., Walter, I.: Do Financial Conglomerates Create or Destroy Economic Value? Journal of Financial Intermediation 18(2), 193–216 (2009)
17. Wu, W.C., Lai, M.C., Huang, H.C.: Rating the Relative Efficiency of Financial Holding Companies in An Emerging Economy: A multiple DEA Approach. Expert Systems with Applications 36(3), 5592–5599 (2009)

Research of Data Resource Management Platform in Smart City

Jing Shao[1,2], Li-na Yang[1,*], Ling Peng[1], Xiao-jing Yao[1], and Xue-Liang Zhao[1,2]

[1] Institute of Remote Sensing and Digital Earth, CAS, 100012 Beijing, China
[2] University of Chinese Academy of Sciences, 100049 Beijing, China
yangln@radi.ac.cn

Abstract. Recently, there stands an upsurge to construct smart city all over China. A higher demand is put forward to real-time and dynamic data. At present, city data is multi-source, inconsistent and hard to match in most region of China, which can't meet the need of smart city construction. This paper focuses on the connotation and techniques of data resource management platform, which is oriented to data acquisition and integration. Based on systematic analysis of city data resource, we discussed the common features from four dimensions of time, space, user and theme, and proposed a multi-dimensional data model. Besides, we designed and implemented the data resource management platform. Key techniques, such as comprehensive data acquisition, data storage and dynamical update, fusion method for multi-source heterogeneous data, full-scale data encoding and spatial-temporal data warehouse, were described. The data resource management platform can provide a data support for comprehensively application and multi-dimensional decision analysis in smart city.

Keywords: Smart City, Data Management, Data Model, System Construction, Data Warehouse.

1 Introduction

Since the concept of "smart planet" was put forward by IBM in November 2008, more and more governments, enterprises and social organizations paid attentions to smart city all over the world. They expect that problems in urban management, economic development and human life can be solved through smart city construction. "Smart city", which is a continuation of digital city, provides intelligent analysis and service on the basis of urban informatization [1], and puts forward a higher demand for real-time, dynamic and integrated data resource. Scholars have pointed out that technique researches in acquisition, storage and analysis for mass data hold important significance for comprehensive management and multi-dimensional analysis in smart city [2].

---

[*] Corresponding author.

F. Bian and Y. Xie (Eds.): GRMSE 2014, CCIS 482, pp. 14–22, 2015.

After ten-year period construction of digital city, a large amount of data about urban management and operation has been accumulated. Most of the data is stored in independent departments (e.g., land department, forest department and electricity department, etc.) [3,4], and is hard to match. With simple manage method and long data update cycle, exiting data resource can't meet the need of smart city construction. The scientific interests of our research focus on 1) analyzing the connotation of city data resource from the perspective of integrated management, 2) designing and implementing the data resources management platform in Sino-Singapore Tianjin eco-city, 3) providing a data support for comprehensively application and multi-dimensional decision analysis in smart city.

2 Connotation of City Data Resource

It is essential to define the content and scope of city data resource, analyze its characteristics and internal relations in city data management. Many decisions can be made based on the connotation of city data resource, such as organizing means of data acquisition and designing methods of data integration.

2.1 Content Classification of City Data

During the construction of smart city, there are a variety of forms and standards of data content classification. Guided by the concept of "people-oriented", this paper improves the traditional data content classification systems and classifies data into three categories: basic geographic information elements, economic entity elements and social event elements (a detailed classification system was shown in Table 1).The proposed content classification system almost covers the entire content in a city, reflects the dynamic characteristics of data, and is easy to enrich and expand.

Table 1. Content classification of city data in smart city

Data classification	*Data content*
Basic geographic information elements	River, Residents and facilities,Traffic,Pipeline,Realm and Administrative Region,Landforms,Vegetation and soil,Indoor Maps,Other geographic elements
Economic entity elements	Individual, Enterprise, government
Social event elements	City environment,Advertising,Construction Management,Emergencies,Street order,City Operations,Facilities regulatory,Other events

Basic Geographic Information Elements. They are the carrier of the spatial positioning and analysis. With the feature of versatility and growing needs of sharing, they are almost applied to all industries related to geographic information. And they are the basis of data integration, analysis and service in smart city.

Economic Entity Elements. Economic entities are the main decision-makers and participants in a city, which includes government, enterprises and individuals. They are the main targets to service and analyze in smart city. Population database, legal person database and other basic database can be built based on these elements.

Social Event Elements. They are events that have happened, is happening and about to happen with the participation of basic geographical environment and economic entities, such as urban disaster, environmental management, emergency response and rescue, population migration, etc. And a variety of thematic databases are formed during the process of urban management.

2.2 Conceptual Model of City Data

In practice, we found that city data has four-dimensional characteristics: multi-space dimension, multi-time dimension, multi-user dimension and multi-theme dimension, as shown in figure 1. Space-time attributes are the basic elements of city data. The feature of space refers to the geometric characteristics of urban geospatial entities and their spatial relations. The feature of time refers to the different forms of object with time passed. City data have different users depending on the different types of business and the different stages of development. And it reflects different features in different application theme. As investment project object focused on the annual investment and funds types in economic thematic, while on the environmental investment, sewage discharge destination, garbage disposal types in environmental thematic.

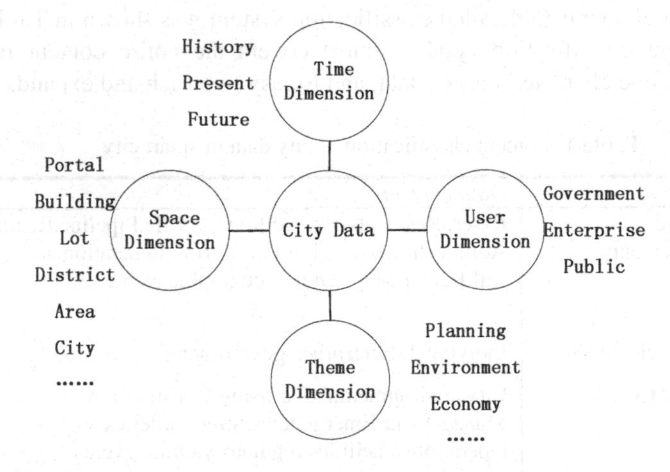

Fig. 1. Four-dimensional characteristics of city data

It possesses important significance to take city data as a set of interconnection organic system and establish the integrated management model for data mining and multi-dimensional decision-making in smart city. We propose a city data conceptual model based on the four-dimensional characteristic of city data, which fuses business information in various types to the space-time information framework, and makes the abilities of data comprehensive services and integrated analysis stronger by associating, organizing and establishing index in accordance with the attribute information and dimensional features. The conceptual model can avoid data redundancy, conflicts and errors, and save time of data transmission, conversion and relevance processing. It can also improve the analysis capability, providing a data foundation of management and multi-dimensional analysis.

3 Key Techniques of Data Resource Management Platform

To manage data effectively and comprehensively, we design the data resource management platform based on conceptual model of city data, including the process of data acquisition, storage, update, fusion and encoding, as shown in figure 2.

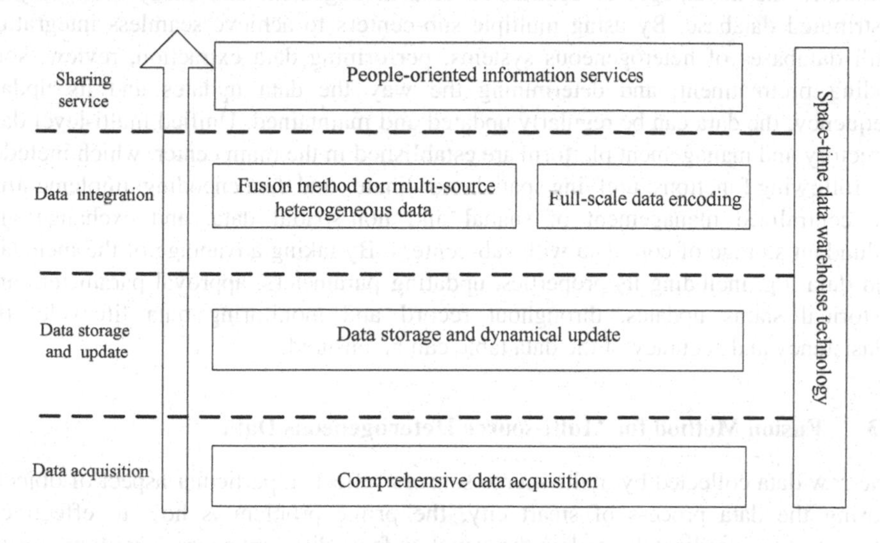

Fig. 2. key techniques of data resource management platform

3.1 Comprehensive Data Acquisition

Comprehensive perception is an essential feature of smart city. The sensor network consist of sensor terminals, such as two-dimensional code recognizers, RFID recognizers, cameras and GPS, provides foundation for object recognition and information collection. The means of city information acquisition is divided into three parts according to different sensors in this paper: urban data collection by the city sensor network, the hot spot information gathering with mobile terminals and the urban resource data acquiring through multi-source remote sensing technique. The urban sensor network, main source of data acquisition, undertakes the vital task of essential data acquisition in urban management and operation. And mobile terminals and multi-source remote sensing technique reinforce the ability to get information.

Besides, the mobile terminals provide information as videos, pictures and position of city hot spots. This method is flexible and effective. Remote sensing technology provides the condition for extracts and updates city data. It plays an important role in information collection of urban land utilization, city dynamic change, urban physical environment, ecological system, environmental pollution, resource management and traffic monitoring.

3.2 Data Storage and Dynamical Update

All aspects of real-time city data are captured by the dense sensor network and aggregate to data center. The techniques in storage, pretreatment and dynamic updates of heterogeneous mass data are major challenges for data management. A distributed heterogeneous database based on data center model is built in this paper, which combines the advantages of centralized data management and easily scalability of distributed database. By using multiple sub-centers to achieve seamless integration with databases of heterogeneous systems, performing data extraction, review, sort, coding pretreatment, and determining the way the data updates and its update frequency, the data can be regularly updated and maintained. Unified multi-level data directory and management platform are established in the main center, which includes the following functions: unifying spatial coordinates and data encoding, implementing the centralized management of spatial and non-spatial data, and exchange and redundant storage of core data with sub-centers. By taking advantage of the metadata and data log, including its properties, updating parameters, approval parameters and historical status updates, throughout record and monitoring data lifecycle, the consistency and accuracy of the data table can be ensured.

3.3 Fusion Method for Multi-source Heterogeneous Data

The raw data collected by multi-source sensors reflects a particular aspect of objects. During the data process of smart city, the prime problem is how to effectively integrate and coordinately explain the raw data from different sources, with properties of different format, frequency, time and space.

By using appropriate data fusion method, the heterogeneous, massive, multi-source data collected by sensor network is fused to form a basic data, which is more credible, thus providing a basis for a comprehensive grasp of intelligent urban conditions. Urban spatial and temporal data fusion in remote sensing applications mainly refers to fusing images with different spatial resolution, time resolution and spectral resolution. In this way, a variety of information resources are complemented, which improves the resolution of the remote sensing images and the accuracy of the comprehensive analysis of remote sensing images. For example, to attain the reliable information of vegetation cover in urban, we can fuse the TM images (high spectral resolution and low spatial resolution) with SPOT images (low spectral resolution and high spatial resolution) to generate new images with both high spectral and spatial resolution.

3.4 Full-Scale Data Encoding

Urban geospatial information is a basis for information storage, management and analysis. In order to integrate the data of demographic, economic, transportation, planning, environmental protection and other themes, it is necessary to break conventional methods based on grid (such as quad tree) for spatial subdivision, using irregular spatial grid with a natural semantic for geospatial multi-level split, increasing adaptation between split space and the corresponding target mapping [5], forming the total element coding for city-level which meets the multi-scale applications requirements.

We proposed a full-scale data encoding model in city-level space, combined by the preceding classification codes, middle spatial location codes and posterior sequential order codes. Preceding classification codes refer to " classification and code for fundamental geographic information elements "standards [6], used to identify coding categories (geographic entities, business entities) ; middle spatial location codes combine the natural attributes of the geographic entity with its management attributes, in accordance with 8 scale levels, "provincial administrative regions - municipal administrative regions - district (county) administrative regions - street-level administrative regions - Community – unit - lot - building", which split urban geographic space and construct a unique and unified space level coding ; posterior sequential order codes identify the elements in order, which are of the same category and with the same location. This encoding system has characteristic of unity, promotion and scalability.

3.5 Space-Time Data Warehouse Technology

Data warehouse is a subject-oriented, integrated, non-volatile, time variant collection of data, used for decision-making support [7]. Heterogeneous data from different sources or databases can be stored, extracted, and maintained in the data warehouse after processing [8]. Spatial-temporal data warehouse is a combination of GIS and data warehouse technology. By introducing temporal dimension data on the basis of the data warehouse, its capabilities for the spatial and temporal data storage, management and analysis has increased.

The data resource management platform involves techniques of automatically updating, spatial index and spatial online analytical processing. Data conversion is a key step that source data can be extracted to data warehouse, where ETL tools, such as IBM Warehouse Manager, Oracle Warehouse Builder, Microsoft DTS and GeoKettle can be used in the processes of data extraction, purification, consistency checks, decompose, merge, time label attachment and so on. The data set stored in the data warehouse is then formed, which is of code consistency, properties uniformity, description uniqueness and key unambiguousness. By creating spatial index , those data can be quickly accessed, thus avoiding massive spatial data scanning. It is an important way to optimize query response speed. In the multi-dimensional view of the data warehouse, it is necessary to create an index for each dimension. Online analytical processing (OLAP) is a technology which is to enable analysts to access information fast, consistently, interactively, and gain a deeper understanding of the data, including operations of select, slice, dice, pivot, drill-down, roll-up.

4 Implementation of Data Resource Management Platform

4.1 Function Design

The functions of city data resource management platform are data management, data processing, operation management, display and so on, as shown in figure 3.

Data Management Subsystem. Its main functions are: distributed management and supervisory control of data center (catalog management, metadata management etc.) to ensure date integrity and consistency; keeping the updating data regularly; calling processing tool set for data extraction , integration, query , statistics and etc.

Operation Management Subsystem. The subsystem, which includes user management, privilege management, log management and early warning of errors, can ensure security of whole system.

Data Processing Toolset. The subsystem contains data type conversion, projection transformation, code generation, data quality checking, creating data index and other basic data processing tools for data management subsystem and display subsystem.

Display Subsystem. The subsystem provides functions of data retrieval, display and analysis for basic geographic and thematic data. It's the visualization of data resource and provides convenient services for the public, uniform query managements for the operator and decision support for the government.

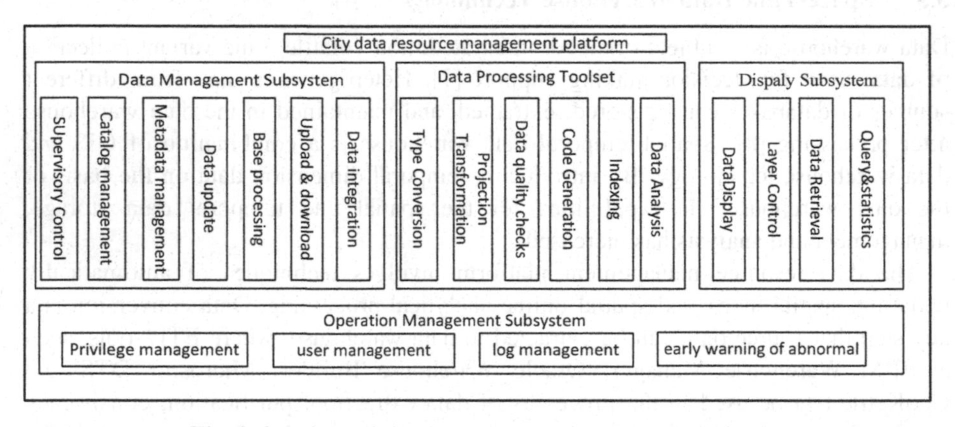

Fig. 3. functions of city data resource management platform

4.2 Application Case

According to the connotation of urban data resource and design of management platform, a prototype system was built with JAVA development language. It used ArcSDE of ArcGIS 10.0 enterprise edition and the database of oracle 10g to save and manage data. And the display subsystem was developed with ArcGIS for FLEX based on B/S model. The platform has been applied in Sino-Singapore Tianjin eco-city.

Practical experience shows that the platform has the ability to support simultaneous access and concurrence transaction of the data covering the whole city by multiple users, and interact with the urban data management well. The main interface of the system was shown as follows: figure 4 showed the updating parameters (fields, means and frequency, etc.) being set when update data between databases, figure 5 showed the query and display of air quality data that extracted from environment database.

Fig. 4. Updating parameters setting

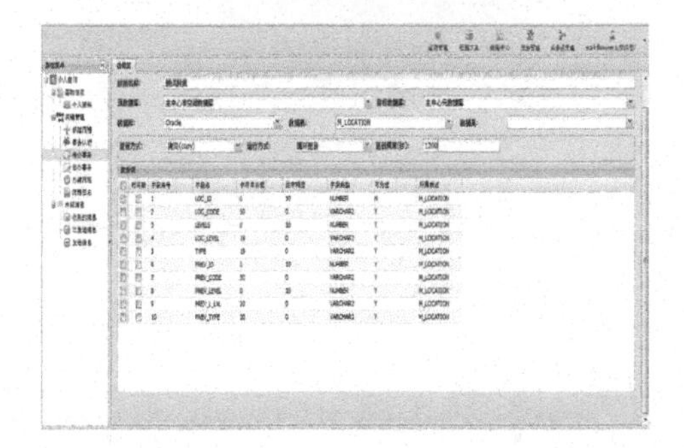

Fig. 5. Query and display of air quality data

5 Conclusion

To meet the higher demand of data resource management in smart city, this paper has analysed the connotation of city data resource, proposed a multi-dimensional data model, designed and implemented data resources management platform. It provides a data foundation for the multi-dimensional decision and comprehensively application analysis in smart city.

References

1. Li, D.R., Shao, Z.F., Yang, X.M.: Theory and Practice from Digital City to Smart City. Geospatial Information 9(6), 1–5 (2011)
2. Wu, H.Q.: Data Management in Smart City. Internet of Things Technologies (11), 11–14 (2012)

3. Chang, Y.F., Wu, H.G., Dong, Z.H., et al.: A National Monitoring and Warning System for Forest Pest Based on Service-Oriented Architecture. Scientia Silvae Sinicae 47(6), 93–100 (2011)
4. Mo, H.Y., Wang, Y.J., Luo, B., et al.: Design and Implementation of Land Planning System: A Case Study in Guangdong Province. Journal of Geo-information Science 12(5), 687–699 (2010)
5. Liu, Y., Peng, Z.F., Liu, J.T.: Study of Key Technologies of Smart Spatial Information Platform and Smart Shenzhen. Bulletin of Surveying and Mapping (6), 78–81 (2013)
6. National Administration of Surveying, Mapping and Geoinfomation of China.: GB/T 13923-2006: Classification and Code for Fundamental Geographic Information elements (2006)
7. William, H.I.: Building the Data Warehouse. Wiley Publishing, Indianapolis (2005)
8. Hu, K., Xia, S.W.: Large Data Warehouse-based Data Mining: a Survey. Journal of Software (1), 54–64 (1998)

Study on Urban Multidimensional Niche of Urban Agglomeration in Northern Slope of Tianshan Mountains

Zuliang Duan[1], Jun Lei[1], and Yaxuan Liu[2]

[1] Xinjiang Institute of Ecology and Geography, CAS,
830011 Urumqi, P.R. China
[2] School of Economics, Xinjiang University of Finance and Economics,
830012 Urumqi, P.R. China
{zuliangduan,LYX1022}@163.com

Abstract. Based on multidimensional niche theory, the paper tried to explore the function of cities and interaction relations among urban agglomerations, so that to provide a new perspective realizing the sustainable development of urban agglomerations. Using the principal component analysis, urban ecological niche breadth model and urban ecological niche differentiation index model on 16 indicators for 10 cities in Northern Slope of Tianshan Mountains (NSTM), ecological niche breadth and differentiation characteristics of the urban agglomeration were measured and calculated. The results are as follows.1) Comprehensive urban niche breadth of primate city is bigger than most cities in NSTM. The polarization is obvious.2) Comprehensive differentiation index of urban niche breadth shows an increasing tendency. The regional development elements gather in regional central city, moreover, the accumulative effect is even enhancing. 3) In general, the niche differentiation index of economy and public service, equitable development and population increasing rises apparently, however, the infrastructure status declines.

Keywords: urban multidimensional niche, niche breadth, principal component analysis, urban agglomeration in northern slope of Tianshan Mountains.

1 Introduction

Under the condition of limited resources and environment, species can coexist without competition on one dimension of multidimensional niche. That is, competition can become cooperation or interdependence to some degree. The cooperation and interdependence of ecological niche has some reference for coordinated development of the urban agglomeration. According to the viewpoint of urban life cycle theory and urban complex ecosystems[1], the city is likened to ecological units, and the city's location and role to urban ecological niche. Then the urban agglomeration can be optimized and integrated on the basis of evaluation of cities according to ecological niche theory.

The concept of ecological niche is from avian population ecology. J. Grinell was the first person to put forward the definition of ecological niche[2-3]. Odling-Smee put forward the concept of ecological niche construction in terms of evolution

F. Bian and Y. Xie (Eds.): GRMSE 2014, CCIS 482, pp. 23–32, 2015.

ecology[4]. Silvertowm pointed up the preoccupation principle and species competition effect of ecological niche[5]. Tilman raised the theory of stochastic ecological niche[6]. Domestic scholars made researches on the definition of ecological niche, extended theory of ecological niche, niche ecostate-ecorole theory, niche breadth, and niche overlap[7-8]. The ecological niche theory, as the basic theory of ecology, is widely used in the ecology field. In recent years, with the development of ecological niche concept, ecological niche theory has been used to make researches on the city. And its meaning is being expanded. The research results focus on the concept of urban ecological niche [9-10], urban ecological niche measurement[11-12], urban planning and sustainable development of ecological niche theory[13-14], and strategic niche administration[15].

The research on urban ecological niche of coastal developed areas is more than that of western areas especially oasis urban agglomeration. And most of the research focuses on explanation of the related concept of urban ecological niche, rather than how to measure urban ecological niche and use related models to measure urban agglomeration. And most of the researchers directly used the concept and method of ecological niche to get society-economy-ecology index, and evaluated the urban ecological niche. The city is not studied as an independent ecological unit, and its location is not taken into account. According to multidimensional niche theory, preliminary quantitative analysis of 16 indicators for 10 cities in Northern Slope of Tianshan Mountains (NSTM) is presented. By the Principal Component Analysis, the niche differentiation index of economy and public service, population, resources and environment, infrastructure status, and space is got. Using urban ecological niche breadth model and urban ecological niche differentiation index model, ecological niche breadth and differentiation characteristics of the urban agglomeration are measured and calculated, Based on the study, some suggestions of coordinated development for the urban agglomeration in NSTM are put forward.

2 Evaluation Model

The urban multidimensional niche refers to the location and function of the city in the urban agglomeration, which includes resource niche, economic niche, social niche, etc. It is the n-dimensional resources space of the city in the urban agglomeration. The urban niche competition is to compete for high quality resources and to expand the access to resources. After the competition, the winning city can acquire more natural resource or social and economic resource, and then has more strategy to choose. Meanwhile, the niche of the inferior city is compressed and it will be at a disadvantage in future competition. The urban niche has the characteristic of expansion compression and subjective [12, 16].

2.1 Breadth Model

The niche breadth of urban present situation depends on the “state” of resources that can be acquired, that are the total amount of human resources, economic resources, etc. And the niche breadth of urban future depends on the “tendency” of resources, which is the change rate of resource stock and time-lapse. The condition of “state” and "tendency" comprehensively indicates the urban niche breadth. Formula is as follows:

$$N_i = \frac{S_i + A_i P_i}{\sum_{j=1}^{n}(S_j + A_j P_j)} \tag{1}$$

In this formula, i, j =1, 2, ..., n. N_i is the niche space breadth of the city i. S_i is the state of the city i, and P_i is the tendency of the city i. S_j is the state of the city j, and P_j is the tendency of the city j. A_i and A_j are dimensional conversion coefficients. And $S_j + A_j P_j$ refers to absolute niche.

2.2 Differentiation Index Model

The urban niche breadth indicates the relative position of the city in this urban agglomeration. In order to find the difference among the urban niche breadth, niche differentiation index is structured, which can be used to find out the change in the structure of the urban niche. Formula is as follows:

$$C = \sqrt{\sum_{i=1}^{n}\left[\left(N_i/\overline{N}\right)-1\right]^2 / n} \tag{2}$$

In this formula, C refers to the niche differentiation index. N_i is the niche space of the i-th city. $\overline{N}$ is the average value of niche space. And n refers to the number of the city.

2.3 Measurement Index System

According to the comprehension of the urban multidimensional niche concept, the index is chosen by these principles as follows: (1) Scientificalness principle. The index should have clear and definite meaning, complete system, and can reflect the niche breadth of every city; (2) Comparability principle. The index should be measured and acquired. (3) Difference principle. The index should be different among cities. As the city is a natural-economic-social complex ecosystem, according to the concept of urban multidimensional niche, city construction, resource and environment, economic development, urbanization, population growth, and social development are taken into account to construct the measurement index system of urban niche with 16 indexes, that is per capita road area X_1, per capita construction land area X_2, per capita daily water consumption X_3, per capita public green space X_4, per capita gross value of production X_5, secondary and tertiary industries added value X_6, general budget revenue of local finance X_7, fixed-asset investment of whole society X_8, total retail sales of consumer goods X_9, non-agricultural population X_{10}, urbanization rate X_{11}, urban population density X_{12}, natural population growth rate X_{13}, post and telecommunications business gross X_{14}, the amount of doctors X_{15}, and middle and primary school enrollments X_{16}.

3 Calculation and Analysis

The urban agglomeration in NSTM lies in the hinterland of Eurasia, on the south margin of Junggar Basin, and at the middle part of the northern slope of Tianshan Mountains, with well natural resources environment and economic development. The urban agglomeration in NSTM is one of great importance in Northwest China, which is built with rapid development of the economic belt, convenient transportation, and preferential policy support. And the core area of the urban agglomeration in Northern Slope of Tianshan Mountains includes Urumqi, Karamay, Shihezi, Changji, Fukang, Miquan, Kuitun, Wusu, Wujiaqu, Hutubi, Manas, and Shawan[17].

3.1 Data Acquisition and Processing

In 2007, Miquan was repealed upon approval by the State Council, and Midong District was founded with Miquan and Dongshan District of Urumqi merged. So the data of Miquan is added to that of Urumqi. And Wujiaqu is not involved in the measurement, for some of its data is missing. According to the measurement index system of urban niche, 10 cities in NSTM are selected. With the statistical data from statistical yearbook of Xinjiang (2000-2010) and statistical year report of cities and counties construction, eigen value, variance contribution rate (Table 1), and factor score coefficient matrix (Table 2) of the urban agglomeration in NSTM are measured and calculated with the Principal Component Analysis.

Table 1. Eigen Value and Variance Contribution Rate

Factor	Eigen Value	Variance Contribution Rate	Cumulative Contribution Rate
F_1	8.227	51.416	51.416
F_2	2.715	16.967	68.383
F_3	1.782	11.138	79.522
F_4	1.013	6.329	85.850

From the cumulative contribution rate in Tab.2, the cumulative contribution rate of the former 4 factors reaches 85.85% (normal if higher than 85%). Thus, the 16 indexes are dimensionally reduced to 4 factors with the Principal Component Analysis. From Tab.2, the first factor refers to economic growth and public service factor (F_1), reflecting the urban economic development and public service level, which is positively related to these indexes, such as secondary and tertiary industries added value, general budget revenue of local finance, fixed-asset investment of whole society, total retail sales of consumer goods, telecommunications business gross, the amount of doctors, and middle and primary school enrollments. The second factor refers to equitable development and population growth factor(F_2), reflecting equitable distribution of water resource among cities during regional development, movement of population between city and countryside, living standard, and balanced population development, which is positively related to per capita daily water consumption, urbanization rate, per capita gross value of production and negatively related to natural population growth rate. The third factor refers to urban infrastructure factor(F_3), reflecting the urban infrastructure level, which is positively related to per capita road area, per capita public green space, and negatively related to urban population density. The fourth factor refers to urban spatial constraint factor(F_4), reflecting urban spatial intensification, which is positively related to per capita construction land area.

Table 2. Factor Score Coefficient Matrix

Indexes	F_1	F_2	F_3	F_4	Indexes	F_1	F_2	F_3	F_4
X_1	-0.212	0.06	0.833	0.147	X_9	0.968	0.04	-0.153	0.076
X_2	0.148	-0.025	0.017	0.857	X_{10}	0.906	0.241	-0.281	-0.071
X_3	0.233	0.768	0.144	-0.008	X_{11}	0.18	0.872	-0.059	0.154
X_4	-0.217	0.283	0.689	-0.007	X_{12}	0.346	0.247	-0.63	0.458
X_5	0.158	0.572	0.396	0.559	X_{13}	-0.248	-0.72	-0.109	0.551
X_6	0.86	0.259	0.023	0.379	X_{14}	0.909	0.13	-0.301	0.005
X_7	0.925	0.139	-0.019	0.245	X_{15}	0.921	0.172	-0.277	-0.073
X_8	0.842	0.355	0.037	0.344	X_{16}	0.927	0.134	-0.285	-0.044

With principal component analysis, the factor score coefficient matrix of the city in the urban agglomeration in NSTM is got by the score coefficient matrix timing matrix after data standardization (the variable after standardization is 0, and the variance is 1). Meanwhile, urban comprehensive development level model of urban agglomeration in NSTM is built with the variance contribution rate as weight. Formula is as follows:

$$F=0.51416\,F_{1i}+0.16967\,F_{2i}+0.11138\,F_{3i}+0.06329\,F_{4i} \tag{3}$$

In this formula, Fi is the score of the four factors of the city i, F is the comprehensive development level. And the comprehensive development level of every city in the urban agglomeration in NSTM is calculated by the formula (3).

The score of some of the cities ranking last is negative. Data is conversed to measure the urban niche. With the minimum-maximum standardization method of Jiawei Han and Micheline Kamber, the original data is converted, maps to a new data region, and forms a new data line. The formula is as follows:

$$v'=\frac{v-A_{min}}{A_{max}-A_{min}}\left(A_{new\text{-}max}-A_{new\text{-}min}\right)+A_{new\text{-}min} \tag{4}$$

In this formula, v' refers to the data after standardization. v refers to the original data. A_{max} and A_{min} respectively refer to the maximum and minimum of the original data line. $A_{new\text{-}max}$ and $A_{new\text{-}min}$ respectively refer to the maximum and minimum of the new data line. In order to make the data converted positive and keep the data in the original order, the score of the factor is converted in the equivalent positive interval region with 0 as the minimum. Thus, urban factor score and comprehensive development level value of the urban agglomeration in NSTM can be seen (Table 3).

Table 3. Urban Comprehensive Development Level of Urban Agglomeration in NSTM from 2000 to 2010

Year	2000	2002	2004	2006	2008	2010
Urumqi	1.028	1.201	1.589	1.822	2.396	3.116
Karamay	0.609	0.63	1.006	1.239	1.203	1.253
Shihezi	0.6	0.642	0.745	0.919	0.982	1.079
Changji	0.015	0	0.045	0.341	0.368	0.458
Fukang	0.103	0.377	0.412	0.495	0.4	0.49
Hutubi	0.159	0.171	0.415	0.697	0.536	0.579
Manas	0.224	0.3	0.44	0.564	0.625	0.668
Kuitun	0.372	0.268	0.321	0.246	0.449	0.568
Wusu	0.076	0.236	0.29	0.417	0.383	0.512
Shawan	0.285	0.321	0.23	0.185	0.308	0.382

3.2 Urban Comprehensive Niche Breadth

The measurement index "state" and "tendency" in the special time refers to the comprehensive development level and average increment. The time is two years, and dimensional conversion coefficient is 0.5. According to formula (1) and the data of Tab.3, the urban comprehensive niche breadth of urban agglomeration is figured out (Table 4).

The comprehensive niche of urban agglomeration in NSTM shows several levels. Firstly, the comprehensive niche breadth of most cities is very small. Among these ten cities, that of only Urumqi, Karamy, and Shihezi is more than 0.1. Secondly, the niche breadth of the first city was larger, and urban comprehensive niche breadth polarized apparently. In 2010, the urban comprehensive niche breadth of Urumqi was 0.329, which was eight times of Shawan, the narrowest niche breadth city. The urban comprehensive niche breadth of the urban agglomeration in NSTM presented three urban economic zone, centering on Urumqi, Karamy, and Shihezi. Thirdly, the urban comprehensive niche breadth of cities in NSTM saw different changing trend. From 2000 to 2010, the urban comprehensive niche breadth of Urumqi, Kuitun, and Shawan showed a first decreasing and then increasing tendency, while that of Karamy, Fukang, Hutubi, Manas, and Wusu first increasing and then decreasing. Changji presented a sustainable growth, while Shihezi continuously decreasing.

Urban agglomeration in NSTM has the embryonic form[18]. Cities researched are built on the oasis, which are different on population, technology, and fund. As the only super-huge type city, Urumqi has the most centralization on region development factor. From 2000 to 2010, the niche breadth of economy and public service factor is 0.396, and equitable development and population growth factor is 0.119. Karamy is one of the two prefecture-level cities in Xinjiang, and is the important oil and petrochemical engineering base and the energy production base. During these eleven years, the niche breadth of equitable development and population growth factor, urban infrastructure factor, and urban spatial expansion factor of Karamy is respectively 0.195, 0.139, and 0.179. Shihezi is the city under dual administration of division and municipality of Xinjiang Production & Construction Corps, which can promote the development of regiments under its administration. During these eleven years, its niche breadth of economy and public service factor, equitable development and population growth factor, and urban infrastructure factor is respectively 0.122, 0.155, and 0.141. In the urban agglomeration in NSTM, the urban comprehensive niche breadth of Urumqi, Karamy, and Shihezi is wider than the other cities.

Table 4. Comprehensive Niche Breadth of Urban Agglomeration in NSTM from 2000 to 2010

Year	2000~2002	2002~2004	2004~2006	2006~2008	2008~2010	Mean Value
Urumqi	0.293	0.289	0.275	0.289	0.329	0.295
Karamay	0.163	0.17	0.181	0.168	0.147	0.165
Shihezi	0.163	0.144	0.134	0.13	0.123	0.139
Changji	0.002	0.005	0.031	0.049	0.05	0.027
Fukang	0.063	0.082	0.073	0.061	0.053	0.066
Hutubi	0.043	0.061	0.09	0.085	0.067	0.069
Manas	0.069	0.077	0.081	0.082	0.077	0.077
Kuitun	0.084	0.061	0.046	0.048	0.061	0.06
Wusu	0.041	0.055	0.057	0.055	0.053	0.052
Shawan	0.08	0.057	0.033	0.034	0.041	0.049

3.3 Urban Niche Breadth of Factors

The four factors score of urban economy and public service, equitable development and population growth, urban infrastructure, and urban spatial constraint, and the average increment of many years respectively refer to the measurement index of state and tendency. The time interval is two years. And the dimensional conversion coefficient is 0.5. According to the formula (1), the urban niche breadth of the factor of the urban agglomeration in NSTM can be calculated.

The niche breadth of economy and public service factor among cities in urban agglomeration in NSTM is quite different. In 2010, Urumqi is the broadest (0.463). And Kuitun is the narrowest (0.009), only 1/51 of Urumqi. Shihezi (0.102) takes the second position. And the other cities are between 0.046 and 0.086. Meanwhile, the factor niche breadth of Urumqi and Changji rises all the time. That of Shihezi, Fukang, Manas, and Kuitun declines obviously. And that of Hutubi and Wusu rises first, and then declines.

As to the niche of the equitable development and population growth factor, the difference among cities in urban agglomeration in NSTM is not so significant. The range is only 0.222, much narrower than the former 0.454. In 2010, that of Kuitun is the widest, which is 0.223. And that of Wusu is narrowest, only 0.01, which is just 1/22 of that of Kuitun. Karamy (0.187) and Shihezi respectively take the second and third position. During the eleven years, the niche breadth of this factor of these cities, such as Kuitun, Shihezi, Fukang, and Hutubi, increased steadily, while that of Urumqi and Wusu decreased. And others had the fluctuation change law.

As to the niche of urban infrastructure factor, there is hardly difference among them. In 2010, Manas was the widest (0.145), then was Shihezi (0.135), Karamy (0.121), Hutubi (0.120), and Wusu (0.108). As to this factor, the niche breadth of half of the cities was more than 0.1. And Changji was the narrowest (0.054). During the eleven years, the niche breadth of Urumqi and Changji was becoming broader, while Karamy, Shihezi, Kuitun, and Shawan declined.

As to the niche of the urban spatial expansion factor, there is difference among cities in the urban agglomeration in NSTM. In 2020, there are 5 cities where the niche breadth of spatial expansion factor was more than 0.1. Karamy was the widest (0.202). Then was Wusu (0.148), Changji (0.122), Urumqi (0.116), and Manas (0.103). Shihezi was the narrowest (0.021). From 2000 to 2010, Karamy and Urumqi saw an obvious increase in the niche breadth of the urban spatial expansion factor. And the urban expansion space was compressed, which was directly related to the per capita construction land area index. The per capita construction land area increased by 20 m^2 in eleven years. There was no evident increase in Shawan and Fukang, while that of Shihezi, Hutubi, and Kuitun became smaller.

3.4 Urban Niche Differentiation Index

According to formula (2) and the number of the niche breadth of the above cities, the urban niche differentiation index of the urban agglomeration in NSTM can be calculated, which is used to compare the difference.

From 2000 to 2010, it is clear that to the comprehensive differentiation index (C) of urban niche breadth, it shows a first descending and then ascending tendency. From 2006 to 2010, the index is ascending. Since the Development of the West Region, Xinjiang has taken the balanced development strategy, and the difference on development level became less. Since 2006, it has taken the unbalanced development strategy, focusing on the economic belt in NSTM, which led to the agglomeration of development factor to the core cities. And the comprehensive niche differentiation index has increasing.

The comprehensive niche differentiation index increases with the increase of niche differentiation index of economy and public service factor (C1), equitable development and population growth factor (C2), and urban spatial expansion factor (C4), but adverse to urban infrastructure factor (C3). And in these four differentiation indexes, the economy and public service factor is the largest with the most increase. It shows that more technology, investment, and public service factors are collected to urban agglomeration in NSTM. The development level among these cities become unbalanced and the above factors begin to flow to larger cities or the central city. Urban infrastructure factor is the smallest, and decreases. With more investment of infrastructure construction, the difference of infrastructure and public service among cities becomes less. With the improvement of living condition and infrastructure, the urban niche of middle-sized and small cities will stop being compressed.

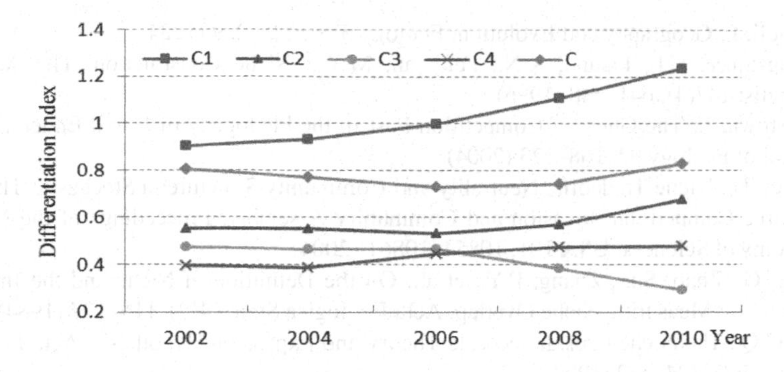

Fig. 1. Niche Differentiation Index of Urban Agglomeration in NSTM from 2000 to 2010

4 Conclusions

The urban agglomeration in NSTM is the typical oasis urban agglomeration. The development of it is restricted by the land and water resources. With the principle of specialization and diversity, it should take "dot-axis" type on spatial development to form the characteristic city clusters, where different cities in each cluster complement each other's advantages.

The urban niche theory is still at the exploration stage. It needs further research especially on quantitative calculation of niche breadth. According to the theory of multidimensional niche, the measurement index system of urban niche is constructed with the urban agglomeration in NSTM as the object of research. By the Principal Component Analysis, the weight of every index is calculated. The four main factors, such as economy and public service, equitable development and population growth, infrastructure status, and space, are got. Thus, the comprehensive niche and factor niche is measured. With the quantitative analysis of development trend of urban niche in NSTM, spatial difference and feature is objectively reflected. In this research, urban comprehensive niche is calculated by the index weight rather than index mean value. The development of urban multidimensional niche theory can provide new angle for research theory and method of urban niche. It can promote the coordinated development of the urban agglomeration, advance urban function, promote the ability of agglomeration, and boost regional social and economic development.

Acknowledgments. This study was financially supported by National Natural Science Foundation of China (No.41101159, No.41101157), and Western Doctor of West Light Foundation of The Chinese Academy of Sciences (No. XBBS201009).

References

1. Ding, S.Y., Li, Z.H.: Niche Pattern Dynamics of Different Functional Modules in Kaifeng City. Acta Geographica Sinica 61(7), 752–762 (2006)
2. Grinnell, J.: The Niche Relationships of the California Thrasher. The Auk 34(4), 427–433 (1917)

3. Grinnell, J.: Geography and Evolution. Ecology 5(3), 225–229 (1924)
4. Odling-Smee, F.J., Laland, K.N., Feldman, M.W.: Niche Construction. The American Naturalist 147(4), 641–648 (1996)
5. Silvertown, J.: The Ghost of Competition Past in the Phylogeny of Island Endemic Pants. Journal of Ecology 92, 168–173 (2004)
6. Tilman, D.: Niche Tradeoffs, Neutrality and Community Structure: a Stochastic Theory of Resource Competition, Invasion and Community Assembly. Proceedings of the National Academy of Sciences, USA 101, 10854–10861 (2004)
7. Wang, G., Zhao, S.L., Zhang, P.Y., et al.: On the Definition of Niche and the Improved Formula for Measuring Niche Overlap. Acta Ecologica Sinica 4(2), 119–127 (1984)
8. Zhu, C.Q.: The Niche Ecostate-ecorole Theory and Expansion Hypothesis. Acta Ecologica Sinica 17(3), 324–332 (1997)
9. Chen, S.Y., Zhang, H.G., Lin, J.P., et al.: City Community Ecology:Ecological Interpretation of Urban Agglomerations. Economic Geography 25(6), 810–813 (2005)
10. Chen, S.Y., Lin, J.P., Yang, L.J., et al.: Study on Urban Competition Strategy Based on the Theory of Niche. Human Geography 21(2), 72–76 (2006)
11. Chen, L., Wang, R.S., Wang, Z.L.: Niche Assessment of China Provincial Social-Economic-Natural Complex Ecosystems in 2003. Chinese Journal of Applied Ecology 08, 1794–1800 (2007)
12. Xiao, Y., Mao, X.Q.: Urban Niche Theory and its Application. China Population, Resources and Environment 18(5), 41–45 (2008)
13. Darchen, S., Tremblay, D.G.: The Development of a High Technology Node in the Inner Suburbs: from a Functionalist to a Niche Approach to Urban Planning. The Case of Saint-Laurent in the Montreal Metropolitan area. Canadian Geographer-Geographe Canadien 53(2), 222–238 (2009)
14. Keirstead, J., Leach, M.: Bridging the Gaps between Theory and Practice: a Service Niche Approach to Urban Sustainability Indicators. Sustainable Development 16(5), 329–340 (2008)
15. Cre, I., Rupprecht, S., Buhrmann, S.: The Development of Local Implementation Scenarios for Innovative Urban Transport Concepts: the Niches plus Approach. In: Papaioannou, P. (ed.) Transport Search Arena 2012, pp. 1324–1335. Elsevier Science BV, Amsterdam (2012)
16. Duan, Z.L., Zhang, X.L., Lei, J.: The Ecologic Research Direction of Urban Agglomerations: Urban Coenology. Environmental Protection of Xinjiang 04, 1–6, 13 (2011)
17. Zhang, Y.F., Yang, D.G., Zhang, X.L., et al.: Research on Regional Structure and Spatial Fractal Characteristics of Urban Agglomerations at Northern Slope of Tianshan Mountains. Journal of Desert Research 28(4), 795–804 (2008)
18. Fang, C.L.: New Structure and New Trend of Formation and Development of Urban Agglomerations in China. Scientia Geographica Sinica 31(9), 1025–1034 (2011)

Localized Spatial Association: A Case Study for Understanding Vegetation Successions in a Typical Grassland Ecosystem

Zongyao Sha[1], Xicheng Tan[1], and Yongfei Bai[2]

[1] Department of Spatial Information and Digital Engineering, Wuhan University, Wuhan, China
{zongyaosha,xctang}@wuhan.edu.cn
[2] Laboratory of Quantitative Vegetation Ecology, Institute of Botany, Chinese Academy of Sciences, Beijing 100093, P.R. China
{yfbai}@ibcas.ac.cn

Abstract. Spatial heterogeneity exists widely in geographical space. We here suggest that spatial association mining should consider spatial variations when designing knowledge-discovering models and applying the models in geographical studies. A Quadtree-based framework was proposed to mine localized spatial associations. Unlike many other approaches, the novelty of this Quadtree-based frame is its suitability in finding strong spatial association rules that are valid in smaller areal patches or "hot spot" area rather than the whole region, and therefore provides domain experts an insight to explore further the associations among different geographical phenomena. The principle underlying the Quadtree-based algorithm is that it employs a Quadtree data structure to explore the multi-level nodes (each representing a patch at that level), recursively. This recursive process is used to check whether the explored nodes satisfy predefined criteria (including support and confidence threshold for association rules, and event density and minimum area for each node), denoting a strong association. Practical application in an ecology study proved that, compared to traditional global association mining, the proposed model and algorithm for mining localized association rules are more meaningful under spatially heterogeneous environment.

Keywords: spatial association, data mining, algorithm, quadtree.

1 Introduction

Data mining from spatial dataset is a hot and interesting topic in the field of geosciences. Due to the explosive volume of spatial data, valuable patterns hidden in large geospatial dataset need to be dug out. Among the patterns, spatial association rules, which usually explain the co-occurrence of certain events or spatial phenomena, can be taken to predict outcome through the "condition and result" sequence patterns. A typical case of association rules is that vegetation usually grows well in regions with sufficient water availability as well as abundant sun-shine. Such an association

F. Bian and Y. Xie (Eds.): GRMSE 2014, CCIS 482, pp. 33–45, 2015.

rule can be expressed as "*sufficient water-supply∩abundant sun-shine ->healthy vegetation*". In this expression, "*sufficient water supply*" and "*abundant sun-shine*" are the condition and "*healthy vegetation*" is the result, while the symbol "->" represents the relationship between the condition the result. By taking use of such rules, it is possible to predict the vegetation growing status based on condition of the natural conditions of any region.

Numerous algorithms have been proposed to extract association rules from spatial dataset (referred to as spatial association rules). The usefulness of those association rules has been proved. However, there is scarce literature dealing with the localization problem of spatial association rules. Such a problem shares similar background as the widely applied geographically weighted regression (GWR) which is used to examine localized effect of one or a set of variables on a particular topic, also known as the dependent variable. However, GWR is only effective in exploring quantitative relationships of numerical variables. For revealing the association relationships (e.g., relationships expressed similar as the above association rules) among any categorical variables, new models must be developed.

2 Problem Definition

An association rule takes the form of X=>Y and thus a set of association rules can be well applied for classification issues with possible X as the antecedent and Y as the consequent which is usually correspondent to a class label. Classification can be defined as the process of determining the labels of spatial objects, given a set of candidate labels. Without loss of generality, suppose an object in a spatial space can be labeled as c_i (i=1, 2, …, n, where n is the number of class labels), we can always label the object a binary value either as c_i or not c_i. Therefore, for the task of labeling an object to one class from the possible n labels, the work can be simplified as assigning the object to a binary label (either 1 indicating belonging to c_i, or 0 indicating the opposite), recursively, through (n-1) times. For simplicity, our work in this study supposes that classification problem is a process of labeling a spatial object to a binary class.

Global association rules are referred to as the rules that hold for the complete dataset or region. Common methods for mining global association rule mining try to generate association rules pertaining to the whole region. Various phenomena tell us that such methods may not be useful or valid to spatially varied space, since associations valid in one area are probably invalid in another. Thus, localized association rules, valid only for a particular region or a group of regions, are more reasonable and useful in helping us understand the nature of the relations between geographical phenomena. To the best of our knowledge, there is very limited literature dealing with mining localized association rules. The main difficulty lies in the lack of effective models to quantify such rules. For traditional association rule mining, at least two parameters, minimum support (*Sup*) and minimum confidence (*Conf*) are needed to indicate the strength of an association rule. However, when the fact of localization is considered, the problem becomes much difficult. For example, a large study area can be divided into smaller regions and the association mining process can then be performed on each small region. However, such a method relies heavily on the practice of area division and thus may be criticized for being subjective. Furthermore, due to the small area of the regions derived from the division of the whole area, randomness can be a critical drawback for the discovered associations, because the number of transactions (spatial features) contained in a sub-region may be too few. Therefore, when localization is considered, a third parameter, minimum patch area that an association rule holds, should be introduced to describe such rules. The objective of localized association rule mining extracts strong association rules not from the whole study area but from the sub-regions. Global association rule mining can be viewed as an average result from the approach dealing with mining localized rules and thus global association rules show general characteristics of the whole area. Therefore, we consider a balanced strategy of mining association rules that have little loss of generality and are able to reveal spatial variation as the same time.

Given a complete (usually large in area) region *R*, localized association rules are only valid in a patch or a set of discrete patches (denoted as r_i and i=1, 2, …, k, where k is the number of patches and $R >= \cup r_i$) in *R* that each patch is contiguous. Any localized association rule must be valid at least in one patch. An association rule is valid in patch r_i only that the transactions spatially located in r_i satisfy the thresholds of both support (*Sup*) and confidence (*Conf*).

Event Density

For a given area *R*, the event density is defined as the frequency of the event happened in that area. An event is the geographical phenomenon or variable such as vegetation succession or soil water availability. Global Event Density (GED) for a whole region and Local Event Density (LED) for any areal patch in that region, are defined. Suppose the overall area of region *R* has an area of A_R, and a subset patch *P* within *R* has an area of A_p, then GED=T_R / A_R, and LED= T_P / A_p, where T_R and T_P are the occurring times of an interested event in the specified area *R* and *P*, respectively. It holds that GED/LED= A_R / A_p.

Example 1

Given vegetation succession on grassland as an example, if *R* covers an area of 100 hectares (R_o) and 1000, 000 samples (T_R) from the area are collected by a regular grid sampling method. Out of the collected samples, 10, 000 showed vegetation succession, therefore, GED = T_R/ A_R = 1000, 000/100 = 10,000. Suppose further that *R* is divided into 100 patches and, on average, each patch will have LED = T_P / A_p = (1000, 000/100) / (100/100) =10,000, where both T_P and A_p are divided by the number of patches (i.e., 100).

Similar to the support threshold defined in transactional association mining to prune frequent itemsets (event sets in geographical space), we define the Support of Event Density (Sup_{ED}) to reflect the occurring frequency of an event. The support of event density reveals the frequency of the event that both antecedent and consequent occur within a given region *R*, and is used as an indicator for the support threshold of association rules. It is assumed that only the frequency of both the antecedent event and consequent events reaches a predefined level, can the discovered patterns from the dataset be regarded as reliable. Only the association rules whose Sup_{ED} is greater than a minimum threshold Min-Sup_{ED} (that is, $Sup_{ED} \geq$Min-Sup_{ED}), can they be further processed as potential strong association rules. We illustrate the study of the vegetation succession as a consequent event and the grazing intensity (labeled as intensive grazing and non-intensive grazing) as an antecedent. Suppose out of all the 1000, 000 samples, the event that both intensive grazing and vegetation succession occurs accounts for 50,000. The support value of the association: X=>Y (indicating vegetation succession is caused by intensive grazing) for the global region *R* is 0.05 (that is, Sup_{ED-R} = 50,000 /1000, 000). Only Sup_{ED} is over a predefined level, it may then be meaningful to analyze this pattern. The support of the event density also applies to any patch *P* (denoted as Sup_{ED-P}) from region *R*. For *R* that is divided into *k* patches, the global support of event density Sup_{ED-R} equals to the area-weighted average of local support of event density Sup_{ED-P}.

The association rules selected by the constraint of minimum support may not be strong and need to be further verified by a confidence test. Suppose that samples in *R* given by Example 1 all show intensive grazing, then the probability of vegetation succession out of those experienced intensive grazing would be very low (that is, 10, 000/1000, 000). Though the pattern that "intensive grazing leads to vegetation succession" satisfies the minimum support constraint, it is not strong due to its low confidence value. We define the Confidence of Event Density ($Conf_{ED}$) to explore localized association rules from spatial dataset. $Conf_{ED-R}$ (R stands for region) is the ratio of consequent event density to the antecedent event density. As illustrated above, since the event density with vegetation succession is 10, 000/100 and suppose the event density with intensive grazing is 1000, 000/100, then the $Conf_{ED-R}$ would be 0.01 for region *R*. Similarly to the support of event density, the confidence of event density also applies to any patch *P* (denoted as $Conf_{ED-P}$). For the region *R* that is divided into *k* patches, the global confidence of event density $Conf_{ED-R}$ equals to the area-weighted averaged confidence of event density $Conf_{ED-P}$ of the patches.

3 A Quadtree-Based Framework

We propose here a quadtree-based frame to mine localized association rules. Unlike many other association rule mining approaches, the novelty of the quadtree-based frame is its capability of finding strong association rules that are only valid in small patches rather than the whole area, and therefore provides domain experts insight to explore interested "hot spot". The key of the framework is a quadtree-based algorithm that employs quadtree data structure to facilitate the computing process. The principle underlying this framework is that the framework takes use of a multi-level quadtree structure that splits the whole area into four child patches (represented by pixels), recursively, until a set of stop conditions are satisfied.

3.1 Indicators for the Framework

This proposed approach starts from a single-event association rule recursively until all events are included. Single-event association rule discovery is defined as a type of association rules that has a single antecedent event and a consequent event labeled by a class name. Given an antecedent *X* with *m* events in a spatial space, at most *m* single-event association rules can be extracted if they satisfy all the conditions discussed below.

Localized association rules are only valid for patches that are large in area. It is necessary to put a minimum area threshold as a baseline to make sure that the association rules are stable and meaningful. The localized association rules also need to meet both frequency and confidence levels. Suppose within a patch that only very few cases supporting an association rule (resulting in low support) and that a great part of those cases satisfying the antecedent do not lead to the consequent class label (meaning low confidence), it would be untrue to claim that the rule holds. Practically, the support and confidence levels are determined subjectively and different practitioners may have different preferences. As a result, one may often find it is hard to set the threshold values for support and confidence. We use the global $Sup_{ED\text{-}R}$ for support threshold and the global $Conf_{ED\text{-}R}$ for confidence threshold. Our method is based on the concept of event density discussed previously. The reason for this selection is twofold. 1) For a given spatial dataset which contains a set of spatial objects with a binary class label and a boundary region R, we calculate the global $Sup_{ED\text{-}R}$ and $Conf_{ED\text{-}R}$. We understand that, for the region R that is divided into *k* patches P_1, P_2, …, P_k ($k \geq 2$), there exists at least one P_i ($1 \leq i \leq k$) satisfying $Sup_{ED\text{-}R} \leq Sup_{ED\text{-}Pi}$, and least one P_j ($1 \leq j \leq k$) satisfying $Conf_{ED\text{-}R} \leq Conf_{ED\text{-}Pj}$. In other words, there is great chance that, potentially, patches meeting the support and confidence constraint can be extracted if the global $Sup_{ED\text{-}R}$ and $Conf_{ED\text{-}R}$ are adopted as the support and confidence thresholds. 2) Another consideration of selecting the global $Sup_{ED\text{-}R}$ and $Conf_{ED\text{-}R}$ is that, different from global association rule discovery, localized association rule discovery is to find spatially varied patterns.

As discussed before, localized association rules valid for spatial patches depend on the patch area in terms of their reliability. Thus, area is an important factor for controlling the computing process. This is an important step for controlling the splitting process based on the quadtree structure. The split process of the quadtree continues until the pixel area reaches or less than the minimum area, leaving leaf pixels at the bottom. Any leaf pixel, if unable to be merged to its adjacent units (at the same level or levels up) in the quadtree, will be removed during the merging process.

The framework also takes support and confidence levels as filter factors to control the process. For any level that the pixel area satisfies the minimum area, support and confidence threshold will be evaluated for the nodes (representing a patch) instead of the whole region. For pixels that both support and confidence satisfy the requirement, they are marked as valid leaf and the splitting process stops. If the current node satisfies the following condition, the node will be marked as invalid,

$$Sup_{ED\text{-}N} < Sup_{ED\text{-}R} \times MinA/Area_N \text{ or } Conf_{ED\text{-}N} < Conf_{ED\text{-}R} \times MinA/Area_N,$$

where $Area_N$ is the area of current node *N*, because it can be proved that no child leaves can satisfy support and confidence threshold and minimum area requirements at the same time even after further division at lower levels. Otherwise, the pixel will be marked as unknown node which will be undergone further splitting at lower levels. This splitting will continue for all the unknown nodes until they do not meet a minimum area threshold and then be either marked as valid or invalid based on its support and confidence value.

3.2 Structure of a Quadtree

A quadtree consists of a root, a set of leaves and a set of middle nodes. The root stands for the whole region R and the leaves are localized regions that an association rule is pertained to, while a middle node is the element that kinks the leaves and the root. For a built quadtree, all leaves must satisfy the minimum area requirement at each splitting step.

Definition: Valid/Invalid Leaf

A valid leaf in a quadtree is a leaf that is labeled as valid if its support and confidence satisfy the corresponding minimum threshold.

A valid leaf may turn into invalid if the standard of the minimum threshold increases. All leaves in a quadtree are labeled either as valid or invalid.

Definition: Valid/Invalid Quadtree

A valid quadtree is the quadtree that has at least one valid leaf.
Similarly, a valid quadtree may become invalid with the increase of minimum support and confidence levels because with such an increase, all valid leaves may become invalid.

In our study, we adopt the global support ($Sup_{ED\text{-}R}$) and global confidence ($Conf_{ED\text{-}R}$) as the minimum threshold values to label nodes. Potentially, this practice can produce a valid quadtree since both the global support and the global confidence are average indexes for the whole region. Due to the possibility of non-intersection of the support and confidence requirement, an invalid quadtree may also be produced. This can be explained by Example 2.

Example 2

The case of vegetation succession on grassland is illustrated. The global region R has 1000, 000 samples (T_R), attributed by grazing intensity as the antecedent event and a binary class label succession/non-succession as the consequent. We suppose 200, 000 show intensive grazing and among the intensive grazing samples, 120, 000 are labeled as vegetation succession; therefore 120,000 show both intensive grazing and vegetation succession. The global support level and confidence level will be: $Sup_{ED\text{-}R}$ =0.12 (120,000/1000, 000) and $Conf_{ED\text{-}R}$ = 0.60 (120, 000/200, 000), respectively. In the first round of region splitting, R is divided into 4 leaves (by quad-division): A, B, C and D. The distribution of samples for the leaves in the form of (total samples; intensive grazing samples; vegetation succession samples from intensive grazing ones; intensive grazing and vegetation succession samples) is listed below for each split patch, A, B, C and D,

A (300,000; 40,000; 25,000; 25,000), B (350,000; 70,000; 35,000; 35,000)
C (200,000; 60,000; 35,000; 35,000), D (150,000; 30,000; 25,000; 25,000)

We have

$Sup_{ED\text{-}A}$ =0.08 and $Conf_{ED\text{-}A}$ = 0.63, $Sup_{ED\text{-}B}$ =0.10 and $Conf_{ED\text{-}B}$ = 0.50
$Sup_{ED\text{-}C}$ =0.18 and $Conf_{ED\text{-}C}$ = 0.58, $Sup_{ED\text{-}D}$ =0.17 and $Conf_{ED\text{-}D}$ = 0.83
based on

$Sup_{ED\text{-}P}$ = (intensive grazing and vegetation succession samples)/(total samples), and $Conf_{ED\text{-}p}$=(vegetation succession samples in intensive grazing ones)/(intensive grazing samples)

Since node C and D satisfy both minimum support and minimum confidence while node A and B do not meet the requirement, the current round produces a valid quadtree with C and D showing localized association rule "intensive grazing->vegetation succession".
For a large region with total area A_t, the splitting process will produce a quadtree with depth *h* that satisfies $A_{pixel} \cdot 2^{2 \cdot h} <= A_t$, where A_{pixel} is the area of a single pixel at the bottom level of the quadtree and h is the depth of the quadtree (Algorithm 1).

Algorithm 1: quadtree-based algorithm for t antecedent events

Input:
spatial objects in Region R attributed by t antecedent events and a binary class label as result
initial splitting height h_{start}; ending splitting height h_{end}; minimum area height h_{min_a}
Output: *t* quadtrees
Steps:
Extract the boundary (BND) of the dataset
calculate Minimum Enclosing Rectangle (MER) of BND
// MER is the tree root which can be viewed as the largest pixel
List lst, tmp_lst
//define two stacks to hold father nodes P and children nodes P_i from P
push in lst the nodes from the quad-divisions of MER
for each event in antecedent
  for each node P in lst
    if P is the first node then empty tmp_lst
    calculate A_P// A_P is the area located within P and BND
    if $Sup_{ED\text{-}P} \geqslant Sup_{ED\text{-}R}$ then
        label node P as valid
        go to next for
    else if ($Sup_{ED\text{-}P} \cdot A_P$ / Min_A < $Sup_{ED\text{-}R}$) then
        label node P as invalid
        go to next for
    quad division of P
    for each node P_i ($1 \leqslant i \leqslant 4$) in P
      push P_i in tmp_lst
    if A_P < Min_{leaf} exit for
    end for
  if P is the last node then lst= tmp_lst
  end for
end for

Multiple-event association rule discovery can be realized by a merging process from multiple single-event quadtrees. A multiple-event association rule is defined as a type of association rules that has more than one antecedent event but with the same consequent event labeled by a class name. Given *m* antecedent events, recursive merging processes are required to produce a multi-event quadtree (Algorithm 2).

Algorithm 2: **Mining association rules with multi-event antecedent by quadtree overlay**

Input: two quadtrees TreeA an TreeB, with depth $h_{TreeA} \geqslant h_{TreeB}$
Output: a quadtree showing association between multi-event antecedent and class label
Steps:
i=1
For each node at level h_{TreeA} -i
while the current level at TreeA is not root level
i=i+1
end while

The output quadtree from algorithm 2 is matched nodes labeled as valid, if both of their father quadtree (TreeA and TreeB) is valid, and therefore the number of the valid leaves is less than any of TreeA and TreeB.

It can be proved that if any of the two quadtree is invalid, the resultant quadtree will be invalid. This feature prevents the need to create association rules with higher antecedent events X if any sub-event contained in X corresponds to an invalid quadtree.

The valide nodes or leaves in the quadtree from both algorithm 1 and 2 may represent small patch (e.g., the leaves at the bottom layer of the quadtree), it is desirable to look for all those leaves that are spatially adjacent (sharing a common edge). Those adjacent nodes, no matter whether they are at the same level, can be merged to make a larger region. The combined region from multiple nodes satisfies the maximized area requirement and is better for any single node showing that any association rules hold locally. Since the combined region comes from at least two adjacent leaves from the quadtree, the support and confidence may differ from any of original nodes. To calculate the support and confidence value for each connected region, we adopt a simple way by area-weighted average of the leaves that consist of the region,

$$Var_N=(area_1 x Var_{Leaf1}+ area_2 x Var_{Leaf2}+\ldots)/(area_1+area_2+\ldots)$$

where Var_X indicates either support or confidence value of region/leaf X.

Let O be the set of a combined region from *n* nodes and each node in O (O_i) the *s* adjacent leaves that consist of O_i, we have $O=\{O_1, O_2, \ldots, O_n\}$ and $O_i=\{L_1, L_2, \ldots, L_s\}$

where all L_j are adjacent leaves from the quadtree.

For any candidate rule to be mined, we use the following steps to merge its valid region based on the quad-tree.

The merging process adopts a bottom-top and area-preceding strategy. Specifically, the merging process starts from the bottom level and grows gradually to the top level to produce a larger area for the interested rule. For any leaf node in the tree, the leaf first finds its adjacent and valid leaves (sharing an edge), if there is any, to merge. Then a single leaf (if no adjacent leaves could be found) or the merged leaves (if merging process happened) will look for a higher level valid leaves (i.e., leaves at the father level) with shared edges to join in. This looping process will continue until all nodes are either merged to form a larger area or not being merged because no adjacent nodes available (Algorithm 3).

Algorithm 3: Calculation of support and confidence of merged nodes by **Adjacent node search**

Input: quadtree representing a localized association rule

Output: set of connected regions O with each element attributed by Sup_{ED-Oi} and $Conf_{ED-Oi}$

Steps:

i=1

For each node at level h_{TreeA} -i

while the current level at TreeA is not root level

i=i+1

end while

4 Application

We applied the localized association rule mining approach to identify the underlying factors that cause vegetation succession in the typical grassland vegetation area in Inner Mongolia, China. The study region is covered by 11 vegetation communities (indicated by climax species) (Sha et al., 2008). The two main vegetation communities, i.e., Stipa grandis (SG) and Leymus chinensis (LC), are widely distributed in the research area. Studies revealed that SG and LC have been partially replaced by other vegetation communities in recent two decades, most of which are Cleistogenes squarrosa (CS) and Artemisia frigida (AF). For example, in the Xier Plain which is located at middle to upper reach of Xilin river, AF and CS had almost dominated the area where the primary climax vegetation is supposed to be SG and LC, indicating an obvious vegetation succession process occurred (Tong et al., 2004). Those successions are regarded as a signal of grassland degradation. However, the reasons underlying the successions and the spatial variation of the observed successions need to be further investigated. In this work, the underlying successions were analyzed by localized spatial association using the vegetation successions as the consequent event and the possible explaining variables as antecedent events extracted from natural and socio-economic dataset. The spatial distributions of the two types of vegetation successions, i.e., from SG to CS/AF and from LC to CS/AF, for the period (1998~2004) were defined, mapping two consequent events (variables), $SUCC_{SG}$ and $SUCC_{LC}$, respectively. Ten antecedent events (variables), listed in Table 1, were selected as predictors for explaining the occurrence of each vegetation succession ($SUCC_{SG}$ and $SUCC_{LC}$). The statistics of each variable is given in Table 2.

Table 1. Included variable in the analysis

Variable	Description
NDVI98	Normalized difference vegetation index in 1998
DVSG_98	Density of vegetation LC in 1998
DVLC_98	Density of vegetation SG in 1998
DS (km)	Distance to village settlement center
DR98 (km)	Density of road in 1998
DW (km)	Distance to water (river) body
SLP (degree)	Slope in degree
ORI (degree)	Orientation from North in degree
ALT (m)	Altitude in meters
AGI98-04 (sheep/km2)	Average grazing intensity during 1998~2004

Table 2. Statistical information for the variables

Variable abbr.	Minimum	Maximum	Mean	Std. dev
NDVI98	-0.21	0.66	0.16	0.06
DVSG_98	0.00	90.00	66.01	20.62
DVLC_98	0.00	80.00	5.84	11.14
DS (km)	0.00	8.10	3.53	0.59
DR98 (km)	0.00	2.05	1.17	1.18
DW (km)	0.00	2.55	0.87	0.73
SLP (degree)	0.00	66.00	4.79	6.30
ORI (degree)	0.00	180.00	76.46	59.00
ALT (m)	902.00	1608.00	1160.81	73.91
AGI98-04 (sheep/km^2)	79.28	170.12	120.69	18.88

Table 3. Discretization of variables

Variable	**Classes (classification threshould)**
$NDVI_{98}$	$NDVI_{high}$ (0.35~0.66), $NDVI_{mid}$ (0.02~0.35), $NDVI_{low}$ (-0.21~0.02)
DV_{SG_98}	DV_{SG_high} (39~90), DV_{SG_low} (0~39)
DV_{LC_98}	DV_{LC_high} (31~80), DV_{SG_high} (0~31)
DS (km)	DS_1(0~2.3), DS_2(2.3~4.5), DS_3(4.5~6.2), DS_4(6.2~8.1)
DR_{98} (km)	DR_{high}(1.7~2.05), DR_{low}(0~1.7)
DW (km)	DW_{high}(1.4~2.55), DW_{low}(0~1.4)
SLP (degree)	SLP_{high}(25~66), SLP_{mid}(9~25), SLP_{low}(0~9)
ORI (degree)	ORI_{high}(120~180), ORI_{mid}(55~120), ORI_{low}(0~55)
ALT (m)	ALT_{high} (1211~1608), ALT_{high} (902~1211)
AGI_{98-04}	AGI_{high}(119.3~170.12), AGI_{mid}(91.2~119.3), AGI_{low}(79.3~91.2)

All the antecedent variables are continuous type, which does not satisfy our proposed model and algorithm. Therefore, those variables were processed to be multi-level classified labels based on the statistical characteristics for each variable. For example, the NDVI value, ranging from -0.21 to 0.66, was converted to three levels as high NDVI ($NDVI_{high}$ with NDVI valued between 0.35~0.66), middle NDVI ($NDVI_{mid}$ with NDVI valued between 0.02~0.35) and low NDVI ($NDVI_{low}$ with NDVI valued between -0.21~0.02). All the other variables were converted in a similar way (table 2).

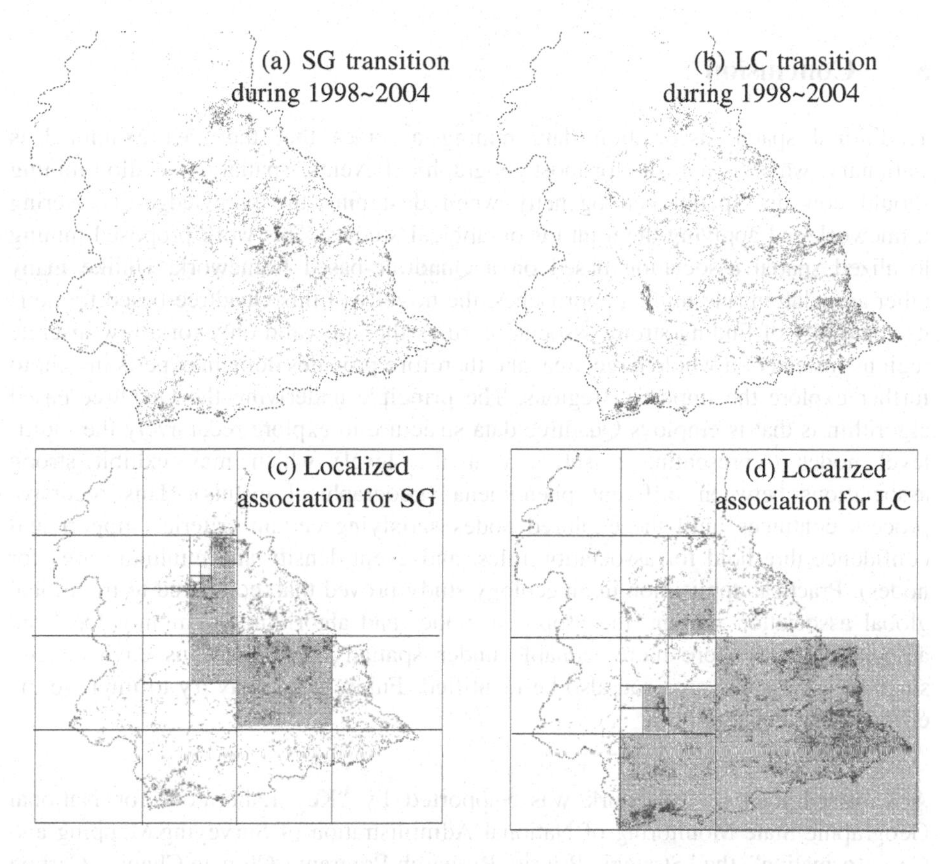

Fig. 1. Spatial distribution of vegetation transition of SG and LC
Red: occurrence of vegetation transitions; white: non-occurrence of vegetation transitions

The prepared data was undergone a systematic analysis using the proposed localized association rule mining and a Quadtree-based framework. A couple of localized association rules relating the explain factors and the occurrence of vegetation succession ($SUCC_{SG}$ and $SUCC_{LC}$) could be discovered. For example, the rules AGI_{high} => $SUCC_{SG}$ and AGI_{high} => $SUCC_{LC}$, which mean that higher grazing intensity could lead to SG and LC successions, were identified and the rules are not globally valid but locally (termed as "hot spot" area, as shown in grey area, Fig. 1). Those "hot spot" area should be key focus when we do research in vegetation successions in the context of grassland grazing. Compared to the traditional global association rule mining, such results may be more meaningful since they reflect the spatial variations. Such localized association rules, combined with illustration of their valid region, serve better understanding for the trajectory of vegetation successions compared to previous study which did not taken spatial variations into account (Xie et al., 2012).

5 Conclusion

Traditional spatial association data mining assumes that the area examined is stationary, which is not true for most geographical events. Spatial association mining should consider spatial heterogeneity when designing the knowledge-discovering framework and applying them into geographical studies. This work proposed mining localized spatial association based on a Quadtree-based framework. Unlike many other association rule mining approaches, the novelty of this Quadtree-based frame is its capability of finding strong association rules that are valid only for a few discrete regions instead of a whole large area, and therefore provides domain experts insight to further explore the important regions. The principle underlying the Quadtree-based algorithm is that is employs Quadtree data structure to explore recursively the multi-level nodes (representing a sub-area at the level) which may exhibit strong associations between different phenomena (geographical events). This recursive process continues until the explored nodes satisfying certain criteria (support and confidence threshold for association rules, and event density and minimum area for nodes). Practical application in an ecology study proved that, compared to traditional global association mining, the proposed model and algorithm for mining localized association rules were more valuable under spatially heterogeneous environment, since the "hot spot" area can also be identified. Future work may try to improve the efficiency of the algorithm.

Acknowledgments. This work was Supported by "Key Laboratory for National Geographic State Monitoring of National Administration of Surveying,Mapping and Geoinformation", the "Strategic Priority Research Program - Climate Change: Carbon Budget and Relevant Issues" of the Chinese Academy of Sciences (No. XDA05050402) and the Natural Science Foundation of China (Grant Nos. 41071249 and 41371371).

References

1. Li, Y., Xue, Y., He, X., Guang, J.: High-resolution aerosol remote sensing retrieval over urban areas by synergetic use of HJ-1 CCD and MODIS data. Atmospheric Environ 46, 173–180 (2012)
2. Xie, Y., Sha, Z.: Quantitative Analysis of Driving Factors of Grassland Degradation: A Case Study in Xilin River Basin, Inner Mongolia. The Scientific World J. 2012 Article ID 169724, 14 pages (2012), doi:10.1100/2012/169724

3. Agrawal, R., Imielinski, T., Swami, A.: Database mining: A performance perspective. IEEE Transactions on Knowledge and Data Engineering 5, 914–925 (1993)
4. Anthony, J.T., Hong, R.W., Ko, W., Tsao, W., Lin, H.: Mining spatial association rules in image databases. Information Sciences 177, 1593–1608 (2007)
5. Lu, W., Han, J., Oci, B.C.: Discovery of general knowledge in large spatial databases. In: Far East Workshop on Geographic Information Systems, Singapore (1993)
6. Sha, Z., Bian, F.: Mining Knowledge from Result Comparison Between Spatial Clustering Themes. Geo-spatial Information 8, 57–64 (2005)
7. Wang, C.Y., Tseng, S.S., Hong, T.P.: Flexible online association rule mining based on multidimensional pattern relations. Information Sciences 176, 1752–1780 (2006)
8. Yee, L., Ma, J.H., Zhang, W.X.: New method for mining regression classes in large data sets. IEEE Transactions on Pattern Analysis and Machine Intelligence 23, 5–21 (2001)
9. Xiaolei, L.: Mining Spatial Association Rules in Spatially Heterogeneous Environment. In: Proceedings of SPIE, vol. 10, pp. 3x-1–3x-10 (2008)
10. Sha, Z., Bai, Y., Xie, Y., Yu, M., Zhang, L.: Using a hybrid fuzzy classifier (HFC) to map typical grassland vegetation in Xilin River Basin, Inner Mongolia, China. International Journal of Remote Sensing 29, 2317–2337 (2008)

Construct Climate Observation Network and Discover Similar Observation Stations

Wei Yu and Hong Zhang*

Faculty of Computer and Information Science, Southwest University, No. 2, Tiansheng Road, Beibei, Chongqing, P.R. China
yuweihao@hotmail.com, zhangh@swu.edu.cn

Abstract. Complex network theory provides a powerful framework to statistically investigate the topology of complex system include both artificial systems and natural systems. We propose a method to construct a climate observation network with climate observation records from automatic weather stations (AWS) in different locations. A link between AWS represents the cross-correlation between them. Apply this method to the climate observation records from the city of Chengdu and find that AWS with edge connected are located very close to. And the area with dense AWS has a significantly higher correlation between AWS compared to the area with exiguous AWS. This work would be helpful for identifying the preferred strategy for location optimization problems / discovering similar observation stations associated with AWS or using this information to complete missing/error values.

Keywords: Complex network, Complex system, Automatic Weather Stations (AWS).

1 Introduction

Meteorological observations and related environmental and geophysical measurements are necessary for a real-time preparation for weather analyses, forecasts and severe weather warnings, for the study of climate, for hydrology and agricultural meteorology, and for research in meteorology and climatology [1].

A weather station is a facility for observing atmospheric conditions with instruments and equipment for observing atmospheric conditions, either on land or sea. The measurements taken include surface air temperature and relative humidity of the air, barometric pressure, solar radiation, wind speed and direction, and precipitation, etc. An automatic weather station (AWS) is an automated version of the traditional weather station, either to save human labour or to enable measurements from remote areas [2]. During the last two decades, the number of AWS has greatly increased throughout the world [3].

Recently, the study of complex networks – that is, networks which exhibit non-trivial topological properties – has permeated numerous fields and disciplines

* Corresponding author.

F. Bian and Y. Xie (Eds.): GRMSE 2014, CCIS 482, pp. 46–55, 2015.

spanning the physical, social, and computational sciences. The analytic capabilities in particular are quite powerful, as networks can uncover structure and patterns at multiple scales, ranging from local properties to global phenomena, and thus help better understand the characteristics of complex systems [4].

In this paper an application of complex network theory to the climate observation records is proposed to construct climate observation network and found unreasonable observation stations/optimize network. Experiments show that those AWS with edge connected are located very close to. The area with dense AWS has a significantly higher correlation between AWS compared to the area with exiguous AWS.

The remainder of the paper is organized as follows: Section 2 describes the data and basic methodology for constructing climate networks; Section 3 experiments with the climate observation records from the city of Chengdu; Section 4 displays previous experimental results in graphics; Section 5 verify the rationality of network we have constructed; some conclusions are drawn in Section 6.

2 Basic Methodology of Network Construction

A network is any set of entities (vertices) with connections (edges) between them. The vertices can represent physical objects, locations, or even abstract concepts. Similarly, the edges can have many interpretations ranging from physical contact to statistical interdependence and conceptual affiliations [4].

2.1 Vertices

Until now, researchers mainly use the data set consists of a regular spatio-temporal grid with time series x_i associated to every spatial grid point i at latitude λ_i and longitude φ_i [4, 8, 9, 10]. Vertices in these networks are uniform distribution.

Vertices in our network represent the geographical location of each AWS, vertices v_i associated to every AWS i at its geographical location. The location of AWS is non-uniform distribution. So vertices in our network are also non-uniform distribution. We have an example shown in Fig 1.

Fig. 1. Example of vertices (Automatic weather station, AWS)

2.2 Edges

Edges are added between pairs of vertices depending on the degree of statistical interdependence between the corresponding pairs of anomaly time series taken from the climate data set [6].

Formally, a network or graph is defined as an ordered pair $G:=(V,E)$ containing a set $V=\{1,\ldots,N\}$ of vertices together with a set E of edges $\{i,j\}$, which are 2-element subsets of V. In this work we consider undirected and unweighted simple networks, where only one edge can exist between a pair of vertices and self-loops of the type $\{i,i\}$ are not allowed. This type of network can be represented by the symmetric adjacency matrix.

$$A_{ij} = \begin{cases} 0 & \{i,j\} \notin E \\ 1 & \{i,j\} \in E. \end{cases} \tag{1}$$

The edge density of a network is given by $\rho=2E/(N(N-1))$, E being the number of edges in the graph [6].

2.3 Data

AWS are controlled by a programmable data logger and are equipped with sensors for measuring surface air temperature and relative humidity of the air, barometric pressure, solar radiation, wind speed and direction, and precipitation, etc. [7]. Data loggers have been programmed to scan the instruments every once in a while--at least once an hour. For consistency, we utilize climate describing data of hourly intervals to construct corresponding network. In a period of time, we get a climate descriptors time series x_i associated to AWS i. Time series of multiple AWS constitutes an original data matrix X.

2.4 Correlation Measures

We quantify the correlation value of the links, which characterizes the interdependence of the different pairs of AWS.

Until now, researchers have mainly used two methods to quantify the degree of statistical interdependence between different vertices in climate network—linear Pearson correlation [8] and nonlinear mutual information [9]. It deserved to note that nonlinear relationships are known to exist within climate, which might suggest the use of a nonlinear correlation measure [4].

Mutual information from information theory is another nonlinear measure now widely applied in many fields of science. Mutual information can be interpreted as the excess amount of information generated by falsely assuming the two time series x_i and x_j to be independent, and is able to detect nonlinear relationships [6, 14]. For numerical-valued variables and time series, the mutual information increases with finer partition depending on the underlying distribution or process and the sample size [10]. However, it's complicated to choose a suitable partitioning scheme. And partitioning all data, whether it is the record of the same time or different time, is unreasonable. Because the date recorded at different time is not comparable. Namely, we want to compare observation data between different AWS which were recorded in the same time only.

We compute correlation measure for each pair of AWS with equation below:

$$S_{ij} = \sum_{k=1}^{n} \Theta\left\langle \omega - 2 \times \frac{\left| x_{ik} - x_{jk} \right|}{(x_{ik} + x_{jk})} \right\rangle \tag{2}$$

Or

$$S_{ij} = \sum_{k=1}^{n} \Theta\left\langle \omega - \left| x_{ik} - x_{jk} \right| \right\rangle \tag{3}$$

$\Theta(x)$ is the unit step function(Heaviside), which

$$\Theta(x) = \begin{cases} 0, & x < 0, \\ 1, & x \geq 0. \end{cases} \tag{4}$$

For large discrete variables such as precipitation, we use equation (2). While for small discrete variables such as surface air temperature or barometric pressure we use equation (3) instead. The parameter ω is the threshold. Note that in equation (2), if x_{ik} equal to x_{jk} equal to 0, $\Theta(x)$ equal to 1.

2.5 Obtaining the Network Adjacency Matrix

We construct the AWS observation network by thresholding the correlation measure matrix S_{ij}, i.e. only pairs of vertices$\{i,j\}$ that satisfy $S_{ij}>\tau$ are regarded as linked. By definition $S_{ij}\geq 0$, $\forall\{i,j\}$. Using the Heaviside function $\Theta(x)$, the adjacency matrix A_{ij} of the AWS network is then given by

$$A_{ij} = \Theta(S_{ij} - \tau) - \delta_{ij} \tag{5}$$

where δ_{ij} is the Kronecker delta. Note that A_{ij} inherits its symmetry from S_{ij} and the resulting AWS observation network is an undirected and unweighted simple graph. One could construct a network with edges $\{i, j\}$ weighted by S_{ij}.

The next step in climate observation network construction is the selection of a threshold τ, above which we consider a pair of vertices to be connected. From a statistical point of view it is desirable to only maintain connections that are statistically significant with respect to some reasonable test and reject those not meeting this criterion.

Construction steps can be summarized as follows:

Input: The original data matrix X, threshold τ

Step 1: Correlation measure is calculated between each row vector of the matrix X with equation (2)/(3), all correlation measure form a correlation measure matrix S.

Step 2: Traverse the correlation measure matrix S. If the element of S no less than threshold τ, the corresponding element of adjacency matrix A is set to 1, otherwise set to 0. The main diagonal elements of the matrix A are set to 0.

Output: Adjacency matrix A

Until now, we can get the adjacency matrix of climate observation network.

3 Construct Climate Observation Network in Chengdu

The data source what we used comes from Chengdu meteorological bureau. A data set consists of a series of AWS with time series x_i associated to every AWS at latitude λ_i and longitude φ_i. The distribution of the AWS shown in Fig 2.

Fig. 2. The distribution of the AWS in our network

There are many types of AWS. Some AWS observe multiple variables: surface air temperature, humidity of the air, barometric pressure, solar radiation, wind speed and direction, and precipitation, etc. While some AWS observe only two properties--surface air temperature and precipitation. We utilize precipitation to construct our network. Observational properties, start and end dates, length of time series N and the number of vertices V of the corresponding network are given in Table 1.

Table 1. Precipitation data sets we used

Observational properties	Precipitation
Temporal coverage	May 1,2011-August 31,2011 & May 1,2012-August 31,2012 & May 1,2013-August 31,2013
N[hours]	3*2952
V	153

Precipitation from May to August in the last three years has been taken as the rainy season. The precipitation of non-rainy season is very small and not conducive to analysis. We take each hour observation data and the length of time series N is 2952(from 0:00 on May 1st to 23:00 on August 31st) for each year. In conclusion, we construct the rainy season precipitation network in Chengdu for the last three years.

Then we calculate the correlation measure between different AWS with equations in Sect. 2.4. We let ω equal to 0.67, because we hope that if x_{ik}, x_{jk} satisfy $1/2 \leq (x_{ik}: x_{jk}) \leq 2, \Theta(x)$ equals to 1.

At this stage, we keep our method simple by studying an unweighted network. We suggest that weight information could help to identify the backbone structure even more clearly, however, would not alter our conclusions below, because we use only a small number of edges with high correlation that dominate the network [5].

The last but nontrivial step in climate observation network construction is the selection of a threshold τ.

Since it is impossible to determine an optimal threshold [11], we must rely on some other selection criterion. For example, Tsonis and Roebber [12] opt for a threshold of Pearson correlation coefficient greater than 0.5 while Donges et al.[5, 10] use a fixed edge density ρ to compare different networks, than to fix τ as it was done in all earlier works. The threshold $\tau=\tau(\rho)$ is thus chosen to yield a prescribed edge density ρ. Obviously, the edge density is strictly monotonic decreasing with τ and induces a one to one correspondence between τ and ρ. For increasing edge density, edges with decreasing correlation measure are added to the network. Consequently, climate networks with a very high edge density $\rho \geq 0.1$ are not expected to contain meaningful information for climate data analysis, because they contain many connections that are not statistically significant, i.e. that are much more likely to arise by chance [6]. We fix the edge density at $\rho=0.005$. The remaining 0.5% of all possible edges correspond to statistically significant and robust relationships.

4 The Experimental Results

After having introduced our methodology for climate network construction, we proceed to the main aim of this study: Construct climate observation network using above correlation measure calculation method and visualize it.

For the precipitation from May 1, 2011 to August 31, 2011, May 1, 2012 to August 31, 2012 and May 1, 2013 to August 31, 2013, construct network with above correlation measure calculation method and edge density. Fig 3, 4 and 5 are the results of 3 temporal coverage which calculate the correlation measure with methods in Sect. 2.4.

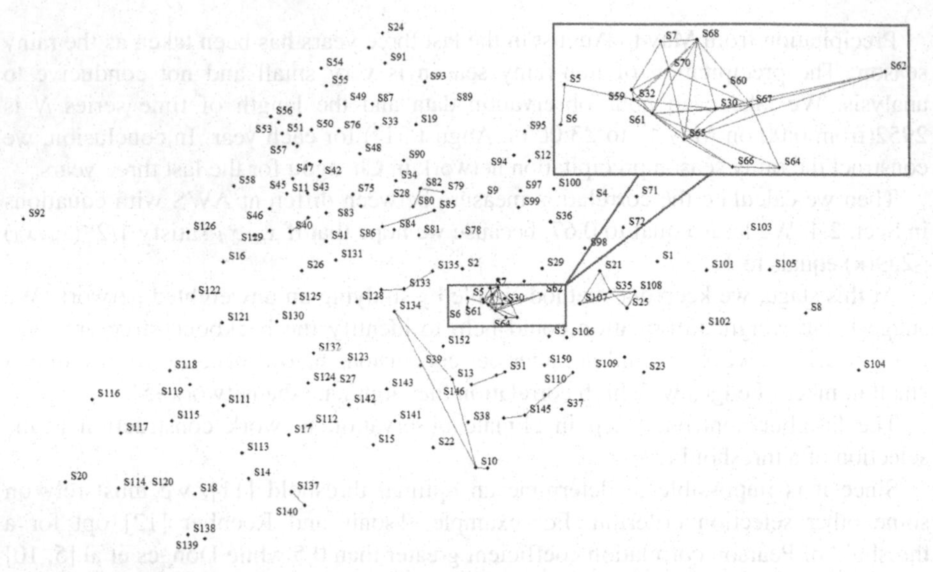

Fig. 3. Precipitation from May 1, 2011 to August 31, 2011, ρ=0.005

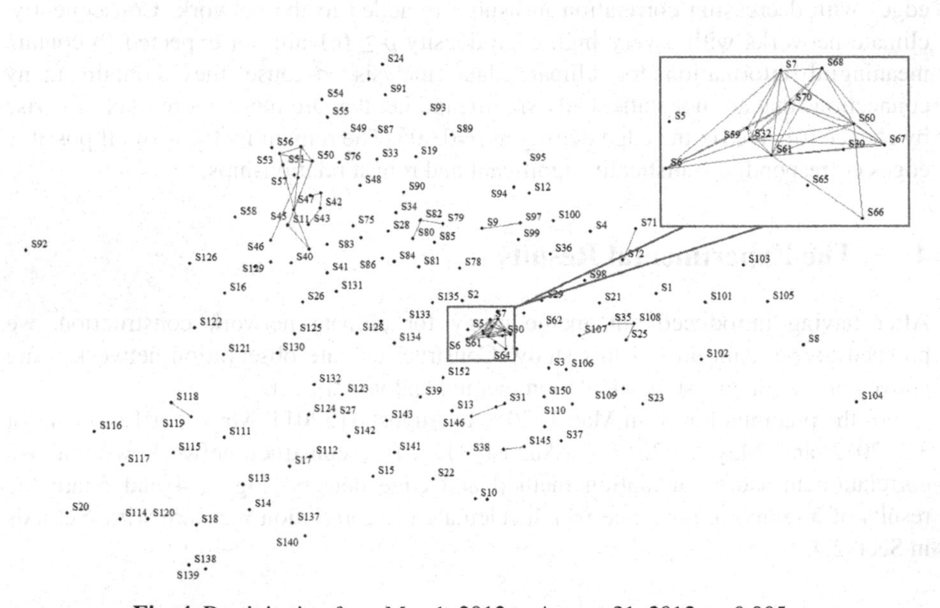

Fig. 4. Precipitation from May 1, 2012 to August 31, 2012, ρ=0.005

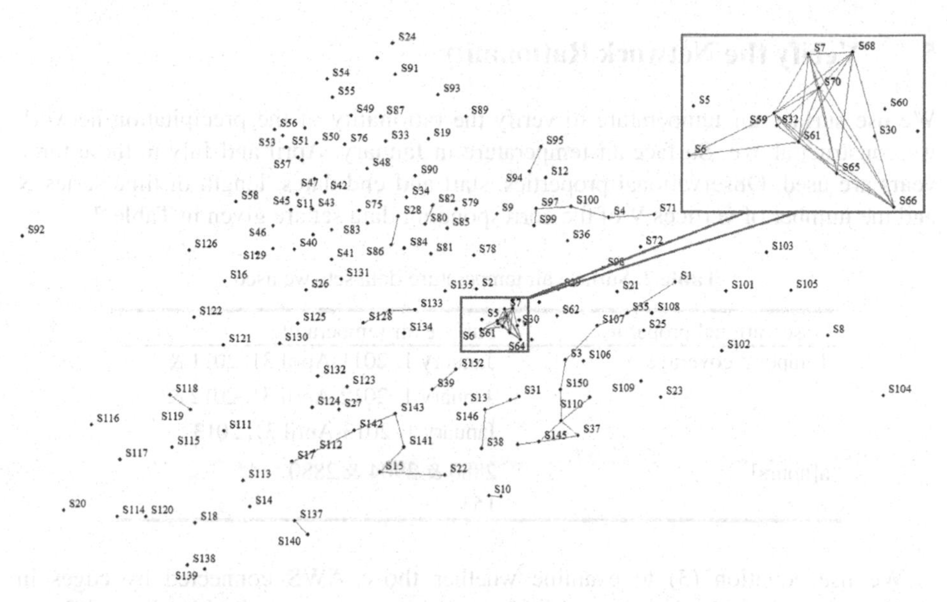

Fig. 5. Precipitation from May 1, 2013 to August 31, 2013, ρ=0.005

Fig 6 shows intersection edges of Fig 3, 4 and 5.

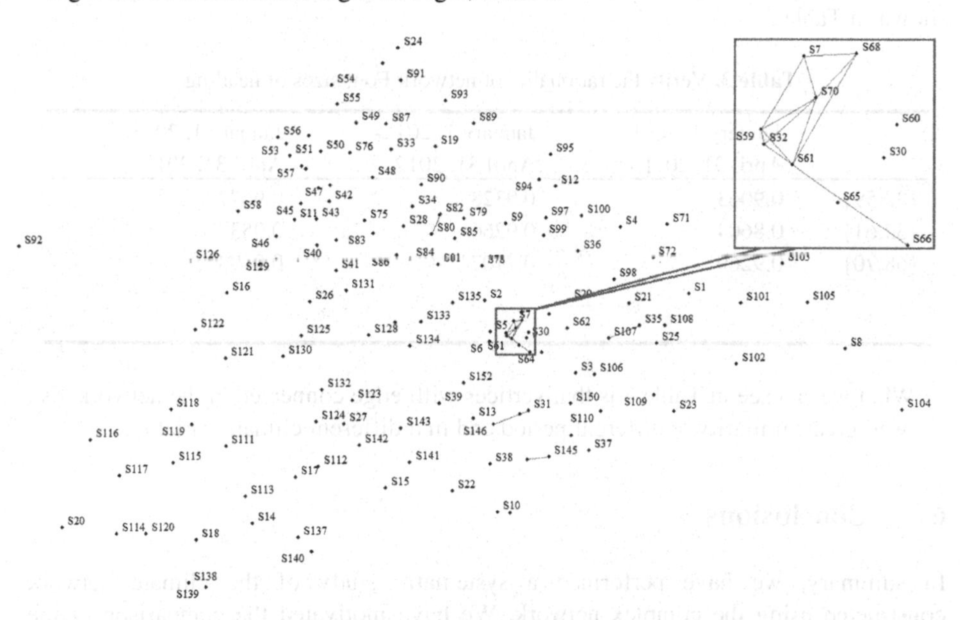

Fig. 6. The intersecting edges of 3 years

We can see that, length of edges in these figures is very short. And there are many edges in the central region where vertices are very dense. Especially, the intersecting edges of 3 years are almost in the central region.

5 Verify the Network Rationality

We use surface air temperature to verify the rationality of the precipitation network we construct above. Surface air temperature in January, April and July in these three years are used. Observational properties, start and end dates, length of time series N and the number of vertices V of the corresponding data set are given in Table 2.

Table 2. Surface air temperature data sets we used

Observational properties	Surface air temperature
Temporal coverage	January 1, 2011-April 31, 2011 & January 1, 2012-April 31, 2012 & January 1, 2013-April 31, 2013
n[hours]	2880 & 2904 & 2880
v	153

We use equation (3) to examine whether those AWS connected by edges in precipitation network also showed surface air temperature similarities in a different period. We set ω equals to 1 here. S_{ij}/n indicates the degree of similarity of two series to verify the rationality of the intersecting edges. We select 3 edges and results are shown in Table 3.

Table 3. Verify the rationality of network Font sizes of headings

	January 1, 2011-April 31, 2011	January 1, 2012-April 31, 2012	January 1, 2013-April 31, 2013
{32,59}	0.9063	0.9328	0.9677
{32,61}	0.8604	0.9266	0.9538
{68,70}	0.9260	0.7707	0.9493
.			
.			

What we can see in Table 3 is that vertices with edge connected in the network also showed great similarity at different period and in a different climate variable.

6 Conclusions

In summary, we have performed a systematic study of the climate network constructed using the complex network. We have motivated the comparison of the precipitation amount of three years' rainy season of networks at equal edge densities with above correlation measure calculate method. We have considered only low edge densities, which were shown to yield networks containing statistically highly significant edges as established on the basis of various significance tests. We find that those vertices with edge connected are located very close to. And the area with dense

AWS have a significantly higher correlation between AWS compared to the area with exiguous AWS. This is consistent with the objective laws.

This work would help policy makers to identify the preferred Strategy for location optimization problems / finding unreasonable observation stations associated with AWS. And it's helpful to use this information to complete the missing/error value.

Acknowledgments. This work is supported by the Fundamental Research Funds for the Central Universities (XDJK2014C143) and the Fundamental Research Funds for the Central Universities (SWU1309117).

References

1. World Meteorological Organization. Guide to Meteorological Instruments and Methods of Observations. WMO-No.8, Geneva, Switzerland (2008)
2. Automatic Weather Stations, `http://www.automaticweatherstation.com/index.html`
3. Miller, P.A., Barth, M.F.: Ingest, integration, quality control, and distribution of observations from state transportation departments using MADIS. In: 19th International Conference on Interactive Information and Processing Systems (2003)
4. Steinhaeuser, K., Chawla, N.V., Ganguly, A.R.: Complex networks in climate science: Progress, opportunities and challenges. In: Proc. Conf. on Intelligent Data Understanding, San Francisco, CA, NASA, pp. 16–26 (2010)
5. Donges, J.F., Zou, Y., Marwan, N., et al.: The backbone of the climate network. EPL (Europhysics Letters) 87(4), 48007 (2009)
6. Donges, J.F., Zou, Y., Marwan, N., et al.: Complex networks in climate dynamics. The European Physical Journal Special Topics 174(1), 157–179 (2009)
7. Estévez, J., Gavilán, P., Giráldez, J.V.: Guidelines on validation procedures for meteorological data from automatic weather stations. Journal of Hydrology 402(1), 144–154 (2011)
8. Zhou, C., Zemanová, L., Zamora-Lopez, G., et al.: Structure–function relationship in complex brain networks expressed by hierarchical synchronization. New Journal of Physics 9(6), 178 (2007)
9. Zamora-López, G.: Linking structure and function of complex cortical networks[D]. Universitätsbibliothek (2009)
10. Papana, A., Kugiumtzis, D.: Evaluation of mutual information estimators on nonlinear dynamic systems. arXiv preprint arXiv:0809.2149 (2008)
11. Serrano, A., Boguna, M., Vespignani, A.: Extracting the multiscale backbone of complex weighted networks. Proceedings of the National Academy of Sciences USA 106(16), 8847–8852 (2009)
12. Tsonis, A.A., Roebber, P.J.: The architecture of the climate network. Physica A 333, 497–504 (2004)

The Design and Application of the Gloud GIS

YongGe Shi[1] and FuLing Bian[2]

[1] School of Resource and Environmental Sciences, Wuhan University, Wuhan 430079, China
[2] International School of Software, Wuhan University, Wuhan 430079, China

Abstract. Compared wth the traditional GIS, the Gloud GIS has many advantages in infrastructure management, such as improving the utilization rate of hardware and the high performance of services, which can solve the existing problems in construction of digital city. Thus the Gloud GIS is studied and implemented. The experimental results showed that the proposed scheme outperforms the traditional GIS methods not only in efficiency but also in accuracy.

Keywords: The Gloud, Resources Pool, Public service platform.

1 Introducation

Gloud is a new technology in the fusion development of Grid and Cloud. Grid and Cloud computing are the new technique in the field of computer, both of which manage infrastructures by virtualization technology. Grid is skilled in task decomposition and parallel computing and Cloud has strong advantages in platform service, both of which have some generality and complementary factors with its strengths. There are some research about the combined application of Cloud and Grid in recent years: The literature[1-8] proposed dynamic architecture and experiments by Cluster technology under the support of the high-performance computing Grid, Cloud, or Hybrid Computing and also there are some further studies in the resource management and load balancing mechanism.

Peng Liu, the professor of PLA University of Science and Technology, the expert about Cloud in China, who first put forward the Gloud (Gloud = Cloud + Grid) concept, discussed the complementary relationship between Cloud and Grid from the unified platform, calculation, data, resource integration and information security, and constantly promoted the Gloud technology research and application[9].

Professor Geoffrey Fox (Indiana University Bloomington) thought that Future Grid was an international testbed modeled on Grid5000 and supported international computer science and computational science research in Cloud, Grid and High Parallel Computing (HPC) in his report in State Key Laboratory for Information Engineering in Surveying Mapping and Remote Sensing(LIESMARS) in Wuhan University on November 19, 2011 and looked forward to the future application of Future Grid[10].

Zhi Zeng in Zhejiang University integrated Grid and Cloud by enhancing internal mechanism of Grid infrastructure by Cloud, built the environment and application of

F. Bian and Y. Xie (Eds.): GRMSE 2014, CCIS 482, pp. 56–67, 2015.

Gloud. In the Gloud environment, it managed resources based on Grid, distributed resource and services in Cloud and realized efficient allocation of resources and services together[11].

Therefore, the paper follows the concept "GLoud" proposed by Professor Peng Liu and continues to study the integration of Cloud and Grid. Then the Gloud is applied to the public service platform of urban GIS, which helps to integrate hardware and spatial information resources and formats the one-stop service of GIS in Gloud.

2 The Gloud Environment Design

Firstly, it is to discuss the coexistence form of Grid and Cloud, and then build the Gloud environment. Based on SOA, the framework of Gloud integrates Grid and Cloud resources through the mode of federations. In the whole system, Grid and Cloud can integrate social idle resources in their own way. it is not only to develop computing ability with ultra large scale and the advantage on dynamic management of resource on Grid but also to develop the storage and computation on cloud data center or cloud services. The platform in GLoud achieves the integration and allocation of resources by building modulation and demodulation server in management center. Alliance form of Grid and Cloud as shown in Fig.1:

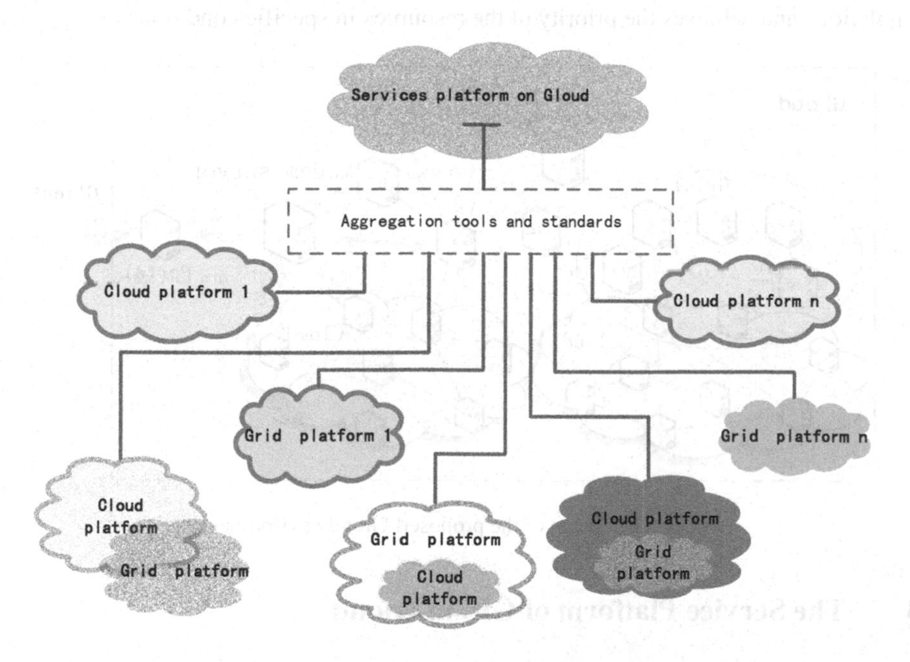

Fig. 1. The alliance mode between Grid and Cloud

The Gloud system is as shown in Fig.2. Clouds include data service cloud, Software application cloud, hardware environment service cloud, map service cloud or combined service cloud, which are perhaps the private cloud, the public cloud or the hybrid cloud. Federations of the private cloud and the private cloud, or the public cloud and the public cloud are called the single system combined cloud. Federation has little difficulty in single system combined cloud. Esri ArcGIS 10.1 inserts plug-in nodes and extends dynamically the cloud platform, which achieves the loose and hotplug framework of GIS in Cloud and federate systems of GIS in a peer-to-peer (P2P) way[12]. But for federations among external heterogeneous cloud platforms such as private, public and hybrid cloud, there is no one application example or universal standards. It is to adopt flexible methods and mechanism to integrate Cloud and Grid platforms: it can be to integrate Grid and Cloud by cloud's strengthening the internal mechanism of Grid infrastructure, or to manage infrastructure to put the Grid resources into cloud framework and provide high performance service on demand by Middleware or other ways. The study about federations of Cloud and Grid is still in the experimental stage and so there is no common alliance standards and general case, therefore, the study is also exploratory.

It is to simplify structure design of the Gloud in the relevant experiment to carry the main functions out. The Gloud environment is designed as shown in Fig.2:

The Gloud establishes a unified coordination mechanism of the resources by modems among Grid and Cloud clusters, manages the Gloud infrastructure and service resources with delivery mechanism of message bus, establishes relevant rules and regulations and achieves the priority of the resources in specific conditions.

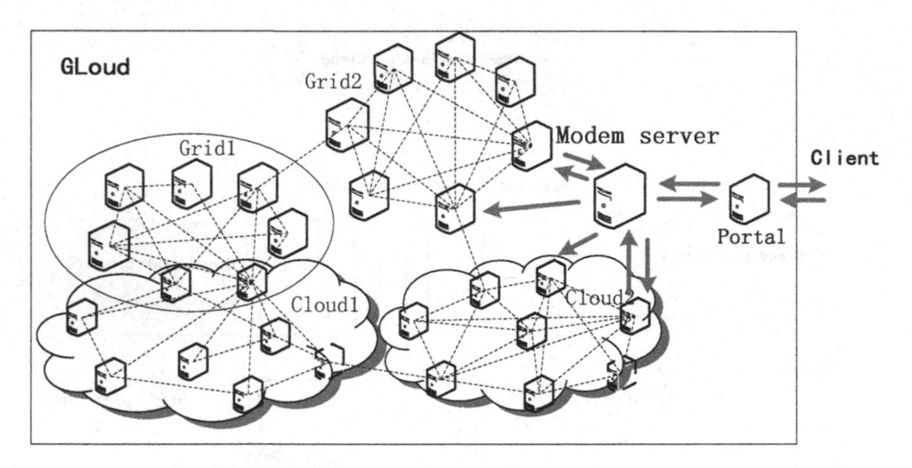

Fig. 2. The design of the proposed Gloud environment

3 The Service Platform of GIS in Gloud

Firstly, it is to build the framework in Gloud, the framework in Cloud is CloudStack and virtual management software is XenServer, and the Grid platform is Gridgain. Cloud and Grid comanage and schedule physical resources, build the Cloud data center and the Grid database. The data center includes basic geographic information database clusters and system database. Grid database is used to store municipal business data.

It is to build the public service platform of spatial information based on data center and Grid database. The public service platform includes C/S and B/S management interface and can realize the management such as map service publishing, users registering and permissions management, order management, the supervision of system and tasks etc. Demonstration projects are supported and implemented on public service platform with its map services and storage services. As shown in Fig.3: it is the running architecture of the public service platform in Gloud:

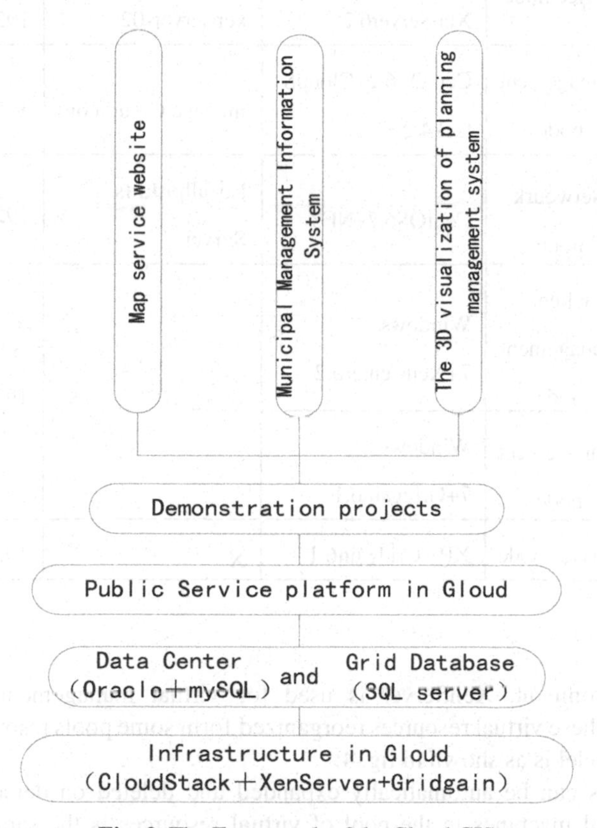

Fig. 3. The Framework of the Gloud GIS

4 Implementation of the Gloud GIS

4.1 The Experimental Environment in Gloud

The experimental system includes 6 server: 2 as management nodes of services in Cloud, in which the operating environment is CentOS6.2 + CloudStack4.2 and CentOS6.2 + NFS; 2 as computing nodes, deployed and installed XenServer6.2, used for management and integration of hardware resources; 1 as server for the Grid computing node and the municipal management information system shared, the rest one as the user management node and the Grid management node shared, experimental environment designed as shown in table 1:

Table 1. The Gloud system environments Design

	Compute node	System and Software	Machines	IP
Cloud enviroment	Service nodes	XenServer6.2	xenserver-01	192.168.120.161
		XenServer6.2	xenserver-02	192.168.120.162
	Management node	CentOS6.2+CloudStack4.2	manage.Cloud.com	192.168.120.168
	Netwourk node	CentOS6.2+NFS	Localhost.nfs Server	192.168.120.167
	Client management node	Windows 7+XenCenter6.2	xc-12	192.168.120.163
Grid enviroment	Management node	Windows 7+Gridgain6.1		
	Service node	XP+ Gridgain6.1	X	192.168.120.164

In Cloud environment, XenServer is used for virtual management of hardware resources, then these virtual resources reorganized form some pools resources, resource management model is as shown in fig. 4:

Physical sites can be automatically expanded and deleted on demand. Installing system in virtual machines in the pool of virtual resources is the same as in actual physical machines. A single physical machine can be divided into a plurality of resource pool and can be installed multiple virtual machines in the performance range allowed of computers. These virtual machines can run at the same time and do not interfere with each other, so the platform improves the efficiency of computer physical resources. Of course, multiple computers with Low allocation can form one or more larger virtual servers with high performance and provide large computing services. At the same time, Cloud and Grid build a coordination mechanism for the management of joint resources through modems, and can realize priority in allocation of resources.

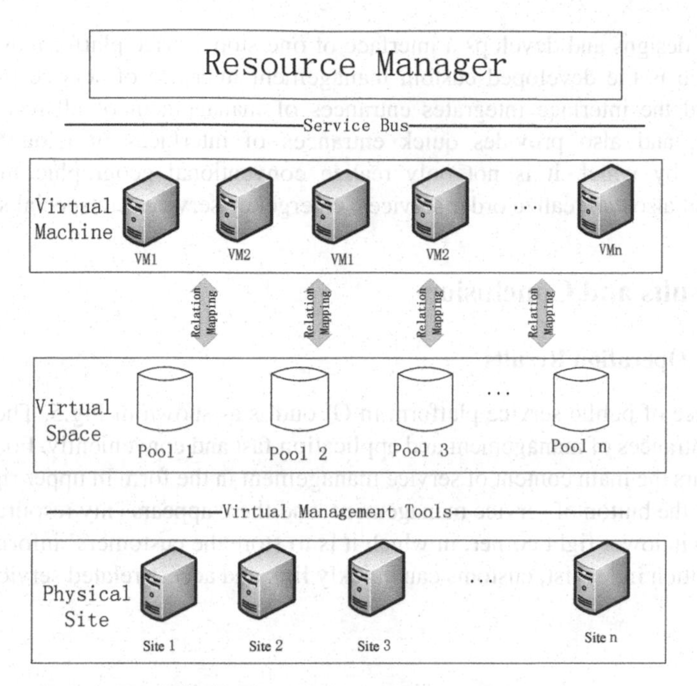

Fig. 4. Resource management mode

4.2 Designing and Deployment of Public Service Platform of the GLoud GIS

It is to build municipal special database and data centers in the Grid server and virtual machines in Cloud, based on which the map server, public service platform of GIS etc are built. the main environment deployment and virtual server resource allocation is as shown in Table 2 and Table 3:

Table 2. The System Environments Design of the GLoud GIS

Compute node	System and Software	network	IP
The database server	Winserver2008 R2 64	Virtual routing	192.168.120.25
The map service server	Winserver2008 R2 64		192.168.120.26
The server of the public service platform in GLoud	Windows XP		192.168.120.27
The 3D visualization of planning management system server	Windows XP		192.168.120.28
The map service websit server	Winserver2008 R2 64		192.168.120.29
Municipal Management Information System	Windows XP		192.168.120.164

The Gloud designs and develops a interface of one-stop service platform as shown in Fig.5, which is the developed custom management interface of service platform in GLoud, and the interface integrates entrances of management of all resources and application, and also provides quick entrances of interfaces of CloudStack and XenCenter, by which it is not only realize conventional geographic information services, but also can realize order services, emergency services etc special services.

5 Results and Conclusion

5.1 The Operation Rcsults

The interface of public service platform in GLoud is as shown in Fig.5. The platform provides entrances of management and application fast and conveniently. For example, there appears the main content of service management in the form in upper right corner by clicking the button of service management. and there appears "my resource library" in the form in lower right corner, in which it is to store the customers' information, by the information index list, customs can quickly link and access related services.

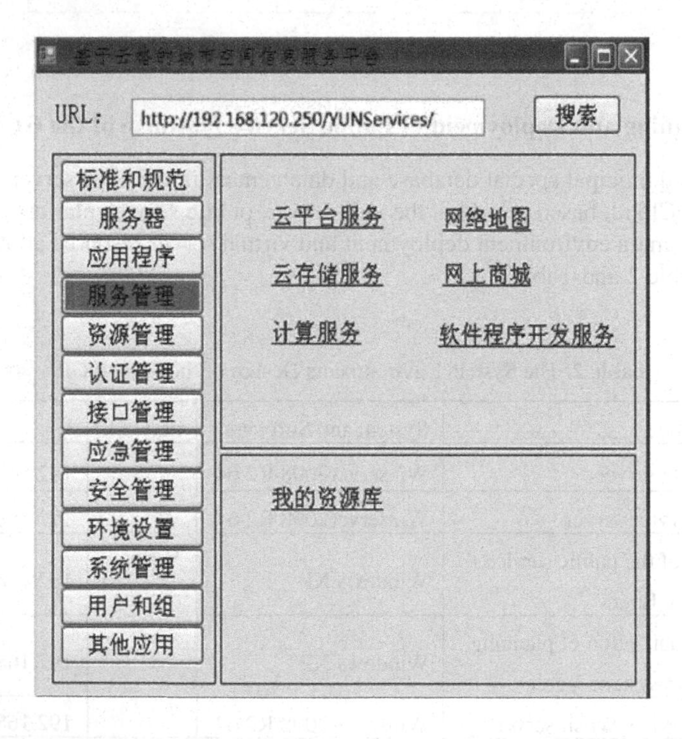

Fig. 5. The General interface of the public service platform of the GLoud GIS

CloudStack and Grid console can realize management of background resources, providing a deployment environment for virtual machines as Fig.6. The CloudStack4.2 management interface and Gridgain6.1 console all sets shortcut entrance in the interface of public service platform in GLoud, It is to realize resource services by the mechanism of resource comanagement.

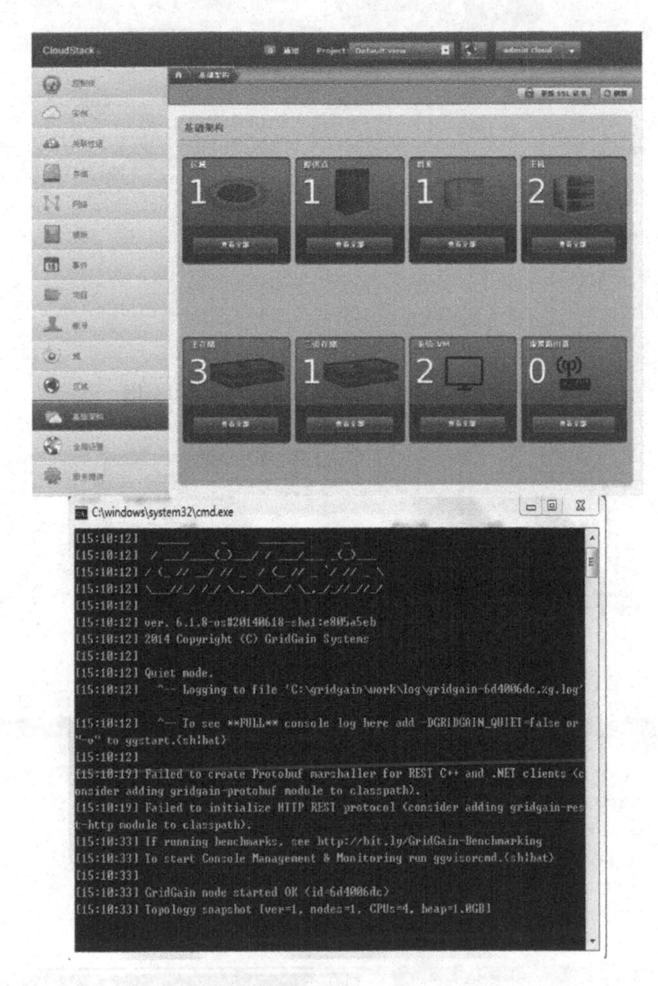

Fig. 6. CloudStack management interface and Gridgain console

The Gloud not only builds one-stop service platform, by which data and map service are exchanged but alsorealizes operations and applications of urban spatial information service platform and the engineering example as shown in Fig.7:The urban spatial information service platform not only includes C/S and B/S interface. At the same time, In order to guarantee the security of the service platform, it is to built ware firewall and system software for real-time monitoring which can timely collect the operation data for the convenience of system management and encrypt system, user, data products and map services.

Fig. 7. The CS/BS interface of the GLoud GIS

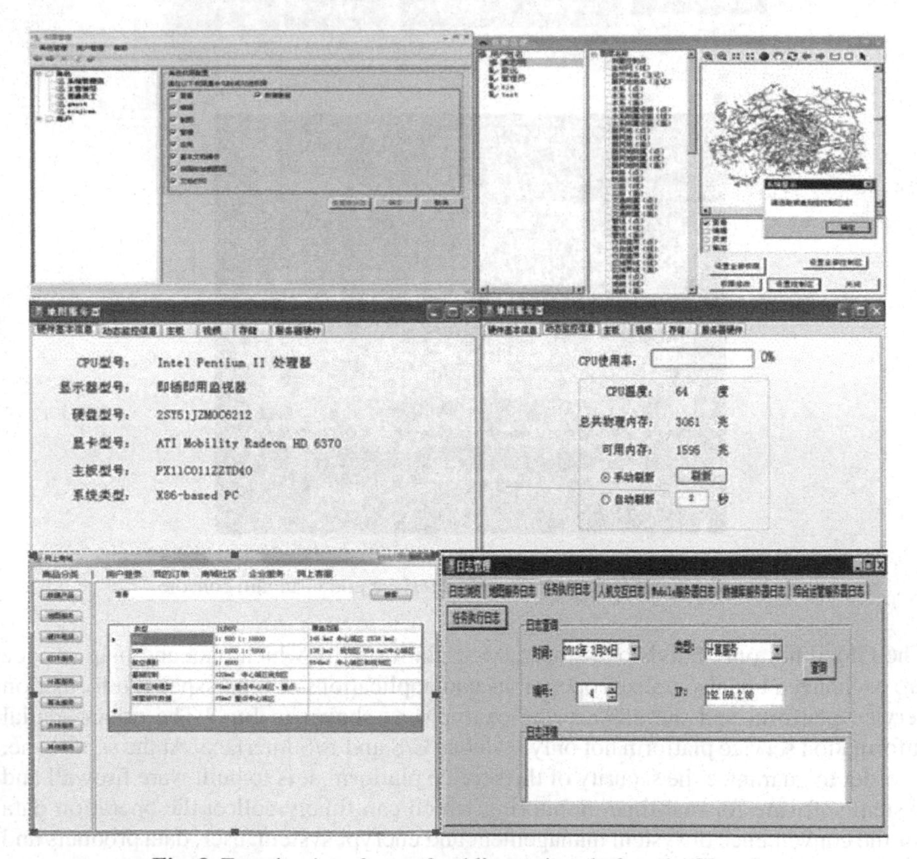

Fig. 8. Function interfaces of public service platform in GLoud

the system logs record real-time running status of servers and clients, human-computer interaction and task implementation in detail. When the system fails, the fault nodes and states can be tracked and task execution status can also be real-time controlled. Function interfaces of the platform in GLoud is as shown in fig.8.

5.2 Analysis and Discussion

The services platform in GLoud not only makes the geo-spatial framework for infrastructure extend flexibly, solves problems about physical resources idle and Low utilization rates of resources, achieves the centralized management of physical resources in background, and improves the efficiency of resources and reduces the waste of energy, but also integrates all resources and improves the parallel computing ability of system in foreground. A set of data following in Table 3 supports the conclusion.

As shown in table 3, each server can only perform assigned tasks because of the distributed task, in traditional environment, when the server is not full load, there are idles in memory and hard disk, due to physical limitation, idle resources cannot be redistributed. Besides, the server running depends on computer hardware devices. when the server is overloaded, the hardware system is unable to expand and the server cannot provide better services, up to the maximum limit of physical resources, users' demands are limited. So idle resources and upper limit of available resources in the traditional environment will affect spatial information services. Moreover, system and data need another backup environment, otherwise once the computer system paralysis or collapse, the data and the system environment perhaps will be damaged and cannot be recovered.

In Gloud environment, apart from CPU and the hardware resources occupied by system management nodes, the resources can be integrated into the virtual resource pool and can be reschedule for using as the same physical environment. As shown in table 3, based on minimum allocation of resources, utilization rate of overall resources (total CPU and total hard disk space) in GLoud has Little differences from that in the traditional environment, with the user increasing gradually, the resources of virtual machines with minimum allocation of resources gradually increase to meet demands no longer, and the system will automatically schedule idle system resources to the virtual machine on demand, then system achieves load balance, besides, in the same physical conditions, utilization rate of the CPU and virtual resource of hardware can increase obviously with more virtual machines and more applications. These advantages are not available in traditional environment. In addition, security is also very high. System files of virtual machines can be made into templates, snapshots or image file and be uploaded to the network at the same time in order to guarantee security of system and services, which can help to isolate a failure and recover resources timely and fast. when a host node fails in Gloud environment, virtual machine can be transferred to anot her host to avoid data loss and ensure the security of system. Systems in GLoud break away from one-to-one limitations of physical machines and really realize virtualization services of physical hardware.

Table 3. The hardware configuration and comparison of utilization rate

Environment		Resource site	CPU	Hard disk space	Total CPU	Utilization rate of total CPU in same physical enviroment	total Hard disk space	Utilization rate of total hard disk space in same physical enviroment
The traditional enviroment		The database server	4G	1T	24G	≥10% There is upper limit	3.5T	7.8% There is upper limit
		The map service server	4G	500G				
		The server of the public service platform in GLoud	4G	500G				
		The 3D visualization of planning management system server	4G	500G				
		The map service websit server	4G	500G				
		Municipal Management Information System	4G	500G				
The Gloud environment		The Management node	4G	50G	24G	≥34% There is no upper limit	3T	≥21.8% There is no upper limit
		Netwourk node (NFS)	4G	500G				
		Client management node	4G Shared	500G				
	Virtual space	The database server	2G	100G				
		The map service server	1G	100G				
		The server of the public service platform in GLoud	1G	100G				
		The 3D visualization of planning management system server	1G	100G				
		The map service websit server	1G	100G				
		Expropriating resources awaited	2G	1.45T		Expropriating resources awaited		Expropriating resources awaited
		Management node (Grid)	4G Shared	500G	Dynamic allocation	Resource allocation	Dynamic allocation	Resource allocation
		Service node (Grid)	4G Shared	500G				
		Municipal Management Information System						

6 Conclusions

To sum up, the Gloud can help to solve the problems such as idle resources and dynamic extension of infrastructures, improve utilization rates of resources and enhance service performances in limited resources and devices. These advantages can make customers to deploy basic geographic information services more conveniently and quickly and reduce effectively the investment and operation cost of projects. Besides, it is to integrate all resources by public service platform in GLoud, and overall coordinate and change the status of development and life in cities with scientific information carried in geospatial information data or service which can help construct of geospatial framework of smart city.

References

1. Gallard, J., Lèbre., A., et al.: Architecture for the next generation system management tools. Future Generation Computer Systems 28, 136–146 (2012)
2. Mateescu, G., Gentzsch, W., Ribbens, C.J.: Hybrid Computing-Where HPC meets Grid and Cloud Computing. Future Generation Computer Systems 27, 440–453 (2011)
3. Vázquez, C., Huedo, E., et al.: On the use of Clouds for Grid resource provision. Future Generation Computer Systems 27, 600–605 (2011)
4. Buyy, R., Ranjan, R.: Special section: Federated resource management in Grid and Cloud computing systems. Future Generation Computer Systems 26, 1189–1191 (2010)
5. Murphy, M.A., Goasguen, S.: Virtual Organization Clusters: Self-provisioned Clouds on the Grid. Future Generation Computer Systems 26, 1271–1281 (2010)
6. Dou, W., Qi, L., et al.: An evaluation method of outsourcing services for developing an elastic Cloud platform. Supercomput (2010)
7. Jha, S., Merzky, A., Fox, G.: Using Clouds to provide Grids with higher levels of abstraction and explicit support for usage modes
8. Abraham, M.A.M.I., Goasguen, M.F.S.: Autonomic Cloud on the Grid. Spriner Science+Business Media B.V (2009)
9. Liu, P.: Cloud Computing. Publishing House of Electronics Industry, Beijing (2011) (chinese)
10. Fox, G.: Geoinformatics and Data Intensive Applications on Clouds. State Key Laboratory of Information Engineering in Surveying Mapping and Remote Sensing. Wuhan University (2011)
11. Zeng, Z., Wang, J., et al.: A mechanism provisioning on computational resource and services with high effectively under Gloud environment. Journal of Zhejiang University (Science Edition) (41), 353–357 (2014) (chinese)
12. ArcGIS Cloud, `http://www.esrichina.com.cn/softwareproduct/technology/ArcGIS_Cloud/`

Monitoring the Dynamic Changes in Urban Lakes Based on Multi-source Remote Sensing Images

Bo Cao[1,2], Ling Kang[1], Shengmei Yang[2], Debao Tan[2], and Xiongfei Wen[2]

[1] College of Hydropower and Information Engineering, Huazhong University of Science and Technology, 1037 Luoyu Road, Wuhan 430074, China

[2] Changjiang River Scientific Research Institute, 23 Huangpu Road, Wuhan 430010, China

sunbathing@163.com, kling@hust.edu.cn,
{liddy533,wxfei19}@gmail.com, tdebao@ 126.com

Abstract. To enhance the urban lake information management, this paper applied the remote sensing technique to timely acquire the lake area reduction and water quality deterioration information of urban lakes. The appropriate classification rules and retrieval model were selected to obtain the important information concerned by the lake management department, such as the lake boundary, land use classification, lake temperature and chlorophyll content based on the multi-source remote sensing images. We found that the remote sensing technique can detect the abnormal change of the urban lake and track the development trend consistently. It highly improves the lake survey efficiency and will also promote the sustainable development of the lake ecosystem.

Keywords: Urban lakes, Remote sensing, Monitoring, Multi-source imagery.

1 Introduction

Water resources are one of the irreplaceable strategic resources for human survival. With the development of social economy and the rapid expansion of city size, the health of urban lakes is threatened by some unsustainable development model in recent years. The problems of the illegal construction around the lake and the sewage discharge have become increasingly serious. Some urban lakes are suffering from such phenomena as the discharge of sewage from living and industry, and the farmland reclaiming from lakes. A number of issues have arisen, such as deterioration of water quality, eutrophication, shrinking of lake area and decrease of water. So, it is urgent to supervise and monitor the urban lake status, so as to improve the environment of the urban living and the water ecosystems.

Currently, the primary investigation method of urban lakes is based on the ground-based survey, which is mainly by the artificial means and usually expenditure of time and effort. In recent years, spatial information technology develops quickly, which is widely used in many fields, such as resources, environment and disaster. As one of the most important spatial information technology, remote sensing (RS) shows some special advantages like a wide range, low cost, ease of long-term dynamic monitoring

F. Bian and Y. Xie (Eds.): GRMSE 2014, CCIS 482, pp. 68–78, 2015.

and so on. The RS technique plays an important role in wetland monitoring [1], flood monitoring [2-3], surface water area estimation [4-6], and water resources management [7-8]. Over three decades, multi-resource remote sensing data have been employed to extract information like water quality and temperature on land surface water bodies [9-11].

This paper applies the RS technique to monitor the dynamic changes of urban lakes. The multi-source remote sensing images, ZY3, GF1, Rapdideye and TM will be employed. The appropriate classification rules and retrieval model will also be constructed to obtain the lake factors, like the lake boundary change, land use classification, lake temperature and chlorophyll, as well as their change trends.

2 Study Area and Materials

2.1 Study Area

In order to better study the topographic feature classification and boundary change of a lake, these typical lakes with a more complete topographic feature category and a greater boundary change should be chosen. In the lake quality and temperature studies, the lakes with a larger area should be selected. The multi-spectral remote sensing imagery usually has a lower spatial resolution. And in those smaller lakes, the significant level change of water quality and temperature cannot be well reflected. In this paper, the Xiwan Lake and the Tangxun Lake in Wuhan were selected to study the lake feature classification and boundary change. Two typical larger lakes, the East Lake in Wuhan city and the Chaohu Lake in Anhui province, were chosen to monitor the change trend of the water temperature and quality.

2.2 Materials

Four kinds of satellite images, ZY3, GF1, Rapideye and Landsat TM were employed in this study. Further information about the specifications of the remote sensing data used is given in Table 1.

Table 1. Specifications of the ZY3, GF1, Rapideye and TM data used in this study

Data	Band	Range of spectrum	Resolution(m)	Amount	Acquisition Date
ZY3	4	B, G, R, NIR	Panchromatic: 2.1 Multi-spectrum: 6	18	From 2012-04 to 2013-12
GF1	4	B, G, R, NIR	Panchromatic: 2 Multi-spectrum: 8	15	From 2013-10 to 2014-05
Rapideye	5	B, G, R, R-edge, NIR	5	3	From 2011-10 to 2012-10
TM	7	B, G, R, NIR, SWIR, LWIR, SWIR	LWIR:120 Others: 30	20	From 1985-06 to 2013-12

2.3 Research Progress

The research progress of this study mainly contains four parts, data collection and pre-processing, lakeside topographic features classification, lake boundary change monitoring and water environment indicator retrieval (Fig. 1). In the data collection stage, several kinds of multi-spectral remote sensing images, and some measured data of lake water quality and temperature are collected. The acquisition time of remote sensing data and measured data should be approximately matched.

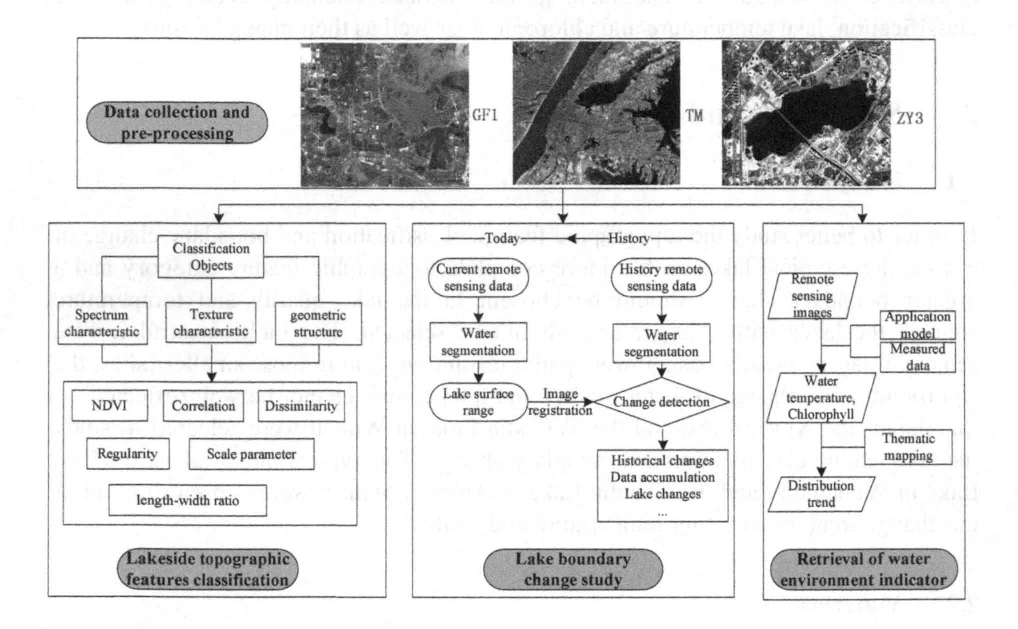

Fig. 1. Research progress of this study

3 Methodology

3.1 High-Resolution Image Classification

The most concerns in the lake management are mainly about the lake area, lake side vegetation, transportation, buildings, construction sites, etc. Therefore, in accordance with the regulatory requirements, the corresponding topographic features are extracted and classified into five categories as water body, vegetation, road, bare soil and building. The traditional supervised and unsupervised classification methods such as maximum likelihood and K-means, both are based on the pixel spectral information. The texture and geometric information of high-resolution remote sensing imagery is underutilized. Based on the spectrum, texture and geometric characteristics of the main features in the lake area, some indicators such as NDVI, correlation, dissimilarity, regularity, length-width ratio and scale parameter are selected to build an object-oriented classification rule of the lake topographic features based on the decision tree theory (Table 2).

Table 2. Classification indicator

Information category	Indicator	Expression	Specification
Spectrum information	NDVI	NDVI=(nir-red)/(nir+red)	Distinguish vegetation and water from other features
Texture information	Correlation	$f=\left[\sum_i\sum_j ijp(i-j)-\mu_1\mu_2\right]/\sigma_1\sigma_2$	Describe the texture repeatability under a window
Texture information	Dissimilarity	$f=-\sum_i\sum_j (i-j)p(i,j)$	Reflect heterogeneity of texture
Geometric information	Regularity	$f=\lVert B\rVert^2/4\pi A$	Reflect the irregular degree of the object shape
Geometric information	Length-width ratio	F= Length/Width	Reflect the object shape
Geometric information	Scale parameter	Object area	Control the minimum area of the classification polygons

In the above table, P (i, j), B, A, μ , σ represent the pixel value, perimeter, area, mean value and mean square deviation respectively.

Normalized difference vegetation index (NDVI) can distinguish water and vegetation well because they have completely different reflection features in the NIR band. Bare soil and construction both have high reflection values on the high spatial resolution satellite images, which lead to it very difficult to classify the bare soil and the construction only using the spectrum information. While bare soil around lakes is an important category concerned by the lake management department, because the big scale of bare soil may be an illegal construction site which is filling the lake. The traditional visual interpretation cannot satisfy the demand of the massive data processing. So, the correlation, regularity and dissimilarity rules, which can take advantage of texture and geometry information contained in the high resolution images, are used to distinguish between the bare soil and the construction automatically. The construction category has more styles in texture and more regular shapes in geometry than the bare soil. Then, the length-width ratio is applied to divide the construction category into the building and the road. The scale parameter can control the minimum area of classification polygons and avoid too tiny polygons. The scale parameter needs to repeatedly adjust to achieve the perfect classification effect.

The decision tree of the classification rule is shown in Fig. 2. The benefit of using the classification decision tree is that, the new classification rule can be added in at any time according to the lake management demand. The broad category also can be subdivided into several smaller categories. For example, some lakes are in a serious condition of land reclaiming and fish breeding. We can expand the decision tree and add two rules, regularity and area to continue subdividing the water category. Then, the fishpond and the paddy field are separated from the water body of the lake.

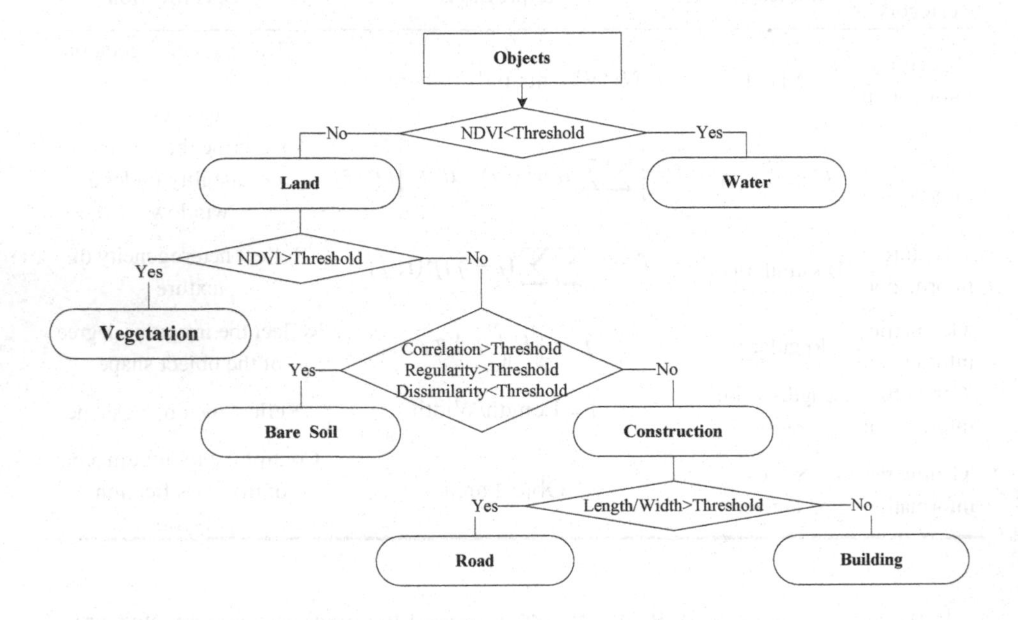

Fig. 2. Decision tree of the classification rule

3.2 Lake Boundary Change Monitoring

The specific monitoring method of the lake boundary change is to overlay and contrast the water classification results of different phases of remote sensing images. In case of sufficient data source, the images should be collected from the similar season of the different years, which guarantees the comparability of images.

3.3 Retrieval of Lake Water Environment Indicators

Lake Chlorophyll Retrieval. In this paper, NDVI is adopted to establish a linear regression model. The model is defined as below.

$$C_{chla} = a_1 \times NDVI + a_2 \quad (1)$$

Where C_{chla} represents the chlorophyll content($\mu g / L$) ; a_1 , a_2 are the regression Coefficients.

Satellite images, ZY3, GF1, Rapideye and TM all have red and near Infrared bands. So the regression model can be generally applied to the four remote sensing images that is the reason why we selected NDVI to retrieve the lake chlorophyll content.

The retrieval principle is to establish a quantitative relationship between the remote sensing images and the chlorophyll content. First we acquire a group of measured data including water quality and spectrum data by the field survey. Then, the regression coefficients a_1 and a_2 are calculated through several groups of NDVI and measured C_{chla} values. Once the regression model established, the entire lake chlorophyll content can be retrieved with the remote sensing images, whose acquisition time is the same or similar as the field survey time.

Lake Surface Temperature Retrieval. The lake surface temperature can be retrieved using the mono-window algorithm. Qin[12] (2001) deduced the mono-window algorithm for the lake surface temperature retrieval from Landsat TM6 data according to thermal radiation transmission equation. The algorithm is defined as below.

$$T_s = \left[a(1-C-D)+(b(1-C-D)+C+D)T_6 - DT_a\right]/C \ . \tag{2}$$

Where a and b are the empirical coefficients (a=-67.355351, b=0.458606). C and D are the intermediate variables, expressed as:

$$C = \varepsilon\tau \ . \tag{3}$$

$$D = (1-\tau)[1+(1-\varepsilon)\tau] \ . \tag{4}$$

In which, τ is the atmospheric transmittance and ε is the ground emissivity. T_s represents the land surface temperature, and T_6 is the brightness temperature of TM6. T_a is the effective mean atmospheric temperature.

4 Results

4.1 Classification Result Based on GF1 Imagery

GF1 satellite image of Xiwan Lake in Wuhan in April 2014 was taken for example of lake topographic features classification. As the aforementioned classification rule and decision tree, a typical area was chosen from the image for training. In order to achieve the best classification effect, the threshold often needs to constantly adjust. Then, the entire image classification can be finished automatically. The classification result of the topographic features in Xiwan Lake is shown in Fig. 3.

Through the visual interpretation, it is found that the classification method of decision tree achieves a high accuracy, which can meet the needs of the routine monitoring and analysis of the lake management.

(a) Xiwan Lake area. (Satellite: GF1; Acquisition time: April 2014)

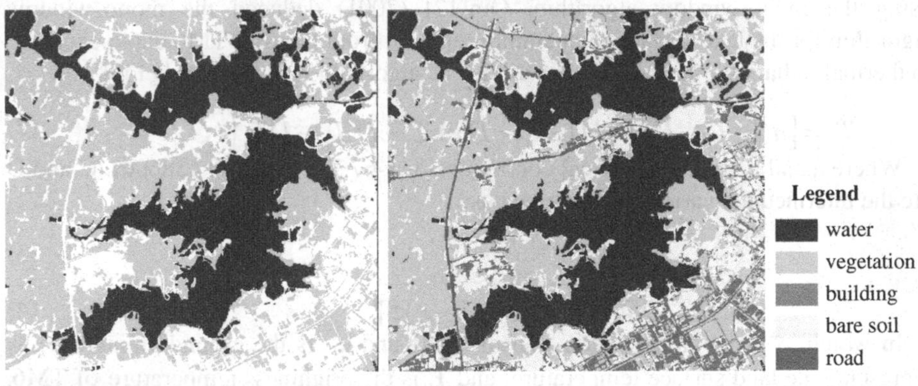

(b) Water and vegetation classification. (c) Water, vegetation, bare soil, road and building result.

Fig. 3. The classification result of Xiwan Lake

4.2 Monitoring Result of Lake Boundary Change

Two periods (April 2012 and February 2013) of ZY3 satellite image of Tangxun Lake were employed to monitor the lake boundary change. Tangxun Lake is the biggest one in Wuhan and locates between urban and rural regions. The fused images with more spatial and spectral information were shown in Fig. 4 (a) and (b). After the water boundaries of the two phase images were obtained, the variable region can be extracted with the image registration, as is shown in Fig. 5. The yellow, red and blue colors represent the increase, decrease and no change water area respectively. From the monitoring result, it is found that the water area of Tangxun Lake has a good recovery, due to the lake protection and sewage discharge prohibition recently.

(a) Tangxun Lake, April 2012. (b) Tangxun Lake, February 2013.

Fig. 4. Tangxun Lake area. (Satellite: ZY3)

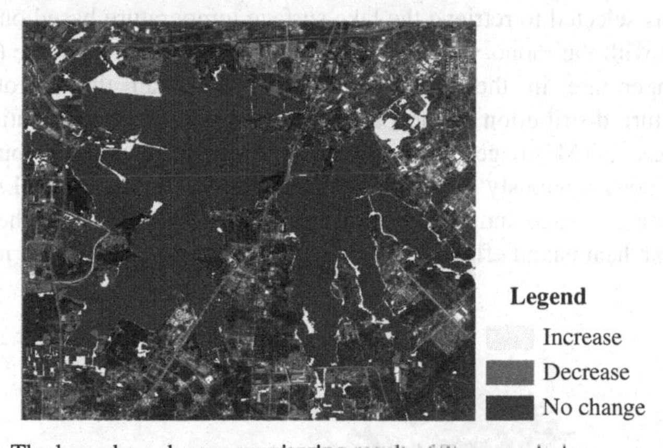

Fig. 5. The boundary change monitoring result of Tangxun Lake

4.3 Retrieval Result of Lake Chlorophyll and Surface Temperature

Retrieval Result of Lake Chlorophyll Based on Rapideye Imagery. Three Rapideye images (shot on October 14, 2011, October 17, 2011 and October 17, 2012) were collected to retrieve the chlorophyll content of Chaohu Lake. The measured data mainly including the chlorophyll content and spectrum information were acquired between October 10, 2012 and October 11, 2012. The sampling sites distribution is shown in Fig.6 (a). The linear regression model coefficients were calculated by the measured data. The regression equation is expressed as below.

$$y = 33.893238\ x + 14.444855\ . \tag{5}$$

Where x=(Band5-Band3)/(Band5+Band3), R^2=0.723. Then the chlorophyll content retrieval result (Fig.6 (b)) can be obtained with the above regression equation. The result shows that the chlorophyll content in the western is higher than the eastern, which can achieve 60 μg/L. Similarly, the Rapideye images shot on October 14, 2011, October 17, 2011 can also be retrieved by the same model coefficients.

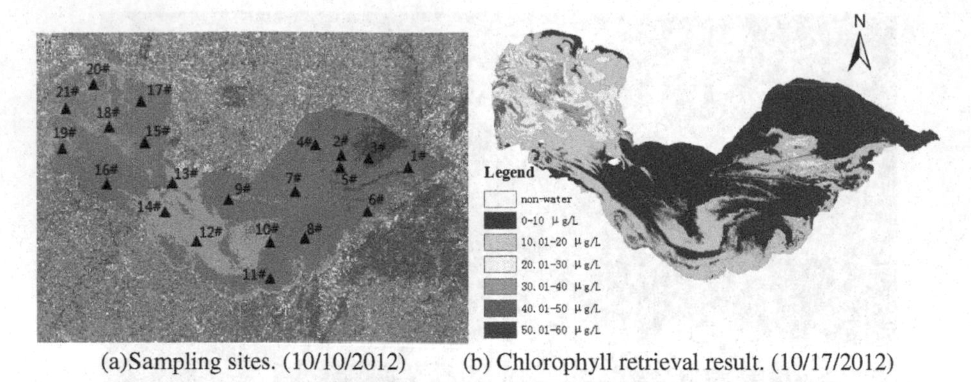

(a)Sampling sites. (10/10/2012) (b) Chlorophyll retrieval result. (10/17/2012)

Fig. 6. The retrieval result of the lake Chlorophyll based on Rapideye image

Retrieval Result of Lake Surface Temperature Based on TM Imagery. The East Lake was selected to retrieve the lake surface temperature based on the band 6 of TM imagery with the mono-window algorithm. The retrieval outcome (Fig. 7) shows that the temperature in the lake center is higher than the surrounding area. The temperature distribution of the East Lake has the similar distribution regularity from the different TM images (since 1985 to 2013). The dense population and human activity does obviously influence the nearby lake temperature. Likewise, the dense vegetation coverage and little land use can effectively decrease the lake temperature. The urban heat island effect also influences the temperature of the nearby water body.

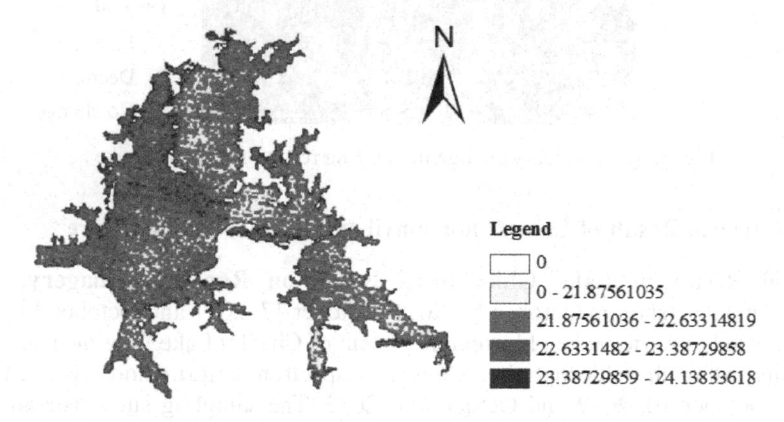

Fig. 7. The retrieval result of the East Lake surface temperature. (10/13/2002, unit: °C)

5 Conclusion

From the application results, we can find that the multi-source remote sensing images can satisfy the primary demands of the lake management. Through establishing the appropriate classification rules and retrieval model, this study can promptly obtain the concerning information about the lake factors, including the lake boundary, land use classification, lake temperature and chlorophyll, as well as their change trends. Remote sensing has more advantages than the traditional methods of urban lake monitoring because it is a low-cost and also reliable information source that is capable of making repeatable observations on lakes. It is benefit to assist the related water management department in monitoring the illegal exploitation, water quality, and other abnormal changes of urban lakes. So, the remote sensing technique can greatly improve the existing lake survey efficiency and promote the sustainable development of the lake ecosystem. Meanwhile, the remote sensing technique will also be the prevalent technique in the future lake management.

Acknowledgments. The authors would like to appreciate the financial supports for this study from the Science and Technology Program of Wuhan (Grant #2014060101010062), the Basic Scientific Research Operating Expenses of Central-Level Public Academies and Institutes (Grant No. CKSF2014032/KJ and CKSF2014035/KJ) and the National Natural Science Foundation of China (Grant #41301435).

References

1. Bortels, L., Chan, J.C.W., Merken, R., et al.: Long-term monitoring of wetlands along the Western-Greek Bird Migration Route using Landsat and ASTER satellite images: Amvrakikos Gulf. Journal for Nature Conservation 19, 215–223 (2011) (Greece)
2. Kuenzer, C., Guo, H., Huth, J., et al.: Flood Mapping and Flood Dynamics of the Mekong Delta: ENVISAT-ASAR-WSM Based Time Series Analyses. Remote Sensing 5, 687–715 (2013)
3. Sheng, Y., Gong, P., Xiao, Q.: Quantitative dynamic flood monitoring with NOAA AVHRR. International Journal of Remote Sensing 22, 1709–1724 (2001)
4. Du, Z., Bin, L., Ling, F., et al.: Estimating surface water area changes using time-series Landsat data in the Qingjiang River Basin, China. Journal of Applied Remote Sensing 6, 063609 (2012)
5. Feyisa, G.L., Meilby, H., Fensholt, R., et al.: Automated water extraction index: a new technique for surface water mapping using Landsat imagery. Remote Sensing of Environment 140, 23–35 (2014)
6. Ding, X., Li, X.: Monitoring of the water-area variations of Lake Dongting in China with ENVISAT ASAR images. International Journal of Applied Earth Observation and Geoinformation 13, 894–901 (2011)
7. Giardino, C., Bresciani, M., Villa, P., et al.: Application of remote sensing in water resource management: the case study of Lake Trasimeno, Italy. Water Resources Management 24, 3885–3899 (2010)
8. Van Dijk, A., Renzullo, L.J.: Water resource monitoring systems and the role of satellite observations. Hydrology and Earth System Sciences 5, 39–55 (2001)
9. Dekker, A.G., Vos, R.T., Peters, S.W.M.: Analytical algorithms for lake water TSM estimation for retrospective analyses of TM and SPOT sensor data. International Journal of Remote Sensing 23(1), 15–30 (2002)

10. Barnes, B.B., Hu, C., Holekamp, K.L., et al.: Use of Landsat data to track historical water quality changes in Florida Keys marine environments. Remote Sensing of Environment 140, 485–496 (2014)
11. Wan., Z.: New refinements and validation of the collection-6 MODIS land-surface temperature/emissivity product. Remote Sensing of Environment 140, 36–45 (2014)
12. Qin, Z., Karnieli, A., Berliner, P.: A mono window algorithm for retrieving land surface temperature from Landsat TM data and its application to the Israel-Egypt border region. International Journal of Remote Sensing 22(18), 3719–3746 (2001)

Automatic Extraction of Building Footprints from LIDAR Using Image Based Methods

ZhenYang Hui*, YouJian Hu, and Peng Xu

Faculty of Information Engineering, China University of Geosciences, Wuhan 430074, P.R. China
huizhenyang2008@163.com

Abstract. This paper proposes a method by which scattered LIDAR point clouds are converted into a two-dimensional image and then building footprints are extracted through the image processing. Firstly point cloud grid is handled to generate georeferenced feature image, and then image threshold segmentation, morphological close operation, connectivity analysis and contour tracking method is used to obtain the final building footprints. Finally, based on the mapping relationship between georeferenced feature image and scattered point clouds, building outline points are obtained. Experimental result shows that this method could extract building footprints very well in plain area, but due to the adoption of single image segmentation method in the georeferenced feature image, it is not suitable for the building footprints extraction in mountainous area.

Keywords: LIDAR georeferenced feature image, image threshold segmentation, morphological close operation, connectivity analysis, automatic building footprints extraction.

1 Introduction

Airborne laser scanner (ALS) is a new technical measure which is developing very rapidly in recent years. Due to its ability to quickly obtain space 3d information, nowadays it is widely used in 3d city modeling and spatial information analysis [1]. In the 3d city modeling, building is the main entity of spatial information. Therefore, it is of great research value to extract the building footprints information from LIDAR point clouds quickly and accurately. Fig. 1 shows the measurement process of airborne LIDAR system [2].

Methods of building footprints extraction have been extensively studied at home and abroad in recent years. Haithcoat, et al. (2001) set different thresholds to extract building footprints from LIDAR point clouds by adopting different height and shape characteristics [3]. Zhang, et al. (2006) used region growing and plane fitting methods to distinguish building from vegetation with the help of normalized DSM (nDSM) [4]. Li, et al. (2008) firstly eliminated lower ground objects through setting threshold, and

* Corresponding author.

F. Bian and Y. Xie (Eds.): GRMSE 2014, CCIS 482, pp. 79–86, 2015.

then used morphological open operation to interrupt the non-building part attached to the building. Finally, he used region growing method to detect the connection area and then clustered buildings [5]. Al-Durgham, et al. (2012) firstly used region growing method to extract plane, and then adopted the improved convex hull algorithm to estimate rough outline of the roof. Finally, he used recursive minimum bounding rectangle method and Boolean operator to produce the final rules and boundaries [6].

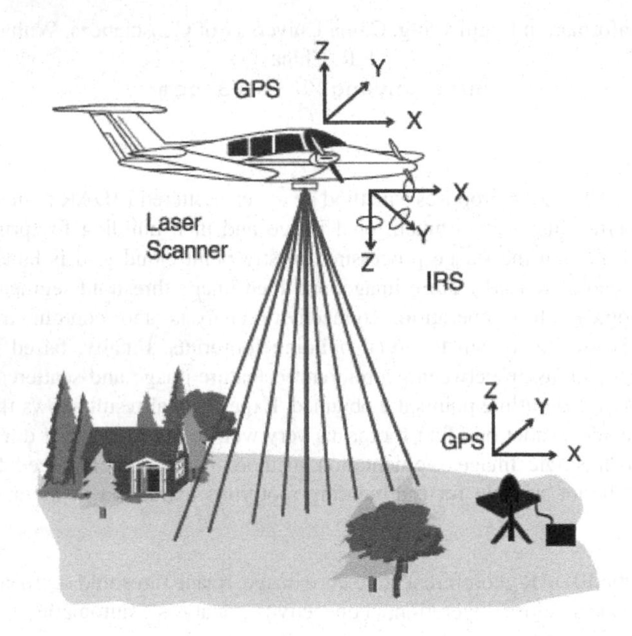

Fig. 1. Sketch map of LIDAR scanning (Flood and Gutelius, 1997)

All methods above can extract building footprints with good effects, but most of them are more complicated. Due to the huge number of LIDAR point clouds, complicated methods always need more complex computer computation and longer computing time, which is against the will of quick spatial information acquired by using airborne radar. Therefore, it is very necessary to design a simple, rapid and effective building footprints extraction method.

2 Building Footprints Extraction

Airborne LIDAR point clouds mainly contain three-dimensional coordinate information, so we cannot conduct point clouds classification and feature extraction directly. The main train of thought of the method proposed by this paper is that elevation values can be converted to grayscale values through grid handling, which can be generated georeferenced feature image. And then through image threshold segmentation, morphological close operation, connectivity analysis and contour tracking, building footprints can be extracted. Fig. 2 elaborates the processing flow chart.

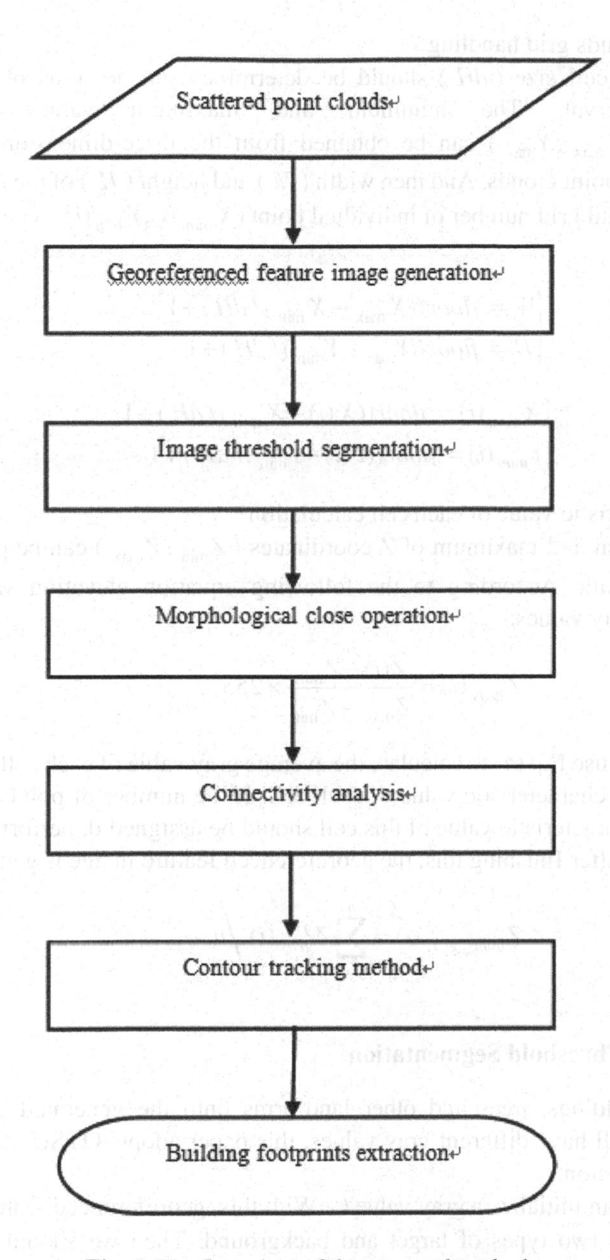

Fig. 2. The flow chart of the proposed method

2.1 Georeferenced Feature Image Generation

To turn scattered point clouds into two-dimensional image, firstly what needs to do is grid handling which can project the scanned points onto X-Y plane. Then according to certain rules elevation values can be converted to gray values. Finally, characteristic value of each cell should be calculated to generate a georeferenced feature image.

(1) Point clouds grid handling

First of all, cell size (dH) should be determined on the basis of point clouds sampling interval. The minimum and maximum values of X, Y (X_{min} , Y_{min} , X_{max} , Y_{max}) can be obtained from the three-dimensional coordinate information of point clouds. And then width (W) and height (H) of the georeferenced feature image and grid number of individual point $(X_{num}(i), Y_{num}(i))$ can be calculated as below.

$$\begin{cases} W = floor((X_{max} - X_{min}) / dH) + 1 \\ H = floor((Y_{max} - Y_{min}) / dH) + 1 \end{cases} . \quad (1)$$

$$\begin{cases} X_{num}(i) = floor((X(i) - X_{min}) / dH) + 1 \\ Y_{num}(i) = floor((Y(i) - Y_{min}) / dH) + 1 \end{cases} . \quad (2)$$

(2) Characteristic value of each cell calculation

The minimum and maximum of Z coordinates (Z_{min} , Z_{max}) can be obtained from point clouds data. According to the following equation elevation values can be converted to gray values.

$$Z_{gray}(i) = \frac{Z(i) - Z_{min}}{Z_{max} - Z_{min}} \times 255 . \quad (3)$$

Then we can use Eq. (4) to calculate the average gray value of each cell which can be regarded as the characteristic value of each cell. If the number of point clouds within the cell is 0, characteristic value of this cell should be assigned 0, performing black on the 2d image. After finishing this, the georeferenced feature image is generated.

$$Z_{num\_gray}(i) = \sum_{j=1}^{n} Z_{gray}^{j}(i) \Big/ n . \quad (4)$$

2.2 Image Threshold Segmentation

In light of buildings, trees and other landforms onto the generated georeferenced feature image all have different gray values, this paper adopts OTSU algorithm to do image segmentation.

Firstly, we can initialize a gray value k . With this, georeferenced feature image can be divided into two types of target and background. Then we should calculate the probability (ω_1 , ω_2) and average (μ_1 , μ_2) of the two kinds. After this, the variance ($\sigma^2(k)$) between the two kinds of classes can be calculated.

$$\sigma^2(k) = \omega_1(\mu_1 - \mu)^2 + \omega_2(\mu_2 - \mu)^2 . \quad (5)$$

Where μ is the mean gray value of georeferenced feature image.

Through constantly iterative calculation, once $\sigma^2(k)$ reaches maximum, at this time the gray value k is requested. The targets whose gray values are greater than k are buildings[7].

2.3 Morphological Close Operation

After image threshold segmentation the georeferenced feature image is performed as binary image, in which buildings are white color. In order to eliminate small hole on the binary image and repair the broken contour line to make it more smooth, we need to do morphological close operation.

Morphological close operation can be defined according to the following formula. That is to say the structural element B to set A morphological close operation is to use B to do first expansion for A, and then use B to do second corrosion for the result.

$$A \bullet B = (A \oplus B) \Theta B \quad (6)$$

Among them, corrosion is to lead the outline to internal contraction, which can help eliminate the outline points. While expansion is to contact with the object of all the background to the object, which can make the outline expand to external [8].

2.4 Connectivity Analysis

Due to the influence of external environment, point clouds data obtained by airborne radar always has gaps, which would cause a same roof of the house to be broken into two or more pieces embodied in the two-dimensional image. To solve this problem, we should do connectivity analysis. Connected in image processing usually refers to four connected or eight connected, which one to chose can be decided according to actual processing effect.

Another function of connectivity analysis is to do the second judge for the target building. After the image threshold segmentation, some high trees, utility poles and high tension lines also would be mistaken for the target building. To avoid this situation, we should set an area threshold to judge the connected area of binary image, the one whose area is less than it should be weeded out.

2.5 Contour Tracing Method

To extract building footprints from connected binary image, a contour tracing method proposed by Pavlidis (1982) can be applied [9]. After finding building footprints in binary image, we can extract building outline points from LIDAR points according to the mapping relationship between image and point clouds data.

3 Experiment and Analysis

The experimental data comes from eight scenes data located in Vaihingen/Enz test field and the center of Stuttgart city which was obtained by ISPRS Commission

Ⅲ Working Group Ⅲ/3 (http://www.itc.nl/isprswgIII-3/filtertest/index.html). To test the effectiveness of the algorithm proposed by this paper, the experiment uses S31 datasets as representative, which contains high density cement roof constructions, ups and downs vegetations and data gaps. The total number of S31 sample point clouds data is 28862. The point spacing is 1-1.5m and point density is 0.67 per square meter [8].

3.1 Georeferenced Feature Image Generation

To make sure that most cells contain at least one point and the georeferenced feature image is without distortion, the cell size here is set to be 2m. After characteristic value of each cell calculation, we can get the final georeferenced feature image. As shown in Fig. 3, cells without point were turned out to be black, while roads and some low plants were characterized by dark color as their gray values were small. The roof of the building as well as some tall trees whose gray values were lager was characterized by bright color. Hence, the georeferenced feature image preserved the original shape information of the building very well.

3.2 Image Threshold Segmentation Outcome

As shown in Fig. 4, after image threshold segmentation the georeferenced feature image performed as binary image. The shape of buildings can be clearly recognized from the image. The reason why the borders are jagged is mainly result of grid handling, which has no effect on the extraction of building footprints. Some small white patches are caused by high trees or poles.

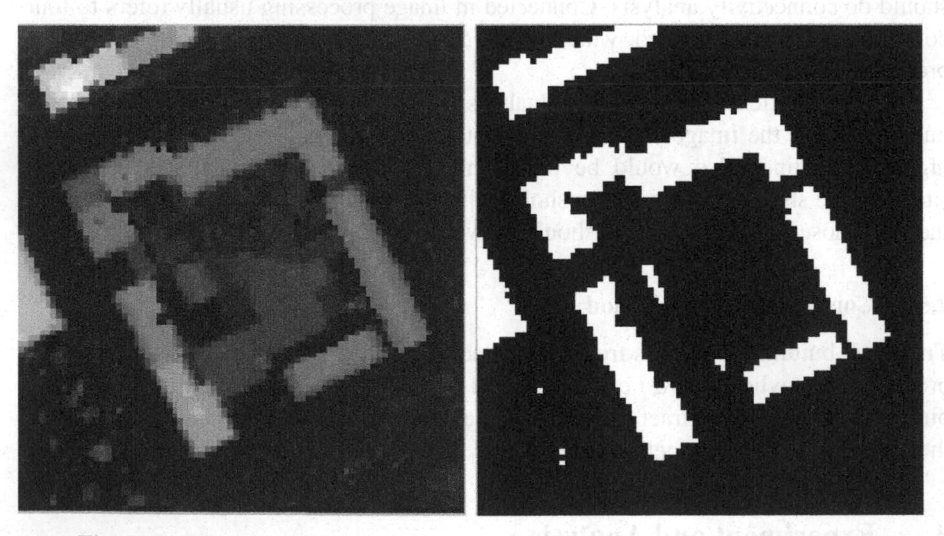

Fig. 3. Georeferenced feature image **Fig. 4.** Binary image

3.3 Connectivity Analysis Outcome

Here we adopted four connected processing method and set the area threshold to be 20 m^2. In order to facilitate subsequent contour tracking, the transform was made for the binary image to turn the building to be black color. As shown in Fig. 5, after connectivity analysis the two houses at the lower right corner were connected together, which was in line with the field situation. In addition, some small patches whose areas were less than threshold were also removed.

3.4 Building Footprints Extraction Outcome

Applied the contour tracing algorithm to the binary image, we can get the final building outline, as shown in Fig. 6. Then according to the mapping relationship between georeferenced feature image and point clouds data, we can get 3d coordinates of building outlines.

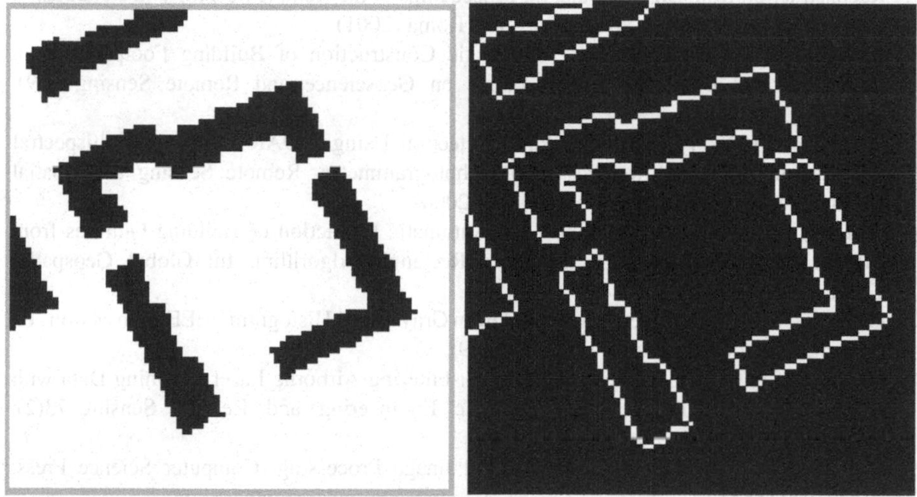

Fig. 5. Building shape after connectivity analysis **Fig. 6.** Final building outlines

4 Conclusion

The method proposed by this paper firstly converts LIDAR point clouds to image information and then do a series of operation to the 2d image. The principle of this method is simple and it needs a small amount of calculation. From the experimental analysis, we can get that this method can extract the building footprints effectively and it can apply to most of plain area for building footprints extraction. As this method adopts single threshold segmentation method, it is prone to miscarriage in the mountainous area [10]. This also is what needs to be improved for this method.

Acknowledgments. The experimental datasets were provided by ISPRS Commission Ⅲ Working Group Ⅲ/3. Thanks very much for their hard work. In addition, the authors would like to thank anonymous reviewers for their insightful suggestions and comments.

References

1. Flood, M., Gutelius, B.: Commercial Implication of Topographic Terrain Mapping Using Scanning Airborne Laser Radar. Photogrammetric Engineering and Remote Sensing 63(4), 327–366 (1997)
2. Vosselman, M.G., Kessels, P., Gorte, B.G.H.: The Utilization of Airborne Laser Scanning for Mapping. International Journal of Applied Earth Observation and Geoinformation 6(3/4), 177–186 (2005)
3. Haithcoat, T.L., Song, W., Hipple, J.D.: Building Footprint Extraction and 3-D Reconstruction from LIDAR Data. In: Proceedings of the Remote Sensing and Data Fusion over Urban Areas, pp. 74–78. IEEE/ISPRS, Roma (2001)
4. Zhang, K., Yan, J., Chen, S.C.: Automatic Construction of Building Footprints from Airborne Lidar Data. IEEE Transactions on Geoscience and Remote Sensing 44(9), 2523–2533 (2006)
5. Li, Y., Wu, H.Y.: Automatic Building Detection Using LIDAR Data and Multispectral Imagery. International Archives of the Photogrammetry, Remote Sensing and Spatial Information Sciences 37(pt. B1), 197–202 (2008)
6. Al-Durgham, M., Kwak, E., Habib, A.: Automatic Extraction of Building Outlines from LiDAR Using the Minimum Bounding Rectangle Algorithm. In: Global Geospatial Conference (2012)
7. Otsu, N.: A Threshold Selection Method from Gray-Level Histogram. IEEE Transactions on Systems Man Cybernetics 9(1), 62–66 (1979)
8. Chen, Q., Gong, P., Baldocchi, D.D., et al.: Filtering Airborne Laser Scanning Data with Morphological Methods. Photogrammetric Engineering and Remote Sensing. 73(2), 175–185 (2007)
9. Pavlidis, T.: Algorithms for Graphics and Image Processing. Computer Science Press. Rockville (1882)
10. Hui, Z.Y., Wu, B.P., Xu, P., Guo, J.X.: Three-dimensional Laser Scanning Topographic Data Acquisition Processing. Science Technology and Engineering 14(18), 1–3 (2014)

Research on Visualization of Ocean Environment Data Using ArcGIS

Xiu Li, Zhuo Jia, and Jin Yu

Graduate School at Shen Zhen, Tsinghua University,
518055 Shenzhen, China
li.xiu@sz.tsinghua.edu.cn,
{jz12,j-yu12}@mails.tsinghua.edu.cn

Abstract. ArcGIS is a series of geographic information system (GIS) software that has been widely applied in many industries in people's daily life. However, there are few applications in the ocean domain. This article discussed the basic concepts and methods of using ArcGIS to visualize ocean environment data and an experiment based on Hong Kong water quality monitoring stations data was conducted. The results show that ArcGIS can be very helpful and powerful when dealing with ocean environment data and it can make a significant contribution to the protection of the ocean ecological environment.

Keywords: ArcGIS, Visualization, Ocean Environment Data.

1 Introduction

Nowadays, due to the rapid development of science and technology, it becomes possible to further observe and study the ocean. The ocean environment data acquired by the ocean observation system usually contains information such as seawater temperature, salinity, turbidity, flow rate, etc. Generally speaking, ocean environment data has some characteristics as follows: long acquisition cycle, strong reusability and close relationship with geospatial information. An effective visualization method of these data can significantly be helpful to describe the ocean environment, and finally help to make decisions. ArcGIS is a suite of geographic information system (GIS) software that provides many useful tools on our research.

2 Related Work

At present, the main aspects of GIS applications in the ocean domain are the reconstruction of the seabed topography, coastal landscape modeling, simulation and analyses of 3D marine elements and so on. However, the visualization work on marine data didn't tightly integrate with geospatial information until ESRI put forward the model of ocean data based on ArcGIS in the 2000s [1]. Since then, ArcGIS has been used to carry out experiments on visualization as well as analyses of marine data.

F. Bian and Y. Xie (Eds.): GRMSE 2014, CCIS 482, pp. 87–94, 2015.

In 2007, Monterey Bay Aquarium Research Institute (MBARI) started to build up a single-node seafloor observatory network system that was completed in November, 2008. This system used ArcGIS in many aspects during construction, for instance, storage, processing, analysis, and also data application. American researchers proposed a concept of MarGIS which is the combination of Marine and GIS. They developed a browser application named MarGIS Viewer using ArcIMS. Users can do some panning, zooming, query, calculation in MarGIS Viewer, besides, they can also check all the layers of different types or download data that they are interested in [2]. China's Fujian Province established a Marine Management System based on ArcGIS Server which allows people access the system through the Internet [3].

GIS is gradually applied in data management [4], however, due to the special ocean environment and the limited technology, there are still some problems to solve when dealing with visualization issues.

1) Organization of data from multiple sources

The ocean observation system [5] contains all kinds of sensors, so the data obtained are packaged in different types. This brings a huge problem on how to organize and visualize these data effectively.

2) Data mining and processing on huge data set

Marine data are characterized by their abundance, hugeness and complexity. Handling this huge data set can bring heavy workload. Data mining is the method which can extract useful or interesting information from large amounts of data [6]. This method is capable to help researchers better understand the data, and thus to analyze and visualize them in an appropriate way.

3) Construction of visual scenes

Ocean environment includes not only environment data such as salinity, but also seabed topography data. Since topography data is kind of related to security issues, it is necessary to simulate the topography by some modeling tools.

Fortunately, ArcGIS can largely help us solve the problems above.

3 Basic Concepts and Methods

3.1 Data Classification

The spatial data are mainly divided into vector data and raster data in ArcGIS [7]. Vector data is designed to express spatial objects by recording their boundaries. Because of this, zooming in or zooming out slightly affects the image. So vector data shows relatively high precision. On the other hand, raster data represents geographic entities by a regular grid. In fact, the structure of raster data is just like a pixel array. Each pixel is located by the array's row number and column number.

Although the accuracy of raster data is not as high as vector data, raster data has many advantages. So it is necessary to consider both types of spatial data when we are faced with GIS problems. Table 1 compares the advantages and disadvantages of the two types of data.

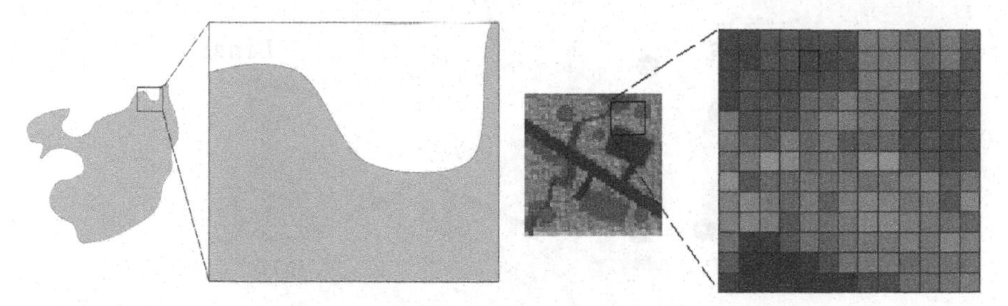

Fig. 1. A comparison between vector data and raster data when they are zoomed in. There is hardly any loss of image information in the vector data.

Table 1. Advantages and disadvantages of vector data and raster data.

	raster data	vector data
Advantage	a) Simple structure b) Easy to overlap and assemble c) Easy to perform spatial analysis d) Low development costs	a) High precision b) Small size of data c) Good appearance of graphical output d) Complete description of graphical topology
Disadvantage	a) Large information loss in higher resolution b) Big data size c) Poor appearance in some resolution d) High time costs when dealing with projection transformation and some other operations	a) Complex structure b) Complex to overlap layers c) Difficult to perform spatial analysis d) High costs to visualize and analyze

The ocean environment data are divided into two categories as vector data and scalar data. Vector data contain information that describes the data's direction, while scalar data don't. In this paper, we mainly focus on scalar data.

3.2 Feature Class

Feature class [8] is a collection of features that have the same spatial representation as well as a common attributes column. For instance, several monitoring sites can be composed of a point feature class. In ArcGIS, the most common used feature classes are points, lines, polygons and labels. Fig.2 shows an example.

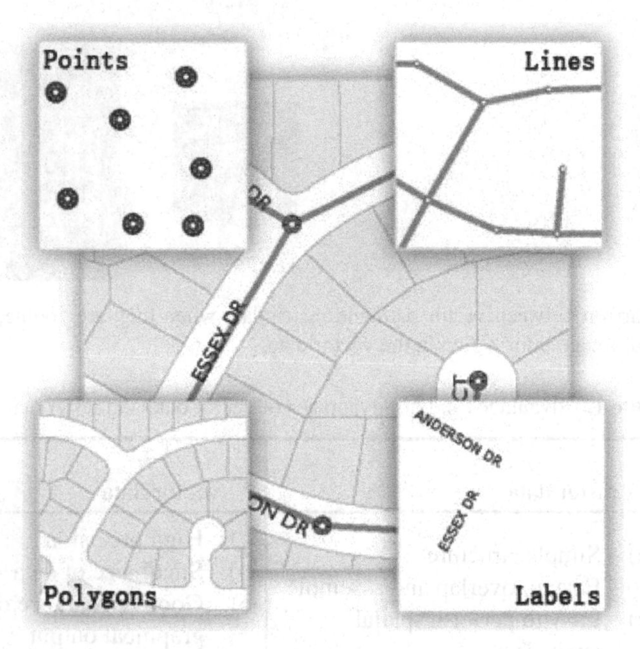

Fig. 2. An example of feature class in ArcGIS

3.3 Spatial Reference

Each point on the map is assigned to a precise geographic coordinate by spatial reference. This helps to reflect its exact position on the earth. The spatial reference of all the data displayed on one map must be consistent. In ArcGIS, each feature has its coordinate system, and we can put different feature classes together on a single map by defining a common spatial reference.

Understanding the concepts above, we can conclude that the basic procedures to visualize ocean environment data are as follows:

For spatial data,

a) Define a spatial reference, and then draw feature classes, feature sets formed by feature classes or raster files on the map as layers.
b) Locate the elements in a) by their x-y coordinates. In this step, we may need to use geo-referencing tool to register a spatial reference system that doesn't exist before in the elements such as a raster file.
c) Modify the appearance to make them more expressive. The simple methods we can probably use are data symbolization, data layer annotation, and adding legend, etc.
d) Conduct spatial analyses or some other operations based on the spatial data, and visualize the results.

In addition, for non-spatial data,

e) Draw statistical charts using ArcGIS, and we can also conduct some statistical analysis on these data.

4 Experiments

In this section I will operate an experiment on visualization of Hong Kong water quality monitoring stations data Using ArcMap.

Hong Kong has about 1651 square kilometers of ocean area with a very long coastline. To protect its ecological environment, Hong Kong Environmental Protection Department launched a marine monitoring program in 1986. Since then, there have been 76 monitoring stations in total built in this area. Ocean environment data are collected monthly by these stations.

Because the base map in hand is a raster file with no coordinate system, it has to be geo-referenced before we add it to ArcMap. Geo-referencing tool in ArcGIS helps a lot with this operation. By entering 15 points whose coordinates are known already, we can obtain a base map that is well spatially referenced after processing.

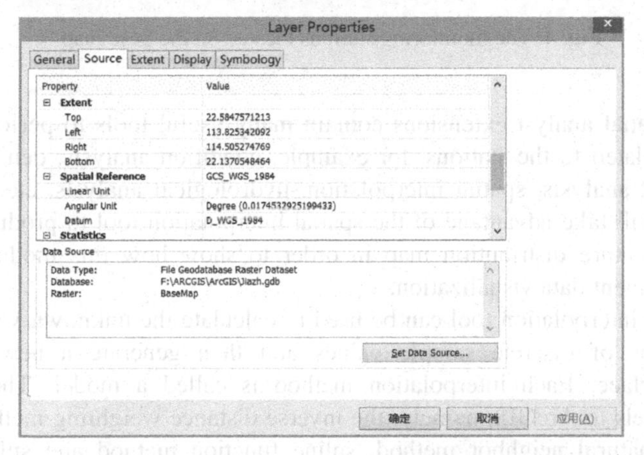

Fig. 3. The base map was spatially referenced with a coordinate system of WGS_1984 and an accurate range of map

The data of the stations which include the longitude and latitude, together with some monitoring data such as temperature and dissolved oxygen are stored in an excel file. What we are going to do is to locate all the stations on the base map. We can achieve this goal with the help of X-Y event layer. Using the longitude and latitude provided, the stations can be positioned exactly where they should be by the coordinate system in ArcMap. However, the original appearance may be simple and information can't be well conveyed, so we must modify the appearance to better represent the stations. Here we can decide not only what symbol and label to use, but also how they will display.

To verify the results, I make a comparison with official site distribution map around Tolo Harbour and Channel and it returns a nearly perfect match.

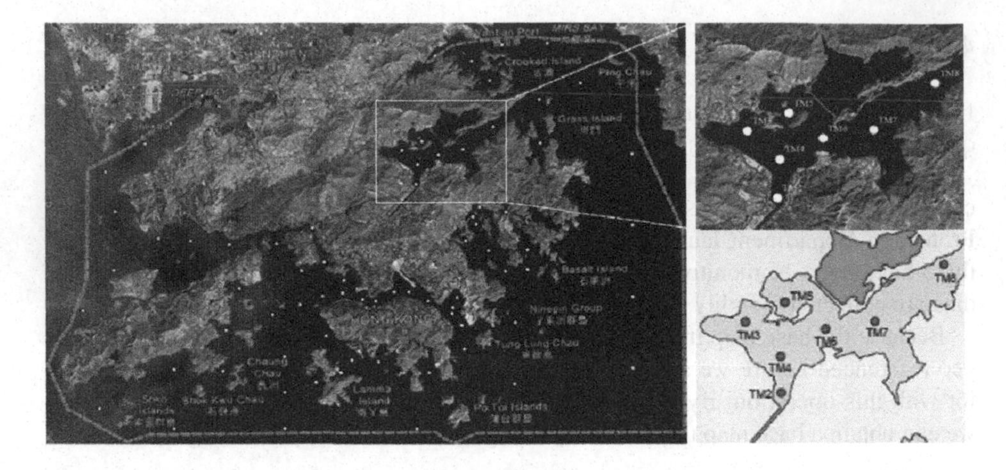

Fig. 4. The monitoring stations located on the base map

ArcGIS spatial analyst extensions contain many useful tools to perform analyses on the data related to the stations, for example, condition analysis, density analysis, extraction and analysis, spatial interpolation, hydrological analysis, etc. In the next paragraph, I will take advantage of the spatial interpolation tool to produce an ocean surface temperature distribution map in order to show how this module works in ocean environment data visualization.

The spatial interpolation tool can be used to calculate the unknown values of each point in terms of discrete sample values and then generate a new continuous prediction surface. Each interpolation method is called a model. There are five available models in ArcGIS, namely: the inverse distance weighting method, Kriging method, the natural neighbor method, spline function method and spline function method with obstacles. In this experiment, the region where monitoring stations locate is filled with staggered land and sea. It is clear that we only need the interpolation surface of the range of water, thus I have to get the mask layer of the ocean and the barrier layer of the land.

Firstly, I make a binary image of the base map, and then apply the vectorization tool to extract the layer of the ocean. Fig.5 is the result.

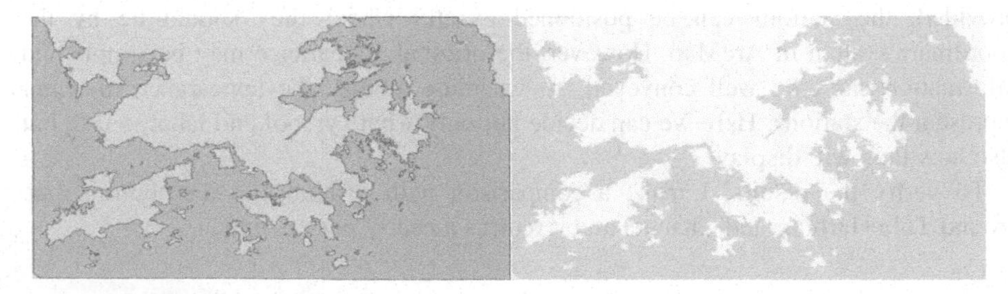

Fig. 5. The mask layer of the ocean and the barrier layer of the land

Now the preparations are completed, we choose the spatial analyst tool to perform interpolation based on the temperature measured by the stations. After entering appropriate parameters and the mask layer of the ocean, the prediction surface is generated as shown in Fig.6.

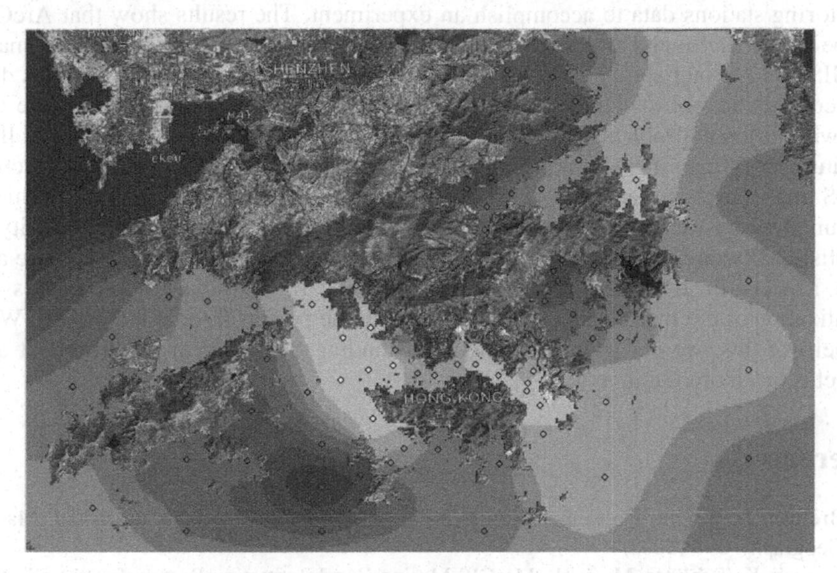

Fig. 6. The prediction of the temperature on the ocean surface using spline function method with obstacles

Now we have got an overall temperature distribution image of this region, however, we need to choose a most suitable interpolation method concerning the characteristics of the data and our own analysis needs in our practical work.

Charts can help to access data more intuitively, so drawing charts is also an effective way to visualize data. In ArcGIS, we can draw a variety of charts for different needs such as bar charts, area charts, bubble charts, scatter plot matrix and so on. Follows are some experiments on ocean environment data.

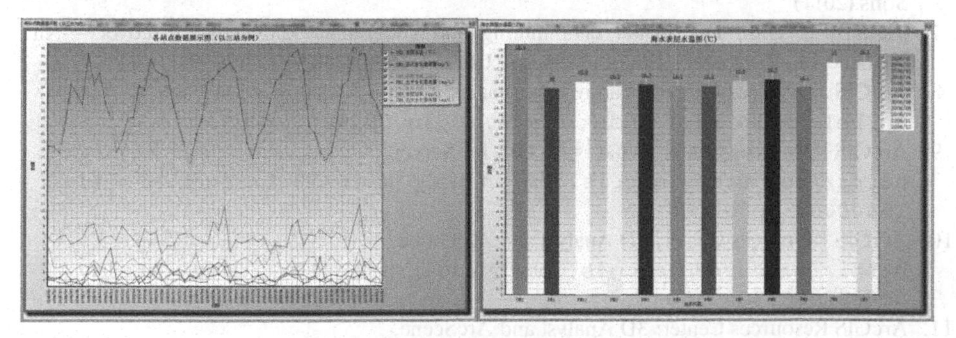

Fig. 7. Two examples of charts to visualize ocean environment data

5 Conclusion

In this article we have described the basic concepts and methods on visualization of ocean environment data Using ArcGIS, and then applied Hong Kong water quality monitoring stations data to accomplish an experiment. The results show that ArcGIS can be a very useful and effective visualization tool in the ocean environment domain. ArcGIS has a strong ability to handle different types of data, and it also provides data type conversion tool to satisfy a variety of needs. Many analysis tools available can deal with different problems with data, spatial or non-spatial, large data set or small.

Future work may include migration of the visualization system from C/S structure to B/S and from 2-D scene to 3-D. Since ArcGIS for Server [9] is capable to achieve full functionality on the server and ArcGlobe [10] or ArcScene [11] has the ability to establish 3-D scene model, we are pretty sure that ArcGIS will be applied to more and more aspects in the ocean environment domain. In addition, more studies on visualization of the marine data will be done with the powerful tools in ArcGIS. With the help of this, we can gain a better understanding of the ocean environment and protect it in a more intuitive way.

References

1. Breman, J., Wright, D., et al.: The inception of the ArcGIS marine data model. Marine Geography, 3–9 (2002)
2. Jerosch, K., Schlüter, M., et al.: MarGIS Marine Geo-Information-System for visualisation and typology of marine geodata (2005)
3. Zhang, R., et al.: Development of the Marine Management System Based on ArcGIS Server Technology. Geo-Information Science (2007)
4. Li, X., Yan, T., Gao, F., Zhou, L., Yu, J., Guo, Z.: Design of Data Management System for Seafloor Observatory Network. In: 2013 International Conference on Service Sciences (ICSS), pp. 147–150. IEEE Press, Shen Zhen (2013)
5. Martin Taylor, S.: Transformative ocean science through the VENUS and NEPTUNE Canada ocean observing systems. Nuclear Instruments and Methods in Physics Research Section A: Accelerators, Spectrometers, Detectors and Associated Equipment, pp. 63–67 (2009)
6. Kantardzic, M.: Data mining: concepts, models, methods, and algorithms. John Wiley & Sons (2011)
7. Mennis, J., Guo, D.: Spatial data mining and geographic knowledge discovery—An introduction. Computers, Environment and Urban Systems, 403–408 (2009)
8. ArcGIS Resources Center: Feature class basics, `http://resources.arcgis.com/en/help/main/10.1/index.html#//003n00000005000000`
9. ArcGIS Resources Center: What is ArcGIS for Server?, `http://resources.arcgis.com/en/help/main/10.1/index.html#//01540000037p000000`
10. ArcGIS Resources Center: 3D Analyst and ArcGlobe, `http://resources.arcgis.com/en/help/main/10.1/index.html#//00q800000053000000`
11. ArcGIS Resources Center: 3D Analyst and ArcScene, `http://resources.arcgis.com/en/help/main/10.1/index.html#//00q8000000p0000000`

The Design of a Collaborative Social Network for Watershed Science

Michael P. McGuire[1] and Martin C. Roberge[2]

[1] Towson University
Department of Computer and Information Sciences
[2] Towson University Department of Geography and Environmental Planning
8000 York Rd
Towson, Maryland 21252, USA
{mmcguire,mroberge}@towson.edu

Abstract. There is a strong and persistent demand amongst scientists, citizen scientists and the general public for hydrologic data such as NEXRAD imagery and stream gauge time-series. Despite this interest, basic analysis tools are available only through specialized scientific software that is accessible to a small cadre of users. Furthermore, hydrologic data, while highly available, has not been integrated in a single system and no system exists to facilitate collaboration for scientists, citizen scientists, and the general public. This paper presents the design of the Watershed Science Network which is a collaborative social network aimed at multiple user groups who are focused on hydrology and watershed science. More specifically, we present a lightweight system that can analyze large datasets quickly and efficiently, while allowing users to interact with one another and perform collaborative analysis. This online gathering spot will allow citizens to post photos of local conditions, data providers to post announcements to users, and field scientists to view station data in the field. Users of the system can subscribe to watersheds of interest and automatically receive updates of recent analysis, visualization, and discussion activity regarding the watershed.

Keywords: hydrology, watershed, social network, collaborative network, visualization, geographic information system.

1 Introduction

Recent advances in data visualization have made it possible for the public to explore large datasets using a simple web browser on their cell phone. These techniques allow terabytes of aerial photography to be quickly accessed through Google Maps, they enable playful, interactive visualizations of election result scenarios [29], and they encourage the user to find patterns in datasets that are complex enough that they otherwise would have seemed too daunting to explore casually [30]. Collectively, these three visualization examples became popular and are useful because they are platform independent, they encourage social interaction through comments and

F. Bian and Y. Xie (Eds.): GRMSE 2014, CCIS 482, pp. 95–106, 2015.

sharing, they do not require users to install specialized software, they work quickly and intuitively, and they employ clean, sleek designs that follow many of the graphical principles of Edward Tufte [35] and others [20] for the display of quantitative information. Several technologies make these visualizations possible, but two in particular lie at their heart: the use of distributed processing and storage to precompute queries to reduce processing times and improve data transfer rates and the direct, client-side manipulation of data within the web browser using streamlined code.

Hydrologic datasets are some of the most highly sought after scientific datasets amongst the general public and the scientific community. NEXRAD precipitation imagery forms the backdrop for TV weather reports and cell phone weather maps. USGS stream gauge data get analyzed by engineers, biologists, and planners in their work, and more casually by environmental groups, fishermen, and kayakers. Straightforward tasks such as hydrograph separation or the calculation of a flow duration curve are not nearly as effortless as they should be. Commonly used software such as HYSEP [31] must be downloaded and installed, creating platform compatibility issues; additional software is needed to format data for input; and the source code is difficult to modify for other purposes. Most scientists find it easier to calculate flow duration curves from scratch in general purpose software such as MS Excel or MATLAB rather than use specialized software; however this "manual" approach prohibits rapid data browsing of many datasets, and again creates barriers to entry for citizen scientists. Differing formats used by providers of hydrologic data such as NOAA and the USGS prevents many users from using these products together in a single graph or in a basic rainfall / runoff analysis. Furthermore, the use of unique formats by different providers hinders search operations and interferes with the interpretation of the observations.

1.1 Background

This paper presents the design of a collaborative infrastructure around which scientists and concerned members of the public will gather to interact with their data and one another. By improving access to hydrologic data and the tools to analyze them, a synergy will be created between citizens and scientists to improve understanding of environmental conditions and increase environmental stewardship. With this in mind, the system is designed in such a way that reduces barriers and encourages social interaction. For example, the use of social media tools to share results of data analysis and visualizations enables users to share irregularities they might discover in the data, or to post results of apps that generate interactive graphs. Furthermore, social media components allow stream gauge operators to post maintenance updates, and for interested users to comment upon recent site conditions, or to post photos of nearby flood conditions.

This project has grown out of the geosciences community in a number of ways. First and foremost, existing methods for data storage have resulted in data silos that make it prohibitive for users to integrate data in spatial and temporal dimensions. Also, existing visualization tools are cumbersome and have a steep learning curve.

Finally, the results of empirical analysis are typically created by a small group of collaborators and only shared in publications and static websites. This system addresses these issues by creating a collaborative environment results in a social network focused on watersheds where users can join a watershed or set of watersheds and jointly participate in data analysis, visualization, and the discussion of results. The collaborative system encourages the existing scientific community as well as the general public to collaborate on watershed research activities to characterize watershed hydrology, determine possible causes to watershed degradation, and ultimately propose interventions to improve watershed quality. Another added benefit of the collaborative system is that it results in a large network of collaborators over a given watershed and therefore will make data collection, analysis, and visualization more efficient.

2 Related Work

Fortunately, many of the problems surrounding hydrologic data have been recognized and steps have been taken to solve them. The CUAHSI Hydrologic Information System has created a vast network of data providers who share hydrologic information in a standardized format using semantic structures that enable searches for specific sites and information. Processed NEXRAD imagery is available as a web tile service [18], enabling mobile users to access real-time data through popular web mapping services such as the Google Maps API. HydroDesktop [3] is a MS Windows-based software application that allows users to search and visualize online hydrologic data services. The WHAT hydrograph separation tool [21] and the web version of the L-THIA hydrologic model [15,13] provide the cross-platform advantages of online analysis, but rely upon server-side processing, which can respond slowly to user input. Furthermore, since the analysis occurs on the server, advanced users do not have access to the code. The design presented in this paper is extensible in that all code is freely available and developers will be encouraged to contribute analysis tools to the network.

There are a plethora of projects in the geosciences community that focus on the integration of large datasets for scientific discovery. Recently the EarthCube [12] and DataOne [27] projects have set out to create cyber infrastructure to integrate data across the geosciences. The LTER Network [6] and Knowledge Network for BioComplexity [5] have established extensive efforts to integrate ecological data across a host of sites. There have also been a number of efforts to integrate hydrologic data. For example, the ArcHydro GIS toolset allows hydrologic analysis within ESRI ArcGIS® [23]. The National Hydrography Dataset (NHD) developed by the United States Geological Survey represents hydrographic features for the entire United States [32] in a standardized GIS database. This improves the spatial analysis of digital elevation models, stream networks, and watersheds. These data models, while very useful, require the user to have expertise in complex GIS systems which creates a major barrier to use. The Consortium of Universities for the Advancement of Hydrologic Sciences has developed a Hydrologic Information System (HIS) consisting of web services and analysis tools that allow users to access a variety of publicly available datasets [24]. The CUAHSI HIS web services also known as WaterML allow research sites that collect data and store the data in the CUAHSI Observations Data Model (ODM) [17] the ability to easily publish their data to make it available through the HydroDesktop interface or as a stand-alone web service interface. The USGS National Water information System (NWIS) has created web services based on the WaterML standard [38]. The most recent effort born out of this project is HydroShare [33]. One thing common to all of the above projects is that they are focused on a small cadre of expert scientists as their user base. This paper extends this work by incorporating a collaborative social network design for scientists, citizen scientists, and the general public who are interested in watershed science.

Recently, many applications have moved to NoSQL databases because of the flexibility that they offer to store structured, semi-structured, and unstructured data. Cloud-based data warehouses such as Hive have been shown to support data on the petabyte scale [34] and it is possible to conduct aggregation queries using the MapReduce framework [8], [9]. The use of cloud architectures in geospatial science has been identified as an emerging technology that will extend cloud computing beyond its current technological boundaries [37]. Recently, there has been research on processing spatial queries using the MapReduce framework [10] and a grid-based architecture has been implemented on Amazon EC2 cloud storage to perform hydrologic simulation [1]. Most recently, a cloud-based architecture built on Hadoop has shown to be effective in querying spatial dimensions of scientific datasets [2] and medical imaging [14]. Document databases such as MongoDB have shown to be very effective in the distributed management of spatial data [16], geoanalytic frameworks [22], and visual analytics [11]. The system architecture presented in this paper incorporates a flexible document oriented schema developed using MongoDB to store a diverse set of data.

Over the years a number of system development efforts have undertaken the need for online visualization and analysis of watershed data. In [19] a stand-alone application was created for the download, analysis, and visualization of hydrologic data. The US Environmental protection Agency has created the "Surf Your Watershed" web mapping application which allows users to input a zip code to receive information about the eight digit hydrologic unit code (HUC) watershed in which they live [36]. The CUAHSI HIS system has a number of tools that are integrated with its HydroServer such as the Time Series Analyst, Observations Data Model Tool, and HydroDesktop [24]. The Water-Hub framework [26] uses Hub-Zero, [25] an NSF-funded scientific collaboration framework to create collaborative data access, sharing, GIS, and data visualization. Finally, HydroShare [33] proposes to use the Hub-Zero architecture to provide access to data from raw form to publication. This paper adds to the existing work by focusing on the design and development of collaborative social networks for watershed science where we focus on a modular design with the aim of providing freely available software that can be easily incorporated into any large project framework. Furthermore, while this paper is focused on data-driven empirical analysis of watersheds, our approach can be extended to any scientific workflow centered on an area or site of interest.

3 System Architecture

The system architecture for the Watershed Science Networkis shown in Fig. 1. At the core of the system is a distributed data warehouse that will store all of the data associated with the site. The data generally consist of standardized data that can be ingested fromgovernmental sources such as the USGS, EPA, and NOAA as well as data from community projects such as the CUAHSI HIS System and user generated data that can be loaded through the web interface. A standardized data collector functions as a daemon that runs periodically to download new data into the data warehouse. Along with the document database that is used to store semi-structured and unstructured data, a relational database system is used to store the data to manage the collaborative social network. This includes user profiles and links to watershed pages, collaborators, and apps. The web interface has a dashboard (Fig.5a and Fig. 5b) that allows users to subscribe to a watershed or sets of watersheds and collaborate with individuals with common watersheds of interest. Included in the dashboard are apps for loading data into and downloading data from the data warehouse and apps for analysis and visualization. Social apps such as a news roll and a discussion board allow users to be aware of recent activity inside their watershed. The web interface will also be extensible in that users will be free to create their own visualization and analysis apps, the results of which are shared by all subscribers of a particular watershed. The web interface employs a responsive design, making it directly available for mobile devices without any additional coding. The code libraries used to create the visualizations are organized as easily reusable objects, and are shared through a public website.

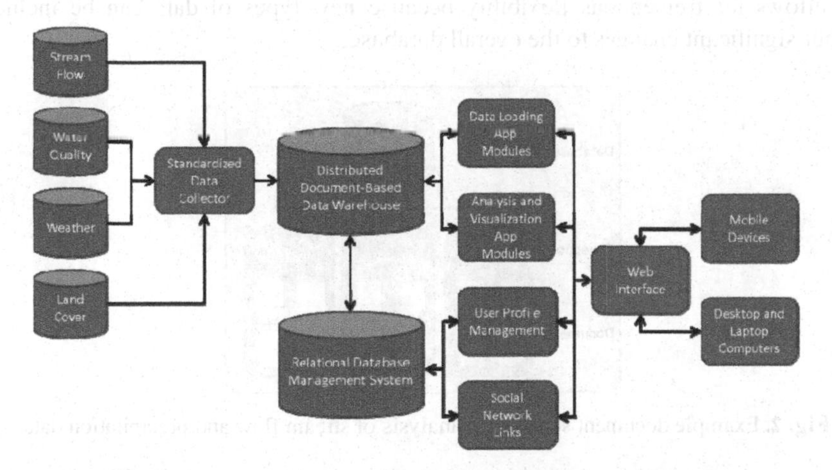

Fig. 1. System architecture for the Watershed Science Network

4 Distributed Data Warehouse Model

The data warehouse model design for the Watershed Science Network employs a document oriented database system to store multi-type data including spatial, timeseries, raster, documents, metadata, and user profile data. This type of data does not easily translate to a conventional RDBMS schema which consists of largely flat tables. The same problem exists for key/value stores such as Hadoop and Cassandra, where a great deal of processing needs to be done to translate the data back into a its original format. For this reason we use MongoDB to model the high dimensional spatio-temporal data. Our choice of a document oriented approach is based on the fact that while this type of NoSQL database is largely schemaless, a structure can be imposed on the data because it can be modeled using JavaScript Object Notation (JSON), which supports hierarchical objects. With this in mind, MongoDB (www.mongodb.org) serves as the back end database for this project because of its ability to scale to rapidly growing datasets using a sharded implementation where a single logical database is distributed across a cluster of machines.

4.1 Data Warehouse Design

MongoDB supports R-tree spatial indices making it a popular NoSQL solution for geospatial data [16]. MongoDB has also been shown to be very effective in the storage and query of time series data [11] as well as a time series of rasters [22]. MongoDB stores data in BSON format which is a binary form of the JavaScript Object Notation (JSON). MongoDB is scalable in that a MongoDB instance can easily be distributed across a cluster of machines using sharding [38]. MongoDB is modeled as a collection of hierarchical objects which includes, from top to bottom, the database, collection, and document. MongoDB is technically a schemaless database in that it doesn't require a rigid schema like a relational database system. This allows for tremendous flexibility because new types of data can be included without significant changes to the overall database.

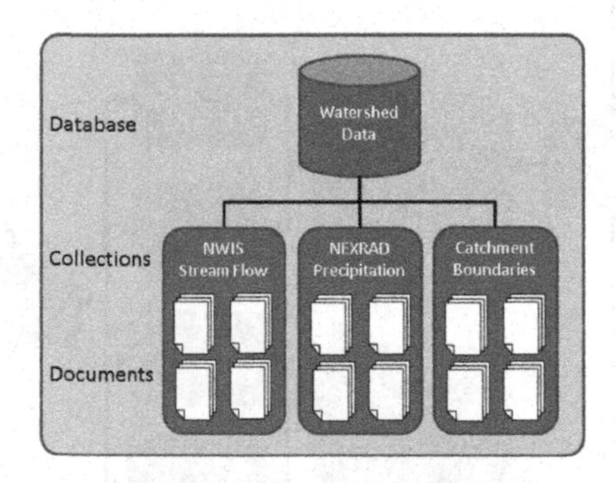

Fig. 2. Example document schema for analysis of stream flow and precipitation data

However, this is not to say that data stored in MongoDB does not need to be organized. With this in mind, we have designed a MongoDB schema to handle multi-source and multi-type data needed for empirical watershed science. The MongoDB schema for the prototype system is shown in Fig. 2. The example schema contains data that is used in the analysis of rainfall and stream flow for gauged catchments in the USGS NWIS System. This schema is used by the watershed mapping and hydrograph visualization apps shown in Fig.5a and Fig. 5b. In this schema, the data is contained in one database and collections exist for NWIS stream flow, NEXRAD precipitation, and catchment boundaries. Documents representing the data elements are then stored within each collection.

Example documents for each collection are shown in Fig. 3. This example shows the flexibility of the document schema to model a) a time series object, b) a raster object, and c) a polygon object. In the schema, time series objects, shown in Fig. 3 a), are stored as a single document for each measurement where the document consists of a data type, site id, variable, units of measure, sample rate, time, and data value. We have chosen to store each measurement as a single document because it is more efficient to insert new measures as new documents. Fig. 3 b) depicts a sample document for a NEXRAD precipitation grid where we have fields for type, site, variable, units, sample rate, time, and data. In this document each grid is stored as an array under the data field. Using this model, it is easy to query grids by time and bring them into apps for visualization and processing. Finally, an example watershed boundary is shown in Fig. 3 c). This document contains field for type, site, area, area units, and geom. The geom field is where the geometry for the watershed polygon is stored as a set of coordinate pairs.

```
{
"type": "NWIS",
"site": "01582500",
"variable": "Discharge",
"units": "CFS",
"sample rate": "5 minutes",
"time": "2012-07-17T17:30:00.000-05:00",
"data": {
   "value": "111"
}
}
```

a) Document Schema for NWIS Data

```
{
"type": "NEXRAD",
"site": "KWLX",
"variable": "Precipitation",
"units": "millimeters",
"sample rate": "15 inutes",
"time": "2012-07-17T17:30:00.000-05:00",
"data": {
   "PrecipitationGrid": "[
        [2.5,2.3,2.4,…],
        [2.3,2.3,2.2,…],
        [2.2,2.1,2.1,…],
        …
   ]"
}
}
```

b) Document Schema for NEXRAD Data

```
{
"type": "Catchment",
"site": "01582500",
"area": "175"
"area units": "hectares"
"geom": "(-74.23995830577357 43.981915607019829,
-74.240323808107561 43.981976665508334,
-74.240689311096773 43.982037722996381,
-74.240774492143913 43.981775734641211,
    )"
}
```

c) Document Schema for catchment

Fig. 3. Example BSON Document Schemas

4.2 Extraction Transformation and Loading of External Data Sources

The data warehouse also includes routines for harvesting external watershed data sources, extracting the data, transforming the data to BSON format and loading it into the data warehouse. This process is also known as extraction transformation and loading (ETL). The ETL workflow for ingesting external watershed data sources is shown in Fig. 4. The ETL workflows are implemented in the Python programming using the PyMongo API for MongoDB. In this example ETL workflow three data sources are being ingested including Stage IV NEXRAD precipitation data available through ftp, USGS stream flow data available through web services, and watershed boundary shape files available through http. The Stage IV NEXRAD precipitation data is in gridded binary (.grib) format and is available through ftp. The ETL process decodes the grib file into three components which include arrays for precipitation, latitude, and longitude values. Each grid array is converted to a document similar to that shown in Fig. 3 b), encoded into BSON format and loaded into the database.

The USGS stream flow data is a time series available for approximately 12,000 watersheds throughout the United States. This data is available through a REST web services interface. The ETL workflow parses the REST response, which is in JSON format, extracts the stream flow values, creates the document shown in Fig. 3 a), and loads them as documents into the data warehouse. The watershed boundaries are downloaded in shape file format and we extract the geometry for each polygon in the watershed shape file. Once the geometry is extracted, a document is created for each polygon similar to that shown in Fig. 3 c). This document is then encoded in BSON format and inserted into the database.

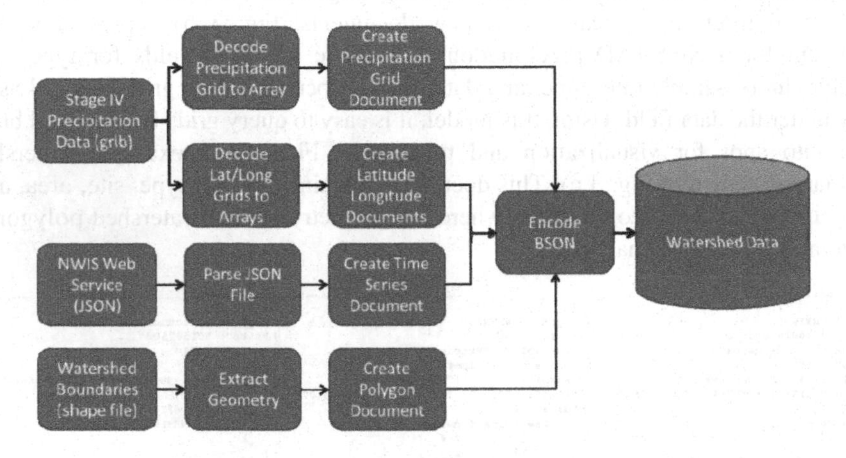

Fig. 4. ETL workflow for harvesting watershed data

5 Social Networking Interface

The interface for the Watershed Science Network is designed so that scientists and the general public can collaboratively visualize and analyze watershed data to promote the discovery of new knowledge. To do this, we have created a social networking-style interface centered on watershed science. The interface design was inspired by social networks such as Facebook, MySpace, and Google+. However, while these networks focus on the individual, this interface adds a spatial dimension. Users will subscribe to a watershed or a set of watersheds, and then use a series of embedded apps that allow data visualization, analysis, and the discussion of results and events. Fig.5 shows a prototype interface for the proposed Watershed Science Network. The interface consists of two major sections: the activity pane on the right, and the dashboard on the left. The activity pane is where users interact with applications designed for watershed science. The dashboard allows the user to customize their personal site by subscribing to watersheds, apps, and friends, as well as keeping track of recent activity in their set of subscribed watersheds.

The app model employed by our design allows users to add different apps to their personal site. The types of apps include mapping and visualization tools, discussion modules, event calendars, raster subsetting and time series conversion. Mapping apps use the mobile-friendly Leaflet JavaScript library and visualization tools use the powerful and lightweight D3 JavaScript library [7]. Mapping and visualization apps are designed to be modular so that code can easily be repurposed and incorporated

Into new apps. The typical user has the ability to customize existing apps to focus on a particular time or place while more advanced users can develop their own apps using a tutorial and available source code. To facilitate this, all source code will be made available on GitHub.

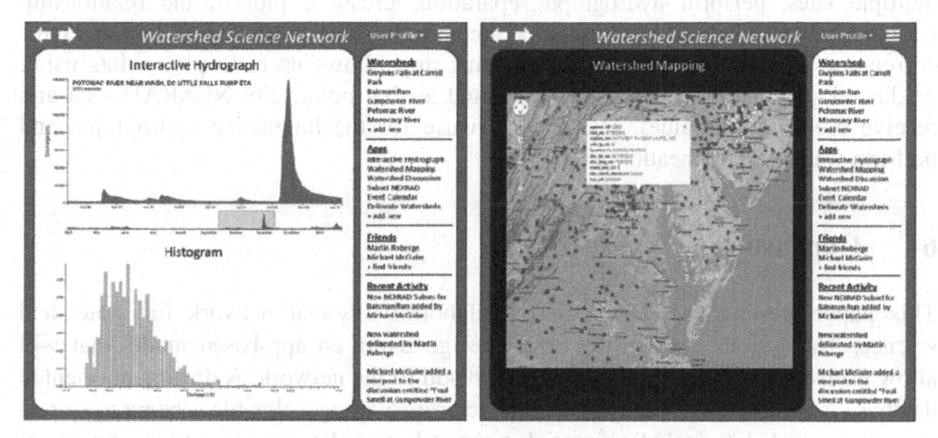

Fig. 5a. Interactive Hydrograph App **Fig. 5b.** Watershed Mapping App

The interface design for the Watershed Science Network is shown in Fig.5. The interface includes an activity pane and a dash board depicted on the left of the figure. Once a user subscribes to a watershed, a link to the watershed page will be added to this section of the dash board. Watershed pages include a map of the watershed; watershed statistics; a discussion forum for each watershed where users can discuss data;and links to visualizations, field visits, reports, and events. Possible discussion items include observations from field visits (e.g. photos of sensors after floods, after damage, of snow conditions), events discovered from sensor visualizations (e.g. a flood of record), or anomalous sensor values. In order to aid this process, we include applications for loading and editing data and creating derivative data products.

The Apps section of the dash board shows the apps that the user has added to their site. Apps that are included in the system include an interactive hydrograph app (Fig.5a), a watershed mapping app (Fig.5b), an app for discussion forums surrounding watersheds, an app for subsetting NEXRAD precipitation data, an event calendar, and an app for delineating watersheds. The app section of the dash board also has a link to add new apps. This link takes the user to an "app store" where the user can search for new apps to add to their dash board.

The dash board also contains links to friends pages where the user can peruse the pages of collaborators. This part of the dash board facilitates social interaction between collaborators. For example, by browsing collaborator's pages the user can see the watersheds that their collaborators are subscribed to, new discussions that the collaborator has joined, and any new analyses or visualizations that the friend has posted to their site. The dash board contains a section where recent activity will be posted. This section will contain a log of new activities in subscribed watersheds as well as new activities by a user's friend network. This will include new analysis results, new discussion posts, and new analyses by collaborators.

Using the app model a number of apps allow users to explore hydrologic data in a number of ways. For example, the Interactive Hydrograph app shown in Fig. 5a allows users to explore time series plots of stream flow and precipitation for a watershed. In its current design, the interactive hydrograph app also allows users to The Watershed Mapping app allows users to generate flow duration curves for multiple sites, perform hydrograph separation, create a plot of the relationship between stage and discharge, and plot precipitation and discharge at two sites for correlation analysis. The watershed mapping app allows users to plot points using graduated symbols based on values collected at that point, plot NEXRAD data and receive precipitation values, select sites to use for the interactive hydrograph, and perform watershed delineation.

6 Conclusion

This paper presents the design of a collaborative social network for watershed science. The system is extensible in its design using an app-based model that will allow users to develop their own modules to add to the network. A document oriented database is at the core of the system architectureproviding a flexible schema to store a disparate set of data including spatial, temporal, metadata, user profiles, and social network related data. The backend database will also store user contributed data including field collected data, figures, and maps. A particularly novel aspect of this design is a user interface which centers on social interaction between collaborators and allowsusers to subscribe to watersheds that they are interested in and post results and comment on the results of others. The vision is to engage scientists, citizen scientists, and the general public to facilitate a better understanding of watershed processes from those living in the watershed. Future steps in this research include testing the design where we will implement the design and test it with a cadre of scientists and citizen scientists from the hydrologic community. We will also test the extensibility of the network by including app developers in the testing where they will be tasked with developing custom apps for the network.

Acknowledgments. This work was supported by the Towson University School of Emerging Technologies.

References

1. Aji, A., Wang, F.: High performance spatial query processing for large scale scientific data. In: Proceedings of the SIGMOD/PODS 2012 PhD Symposium - PhD 2012, p. 9 (2012)
2. Aji, A., Wang, F.: Towards Building a High Performance Spatial Query System for Large Scale Medical Imaging. In: Proceedings of the 20th ACM SIGSPATIAL International Conference on Advances in Geographic Information Systems (2012)

3. Ames, D.P., Horsburgh, J.S., Cao, Y., Kadlec, J., Whiteaker, T., Valentine, D.: HydroDesktop: Web services-based software for hydrologic data discovery, download, visualization, and analysis. Environmental Modelling& Software 37, 146–156 (2012)
4. Amazon Elastic Compute Cloud (EC2) (2011), `http://aws.amazon.com/ec2/` (accessed online June 30, 2013)
5. Andelman, S.J., Bowles, C.M., Willig, M.R., Waide, R.B.: Understanding environmental complexity through a distributed knowledge network. BioScience 54(3), 240–246 (2004)
6. Baker, K.S., Benson, B.J., Henshaw, D.L., Blodgett, D., Porter, J.H., Stafford, S.G.: Evolution of a multisite network information system: the LTER information management paradigm. BioScience 50(11), 963–978 (2000)
7. Bostock, M.: Data-Driven Documents D3.js. JavaScript library (2012), `http://d3js.org/` (accessed online: June 30, 2013)
8. Brezany, P., Zhang, Y., Janciak, I., Chen, P., Ye, S.: An Elastic OLAP Cloud Platform. In: 2011 IEEE Ninth International Conference on Dependable, Autonomic and Secure Computing, pp. 356–363 (2011)
9. Cao, Y., Chen, C., Guo, F., Jiang, D., Lin, Y., Ooi, B., Tam, H., Wu, S., Xu, Q.: Es2: A cloud data storage system for supporting both OLTP and OLAP. In: 2011 IEEE 27th International Conference on Data Engineering (ICDE), pp. 291–302 (2011)
10. Chiang, G.T., Dove, M.T., Bovolo, C.I., Ewen, J.: Implementing a Grid/Cloud eScience Infrastructure for Hydrological Sciences. In: Yang, X., Wang, L., Jie, W. (eds.) Guide to e-Science, pp. 3–28. Springer, London (2011)
11. Cube Time Series Data Collection and Analysis (2012), `http://square.github.com/cube/` (accessed online: June 30, 2013)
12. Di, L., Yue, P.: Provenance in earth science cyberinfrastructure. A White Paper for NSF EarthCube (2011)
13. Engel, B.A., Choi, J.Y., Harbor, J., Pandey, S.: Web-based DSS for hydrologic impact evaluation of small watershed land use changes. Computers and Electronics in Agriculture 39(3), 241–249 (2003)
14. Han, Y., Park, D.-S., Jia, W., Yeo, S.-S. (eds.): Ubiquitous Information Technologies and Applications, vol. 214. Springer, Dordrecht (2013)
15. Harbor, J.M.: A practical method for estimating the impact of land-use change on surface runoff, groundwater recharge and wetland hydrology. Journal of the American Planning Association 60(1), 95–108 (1994)
16. Heard, J.R.: Geoanalytics. Technical Report TR-11-03, Renaissance Computing Institute (2011), `http://www.renci.org/wp-content/uploads/2011/03/TR-11-03.pdf` (accessed: December 26, 2012)
17. Horsburgh, J.S., Tarboton, D.G., Maidment, D.R., Zaslavsky, I.: A relational model for environmental and water resources data. Water Resources Research 44(5) (2008)
18. Mesonet, I.: Iowa ag climate network (2008), `http://mesonet.agron.iastate.edu/agclimate/index.php` (accessed online: June 30, 2013)
19. Jeong, S., Liang, Y., Liang, X.: Design of an integrated data retrieval, analysis, and visualization system: application in the hydrology domain. Environmental Modelling & Software 21(12), 1722–1740 (2006)
20. Kelleher, C., Wagener, T.: Ten guidelines for effective data visualization in scientific publications. Environmental Modelling & Software 26(6), 822–827 (2011)
21. Lim, K.J., Engel, B.A., Tang, Z., Choi, J., Kim, K.S., Muthukrishnan, S., Tripathy, D.: Automated Web GIS Based Hydrograph Analysis Tool, WHAT. JAWRA Journal of the American Water Resources Association 41(6), 1407–1416 (2005)

22. Lu, S., et al.: A Framework for Cloud-Based Large-Scale Data Analytics and Visualization: Case Study on Multiscale Climate Data. In: 2011 IEEE Third International Conference on Cloud Computing Technology and Science, pp. 618–622 (2011)
23. Maidment, D.R.: ArcHydro: GIS for water resources. ESRI, Inc. (2002)
24. Maidment, D.R.: Bringing water data together. Journal of Water Resources Planning and Management 134(2), 95–96 (2008)
25. McLennan, M., Kennell, R.: HUBzero: a platform for dissemination and collaboration in computational science and engineering. Computing in Science & Engineering 12(2), 48–53 (2010)
26. Merwade, V., Ruddell, B., Song, C., Brophy, S., Mohtar, R., Yerrammilli, A.: Water-HUB-A community cyberinfrastructure for hydrology education and research. AGU Fall Meeting Abstracts (2010)
27. Michener, W., Vieglais, D., Vision, T., Kunze, J., Cruse, P., Janée, G.: DataONE: Data Observation Network for Earth-preserving data and enabling innovation. In: Paths to the White House - Election 2012, NYTimes.com (2012), http://elections.nytimes.com/2012/results/president/scenarios (accessed online: June 30, 2013)
28. MongoDBSharding Overview (2012), http://docs.mongodb.org/manual/core/sharding/#sharding-overview (accessed: December 26, 2012)
29. Paths to the White House - Election 2012 - NYTimes.com (2012), http://elections.nytimes.com/2012/results/president/scenarios (accessed online: June 30, 2013)
30. Rosling, H.A.: Visual technology unveils the beauty of statistics and swaps policy from dissemination to access. Statistical Journal of the IAOS: Journal of the International Association for Official Statistics 24(1), 103–104 (2007), http://www.gapminder.org/
31. Sloto, R., Crouse, M.: HYSEP: A Computer Program for Streamflow Hydrograph Separation and Analysis: US Geological Survey Water-Resources Investigations Report 96-4040. 46 (1996)
32. Simley, J.D., Carswell Jr., W.J.: The national map—hydrography. US Geological Survey Fact Sheet 3054(4) (2009)
33. Tarboton, D., Idaszak, R., Ames, D., Horsburgh, J., Goodall, J., Band, L., Merwade, V., Couch, A., Arrigo, J., Hooper, R., Valentine, D.: HydroShare: An online, collaborative environment for the sharing of hydrologic data and models. In: Fifty Years of Watershed Modeling - Past, Present, and Future, Boulder, CO, September 24-26 (2012)
34. Thusoo, A., Sarma, J.S., Jain, N., Shao, Z., Chakka, P., Anthony, S., et al.: Hive: a warehousing solution over a map-reduce framework. Proceedings of the VLDB Endowment 2(2), 1626–1629 (2009)
35. Tufte, E.R.: The visual display of quantitative information, vol. 2. Graphics Press, Cheshire (1983)
36. US Environmental Protection Agency Surf Your Watershed (2004), http://cfpub.epa.gov/surf/locate/index.cfm (accessed Online June 30, 2013)
37. Yang, C., et al.: Spatial cloud computing: how can the geospatial sciences use and help shape cloud computing?. International Journal of Digital Earth 4(4), 305–329 (2011), Akdogan, A., Demiryurek, U., Banaei-Kashani, F., Shahabi, C.: Voronoi-Based Geospatial Query Processing with MapReduce. In: 2010 IEEE Second International Conference on Cloud Computing Technology and Science, pp. 9–16 (2010)
38. Zaslavsky, I., Valentine, D., Whiteaker, T.: CUAHSI WaterML v0. 3.0. OGC07-041r1, 76 p. Open Geospatial Consortium, Inc. (2007)

Construction Land Layout in Qi River Ecological District of Hebi City Based on GIS

Xi Wang [1], Xiaolei Wu[2], and Weixing Mao[1]

[1] The College of Environment and planning, Henan University, Kaifeng 475004, Henan, China
[2] Henan Mechanical and Electrical Engineering College, Xinxiang, 453001, Henan, China

Abstract. This paper explored the rational construction land layout in Qi river ecological district of Hebi city so as to provide references for the construction land layout in other similar domestic area. We took the geographical information system software (ArcGIS) and statistical software (SPSS) as technical support. As research methods, combining Qualitative analysis and quantitative calculation, data analysis with graphical analysis were adopted in the study. The paper evaluated the suitability and the ecological sensitivity of the construction land in Qi River ecological area, then analyzed the evaluation results supported by ArcGIS, finally got the suitable part of construction land in the ecological region. According to research, rational layout of construction land in ecological region is not only relates to suitability evaluation but also sensitivity evaluation; There are strong correlations among ecological sensitivity and water system, land cover type, elevation, special value.

Keywords: Geographic information system (GIS), Construction land layout, Ecological sensitivity assessment, Land suitability evaluation, Rational layout.

1 Introduction

Healthy and stable ecological environment is the premise of survival and development of human society, and the layout of the construction land is an important regional economic foundation. Hebi is a both resource and tourist city. At present, the city is in the accelerated development period of industrialization, urbanization and modernization. Due to coal mining, environmental, pollution of Hebi spread from the point to plane. The resource destruction is becoming more and more serious, which is a serious threat to the sustainable development of social economy. Therefore, the study on reasonable layout of construction land in Hebi ecological area has a strong practical significance in the rational utilization of ecological natural resources, human resources and tourism resources

2 Research Methodologies

Taking the sustainable development and recycling economy theory as the guiding theory of ecological construction, the ecological zone land is evaluated under the

F. Bian and Y. Xie (Eds.): GRMSE 2014, CCIS 482, pp. 107–116, 2015.

theory of construction land suitability assessment and the theory of ecological sensitivity ecological, and then the two evaluation results are superimposed in support of ARCGIS software to obtain the construction land layout suitable area of ecological zone [1, 2].

3 Empirical Analyses

3.1 Full-Sized Camera-Ready (CR) Copy Ecological Area Construction Land Analysis of Qi River in Hebi

Qi River ecological zone was established in 2007, approved by Hebi People's Government, with a total area of 37 square kilometers and a population of about 16000 people; Qi River flows from northwest to southeast, through the length of about 18 kilometers, and the water quality is good. According to the map of ecological region present land use of Qi River in Hebi, the current land use situation of ecological zone: mountain and forest land area is large, the forest vegetation coverage is high; the town, rural residential land is relatively dispersive, local population density is small; traffic land, industrial and mining land, scenic area land use types are available, especially scenic spot, with rich tourism resources [3].

Ecological problems summed up in the following aspects: the infrastructure was weak, the existing road grade mixed with poor quality; construction land was dispersive, low using efficiency and adverse to the ecological environmental; municipal infrastructure corridor across the region and the region was divided into pieces of land, with a high degree of landscape fragmentation; unauthorized reclamation and unauthorized construction in the ecological region did big harm to the wetland resource of Qi River.

3.2 Ecological Area Construction Layout of Land Suitability Evaluation

According to the condition and features of Qi River ecological zone and combined with some recognized indicators of construction land adaptability evaluation, it built Qi River ecological zone construction land suitability evaluation system from 5 aspects(containing 15 impact factors): the engineering geological conditions, terrain conditions, geographical conditions, natural disasters and fundamental condition [4,5,6]. Analytic hierarchy process was used to determine the index weight of Qi River ecological zone construction land suitability evaluation. The general steps of the analytic hierarchy process are in the below. First of all, set up target layer, criterion layer and project layer, and then construct the judgment matrix, the first step to construct the level indicators of judgment matrix, the second step to construct the two level indicators of judgment matrix. Let A, B, C, D, E respectively represent the terrain conditions, engineering geological conditions, geographical conditions, natural disasters and basic conditions. Using 9 numerical scaling method, "1" means two factors are of equal importance, "3" means the former is obviously important, " 5 " means very important, "7" means the former is very obviously important, "9" is extremely important; numerical 2, 4, 6, 8 is in the intermediate value of the above-mentioned judgment. Using the above method to calculate each impact factor of the corresponding factors influencing proportion, the formula was used again to calculate the weight of various influencing factors on Qi River ecological zone construction land.

After using this method to calculate each influencing factor weight value, the expert scoring method was used to amend the weight (Table.1).

Table 1. Index weight of influence factor for suitable assessment system

Influence factor	Weight	Specific factor	Weight	Index Weight
Terrain conditions	0.150	Slope	0.200	0.030
		Aspect	0.150	0.023
		Elevation	0.350	0.053
		Topography types	0.300	0.045
Engineering geological conditions	0.100	Components of the earth's surface	0.300	0.030
		Bearing capacity of foundation (t/m2)	0.250	0.025
		Groundwater's depth	0.350	0.035
		Water and soil erosion	0.100	0.010
Geographical conditions	0.450	Hebi city circle radiation zone(Km)	0.200	0.090
		Radiation of ecological region and nearby Towns (m)	0.450	0.203
		Traffic location (m)	0.350	0.158
Natural disaster	0.050	Flood buffer and water ecological isolation zone	0.400	0.020
		Geological stability	0.600	0.030
Fundamental condition	0.250	Land utilization	0.700	0.175
		Communication, electric and water conditions	0.300	0.075

Using ARCTOOLBOX Union and Buffer functions for data spatial analysis in ARCMAP, the thematic map of impact factors was obtained.

According to the respective weight of each topography factor in the factor graph based on slope factor, aspect factor, elevation factor and topography types factor, four maps were superimposed to obtain topography factors influence graph.

According to each factor respective weight of the overall goal in the topography factors influence graph, flood buffer and water ecological isolation zone factor graph, traffic location factor graph, land utilization factor, ecological region and nearby towns' radiation factor graph, construction land suitability assessment graph was obtained in the ARCGIS software. By the construction land suitability assessment graph of Qi River ecological zone, the appropriate and suitable for the arrangement of the construction land area was obtained. Number 1 to 5 in figure 1 represents suitable degree of ecological area from low to high (Fig.1).

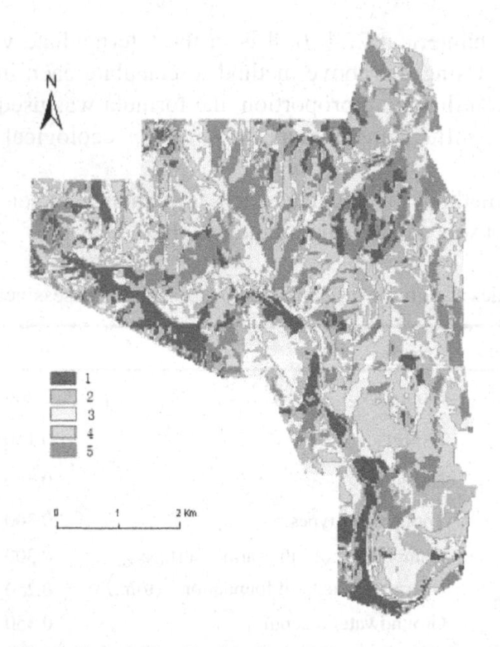

Fig. 1. Suitability Evaluation of Construction Land

3.3 Ecological Sensitivity Assessment of the Layout of Ecological Zone Construction Land

On the analysis of current situation of Qi River ecological zone in Hebi, based on the characteristics of water factor, natural landscape, the cultural landscape of special value factor, height factor, slope factor, land cover types factor 5 aspects to build the construction land ecological sensitivity assessment system of Qi River ecological zone [7,8,9,10].

The Delphi method was used to determine the sensitivity assessment value of the internal components of single factor, and the assessment standards is divided into five levels,respectively1, 2, 3, 4, 5,representative significance is as follows:1, non-sensitive index, the corresponding regional non-sensitive area;2, low sensitivity, its index corresponding to the area for the low sensitive area;3, middle sensitive, its index corresponding to the area of the middle sensitive area;4,sensitive,its index corresponding to the area of sensitive area;5,high sensitive, its index corresponding to the region of high sensitive area. The higher ecological sensitivity of the area there is, the fewer suits for the construction land layout, and vice versa (Table.2).

Table 2. Specific assignment of ecological sensitivity evaluation standard

Factor	Classification	Index
Elevation	90——120	1
	120——150	2
	150——180	3
	180——210	4
	210——495	5
Slope	Land of 0°——5°	1
	Land of 5°——10°	2
	Land of 10°——15°	3
	Land of 15°——20°	4
	Land of >20°	5
Water	200 meters outside of the river buffer	1
	150——200 meters outside of the river buffer	2
	100——150 meters outside of the river buffer	3
	50——100meters outside of the river buffer	4
	50 meters of water around	5
Land and vegetation	Current construction land	1
	Farmland	2
	Woodland	3
	Natural scenery protection areas, rivers and wetlands	5
Special value	Tai chi natural scenic area	5
	Qi River coast and core wetland area	5
	Jinshan Temple Scenic Area, Luo Guanzhong Literature Research Institute	5

In the ecological sensitivity assessment of Qi River in Hebi, the analytic hierarchy process method was used again to determine the weight of every influence factor value. After calculating the weight of every influence factor, the expert scoring method was used to amend the weight, finally the assessment system weight table was obtained (Table.3).

Table 3. Index weight of Ecological sensitivity evaluation system impact factor

Factor	Water	Elevation	Slope	Land cover types	Special value
Weight	0.3	0.20	0.10	0.2	0.2

Again using ARCTOOLBOX Union and Buffer functions for data spatial analysis in ARCMAP, the thematic map of impact factors was obtained.

According to the respective weight of each topography factor in the factor graph based on water factor, special value factor, elevation factor, slope factor, land cover types factor and topography types, four maps were superimposed to obtain ecological sensitivity factors influence graph. Number 1 to 5 in Fig.2 represents sensitive degree of ecological area from low to high.

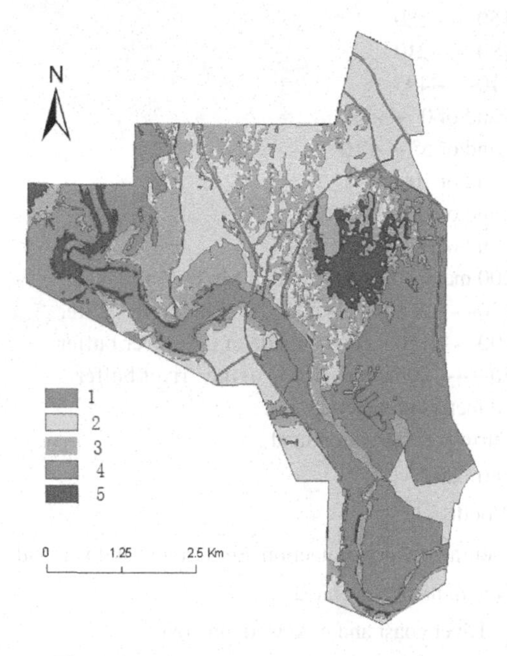

Fig. 2. Ecological Sensitivity Evaluation

3.4 Comprehensive Assessment of the Construction Land in Ecological Area

The suitability assessment and ecological sensitivity assessment graph of Qi River ecological zone land construction in Hebi was superimposed in support of ARCGIS software, and the comprehensive assessment of construction land in the ecological zone was obtained. The main factor of suitability assessment of construction land and ecological sensitivity accounted for the influence of weight 0.3 and 0.7 (by the expert consulting method), then the comprehensive overlay analysis graph of ecological sensitivity and the construction land suitability was obtained. Number 1 to 5 in Fig.3 represents suitable degree of ecological area from low to high.

Number 1 to 5 in figure 3 represents suitable degree of ecological area from low to high. Number 4 and 5 represents very appropriate and suitable distribution of Qi River ecological zone construction land layout in Hebi, mainly for the existing urban construction land, the original rural residential, industrial and mining land distribution. These areas can withstand a certain degree of human interference, but serious interference would lead to the soil erosion and other related natural disasters,

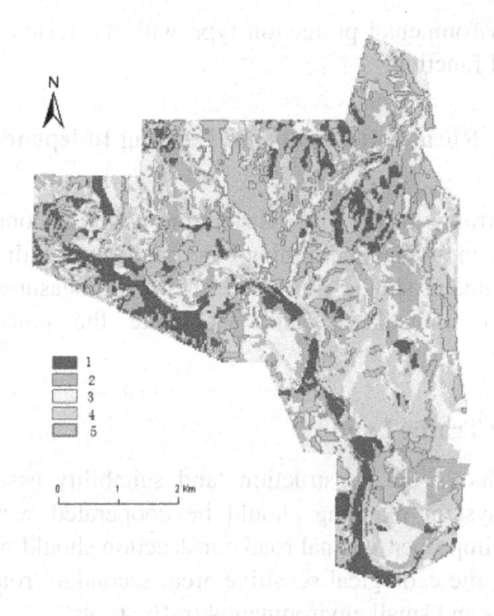

Fig. 3. The comprehensive map of ecological sensitivity and suitability assessment of construction land

with slower ecological restoration; It is unsuitable and not appropriate for the distribution area of construction land, as the area with fragile ecological environment, vulnerable to human disturbance, resulting in the ecosystem instability, mainly along the river 50 meters vertically away from the strip, and some destroyed ecological vegetation which is being ecological restoration construction; the rest area for construction land distribution is not suitable distribution area, mainly for steep woodland distribution.

4 Suggestion of the Reasonable Layout of Qi River Construction Land in Ecological Zone

4.1 Suggestion for Urban Construction Land

The urban construction land in original ecological zone is very suitable, so the area should be expanded in the present situation, and should improve the urban population density and land utilization to make eco-town township form a centralized sheet pattern based on the original building area.

4.2 Suggestion for Rural Settlements

We should merge the existing rural settlements, relocate to the very suitable 4 resident areas gradually, narrow the existing area scale and improve the land use intensive degree; the village construction should respect ecological resources and pay equal attention to the protection and development, to build the rural ecological community, which is of green environmental protection type with the leisure and tourism resort agricultural ecological function.

4.3 Suggestion for Rural Settlements the Existing Independent Industrial and Mining

We can continue to arrange in situ site distribution, and take control of the area size and number. For those industrial and mining enterprises, which did serious damage to the ecological environment, should be closed, then take measures to reconstruct the ecological restoration immediately and accelerate the construction of water conservation forest.

4.4 Suggestion for Traffic Land

Considering the results of the construction land suitability assessment, ecological zone road skeleton system planning should be cooperated with the urban space development, and the improper original road construction should be adjusted, the road should avoid crossing the ecological sensitive area; secondary road should set traffic road for walk, bicycles and small environmental traffic tools.

4.5 Suggestion for Waters and Water Conservancy Facilities

The area along Qi River is not suitable for construction land, so it should be defined strictly protected areas, the establishment of banning digging, banning mining, banning lumbering, forbidden grazing area, banning reclamation area, phosphorus prohibition area, banned building the contaminative and poisonous industry in the area .But some layout of the construction land is mainly for the purpose of better water resources protection and wetland development, such as wetland science base, scientific research and observation point, should be arranged along the Qi River ecological protection buffer zone.

4.6 Suggestion for Scenic Spots and Special Use Land

Unsuitable areas should be prohibited the construction activity. We should carry out the strict measures of biological species, ecological environment and natural landscape to protect the whole regional ecological safety. The core area of land is limited to scientific research and observation using. The land out of the core area is proper to develop the tourism and health as the main content of the catering, accommodation, medical and other industries to improve the economic value of land.

5 Conclusion

The layout and assessment of construction land in ecological zone is a comprehensive, highly relevant research topic. The following conclusion was obtained through this study:

Study on the reasonable layout of construction land in ecological zone should not only assess the ecological sensitivity, but also pay attention to combine with the ecological suitability assessment of construction land, so as to characterize the suitable layout of the construction land in ecological zone fully and arrange the layout of construction land reasonably.

Ecological sensitivity is great influenced by water system, land cover types, elevation and special value. Study on the ecological sensitivity assessment shown the weight of water system, land cover type, elevation and special value factors are higher. Therefore, their influence on ecological sensitivity; we should strengthen the development and protection on these factors reasonably in order to promote the sustainable development of Qi River ecological area.

Acknowledgments. Supported by the National Natural Science Foundation of China (No. 41171438), Major Project for Humanistic and Social Science Base of Ministry of Education, China (No. 10JJDZONGHE015) and Project for Humanistic and Social Science of the Education Department Henan province (No. 2012-JD-013).

References

1. Zhang, L., Zong, Y.G., Yang, W.: Ecological Suitability Assessment of Urban Construction Land Use Based on GIS——the Case Study of Liancheng County of Fujian Pprovince. Journal of Shandong Normal University (Natural Sciences) 23(9), 95–98 (2008)
2. Zhou, J.F., Zeng, G.M., Huang, G.H.: The Ecological Suitability Evaluation on Urban Expansion Land Based on Uncertainties. Acta Ecologica Sinica 2(2), 774–781 (2007)
3. Wang, C.G., Zong, Y.G.: GIS-based Ecological Suitability Evaluation for Town Development Used-land in Dalian City. Journal of Zhejiang Normal University (Natural Sciences) 30(1), 109–114 (2007)
4. Chen, Y.F., Du, P.F., Zheng, X.J.: Evaluation on Ecological Applicability of Land Construction in Nanning City Based on GIS. Journal of Tsinghua University (Science and Technology) 46(6), 801–804 (2006)
5. González, A., Gilmer, A., Foley, R., Sweeney, J., Fry, J.: Applying Geographic Information Systems to Support Strategic Environmental Assessment: Opportunities and Limitations in the Context of Irish Land-use Plans. Environmental Impact Assessment Review 31(3), 368–381 (2011)
6. Raizada, A., Dogra, P., Dhyani, B.L.: Assessment of a Multi-objective Decision Support System Generated Land Use Plan on Forest Fodder Dependency in a Himalayan Watershed. Environmental Modeling & Software 23(9), 1171–1181 (2008)
7. Liu, J., Ye, J., Yang, W., Yu, S.X.: Environmental Impact Assessment of Land Use Planning in Wuhan City Based on Ecological Suitability Analysis. Procedia Environmental Sciences 2(1), 185–191 (2010)
8. Barral, M.P., Oscar, M.N.: Land-use Planning Based on Ecosystem Service Assessment: A Case Study in the Southeast Pampas of Argentina, Agriculture. Ecosystems & Environment 154(7), 34–43 (2012)

9. Santé-Riveira, I., Crecente-Maseda, R., Miranda-Barrós, D.: GIS-based Planning Support System for Rural Land-use Allocation. Computers and Electronics in Agriculture 63(2), 257–273 (2008)
10. Nuissl, H., Haase, D., Lanzendorf, M., Wittmer, H.: Environmental impact assessment of urban land use transitions—A context-sensitive approach. Land Use Policy 26(2), 414–424 (2009)

Accelerated Extraction Technology Research on Damaged Building Information by Earthquake Based on LiDAR Image

Xiang Wen, Chonggang Miao*, Lijuan Lu, Hua Zhang, Fan Zhang, and Xirong Bi

Earthquake Bureau of the Guangxi Zhuang Autonomous Region, Nanning 530022, China
{yaya997,huazhang1222,bixirong1989}@163.com, miaocg@gmail.com,
65616555@qq.com, zhangfan530@126.com

Abstract. To obtain the damaged building information rapidly and accurately can provide support for the disaster-relief work after the earthquake, and it is also an important part of the evaluation of the earthquake disaster losses. The high-resolution remote sensing image and the satellite radar data are important technical means for the disaster monitoring, but the accuracy of the automatic extraction of information is subject to certain restrictions. However in recent years, the newly-presented LiDAR technology can provide the elevation information for ground targets, which can be applied to rapidly and accurately extract the earthquake disaster information under the circumstances that the remote-sensing image is absent before the earthquake. Adopting the LiDAR data and remote sensing data in the Yushu disaster area, the damaged building information after the earthquake is automatically extracted through the Arcmap collection category samples and by using the object-oriented SVM classification, the overall accuracy can reach 80.96% in the research.

Keywords: LiDAR, ARCMAP, SVM, Damaged Buildings, Information Extraction.

1 Introduction

The earthquake disaster has become one of the frequent natural disasters; the building damage is a very important index among the disaster evaluation works after the earthquake. At present, the extraction researches on damaged building information with high-resolution satellite, aerial remote sensing image and satellite radar data are relatively more. Kaya S et al, Wen Chunjing et al, Sakamoto M et al extract the collapsed buildings and other disaster information by comparing the changed regions of buildings in high-resolution remote sensing image before and after the earthquake; some researchers apply the DEM 3D information extracted from the prior knowledge in GIS vector and aerial or satellite stereo image, as well as the satellite radar data and so on to the building information extraction before and after the earthquake. However, the accuracy of the high-resolution remote sensing image and satellite radar data durin the

---

* Corresponding author.

F. Bian and Y. Xie (Eds.): GRMSE 2014, CCIS 482, pp. 117–132, 2015.

information extraction is subject to certain restrictions, and in reality, it often fails to accurately extract the changed information of buildings due to the lack of pre-earthquake seismic high-resolution remote sensing data or accurate prior knowledge in vector. As a new remote sensing technology and integrating GPS technology, INS technology and laser scanning technology and computer technology as a whole, the Light Detection and Ran-ging (LiDAR) can automatically acquire 3D point cloud data in the earth's surface of high accuracy and high density, and establish the Digital Terrain Model (DTM)to apply to extract the damaged building information, which is a research hotspot. Taking the Qinghai Yushu earthquake of magnitude 7.1 happened on April 14, 2010 as an example, the paper completes the automatic extraction of damaged building information after the earthquake by combining the LiDAR point cloud data with the aerial image data, using the object-oriented classification technology to segment the image as the object, analyzing the features of the multi-space data, spectral and texture, collecting samples by Arcmap, and using the classification methods of the Support Vector Machine (SVM). This research has important practical and realistic significance to rapidly and accurately extract the earthquake disaster information under the circumstances that the remote-sensing image is absent before the earthquake.

2 Data Acquisition and Preprocessing

The data used in this research is collected by School of Remoter Sensing and Information Engineerig of Wuhan University, among which QB remote sensing images with spatial resolution of 0.61 m of the earthquake affected areas and LiDAR data has been obtained; the study area is Xianfeng Village Yushu County in Qinghai Province (North Latitude33°00′30″-33°04′00″, East Longitude 97°01′15″-97°05′15″), the acquisition time is April 20 th-April 26th 2011. The study areas include buildings that not yet collapsed (being made up of the intact buildings and partially destroyed buildings), completely collapsed buildings, vegetation, bare land, and massif and other typical information of surface features (Fig.1).

Pre-process the LiDAR point cloud data of the study area by using the data preprocessing method proposed by Alexander B et al, with the steps as follows:①calculate the 3D coordinates of each point cloud data by combining the GPS with the observation data of inertial surveying system, and the height precision of the data acquired is 11 cm, while the horizontal precision is 15 cm; ②detect and clear the system errors and gross errors, and filter and classify the point cloud data by using Adaptive TIN (ATIN) to separate the ground points and non-ground points;③set up the Ground Model based on the Kriging interpolation method, establish the Digital Surface Model (DSM) by using the first time returned data of LiDAR simultaneously; ④the data of Normalized Digital Surface Model DSM (nDSM) is obtain by DSM minus DEM (Fig.2), used for extracting information of buildings that not yet collapsed, vegetation , massif and surface features with a certain elevation information.

Fig. 1. Aerial images of study area

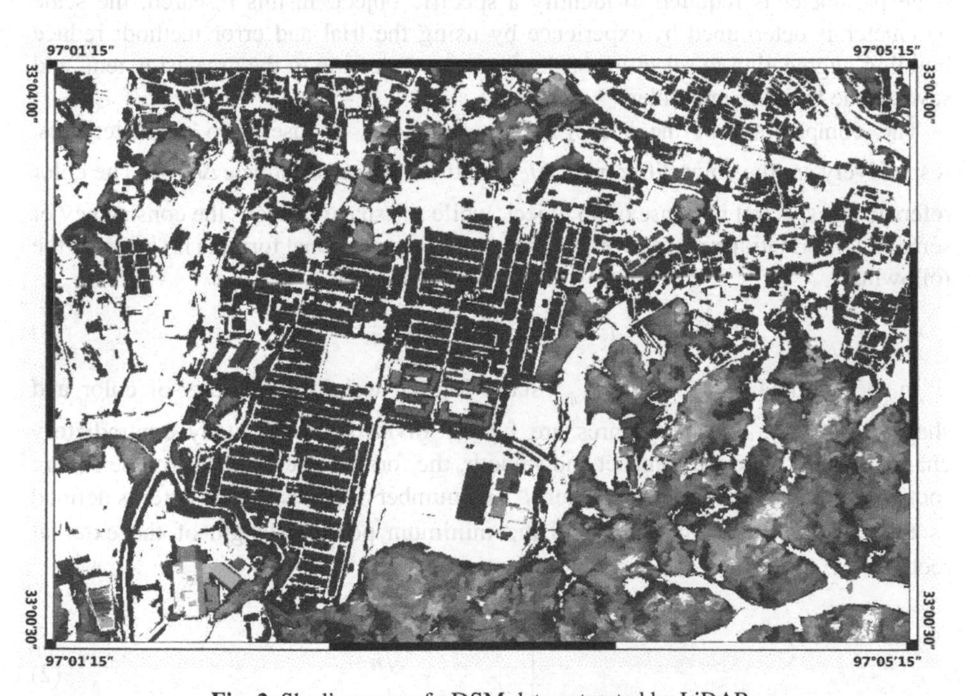

Fig. 2. Shading map of nDSM data extracted by LiDAR

3 Extraction of Object-Oriented Damaged Building Information

Object-oriented Classification Method is the method that composite the pixels with the same characteristics as objects based on the pixel shape, texture, spectral characteristics, and classify them according to the characteristics of objects. The basic idea for the object-oriented information extraction of the damaged buildings describes as below: multi-scale segment the QB image after the earthquake to generate a image object; construct a feature space, and analyze and combine the various features of the image objects, including nDSM, spectral, shape, and texture; extract the damaged building information through the Arcmap collection category samples and by using the SVM classification method.

3.1 Multi-scale Segmentation

Combine the adjacent regions by using the method that control the scale parameter and the homogeneous parameters proposed by Baatz M et al, to obtain the object segmentation algorithm for image segmentation.

The Scale refers to the minimum size required to identify specific objectives, which closely relates to the image resolution. The higher the image resolution, the larger the scale parameter is required to identify a specific object. In this research, the scale parameter is determined by experience by using the trial and error method; reduce building aggregation to select the scale parameter selection to the greatest extent, and save all the building integrities.

The computation of the homogeneous parameter f uses two characteristics, respectively are the color difference Δh_{color} and the shape difference Δh_{shape}. The color refers to the spectral response of an object, while the shape conveys the consistency of semantic information related to object space; the computational formula is shown as the following:

$$f = w_{color} \cdot \Delta h_{color} + w_{shape} \cdot \Delta h_{shape} \tag{1}$$

In the formula, w_{color} and w_{shape} stand for the weight parameters of color and shape difference. Shape features are further divided into mutually contradictory characteristics, there are compact and smooth, the former is represented by the square root of the ratio of the object perimeter l and number of pixels n the latter is defined as the ratio of the perimeter l and the minimum possible length of the external rectangle b ,that is:

$$\begin{cases} compact = \dfrac{l}{\sqrt{n}} \\ smooth = \dfrac{l}{b} \end{cases} \tag{2}$$

The ideal object segmentation result for a particular object is that, after the segmentation, the polygon can either clearly express object boundary, or it can be distinguished from other objects; neither too general, nor too much broken, in case of reducing the subsequent object extraction accuracy. Looking from the image, the buildings in the study area are more dense, hence, it is very necessary for guaranteeing the building integrity and making sure they are not be much broken in the segmentation process, and to handle the scale is not easy that requires for repeated attempts. This paper determines the final scale parameter is 25, the shape factor is 0.1, and compactness factor is 0.5 (Table 1). As you can see from Figure 3, the size of the segmented building polygon is basically the same as the image, which ensure the integrity of the building outline so as to lay a good foundation for the further determination of the damaged buildings.

Table 1. Segmentation Parameters of Image

Scale Parameter	Tone	Shape	Compactness	Smoothness
25	0.9	0.1	0.5	0.5

Fig. 3. Segmentation Result of Image(Blue line stands for polygon boundary of object)

3.2 Construction of Feature Space

Both for the target recognition and classification, the characteristics are the decisive factors of the processing results. The specific object is always associated with the corresponding features or feature combinations, thus only selects the appropriate features or feature combinations to distinguish a certain object with other objects. The buildings that not yet collapsed, vegetation, massif and earth objects, relatively bare land, completely collapsed buildings in the study area has a certain elevation information, therefore, firstly apply the nDSM feature to extract the mixed elevation features, then further use the image standard deviation (δ_L), mean value ($\overline{C_l}$), shape index shape (S), length-width ratio (γ), angular second moment (ASM) in the gray level co-occurrence matrix, and entropy and other combination features to classify the mixed ground objects as basic intact buildings, partially destroyed buildings, vegetation and massif. From the analysis of the image, the texture differences between the barc land and completely collapsed buildings are more obvious, and the brightness features are different, so the texture and brightness feature can be used to identify the completely collapsed buildings; construct the feature space by selecting the brightness value (B) and the homogeneity index of the gray level co-occurrence matrix (Hom), referring to the previous relevant researches. The specific formulation of each feature parameter is as follows:

(1) δ_L is calculated by all n pixel values of layers that constitute an image object, and the range of its characteristic value is [0, determined by the data bits].

$$\delta_L = \sqrt{\frac{1}{n-1}\sum_{i=1}^{n}(c_{L_i} - \overline{c_L})^2} \tag{3}$$

In the formula, n stands for the number of object pixels, C_{L_i} is the pixel value of the i pixel within the object in the L wave band, C_{L_i} is the pixel value within the object in the L wave band.

(2) $\overline{C_l}$: the mean value that is used for calculating the spectral values of the pixels that constitute as the objects of each image in each wave band.

$$\overline{C_l} = \frac{1}{n}\sum_{n-1}^{n} C_{L_i} \tag{4}$$

In the formula, L stands for the band number, C_{L_i} is the pixel value of the i pixel within the object in the L wave band, the range is [0,255], n is the number of pixels within the object.

(3) S : the mathematical shape index refers to the number that 4 times of the boundary length of the image object divide the square root of the area.

$$S = \frac{e}{4\sqrt{A}} \tag{5}$$

In the formula, e expresses as the boundary length of the object, A expresses as the area of the object. Use the shape index S to describe the boundary smoothness of the image object. The more broken the image object is, the larger the shape index is. The range of the characteristic value is [1, decided based on the image object shape].

(4) γ : expressed by the ratio of the length(l) and width (w) of the bounding oval, The range of the characteristic value is [0,1].

$$\gamma = \frac{l}{w} \tag{6}$$

(5) B : express the spectrum mean of the object in all bands.

$$B = \frac{1}{n_L} \sum_{i=1}^{n_L} \overline{C_i} \tag{7}$$

In the formula, $\overline{C_i}$ is the spectrum mean of the object in the layer i , n_L is the total number of layers.

(6) Hom : reflect the homogeneity of the image texture to measure the its number of the local changes. Its value is calculated by using the gray level co-occurrence matrix, the big value explains that there are lacks of changes between the different regions of the image texture and the local is very uniform.

$$HOM = \sum_{i=0}^{n-1} \sum_{j=0}^{n-1} \frac{p_{i,j}}{1+(i-j)^2} \tag{8}$$

In the formula, n is the order of the co-occurrence matrix; j is the coordinate of the co-occurrence matrix; p is the numerical value of the co-occurrence matrix in(i , j).

(7) Angular second moment (ASM): the quadratic sum of each element of the gray level co-occurrence matrix, also known as energy. It is a measure of the gray change uniform of the image texture, reflecting the image gray distribution uniformity and texture coarseness.

$$ASM = \sum_i \sum_j p(i,j)^2 \tag{9}$$

In the formula, i and j is the coordinate of the co-occurrence matrix; p is the numerical value of the of the co-occurrence matrix in(i , j).

(8) Entropy: a random variable, expressing the amount of the image information. It reflects the heterogeneity degree or complexity of the texture. The higher the complexity is, the bigger the entropy value is; the lower the complexity, the smaller the entropy value or 0.

$$ENT = -\sum_i \sum_j p(i,j) log(p(i,j)) \tag{10}$$

In the formula, i and j is the coordinate of the co-occurrence matrix; p is the numerical value of the of the co-occurrence matrix in(i , j).

A certain size of the sliding window shall be selected to extract the texture features for the gray level co-occurrence matrix; the different window size restricts the accuracy of the image classification. Therefore, the optimal window size of the texture extraction shall be determined under the highest classification accuracy before the classification.

In this paper, the textures are extracted respectively in the window sizes of 3×3,5×5,7×7,9×9,11×11,and 13×13 to classify the image, the accuracy of the classification results is shown in Table 2 and Fig.4, the accuracy of the classification results has a trend of gradual decrease along with the increasing of the texture window. The highest accuracy of image classification is achieved when using the texture features in the window size of 3×3, the total accuracy is 76.78%, and *Kappa* is 0.661.

Table 2. Effect of window size on classification accuracy of texture feature

Window size	Total accuracy (%)	*Kappa*
3×3	76.78	0.661
5×5	74.51	0.643
7×7	73.14	0.627
9×9	71.37	0.611
11×11	66.52	0.554
13×13	64.48	0.547

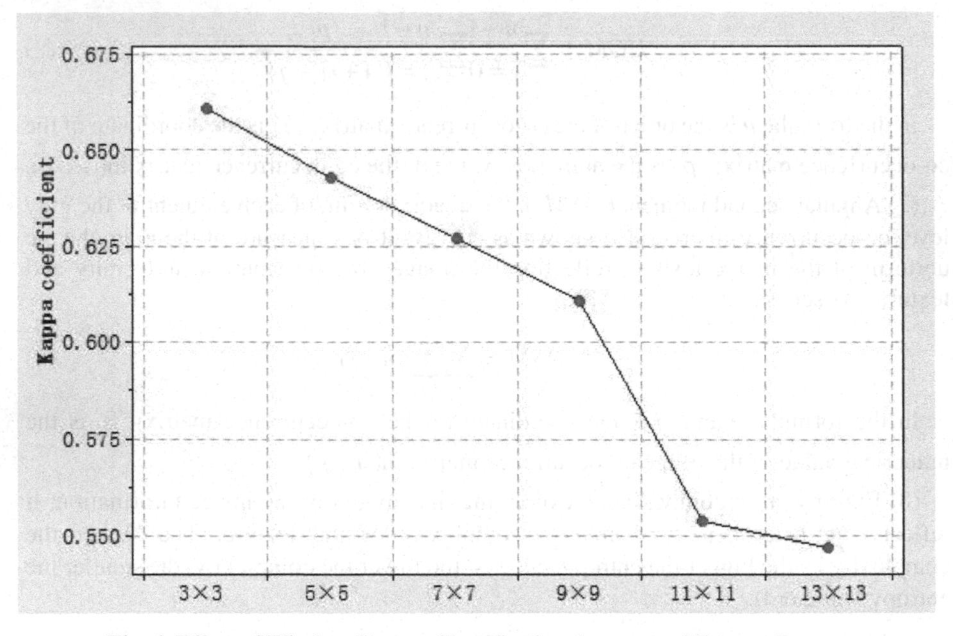

Fig. 4. Effect of Window Size on Classification Accuracy of Texture Feature

3.3 Object-Oriented Image's Classification

Classification of Basically Intact Buildings and Partially Destroyed Buildings. Use the nDSM feature in QB image (Table 3) to extract the mixed surface features of the buildings that not yet collapsed, vegetation and massif (Fig.5); and divide the mixed surface features into buildings that not yet collapsed, vegetation and massif through the

Arcmap collection class samples and by using the SVM method; finally, subdivide the buildings that not yet collapsed into basically intact buildings and partially destroyed buildings by using the combined features between *ASM* and *entropy*.

Table 3. Types and Rules of Image's Classification and Extraction

Level	Extracting information	Member function based on classification
1	Buildings that not yet collapsed, massif, vegetation	nDSM>1.5
2	Bare land, completely collapsed buildings	Unclassified objects

Fig. 5. Mixed surface features of buildings that not yet collapsed, vegetation and massif (purple red surface features)

The SVM classification is a proposed new machine learning method based on the VC dimension theory and structural risk minimization principle, which finds the best compromise between the model complexity and learning ability according to the limited sample information, in order to obtain the best generalization ability. The main problem of the SVM is to determine the optimal classification hyperplane according to the limited samples, to gain the maximum classification interval and the minimum expected risk when predicting the position of samples. Based on the known training samples categories and their characteristic attributes, the SVM obtains the relationships between the training samples and samples or characteristic attributes; and classify the

Fig. 6. Sample Collection(Blue circle represents as the collected samples of buildings that not yet collapsed, yellow circle represents as the collected samples of massif, while the green circle represents as the collected samples of vegetation)

Fig. 7. Sample Creation(Purple features stand for the created samples of buildings that not yet collapsed, yellow features stand for the created samples of massif, and green **features** stand for the created samples of vegetation)

training samples by categories or attributes; and finally forecast the unknown sample categories, attributes and their distributions. The classification procedures of the SVM are listed as follows:

(1) Acquisition and sample creation: determine the classification of the required samples of the buildings that not yet collapsed massif and vegetation in the Arcmap, used for the establishment of SVM classification model. When choosing the samples, all samples shall be distributed evenly over the whole image area (Fig.6); after the sample collection completes, create the samples of the buildings that not yet collapsed, massif and vegetation on the image (Fig.7).

(2) Training samples and classification: Train the samples of buildings that not yet collapsed, massif and vegetation by using SVM classifier, select the δ_L , $\overline{C_l}$, S , γ as classification feature, and classify image by using classifier algorithm (Fig.8).

Fig. 8. SVM Classification Results(Blue features represent as buildings that not yet collapsed, yellow features represent as massif, and green features represent as vegetation)

(3) Classification of buildings that not yet collapsed: The texture difference between basically intact buildings and partially destroyed buildings is obvious. Classify the buildings that not yet collapsed by using the combined features between ASM and entropy and adopting the classifier algorithm (Fig.9).

Fig. 9. Classification results of buildings that not yet collapsed(Blue features present basically intact buildings, pink features present partially destroyed buildings)

Extraction of Completely Collapsed Buildings. The brightness and texture differences between the completely collapsed buildings and the bare land is obvious, therefore, the B and Hom can be relied on building the space characteristics (Table 4), to extract the information of the completely collapsed buildings.

Table 4. Types and rules of image's classification and extraction

Layer	Extracting information	Member function based on fuzzy classification
1	Completely collapsed buildings	B>150,Hom<0.55
2	Bare land	Unclassified objects

Fig. 10. Completely collapsed buildings (red features)

Table 5. Accuracy of image classification after the earthquake

	massif	Partially destroyed buildings	Basically intact buildings	Vegetation	Completely collapsed buildings	Column precision /%
massif	28542	345	1587	622	347	90.77
Partially destroyed buildings	435	3556	383	156	145	76.06
Basically intact buildings	1456	491	47687	1314	245	93.15
Vegetation	5868	453	1579	31656	317	79.39
Completely collapsed buildings	756	56	1282	234	3478	59.90
Line precision /%	77.02	72.56	90.80	93.16	76.74	

It can be seen from the table 5 that the classification accuracy of the basically intact buildings is the highest one after the earthquake, up to 90.80%, while the classification accuracy of the partially destroyed buildings and completely collapsed buildings is completely ideal, respectively reached 72.56% and 76.74%.

4 Evaluation of Classification Accuracy

Evaluate the classification accuracy of SVM by selecting the building area of interest. The overall expression formula for classification accuracy is as follows:

$$p_c = \sum_{k=1}^{n} p_{kk} / p \tag{11}$$

In the formula, n stands for the number of samples, p stands for types of the categories. The expression formula for accuracy evaluation of *Kappa* statistics is as follows:

$$K_{hat} = \frac{N\sum_{i=1}^{r} x_{ii} - \sum_{i=1}^{r}(x_{i+} + x_{+i})}{N^2 - \sum_{i=1}^{r}(x_{i+} + x_{+i})} \tag{12}$$

In the formula, N represents the total number of samples, r represents the total number of categories, x_{ii} represents the correct number of the classification, x_{+i} and x_{i+} respectively are the total sample quantity of the row i and the column i.

It can be seen from the table 6 and figure 10 that the overall classification accuracy of SVM reaches 80.96%, *Kappa* statistics is 0.6981, the classification results of the damaged buildings and other surface features are clear and intact.

Table 6. Accuracy evaluation of SVM classification

Classification methods	Overall accuracy	*Kappa* coefficient	Performance period
SVM	80.96%	0.6981	35.7s

5 Conclusion

Taking the high-resolution QB remote-sensing image and LiDAR data in Xianfeng Village Yushu County after the earthquake as an example, the author segments the image as the objects by using object-oriented method, analyzes the nDSM, spectral and texture features, realizes the automatic extraction of damaged buildings information after the earthquake through the Arcmap collection category samples and by using SVM classification in this paper. Through analyzing the classification results and precision, the author draws the following conclusions: with regard to the high-resolution LiDAR data and aerial image data after the earthquake, it is valid to combine the object-oriented application with SVM method to extract the damaged building information, which can provide support for earthquake relief and disaster evaluation and so on.

Acknowledgements. Fund Project: This work was supported by the science-technology plan of Guangxi (Project number: 12426001).

References

1. Guo, H.-D., Lu, L.-L., Ma, J.-W., et al.: An improved automatic detection method for earthquake-collapsed buildings from image. Chinese Sci. Bull. 54(17), 2581–2585 (2009)
2. Kaya, S., Curran, P.J., Llewellyn, G.: Post-earthquake Building Collapse; A comparison of government statisties and estimates delved from SPOT HRVIR data. Int. J. Remote Sens. 26(3), 2731–2740 (2005)
3. Wen, C.-J., Zhao, S.-H., Li, H., et al.: One object–oriented change detection method on the high resolution images. Journal of Shandong Normal University 12(1), 126–129 (2010)
4. Sakamoto, M., Takasago, Y., Uto, K., et al.: Automatic detection of damaged area of iran earthquake by high-solution satellite imagery. In: Proceedings of IGARSS 2004, Alaska, pp. 1418–1421 (2004)
5. Bi, L.-S., He, H.-L., Xu, Y.-R., et al.: The extraction of knickpoint series based on the high resolution dem data and the identification of paleo-earthquake series-a case study of the huoshan mts. piedmont fault. Seismic Geology 33(4), 963–977 (2011)
6. Liu, H.-G., Ran, Y.-K., Li, A., et al.: Attitude extraction of shallow stratum based on P5 stereo images and geoeye-1 image. Seismic Geology 33(4), 951–962 (2011)
7. Shi, J.-N., Zhang, F.: Research on methods of extracting DEM by using SPOT5 HRS/HRG stereoscopic image. In: The Thirty-Third volume Sixth Issue of Surveying and Mapping, pp. 263–265 (2010)
8. Turker, M., Cetinkaya, B.: Automatic detection of earthquake-damaged buildings using DEM created from Pre-and Post-earthquake stereo aerial photographs. Int. J. Bemote. Sens. 26(4), 823–832 (2005)
9. Li, X., Shu, N., Li, L., et al.: Research of change detection using gis auxiliary data and samples. Computer Engineering and Applications 46(14), 215–217 (2010)
10. Turker, M., San, B.T.: Detection of collapsed buildings caused by the 1999 Izmit, Turkey Earthquake Through Digital Analysis of Post-event Aerial Photographs[J]. Int. J. Remote. Sens. 25(21), 4701–4714 (2004)
11. Yang, J., Liao, M.-S., Jiang, W.-S., et al.: Extraction of the elevation from single SAR image. Journal of Wuhan University of Surveying and Mapping Technology 25(6), 121–127 (2000)
12. Garnba, P., Dell Acqua, F., Trianni, G.: Rapid damage detection in the bam area using multitemporal SAR and exploiting ancillary data. IEEE Transactions on Geoscience & Remote Sensing 45(6), 1582–1589 (2007)
13. Yu, H.-Y., Cheng, G., Zhang, Y.-M., et al.: The detection of earthquake-caused collapsed building information from LiDAR data and aerophotograph. Remote Sensing for Land & Resources 3(90), 77–81 (2011)

14. Alexander, B., Christian, H., Goepfet, J., et al.: Aspects of generating precise digital terrain models in the wadden Sea from Lidar-water classification and structure line extraction. ISPRS Journal of Photogrammetry & Remote Sensing 63(5), 510–528 (2008)
15. Baatz, M., Schape, A.: Multiresolution segmentation-an optimization approach for high quality multi-scale image segmentation. In: Strobl, J., Blaschke, T., Griesebner, G. (eds.) Angewandte Geographische Informations-Verarbeitung XII, pp. 12–23. Wichmann Verlag, Karlsruhe (2000)
16. Tian, X.-G.: Information extraction of object-oriented high resolution remote sensing images, vol. 22(5), pp. 66–72. Chinese Academy of Surveying and Mapping, Beijing (2007)
17. Wang, H.-M., Li, Y.: Object-oriented damage building extraction, remote sensing application 23(5), 80–85 (2011)
18. Peng, L., Yang, W.-N., Li, X.-D., et al.: Information extraction of geological hazards in an object-oriented approach—a case study in Wenchuan earthquake. Journal of Southwestern Normal University 36(3), 77–82 (2011)
19. Zhang, Q., Wang, J., Gong, P., et al.: Study of Urban Spatial Patterns from SPOT Panchromatic Imagery Using Textural Analysis. International Journal of Remote Sensing 24(21), 4137–4160 (2003)
20. Du, P.-J.: The classification review of SVM-based high spectrum remote sensing. Bulletin of Surveying and Mapping 12(3), 37–40 (2006)
21. Wang, X.-Q., Wei, C.-J., Miao, C.-G.: Research of hazard remote sensing fast extraction. Earth Science Frontiers 10(8), 285–291 (2003)
22. Chen, W.-K.: Remote sensing technology research of hazard-oriented assessment. Lanzhou Earthquake Research Institute, China Earthquake Administration 22(4), 56–60 (2007)
23. Yu, X.-C., An, W.-J., He, H.: Classification of remote sensing image based on object oriented and class rules. Progress in Geophysics 27(2), 120–126 (2012)

Finger-Screen Interaction Mechanism on 3D Virtual Globe for Mobile Terminals

Jiawei Li and Jing Chen

State Key Laboratory of Information Engineering in Surveying, Mapping and Remote Sensing, Wuhan University, 430079, Wuhan, China
{lijiawei,jchen}@whu.edu.cn

Abstract. Virtual Globe technology has become a cutting-edge technology of geographic information industry in recent years. In view of the touch operations of mobile terminals and the interactive requirement in Virtual Globe, we design a set of finger-screen interactive control mechanism, which consists of a series of multi-touch gestures such as tap, slide, pinch, swipe, double click and so on, to handle the multi-scale interactive browse of 3D Virtual Globe in mobile terminals. Pixel threshold and time threshold are introduced into error control, which solves the problem of unstable disturbances and finite precision while operating the touch-screen interface. To demonstrate thc fcasibility and effectiveness of this mechanism, experiments were carried out on the platform of Android. The results showed that the shaking screen phenomenon and the error responses were eliminated. It achieves highly effective seamless browse of 3D Virtual Globe in mobile terminals.

Keywords: Virtual Globe, Mobile GIS, Interactive operation, Error control.

1 Introduction

3D Virtual Globe is a worldwide 3D Geographic Information System which combines multi-scale spatial data such as satellite imagery, aerial imagery and digital elevation models together for integrated management and rendering [1]. It provides a platform for the collaborative service, online sharing, rapid display and efficient retrieval of 3D geographic information [2]. In recent years, with the fast integration of mobile communication technology and GIS applications, a new generation of mobile geographic information services based on IOS and Android has shown a broad development prospect [3-4]. As an integrative product of mobile GIS technology [5] and 3D GIS technology [6], 3D Virtual Globe based on mobile terminals achieves the acquisition, storage, retrieval, analysis and display of 3D spatial information in a 4A (Anytime, Anywhere, Anybody, Anything) mode [7].

Compared with the traditional mobile 2D map service, mobile 3D Virtual Globes will realistically reproduce the earth scenario in different perspectives, especially from the sphere to the plane and from the macro to the micro [8]. Diverse from the desktop terminal that takes the mouse-keyboard input as the main interaction mode, touch operations on mobile terminals greatly shorten the interactive distance between

F. Bian and Y. Xie (Eds.): GRMSE 2014, CCIS 482, pp. 133–142, 2015.

the user and the equipment, which will bring about immersive interaction experience and make it possible to play the virtual earth with your fingertips. However, finger-screen interaction has limited precision compared with the mouse-keyboard operations, and it easily causes error responses especially when the target is a very tiny spot [9]. In addition, the interactive mode of mobile terminals lacks multiple key-press response, and there is inevitable ripple effect among different gestures, such as multiple finger gestures always trigger single finger gestures and tap gestures always trigger slide gestures. Therefore, how to eliminate invalid interference from various screen-touching messages, precisely control the parameter changes of the Virtual Globe camera, avoid the jitter and jump of the viewpoint and guarantee a stable and smooth browse of the 3D scene is the technological difficulty to the 3D Virtual Globe on mobile terminals.

2 Finger-Screen Interaction Mechanism

2.1 Virtual Camera Principle Based on Quaternion in Virtual Globe

The perspective principle in 3D Virtual Globe is to simulate the situation of taking photographs with camera in a standard way, which means the virtual camera is responsible for managing the position, orientation and rotation of the viewpoint and dynamically scheduling the spatial data in the observation range, such as imagery, terrain, models, vector and so on. The construction of the view camera requires frequent vector rotation, coordinate transformation, angular displacement calculation and smooth orientation interpolation. Compared with rotation method based on Euler angle, rotation based on quaternion will avoid the gimbal lock problem effectively. At the same time, it's more convenient to transform the quaternion to the rotation matrix.

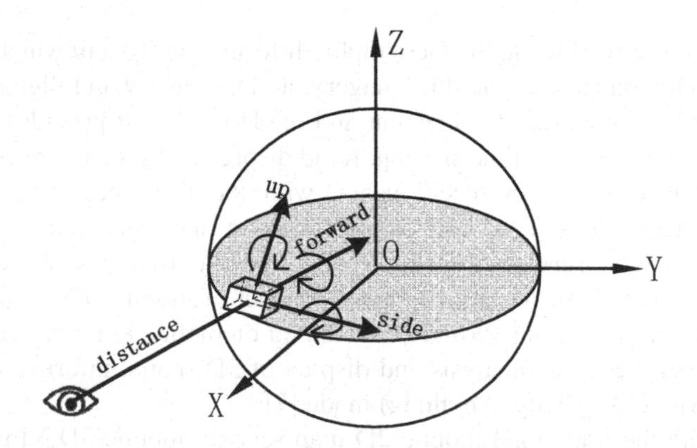

Fig. 1. The virtual camera theory based on quaternion in 3D Virtual Globe

Moreover, the Slerp algorithm has been proposed to support smooth interpolation method for rotation transform [10], which is widely used in the construction of virtual camera.

In view of that the coordinate value is usually very large under the global scenario, even tiny scene interaction also need a vast number of coordinate transformation calculations [11], which will definitely causes certain accuracy loss due to the approximate interception of numbers in the process of calculation. For this purpose we introduce the concept of ground reference point, which refers to the intersection point of the sight line and the ground plane, decided by the location of viewpoint and the sight distance. The ground reference point defines the origin of coordinates to establish a local coordinate system, while the normal direction of the reference point determines the camera aiming axial vector $\overrightarrow{forward}$, the parallel direction from west to east determines the camera side axial vector $\overrightarrow{side}$, and the meridian direction from south to north determines the camera vertical axial vector $\overrightarrow{up}$. The local coordinate system finally determines the basic pose of the virtual camera at the reference point. On this basis, the virtual camera rotates around the three-axis of its basic pose, which determines the relative pose of the virtual camera in the local coordinate system. The final absolute pose of the virtual camera expressed in quaternion is decided by the basic pose and the relative pose, as is shown in Figure 1.

2.2 Basic Process of the Interaction Mechanism

Due to the lack of key-press response mechanism on mobile terminals, the user interactions with the three dimensional earth scene entirely depend on the finger-screen touching gestures. And the so-called "Gesture" refers to all kinds of finger actions and touching messages from the moment that finger touches the screen until the moment that finger is lifted from the screen [12]. The touch parameters of gestures consist of the finger number, the touching position, the action type and so on. Diverse from the desktop terminals, where the left mouse button controls translation, the middle mouse button controls scaling, and the right mouse button controls rotation, we can't figure out the accurate interaction purpose immediately even after the finger has already been put on the screen. We have to dynamically identify the gesture type by real-time monitoring and calculating the finger trajectory in the process of the gesture interactions.

The main idea of the interactive process in Virtual Globes is as follows: Extract the finger trajectory from the touching parameters; Eliminate the interference information from the chain gestures; Abstract separate gesture types from the finger trajectory and map them into key-press response based on mouse-keyboard input on desktop terminals; Match the interactive messages with corresponding event handler; Handle the interactive missions in Virtual Globes such as querying feature properties, editing feature elements, modifying camera parameters and so on. The flow chart is shown in Figure 2.

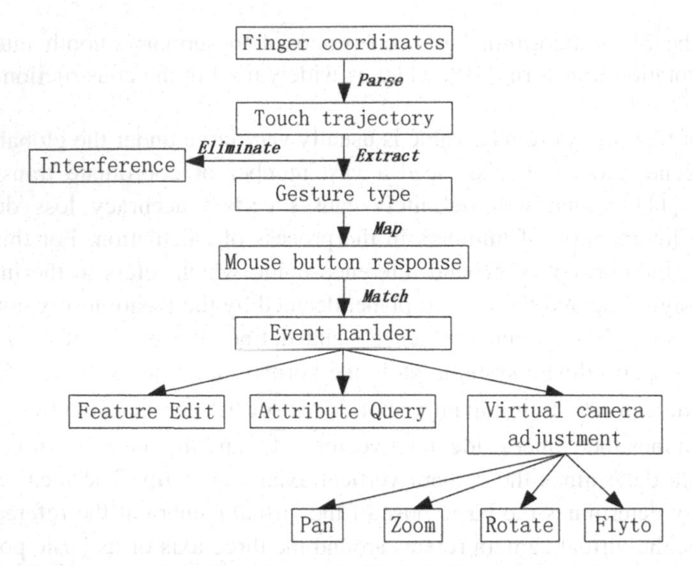

Fig. 2. The interaction procedure in 3D Virtual Globes for mobile terminals

2.3 Gesture Simulation and Scene Interaction

The main gestures designed for scene interaction in Virtual Globes are as follows: one finger tap, one finger slide, double finger pinch, double finger swipe and double click. The tap gesture means that once the finger touches the screen, lift it up immediately without any finger movement. The slide gesture means that move the finger on the screen with steady contact state. The pinch gesture means that move two fingers along the opposite directions with steady contact state, including moving towards and moving away from each other. The swipe gesture means that move two fingers along the same direction with steady contact state and steady finger clearance. Double click gesture means that tap the screen with the same finger twice in rapid succession.

The tap gesture of mobile terminal can be mapped into the left mouse button click of desktop terminal. By matching the mouse click event handler, the interaction messages of the tap gesture can be interpreted, which can be used for feature edit or feature pick-up in Virtual Globes. The core calculations are based on the translation of coordinates from 2D screen space to 3D geometry space with the formula below:

$$(X, Y, Z) = (X_{down}, Y_{down}, Z_{buffer}) \times Inv(M_{view} \times M_{project} \times M_{window}). \quad (1)$$

Where X_{down}, Y_{down}, Z_{buffer} represent the touch screen coordinates and the corresponding depth buffer value, respectively. M_{view}, $M_{project}$, M_{window} represent the model view matrix, projection matrix and viewport matrix in the rendering pipeline, respectively. Inv represents the matrix inverse operation.

The one-finger slide gesture of mobile terminal can be mapped into the mouse move of desktop terminal. By matching the mouse move event handler, the interaction messages of slide gesture can be interpreted to control panning the virtual camera. The position offset of the reference point in the Virtual Globe is recalculated according to the finger slide amplitude (dx, dy), as is shown below:

$$\overrightarrow{\Delta Center} = \overrightarrow{V_{Up}} * (dy \times f_{pan}) + \overrightarrow{V_{Side}} * (dx \times f_{pan}) . \quad (2)$$

Where $\overrightarrow{\Delta Center}$ represents the position offset vector of the reference point in the Virtual Globe. $\overrightarrow{V_{Up}}$ represents the normalized vector of the camera's vertical upward direction in the global coordinate system. $\overrightarrow{V_{Side}}$ represents the normalized vector of the camera's sideward direction in the global coordinate system. f_{pan} represents the pan factor of the virtual camera, which is used to adjust the touch sensitivity of controlling the gradual movement of the viewpoint. Generally the value of this factor is set to be proportional to the distance between the position of the virtual camera and the reference point on the ground surface. The schematic diagram is shown in Figure 3.

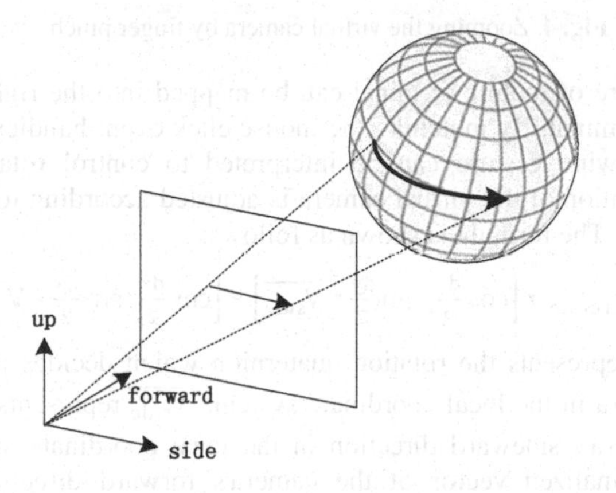

Fig. 3. Panning the virtual camera by finger slide

The pinch gesture of mobile terminal can be mapped into the middle mouse button click of desktop terminal. By matching the mouse roller event handler, the interaction messages of the pinch gesture can be interpreted to control zooming in and out the 3D earth scene. The distance D_{view} between the virtual camera and the reference point is adjusted by the scale ratio r_z with the formula below:

$$r_z = \left(\sqrt{(X_1 - X_2)^2 + (Y_1 - Y_2)^2} - \sqrt{(X'_1 - X'_2)^2 + (Y'_1 - Y'_2)^2}\right) \times f_z . \quad (3)$$

$$D_{view} = D_{view} \times (1 + r_z) . \quad (4)$$

Where f_z represents the zoom factor which is used to adjust the touch sensitivity of controlling the scale of the 3D scene. (X_1, Y_1),(X_2, Y_2) respectively represents the screen coordinates the moment that the two fingers touch the screen. (X'_1, Y'_1), (X'_2, Y'_2) respectively represents the real-time screen coordinates when the two fingers slide across the screen. The schematic diagram is shown in Figure 4.

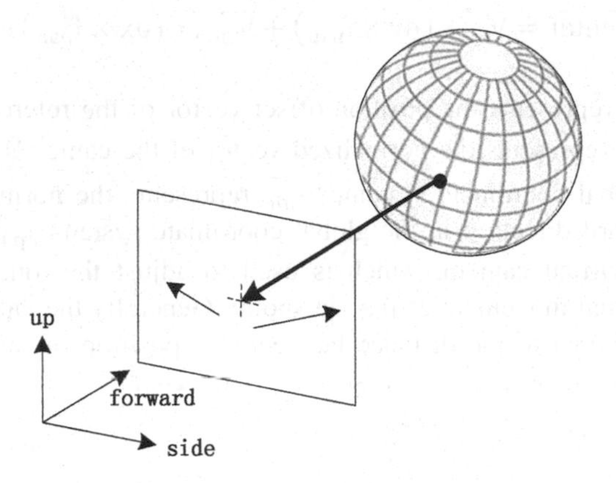

Fig. 4. Zooming the virtual camera by finger pinch

The swipe gesture of mobile terminal can be mapped into the right mouse button click of desktop terminal. By matching the mouse click event handler, the interaction messages of the swipe gesture can be interpreted to control rotating the virtual camera. The orientation of the virtual camera is adjusted according to the finger slide amplitude (dx, dy). The formula is shown as follows:

$$Q_{rotate} = Q_{rotate} * \left[\cos\frac{dy}{2}, \sin\frac{dy}{2} * \overrightarrow{V_{side}}\right] * \left[\cos\frac{dx}{2}, \sin\frac{dx}{2} * \overrightarrow{V_{forward}}\right]. \quad \textbf{(5)}$$

Where Q_{rotate} represents the rotation quaternion which decides the relative pose of the virtual camera in the local coordinate system. $\overrightarrow{V_{side}}$ represents the normalized vector of the camera's sideward direction in the local coordinate system. $\overrightarrow{V_{forward}}$ represents the normalized vector of the camera's forward direction in the local coordinate system. The schematic diagram is shown in Figure 5.

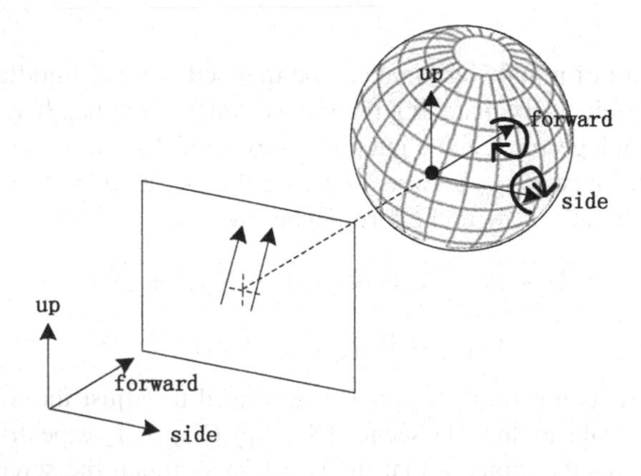

Fig. 5. Rotating the virtual camera by finger swipe

The double tapping gesture of mobile terminal can be mapped into the mouse double click of desktop terminal. By matching the mouse click event handler, the interaction messages of the double tapping gesture can be interpreted to control navigating to the specified location in Virtual Globe. The screen coordinates captured by the finger will be used for calculating the corresponding actual coordinates and replacing the original reference point. At the same time, the distance between the virtual camera and the reference point will be decreased to the specified rate.

Taken together, all these finger gestures are combined to achieve continuous seamless roaming in Virtual Globe of mobile terminals.

2.4 Error Control and Interference Eliminating

Since the accuracy of finger-screen interaction is limited, tiny unstable disturbances are inevitably triggered no matter how fast you lift your finger up after tapping the screen. The key point of distinguishing the tap gesture and slide gesture is to record the original screen coordinates (X_{down}, Y_{down}) where the finger touches the screen and the real-time screen coordinates (X_{move}, Y_{move}) where the finger slide across the screen. The finger's offset distance from the original position is monitored in real time. If the distance is under the pixel threshold e_{pixel} all the time, it indicates that the disturbance information is caused by ripple effects and need to be eliminated. At the same time, the displacement difference between finger down and finger up should also be ignored and only the finger down coordinates will be preserved to participate in the subsequent calculations. In this way, the jitter and jump of the rendered frame can be avoided when touching the screen causes additional slide messages. Otherwise, if the distance exceeds the pixel threshold, it indicates that the disturbance is beyond the tolerance and the interaction needs to be considered as finger slide gesture. Besides, in order to avoid the interference of long press gesture, the time threshold e_{time} is introduced to restrict the interval between finger down and finger up.

The criterion conditions for tap gestures with error control are shown as follows:

$$\sqrt{(X_{move} - X_{down})^2 + (Y_{move} - Y_{down})^2} \leq e_{pixel} \,. \tag{6}$$

$$T_{up} - T_{down} \leq e_{time} \,. \tag{7}$$

The criterion condition for slide gestures with error control is shown as follows:

$$\sqrt{(X_{move} - X_{down})^2 + (Y_{move} - Y_{down})^2} \geq e_{pixel} \,. \tag{8}$$

Different from the single finger interaction, the multi-touch interaction cannot guarantee the timings of touching screen with diverse fingers absolutely identical. The time difference will lead to the problem that multiple finger gestures always trigger single finger message response by mistake. Thus, on the one hand, the slide message response between the interval when the two fingers successively fall onto the screen needs to be eliminated. On the other hand, the tap operation of the first finger should be released before the second finger touches the screen, which prevents the jitter and jump of the rendered frame and the viewpoint.

When it comes to the recognition of the double finger gestures, it not only demands real-time monitoring the touch trajectory for comparing the two fingers' relative movement, but also demands real-time monitoring the distance change of the two fingers, which can be used to set the scale ratio and rotate angle of the virtual camera in the 3D earth scene. In order to avoid the unstable disturbances affecting the judgment of movement trend, the pixel threshold e_{pixel} is introduced to restrict the error response. The double finger gestures should satisfy the following conditions:

$$\sqrt{(X_1 - X'_1)^2 + (Y_1 - Y'_1)^2} \geq e_{pixel} \,. \tag{9}$$

$$\sqrt{(X_2 - X'_2)^2 + (Y_2 - Y'_2)^2} \geq e_{pixel} \,. \tag{10}$$

Based on the above formulas, the criterion condition for pinch gestures with error control is shown as follows:

$$(X_1 - X'_1) \times (X_2 - X'_2) + (Y_1 - Y'_1) \times (Y_2 - Y'_2) \leq 0 \,. \tag{11}$$

The criterion condition for swipe gestures with error control is shown as follows:

$$(X_1 - X'_1) \times (X_2 - X'_2) + (Y_1 - Y'_1) \times (Y_2 - Y'_2) > 0 \,. \tag{12}$$

(X_1, Y_1),(X_2, Y_2) respectively represents the screen coordinates the moment that the two fingers touch the screen. (X'_1, Y'_1), (X'_2, Y'_2) respectively represent the real-time screen coordinates when the two fingers slide across the screen.

3 Experiment

According to the foregoing interaction mechanism of 3D Virtual Globe on mobile terminals, we chose OpenGL ES1 as the underlying 3D graphics library and transplanted GeoGlobe, namely the open virtual globe-based interaction and sharing service platform [13], into the Android 4.0 system. Users use clients to access the one-stop geographic information services such as 256*256 pixels imagery tile services and 32*32 pixels terrain tile services, from the "National Platform for Common Geospatial Information Services" website. Via applying the foregoing finger-screen interactive operations to the 3D earth scene, we finally achieved a stable and smooth browsing speed at an average 40fps frame rate. The 3D earth scene in different perspectives is shown as follows:

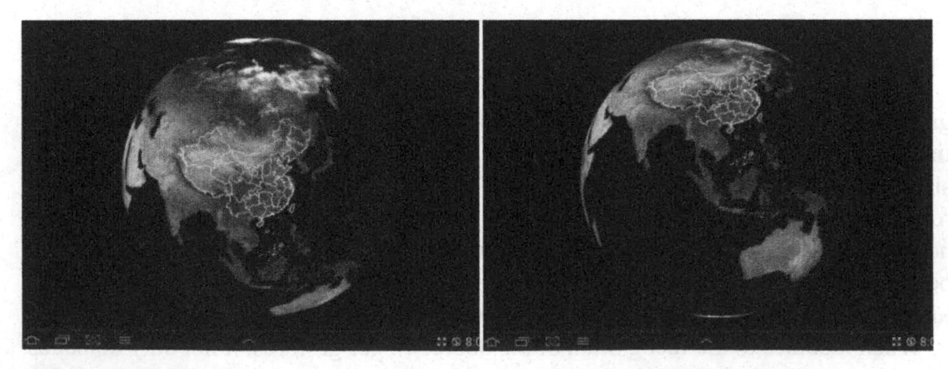

Fig. 6. Virtual earth scenarios pan controlled by finger slide interaction

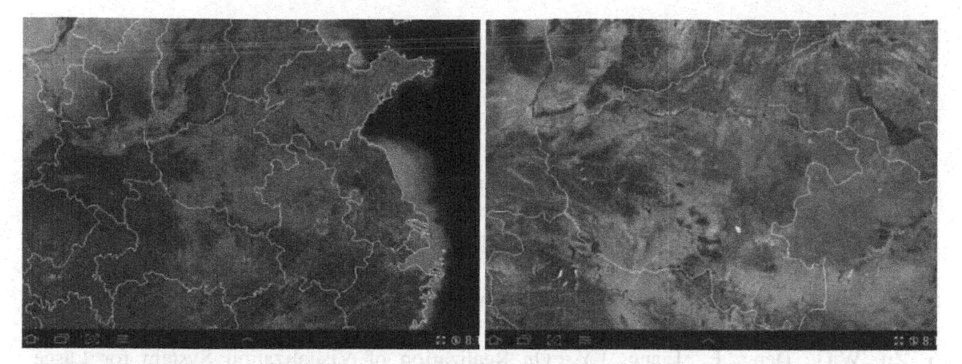

Fig. 7. Virtual earth scenarios zoom controlled by finger pinch interaction

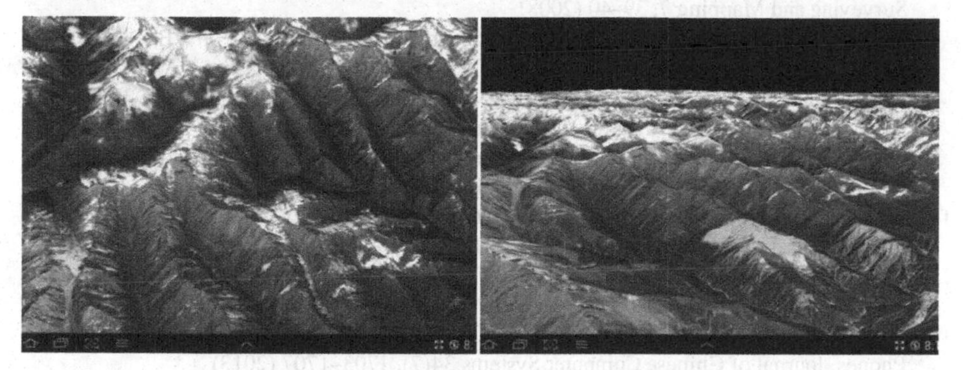

Fig. 8. Virtual earth scenarios rotation controlled by finger swipe interaction

4 Conclusion

Gesture simulation and quantitative calculation for multiple interactive response in mobile terminals have been applied in this paper, which realizes the seamless transformation from mobile finger trajectories to 3D virtual camera parameters. By introducing error control conditions into this approach, the accuracy and stability of virtual reality interaction in the mobile Virtual Globes have been improved. With the continuous development of touch screen technology in mobile terminals, more concise and practical multi-finger gestures will fully replace the external mouse due to its better interaction characteristics, which will contribute to more complex 3D interaction functions and extend the application domains of Virtual Globes in the near future, such as surveying and mapping outdoors, feature query, spatial analysis and so on.

References

1. Gong, J.Y.: The Development and Application of 3-D Virtual Earth Technology. Geomatics World 09(2), 15–18 (2011)
2. Gong, J.Y., Chen, J., Xiang, L.G.: GeoGlobe: Geo-spatial Information Sharing Platform as Open Virtual Earth. Acta Geodaetica et Cartographica Sinica 39(6), 551–553 (2010)

3. Suarez, J.P., Trujillo, A., De La Calle, M., Gomez-Deck, D.D., Santana, J.M.: An Open Source Virtual Globe Framework for IOS, Android and WebGL Compliant Browser. In: 3rd International Conference and Exhibition on Computing for Geospatial Research and Application, COM, Geo 2012, July 1-3. Association for Computing Machinery, Washington (2012)
4. Sterk, M., Agustin, M., Palacio, C.: Virtual Globe on the Android - Remote vs. Local Rendering. In: 6th International Conference on Information Technology: New Generations, ITNG 2009, April 27-29, pp. 634–639. IEEE Computer Society, Las Vegas (2009)
5. Kang, M.D., Peng, Y.Q.: The Key Technology and application of Mobile GIS. Bulletin of Surveying and Mapping 9, 50–53 (2008)
6. Zu, W.G., Deng, F., Liang, J.Y.: On Realization of Visualization System for Large Volume 3D GIS Data. Identification of Common Molecular Subsequences. Bulletin of Surveying and Mapping 7, 39–40 (2008)
7. Chen, F.X., Yang, C.J., Shen, S.L.: Research on Mobile GIS Based on LBS. Computer Engineering and Applications 2, 200–202 (2006)
8. Ding, J., Qin, L.J.: Research on Real-time Rendering Technology of Mobile 3D Scene. Urban Geotechnical Investigation & Surveying 6, 18–22 (2011)
9. Shi, J., Bai, R.L., Zou, J.Y.: Android Multi-touch Screen Input System Design and Realization. Computer Engineering and Applications 48(28), 66–70 (2012)
10. Du, J., Kai, Q.J.: The Display and Free Rotation of 3D Model in Virtual Assembly System Based on ACT. Advanced Materials Research, 588–589, 1178–1183 (2012)
11. Zhu, J., Wang, J.H.: Interactive Virtual Globe Services System Based on OsgEarth. Applied Mechanics and Materials 340, 680–684 (2013)
12. Ma, J.P., Pan, J.Q., Chen, B.: Self-adaptive Gesture Recognition Method of Android Smart Phones. Journal of Chinese Computer Systems 34(7), 1703–1707 (2013)
13. Gong, J.Y., Xiang, L.G., Chen, J., Yue, P.: Multi-source Geospatial Information Intergration and Sharing in Virtual Globes. Science China Technological Sciences 53, 1–6 (2010)

GIS Spatial Data Updating Algorithm Based on Digital Watermarking Technology

Na Ren, Qisheng Wang, and Changqing Zhu

Key Laboratory of Virtual Geographical Environment, Ministry of Education,
Nanjing Normal University, Nanjing 210023, China
rena1026@163.com

Abstract. The need for updating GIS spatial data automatically and effectively, especially ensuring data security which motivated us to propose a scheme based on digital watermark technology that detects and locates modification data with high accuracy while guarantees exact recovery of the original content and the updated data. In this paper, the data updating algorithm is explored based on digital watermark technology. The copyright and version information are embedded in the process of data distribution. Then, the algorithm of removing watermark information is designed after updating data. Experimental results show that the proposed scheme has good performance in invisibility, and can guarantee that the data is used legally and security. Thus, the authority and effectiveness of data updating can be maintained, and the requirements of storage the qualified data in database is satisfied.

Keywords: Digital watermark, GIS spatial data, data updating.

1 Introduction

With the development of global information technology and Internet, the GIS spatial data with time and space features have become the important foundation and strategic resources of national economy and social development. As the foundation for many other social economy and human statistics, the GIS spatial data has been widely used in various fields around us. Meanwhile, the security of GIS spatial data is facing great challenges, and the encryption already cannot satisfy the security protection of vector geographic data. As an effective security protection technology, digital watermarking is playing more important roles [1-5].

With more and more use of GIS spatial data, in particular for its use in internet, the requirements in authority of GIS spatial data is increasing [6-8]. Data authentication thus requires to solve some issues, such as the protecting security of GIS spatial data during transfer, detecting destructive information effectively, and discovering and locating the updated information [9-10].

The work is organized as follows. In section 2, the process of GIS spatial data updating is discussed. In section 3, the data updating algorithm based on digital watermark technology is discussed, by paying attention to introduce the steps of the updating algorithm. Section 4 is devoted to the presentation of experimental results. Finally, some conclusions are drawn in section 5.

F. Bian and Y. Xie (Eds.): GRMSE 2014, CCIS 482, pp. 143–150, 2015.

2 Process of GIS Spatial Data Updating

During updating GIS spatial data, the data is distributed to various producing departments by department of data management. The producing departments produce and update the data. The updated data is passed to the examining department for checking. The examining department will examine the integrity of updated data and original data by content checking, and obtain the qualified data. Finally, the qualified data is stored in database and submitted to the department of data management. The process is shown in Fig. 1.

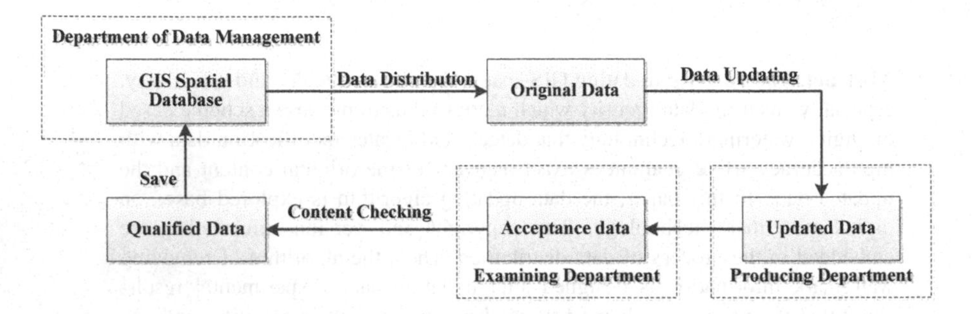

Fig. 1. The process of GIS spatial data updating

During updating the GIS spatial data, the security of data usage should be guaranteed in the whole process of data distribution, use and submission. After producing and passing the data by the producing departments, the examining department will check the content of data. One important aspect is to judge whether the data updating is based on the distributed data. However, it is a very difficult problem to implement this point. By using digital watermarking technique, the distribution instant is embedded in the watermarking information while distributing data. The watermarking information can be extracted while submitting the data, and data distribution instant embedded before can be used to solve this problem.

During this process, two points should be noted: One is that GIS spatial data are all national fundamental data. They should be used based on strict principles and cannot be given away. Second is that the data submitted should be the newly updated data since the data may be update many time in the same district. Thus, the data in the database is always newest version.

3 Data Updating Algorithm Based on Digital Watermarking Technology

3.1 The Data Updating Process Based on Watermark Technology

The application of digital watermarking technique can solve the security in updating GIS spatial data effectively, and supervise the data updating better. Hence, the effectiveness of topography database can be guaranteed. The flowchart of updating GIS spatial data with digital watermarking technique is given in Fig. 2.

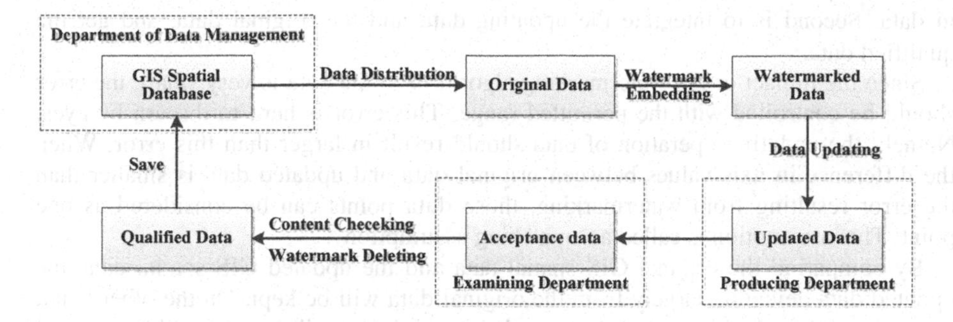

Fig. 2. The process of data updating based on watermark technology

During updating, the department of data management is to embed watermark in original data, and the producing department obtains the data with watermark. Then, the vector data is updated by data producing department, and the updated data is submitted to the examining department, where the submitted data is detected for the watermarking information. By checking the watermarking information, the updated information in producing departments is determined new or not. After confirming the new data, the watermark removing operation is carried by investigating the original data and the submitted data, and the qualified data is obtained. Finally, the qualified data is stored in database.

3.2 Watermark Embedding and Detecting

During updating data by producing departments, the data may be attacked, in particular for operations such as addition, deletion, etc. This puts forward high requirements in anti-attacking abilities for digital watermarking algorithm. Namely, the digital watermarking information can be extracted even when the watermarking data is maintained less.

The watermark is embedded in the point and linear data of GIS spatial data. Thus, such embedding method has been verified to implement the watermarking effect and reduce the algorithm's complexity in practice. Considering the use of GIS spatial data, the graphical deform should be controlled in watermark embedding process. Although the data is changed slightly by watermarking algorithms, the large deform may be resulted in for some related data points. At this time, the control should be carried from the aspects of distances and angles. So, it will not affect the use of GIS spatial data. After detecting the watermarking information, the copyright of data such as user's information and distribution instant will be recovered.

3.3 Watermark Removing

When the department of data management requires to store the updating data, the watermark removing operation is still in need due to the more strict requirements of data storage. The original data after updating can thus be stored. So, the following work is needed for the qualified data. One is to remove the watermarking information in data. Second is to integrate the updating data and the original data, and get the qualified data.

Since the impact of the watermarking algorithm to the data is very small, the error should be controlled with the permitted range. This error is hard to discern by eyes. Namely the updating operation of data should result in larger than this error. When the difference in axis values between original data and updated data is smaller than the error resulting from watermarking, those data points can be considered as one point. This assumption is called as "matching assumption".

By comparing the original GIS spatial data and the updated GIS spatial data, the updated data deviating largely from the original data will be kept. On the other hand, the updated data deviating slightly from the original data will be replaced by original data. Thus, the combination of updated data and non-updated data without watermark can be obtained and stored in the database.

The detailed process is as follows.

Read Original Data and Updated Data, Respectively. According to the data structure and embedding location of watermarks, the point sequences and line sequence are extracted and stored as below.

$$\begin{cases} Op = \{x_i, y_i \mid i = 1, 2, ..., n\} \\ Ol = \{l_i \mid i = 1, 2, ..., m\} \\ Vp = \{x_i, y_i \mid i = 1, 2, ..., p\} \\ Vl = \{l_i \mid i = 1, 2, ..., q\} \end{cases} \tag{1}$$

where, n is the number of solid dots in original data, m is the number of solid lines in original data, p is the number of solid dots in updated data, and q is the number of solid dots in updated data), respectively.

Among them, each line in the line sequence Ol and Vl are still composed of dot sequences, and marked as follows.

$$\begin{cases} Ol_j p = \{x_i, y_i \mid i = 1, 2, ..., r\} \\ Vl_k p = \{x_i, y_i \mid i = 1, 2, ..., s\} \end{cases} \tag{2}$$

where, r is the dot number in solid lines Ol_j of original data, $j = 1, 2, ..., m$, and s is the dot number in solid lines Vl_k of updated data, $k = 1, 2, ..., q$.

For Op, $Ol_j p$, Vp and $Vl_k p$, the mark is added to each point, to mark whether the matching is realized or not.

Compare Each Point Between the Original Data and the Updated Data. First choose a point Vp_1 in Vp according to the storage sequence, and compare it with all the points in Op. Firstly, the mark of $Op_i (i = 1, 2, ..., n)$ is judged whether the point matches successfully. If yes, Op_i is passed. Otherwise, it should be compared.

According to "matching assumption", if there is a point Op_t in Op and the difference between Op_t and Vp_1 is smaller than the error due to watermark algorithm, the point Vp_1 is considered as a watermarking point. Thus, Vp_1 is replaced by Op_t, and Op_t is marked matching successfully.

If there is no point in Op deviating from Vp_1 smaller than the error, Vp_1 is considered as the updated point, and it will be maintained without change.

According to the storing sequence in Vp, the points $Vp_2, Vp_3, \ldots, Vp_q$ are got, and compare each other with all points in Op. Then, repeat the process until all the points in Vp have been compared. Therefore, original points and updated points are obtained, and Vp' can be obtained, which is the dot sequence of combined data.

Obtain the Qualified Data. The updated data is considered as the updated content and kept directly. They together with new dot sequence Vp' and line sequence Vl' consist of the combined data, which are qualified data to be stored in database.

By using the method above, the watermark can be removed from the updated data, and the qualified data after updating can be obtained and stored. By using the matching mark, the calculating burden is reduced greatly, and the calculating work is decreased gradually.

4 Experiments and Analysis

Since the large-ratio GIS spatial data of cities are mainly dependent of AutoCAD, a 1:500 GIS spatial data in "dwg" format is exemplified as shown in Fig. 3. The experimental verification is given for the proposed scheme.

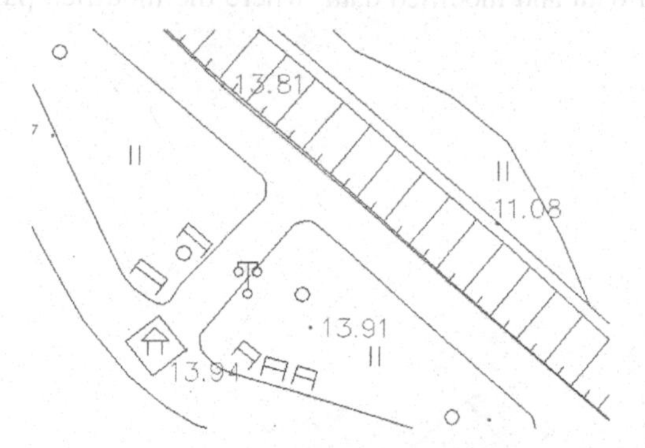

Fig. 3. Part of GIS spatial data

4.1 Experimental Analysis of Watermarking Embedding

The watermark is embedded in the data with watermarking algorithm. The intension of watermark is controlled so that the embedded watermark will not affect the accuracy of data. The embedded watermark exhibits not distinct difference with original data. The overlapping map with original data and watermarked data is shown in Fig. 4.

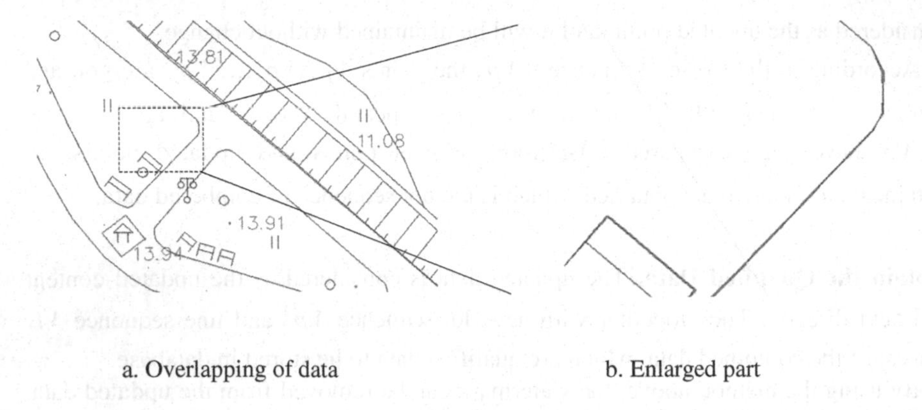

a. Overlapping of data b. Enlarged part

Fig. 4. Overlapping map of original data and watermarked data of GIS spatial data

It can be observed that there is slight difference between the original data and watermarked data in the enlarged part in figure 4(b). As we known, the watermarked data will not destroy the accuracy of data, and will not affect the use of data later.

4.2 Experimental Analysis of Data Updating

The data updating is also be simulated, Fig. 5 shows the enlarged figure after overlapping original data and modified data, where the modified part is marked with dashed line.

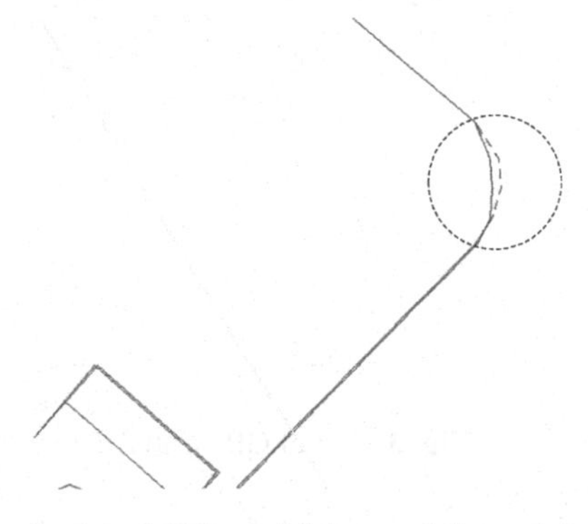

Fig. 5. Overlapping of original GIS spatial data and the updated data (part)

It can be observed that there is large difference in the updated part, and the non-updated part contains a certain small difference due to the process of embedding watermark from figure 5.

4.3 Experimental Analysis of Watermarking Detection

The watermark information is detected from updated data, which can be obtained before updating. The detected information shows the distribution information of data, and thus control the effectiveness and use of updated data.

4.4 Experimental Analysis of Watermarking Removing

The watermarking removing is carried on the updated data. The processed data is combined with original data, and the qualified data is got. Fig. 6 shows the enlarged data of overlapping qualified data with original data, and the updated part is presented by dashed line.

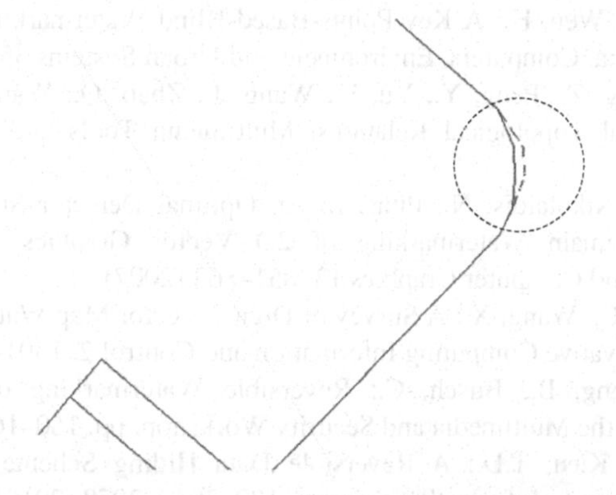

Fig. 6. Overlapping of qualified data with original data

From Fig. 6, it can be seen that the embedded watermark has been removed, and the corresponding place matches the original data completely. The updated part is kept, and replaces the original data.

From the experiments above, the proposed method of data updating in this paper can improve the security, effectiveness of updating operation, which satisfies the requirements in process of GIS spatial data updating.

5 Conclusions

To solve the security issues and find proper techniques to guarantee effectiveness in updating GIS spatial data, a secure data updating method is proposed and designed in this paper by using the watermark technology. The corresponding watermarking removing algorithm is proposed to satisfy the requirements to store qualified data.

The experimental analysis shows the proposed method can realize the illegal use of data and control the use of data. The newly distributed data can be used for updating and submission, and the watermark information can be removed from data effectively, which could satisfy the requirements in data storage.

Acknowledgements. The work was supported by the National Natural Science Foundation of China (Grant No. 41301413), Natural Science Foundation of Jiangsu Province (Grant No. BK20130903, BK20140066), Open Foundation of State Key Laboratory of Information Engineering in Surveying, Mapping and Remote Sensing of Wuhan University (Grant No. 12I01).

References

1. Wang, N., Zhang, H., Men, C.-G.: A High Capacity Reversible Data Hiding Method for 2D Vector Maps Based on Virtual Coordinates. Computer-Aided Design 47, 108–117 (2014)
2. Yan, H., Li, J., Wen, H.: A Key Points-Based Blind Watermarking Approach for Vector Geo-Spatial Data. Computers, Environment and Urban Systems 35, 485–492 (2011)
3. Wang, C., Peng, Z., Peng, Y., Yu, L., Wang, J., Zhao, Q.: Watermarking Geographical Data on Spatial Topological Relations. Multimedia Tools and Applications 57, 67–89 (2010)
4. Doncel, V.R., Nikolaidis, N., Pitas, I.: An Optimal Detector Structure for The Fourier Descriptors Domain Watermarking of 2D Vector Graphics. IEEE Transactions on Visualization and Computer Graphics 13, 851–863 (2007)
5. Niu, X., Shao, C., Wang, X.: A Survey of Digital Vector Map Watermarking. International Journal of Innovative Computing Information and Control 2, 1301–1316 (2006)
6. Voigt, M., Yang, B., Busch, C.: Reversible Watermarking of 2D-Vector Data. In: Proceedings of the Multimedia and Security Workshop, pp. 160–165 (2004)
7. Chang, C.-C., Kieu, T.D.: A Reversible Data Hiding Scheme Using Complementary Embedding Strategy. Information Sciences 180, 3045–3058 (2010)
8. Huang, H.-C., Fang, W.-C.: Authenticity Preservation with Histogram-Based Reversible Data Hiding and Quadtree Concepts. Sensors 11, 9717–9731 (2011)
9. Peng, F., Guo, R.-S., Li, C.-T., Long, M.: A Semi-Fragile Watermarking Algorithm for Authenticating2D CAD Engineering Graphics Based on Log-Polar Transformation. Computer-Aided Design 42, 1207–1216 (2010)
10. Shao, C., Wang, X., Xu, X.: Security Issues of Vector Maps and Reversible Authentication Scheme. Doctoral Forum of China 2005, 326–331 (2005)

Hyperspectral Satellite Remote Sensing of Dust Aerosol Based on SVD Method

Ruiling Lv, Xiaobo Deng, Jilie Ding, Hailei Liu, and Qihong Huang

Chengdu University of Information and Technology, Chengdu, 610225
{Ruiling Lv, ruilinglv}@163.com,
{Xiaobo Deng, dxb}@cuit.edu.cn

Abstract. Satellite remote sensing of dust aerosol depth is quite significant for practical application. In this paper, airborne dust AOD is retrieved from the hyperspectral observed data of the Atmospheric Infra-Red Sounder (AIRS) by using Singular Value Decomposition (SVD) method which is first proposed by L Kuser in 2011. According to the analysis, 8.8-12 infrared observation can be used for dust aerosol retrieval. This method took advantage of the spectral shape of dust extinction and surface and atmospheric influence over the total 8.8–12μm window band. Though the proper linear combination of the singular vectors, dust signal was finally distinguish from the influence of surface emissivity and gas absorption. Then dust AOD of Beijing areas was retrieved to validate this method. As a result, the inversion by using SVD is good with ground-based observations of Aerosol Observation Network (AERONET) data, where their correlation coefficient is 0.9891. In contrast to the traditional physical methods, this method takes advantage of the statistics without losing the physical meaning.

Keywords: dust, AOD, hyperspectral, SVD.

1 Introduction

Dust aerosol is the main component of tropospheric aerosols, accounting for 30% of the atmospheric aerosol content, which is also an important factor influencing the Earth's climate, because it can simultaneously absorb and scatter both solar and infrared radiation, thereby heating and cooling of the climate effect under different conditions. Besides, dust aerosol has a significant impact on the atmospheric environment. When the wind goes through, the regional pollution index and the atmospheric AOD increase significantly. As the recent massive dust severe weather caused by human activities is appearing frequently, how to effectively control dust and forecast meteorological research activities timely and accurately becomes the focus of disaster prevention and mitigation.

Satellite remote sensing inversion theory is quite well established over the ocean areas [1]. However, the traditional inversion method can not get dust aerosol AOD for the bright surface such as desert region and so on. Scientists have been working on satellite remote sensing research of aerosols over the bright surface for a long time, Hsu et al [2] proposed Deep Blue algorithm for Moderate Resolution Imaging Spectrometer MODIS

F. Bian and Y. Xie (Eds.): GRMSE 2014, CCIS 482, pp. 151–161, 2015.

(The Moderate resolution Imaging Spectral-radiometer), and analyzes the Asian dust aerosols optical properties. The multi-angle observations of the Multi-angle Imaging Spectral-radiometer [3] (MISR) can detect surface reflectance and also inverse dust desert aerosol optical properties. Lean and Sayers [4] improve the ORAC (Oxford-RAL Aerosol and Cloud) algorithm, a new micro- physical model of dust aerosols can be designed for remote sensing of desert dust aerosols and so on. However, these methods are not commendable for separating dust AOD signal from the surface reflectance information, thus the surface reflectance and atmospheric conditions are indispensable (temperature and humidity profiles) and a reliable estimate estimation should be given.

Due to large sizes of dust particles, dust aerosols have strong absorbency in the infrared bands, so it is achievable remote sensing dust AOD over the thermal infrared bands. AIRS [5] has 2378 channels covering the spectral range from 650cm-1 to 2700cm-1 , the spectral resolution is up to 1200 , whose data can be used to invert the dust AOD. Thus this paper presents a new method base on SVD, which can retrieve dust AOD without the priori assumptions of the atmospheric state and the surface emissivity. First, the basic theory about SVD is introduced in section 2, then the retrieval route is described in section 3 and details in section 4, 5 and 6, finally an example of retrieval in Beijing areas is explored in section 7.

2 Algorithm Theoretical Basis [6]

Principal component analysis of the data field is decomposed into a primary method of independent uncorrelated signal decomposition was independent uncorrelated signals form the principal component analysis of the main ingredients. Principal component analysis is mainly used in complex and redundant data, through finding the principal component of the data field to achieve the purpose.

Principal component analysis is a classical mathematical statistical analysis method. It is mainly used to explore the extraction of the uncorrelated and independent variables and concentrate attention only on main (principal) components which should be retrieved.

For a non- symmetric matrix, Singular Vector Decomposition (SVD) can be mathematically regarded as solving a non-symmetric eigenvalue problem. Thus SVD can be seen as principle component analysis of the original data matrix instead of the covariance matrix. For a non-symmetric m×n data matrix X, there certainly exists two orthogonal matrixes Um×m and Vn×n which make the singular value decomposition

$$X=U\Sigma V^T \tag{1}$$

$$\Sigma = \underset{\begin{matrix} r & n-r \end{matrix}}{\begin{bmatrix} \Sigma_r & 0 \\ 0 & 0 \end{bmatrix}} \begin{matrix} r \\ m-r \end{matrix} \tag{2}$$

Σ is a m × n diagonal matrix, in which Σr contains singular values of X, U and V are the left and right singular vectors, which are the basic vectors for the observations space M, including the data matrix X, so SVD can be applied to the observation field directly and the corresponding singular vectors form a basis of the observation space.

3 Inversion Method [7]

For hyperspectral infrared atmospheric sounder AIRS channel performance indicators such as response function , sensitivity, channel center wave number and width and so on, and because of the impact of surface and gas absorption , data preprocessing is first made, for filtering pixels that are not capable of retrieval. Then SVD is applied to retrieve dust AOD over the pixels left after screening. The specific operation route is shown in Figure 1.

4 Data Preprocessing

AIRS hyperspectral data provides much information that can greatly improve the accuracy of remote sensing of atmospheric parameters, but such high-dimensional data itself also brings the inevitable question to the high spectral resolution infrared remote sensing inversion. Therefore, this study section is introduced for dimensionality reduction and reducing the impact of narrowband gas absorption and removal of cloud.

4.1 Narrowband Gas Absorption

8-12um is a TIR window, where gas absorption significantly reduced, or even can not be considered. At the same time, the dust aerosol extinction spectra in this region are for V-shaped, which can be used in this inversion, this article select AIRS 672 channels data of 8.8-12um band for inversion dust AOD. In order to avoid the influence of narrowband gas absorption, these 672 channels are equally divided into 67 bins with 10 channels per group, which form a new brightness temperature spectrum. The baseline brightness temperature is determined as the 10 channels maximum brightness temperature, the hyperspectral sampling of AIRS can finally avoid the narrowband gas absorption line in this region(status level S_0), but water vapor absorption at 9.6um still can not be filtered out.

4.2 High Clouds and Inhomogeneous Field

Cloud existed in high space makes it more complex and difficult to detect the dust AOD, so the next step is cloud monitoring and a partial cloud filtering of observations.

In order to remove the high clouds, inspired by the ice clouds threshold of 238K [8], we assume that pixels with the baseline brightness temperature below 240K are full of high clouds, dust AOD in this region is not suitable for inversion (status level S1).

Due to the high spatial homogeneity requirements for the characteristics of mineral dust, and the heat in a non-uniformity of AIRS spectral radiance values would cause a considerable impact, Only if the variance of field of view brightness temperature is sufficiently low (σ(FOV)<8K), the pixel of nine squares can be identified to fit for dust aerosol inversion (status level S2).

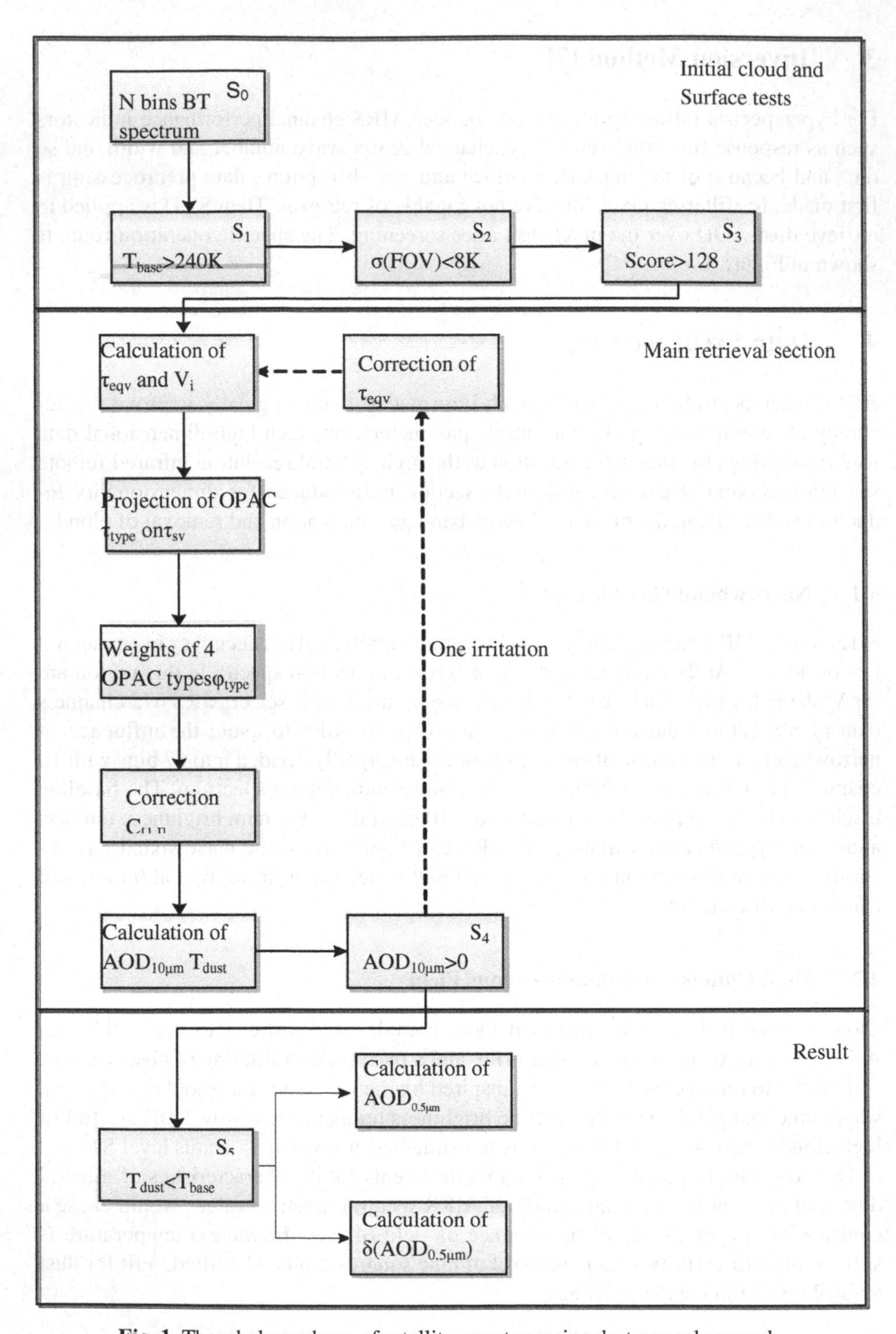

Fig. 1. The whole roadmap of satellite remote sensing dust aerosol research

4.3 Surface Tests

Calculate the equivalent optical thickness, using the following formula:

$$L_{obs}(\lambda) = e^{-\tau_{eqv}(\lambda)/\cos(\Theta_v)} B_\lambda(T_{base}) \tag{3}$$

Where L_{obs} is the radiance whose brightness temperature is the maximum value in each bin, Θ represents zenith angle. As the equivalent optical thickness τ_{eqv} still contains large amounts of information of water vapor and ozone absorption, table 1 provides a set of test conditions. If the test score is too low, it means the dust concentration is too low and the impact of surface emissivity becomes too significant that it is difficult to separate out dust signals [10] from other signals. When the sum of four tests is larger than 128, it means that pixels tested meet the inversion conditions (status level S3).

Table 1. Four test conditions for surface tests

Test No.	Test condition	Score
1	τ_{eqv} (10.00μm)/ τ_{eqv} (10.63μm)>2	128
2	τ_{eqv} (10.10μm)/ τ_{eqv} (10.87μm)>1	64
3	τ_{eqv} (10.10μm)/ τ_{eqv} (11.49μm)>1	64
4	(λ_x >11μm or λ_x <9μm) and τ_{eqv} (10.10μm)/ τ_{eqv} (11.49μm)>2	64

5 Dust Aerosol Model

OPAC [11] is the short of Optical Properties of Aerosol and Clouds; it's a software package which can provide 10 aerosol particles modes in 61 wavelengths through 0.25 to 40 micron. In this paper, we just use four of the modes, namely mineral nucleation mode particles MINM, mineral accumulation mode particles MIAM, crude modal transfer type mineral particles and mineral particles MICM MITR.

As we all know, with no impacts of the water vapor, aerosol particles meet Jung spectral distribution, and then the AOD dependent on the wavelength can be written by Angstrom relationship [12]:

$$\tau_a(\lambda) = \beta\lambda^{-\alpha} \tag{4}$$

Among the formula, α means Angstrom wavelength exponent, β is Angstrom turbidity coefficient. If AOD at λ_1 and λ_2 is known, OPAC dust AOD at some wavelengths with little impacted by water vapor can be calculated by

$$\tau_a(\lambda) = \tau_a(\lambda_1)\left(\frac{\lambda}{\lambda_1}\right)^{-\alpha_{\lambda_1/\lambda_2}} \tag{5}$$

Then using cubic spline interpolation, we get a smooth curve explored in the below:

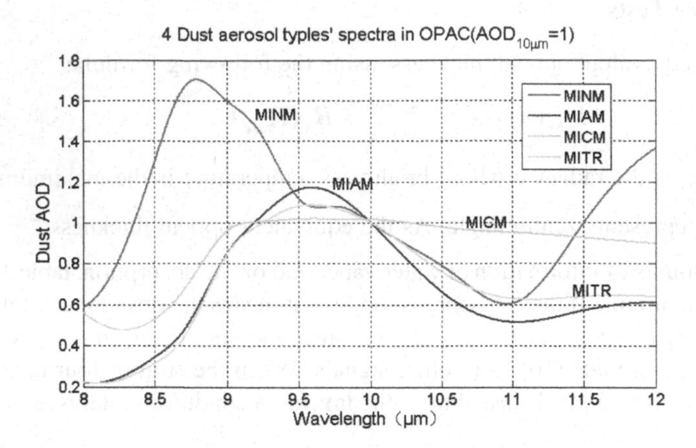

Fig. 2. OPAC dust AOD spectra

6 Dust Aerosol Optical Thickness Retrieval

6.1 The Singular Value Decomposition of Equivalent Optical Thickness

After the singular value decomposition of the equivalent optical thickness calculated by the formula 2, a set of n=67 right singular vectors v_i are received, which form the basis for a new space. As the proportion of the first 15 singular vectors is more then 91%, so we take the first 15 primary singular vectors ($v_1, v_2, \ldots, v_{14}, v_{15}$) constitute 15 basic vectors used for dust aerosol retrieval. The weight of the singular vector ω_i is calculated by

$$w_i = \frac{\sum_\lambda v_i(\lambda)\tau_{eqv}(\lambda)}{\sqrt{\sum_\lambda \tau_{eqv}(\lambda)^2}} \tag{6}$$

We take the first I singular vectors represent the signals of the surface emissivity and the absorption information of the ozone and water vapor:

$$\tau_{[1,I]}(\lambda) = \sum_{j=1}^{I} w_j \cdot v_j(\lambda) \tag{7}$$

And the signal of dust aerosols is

$$\tau_{sv}(\lambda) = \sum_{j=I}^{15} w_j \cdot v_j(\lambda) \tag{8}$$

I is decided by the correlation between $\tau_{[1,I]}$ and τ_{sv}.

6.2 Preliminary Calculations Dust Aerosol Optical Thickness

The projection of τ_{sv} on AOD spectral of the 4 dust aerosols in the software package OPAC at the Uniform conditions: $AOD_{10\mu m}=1$ is computed by the inner product of them.

$$AOD_{type} = \sum_{\lambda}[AOD_{opac,type}(\lambda)\cdot\tau_{sv}(\lambda)] \tag{9}$$

The weight φ_{type} of the 4 OPAC types in the inversion can be obtained by the following formula:

$$\varphi_{type}^{-1} = \arccos\left(\frac{AOD_{type}}{\sqrt{\sum_{\lambda}(AOD_{opac,type}(\lambda))^2}\sqrt{\sum_{\lambda}(\tau_{sv}(\lambda))^2}}\right) \tag{10}$$

Therefore, with a variety of known of the 4 OPAC types proportion, the spectrum of OPAC dust aerosol optical thickness can be calculated from the model:

$$AOD_{opac}(\lambda) = \sum_{type}\left[\varphi_{type}\cdot AOD_{opac,type}(\lambda)\right] \tag{11}$$

Because first I vectors may contain dust aerosols near the ground information, therefore we gives the correction term:

$$c_{[1,I]} = s_{\text{var}}\cdot\sum_{type}\left(\varphi_{type}\cdot P_w\cdot AOD_{type}\right) \tag{12}$$

Where, P_w donates the standard projection of the first I vectors on the left vectors and s_{var} is the signed squared linear correlation between AOD_{type} and $\tau_{[1\,,\,I]}$.

So far, the AOD of dust aerosol at 10μm wavelength is arrived:

$$AOD_{10\mu m} = \sum_{type}\left(\varphi_{type}\cdot AOD_{type}\right)+c_{[1,I]} \tag{13}$$

6.3 Iteration Principal

If the $AOD_{10\mu m}$ is positive (status level S4), the AOD of dust aerosol at the whole wavelengths can be detected. On the other side, it means the image cannot be reviewed.

$$AOD_{\lambda} = \sum_{type}\left[AOD_{10\mu m}\cdot\varphi_{type}AOD_{opac,type}(\lambda)\right] \tag{14}$$

Then bring into the below formula for the correction and estimating the equivalent emission of dust aerosols.

$$L_{\uparrow}(\lambda) = e^{-AOD_{\lambda}/\cos(\Theta)}B_{\lambda}(T_{surface}) + (1-e^{-AOD_{\lambda}/\cos(\Theta)})B_{\lambda}(T_{dust}) \tag{15}$$

Because of the $T_{surface}$ unknown, we use T_{dust} as the initial value instead of it. If $T_{surface} > T_{base}$, using T_{base} instead of T_{dust} repeating formula (3) to (14). Until $T_{dust} < T_{base}$ (status level S_5), the iteration is stopped. Finally, we get the results we want.

7 Examples

This paper selects the April 2012 AIRS data covering of Beijing area for inversion, and compares with AERONET AOD values corresponding three sites measured. Figure 3 show the partial results.

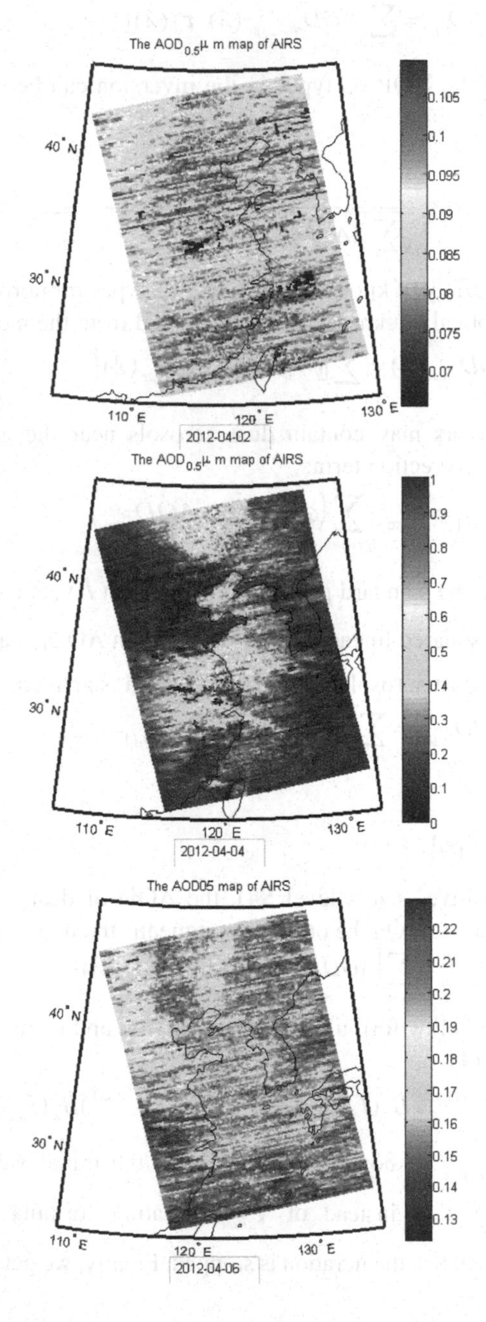

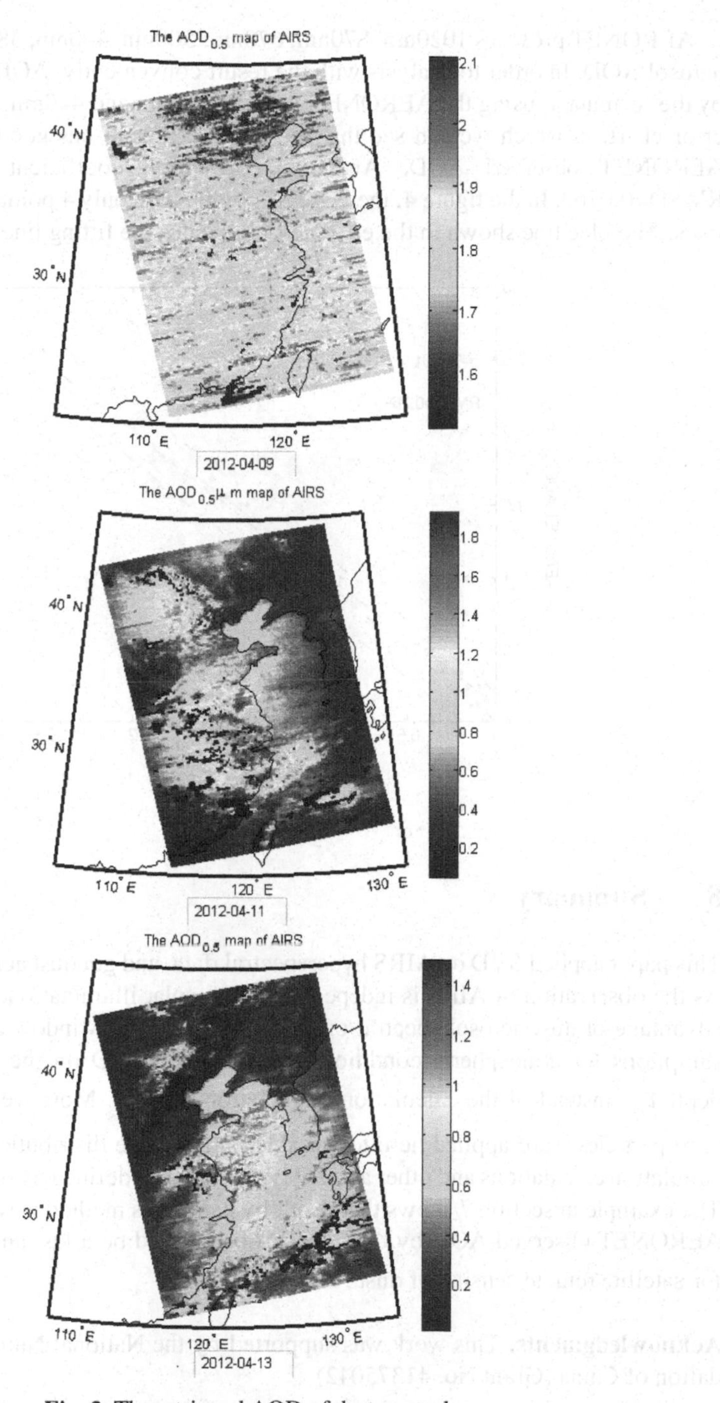

Fig. 3. The retrieved AOD of dust aerosols

AERONET provides 1020nm, 870nm, 670nm, 500nm, 440nm, 380nm, 340nm band aerosol AOD. In order to analysis with the result conveniently, AOD_{500nm} is calculated by the formula 4, using the AERONET AOD at 670nm and 440nm. Finally we get the error chart, in which we can see that the retrieved result has good consistency with AERONET observed AOD. And the correlation coefficient reached 0.9891, RMSD=0.0769. In the figure 4, there are 33 points, with only 4 points out of the dashed lines. The blue line shown in the error chart indicates the fitting line of all points.

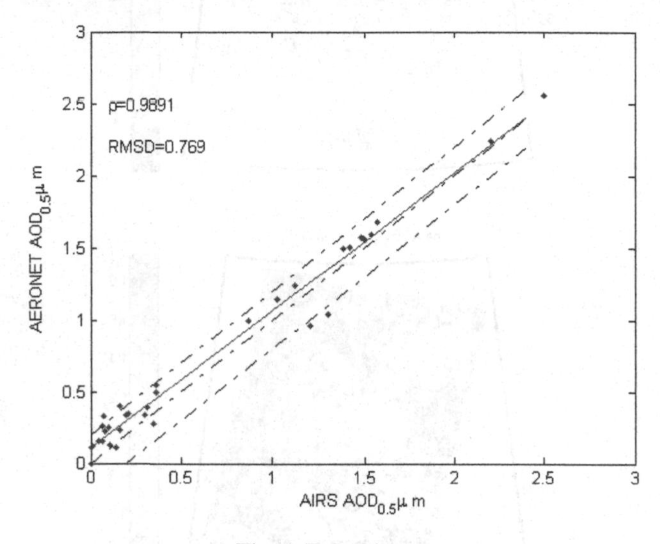

Fig. 4. Error chart

8 Summary

This paper applied SVD on AIRS hyperspectral data, and got dust aerosol depth finally. As the observation of AIRS is independent from solar illumination, This method took advantage of dust aerosol spectral on the 8.8-12 microns window avoiding priori assumptions for atmospheric conditions, then applied SVD on the equivalent optical depth τ_{eqv} instead of the calculation of radiation transfer. Moreover four OPAC dust base particles ware applied here to variable particle size distribution of dust in priori simulation calculations and other aerosol types could be defined by deferent conditions. The example in section 7 shows that result by using this method is well correlated with AERONET observed AOD by ρ =0.9891. So it would be a fast and efficient method for satellite remote sensing of dust AOD.

Acknowledgments. This work was supported by the National Natural Science Foundation of China (Grant No. 41375042).

References

1. King, M.D., Kaufman, Y.J., Tanre, D., Nakajima, T.: Remote Sensing of Tropospheric Aerosols from Space: Past, Present and Future. Bulletin of American Meteorol. Soc. 80, 2229–2259 (1999)
2. Hsu, N.C., Tsay, S.-C., King, M.D., Herman, J.R.: Aerosol Properties Over Bright-Reflecting Source Regions. IEEE Transactions on Geoscience and Remote Sensing 42(3), 557–569 (2004)
3. Diner, D.J., Beckert, J.C., Reilly, T.H.: Multi-angle Imaging SpectroRadiometer (MISR) Instrument Description and Experiment Overview. IEEE Transactions on Geoscience and Remote Sensing 36(4), 1072–1087 (1998)
4. Lean, K.: Empirical methods for detecting atmospheric aerosol events from satellite measurements. Technical report, University of Oxford (2009)
5. Aumann, H.H., Chahine, M.T., Gautier, C., Goldberg, M.D.: AIRS/AMSU/HSB on the Aqua mission: design, science objectives, data products, and processing systems. IEEE Transactions on Geoscience and Remote Sensing 41, 253–264 (2003)
6. Rodgers, C.D.: Inverse Methods for Atmospheric Sounding, Theory and Practice. Series on Atmospheric, Oceanic and Planetary Physics, vol. 2. World Scientific Publishing, Singapore (2000)
7. Kluser, L., Martynenko, D., Holzer-Popp, T.: Thermal Infrared Remote Sensing of Mineral Dust over Land and Ocean: a Spectral SVD based Retrieval Approach for IASI. Atmos. Meas. Tech. 4, 757–773 (2011)
8. Chylek, P., Robinson, S., Dubey, M.K., et al.: Comparison of Near-infrared and Thermal Infrared Cloud Phase Detections. Journal of Geophysical Research 111, D20203 (2006)
9. Evan, A.T., Heidinger, A.K., Pavolonis, M.J.: Development of a New Over-water Advanced Very High Resolution Radiometer Dust Detection Algorithm. International Journal of Remote Sensing 27(18), 3903–3924 (2006)
10. Ogawa, K., Schmugge, T., Jacob, F., et al.: Estimation of Land Surface Window (8–12μm) Emissivity from Multi-spectral Thermal Infrared Remote Sensing-A Case Study in a Part of Sahara Desert. Geophysical Research Letters 30 (2003)
11. Hess, M., Koepke, P., Schult, I.: Optical Properties of Aerosols and Clouds: The Software Package OPAC. Bulletin of the American Meteorological Society 79(5), 831–844 (1998)
12. Alexander, A.: Kokhanovsky. Aerosol Optics: Light Absorption and Scatting by Particles in the Atmosphere. Springer, New York (2008)

The Study of Complex Network of Search Keyword

Chengguang Wei[1], Di Chen[2], Xinyue Gu[3], and Soosang Lee[4,*]

[1] Library of Tsinghua University, 100084, Beijing, China
[2] School of Software, Tsinghua University, 100084, Beijing, China
[3] School of Social Sciences, Tsinghua University, 100084, Beijing, China
weichg@lib.tsinghua.edu.cn,
{chend14,guxy11}@mails.tsinghua.edu.cn
[4] Department of Library, Archives and Information Studies of Pusan National University, 609735, Busan, Korea
sslee@pusan.ac.kr

Abstract. In the era of big data, complex network theory has been applied to the field of internet information mining and information research. This study chose the keywords from query term log data of the Korea NDSL (National Digital Science library), and analyzed the complexity properties of keyword network. The result showed that the keyword network had the character of free-scale complex network, and that the distribution of link degree accords power law distribution. Meanwhile, we proved that the key word network has growth and self-organization attributions. In digital library, there are some other big datasets which are similar with the real information search query data, so we can utilize the theory to other keyword network research directly.

Keywords: user query, complex network, information research, network analysis, digital library.

1 Introduction

1.1 Background and Necessity of the Research

With current technologies, knowledge society doesn't to need be constrained by geographic proximity and current technologies offer much more possibilities for sharing, archiving and retrieving knowledge. A web search query is a query that users enter keywords into the web search engine to satisfy their query needs[1]. Their keywords cover a broad topic for which there may be thousands of relevant results. Each of the keywords is directly entered by users; these keywords formed a network by co-occurrence.

In past few years, increasing happens in researching query term logs. However in web 2.0, such research enable users to find more of what they are looking for, it comes at a social cost. But they all ignore that it has social network properties by

* Corresponding author.

F. Bian and Y. Xie (Eds.): GRMSE 2014, CCIS 482, pp. 162–169, 2015.

connection relation between the keywords. It is urgent for us to provide a method to explore characters about the keyword network. Meanwhile, a variety of complex networks and complexity theories appear.[2] Nowadays the complexity theories are widely applied to solve a dispersed and no-scale network in the field of internet, physics, biology, math and social science etc. Based on the network comprised of 111,444 keywords of library and information science that are extracted from Scopus, and taken into consideration the major properties of average distance and clustering coefficients, Zhu(2013) with the knowledge of complex network and by means of calculation, reveal the small-world effect of the keywords network.[3]

1.2 Content of Research

NDSL is one of the most important nationwide information services for the users in Korea, according to the results of the survey, annual visits to the web-site portal of the library are now increasing. This article focuses on the complexity analysis of NDSL user's search keyword network. It analyzed its structure complexity, self-organization and growth of network of query term below. The way of this study is based on the complexity theory.

1.3 Experimental Progress

An experimental analysis involves the following five major stages as Fig.1 below:

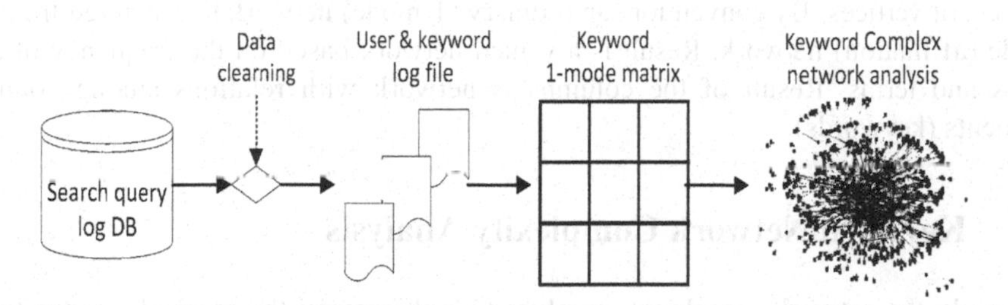

Fig. 1. Process of the experiment

2 Data Preprocessing

2.1 Data Cleaning

Korea NDSL query log files were made by each of the search request sessions in a certain period by users. Each session record item usually has 16 fields, which are "seqno, inputdate, inputtime, loginyn, keyvalue, gubun, contents, searchcnt, loginipsort, local_opacid, gubun_option, libid, libname, userposition, username, usertype". Fig.2 is the sample of query log record, and its data structure is shown as below:

```
seqno,inputdate,inputtime,loginyn,keyvalue,gubun,contents,searchcnt,loginip sort,local_opacid,gubun_option,li
bid,libname,userposition,username,usertype
49850056,20080512,173330,1,mhhyun,0101,(BI:정보처리학회) AND
(BI:data),289,203.250.227.098,NDSL,1,00027,한국과학기술정보연구원,,현미환,1
49841585,20080512,133705,0,203.250.227.222,0101,(TI:science),169330,203.250.227.222,NDSL,1,00027,한
국과학기술정보연구원,0,,
49841664,20080512,133838,0,apple01,0201,(BI:online),5560,203.250.227.222,NDSL,1,00027,한국과학기술정
보연구원,0,,49859878,20080512,234248,1,arirangche,0101,(BI:UV),144542,122.042.044.204,NDSL,1,,,0,,
```

Fig. 2. Query log sample of the Korea NDSL

This experiment was performed on a twelve-month query term log sample of the Korea NDSL (National Digital Science library) from 15th April 2008 to 31st May 2009. The "keyvalue" field records unique id of a user, and access point is recorded in "searchcht" field in each search. User's search keywords were written down in the field of "content". According to the two field of "keyvalue" and "searchcht", we extracted the 24241 login personal searchers and their 104011 search query terms by access point of “BI(all)” from huge records data on the server by data cleaning.

2.2 Matrix Affiliation

After the process of data cleaning, we abstracted users and theirs search keyword sessions, and made a relation matrix. As it consists of two sets of vertices, so it is called user & keyword two-mode matrix.

According to this two-mode matrix, we used "PAJEK" tool's command "Net> Transform>2-Mode to 1-Mode" to create to a one-mode network on each of the user subsets of vertices. By convention, an ordinary (1-mode) network is generated from 2-mode (affiliation) network. Result is a valued network based on the frequency of the users and terms. Result of the columns is network with relations among column elements (keywords).

3 Keyword Network Complexity Analysis

Normally the network complexity involves two characters: the research of structural complexity and the network node complexity. In this article, we analyzed not only the structural complexity but also the node complexity of two real networks. As going deeper into the experimental, we also studied other properties about network, such as self-organization, growth additionally.

Korea NDSL portal are millions of experts or regular persons in the domains where they are working, such as researchers, students, teachers, ordinary information seekers etc. As we known, there are a variety of backgrounds, the context of search keyword in query logs is a record of the users’ interests, according to these keywords, and we can judge his /her interests. Except for a few experts, most of the users often don’t have property vocabularies of a scientific sub-discipline to express their information needs.

Keyword network is not only a simple word net, every word has its own meaning that is used to express the users' information needs or requests. They connect with each other, and have a number of relations, such as homophony relation, synonym relation, antonyms relation, antonyms relation etc. In the interests of users search for the parent category as well as meaningful categories, sub-categories, synonyms, thesaurus and related words together including the meaning which have complex semantic relationships between the users' search query terms. So there could be many different kinds of nodes in the network showing a chaos world to us in its internal or between query terms. Links between nodes could have different weights, directions and signs. Synapses in the nervous system can be strong or weak, inhibitory or excitatory.

3.1 Structure Property of the Query Network

Structures always affect functions. Many researchers have shown that the complex network distributions conform to the power law, or long tail distribution curves. After the theory about the Watts and Strogatz's small world phenomenon of complex network and Albert-László Barabási's no-scale property of the complex network, more scholars are much more interested in the study of real-world network. [4]

Obeying the power law degree contribution is an important criterion to test whether the network is a complex networks or not. In this study, properties search keyword network is laid out in Table1.

Table 1. General Information of Query Network

Direction	Indirection
Density	0.0050825
Number of vertices	104011
Total number of lines	27497238
Power law coefficient(γ)	2.1322

It is an undirected network, its density is 0.0050825. The network includes 104011 numbers of vertices, 27497238 numbers of edges, and its power coefficient (γ) is 2.1322. Fig.3 appears that the keyword network degree distribution has truncated classic power-law type:

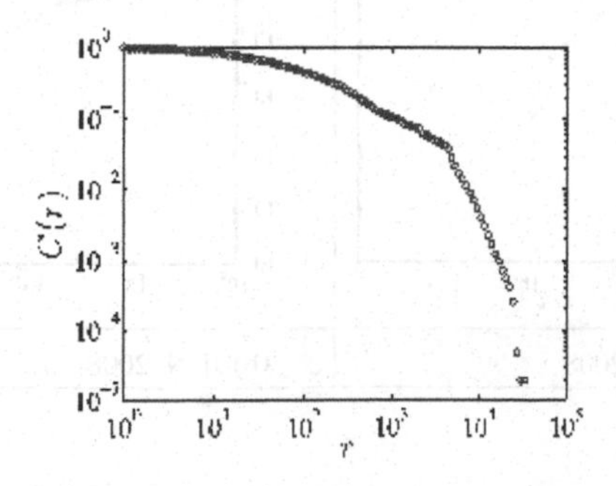

Fig. 3. Power law of degree distribution

3.2 Growth of the Complexity Network

Users and their information needs are changing and show the phenomenon of development. According to the change of users and their variable information needs, search query also has rapid growth on the World-Wide Web. Users and queries that they used are created and lost every month along with their changing interests or other information needs. In this paper we analyzed the keyword network of the search query term monthly from May. 2008 to Apr. 2009 to find whether it has growth attribute or not. Analysis result data is shown in the Table 2 as below:

Table 2. Character of Monthly Query Network

Month	Number of vertices	Average Degree $\langle k \rangle$	γ	Total number of lines
MAY. 2008	1640	14.695	2.430	12050
JUN. 2008	5944	71.322	2.438	211970
JUL. 2008	9019	89.209	2.770	402289
AUG. 2008	18160	82.398	1.924	748182
SEP. 2008	20294	73.382	2.235	744617
OCT. 2008	21415	85.076	1.770	910955
NOV. 2008	22957	92.874	2.193	1066063
DEC. 2008	20935	92.838	1.797	971785
JAN. 2009	20099	123.322	3.508	1239332
FEB. 2009	18694	295.849	4.197	2765303
MAR. 2009	20248	211.784	4.043	2144104
APR. 2009	16844	145.727	4.912	1227318

According to the Table 2, we get the monthly search keyword degree attribution curve as the Fig.4:

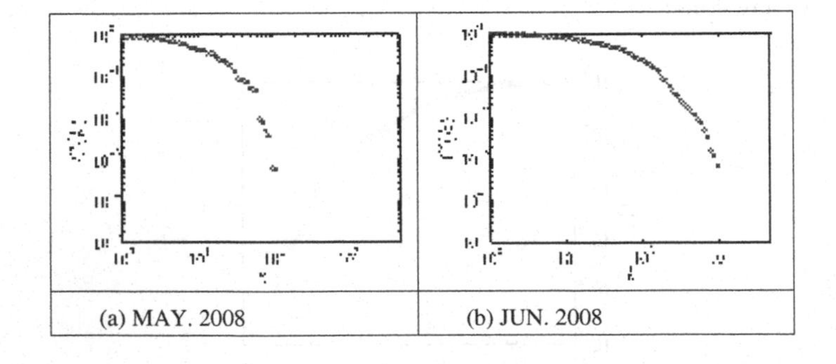

(a) MAY. 2008 (b) JUN. 2008

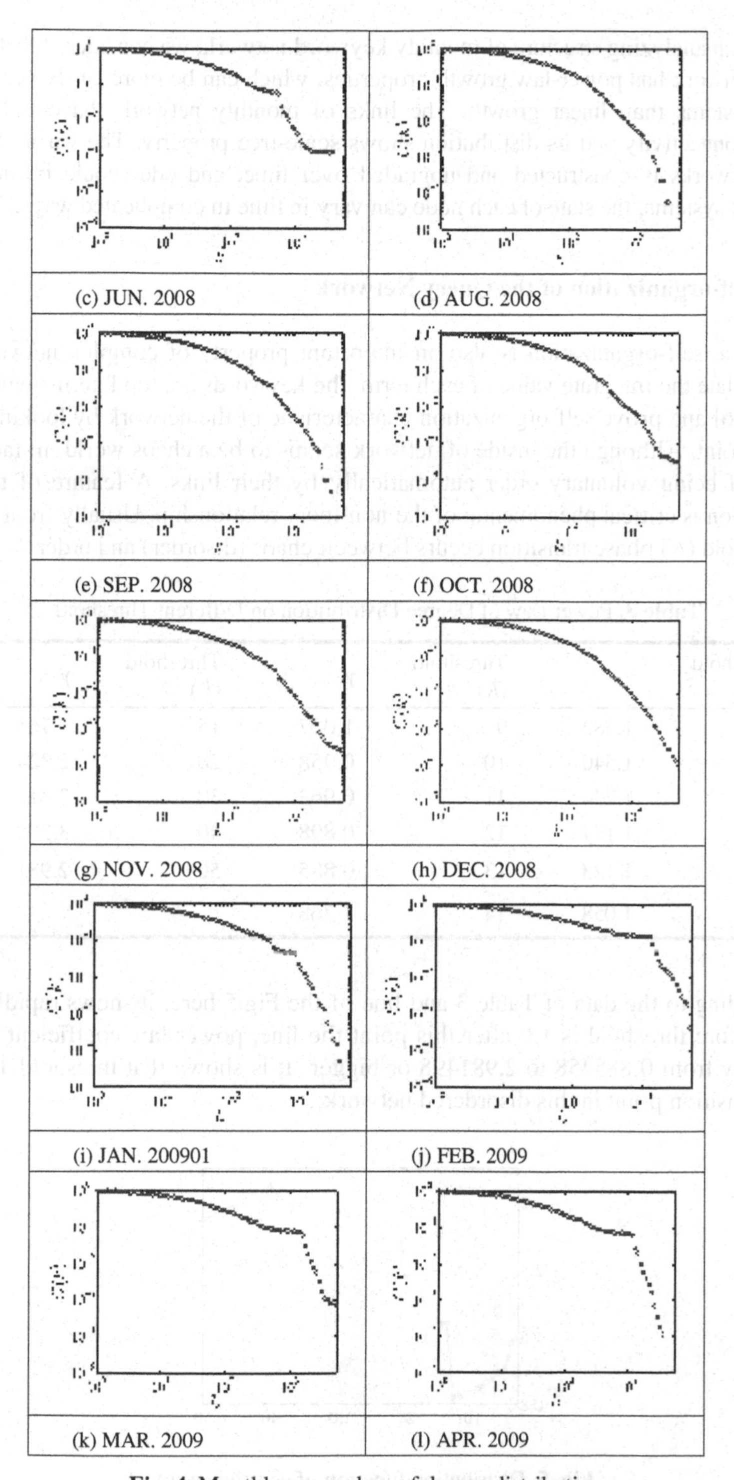

Fig. 4. Monthly power law of degree distribution

Based on analyzing structure of monthly keyword network, we considered that growth of each network has power-law growth properties, which can be more easily generalized to real systems than linear growth. The links of monthly networks have indirection, growth, connectivity and its distribution shows scale-free property. The wiring diagram of the networks is constructed and upgraded over time, and odes could be nonlinear dynamical systems, the state of each node can vary in time in complicated ways.[5]

3.3 Self-organization of the Query Network

In addition, self-organization is also an important property of complex network. We will calculate the integrate value of each term, the keywords are top k terms with greatest value[6] and prove self-organization characteristic of the network by looking phase transfer point. Although the inside of network seems to be a chaos world, in fact it has a trend of being voluntary order automatically by their links. A feature of the self-organization is critical phenomenon of the nonlinear relationship. Usually front back of the threshold (*K*) phase transition occurs between chaos (disorder) and order

Table 3. Power Law of Degree Distribution on Different Threshold

Threshold (*K*)	γ	Threshold (*K*)	γ	Threshold (*K*)	γ
3	1.582	9	1.007	15	1.761
4	1.540	10	0.958	20	2.324
5	1.234	11	0.963	30	2.417
6	1.174	12	0.898	40	3.388
7	1.123	13	0.885	50	2.981
8	1.058	14	1.768		

According to the data of Table 3 and line of the Fig.5 here, it shows rapidly down trends before threshold is 13, after this point the line, power law coefficient (γ) rose up quickly from 0.885458 to 2.981498 or bigger. It is shown that threshold 13 is the phase transition point in this disordered network.

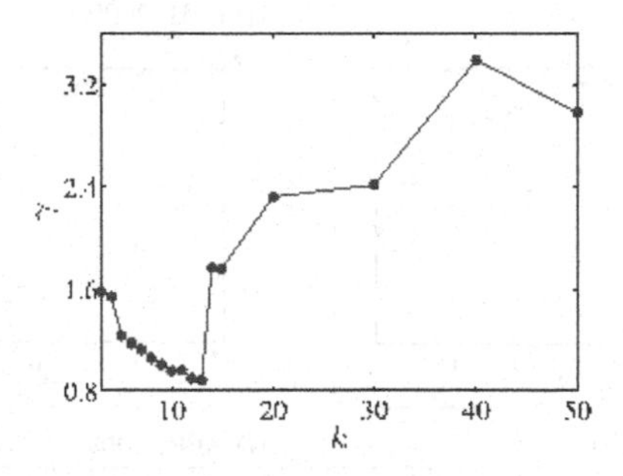

Fig. 5. Distribution function of coefficient (γ)

4 Conclusion

Power law is an important theory in solving and explaining the chaos in real world. In this article, we made query term-query term community. The complexity of structure was defined in the query network's terms of size and number of connections. That is to say, quantity of query terms is big, link directionality and linkage is irregular between the nodes. It shows a chaotic state in the internal structures of the network. Moreover, connections of the nodes are constant changing with time elapsing, it often shows a growth and self-organization trend and tends to evolve more and more complex structure network. By analyzing the properties of query term network, we obtained that:

By analyzing the search query log analysis of the Korea NDSL (National Digital Science library), we know that internal network connection of query term is complex and disorder, and identified its' complexity characters in the field of the information research.

According to this study, we find out the threshold, which is weight of phase transition point of the keyword network. By the cut point, we can extract subnetwork whose link degree is bigger than the threshold, which is a small portion of the terms observed in a large query log used most often, instead of full network analysis.

References

1. White paper of Accessing Web Site Usability from Server Log Files, Tec-ed, Inc. (1999)
2. Murray, G.C., Teevan, J.: Murray., Jaime T.: Query Log Analysis: Social and Technological Challenges. In: WWW 2007 Workshop on ACM SIGIR Forum Archive, vol. 41(2) (2007)
3. Zhu, D.H., Wang, D.B., Hassan, S.U., Haddawy, P.: Small-world phenomenon of keywords network based on complex network. Scientometrics 97(2), 435–442 (2013)
4. Redner, S.: How popular is your paper? An empirical study of the citation distribution. Eur. Phys. J. B 4(2), 131–134 (1998)
5. Dorogovtsev, S.N., Mendes, J.F.F.: Accelerated growth of networks, Wiley-VCH, Berlin (2002)
6. Zhan, Z.J., Lin, F., Yang, X.P.: Keyword Extraction of Document Based on Weighted Complex Network. Mems, Nano and Smart Systems. PTS 1-6, 403–408, 2146–2151 (2012)

Fast MAP-Based Super-Resolution Image Reconstruction on GPU-CUDA

Zhijun Song*, Zhisong Chen, and Rongrong Shi

The 28th Research Institute of China Electronics Technology Group Corporation, Nanjing, China
jason@xmu.edu.cn, zchen@stevens.edu, banirab0601@163.com

Abstract. The traditional super-resolution image reconstruction methods for optimization and implementation are designed for common processor (CPU). According to the parallel computing capability of GPU-CUDA, a fast super-resolution image reconstruction method is presented based on GPU-CUDA. Additionally, we proposed the MAP framework that can allocate sub-pixel displacement information of low-resolution images to the unified super-resolution image grid. On the basis of the parallel architecture and hardware characteristic of GPU, the acceleration method use CUDA programmable parallel framework to optimize the data storage structure, improve the efficiency of data access and reduce the complexity of the algorithm. The experiment expressed that we could get an over ten times speed effect by this method than traditional super-resolution image reconstruction methods.

Keywords: MAP framework, super resolution, GPU-CUDA, image reconstruction.

1 Introduction

With the rapid development of graphics processor GPU (Graphics Processing Unit) technology, today's leading GPU already has strong parallel computing capabilities and floating-point capabilities in many MathWorks products. Meanwhile, serializable in image processing got improved by the parallelism, high-speed features of GPU and programmable GPU introduced in recent years. All those made it possible for the implementation of Super-Resolution Image Reconstruction by GPU parallel computing. It has already became a popular topic today: How to accelerate the computation of some complex algorithm by using GPU parallel computing.

The CUDA [1] (Compute Unified Device Architecture, CUDA) is a general purpose parallel computing architecture proposed by NVIDIA Corporation. The architecture enables developers to build a more efficient data-intensive computing solutions based on the powerful computing ability of GPU. CUDA core is composed of three important

---

* Zhijun Song, Male, PhD, Engineer of the 28th Research Institute of China Electronics Technology Group Corporation, Nanjing, China. He received his PhD from Xiamen University in 2013. His research interests lie in the areas of artificial intelligence.

F. Bian and Y. Xie (Eds.): GRMSE 2014, CCIS 482, pp. 170–178, 2015.

parts: thread processor, shared memory and synchronization boundaries. These three parts provide fine-grained data parallelism, thread parallelism, coarse-grained data parallelism and task parallelism, which can the whole problem be split to the several coarse-grained sub-problems and each sub-problem be processed independently in parallel. Then, these coarse-grained sub-problems can be split into several finer-grained sub-tasks, which can be processed in parallel on any available processor that can achieve the ability to rapidly scale computing.

The application and research of super-resolution image reconstruction can fully exploit the potential of existing image data, and do a great contribution to the development of the biomedical imaging, satellite remote sensing imaging, military reconnaissance and positioning, surveillance video enhancement and restoration, public security surveillance, and other image-processing areas. The traditional super-resolution image reconstruction methods are mainly divided into two main categories: learning-based methods [13,14,22-28] and constraint-based reconstruction methods[2-12,18] which includes frequency domain methods and spatial domain methods. The basic idea of learning-based methods is to study the relationship between low-resolution images and high-resolution images, and the use of this relationship to guide the image super-resolution. These super-resolution image reconstruction method has a high computational complexity, it is difficult to meet the real-time requirements of military, video surveillance and other image-processing areas. Frequency domain methods are earlier super-resolution methods, they can only deal with image sequences that only translational motion are allowed. Spatial methods use general observation models, they only be applied to the case of a global translation and linear space invariant degradation model that lack of a prior spatial ability and flexibility.

In this paper, we proposed a fast super-resolution image reconstruction method that utilizes the parallel architecture of GPU and CUDA programmable parallel framework. Moreover, we built the MAP framework which can allocate sub-pixel displacement information of low-resolution images to the unified super-resolution image grid. Experimental results show that the proposed super-resolution image method is more efficient and practical than traditional methods.

The rest of the paper is organized as follows. Section 2 introduce the MAP-based super-resolution image reconstruction method on GPU-CUDA. The experimental results and discussions are given in Section 3. Finally, the conclusions are presented in Section 4.

2 MAP-Based Method on GPU-CUDA

The aim of super-resolution image reconstruction is to generate a higher resolution image from lower resolution images. High resolution image offers a high pixel density and thereby more details about the original scene. The framework of the fast MAP-based super-resolution image reconstruction on GPU-CUDA is shown in fig.1.

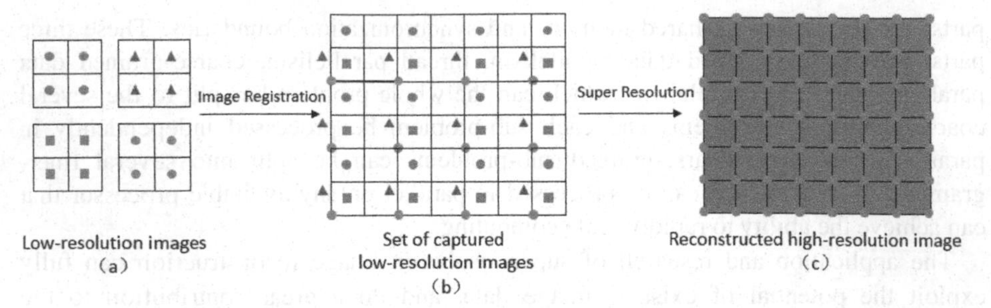

Fig. 1. The stages in super-resolution process: First, we set up an image acquisition model relating the ideal high-resolution image to a set of observed low-resolution images. Then, we build the MAP framework that can allocate sub-pixel displacement information of low-resolution images to the unified super-resolution image grid. Finally, we implement our method on GPU-DUDA.

2.1 Image Acquisition Model

The multiple low-resolution images can be captured from different view-points of the same scene and image registration deal with relating the high-resolution image x and the corresponding low-resolution observation sequence y_k. So, we should first build the acquisition model that can be mathematically expressed as [7]:

$$y_k = HB_k M_k x + n_k \quad k = 1,\ldots,p \tag{1}$$

where y_k is the captured frame k^{th} low-resolution image with the size of $m \times n$, n_k stands for noise in the observation y_k, x is the ideal high-resolution image, H is the down-sampling operator, B_k is the blurring operator, M_k is the warping operator due to the relative motion between the scene and the camera. The above acquisition model depicts the process from an original high-resolution image to multiple observed low-resolution images, which includes the degradation caused by warping, blurring and down-sampling.

2.2 Image Registration on GPU-CUDA

Image registration is used in super-resolution to register low-resolution image frames. Therefore, the first step in the proposed method is to find an estimate of the actual image based on a sub-pixel image registration procedure. Figure 1(a) shows the case for four low-resolution images with different sub-pixel displacement from each other, where the images are aligned following a reference frame, as shown in fig.1 (b). The SIFT (Scale-Invariant Feature Transform) algorithm [16] is one the most popular feature-based image registration methods often used various application. It has the some disadvantages such as low executive efficiency and poor real-time capability, however, so we proposed the parallel SIFT implementation on the basis of the GPU-CUDA technology for the image registration.

The SIFT algorithm [16] procedures is divided into four main steps:keypoints detection, keypoints location, orientation assignment, and keypoints descriptor generation. All the steps of SIFT were specifically distributed and implemented by GPU-CUDA, such as memory allocation, the octaves and sublevels processing, the orientations assigning, and the descriptors generating.

2.3 Super-Resolution Based on MAP

In order to achieve a stable solution of formula (1), we utilize a maximum a posteriori (MAP) probability approach to solve super-resolution image reconstruction. That is, the proposed approach is to estimate the unknown high-resolution image x via its max a posterior (MAP) estimation $\hat{x}$. The maximum likelihood estimate of x is given by:

$$\hat{x} = \underset{x}{\operatorname{argmax}} \Pr\left(x \middle| y_k\right) \tag{2}$$

Based on the low-resolution images y_k, the ideal high-resolution image x can be inferred by the Bayesian law, $\Pr\left(x \middle| y_k\right)$ can be obtained by:

$$\begin{aligned} \Pr\left(x \middle| y_k\right) &= \frac{\Pr(y_k|x)\Pr(x)}{\Pr(y_k)} \\ &\propto \Pr(y_k|x)\Pr(x) \end{aligned} \tag{3}$$

which is a MAP estimation problem and can be modeled by MRFs. Based on the independent identical data distribution, $\Pr\left(y_k \middle| x\right)$ can be defined as

$$\Pr(y_k|x) \propto \Pi \exp(-D(y_k, x)) \tag{4}$$

where $D(y_k, x)$ is the data penalty function which places the penalty of low-resolution images y_k with a high-resolution image x.

Besides, the $\Pr(x)$ in Eq. (2) can be restricted to MRF which involves pairs of low-resolution images as

$$\Pr(x) \propto \exp\left(-\sum_{y_i \in Y} \sum_{y_j \in N(y_i)} V_{y_i, y_j}\left(\mathbf{x}_{y_i}, \mathbf{x}_{y_j}\right) \right) \tag{5}$$

where $N(y_i)$ is the neighborhood of low-resolution image y_i. $V_{y_i, y_j}\left(\mathbf{x}_{y_i}, \mathbf{x}_{y_j}\right)$ named the smoothness penalty function, is a clique potential function. Here, we adopt the generalized Potts model to define the smoothness penalty function as

$$V_{y_i,y_j}\left(x_{y_i},\mathrm{x}_{y_j}\right)=\lambda\cdot\exp\left(\frac{-\Delta\left(y_i,y_j\right)}{\alpha}\right)\cdot T$$
$$=\lambda\cdot\exp\left(\frac{-\left|\mathrm{I}\left(y_i\right)-\mathrm{I}\left(y_j\right)\right|}{\alpha}\right)\cdot T \tag{6}$$

where $\lambda>0$ is a smoothness factor, $\alpha>0$ is used to control the contribution of to the penalty, and T is 1 if its argument is true and 0 otherwise.

From Eq. (3), (4) and (5), we can obtain

$$\Pr\left(x\middle|y_i\right)\propto\prod_{y_i\in Y}\exp\left(-D\left(x,y_i\right)\right)\cdot\exp\left(-\sum_{y_i,y_j\in Y}V_{y_i,y_j}\left(x_{y_i},\mathrm{x}_{y_j}\right)\right). \tag{7}$$

The following energy function can be obtained by taking the logarithm of Eq. (7):

$$E(x)=\sum_{y_i\in P}D(y_i,\mathrm{x})+\sum_{y_i,y_j\in Y}V_{y_i,y_j}\left(x_{y_i},\mathrm{x}_{y_j}\right) \tag{8}$$

where $E(x)\propto-\log\Pr\left(x\middle|y_i\right)$. It includes two parts: the data term

$$E_{data}=\sum_{y_i\in Y}D(\mathrm{x},\mathrm{y}_\mathrm{i}) \tag{9}$$

and the smoothness term

$$E_{\mathrm{smooth}}=\sum_{y_i,y_j\in Y}V_{y_i,y_j}\left(x_{y_i},\mathrm{x}_{y_j}\right) \tag{10}$$

From Eq. (8), we can find that maximizing $\Pr\left(x\middle|y_i\right)$ is equivalent to minimizing the Markov energy $E(x)$. The graph cuts method [17] can be adopted to solve Eq. (8). And we implement the graph cuts on the GPU using CUDA.

3 Experimental Results and Analysis

In this section, we present experimental results on MAP-based super-resolution image reconstruction on GPU-CUDA. Besides, we conduct a comprehensive analysis to compare our algorithm with other matching algorithms.

In order to examine the performance of the proposed method and to examine the possibility of real-time processing, we conducted experiments using some images, as shown in fig.2.

Fig. 2. The original low-resolution images for experiments

Experiments are conducted to compare the proposed approach with NUI [19] (Non-uniform interpolation) and POCS [20] (Projection onto convex sets). In the first experiment, the above super-resolution approaches are implemented using the Matlab programming language and run on PC with a Core2 CPU and a 2G RAM. For comparison, we showed the results obtained by NUI [19], POCS [20] and MAP (Maximum a posteriori), respectively, as shown in fig.3, 4, 5 and 6. Their run time are presented in fig.3. In fig.4, 5 and 6, we give the proposed approach and conventional approaches to obtain super-resolution image. (a) is the result of NUI, (b) is the result of POCS and (c) is the result of our algorithm. From the comparison of these images, it can be clearly seen that our algorithm outperforms the traditional NUI and POCS algorithm in visual quality of the reconstructed image.

As POCS algorithm and NUI algorithm itself does not take into account the ideas of parallel computing, there is not a significant difference in performance between running on CPU and GPU. So, the second experiment is to compare the running times of two different implementations: a CPU implementation of the proposed method and a GPU-CUDA implementation of the proposed method. The table 1 shows the speed up of the GPU implementation of the proposed algorithm when compared to the faster of the CPU implementations. The speedups gained are in the range 38-178, depending on the image size.

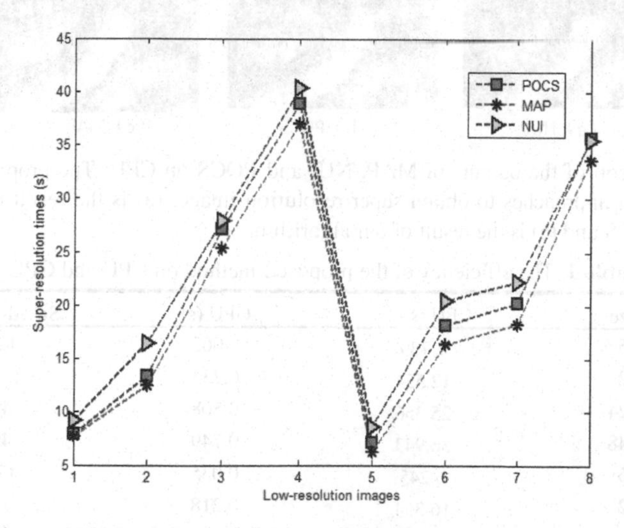

Fig. 3. Comparison of MAP, NUI and POCS on CPU. The proposed MAP approach is slightly lower than that of conventional approaches.

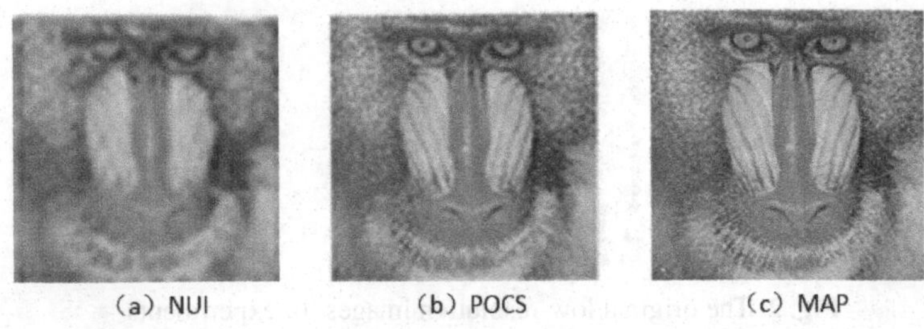

Fig. 4. Comparison of the outputs of MAP, NUI and POCS on CPU: The proposed approach and conventional approaches to obtain super-resolution image. (a) is the result of NUI, (b) is the result of POCS and (c) is the result of our algorithm.

Fig. 5. Comparison of the outputs of MAP, NUI and POCS on CPU: The proposed approach and conventional approaches to obtain super-resolution image. (a) is the result of NUI, (b) is the result of POCS and (c) is the result of our algorithm.

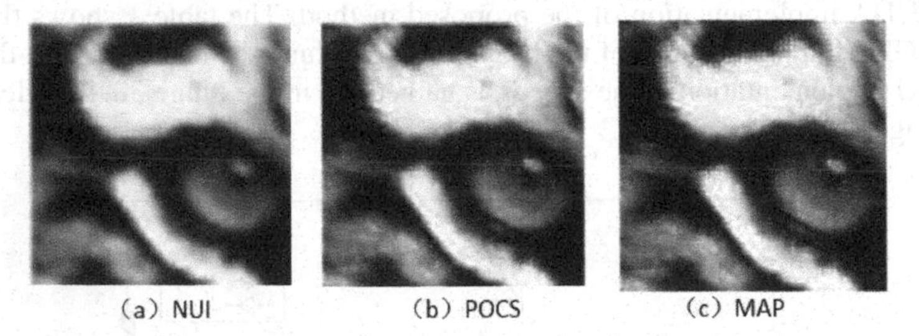

Fig. 6. Comparison of the outputs of MAP, NUI and POCS on CPU: The proposed approach and conventional approaches to obtain super-resolution image. (a) is the result of NUI, (b) is the result of POCS and (c) is the result of our algorithm.

Table 1. The efficiency of the proposed method on CPU and GPU

Image Size	CPU (s)	GPU (s)	Speed-up Ratio
256×256	7.739	0.062	124.8
512×512	12.524	0.235	35.3
1024×1024	25.356	0.508	49.9
2048×2048	36.943	0.749	49.3
256×256	6.245	0.035	178.4
512×512	16.354	0.218	75.0
1024×1024	18.245	0.474	38.5
2048×2048	33.578	0.698	48.1

As shown in fig.7, we can see that the running times of the two implementations with different image sizes.

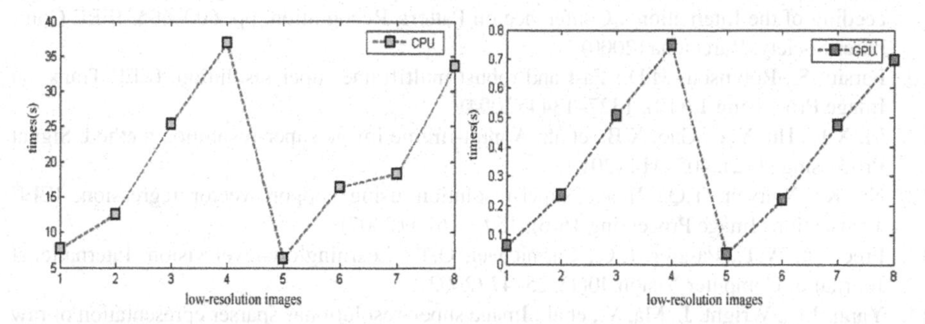

Fig. 7. Comparison of the running time on CPU and GPU. The implementation of the proposed algorithm on CPU is not always faster than the implementation of the proposed algorithm on GPU.

4 Conclusions

In this paper, a novel super-resolution image estimation based on MAP is proposed. The proposed algorithm achieves better performance than the existing resolution enhancement approaches and is successful in real-time performance, as verified by our experimental results. Experimental result shows that the algorithm is feasible and could achieve better result. The proposed algorithm can be used in the biomedical imaging, satellite remote sensing imaging, military reconnaissance and positioning, surveillance video enhancement and restoration, public security surveillance, and other image-processing areas. In future works, we intend to make additional experiments in order to verify the accuracy of the proposed method when compared with the traditional algorithms, and develop the practical applications.

References

1. http://www.nvidia.cn/page/home.html
2. Farsiu, S., Robinson, D., Elad, M., Milanfar, P.: Fast and Robust Multi-frame Super-resolution. IEEE Transactions on Image Processing 13(10), 1327–1344 (2004)
3. Schultz, R.R., Stevenson, R.L.: Extraction of high-resolution frames from video sequences. IEEE Transactions Image Processing 5(6), 996–1011 (1996)
4. Irani, M., Peleg, S.: Improving resolution by image registration. CVGIP 53(3), 231–239 (1991)
5. Elad, M., Feuer, A.: Super resolution restoration of an image sequence: adaptive filtering approach. IEEE Trans. Image Processing 8(3), 387–395 (1999)
6. Hardie, R.: A fast image super-resolution algorithm usingan adaptive Wiener filter. IEEE Trans. Image Processing 16(12), 2953–2964 (2007)
7. Park, S.C., Park, M.K., Kang, M.G.: Super-Resolution Image Reconstruction: A technical Overview. IEEE Signal Processing Magazine, 21–36 (May 2003)

8. Marcelo, V.W.Z., Fermin, S.V.B., Joceji, M.: Determining the regularization parameters for super-resolution problem. Signal Processing 88(12), 2890–2901 (2008)
9. Capel, D., Zisserman, A.: Super-resolution enhancement of text image sequences. In: Proceeding of the International Conference on Pattern Recognition, pp. 600–605. IEEE Computer Society, Barcelona (2000)
10. Farsiu, S., Robinson, M.D.: Fast and robust multiframe super resolution. IEEE Trans. on Image Processing 13(10), 1327–1344 (2004)
11. Li, X.L., Hu, Y.T., Gao, X.B., et al.: A mult-iframe image super-resolution method. Signal Processing 90(2), 405–414 (2010)
12. Ni, K., Nguyen, T.Q.: Image super-resolution using support vector regression. IEEE Transactions Image Processing 16(6), 1596–1610 (2007)
13. Freeman, W.T., Pasztor, E.C., Carmichael, O.T.: Learninglow-level vision. International Journal of Computer Vision 40(1), 25–47 (2002)
14. Yang, J.C., Wright, J., Ma, Y., et al.: Image super-resolutionas sparser epresentation of raw image patches. In: CVPR, pp. 1–8. IEEE Computer Society Press, Anchorage (2008)
15. Lowe, D.G.: Object recognition from local scale-invariant features. In: The Proceedings of the Seventh IEEE International Conference on Computer Vision, vol. 2, pp. 1150–1157 (1999)
16. Boykov, Y., Veksler, O., Zabih, R.: Fast approximate energy minimization via graph cuts. IEEE Transactions on Pattern Analysis and Machine Intelligence 23(11), 1222–1239 (2001)
17. Boykov, Y., Veksler, O., Zabih, R.: Fast Approximate Energy Minimization via Graph Cuts. IEEE TPAMI 20(12), 1222–1239 (2001)
18. Tsai, R.Y., Huang, T.S.: Multiframe image restoration and registration. Advances in Computer Vision and Image Processing 1, 317–339 (1984)
19. Lertrattanapanich, S., Bose, N.K.: High resolution image formation from low resolution frames using delaunay triangulation. IEEE Transactions on Image Processing 11(12), 1427–1441 (2002)
20. Banham, M.R., Katsaggelos, A.K.: Digital image restoration. IEEE Signal Processing Magazine 14(2), 24–41 (1997)
21. Farsiu, S., Robinson, M.D., Elad, M., Milanfar, P.: Fast and robust multiframe super resolution. IEEE Transactions on Image Processing 13(10), 1327–1344 (2004)
22. Freeman, W.T., Jones, T.R., Paztor, E.C.: Example based superresolution. IEEE Computer Graphics and Applications 22(2), 56–65 (2002)
23. Sun, J., Zheng, N.N., Tao, H., Shum, H.Y.: Image hallucination with primal sketch priors. In: IEEE Computer Society Conference on Computer Vision and Pattern Recognition (CVPR 2003), vol. 2, pp. 729–736 (June 2003)
24. Cang, H., Oung, D.Y.Y., Xiong, Y.: Super-resolution through neighbor embedding. In: IEEE Computer Society Conference on Computer Vision and Pattern Recognition (CVPR 2004), vol. 1, pp. 275–282 (July 2004)
25. Freeman, W.T., Pasztor, E.C., Carmichael, O.T.: Learning lowlevel vision. International Journal of Computer Vision 40(1), 25–47 (2004)
26. Dai, S., Han, M., Xu, W., Wu, Y., Gong, Y.: Soft edge smoothness prior for alpha channel super resolution. In: IEEE Computer Vision and Pattern Recognition (CVPR 2007), pp. 1–8 (June 2007)
27. Yang, J., Wright, J., Ma, Y., Huang, T.: Image super-resolution as sparse representation of raw image patches. In: IEEE Computer Vision and Pattern Recognition (CVPR 2008), pp. 1–8 (August 2008)

Remote Sensing Image Segmentation Based on Mean Shift Algorithm with Adaptive Bandwidth

Chongjing Deng, Shuang Li*, Fuling Bian, and Yingping Yang

International School of Software, Wuhan University
sli@whu.edu.cn

Abstract. Image segmentation is an important step in bridging the semantic gap between low level image interpretation and high level information extraction. Many image segmentation algorithms are available, i.e. active contour method, watersheds method, edge based method, threshold method, etc. Most of these algorithms are parametric and require the image with strong gradient. Mean shift algorithm is a non-parametric density estimation algorithm, which is popularly used in image segmentation recently. However, one bottleneck of the mean shift procedure is that the results of segmentation rely highly on selection of bandwidth. We present an improved mean shift algorithm with adaptive bandwidth for remote sensing images. The bandwidth of each pixel is adaptively adjusted according to the corresponding probability distribution. Compared with traditional fixed bandwidth, our proposed algorithm is both with high efficient and accurate in segmentation of high resolution remote sensing image.

Keywords: image segmentation, mean shift algorithm, adaptive bandwidth.

1 Introduction

Image segmentation refers to the process of partitioning an image into non-overlapping homogeneous regions. Thus, the result of segmentation is the division of the image into a set of regions. Each region represents some area of interest. Remote sensing image segmentation is the basic step in landscape change detection or land use/cover classification, which bridging the semantic gap between low level image interpretation and high level information extraction [1-2].

Many image segmentation algorithms are available, i.e. active contour method, watersheds method, edge based method, threshold method, etc. Most of these algorithms are parametric and require the image with strong gradient [3]. Mean shift algorithm was firstly proposed by Fukunaga and Hostetler (1975) [4], which is a non-parametric density estimation algorithm and converges rapidly on the local maximum of the probability density function through iterations. Then it is applied to image segmentation successfully by Comaniciu and Meer (1999) [5], with the advantage of not requiring prior knowledge of the number of clusters and not constraining the shape of the clusters. In recent

* Corresponding author.

F. Bian and Y. Xie (Eds.): GRMSE 2014, CCIS 482, pp. 179–185, 2015.

years, the mean shift algorithm applied in remote sensing image segmentation is also promising, especially in segmentation of high spatial resolution images [6].

As an iterative algorithm, mean shift segmentation is affected much by the selection of bandwidth, which became a bottleneck when applied in remote sensing images with high spatial resolution. In this paper, we have proposed an adaptive mean shift segmentation which addresses the problem of the bandwidth. The histogram is used to make the band width adaptively. Experiments on Qucikbird-2 multi-spectral image have confirmed the affectivity of this proposed approach in comparison to the algorithm without adaptive bandwidth selection.

The paper is organized as follows. The basic idea of mean shift segmentation is described in section 2. The modification of thc basic mean shift segmentation is proposed in section 3. Section 4 gives the experimental results. The conclusions are made in section 5.

2 Basic Idea of Mean Shift Algorithm

Given a set of n data points $x_i, i = 1, 2, \ldots, n$ in the d dimensional feature space R^d, the kernel density estimator at a given location point x can be defined as

$$\hat{f}(x) = \frac{1}{nh^d} \sum_{i=1}^{n} k\left(\frac{x - x_i}{h_i}\right) \tag{1}$$

where h_i is the band width parameter, k is the kernel function which is defined as follows.

$$k(x) = ck\left(\|x\|^2\right) \tag{2}$$

where c is the normalization constant, which ensures $k(x)$ integrates to 1. Mean shift is considered to be based on gradient of the density estimator as follows:

$$\begin{aligned} f'(x) &= \frac{2c}{nh^{d+2}} \sum_{i=1}^{n} (x_i - x)\, g\left(\left\|\frac{x - x_i}{h}\right\|^2\right) \\ &= \frac{2c}{nh^{d+2}} \left[\sum_{i=1}^{n} g\left(\left\|\frac{x - x_i}{h}\right\|^2\right)\right] \left[\frac{\sum_{i=1}^{n} x_i g\left(\left\|\frac{x - x_i}{h}\right\|^2\right)}{\sum_{i=1}^{n} g\left(\left\|\frac{x - x_i}{h}\right\|^2\right)} - x\right] \end{aligned} \tag{3}$$

where $g(x) = -k'(x)$. The first term is proportional to the density estimate and the second term is the mean shift vector.

$$m_h(x)=\frac{\sum_{i=1}^{n}x_i g\left(\left\|\frac{x-x_i}{h}\right\|^2\right)}{\sum_{i=1}^{n}g\left(\left\|\frac{x-x_i}{h}\right\|^2\right)}-x \tag{4}$$

Thus, the mean shift vector points towards the direction of maximum increase in the density. This procedure is obtained by successive computation of the mean shift vector and translation of the kernel $g(x)$ by the corresponding vector.

If the image dimension is p, when the position vector and the color vector integrate into the "space-color" field, the dimension is p+2. Then, each pixel corresponds to a p+2 dimensional vector. $x=(x^s,x^r)$ as the radial symmetry kernel and Euclidean kernel is expressed as:

$$K_{h_s,h_r}=\frac{C}{h_s^2 h_r^p}k\left(\left\|\frac{x^s}{h_s}\right\|^2\right)k\left(\left\|\frac{x^r}{h_r}\right\|^2\right) \tag{5}$$

Where x^s denotes the coordinate of pixel, x^r denotes the p-dimensional vector of corresponding point. C is the normalization constant, h_s, h_r are the bandwidth of the kernel.

3 Improved Mean Shift Algorithm with Adaptive Bandwidth

In image clustering, the bandwidth h determines the influence range of kernel function. If the h is small, the density function $f(x)$ is equivalent to overlay of a number of n data center. Thus, the density function around each pixel is small and many clusters are generated. If the h is large, the density function $f(x)$ is equivalent to overlay of a number of n basic functions with large width. Thus, the density function around each pixel is close to each other and few clusters are generated. We have to optimize the h parameter in mean shift algorithm for image segmentation.

Let the bandwidth $h=h(x_i)$, that's to say, the bandwidth of each pixel is different. We use the histogram to adjust the bandwidth. In high density region, we use small bandwidth. On the contrary, in low density region, we use large bandwidth. That's to say, the distribution of pixels is inversely proportional to the bandwidth. Let

$$h(x_i)=\frac{\alpha h_0}{f(x_i)} \tag{6}$$

where, h_0 is the initial bandwidth, $f(x_i)$ is the two-dimensional histogram, α is the constant. In equation, the constant α is calculated as follows:

$$\log \alpha = \frac{\sum \log f(x_i)}{n} \tag{7}$$

We use the adaptive bandwidth $h(x_i)$ in equation, the new mean shift vector is:

$$m_h(x) = \frac{\sum_{i=1}^{n} x_i g\left(\left\|\frac{x - x_i}{h(x_i)}\right\|^2\right)}{\sum_{i=1}^{n} g\left(\left\|\frac{x - x_i}{h(x_i)}\right\|^2\right)} - x \tag{8}$$

4 Experimental Results

We used Quickbird-2 multi-spectral image with the size of 500*500 for experiment. The spatial resolution of the test image is 2.4 meter. It includes four bands, i.e. blue band (band 1), green band (band 2), red band (band 3) and the near infrared band (band 4). The image is displayed in band 4, band 3 and band 2 in RGB channels, which is shown in figure 1. The land cover includes farmland, pond, river, building, road, trees and grass.

Fig. 1. Original Quickbird-2 images with complex land covers

The histograms of the original image in different bands are shown in figure 2. The bandwidth is calculated according to the two-dimensional histogram.

The results of image segmentation with different bandwidth are shown in figure 3. Figure 3(a) to figure 3(d) are the results of mean shift segmentation with fixed bandwidth. In the fixed bandwidth experiment, the spatial width hs and the color width hr are (10,1), (10,5), (20,10) and (20,20), respectively. Figure 3(e) is the result of

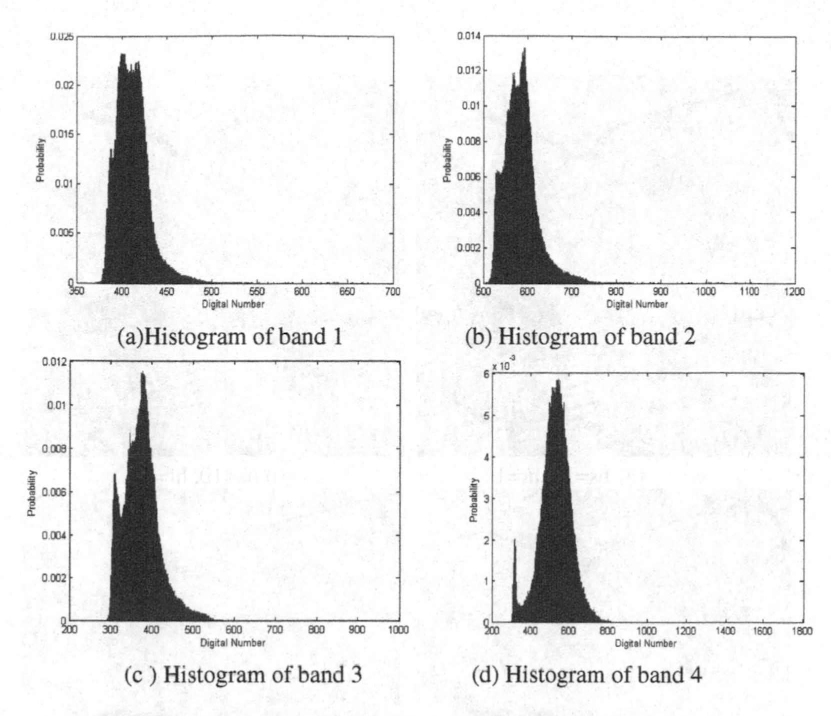

(a)Histogram of band 1 (b) Histogram of band 2

(c) Histogram of band 3 (d) Histogram of band 4

Fig. 2. The histograms of the Quickbird-2 image in different bands

proposed segmentation algorithm. The corresponding outline is shown in figure 3(f). Form figure 3(a) and figure 3(b), we can see that the image is fully segmented. The results of figure 3(a) and figure 3(b) are very close to original image (shown in figure 1). From figure 3(c) and figure 3(d), we can see that the image is not fully segmented. The regions arc too large to distinguish the ridge of farmland. Figure 3(e) shows good segmentation performance. The regions can be distinguished obviously. Figure 3(f) further proved that the segmentation result with adaptive bandwidth is proper.

To further prove the effectiveness of the proposed algorithm, the number of regions and the computation time are recorded in table 1. From table 1, we can see that the number of regions in figure 1(a) is 215650. It suggests that the segmentation is invalid as most of the objects are single pixel. The number of regions decreases as the hr increases. The computation time increases as the hr increases. The hr has more effect on the segmentation results than hs does. Form figure 3 and table 1, it can be seen that the image segmentation results are affected much by the bandwidth selection, especially for the high detail urban area. Our proposed adaptive mean shift segementation outperms the trandistional mean shift algorithm with fixed bandwidth. More importanly, our proposed algorithm is easy implemented and with high efficiency.

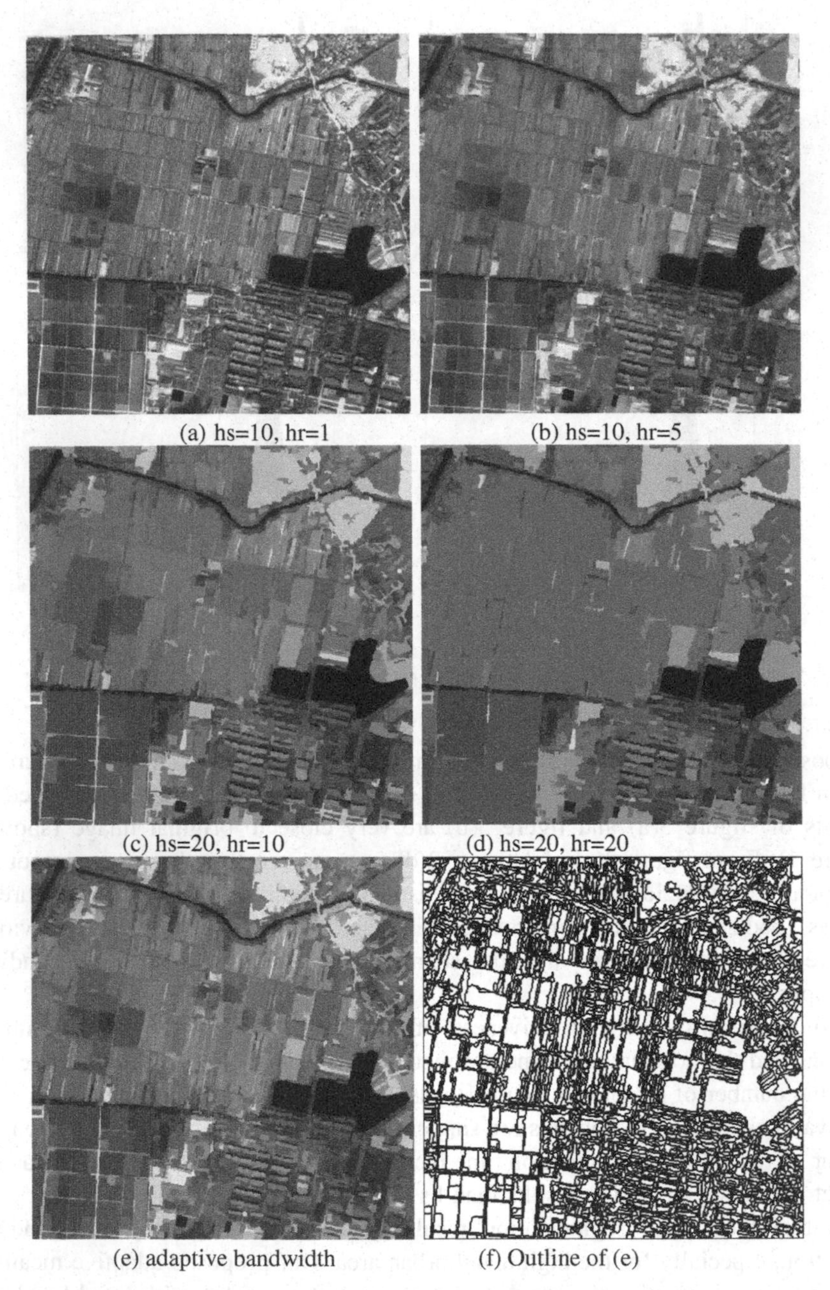

(a) hs=10, hr=1 (b) hs=10, hr=5

(c) hs=20, hr=10 (d) hs=20, hr=20

(e) adaptive bandwidth (f) Outline of (e)

Fig. 3. Results of Image segmentation with different bandwidth

Table 1. The statistical results of segmentation with different bandwidth

	Bandwidth	Number of regions	Time (s)
Fig. 2(a)	(10,1)	215650	19.5
Fig. 2(b)	(10,5)	10618	24
Fig. 2(c)	(20,10)	2098	15
Fig. 2(d)	(20,20)	1290	31.5
Fig. 2(e)	Adaptive	3000	18.1

5 Conclusions

Mean shift algorithm is a non-parametric density estimation algorithm, which is popularly used in image segmentation recently. However, the results of segmentation rely highly on selection of bandwidth. To overcome the shortcoming of basic mean shift algorithm, we proposed an adaptive mean shift algorithm based on probability density, which is calculated by using histogram. The bandwidth of each pixel is adaptively adjusted according to the corresponding probability distribution. Compared with traditional fixed bandwidth, our proposed algorithm is both with high efficient and accurate in segmentation of high resolution remote sensing image.

Acknowledgments. This research was supported by the Natural Science Foundation of Hubei Province (2013CFB285), the Fundamental Research Funds for the Central Universities, Wuhan University (2042014KF0054) and the State 973 project (2012 CB719901) in China.

References

1. Bezdek, J.C., Hall, L., Clarke, L.P.: Review of MR image segmentation techniques using pattern recognition. Medical Physics 20(4), 1033–1048 (1992)
2. Banerjee, B., Surender, V., Buddhiraju, K.M.: Satellite image segmentation: A novel adaptive mean-shift clustering based approach. In: IEEE International Geoscience and Remote Sensing Symposium, IGARSS (2012)
3. Aly, A.A., Deris, S.B., Zaki, N.: Research review for digital image segmentation techniques. International Journal of Computer Science & Information Technology 3(5), 99–106 (2011)
4. Fukunaga, K., Hostetler, L.: The estimation of the gradient of a density function, with applications in pattern recognition. IEEE Transactions on Information Theory 21(1), 32–40 (1975)
5. Comaniciu, D., Meer, P.: Mean shift analysis and applications. In: The Proceedings of the Seventh International Conference on Computer Vision (1999)
6. Wan, F., Deng, F.: Remote sensing image segmentation using mean shift method. In: Lin, S., Huang, X. (eds.) Advanced Research on Computer Education, Simulation and Modeling. CCIS, vol. 176, pp. 86–90. Springer, Heidelberg (2011)

Watermarking Algorithm Based on Data Feature for Tile Map

Bo Wang[1], Na Ren[2,*], and Changqing Zhu[2]

[1] Geological Exploration Technical Institute of Jiangsu Province, Nanjing, China
[2] Key Laboratory of Virtual Geographical Environment, Ministry of Education, Nanjing Normal University, Nanjing, China
rena1026@163.com

Abstract. Watermarking is widely being explored as an effective means of providing copyright protection for digital multimedia data, and there has been increasing interest in applying watermark to digital raster map for the same purpose. Tile map, however, subjected to a very different feature from other multimedia data. Thus, the watermark scheme designed for image can not be directly applicable to tile map. The purpose of this paper is to propose a novel feature based watermarking algorithm for tile map. First of all, characteristics of tile map have been analyzed and corresponding requirements of watermarking algorithm has been discussed. Then, the watermark embedding and detection algorithm concerning features for tile map have been proposed. Finally, experiments have been conducted to verify the proposed algorithm. The experimental results have shown that the algorithm has excellent invisibility, and it is able to resist various attack behavior such as adding noise, compression, stitching and format conversion.

Keywords: Digital watermark, Tile map, Robust.

1 Introduction

"TIANDITU" is the public version of National Geographic Information Public Service Platform in China. It has made a fundamental change of mapping and geographic information department in the way from providing maps and data offline to providing online information services, which have caused great repercussions all over the world [1]. With the situation that builders of "TIANDITU" take much more effort in enriching data and improving the capabilities and levels of services, they have been faced with serious problems, including downloading and using maps and data illegally. These problems have made serious damage to the interests of data owners, and the copyright of data cannot be guaranteed effectively. It became highly essential to watermark the tile map in order to identify the ownership and the copyright holder and to trace out the unauthorized copies [2-3].

Currently, copyright protection has largely been addressed in the field of multimedia data by resorting to watermarking technology, which consists in permanently embedding

* Corresponding author.

F. Bian and Y. Xie (Eds.): GRMSE 2014, CCIS 482, pp. 186–193, 2015.

a digital mark in the original data, such as digital map, remote sensing image, carrying information of ownership and user license rights. In the last decade, some research about digital watermarking in raster maps and remote sensing images have been studied [4-8]. However, there are not much research has been conducted in digital watermarking specifically for tile map [9-10]. This paper is concerned with the characteristics of the tile map and requirements of digital watermarking algorithm, in the case of copyright protection. The desired functionalities of watermarking techniques are discussed, and the possible design options of a watermarking algorithm are evaluated in terms of tile map issues.

This work is organized as follows. In Section 2, a brief overview of the watermark for tile map are analyzed, by paying attention to introduce some characteristics and requirements of tile map that will be used in the following sections. Section 3 is devoted to describe the proposed watermark embedding and detection algorithm. Section 4 discusses the impact of the watermark on robustness in the face of attacks such as adding noise, compression, stitching and format conversion, etc. Finally, some conclusions are drawn in Section 5.

2 Characteristic Analysis of Tile Map and Its Requirements on Watermarking Algorithm

There are some common features between tile map and ordinary digital raster map. However, the size of raster map is often variable but the size of each tile map is fixed. In addition, in terms of anti-attacking ability, these two kinds of maps are much more different. After cutting operation, the raster map still has a certain commercial values and use value. But one single tile map does not have practical use. Therefore, while studying watermarking algorithm for tile map, it's not practical to copy the watermarking algorithm for raster map simply. The watermarking algorithm should be designed according to the characteristics of tile map.

After analyzing the tile map, its characteristics and requirements for watermarking algorithm mainly lie in the following aspects:

(*a*) In order to save storage space effectively, it's common to use index mechanism to store tile map data, which is typically in the format of "png". This method of storage can also carry the transparent channel, which brings more benefits in the expression of tile map.

(*b*) The chromaticity of tile map is not rich, so the number of color index in use is limited, which remains substantially in the range of 20-30 kinds. Therefore, the content of watermark information carried by tile map is very small.

(*c*) The size of tile map is unified. The size of each tile map is 128*128 pixels or 256*256 pixels. Thus more targeted research in accordance with these map sizes can be studied.

(*d*) The features of tile map, especially for the highlighted lines are very obvious, and the blank area is larger, which means tile map is of high brightness and low saturation. So the watermarking algorithm for tile map can use this feature to embed watermark effectively.

(*e*) In the application of tile map, it's often to load tile map according to the naming rules. It's not much possible to conduct different attacks such as cutting, rotation and so on to a single tile map. However, in the process of designing watermarking algorithm, it's still necessary to consider the situation that a single tile map may suffer from noise attack, compression. These kinds of attack behaviors should not affect the use of tile map.

(*f*) When criminals download tile map from the Internet, they usually need to customize the format of tile map as png or jpg according to their own needs. Therefore, the watermarking algorithms for the tile map should be robust to format conversion.

It can be seen from the characteristic analysis of tile map above that there is need to follow its unique characteristics and algorithm requirements in the research of watermarking algorithm for tile map. The data hidden in pixel values and index values are very limited.

3 The Embedding and Detection Algorithm of Watermarking for Tile Map

Based on the research of the characteristic analysis of tile map and it requirements for watermarking algorithm requirements, a feature intensive embedding and detection algorithm of watermarking for tile map is proposed in this paper. As non-blind watermarking algorithms for tile data are completely unrealistic in practical applications, the watermarking algorithm proposed in this paper belongs to blind watermarking algorithm.

3.1 The Generation and Embedding Algorithm of Watermarking Information

The detailed steps of the algorithm are as follows:

(*a*) Generate the binary watermarking sequence $W = (w_0, w_1, \cdots, w_{L-1})$, $W = G(key)$ from the watermarking information to be embedded or the key by using the pseudo-random sequence, where L is the length of watermarking sequence, $w = \{1, -1\}$, G is the embedding algorithm of watermarking information, and key is the set of keys.

(*b*) Transform the tile map into the 2-layer digital wavelet transform (DWT) domain, and choose the low frequency sub-band.

(*c*) Divide the low frequency sub-band into uniform blocks of 8×8, in such a way that non-overlapping blocks are obtained. As the tile map has high brightness and low saturation, the low-frequency mean value of each sub-block is an important characteristic value, which has a strong invariance. Calculate the low-frequency mean value of the current block and its neighboring blocks, and record them as ave_m and ave_{m+1} separately.

(*d*) Construct the mapping function based on the relationship between ave_m and ave_{m+1}, then determine the embedding bit of the watermarking information. The mapping function is constructed as follows:

$$Index_w = f(ave_m, ave_{m+1}) \tag{1}$$

(*e*) For specific low frequency blocks, according to the corresponding watermarking bit based on the mapping function, embed the watermarking information into the corresponding low frequency block based on the quantization rule. The specific embedding rule is given as the Formula (2) below.

$$I'(i,j) = round\left(\frac{I(i,j)}{5\delta}\right) \times 5\delta + \delta \times w_{Index} \tag{2}$$

where, $round(\bullet)$ is rounding function, δ is the quantization step and $Index = \{0,1,2,\cdots,L-1\}$.

3.2 The Watermarking Detection Algorithm

The watermarking detection algorithm is the inverse process of the embedding algorithm. Using the same mapping function and quantization step size of the embedding algorithm, the detection rules is given as follows:

$$w_i = \mathrm{mod}\left(round\left(I(i,j)/\delta\right), 5\right) \tag{3}$$

where $\mathrm{mod}(*,5)$ is a function of modulus 5. Confirm the watermarking information using the majority rule.

The false alarm detection often occurs in the watermarking detection process. Namely, the watermarking information is detected from the data which has not been embedded watermarking information. In order to reduce the probability of false alarm detection, it is necessary to compare correlation between the original and the detected watermarking information. It can be confirmed that there is watermarking information in the detected data when the correlation coefficient is greater than the preset threshold value. To evaluate the similarity between the original and the detected watermarking information, the similarity calculating formula is given as follows:

$$NC = \sum_i b_i / L \tag{4}$$

where $b_i = 1 - XOR(w_i, w_i')$, w_i and w_i' represents the binary sequence of original and detected watermarking separately, and L is the length of the watermarking sequence.

4 Experiments and Analysis

In this section, some experimental results are given to evaluate the performance of the proposed algorithm. Two tile maps stored in the format of "jpg" and "png" are used in the following experiments. They are all in the size of 256*256, as depicted in Fig. 1(a) and Fig. 1(b).

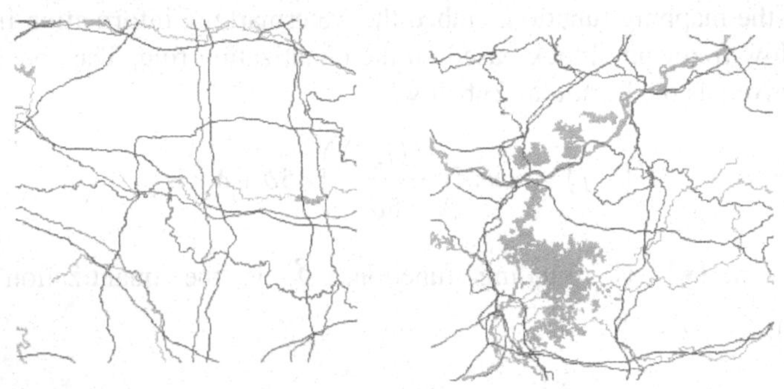

(*a*) a piece of tile map in png format (*b*) a piece of tile map in jpg format

Fig. 1. Original tile maps

4.1 Invisibility

The watermarked tile maps are shown in the Fig. 2.

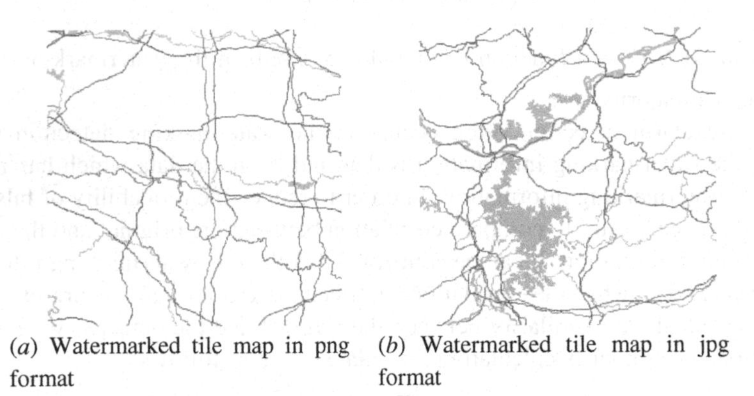

(*a*) Watermarked tile map in png format (*b*) Watermarked tile map in jpg format

Fig. 2. Watermarked tile maps

The comparison of peak signal to noise ratio (*PSNR*) between the original and the watermarked tile map is used to evaluate the invisibility of the proposed algorithm. The *PSNR* is expressed mathematically in the form as follows.

$$PSNR = 10 \times \log_{10} \frac{(M \times M) \times [\max(I) - \min(I)]^2}{\sum_{i=1}^{M} \sum_{j=1}^{N} [I(i,j) - I'(i,j)]^2} \tag{5}$$

where, $M \times M$ is the size of tile map, I is the original tile map, and I' is the watermarked tile map. The comparison of PSNR between the original and the watermarked tile map, and the normalized correlation coefficient (*NC*) between the original and detected watermark are presented in Table 1.

Table 1. The results of *PSNR* and *NC*

Tile map	*PSNR*	*NC*
Fig1(a) and Fig2(a)	39.4766	1.0
Fig1(b) and Fig2(b)	42.1136	1.0

Form the aspect of the subjective vision, it is difficult to perceive the difference between the original and the watermarked tile map. From the objective indicator, the PSNR of two tile maps are both high enough. Furthermore, the NC of two tile maps without any attack are both 1.0. Therefore, the invisibility of this algorithm is verified well.

4.2 Robustness

The embedded tile maps are attacked by noising, compressing and splicing to verify the robustness of the proposed algorithm. There are two ways of splicing tiles. One is that two watermarked tiles are spliced. The other is that a tile embedded with watermark and one without watermark are spliced. In this paper, the threshold of NC for successful detection is set as 0.5.

The results of experiments is as shown in the Table 2.

Table 2. The results after attacking

Tile map	The way of attacking	*NC*	Detecting result
(a1) (b1)	Salt-and-pepper noise (0.01)	0.99 0.95	Successful Successful
(a1) (b1)	Salt-and-pepper noise (0.02)	0.83 0.77	Successful Successful
(a1) (b1)	Salt-and-pepper noise(0.05)	0.73 0.70	Successful Successful
(a1) (b1)	Compression (compression ratio 90)	1.00 1.00	Successful Successful
(a1) (b1)	Compression (compression ratio 80)	0.88 0.81	Successful Successful
(a1) (b1)	Compression (compression ratio 70)	0.70 0.66	Successful Successful
(a1) (b1)	Splice(both tiles is watermarked)	1.00 1.00	Successful Successful

Table 2. (*Continued*)

(a1)	Splice(only one tile is	1.00	Successful
(b1)	watermarked)	1.00	Successful
(a1)	Format conversion	0.95	Successful
(b1)	jpg -> png	1.00	Successful
(a1)	Format conversion	0.96	Successful
(b1)	png->jpg	1.00	Successful

As we can see from Table 2, after attacking a single tile with noise, compression and so on, the detected NC is higher than the threshold of 0.5 which indicates that watermark is detected successfully. In the designed attack of splicing, the spliced tiles containing two watermarked tiles and those containing an original tile and a watermarked one can be both detected successfully because the watermark can be detected from the spliced tiles even the watermark is detected from one tile. The tile map is stored in the formats of "png" and "jpg" in general. After conversions between the two formats, the detected NC is usually higher than 0.9, which indicates that the proposed algorithm is capable of assisting the attack of conversions of data formats. So the algorithm proposed in the chapter is verified to be robust.

5 Conclusions

This paper proposes a watermarking algorithm for tile map based on the characteristics of the tile map. The experimental results verify that the proposed algorithm has good invisibility and robustness, which can satisfy the requirements of copyright for tile maps. Furthermore, the proposed algorithm will play very important role in the application of digital watermark in website of "TIANDITU", security protection of tile map, etc.

Acknowledgements. The work was supported by the National Natural Science Foundation of China (Grant No. 41301413), Natural Science Foundation of Jiangsu Province (Grant No. BK20130903), the University Science Research Project of Jiangsu Province (Grant No. 12KJB420002). A Project Funded by the Priority Academic Program Development of Jiangsu Higher Education Institutions.

References

1. Li, Z.-G., Jiang, J., et al.: "TIANDITU" Technical architecture for supporting distributed service aggregation. Journal of Geomatics 37, 13–15 (2012)
2. Zhu, C.-Q., Yang, C.-S., Ren, N.: Application of Digital Watermarking to Geospatial Data Security. Bulletin of Surveying and Mapping 10, 1–3 (2010)
3. Kbaier, I., Belhadj, Z.: A Novel Content Preserving Watermarking Scheme For Multispectral Images. Information and Communication Technologies 16, 243–247 (2006)
4. Tripathi, S., Ramesh, N., Bernito, A., Neeraj, K.J.: A DWT based Dual Image Watermarking Technique for Authenticity and Watermark Protection. Signal & Image Processing 1, 33–45 (2010)

5. Fu, H.-J., Zhu, C.-Q.: Multipurpose Watermarking Algorithm for Digital Raster Map Based on Wavelet Transformation. Acta Geodaetica et Cartographica Sinica 40, 397–400 (2011)
6. Zhu, C.-Q., Fu, H.-J., Yang, C.-S., et al.: Watermarking Algorithm for Digital Grid Map Based on Integer Wavelet Transformation. Geomatics and Information Science of Wuhan University 34, 619–621 (2009)
7. Rui, Z., Gangwu, J., Shugen, W.: Content-based multipurpose watermarking for remote sensing images. In: Proceedings of the SPIE, pp. 6752–6756 (2007)
8. Wang, Z.-W., Zhu, C.-Q., et al.: An Adaptive Watermarking Algorithm for Raster Map Based on HVS and DFT. Geomatics and Information Science of Wuhan University 36, 351–354 (2011)
9. Ren, N., Zhu, C.-Q., et al.: Research on Methods and Applications of Copyright Protection for Tile Data. Journal of Geo-Information Science 14, 491–493 (2012)
10. Ren, N., Zhu, C.-Q., et al.: A Digital Watermark Algorithm for Tile Map Stored by Indexing Mechanism. In: Cartography from Pole to Pole, pp. 79–86 (2014)

Observation Scheduling Problem for Multi-task with Complex Constraints

Fanyu Zhao[1,2], Rui Xu[1,2], and Pingyuan Cui[1,2]

[1] Institute of Deep Space Exploration Technology, Beijing Institute of Technology, Beijing 100081, China

[2] Key Laboratory of Dynamics and Control of Flight Vehicle, Ministry of Education, Beijing 100081, China

{zfybit,xurui,cuipy}@bit.edu.cn

Abstract. The observation scheduling problem with multi-task and multi-resource for multi-satellites is studied in this paper. Firstly, the time constraints and resource constraints remain in the scheduling process are analyzed, and the mathematical model for the scheduling problem is established. Secondly, the ways of checking the time and resource constraints are respectively given, especially for the resource with both producing and consuming functions. Finally an improved ACO is proposed, combining with the task priorities and resource constraint information, proper transition controlling strategy and pheromone updating mechanism. The simulation proves the validity of the model and the algorithm.

Keywords: Earth observation, scheduling problem, multi-task, complex constraints, improved ACO.

1 Introduction

The space scheduling system with the core of a real-time disperse event dynamic system (DEDS) is responsible for scheduling of TT&C (Tracking Telemetry and Command) resource and supports the normal work for spacecraft. The nature of DEDS is to solve a large-scale optimization problem with complex constraints and multiple resources. In recent years, an increasing number and types of spacecraft are operating and developed. Because of various limits, TT&C resource can not be substantially increased. Therefore, the contravention between large number of spacecraft and a few ground stations is even severe especially for a great amount of low orbiting satellites whose transit time is short at the same time. Many countries attach great importance to this problem and scholars have made multi-angle studies combined with their own national situation.

The early study of Burrowbridge for this problem is representative[1]. In his study, he used the maximizing TT&C task number as the objective function and introduced polynomial time algorithm for TT&C resource scheduling problems. After that, Barbulescu[2] proved in detail that TT&C resource scheduling problems are NPC problems and compared several common optimization algorithms using the maximizing

F. Bian and Y. Xie (Eds.): GRMSE 2014, CCIS 482, pp. 194–203, 2015.

TT&C task number as the objective function. Tapan[3] used the maximizing observation time as the objective function and studied the application of genetic algorithm to TT&C problems of low orbiting satellites. The results of this method were good. Zhang Na[4] built a new kind of composite independent set model which was a meaningful exploration in modeling field.

At present, most of studies use the maximizing TT&C task number as the objective function. However, in actual TT&C tasks, we must consider the influence of the priority of tasks on scheduling results. Although some scholars have done some research on this problem, the studies are unbalanced. For example, He Renjie and Li Yuanxin [5] paid attention to TT&C scheduling problems in single resource system. For multi resource systems, Ling Xiaodong [6] designed tabu search genetic algorithm (TSGA) to solve a kind of TT&C scheduling problem. He assigned different priorities to different spacecraft and used the maximizing sum of priorities as the objective function. Li Yuqing[7] studied multi-spacecraft scheduling problems with multiple kinds of resource. In this kind of scheduling problems, the periods of time of TT&C have priority constraints. Qiu Dishan[8] researched on multi-spacecraft intensive observation scheduling problems with the limits of power and memory capacity.

In conclusion, observation activity of spacecraft usually has multiple targets and there must be temporal constraints for scheduling of observing strip targets. At the same time, scheduling process is constrained by the upper limits of resource. Sometimes, the situation is more complicated that scheduling objects consume resource as well as supply resource when there are multiple tasks. For example, in every lap of the track, the total capacity of power is constant, and scheduling objects only consume resource. That is to say the supplement of resource is not included in scheduling tasks. In general, ground stations have the same regions with observing targets, so downloading data to ground stations and observing targets are synchronous. In the meantime, downloading data can increase the available memory capacity.

This paper focus on large-scale scheduling problems of Earth Observing Satellites in the low earth orbit with multiple satellites, tasks and resource. In these problems, we consider that observing targets have priority and temporal constraints and multiple resources can be consumed as well as supplied. Based on existing research results, we use ant colony algorithm to solve scheduling problems in which every circuit of orbit is independent. Meanwhile, we use task priority, temporal constraints and resource information to improve algorithm.

2 Multi-Target and Multi-Resource Scheduling

The scheduling process for the observations of LEOS deals with the constraints from the multi-target and multi-resource, which could be called as MTRS(Multi-Target and Multi-Resource Scheduling). A scheduling problem is studied in this paper, handling targets with different priorities and resources with different types of characters.

Constraints existing in the scheduling process:

(1) Priorities: Different targets possess different priorities. In the real application, targets corresponding to the requests from the events on earth or on-board a satellite, and the events usually have different science or engineering values.

(2) The constraints existing should be checked along with every step in the scheduling process, and should be satisfied as the scheduling goes on[9].
(3) The time constraints appear as the sequential constraint between tasks, and they are checked using the time variables of the tasks.
(4) The resource constraints are caused by the limit of the resource on-board, appearing as any type resource consumptions cannot exceed the resource capacity. Two types of resource are considered in this paper: resource with only consumers & resource with both consumers and producers. Tasks are selected in no strict time order in the global optimal scheduling, which leads to that, the two types of resource cannot share the same checking mechanism. For the resource with only consumers, only checking on total amount composed by all the selected tasks' consumers in the scheduling time horizon will be enough. However, for the resource with both consumers and producers, as well as the checking on total amount, the current consumption composed by the tasks, which will be executed before the current task, should be checked. A scheduling result could be generated, with a sequential part exceeding the limit, after which the producers could make the total amount constraint being satisfied. The two kinds of checking mechanisms could guarantee to exclude the above results.
(5) Observation targets could be selected only once, while the ground station could be selected repeatedly selected. Ground station could be selected by only one satellite at a specific time.

For the convenience of calculation, combining the real application, assumptions are made as follow:
(1) In the observation of LEO satellites, the targets are set to be strips which parallel with the satellites' track;
(2) If a task has been selected, the execution time for the task occupies the time window for the corresponding target or ground station;
(3) The energy is reset to be a constant at the beginning of each lap, and the progress for out the scheduling horizon is ignored;

2.1 Definitions of the Parameters

A Multi-Target and Multi-Resource scheduling problem could be described using a tuple with seven elements:

$$MTRS = \{S, T, P, G, Tar, R, W\} \tag{1}$$

Where S represents the set of satellites with amounts of N_S, $S = \{s_1, s_2, \cdots, s_{N_S}\}$.

T represents all the candidate tasks that could be available in the scheduling horizon. The observation candidate tasks is N_T, $Tp = \{tp_1, tp_2, \cdots, tp_{N_T}\}$, while the quantity of data downloading tasks is N_D, $Td = \{td_1, td_2, \cdots, td_{N_D}\}$. Tp_{before} represents the observation tasks before the current task, and Td_{before} represents the data downloading tasks before the current task.

P represents the set of the priorities of the tasks, $P=\{P_1,P_2,\cdots,P_{N_T}\}\cup\{P_1,P_2,\cdots,P_{N_D}\}$; $P_i=\{p_1,p_2,\cdots,p_{N_S}\}$ represents the priorities of task t_i for the different satellites.

G represents the set of ground stations, one of which should be occupied when a data downloading task is selected. Assume the quantity of stations is N_G, $G=\{g_1,g_2,\cdots,g_{N_G}\}$; Tar represents the observation targets, each target corresponds to one candidate observation task. R represents the set of the resource constraints existing in the scheduling process, $R=\{E,C\}$; W represents the set of time windows for tasks。 $O^i=\{o_1^i,o_2^i,\cdots,o_{N_O}^i\}$ represents the set of laps of the satellite s_i, where N_O represents the quantity of the laps.

Time parameters: d_i represents the duration for task t_i, $tac_{i,j}$ represents the duration for attitude maneuver from task t_i to task t_j.

Priority parameters: w_i represents the priority for task t_i.

RAM parameters: C_s^p represents the feasible total amount of RAM for satellite s_p; c_i represents the RAM consumption of task t_i; $c_{\max}=\max(c_i)$ represents the max consumption in all the observation tasks.

Energy parameters: E_s^p represents the total amount of energy in one single lap of satellite s_p; e_i represents the energy consumption of observation task t_i; e_i^d represents the energy consumption of data downloading task td_i; e_M represents the energy consumption of maneuvering for a unit angle. e_S represents the constant energy consumption for the attitude stabilization.

The decision variable in the scheduling process:

$$c_{mi}^P=\begin{cases}1 & tp_i\, is\, selected\, in\, lap\, ot_m^p\, of\, sat\, s_P\\ 0 & tp_i\, is\, not\, selected\, in\, lap\, ot_m^p\, of\, sat\, s_P\end{cases} \tag{2}$$

$$r_{mi}^P=\begin{cases}1 & td_i\, is\, selected\, in\, lap\, ot_m^p\, of\, sat\, s_P\\ 0 & td_i\, is\, not\, selected\, in\, lap\, ot_m^p\, of\, sat\, s_P\end{cases} \tag{3}$$

$$c_{ij}^{Pm}=\begin{cases}1 & t_j\, succeeds\, t_i\, in\, lap\, ot_m^p\, of\, sat\, s_P t_i\\ 0 & t_j\, does\, not\, succeeds\, t_i\, in\, lap\, ot_m^p\, of\, sat\, s_P t_i\end{cases} \tag{4}$$

2.2 Constraints in the Model

Considering the different kinds of constraints, the model is established.

The objective function is defined as follow, which is the sum of the product of priorities and the RAM consumptions of the tasks.

$$P=\sum_{i=1}^{N_T}\sum_{m=1}^{N_O}\sum_{p=1}^{N_S} w_i c_{m_i}^P c_i \tag{5}$$

Constraints:

$$\sum_{i=1}^{N_T}\sum_{m=1}^{N_O}\sum_{p=1}^{N_S} ct_{ij}^{pm} = \sum_{r=1}^{N_T}\sum_{m=1}^{N_O}\sum_{p=1}^{N_S} ct_{jr}^{pm} = \sum_{m=1}^{N_O}\sum_{p=1}^{N_S} c_{mj}^{p} \tag{6}$$

$$c_{ij}^{Pm}=1, obt_{mj}^{p} \geq obt_{mi}^{p} + d_i + tac_{ij} \tag{7}$$

$$\sum_{tp_i \in T} e_i c_{mi}^p + \sum_{td_j \in D} e_j^d r_{mj}^P + \sum_{t_i \in (T\cup D)} (e_m tac_{ij} + e_S) ct_{ij}^{pm} \leq E_s^p \tag{8}$$

$$\sum_{m=1}^{N_O}\sum_{i=1}^{N_T} c_i c_{mi}^p + \sum_{m=1}^{N_O}\sum_{j=1}^{N_D} c_j r_{mj}^{Pm} \leq C_s^p \tag{9}$$

$$\sum_{t_i \in Tp_{before}} c_i c_{mi}^p + \sum_{t_i \in Td_{before}} c_j r_{mj}^{Pm} \leq C_s^p \tag{10}$$

$$\sum_{m=1}^{N_O}\sum_{i=1}^{N_T} c_i c_{mi}^p \geq \sum_{m=1}^{N_O}\sum_{j=1}^{N_D} c_j r_{mj}^{Pm} \tag{11}$$

Where (6) represents the uniqueness constraints of observation tasks; (7) represents the time constraints between tasks in a single lap; (8) represents the energy constraints in a single lap; (9) , (10) and (11) represent the RAM capacity constraints: (9) represents the constraints of total usage; (10) represents the constraints of the current usage; (11) represents the constraints of feasible data downloading task.

3 Scheduling Process Based on Improved ACO Algorithm

Considering the characters of the scheduling problem, an improved ACO algorithm is proposed. The basic ACS and MMAS are referenced. The state transition principle and pheromone concentration updating are newly designed combining the priorities, time constraints and resource constraints to improve the ACO.

3.1 Transition Principle

$$j=\begin{cases} \max\limits_{j\in allowed_k(t_i)} \{(\tau_{ij})^{\alpha}(\eta_{ij})^{\beta}(\omega_j)^{\gamma}\cdot(\vartheta_j)^{\varsigma}\} & q \leq q_0 \\ S & q > q_0 \end{cases} \tag{12}$$

Where $allowed_k(t_i) = T - \sum_{i=1}^{N_O N_S} tabu_i - t\_violate_k$ is the set of feasible tasks? $\sum_{i=1}^{N_O N_S} tabu_i$ is a global variable, which represents the tasks that have been explored by all ants. In the orbit, the positions of a satellite is a value of function with the variable of time elements, and the time constraints between tasks should be considered.

$t\_violate_k$ represents the tasks that cannot satisfied the time constrains. The heuristic method of checking time constraints is based on the follow principles[10]:

(1) If a conflict occurs between two observation tasks, and the tasks have different priorities, the task with lower priority would be excluded.
(2) If the two tasks with conflict have same priorities, the task with more feasible time window in the current lap would be excluded.
(3) If the two tasks with conflict have same priorities and feasible time window number, the task with more conflicts with other tasks would be excluded.
(4) If all the above three are same, a task would be excluded randomly.

q_0 represents the parameters controlling the transition, which is a constant between the interval of [0,1]; q represents a random variable equally distributed in the interval of [0,1]. S is determined by the following equation:

$$p_{ij}^k = \begin{cases} \dfrac{(\tau_{ij})^\alpha (\eta_{ij})^\beta (\omega_j)^\gamma (\vartheta_j)^\varsigma}{\sum\limits_{l\in allowed_k(t_i)} \tau_{il} (\eta_{il})^\beta (\omega_l)^\gamma (\vartheta_l)^\varsigma} & j \in allowed_k(t_i) \\ 0 & otherwise \end{cases} \tag{13}$$

Where, τ_{ij} represents the pheromone concentration between task t_i and task t_j; $\eta_{ij} = \dfrac{1}{\xi_{ij}}$, $\xi_{ij} = tac_{i,j} + e_j$ represents the influence of energy request between task t_i and task t_j; ω_j represents the influence of the priority of task t_i.

$$\vartheta_j = \begin{cases} \inf & \sum\limits_{m=1}^{N_O} \sum\limits_{i=1}^{N_T} c_i c_{mi}^p - \sum\limits_{m=1}^{N_O} \sum\limits_{j=1}^{N_D} c_j r_{mj}^{Pm} \le c_{\max} \\ 1 & else \end{cases} \tag{14}$$

ϑ_j represents the influence of the request of random capacity of task t_j.

3.2 Pheromone Concentration Updating

The ACO algorithm proposed in this paper used the transition mode of all ants explores in a random order. Firstly, an ant is randomly allocated with an initial task. The ant select next tasks until the constraints could not be satisfied, when the feasible task sequence t_f_i could be returned. All ants randomly begin selecting, and the feasible task sequences TF could be returned after all ants finish selecting.

$$TF = \bigcup_{i=1}^{N_O N_S} t\_f_i \tag{15}$$

The profits of the TF will be compared with the current best sequences TF_{Best}, to get a new better one.

In the updating of pheromone concentration, the way used by Qiu[8] is borrowed with the new parameters in our model. If the profits P_{TF} of TF, which means the objective function value of all the tasks have been explored by ants, is not greater than the profit P_{Best} of TF_{Best}, then the pheromone concentration of all the paths contained by TF volatilizes. After the volatilization, the new pheromone concentration could be obtained by follow:

$$\tau_{ij}^{new} = (1-\rho)\tau_{ij}^{old} \tag{16}$$

Where ρ represents the pheromones volatility, τ_{ij}^{old} represents the old pheromone concentration between task t_i and t_j.

If the profits P_{TF} is greater than the profits P_{Best}, then the global best solution will be updated as $TF_{Best} = TF$. To accelerate convergence, after each iteration, only the pheromone concentration of current best solution will be updated:

$$\tau_{ij}^{new} = (1-\rho)\tau_{ij}^{old} + \Delta\tau_{ij} \tag{17}$$

Where $\Delta\tau_{ij} = \dfrac{Q}{L_{Best}}$, Q represents a pre-fixed parameter, L_{Best} represents the influence of the resource parameters in the current best solution:

$$L_{Best} = \sum_{t_i, t_j \in TF_{Best}} \xi_{ij} \tag{18}$$

To avoid the algorithm converges early at the locally optimal solution, the pheromone concentration was limited to a interval of [$\tau_{\min}, \tau_{\max}$], and $\tau_{\max} = \dfrac{1}{(1-\rho)L_{Best}}$, $\tau_{\min} = \dfrac{\tau_{\max}}{50}$.

3.3 Terminating Conditions

For the ACO algorithms, a simple but practical principle is to limit a max number of iterations. The algorithm runs until the max number is reached and the results could be obtained. The simulation in this paper uses just this kind of principle. The max number of iterations in this paper was set as $N_{\max}$.

3.4 Realization of the Algorithm

Both the time constraints and resource constraints should be checked every step of the algorithm. The time constraints were checked to generate the candidate missions and when a candidate was selected the resource constraints should be checked. In this way, the validity of the results could be guaranteed. The detailed calculating flow of the algorithm is represented in Fig 1.

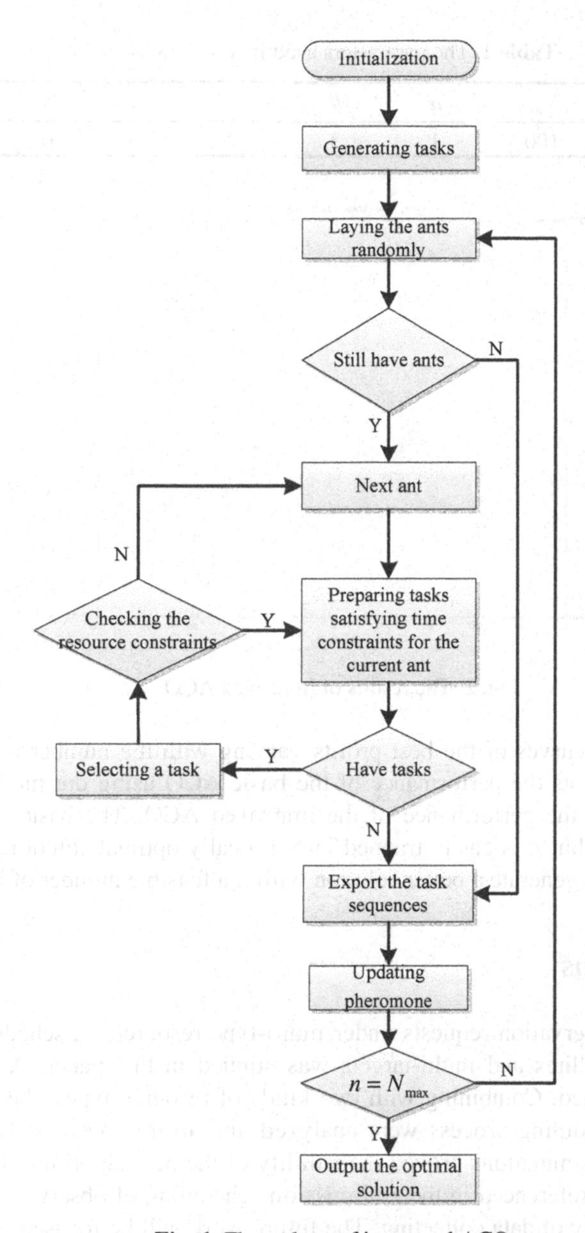

Fig. 1. Flow chart of improved ACO

4 Simulations

In the simulations, observation targets were randomly generated in a range of (60°E~150°E,0°N~50°N). The priorities of the targets were respectively allocated with random integers from 1 to 10, while the ground station' priorities are unified to be zero. The algorithm was completed on Matlab 2010b and runs on a PC with a CPU of 2.6GHz and a 4GB RAM. The parameters used in the simulations are listed in Table 1:

Table 1. The parameters used in simulations

N_T	N_G	$N_{\max}$	α	β	λ	φ	q_0	ρ
100	8	100	1	3	2	1	0.5	0.1

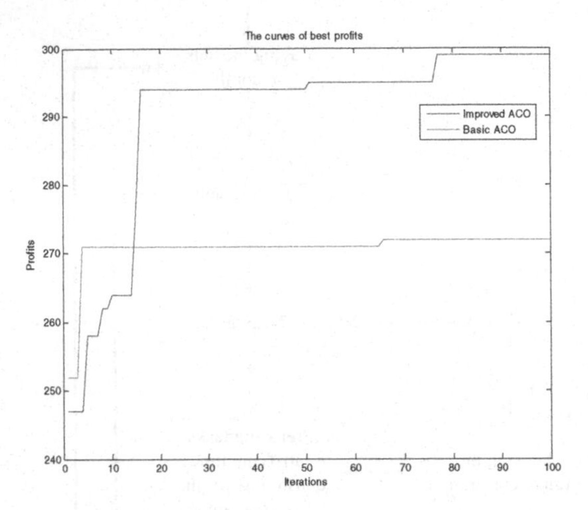

Fig. 2. The results of improved ACO

Fig 2 shows the curves of the best profits varying with the number of iterations. The red line represents the performance of the basic ACO using our model, and the blue one represents the performance of the improved ACO. The basic ACO could convergence faster, but it is easily trapped into a locally optimal solution. While the improve ACO could generate a better solution within a feasible number of iterations.

5 Conclusions

Considering the observation requests under multi-type resources, a scheduling problem with multi-satellites and multi-targets was studied in this paper. A scheduling model was established. Combining with two kinds of resource types, the constraints existing in the scheduling process were analyzed and an improved ACO algorithm was proposed. The simulations proved the validity of the new algorithm. Research in this paper could be referenced in the real mission scheduling of observation satellites to improve the ability of data collecting. The future work will be focused on the scheduling problem with massive targets and more complex constraints.

Acknowledgements. This work was supported in part by the National Basic Research Program of China (973 Program) 2012CB720000, the National Natural Science Foundation of China (61374216, 61304248, 61304226), the Research Fund for the Doctoral Program of Higher Education of China 20121101120006.

References

1. Burrowbridge, S.E.: Optimal allocation of satellite network resources. Masters Thesis. Virginia Polytechnic Institute and State University, Virginia (1999)
2. Barbulescu, L., Watson, J.P., Whitley, L.D., et al.: Scheduling space ground communications for the air force satellite control network. Journal of Scheduling 7, 7–34 (2004)
3. Tapan, P.B.: Near optimal ground support in multi-spacecraft missions: a GA model and its results. IEEE Transactions on Aerospace and Electronic Systems 45(3), 950–964 (2009)
4. Zhang, N., Ke, L.-J., Feng, Z.-R.: A new model for satellite TT&C resource scheduling and its solution algorithm. Journal of Astronautics 30(5), 2140–2145 (2009)
5. He, R.-J., Tan, Y.-J.: Apply constraint satisfaction to optimal allocation of satellite ground station resource. Computer Engineering and Applications 18, 229–232 (2004)
6. Ling, X.-D., Wu, X.-Y., Liu, Q.: Study of GATS algorithm for multisatellite TT&C scheduling problem. Journal of Astronautics 30(5), 2133–2139 (2009)
7. Li, Y.-Q., Wang, R.-X., Xu, M.-Q.: An Improved Genetic Algorithm for a Class of Multi-Resource Range Scheduling Problem. Journal of Astronautics 33(1), 85–90 (2012)
8. Qiu, D.-S., Guo, H., He, C., et al.: Intensive Task Scheduling Method for Multi-agile Imaging Satellites. Acta Aeronautic et Astronautica Sinica 34(4), 882–889 (2013)
9. Xu, R., Xu, X.-F., Cui, P.-Y.: Dynamic Planning and Scheduling Algorithm Based on Temporal Constraint Network. Computer Integrated Manufacturing Systems 10(2), 188–194 (2004)
10. Yang, Y.-A., Fan, H.-H., Feng, Z.-R., et al.: Simulation & realization of satellite TT&G resources scheduling based on event scheduling method. Journal of System Simulation 17(4), 982–985 (2005)

Detection of Trajectory Patterns and Visualization of Spatio-temporal Information Based on Data Stream Approaches

Yicong Wang[1], Kazuhiro Seki[2], and Kuniaki Uehara[1]

[1] Graduate School of System Informatics, Kobe University
1-1 Rokkodai-cho, Nada-ku, Kobe 657-8501, Japan
{wang,uehara}@ai.cs.kobe-u.ac.jp

[2] Faculty of Intelligence and Informatics, Konan University
8-9-1 Okamoto, Higashinada, Kobe 658-8501, Japan
seki@konan-u.ac.jp

Abstract. With the rapid increase of the number of mobile GPS devices including smartphones, it is becoming more and more important to develop efficient and effective algorithms to analyze massive trajectory data streams generated through those devices. Although there are many algorithms that can find patterns from massive trajectory data stream by batch processes, what we need now is a new algorithm that can deal with massive data streams with limited resources by online processes. This study aims at developing such an algorithm and attempts to discover the places at which people often stop when they are walking or driving, or the places which are becoming crowded by analyzing massive trajectory data streams.

Keywords: trajectory, data stream, data mining.

1 Introduction

During the development of data mining technologies in the last decade, what has been changing is not only the scale we are facing but also the object we are dealing with. One aspect is the increase of the number of embedded systems and sensors in the world. For instance, the percentage of cell phone subscribers in the world was 19% in 2002, which rapidly increased to 67% in 2010; most of them are smartphones with GPS sensors. This indicates that there is a new kind of data that we can analyze, i.e., data streams. Data arrives in a stream or streams and if it is not processed immediately, it will be lost permanently. Analyzing real time data streams rapidly plays a critical role in big scale traffic management and will gain potential resource reduction.

Despite the volume and velocity of data streams, the basic requirement for a data stream algorithm is that it must finish the process on the current data before the next data arrive, or push it into a buffer and keep the buffer from overflowing. In this paper, we attempt to deal with a specific kind of data stream from GPS devices. There is an engaging project at Tokyo University Center for Spatial Information Science,

F. Bian and Y. Xie (Eds.): GRMSE 2014, CCIS 482, pp. 204–214, 2015.

called the "People Flow Project" [1]. This project provides the trajectory data for people's positions and time for research purposes. All experiments in this paper are based on the trajectory data in October 1st, 2008 provided by the People Flow Project. The data contain about 600,000 of people's trajectories.All the data were collected by questionnaire, and thus, there is no need to filter out signal noises that would exist in real data streams.

Here is a sample of the data:

126623,5,5,2008/10/01 11:29:00,139.70711,35.69092,2,7, 00007202,14,99,38,0,97
126623,5,5,2008/10/01 11:30:00,139.70682,35.69000,2,7, 00007202,14,99,38,0,97
126623,5,5,2008/10/01 11:31:00,139.70775,35.68961,2,7, 00007202,14,99,38,0,97

The first number is Person ID, followed by Trip Number, Sub-trip Number, time stamp, longitude, and latitude. The remaining numbers encode sex, age, address, occupation, expansion factor 1&2, and means of transportation. These data can be considered as data streams from about 600,000 people. Fig. 1 shows that person A moved from point X to Y to Z in time sequence according to the sample data above. The other polyline in Fig. 1 shows person B moved from O to P to Q according to the same kind of trajectory data.

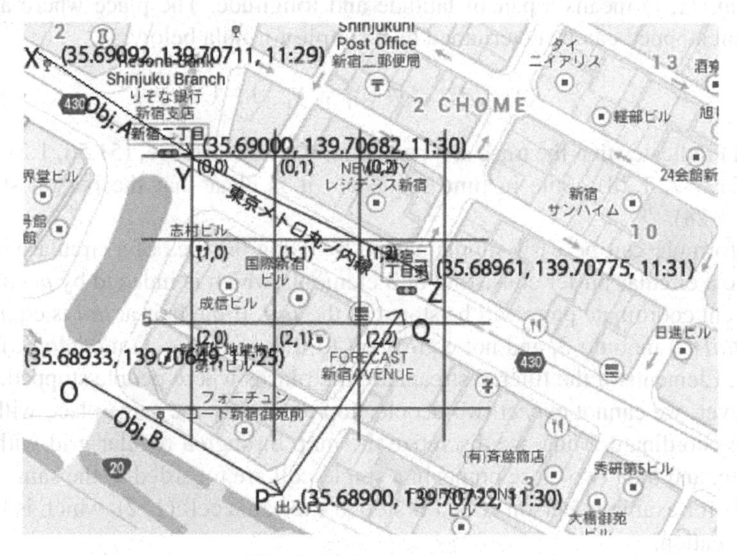

Fig. 1. Trajectory data

2 Background

Although there are many works related to trajectory pattern mining from static data stored in a storage or an active database [2], they cannot handle data stream input. Also those algorithms typically assume that all the data can be loaded into memory, which is often invalid for real-world trajectory data. For example, the data set we use in this study covering five prefectures in Japan for only one day (Oct 1st, 2008) consists of 600,000 samples, amounting to 70 GB of data.

For analyzing data streams, the University of Waikato developed a framework called Massive Online Analysis (MOA) [3]. It provides a few common machine learning algorithms, such as naive Bayes classifiers, StreamKM++ (based on K-means), Den-Stream based on DBscan or density-based spatial clustering. Particularly, Den-Stream can find clusters that have a considerably higher density of points than outside of the cluster. Den-Stream can deal with trajectory data streams, but analyzing these data by Den-Stream only extract relatively crowded regions, not fast changing regions. This means that the patterns found by Den-Stream are no more than crowd of people with density above the threshold. Therefore, even crowd of people with high density which will not cause traffic problems like the audience in a stadium will be detected as clusters. Instead of finding a large crowd of people, we aim at detecting places where people are just starting to gather. In other words, we attempt to find places where there will be a traffic jam before it actually occurs.

How can we find such places where people are gathering? People gathering means there are many people stop at the same place in a short period of time. In other words, we have to find the place where the density is growing. Therefore, determining the stop point from a trajectory data stream is the first step. Formally, a trajectory data stream is a sequence of elements (x_1, y_1), (x_2, y_2), whose length might be infinite. More that, (x, y) means a pair of latitude and longitude. The place where a trace of movement stopped can be determined by a simple formula below:

$$(x_{t-2}, y_{t-2}) \neq (x_{t-1}, y_{t-1}) = (x_t, y_t) \tag{1}$$

Eq. (1) indicates that the trace stopped at (x_t, y_t). For instance, (51.56, 1.2), (51.557, 1.28), (51.557, 1.28) come in time sequence, it is clear that the person stopped at (51.557, 1.28).

This formula can be easily implemented by using the idea of a circular buffer. We can make a circular buffer consisting of 3 elements which is indexed by *head* and *tail*. The current coordinate pairs will be stored at the *tail*. If the pair at *tail* is equivalent to that at (*tail*+2) modulo 3, and not equivalent to that at (*tail* + 1) modulo 3, Eq. (1) is satisfied. Elements of the filtered stream are the places where people stopped.

However, we cannot expect two people stop at exactly the same place with exactly the same coordinate. Thus, we discretize the map through a regular grid with cells of small size, and then two close points in a single cell are regarded as the same location. In Fig. 1, for example, person A and B stop at the same cell (1, 2), which is treated as a same location.

Next, if we can quantify the popularity of each element in the filtered stream, we know where people always stop. However, we cannot focus on a single stream. Every GPS device is streaming its own data, which means we are filtering multiple streams. We will gather all the filtered streams to a single stream. For instance, suppose there are two streams:

Stream from device A: ...(35.667, 139.85, 13:02) ...

Stream from device B: ...(35.686, 139.877, 13:02), (35.241, 139.2413, 13:03),...

Although the filtered streams contain data with the same time stamp, as they come in time sequence, we can gather them to a single stream,

Filtered stream: ...(35.667, 139.85, 13:02), (35.686, 139.877, 13:02), (35.241, 139.2413, 13:03)...

Because we are trying to figure out the how many people stopped at a particular place recently, which device stopped is not of importance. Thus, the device information can be removed. This stream will be the input for the approaches described in the following sections.

3 Decaying Window Approach

Now we will introduce a simple method which can find the most common recent elements in a data stream by a decaying window [4]. The decaying window here means that the older data are given less weights than newer data. For example, assume that the filtered stream is {(35.667, 139.8998, 13:02), (35.686, 139.877, 13:02), (35.241, 139.2413, 13:03), (35.6857, 139.8752, 13:03), (35.2405, 139.2421, 13:03) ...}. Fitting the coordinate to cell, the stream become {cbaba...}, where "a", "b" and "c" denote the cell IDs. In addition, the current data is "a". We can define a sub-stream for each element appeared in the stream. For the element "a" appeared in the stream, if the *i*-th element is "a", we make the *i*-th bit of sub-stream "a" 1, and make it 0 otherwise. Then we get:

substream a=......00101
substream b=......01010
substream c=......10000

Based on this representation, we can calculate the popularity for each element defined by the formula below

$$Popularity_a = \sum_{i=0}^{t} a_i \delta^{t-i} \tag{2}$$

In Eq. (2), a_i becomes 1 if the *i*-th element in the input stream is "a", and 0 otherwise. δ means a constant very close to but less than 1, e.g., $1\text{-}10^{-5}$. Then we get

$$Popularity_a = \cdots 0 \cdot \delta^4 + 0 \cdot \delta^3 + 1 \cdot \delta^2 + 0 \cdot \delta^1 + 1 \cdot \delta^0$$

$$Popularity_b = \cdots 0 \cdot \delta^4 + 1 \cdot \delta^3 + 0 \cdot \delta^2 + 1 \cdot \delta^1 + 1 \cdot \delta^0$$

$$Popularity_c = \cdots 1 \cdot \delta^4 + 0 \cdot \delta^3 + 0 \cdot \delta^2 + 0 \cdot \delta^1 + 0 \cdot \delta^0$$

$$Popularity_a > Popularity_b > Popularity_c$$

This is exactly what we expected. Element "c" should be given the lowest popularity, because element "c" only appeared once, less than both "a" and "b". Although the element "a" and "b" both appeared twice, "a" appeared later than "b" and has higher popularity than "b". That is, current data are given higher weights than the older data.

Mathematically, we indeed need all the data to calculate the summation (Eq. (2)). However, it is computationally unnecessary to keep all the data in memory because the data are coming as a stream in sequence. Precisely, when a new element comes from the stream, we perform the following:

1. For each popularity we are holding, multiply it by δ.
2. Suppose the new element is “a”. If the popularity of “a” already exists, add one. If it does not exist, create the variable and define it as 1.

The decaying window algorithm can be summarized as the following:

```
Input: element e from filtered stream
list := [];
for all element in list do:
        element.popularity*=δ;
        if e exist in list then
                    e.pupularity+=1;
        else
                    list ← e with popularity 1;
        end if
end for
```

Algorithm 1. Decaying Window

In this way, we can calculate the popularity of each cell without loading all the preceding data previously.

However, this algorithm has several problems. First, a threshold is necessary, or each element will be assigned a popularity which will lead to insufficient memory. Fortunately, we have an interesting observation of the cells where people stop. Fig. 2 illustrates the distribution of the popularity of cells, where x axis means popularity and y axis means the number of cells. This indicates that almost 95% of cells do not appear frequently, which means that they should be removed to save memory.

Even if the memory is no longer a problem, we are still facing another problem. That is, when the popularity of a particular cell was assigned is unknown. Imagine that if a traffic accident happened at a crossing and a heavy traffic jam occurred. Hundreds of vehicles were stuck at the crossing and doubtlessly this will be detected

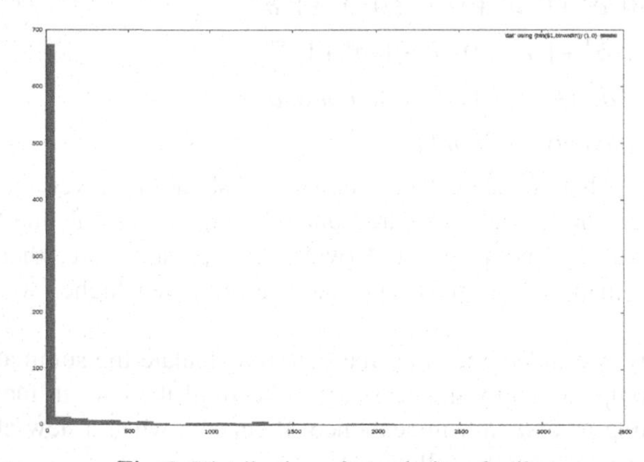

Fig. 2. Distribution of popularity of cells

by the decay window method. However, the problem is that there is no difference between such a special accident and a commonly busy train station by the decaying window method. That is why we need to use a lossy counting [5] to calculate the popularity. Lossy counting is an algorithm that counts the occurrence of each element in a data stream. Meanwhile, it will periodically remove infrequent elements.

4 Lossy Counting with Decaying Window

We will introduce an algorithm called lossy counting (LC) [5] and expand it to lossy counting method with decaying window. Unlike decaying window method, LC does not maintain the entry that represents the popularity of all the elements and determines whether it should be removed by a constant threshold. In advance, we call the memory space where the algorithm records the popularity of each element in stream "the popularity table". For input, LC cuts input data stream into fixed-size windows. For instance, each window has 1,000 elements from the input data stream. We assign the fixed size with ε. However, the problem is that trajectory data are aligned with time. Thus, the quantity of the coming data per unit time may change, which means that the original LC method is not proper for trajectory data.

Fortunately, we can extend the concept such that each window contains the coming data in ε time. Each window is indexed with a window stamp t which represents the window number. For example, if we set ε to 5 minutes, the first window will contain the data in 5 minutes since the algorithm is executed, and the second window will contain the data for the next 5 minutes. Then the algorithm will process windows sequentially, and every element will be checked. If an element has no record in the popularity table, a structure (e, f, Δ) will be created to record the occurrence of this element where e is a label of the element being processed (in the case of map, it is the cell ID), f is the number of occurrence, and Δ is the window number for which e appeared first. If a record already exists, we increment f. After processing each element in a window, the algorithm will traverse the popularity table and check each (e, f, Δ) by the formula below:

$$\frac{f}{current\_window\_number - \Delta + 1} \leq 1 \tag{3}$$

$$f = \sum_{i=j}^{k} n_i \tag{4}$$

In Eq. (4), j denotes Δ in (e, f, Δ), which represents the window number when it first appeared, k is the current window number, and n_i is the number of occurrences of element e in the i-th window.

The denominator of Eq. (3) means how many windows have passed since the entry was assigned in the popularity table. Thus, the left part of the formula becomes the average occurrences per window (ε time). Any entry satisfying Eq. (3) will be detected as an infrequent element and will be removed, which means we define an element as a frequent element when it occurred more than once in ε time on average.

After removing all the infrequent entries, the algorithm will process the next windows and repeat the procedure above. The following shows an example.

In Figure 3, a particular element *e* occurred in window 1, 17 times and in window 2, 4 times, respectively. At the end of window 1, the entry of *e* is (e, 17, 1), because *e* first appeared at window 1 and occurred 17 times. At the end of window 2, *e* occurred 21 times in total since it appeared at window 1, thus it occurred 10.5 times per window on average. If we assume that element *e* never appeared after the second window, finally at the end of window 21, *e* occurred once on average per window, and satisfies the condition of Eq. (3). Being determined as an infrequent element, element *e* will be removed at the end of window 21. In this way, infrequent elements will be removed periodically.

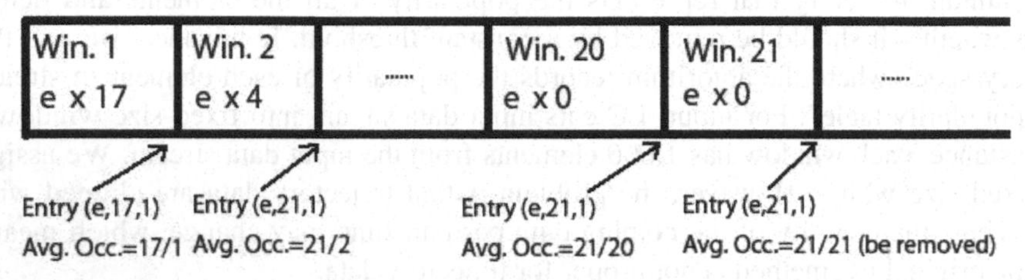

Fig. 3. Example of determining unfrequented elements

This algorithm can be summarized below,

```
1. input: elements e from stream, parameter epsilon
2. D:= [];previous_time := current time;delta:=0;
3. for all e do
4. if entry of e exist in D then
5.        D ← (e, f+1. Delta);
6. else
7.        D ← (e, 1, delta);
8. end if
9. if previous_time + epsilon == current time then
10.       delta := current window_id;
11.       for all (ee, f, delta) in D do
12.               if f + delta <= current window_id then
13.                       remove (ee, f, delta)
14.               end if
15.       end for
16. previous_time := current time;
17. end if
18. end for
```

Algorithm 2. Lossy Counting

Furthermore, this algorithm can be enhanced by the decay window idea. In this way, the current data will have more weight than older data. Thus, the frequent element appeared recently will get a higher popularity than those appeared earlier, even they have the same occurrence. In addition, we can change the decaying speed by modifying the parameter δ. The pseudocode is the same as Algorithm 3 with the following loop inserted between line 8 and 9.

```
for all ee in D except e
        D ←(ee, f*δ, delta);
end for
```

5 Experiment

In this section, we will compare the result of proposed method with that of Den-Stream. In addition, we will evaluate the memory cost between the proposed method and the original decaying window method. At last, we will zoom out the center of Tokyo to the whole Tokyo area to see if there is something out of expectation.

5.1 Comparison with Den-Stream

Fig. 4 and Fig. 5 indicate that our proposed method can find some patterns that cannot be found by Den-Stream algorithm, like somewhere people are gathering. Vice versa, our proposed method omits somewhere being recognized as a cluster by Den-Stream algorithm, such as running trains. As shown in Fig. 4, our proposed method in the left part detected two patterns around the Shinbashi station at 12 a.m; 384 passengers stopped at the station and 83 passengers stopped at the bottom left corner of the figure during 10 minutes. The pattern at the train station is also detected by Den-Stream algorithm in the right part because the density exceeded its threshold (4 meters in this experiment) which means the distance between two people neighbored is no more than 4 meters. However, the pattern in bottom left is not detected by Den-Stream because of low density, although it might be becoming a crowded region.

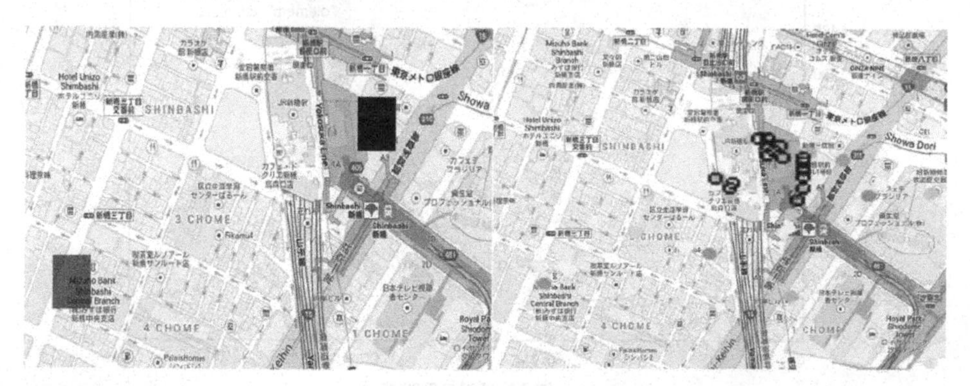

Fig. 4. Comparison b/w our approach (left) and Den-Stream (right) around Shinbashi Station

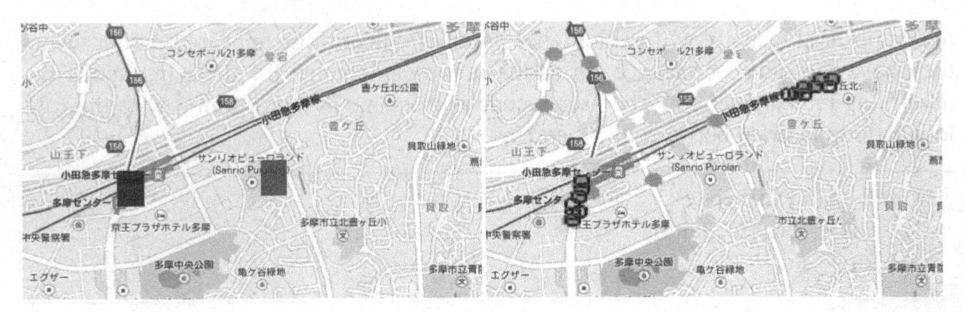

Fig. 5. Comparison b/w our approach (left) and Den-Stream (right) around Tama-center Station

In the right part of Fig. 5, there is a cluster detected in the east of Tama-center Station. In fact, it is a running train. Metros in the rush hour are always the places people stand shoulder to shoulder, with extremely high people density far over the threshold. As designed, this cluster was not detected by our proposed method in the left part because the train did not stop.

5.2 Memory Cost

Fig. 6 shows the number of entries under different window size which can reflect memory consumption directly during 24 hours. The green, blue, red curves represent ε equals 20, 10, and 5 minutes, respectively. The two peaks appeared in the typical rush hour of the day. The pink curve represents the number of entries with the old decaying window method. The number of entries kept increasing because none of them has been removed. Meanwhile, as the number of cells is limited, the pink line is converged. As shown in the figure, at the end of the day, about 75% memory has been recycled.

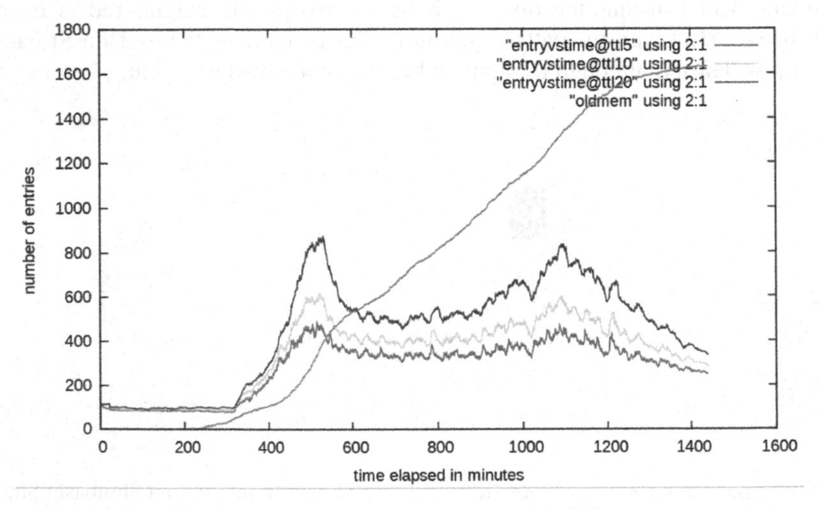

Fig. 6. Number of entries against time at different ε

5.3 Exploring the Big Data

Furthermore, if we zoom out to the whole Tokyo area from the area around Tokyo Station with a 6.33 km radius, we can make some interesting observation. It is said that the Shinjuku Station is the busiest train station in the world [6]. However the Shinjuku Station is not the busiest station in Tokyo Area before 08:05. The busiest train stations before 08:05 are some other stations in the suburbs and exurbs around Tokyo. The Tamaplaza, Aobadai, Chiharadai, Hiyoshi, Senzokuike, Makuhari, Otsuka, Kasai stations around the center of Tokyo City; all have an average throughout between 500 and 600 people per hour before 08:05, far more than 60 people per hour at Shinjuku Station. This might be a discovery out of expectation. What should be noted is that the number is based on the survey data, just a part of the real quantity. All these stations are shown as the red circle in Fig 7.

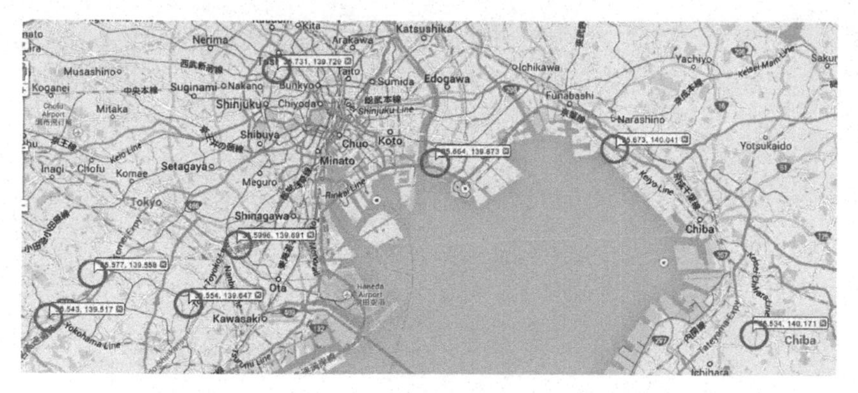

Fig. 7. Busiest train station in the rush hour of morning before 08:05

6 Future work

Although Lossy Counting reduced the memory cost, if we are facing an extraordinarily big map, the memory still might be insufficient. Therefore we are attempting to apply another algorithm called the Probabilistic Lossy Counting (PLC) [7], which can save 30% or more memory than lossy counting. The difference between PLC and Lossy Counting is the way of calculating Δ in (e, f, Δ). The problem is that PLC has a small probability of making a false negative, which means PLC may lose a frequent element. Therefore, some mathematical modifications are needed to avoid the false negatives and to make PLC suitable to find a recently frequent element more easily.

References

1. Peoples Flow Project. Center for Spatial Information Science, Tokyo University, `http://pflow.csis.u-tokyo.ac.jp/`
2. Giannotti, F., Nanni, M., Pinelli, F., Pedreschi, D.: Trajectory pattern mining. In: The 13th ACM SIGKDD International Conference on Knowledge Discovery and Data Mining (KDD 2007), pp. 330–339 (2007)

3. Massive online analysis. Waikato University, http://moa.cms.waikato.ac.nz
4. Rajaraman, A., Ullman, J.D.: Introduction to Information Retrieval. Cambridge University Press (2001)
5. Motwani, R., Manku, G.S.: Approximate Frequency Counts Over Data Streams. In: The 28th International Conference on Very Large Data Bases, pp. 346–357 (2002)
6. The Shinjuku Station, http://en.wikipedia.org/wiki/shinjuku_station
7. Dimitropoulos, X., Hurley, P., Kind, A.: Probabilistic Lossy Count-ing: An Efficient Algorithm for Finding Heavy Hitters. SIGCOMM Computation and Communication 38(1), 5–5 (2008)

Investigating and Comparing Spatial Accuracy and Precision of GPS-Enabled Devices in Middle Tennessee

Leong Lee[1,5], Matthew Jones[2,5], Gregory S. Ridenour[3,5], Maurice P. Testa[3,5], and Michael J. Wilson[4,5]

[1]Department of Computer Science, [2]Department of Mathematics and Statistics, [3]Department of Geosciences, and [4]GIS Center, [5]Austin Peay State University, Clarksville, Tennessee, USA
{leel,jonesmatt,ridenourg,wilsonm}@apsu.edu

Abstract. GPS-enabled mobile devices are extremely popular today. Billions of such devices are currently in use. The application and research potentials of these devices are limitless, but how accurate are these devices? The research team used Average Euclidean Error (*AEE*), Root Mean Square Error (*RMSE*), and Central Error (*CE*) to define and calculate the accuracy and precision of twelve popular GPS-enabled mobiles devices in two different geographical regions in Middle Tennessee. Field data were collected, and the results were ranked and compared. A website and related algorithm were developed to facilitate potential future research. In this preliminary study, it was discovered that various mobile devices performed differently in terms of *AEE*, *RMSE*, and *CE*. Their performance also varied in different geographical regions in terms of both (*AEE*, *RMSE*, and *CE*) values and ranking.

Keywords: spatial accuracy, spatial precision, GPS-enabled devices.

1 Introduction

The Global Positioning System (GPS) has been incorporated into every aspect of civilian life. Billions of GPS-enabled mobile devices such as cellphones and tablets are being widely used by millions today. GPS-enabled mobile handsets (cellphones) sales alone are estimated to reach up to 960 million units in 2014 [1]. Due to its wide availability and dynamic technologies (usually a combination of satellite based, cell tower based, wi-fi based, and/or other local positioning system signals), GPS-enabled mobile devices have limitless research and application potential. Research and application areas include but not limited to traffic control, mobile networks, real-time data management, location mining, disaster relief / damage assessment and geo-informatics.

An important aspect of using a GPS-enabled mobile device is to have accurate and precise determinations of location. It is beyond doubt that accurate and precise location measurements are important, but how exactly are accuracy and precision defined and how do we measure them? How accurate and precise are the common GPS-enabled mobile devices currently available on the consumer market? How accurate and precise are these devices in Middle Tennessee? Our research team presents data to answer these questions in this paper.

F. Bian and Y. Xie (Eds.): GRMSE 2014, CCIS 482, pp. 215–224, 2015.

In physical science and statistics, accuracy is a quantification of how close measurements are to the "target" or "true" value (the value accepted as being correct) [2, 3]. Precision is a quantification of how close replicate measurements are to each other [2, 3]. Consider Fig. 1, which is commonly used to elucidate the conceptual differences between accuracy and precision.

Measurements are said to be accurate if they cluster around the true location. Measurements are said to be precise if they cluster close to each other. For example, upper-right diagram in Fig. 1 shows inaccurate, but precise measurements; lower-left diagram shows accurate, but imprecise measurements.

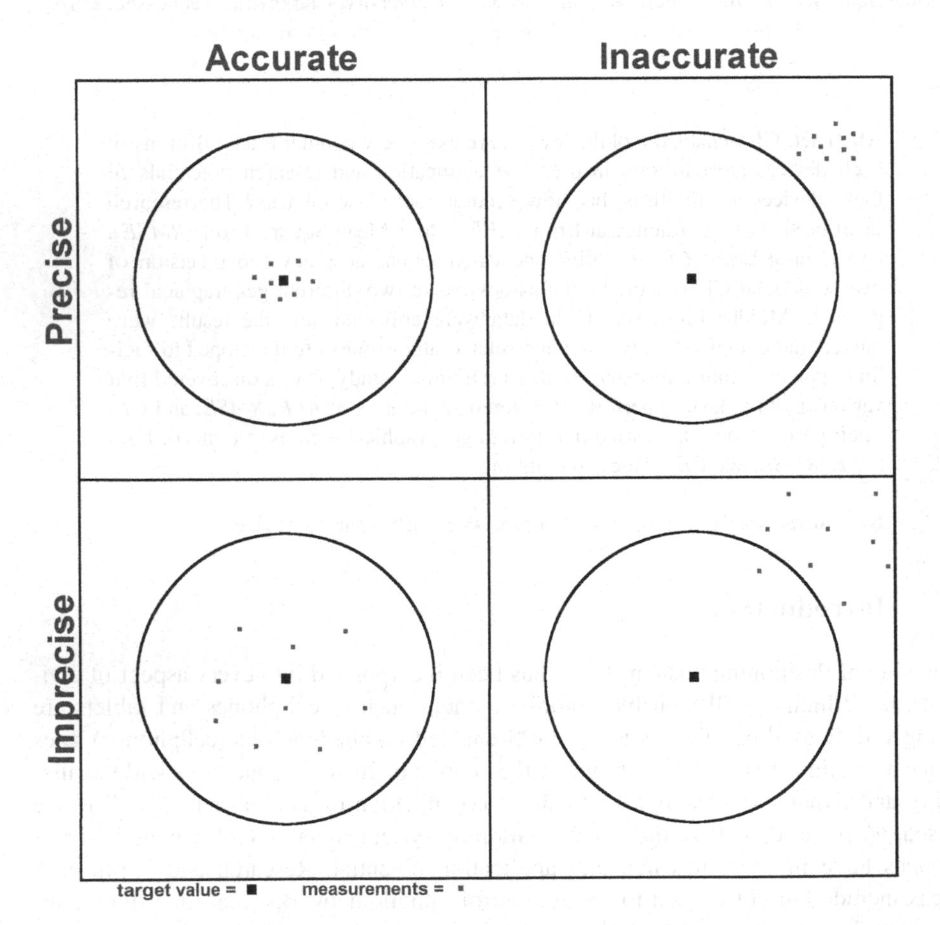

Fig. 1. Comparison of accuracy and precision

Mathematically and statistically, both accuracy and precision are often measured using average Euclidean error (*AEE*) [4] and root mean square error (*RMSE*). It can be shown that root mean square error partitions as a sum of an accuracy measurement (which we call central error, *CE*) and a precision measurement.

The research team enlisted student volunteers to collect field data of spatial coordinates (latitude and longitude) returned by several GPS-enabled mobile devices within two different regions in Middle Tennessee. The accuracy and precision measures *AEE*, *RMSE*, and *CE* calculated from the data. Dixon's Q test [5] was utilized to determine if improvement can be made in accuracy and precision by rejecting statistical outliers from replicate measurements. The results and rankings of the *AEE*, *RMSE* and *CE* are presented in this paper. To facilitate future computation of *AEE*, *RMSE* and *CE* for different GPS-enabled mobile devices and/or for different geographical regions, a website was also built and made available [6]. The algorithm is presented in this paper.

2 Background

Accuracy and precision can be quantified using at least two common measures: average Euclidean error (*AEE*) (also referred to as mean Euclidean distance, *MED* [4]), and root mean square error (*RMSE*). *RMSE* can be shown to partition into a sum of an accuracy measurement (which we call central error, *CE*) and a precision measurement.

2.1 Average Euclidean Error (*AEE*) for Accuracy and Precision Calculation

AEE is the mean of the Pythagorean distances between *n* control points (whose coordinates are known) and their associated field measurements.

$$AEE = \frac{1}{n}\sum_{i=1}^{n}\sqrt{(\Delta lat)^2 + (\Delta long)^2} \quad (1)$$

Where Δlat and Δlong are the differences between coordinates of the control points and the field measurement average.

2.2 Root Mean Square Error (*RMSE*) for Accuracy and Precision Calculation

Accuracy and precision were measured using the root mean square error procedure recommended by the Federal Geographic Data Committee [7]. The horizontal (i.e., within the plane of the earth's surface) $RMSE_r$ for a region is given by

$$RMSE_r = \sqrt{\frac{1}{n}\sum_{i=1}^{n}\left((\bar{x}_{f,i} - x_{c,i})^2 + (\bar{y}_{f,i} - y_{c,i})^2\right)} = \sqrt{(RMSE_x)^2 + (RMSE_y)^2} \quad (2)$$

Where

$$RMSE_x = \sqrt{\frac{1}{n}\sum_{i=1}^{n}(\bar{x}_{f,i} - x_{c,i})^2} \quad (3)$$

And

$$RMSE_y = \sqrt{\frac{1}{n}\sum_{i=1}^{n}(\bar{y}_{f,i} - y_{o,i})^2} \tag{4}$$

Are the root mean square errors in the x- and y- directions, respectively.
Here,

- $\bar{x}_{f,i}$ denotes the average x-coordinates of the ith set of field measurements;
- $x_{c,i}$ denotes the x-coordinate of the ith control point;
- $\bar{y}_{f,i}$ denotes the average y-coordinates of the ith set of field measurements;
- $y_{o,i}$ denotes the y-coordinate of the ith control point.

It is a simple algebra exercise to show $(RMSE_r)^2$ partitions conveniently as the sum of the accuracy measurement CE (defined in the following subsection) and a precision measurement based on the sample variances in the x- and y-coordinates:

$$(RMSE_r)^2 = (CE)^2 + \frac{n-1}{n}(s_x^2 + s_y^2) \tag{5}$$

Accuracy (and incorporated precision) was computed using one of two algorithms depending on whether or not the root mean square error in the longitudinal direction (***RMSE$_x$***) was approximately the same as that in the latitudinal direction (***RMSE$_y$***).

Under the assumptions that (i) the observations come from a bivariate normal distribution centered at the control points, and (ii) the longitudinal and latitudinal position variables are uncorrelated and $RMSE_x \approx RMSE_y$, we estimate 95% of future averages of all the field measurements to be within a circle centered at any true location in that particular region, with radius of $Accuracy_r = 1.7308(RMSE_r)$. When $RMSE_x \neq RMSE_y$ but $0.6 < RMSE_x / RMSE_y < 5/3$, $Accuracy_r \approx 1.22385(RMSE_x + RMSE_y)$.

In other words, there are two cases:

1. When $RMSE_x = RMSE_y$, $Accuracy_r = 1.7308(RMSE_r)$ (6)
2. When $0.6 < RMSE_x / RMSE_y < 5/3$, $Accuracy_r \approx 1.22385(RMSE_x + RMSE_y)$ (7)

2.3 Central Error (*CE*) for Accuracy Calculation

In order to delineate between the notions of accuracy and precision, especially in cases when measurements are not centered at the true locations, an additional measure of accuracy, central error (*CE*), based on a metric advocated by electrical engineers Li and Zhao [8], was proposed for our project.

$$CE = \sqrt{\left(\frac{1}{m}\sum_{i=1}^{m}(x_{o,i} - \bar{x}_{f,i})\right)^2 + \left(\frac{1}{m}\sum_{i=1}^{m}(y_{o,i} - \bar{y}_{f,i})\right)^2} \tag{8}$$

Intuitively, central error measures the distance between the center of mass of the data points and the target location.

2.4 Dixon's Q Test

Dixon's Q test [5] was set up in a spreadsheet to determine if potential improvement could be made in accuracy and precision by rejecting statistical outliers from replicate measurements of spatial coordinates. Dixon's Q test was designed for identification and rejection of outliers from **small data sets** (three to ten replicate measurements). For our study, a data set from which outliers might be rejected consisted of five replicate measurements of the coordinates for a given field location (five is the number of measurements taken for each control point in our field data collection, to be introduced in the next section). After arranging five replicate measurements of latitude or longitude in order of increasing value, the test statistic Q was computed as

$$Q = gap / range \tag{9}$$

where *range* is the difference between extreme (maximum and minimum) values of replicate measurements and *gap* is the difference between a suspected outlier (maximum or minimum) and its closest neighbor within the dataset. Two Q statistics were computed for each set of five coordinates, one for latitude, and one for longitude.

If a Q statistic exceeded Q_{table} (a value from a table whose values are a function of sample size and confidence level), the outlying value (the extreme with the larger gap) was eliminated from the data set. Thus, the measured location from a particular device was taken to have latitude as the average of either four or five latitude measurements, and longitude as the average of either four or five longitude measurements.

3 Methods

3.1 Field Measurements Collection and Accuracy and Precision Computation

Four student volunteers collected the spatial data consisting of latitude and longitude coordinates of positions reported by several GPS-enabled mobile devices. Five latitude and five longitude measurements were taken at each of fifteen locations, and with several devices. For each location and device, the four or five latitudes were averaged and the four or five longitudes were averaged (after applying Dixon's Q test). The fifteen locations were taken across two regions with different landscapes:

- Land Between the Lakes, where there is no cell phone reception,
- Downtown Nashville, where tall buildings might obstruct reception of satellite signals.

The field data were compared with control points ("true" values) whose coordinates were determined by placing a computer's mouse cursor over the field locations on an electronic United States Geological Survey base map and reading the onscreen display of coordinates. Because of the difference in units in which the devices measure spatial coordinates, a spreadsheet was developed that allowed either direct entry of the coordinates of the field measurements in decimal degrees, or that converted data entered in degrees, minutes, and seconds to decimal degrees. Averages of the coordinates for five field measurements were then computed. Accuracy and precision were also computed using two measures:

1. *AEE* according to equation (1) and
2. *RMSE* (and hence $Accuracy_r$) according to equations (2, 3, 4, 5, 6, 7)

To delineate between the notions of accuracy and precision, *CE* was also calculated:

3. *CE* according to equation (8)

The results of the field tests are shown in section 4.

3.2 Online Spatial Accuracy Calculator and Algorithm

A website which contains an online Spatial Accuracy Calculator was developed to facilitate future calculation of field data for its *AEE*, *RMSE*, and *CE* values. This website would allow the research team to easily calculate spatial accuracy and precision for other geographical regions based on *AEE*, *RMSE*, and *CE* equations listed above. The main programming technologies used for building the website are PHP, JavaScript, and AJAX. The algorithm of the Spatial Accuracy Calculator is listed here. The time complexity of the algorithm is polynomial with respect to the number of control points, and the maximum number of field measurements for each control point.

Algorithm 1. *AEE*, *RMSE*, and *CE* calculation (Spatial Accuracy Calculator)
begin
    user to enter the number of control points, save to n;
    for $i = 1$ to n do
        user to enter the number of spatial coordinates (field measurements) for control point i, save the number to m;
        for $j = 1$ to m do
            user to enter spatial coordinate (latitude x and longitude y) for control point i, measurement j, save coordinator x_{ij}, y_{ij} to 2D associative array;
            spatial coordinates can be entered in Decimal Degrees (DD) format or Degrees-Minutes-Seconds (DMS) format;
            if user chooses to enter in DMS format
                convert x_{ij}, y_{ij} to DD format;
            end-if;
        end-for;
        user to enter spatial coordinate for control point i, save to xc_i, yc_i;
        if user chooses to enter in DMS format
                convert xc_i, yc_i to DD format;
        end-if;
    end-for;
    for $i = 1$ to n do
        retrieve the number of measurements for control point i from 2D associative array, save to m;
        retrieve all x and y coordinates of m measurements from 2D associative array;
        calculate average of all x coordinates of m measurements, save to xd_i;
        calculate average of all y coordinates of m measurements, save to yd_i;
        $\Delta x_i = xc_i - xd_i$;

$\Delta y_i = yc_i - yd_i$;
$SS_i = (\Delta x_i)^2 + (\Delta y_i)^2$;
save $(SS_i)^{½}$ SS_i, Δx_i, Δy_i, $(\Delta x_i)^2$, $(\Delta y_i)^2$ to 1D associative array;
end-for;
retrieve all values from 1D associative array;
AEE = average of all $(SS)^{½}$ values;
CE = ((average of all Δx values)^2 + ((average of all Δy values)^2) ^{½};
$RMSE_x$ = (average all $(\Delta x_i)^2$ values)^{½};
$RMSE_y$ = (average all $(\Delta y_i)^2$ values)^{½};
$RMSE_r$ = (average all SS values)^{½};
if $RMSE_x = RMSE_y$
$Accuracy_r = 1.7308 * RMSE_r$;
else
$Accuracy_r = 1.22385 * (RMSE_x + RMSE_y)$;
end-if;
$RMSE = Accuracy_r$;
return final output AEE, $RMSE$, CE;
end

4 Results

The results of the comparative field tests for various GPS-enabled devices are summarized in the following tables. The units for all three measures of accuracy (*AEE*, *RMSE*, and *CE*) are in decimal degrees. The focus of this field study is to compare the accuracy and precision of different devices (in two different regions in Middle Tennessee) based on *AEE*, *RMSE*, and *CE*. However, accuracy and precision based on physical distance can be easily deduced based on our data, since the changes in latitude and longitude are very small.

Table 1. Average of Rankings by Root Mean Square Error (*RMSE*), Central Error (*CE*), and Average Euclidean Error (*AEE*) for the **Land Between the Lakes** (LBL) area

Device	RMSE	Rank	CE	Rank	AEE	Rank	Avg. Rank
Apple iPhone 4	1.43198	10.0	1.42339	10.0	1.42767	10.0	10.0
Apple Ipad	1.43013	1.0	1.42188	1.0	1.42588	1.0	2.0
Asus Transformer	1.43016	3.0	1.42191	3.0	1.42591	3.0	4.0
Garmin 60CSX	1.43018	5.0	1.42192	5.0	1.42593	5.0	6.0
HTC Thunderbolt	1.43906	12.0	1.43027	12.0	1.43421	11.0	12.3
Motorola Bionic	1.43019	6.0	1.42194	6.0	1.42594	6.0	7.0
Motorola X2	1.43037	9.0	1.42208	9.0	1.42610	9.0	10.0
Motorola Xoom	1.43022	8.0	1.42197	8.0	1.42597	8.0	9.0
PC Dongle	1.43016	2.0	1.42191	3.0	1.42591	3.0	3.7
Samsung Galaxy Note	1.43016	4.0	1.42191	3.0	1.42591	3.0	4.3
Samsung Nexus	1.43881	11.0	1.43018	11.0	1.43425	12.0	12.0
Samsung S3	1.43021	7.0	1.42195	7.0	1.42595	7.0	8.0

Table 2. Average of Rankings by Root Mean Square Error (*RMSE*), Central Error (*CE*), and Average Euclidean Error (*AEE*) for the **Downtown Nashville** (Nashville) area

Device	RMSE	Rank	CE	Rank	AEE	Rank	Avg. Rank
Apple iPhone 4	0.00106	7.0	0.00033	7.0	0.00057	7.0	7.0
Apple Ipad	0.02374	12.0	0.00621	12.0	0.00640	12.0	12.0
Asus Transformer	0.00599	10.0	0.00075	9.5	0.00438	11.0	10.2
Garmin 60CSX	0.00835	11.0	0.00241	11.0	0.00293	10.0	10.7
HTC Thunderbolt	0.00047	2.0	0.00011	1.0	0.00038	3.0	2.0
Motorola Bionic	0.00050	3.0	0.00019	5.0	0.00040	4.5	4.2
Motorola X2	0.00211	8.0	0.00048	8.0	0.00083	8.0	8.0
Motorola Xoom	0.00068	5.0	0.00016	4.0	0.00040	4.5	4.5
PC Dongle	0.00088	6.0	0.00023	6.0	0.00052	6.0	6.0
Samsung Galaxy Note	0.00053	4.0	0.00014	2.0	0.00037	2.0	2.7
Samsung Nexus	0.00271	9.0	0.00075	9.5	0.00130	9.0	9.2
Samsung S3	0.00038	1.0	0.00015	3.0	0.00030	1.0	1.7

By inspecting the results of this preliminary field study, a few basic observations were made:

- Physical Distance
  - Spatial accuracy and precision based on physical distance can be easily deduced based on field data collected.
- *AEE*, *RMSE*, and *CE* comparison between devices
  - The spatial accuracy and precision (based on *AEE*, *RMSE*, and *CE*) are not the same for different devices. Based on the ranking and the values for *AEE*, *RMSE* and *CE*, it is shown that the differences in values can be significant. The ranking analysis and physical distance analysis could be an interesting future research topic.
  - The rankings based on *AEE*, *RMSE*, and *CE* for different devices are not the same, although similar. Further analysis using statistical models would be of interest.
- *AEE*, *RMSE*, and *CE* comparison between the two geographical areas
  - The values of *AEE*, *RMSE*, and *CE* for the Land Between the Lakes (LBL) area are significantly different from those for the Downtown Nashville (Nashville) area. The difference may be due to the fact that there was no cell phone reception in LBL area. Further investigation of the actual cause of this difference is warranted.
  - The values of *AEE*, *RMSE*, and *CE* in LBL area are uniformly larger than values in the Nashville area, meaning the devices are less accurate and less precise in LBL area based on our definitions of accuracy and precision.
  - The ranking of the different devices for these two areas are not the same. For example, Apple Ipad's ranking was 1 in LBL area for all three categories (*AEE*, *RMSE* and *CE*), but its ranking was 12 in the Nashville area for all

three categories. Once again, it is an interesting discovery, deserving follow-up investigations.

- o The average rankings of the devices between LBL area and Nashville area are not the same.

5 Conclusion

In this paper, spatial accuracy and precision for GPS-enabled mobile devices were defined based on average euclidean error (*AEE*), root mean square error (*RMSE*), and central error (*CE*). Field data of spatial coordinates (latitude and longitude) returned by several GPS-enabled mobile devices were collected within two different regions (LBL area, which has no cell phone reception, and Nashville) are in Middle Tennessee. The values of *AEE*, *RMSE* and *CE* of these devices in these two areas were ranked and compared. The website with an online Spatial Accuracy Calculator was developed to enable the research team and the research community to calculate *AEE*, *RMSE* and *CE* values for future field studies. The algorithm of the Spatial Accuracy Calculator was presented in this paper. From this preliminary field study, it was observed that the *AEE*, *RMSE* and *CE* values were not the same for different devices. The values were also significantly different between the two geographical regions. The ranking of the devices were not the same in the two geographical regions, meaning a device performing well comparatively in one area may not do well in the other. These preliminary observations also enabled the research team to identify future research topics, such as analysis of data and ranking based on accuracy and precision in physical distance, and analysis and comparison of ranking of devices between different geographical areas.

Acknowledgments. This work was supported by DHS/Southeast Region Research Initiative (SERRI) under funded project number 4000112222 (Disaster Mitigation & Recovery Kit). Field data of spatial coordinates were collected by Maurice Testa, Patrick Robbins, Eric Whitaker, and James Martin, students in the Department of Geosciences, Austin Peay State University (APSU), Tennessee under the supervision of Michael J. Wilson, Director of the Geographic Information Systems Center at APSU. The figure used in this paper was created by Tabitha S. Y. Lee, student of Clarksville Academy, Tennessee. The online Spatial Accuracy Calculator was programmed by Brian N. Rivers, student in the Department of Computer Science, APSU.

References

1. GPS and Mobile Handsets. LBS Research Series (2010)
2. JCGM 200: 2008 International vocabulary of metrology — Basic and general concepts and associated terms (VIM) (2008)
3. Taylor, J.R.: The Grid: An Introduction to Error Analysis: The Study of Uncertainties in Physical Measurements. University Science Books (1999)

4. Zandbergen, P.A.: Positional Accuracy of Spatial Data: Non-Normal Distributions and a Critique of the National Standard for Spatial Data Accuracy. Transactions in GIS 12(1), 103–130 (2008)
5. Dean, R.B., Dixon, W.J.: Simplified Statistics for Small Numbers of Observations. Analytical Chemistry 23(4), 636–638 (1951)
6. Spatial Accuracy and Precision, `http://www.leeleong.com/spatial/`
7. Geospatial Positioning Accuracy Standards Part 3: National Standard for Spatial Data Accuracy. Subcommittee for Base Cartographic Data, Federal Geographic Data Committee, FGDC-STD-007.3-1998 (1998)
8. Li, X.R., Zhao, Z.: Measures of Performance for Evaluation of Estimators and Filters. In: Proc. SPIE 4473, Signal and Data Processing of Small Targets 2001, 530 (2001)

Spatial Multi-resolution Terrain Rendering Based on the Improved Strategy for Terrain Dispatching and Pre-reading

Haibo Wang, Lin Zhang, Jingeng Mai, and Fei Tao

School of Automation Science and Electrical Engineering, Beihang University, Xueyuan Road No.37, Haidian District, Beijing, China.100191
wangyu081189@126.com

Abstract. In this paper, we propose an improved strategy of terrain dispatching and pre-reading during the process of spatial multi-resolution terrain rendering. The improved strategy is adopted to render the 3Dterrain. The fundamental step of the procedure is the terrain data processing including three parts of terrain data source, terrain data block and terrain texture generation. The ultimate goal is to render the terrain successfully via the proposed strategy of terrain dispatching and pre-reading. The displayed rendering effect demonstrates that the improved strategy is feasible and achievable.

Keywords: improved strategy, spatial multi-resolution, terrain data processing, terrain rendering.

1 Introduction

Terrain rendering has significant applications both in the aspects of military and civil, it is also an important branch in computer graphics and widely applied in the area of Geographic Information System (GIS), Virtual Reality (VR), Synthetic Natural Environment (SNE), flight simulators, and even in the Vehicle Terrain Measurement System(VTMS) andvisualization of meteorological data and pollution data, etc.[1,2,3,4]. Many research scholars from home and abroad invest a great deal of concern on the terrain rendering.

Timothy Butler presented three approaches to terrain rendering in 1991, but he only made a simple comparison of the three methods which had little relation to a specific application environment [5].Wang J.J, He X.Het al. compared several methods of modeling and deeply studied the octrees deposing method to realize the visualization of 3D model, but it only elaborated the theory and lacked the factual experimental effects [6].Ricardo Olanda et al proposed a novel wavelet-tiled pyramid to realize the compression of terrain data, but it also lacked further applications in 3D terrain visualizations [7]. Wu L.D et al presented that parallel rendering was an important fast visualization method of large-scale 3D scalar field, but she also only elaborated the two classes of main methods: surface rendering and direct volume rendering with their respective development trends [8], and lacked experimental results. MarkoKuder,

F. Bian and Y. Xie (Eds.): GRMSE 2014, CCIS 482, pp. 225–234, 2015.

MarjanŠterk andBorutŽalik presented a new method for high quality rendering of large LiDAR based on terrain data which mainly used render-to-texture to generate color textures, but this method required more detailed data parameters such as coordinates, normals,colors and or any other LiDAR data fields which would increase the complexity and difficulty of data processing as well as waste time and memory resources [9]. PorwalSudhir presented a uniform scheme for efficient quad-tree based on LOD (Level-of-Detail) to meet the requirement of fast frame rate rendering, but its restricted applicationledto this scheme demanding further research [10]. Wei Xiong et al adopted the method of ROAM (Real-time Optimally Adapting Meshes) which used regular meshes to render by CPU and GPU. Although GPU could implement the algorithm effectively, it also brought thc limit ofrendering velocity between CPU and GPU as well as the memory overload [11,12,13,14].

In this paper, we propose an improved strategy of terrain dispatching and pre-reading during the process of terrain rendering using the Semi-CLOD (Semi Continuous Level of Detail) algorithm. The proposed strategy could be widely applied into environmental protection, industry, meteorology and other 3D visualization fields. The terrain data should be processed according to the given steps. The corresponding textures are generated according to their characteristics and properties by the specific formula. The proposed strategy is used to carry out terrain rending, and thedisplayed rendering effect demonstratesthat the proposed strategy is feasible and achievable.

The rest of the paper is organized as follows. The Terrain Data Processing including their source, block and texture generation is described in Section 2, and the terrain rendering is shown in Section 3. In Section 4, the rendering effect is displayed. We conclude in Section 5.

2 Terrain Data Processing

2.1 Terrain Data Source

Terrain data of the constructed system mainly consists of SRTM (Shuttle Radar Topography Mission) terrain data, including the topographic data from 60 degrees north latitude to 60 degrees south latitude which covers more than 80 percent of Earth's land surface and the resolution of whichis 90 meters [15]. SRTM terrain data has divided the global terrain into 72×24 terrain blocks according to 5×5 degrees of latitude and longitude when it is stored. Meanwhile, each block includes the number of 6000×6000 evaluation data points when it is downloaded in the form of tiff.

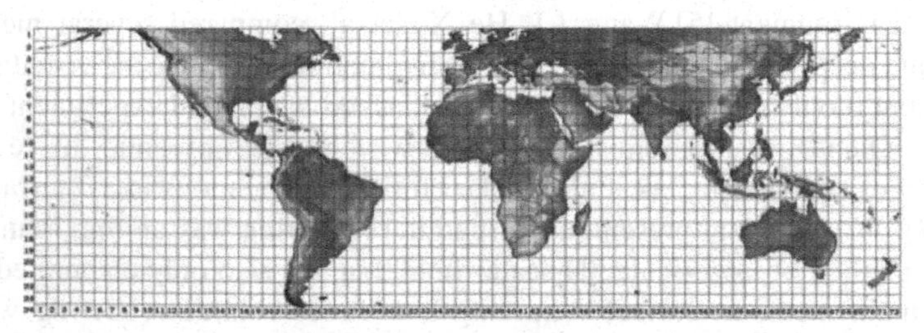

Fig. 1. SRTM data zoning schematic

2.2 Terrain Data Block

Based on the divided SRTM terrain blocks, we constructa multi-resolution terrain data model for each sub-block.The constructed model is divided into 15 levelsconsidering the resolution levels of satellite images which taken as the texture. And the resolution varies from low to high as the level changing from 1 to15. In level 1 to 4, it is unnecessary to add the high-level terrain information when displaying, we could take the world-map as the display navigation map. Whereas, we should carry out sampling step by step forlevel 8 to 15, and the level 15 ought to use the original data. The specific steps are as follows:

1. The level 5 to 7 use the same vertex data as level 8;
2. The level 9 and its above levels make interlaced sampling interval every line and column, the specific effect is shown in Fig.2

Through the above steps, we could fulfill the terrain blocks further divided forming the quad-tree terrain blocks divisions. Similarly, terrain textures also need to be cut to achieve texture mapping.Satellite images of the same terrain blocksshould be corresponding to each other.The number of vertex and sub blocks of each terrain block on the different levels are shown in Table1.

Table 1. Vertex and sub blocks of each terrain block

Level	Total Vertex	Sub Blocks
5	47*47	1*1
6	47*47	1*1
7	47*47	1*1
8	47*47	1*1
9	94*94	2*2
10	188*188	4*4
11	375*375	8*8
12	750*750	16*16
13	1500*1500	32*32
14	3000*3000	64*64
15	6000*6000	128*128

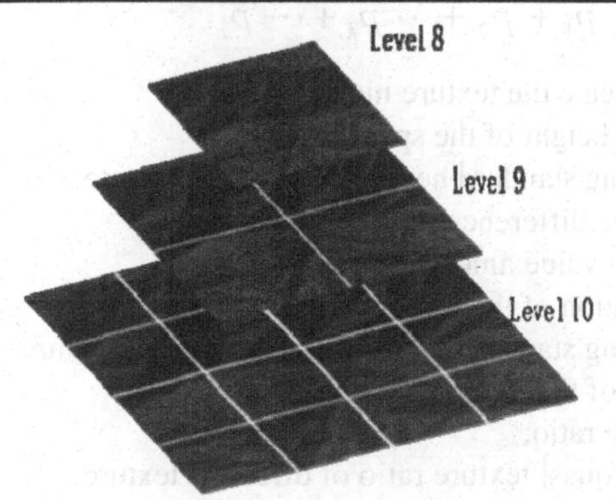

Fig. 2. The effect of quad-tree division

2.3 Terrain Texture Generation

As mentioned above, the terrain textures used in this paper are obtained via two ways: using the satellite images as textures or through texture material to blend and render the terrain texture. This section will focus on the latter one. According to the terrain data obtained by the DEM data, we could utilize the texture material and combine the terrain height with calculated inclination to generate the hybrid texture. The hybrid texture could not only show the changes of terrain height, but also make terrain more realistic.

Texture Material Selection. According to different altitude and topography, natural landscape features generally include water, soil, grass, bushes, desert and rocks, etc. Thus,we choose six kinds of textures to render. The textures are respectively grass (or water), dark rock, bushes, soil, grove and light-colored rocksbased on the actual terrain trend from low to high, corresponding to the numbers 1-6

The concrete method could be divided into 4 steps. Firstly, texture hybrid ratio at eachpoint is respectively calculated according to terrain height and inclination. Secondly, the texture hybrid color values are calculated based on the texture hybrid ratio. Then, the texture ratio of each material could be obtained, using the R, G and B value of the corresponding point to multiply each texture ratio to get the rendered R, G and Bvalue of the texture color. Finally, the final texture will cover to the surface of the 3D terrain model to achieve the texture rendering of the corresponding terrain.

Texture Hybrid Ratio Calculation. The determined texture material needs to be processed to generate the same resolution to facilitate the texture hybrid of the corresponding pixel. Furthermore, the texture hybrid ratio of each point also needs to be calculated separately so that we can further get the blended texture color values. The used formulas are as follows[16]:

$$h_k^{''} = (h - h_k^{'})^2 \quad (1)$$

$$p_k^{'} = (2h_{\min}^{''} - h_k^{''})(1 + | g - g_k^{'} |) \left(g_k^{'} = \begin{cases} 0 & (k: even) \\ g_{\max} & (k: odd) \end{cases} \right) \quad (2)$$

$$p_k = \frac{p_k^{'}}{p_1^{'} + p_2^{'} + \cdots p_k^{'} + \cdots p_n^{'}} \quad (3)$$

k:The subscript indicate the texture number;
h:The actual terrain height of the specific point;
$h_k^{'}$:The corresponding standard height of different texture;
$h_k^{''}$:The square of the difference between h and $h_k^{'}$;
$h_{min}^{''}$: The minimum value among all the $h_k^{''}$;
g:The actual inclination of the terrain at a specific point;
$g_k^{'}$:The corresponding standard inclination of different texture;
g_{max}:The maximum of the inclination among all the $g_k^{'}$;
$p_k^{'}$:The quasi texture ratio;
$p_i^{'}$,(i=1,2,...,n):The quasi texture ratio of different texture;
p_k:The texture raio.

When the calculated texture ratiop_k< 0, the texture ratio of this material is 0 at the-point.

Subscript$k$indicates the texture number, corresponding to the 1-6mentioned above.In the formula (1), $h$indicates the actual terrain height of the point,h_k' indicates the specific texture standard height, and h_{max}indicates the terrain height difference between the maximum and minimum value.After the actual rendering practice, we find that it could achieve a good effect when wetake 1/1000$h_{max}$for grass or water, 1/20$h_{max}$for dark rock, 1/10$h_{max}$for bush, 1/5 $h_{max}$for soil, 1/4$h_{max}$for Grove and 4/5h_{max} light-colored rock as the standard height.Whereas, when the actual altitude is less than 0, we adopt water texture for texture rendering.

In the formula (2),h_{min}' is the minimum one among all theh_k' ,the texture ratio$p_k$in the formula (3) is corresponding to the texture ratiop_k' in the formula (2), andp_k is the prospective texture ratio. As can be seen from formula (1) and (2), the smaller h_k' is,the greater the p_k' becomes, subsequently, the texture ratio $p_k$become greater and the texture will get the maximum texture ratio corresponded by the minimum value amongh_k' . When h_k' is less than 2h_{min}' , we could get the results that are h_{min}' -h_k'<0 andp_k<0.The results indicate that the actual ground height$h$is far from the standard height h_k' of the texture, the texture ration of such material is 0 at the point.$g$indicates the terrain actual inclination of the specific point, g_k' indicatesthe corresponding standard inclination of various textures, and g_{max}indicates the maximum inclination of all points. Actually, the vegetation is generally grown in relatively flat areas, while the steep places are generally bare rock or soil. Thus, when the texture number$k$is odd, the standard inclinationg_k' of the corresponding grass, bushes, grove are$g_k' = g_{max}$,meaning the greater the inclination$g$is, the smaller the difference |g-g_k' |becomes. Meanwhile, the pseudo-texture ratio and the texture ratio also become smaller. On the contrary, when the texture number$k$iseven, the standard inclinationg_k' of the corresponding dark rock,soil and light-colored rock areg_k' =0, meaning the greater the inclination$g$of the specific point, the greaterthe texture ratio becomes.

As can be seen from the formula (2), the ground height dominates among all the impact of various textureratiosp_k, meanwhile,the inclination also has an effect on the texture ratio.

3 Terrain Rendering

In order to guarantee a smooth drawing, it is essential toaccelerate the rendering speedof each frame and make a reasonable dispatching for satellite images and terrain textures [17, 18, 19, 20, 21]. In addition, it is also necessary to pre-read the terrain blocks to realize the terrain's pre-loading, which could facilitate to accelerate the reading speed of terrain blocks from external memory.

3.1 Terrain Dispatching Strategy

The resolution level of blocked terrain data matches with the satellite images which taken as textures that also divided into 15 levels. And its resolution varies from low to high according to level 1-15. Considering thatuser first needs to select the observation area, the constructed terrain model is built to support multiple height levels to observe. When the observer moves, the terrain blocks which need to be rendered change with the changing of height, longitude and latitude. And the terrain blocks that needed to be rendered are determined by the coordinates and height of the observation point when rendering [22]. Specifically, it is critical to determine the key terrain called-KeyTerrainof the observer's location and calculate the level*X* according to the observation height.

When the level*X*is less than 8(*X*<*8*) as shown in Figure3, the texture level *X*of-KeyTerrainis consistent with its adjacent eight terrain blocks. Similarly, the level of-terrain data and texture are descending in order.

When the level *X*is greater than 8(*X*>*8*)as shown in Figure 4, the concrete steps are as follows:

Step 1: Determine the position of the KeyTerrain, and calculate its level *X*;

Step2: Get the eight terrain blocks which are adjacent to theKeyTerrainand guarantee the level of texture and terrain blocks are consistent with KeyTerrain;

Step 3: Make the level of the three terrain sub-blocks included bynext level terrain blocks be consistent with the level of the eight blocks mentioned above;

Step 4: It could form the parent block composed of each four sub-blocks, and the level of the terrain block adjacent to the parent block will lower a level;

Step 5: Repeat the steps 3 and 4 of the above steps until the level of terrain block is consistent with the level of the overall texture called world-map.

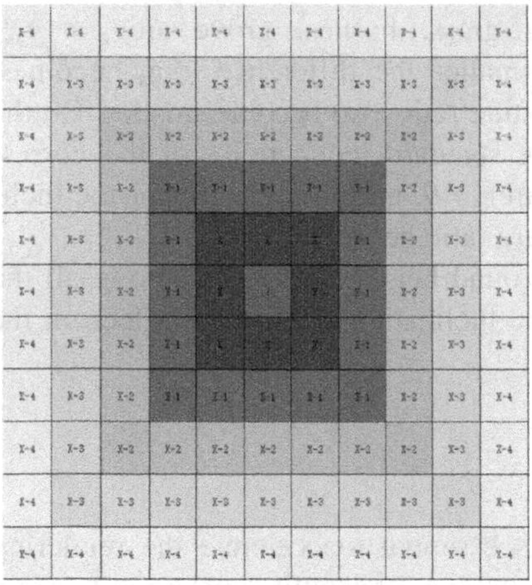

Fig. 3. The level of KeyTerrain less than 8

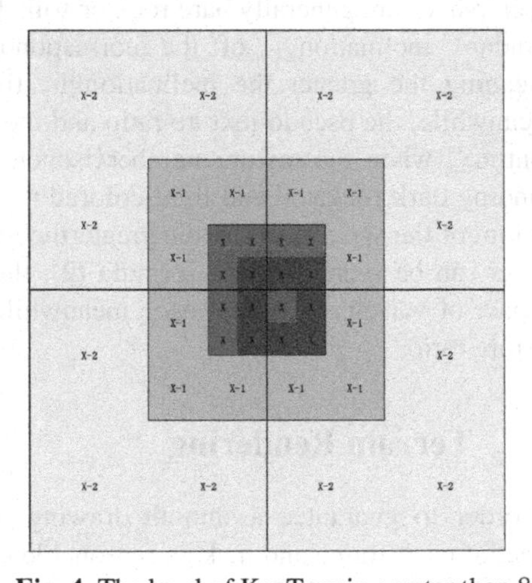

Fig. 4. The level of KeyTerrain greater than 8

3.2 Terrain Pre-reading Strategy

Actually, it is essential to render every frame of the terrain blocks depending on the current position when roaming in the scence. If every terrain block is re-read from the external memory, it will cost a lot of time and make the roaming become choppy. As it will take some time to load the terrain block into buffer,the terrain needs to be pre-read to facilitate the rendering process. The pre-reading of terrain use the strategy described in previous section. The key point is to compare the level of current block-which taken as the KeyTerrain with the level of the former frame. If the two values are same, it indicates that the blocks which are being rendering are unnecessary to update. Then the terrain pre-reading could begin. We must take the two factors of terrain blocks level and key terrain (KeyTerrain) number into account when pre-reading. As long as the terrain blocks level or the key terrain (KeyTerrain) number changes, the pre-reading should be carried out. The concrete process is as the following procedure in table 2.

Table 2. The procedure terrain pre-reading

```
if(KeyTerrain level = = the former frame level)
        {
// Unnecessary to update the terrain block rendering;
        //Carry out pre-reading strategy(According to the terrain dispatching strategy):
                        if( (the KeyTerrain's number unchange and level
                        changes) ||
                (theKeyTerrain's number changes and level unchanges))
                        {
// Carry out rendering;
if( the terrain doesn't exist in the terrainlist)
create one and pre-load it;
 }
else
No operations;
 }
```

3.3 Terrain Memory Management

Two lists are constructed to store the rendered terrain to realize the management of the read terrain data.One is TerrainList mentioned above which will be put into cache to render, the other one is Terrian Candidate List. The Terrain Candidate Listis a candidate list which stores terrain data and textures that could also store the uninstalled or pre-read terrain blocks. In addition, another list Terrain infor List is also constructed to store calculated information of the current terrain blocks which are in need for rendering, meanwhile, it will also choose the existent terrain blocks from the Terrain List and Terrain Cadidat List depending on the TerrainInforList and create the nonexistentterrain blocks to construct a new TerrainList for the current rendering terrain blocks.

In practice, the Terrain infor List of the terrain blocks which are in need for rendering should be calculated firstly when rendering. As the TerrainInforList just records the information ofterrain blocks number and level, the real terrain data and texture are not stored in it. It is essential to traverse the TerrainInforList. After that, we should continue to process terrain blocks according to the following steps as shown in Fig.5.

Step 1:If the terrain blocksthemselves exist in TerrainList, they needn't make changes;

Step2:Put theterrain blocks into theTerrainListwhich exist in both Terrain Candidat List and Terrain Infor List.

Step 3: Make a new strip in TerrainList to store the terrain blocks which only exist in Terrain Infor List;

Step4: Put the terrain blocks into Terrain Candidat List which exist in Terrain List but do not exist in Terrain Infor List. When the terrain blocks in Terrain Candidat List exceeds the given threshold value, the first entered terrain blocks will be deleted.

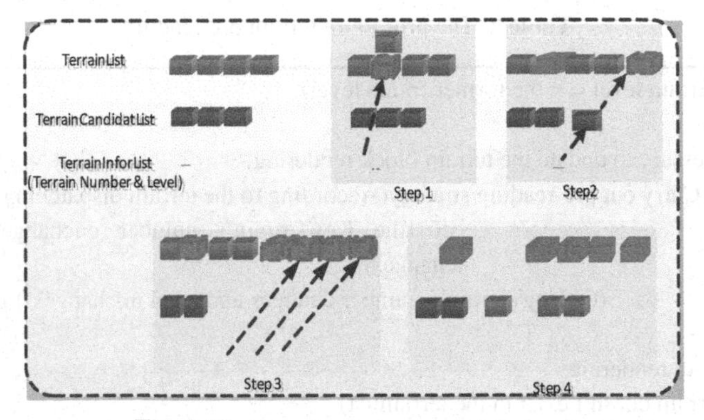

Fig. 5. The memory management steps

4 The Rendering Effect

Combine the specific terrain data and terrain texture, divide the terrain data into the proper format according to the method describing in the section 2.2 and generate the texture of different level via the formula in section 2.3, then render the 3D terrain on the basis of spatial multi-solution. When rendering, we adopt the improved strategy of terrain dispatching and pre-reading and use theSemi-CLODalgorithmby the techniques of DirectX and C++ based on the tool of the Visual Studio 2010. The concrete effects are shown as follows in Fig.6.

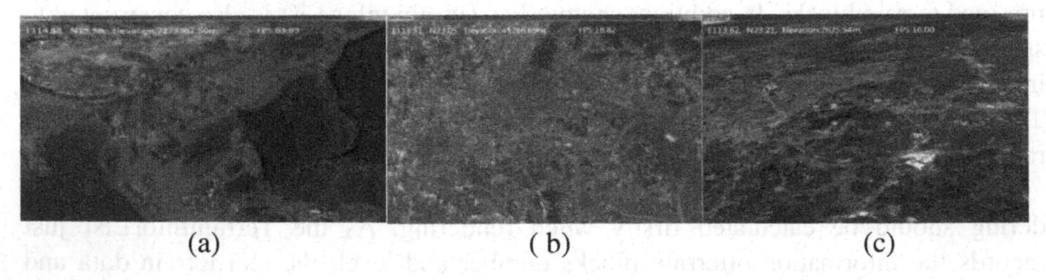

(a) (b) (c)

Fig. 6. The effect of terrain rendering

The above three figures respectively display the rendering speed by the FPS (Frame Per Second), the resolution from (a) to (b) in Fig.6 are 94×94, 375×375 and 750×750. As shown in Fig.6, the average FPS is 32.62 in the cases using the method in this paper. While the average FPS is 24.8 in the cases of different resolutions of 150×150 and 512×512 via the pyramid model in the literature [23]. Both of the two methods are realized on the same software environment of Visual Studio 2010 using C++ and almost the same hardware environment of core i3 processor with 4GB memory. Although the compared resolutions are different, the effect brought by the difference has no great impact on the average FPS. The compared result demonstrates that the average FPS is enhanced by 31.5%. The result further illustrates that the proposed strategy is effective and feasible as well as efficient.

5 Conclusions

In this paper, we propose an improved strategy of terrain dispatching and pre-reading during the process of spatial multi-resolution terrain rendering. The main idea of the method is summarized as follows:

1. Deal with the original terrain data including data blocks division based on the Quad-tree and texture generation according to the relevant formulate.
2. Carry out the terrain rendering based on the proposed strategy of dispatching and pre-reading.
3. Display the terrain rendering effect in 3D view and discuss the results.

The further work is to research combination of GPU and Semi-CLOD algorithm. In addition, it will also be an important point to make another research on the realized platform.

Acknowledgement. This work was financially supported by the National Key Technology Research and Development Program (No. 2011BAK16B03) in China.

References

1. Fan, M., Tang, M., Dong, J.: A Review of Real-time Terrain Rendering Techniques. In: The 8th International Conference on Computer Supported Cooperative Work in Design Proceedings (CSCWD 2004), Xiamen, China, vol. 1, pp. 685–691. IEEE (2004)
2. Wang, H., Mai, J., Song, Y., Wang, C., Zhang, L., Tao, F., Wang, Q.: A 3D visualization framework for real-time distribution and situation forecast of atmospheric chemical pollution. In: Tan, G., Yeo, G.K., Turner, S.J., Teo, Y.M. (eds.) AsiaSim 2013. CCIS, vol. 402, pp. 415–420. Springer, Heidelberg (2013)
3. Chung, H., North, C., Ferris, J.: Developing Large High-Resolution Display Visualizations of High-Fidelity Terrain Data. Journal of Computing and Information Science in Engineering 13, 1–7 (2013)
4. Liu, W., Wang, X., Li, N.: Modeling and Simulation of Synthetic Natural Environment. Journal of System Simulation 16, 2631–2635 (2004)

5. Timothy, B.: Three Approaches to Terrain Rendering. In: Proceedings of the IEEE 1991 National Aerospace and Electronics Conference (NAECON 1991), Dayton, OH, pp. 926–932. IEEE (1991)
6. Wang, J., He, X., Long, M.: Visualization of Disordered Space Data Based on Octrees Decomposing Method. Microcomputer Development 15, 155–160 (2005)
7. Ricardo, O., Mariano, P., et al.: Terrain Data Compression Using Wavelet-tiled Pyramids for Online 3D Terrain Visualization. International Journal of Geographical Information Science 28, 407–425 (2014)
8. Wu, L., Yu, R., Qu, S.: Survey on Parallel Rendering of Large-Scale 3D Scalar Field. Journal of System Simulation 24, 12–16 (2012)
9. Kuder, M., Sterk, M., Zalik, B.: Point-based Rendering Optimization with Textured Meshes for Fast LiDARVisualization. Computers & Geosciences 59, 181–190 (2013)
10. Porwal, S.: Quad Tree-based Level-of-Details Representation of Digital Globe. Defence Science Journal 63, 89–92 (2013)
11. Xiong, W., Wang, X., Zhu, M.: Study of LOD Terrain Rendering Algorithm Based on GPU. In: International Communication Conference on Wireless Mobile and Computing (CCWMC 2011), Shanghai, China, pp. 476–481. IET (2011)
12. Lee, I., Kang, K.-K., Lee, J.-W., et al.: Real-time Rendering for Massive Terrain Data using GPUs. In: 2012 International Conference on Cloud Computing and Social Networking (ICCCSN 2012), Bandung, West Java, Indonesian, pp. 1–3. IEEE (2012)
13. Shen, M., Wang, H.: A Large Scale Terrain Rendering Algorithm based on GPU. Information Technology 6, 176–179 (2013)
14. Liao, X., Wang, R., et al.: Digital Battlefield Visualization Technologies and Applications. Defense Industry Press, Beijing (2010)
15. Pattathal, V.A.: A Comparative Analysis of Different DEM Interpolation Methods. Geodesy and Cartography 39, 171–177 (2013)
16. Yang, G., Wang, M.: The Seamless Joining and Visual Checking of DEM. Journal of Institute of Survey and Mapping 20, 279–281 (2003)
17. Hugues, H.: Smooth View-Dependent Level-of-Detail Control and Its Application to Terrain Rendering. In: Proceedings of the IEEE Visualization 1998, Research Triangle Park, NC, USA, pp. 35–42. IEEE (1998)
18. Grabner, M.: Smooth High-quality Interactive Visualization. In: 17th Spring Conference on Computer Graphics (SCCG 2001), Budmerice, Slovakia, pp. 87–94. IEEE (2001)
19. Chen, B., Swan, J.E., Kuo, E., Kaufman, A.: LOD-Sprite Technique for Accelerated Terrain Rendering. In: Proceedings of the IEEE Visualization 1999, San Francisco, CA, USA, pp. 291–536. IEEE (1999)
20. Tang, S.: Wang, l., Hao, A.:A Method for Terrain Rendering without Seams Based on Image. In: 2007 10th International Conference on Computer-Aided Design and Computer Graphic (CADCG 2007), Beijing, China, pp. 481–484. IEEE (2007)
21. Valdetaro, A., Nunes, G., et al.: LOD Terrain Rendering by Local Parallel Processing on GPU. In: 2010 Brazilian Symposium on Games and Digital Entertainment (SBGAMES 2010), Florianopolis, Brazil, pp. 182–188. IEEE (2010)
22. Mu, L.: Research on Spatial Radiation Field Calculation and Synchronous Visualization Technology. Beijing (2012)
23. Huo, L., Yang, Y., et al.: Rearch and Practice of Tiles Pyramid Model Technology. Science of Surveying and Mapping 37, 144–146 (2012)

Research and Implementation of Smoke Diffusion Parallel Rendering Based on Memory Mapping and Billboard

Yuxun He, Jingeng Mai, Fei Tao, and Lin Zhang

School of Automation Science and Electrical Engineering,
Beihang University, Beijing 100191, China
tashaxing123@163.com

Abstract. Aiming at enhancing the rendering performance of complex natural phenomena such as smoke diffusion, a rendering approach based on Compute Unified Device Architecture (CUDA) and particle system is designed and implemented in this paper. The strategy that host memory resource is mapped to the CUDA memory address, and the technique which uses two-dimensional textured planar graph to simulate the three-dimensional effect combined with CUDA parallel computing (the parallel billboard technique) are integrated applied. The simulation results demonstrate that this proposed approach can effectively accelerate the rendering process, save memory usage, and achieve impressive visual effects.

Keywords: rendering, CUDA, particle, billboard.

1 Introduction

In the Computer Graphics Simulation area, the fog, rain, snow and fireworks are generally realized through particle system, but particle system requires the computer has high computing speed and large memoryin order to be simulated smoothly. The computing power of a single computer is often not qualified when dealing with the massive and complex particle system rendering like the smoke diffusion. However, with the rapid development of multi-core graphics processors (GPU), more and more researchers begin to use Compute Unified Device Architecture (CUDA) programming model and multi-stream processors, which is based on NVIDIA's GPU, to render the particle system parallel. As a result, the rendering performance has been improved and the burden of CPU has been reduced.

Recently, many researchers have used CUDA in the related field. For example, Chen etc.[1],using CUDA technique, achieves the fireworks simulation and significantly improve rendering frame rate in contrast to serial rendering method. Goldsworthy [2] also uses CUDA technique to optimize the rendering of gas diffusion, resulting in prominent rendering speedup. But these two methods are both based on an approach that successively copies the vertex data and index data of particle system from host memory to device memory, then copies the data back to host memory after CUDA

F. Bian and Y. Xie (Eds.): GRMSE 2014, CCIS 482, pp. 235–243, 2015.

computing. The copy process will consume a lot of time, therefore these methods remain to be optimized. Yilmazetc.[3] implements a CUDA-based rendering method of large-scale audience in a football match. In order to reduce the total copy time, the author writes the program in a new way, in which the program copies all the vertex data from host to device only once when the program begins to run, then copies back to host when the rendering process ends. To some extent, this strategy solves the problem of copy-time waste. However, because the thousands of spectators are displayed with three-dimensional mesh entities, resource consumption is too high, making the efficiency of the program declined.

This paper proposes an alternative approach, which is based on CUDA, to render the smoke diffusion parallel. Through memory mapping strategy, the whole program avoids data copy between host memory and device memory, making the CPU side program and CUDA kernel function program can access the same piece of memory, and effectively improves rendering efficiency. Meanwhile, the program attaches a new billboard technique combined with CUDA, which uses thousands of two-dimensional pictures to simulate three-dimensional effect, so as to reduce the resource that a single particle occupies. Moreover, the program computes billboard matrix parallel, further optimizes memory usage, without affecting the visual effect of the simulation.

2 Physical Model of Smoke Diffusion

There are many physical model that can describe the dynamic spread of smoke, such as the classic Euler model, Gaussian model, and Lagrangian model. However, the Gaussian model is only suitable for in-door or small-scale gas diffusion, Euler model is a mathematical description based on macroscopic and statistical method which is not suitable for the computer display of smoke diffusion. In contrast ,the Lagrangian model can be refined to the motion state of individual particles, suitable for the programming model adopted in this paper. Therefore, the Lagrangian dispersion model[4] with random perturbations is selected in this paper as the theoretical basis for the rendering of particle system

In this model, the smoke is considered composed of numerous smoke particles, each particle is emitted from a diffusion source, with quality, spatial coordinates, velocity vectors and other attributes. Under the wind field and disturbing field, the particles will move along its velocity vector, their velocity vectors and coordinates will change as the time pass by. As a result, the microscopic motion of particles constitute the macroscopic diffusion of smoke[5] as a whole. For a single particle, the spatial coordinates at a certain time can be represented by the following formulas:

$$X_{k+1} = X_k + V_x \Delta t + X_{rand} \tag{1}$$

$$Y_{k+1} = Y_k + V_y \Delta t + Y_{rand} \tag{2}$$

$$Z_{k+1} = Z_k + V_z \Delta t + Z_{rand} \tag{3}$$

Wherein X_k and X_{k+1} are the coordinates in the X direction before and after a certain moment,Δt is the time interval,V_x is the velocity in the X direction at time

K,X_{rand} is the disturbing level in X direction at time K. The other two directions' parameters can be described similarly. In X direction, for example, disturbance can be calculated according to the following formula:

$$X_{rand} = N_{rand}\sqrt{2M_k \Delta t} \tag{4}$$

Wherein the random number N_{rand} is generated by standard Mersenne Twister Generator[6],M_k is a disturbance coefficient in X direction (different val-ues in different directions).

In real diffusion phenomenon, the air humidity and the sedimentation of smoke must be considered. Since this article focuses only on realization of computer renderings,for the convenience of programming, these additional parameters are not taken into account. Comprehensively speaking, this diffusion model is discrete in space, sequential in time, and the wind field takes the dominate role. After simplified in this article, this model become convenient for C++ programming.

3 CUDA Parallel Rendering for Smoke Particle System

3.1 CUDA Programming Model

NVIDA's GPU-based CUDA technique [7],can reach good performance of multi-core GPU with parallel computing technique, simplify programming, and optimize logical management of CPU. In the CUDA programming model, a large-scale computational problem with huge number of data can be assigned to a thread block, and each thread block can be divided into multiple threads, each thread bear a separate calculation process. Multiple threads do the parallel computing together, and finally complete the large-scale computing task. It is the kernel function that perform calculation in each GPU thread[8], therefore, a complete[5] CUDA program contains serial processing part in the GPU plus parallel processing part in the GPU, which means the kernel functions.

CUDA is perfect for those computing tasks with large amount of data and can be divided into blocks for parallel programming. The smoke particle system to be achieved in this article just meet this feature: every particle is independent from other particles, each particle's generation, rendering, update and die all correspond to one single thread. So, here the entire process uses CUDA to accelerate the rendering of particle system, in order to achieve better optimization results.

3.2 Parallel Programming Design for Smoke Particle System Rendering

Particle System Design. The concept of particle system was proposed by Reeves[5]in 1983. He thought that the moving and blurred objects can be seen as a collection of flowing particles which has limited number and specific attributes[9],suitable for simulation of complex natural phenomena such as fireworks[10],snow, rain and gas. In practical, the specific type of a particle is determined by its properties like the quality, speed, color, shape and so on. In general, the particle system simulation requires four steps:

- Particle generation
- Particle update
- Particle rendering
- Particle die

The traditional process of programming performs these four steps on CPU, but it will result in heavy burden for CPU, which is not conducive for the program to run on some low-configured devices, and cannot reach smooth rendering effect. As the particle system is easy to be computed parallel, it is considered here to carry these four steps on GPU(realize with CUDA programming, shown in **Fig.1**).Each thread is responsible for one particle's generation, update, rendering and die, all the threads finally achieve large-scale particle system's simulation, and effectively enhance the computing performance.

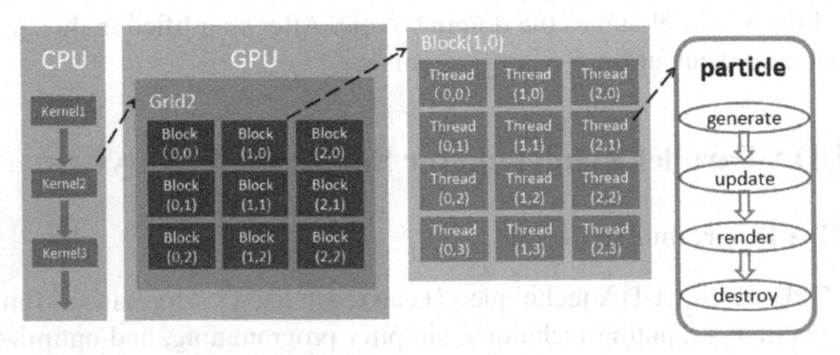

Fig. 1. CUDA programming model combined with particle system rendering procedure

Program Procedure. In order to test the practical effect of CUDA-based parallel rendering, using previously established smoke diffusion model, the C++ program is written to simulate the diffusion of smoke. The graphics programming interface[11] used in this paper is directx3D 9.0c.

The particle system here is constructed, with particle properties specified. The data structures defined in C++ code are as follows:

```
//Piece particle definition
struct Particle
{
floatx,y,z;          //x,y,z position
floatvx,vy,vz;       //x,y,z direction speed
floatxrand;        //x disturb
floatyrand;          //y disturb
floatzrand;        //z disturb
};
```

Each particle has the follow properties: three-dimensional space coordinates x,y,z (the basic unit is pixel),the moving speed in three directions vx, vy, vz, and random perturbation components in three directions, xr and, yr and, zr and. These properties will update[12] constantly in the rendering process. For practical display of individual particles, a rectangular patch affixed with ball-type texture plus alpha channel is used(due only to test the efficiency of the parallel rendering, there is no realistic texture and motion blur).

CUDA-based simulation of smoke diffusion's main process is as follows:

(1) **The initialization.** It includes Directx3D initialization and CUDA device initialization. In the meantime, particles' vertex buffer in the memory will be attached to the rendering pipeline.

(2) **Memorymapping.** Traditional CUDA parallel computing programs need to call the cudaMemcpy function to copy data between host and device memory, but this mechanism is only applicable to large-scale computing programs in which data copy takes only a small portion of the total computing time. This mechanism does not work for the particle system rendering in this paper, because the rendering process requires frequent updating operation on the vertex buffer,which will accumulate the data copy time, thus slowing down the program's running speed. In this paper, the memory mapping method is applied to map Direct3D resources to CUDA address, so that the data can be directly read and written in both Directx3D and CUDA, and data copy process are eliminated. Before using this memory mapping mechanism, the cudaD3D9RegisterResource function [7] must be called to register Direct3D vertex bufferresource, and then mapped it to CUDA. At the end of the program, the cudaD3D9UnregisterResource function is called to deregister the resource. After registering resources to CUDA, function cudaD3D9Map Resources and cudaD3D9UnmapResources can be called to map and unmap.

(3)**Constructing kernel function.** The program needs to allocate the CUDA programming grids and blocks, each particle corresponding to one thread. Then the kernel functions for updating of individual particles should be called, and finally the cudaD3DResourceGetMappedPointer function will be called to access the CPU address in order to invoke CUDA kernel function for parallel rendering.

(4) **Rendering particle system.** After updating the particles in each frame, the particle system is combined with billboard technique to finish the whole rendering.

This improved parallel rendering program's procedure is shown in Fig.2.c,while the Fig.2.a shows the serial rendering procedure and Fig.2.b shows the procedure of the old

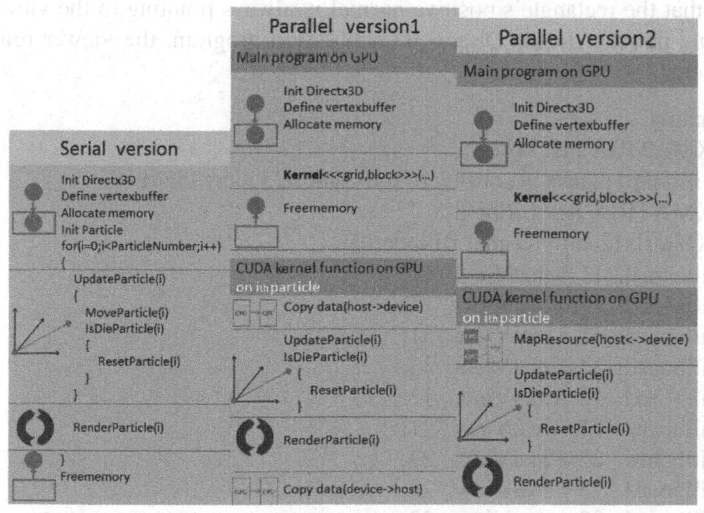

a Serial b Parallel 1 c Parallel 2

Fig. 2. Program flow comparison

parallel rendering method which needs data copy. Some pseudo-codes are also displayed as follows in the pictures.

3.3 Parallel Billboard Technique

Usually,a single particle is displayed with a spherical mesh[13](or any other irregular mesh) which has specific materials that can reflect specific lights, as in Figure3,but this method costs a large memory. For instance,a spherical mesh with 25×25 segments, has a number of 1250(25×25×2=1250) basic triangle primitives. In the practical rendering process, there are thousands of such particles, as a consequence,a lot of buffer space will be occupied. This paper uses a technique called billboard[14] combined with CUDA parallel programming to optimize the rendering process, shown in Fig.3.For a single particle, a rectangular patch with two basic triangle primitives labeled with the corresponding three-dimensional texture[15] is used to simulate three-dimensional effects, eventually saves the vertex buffer space dramatically.

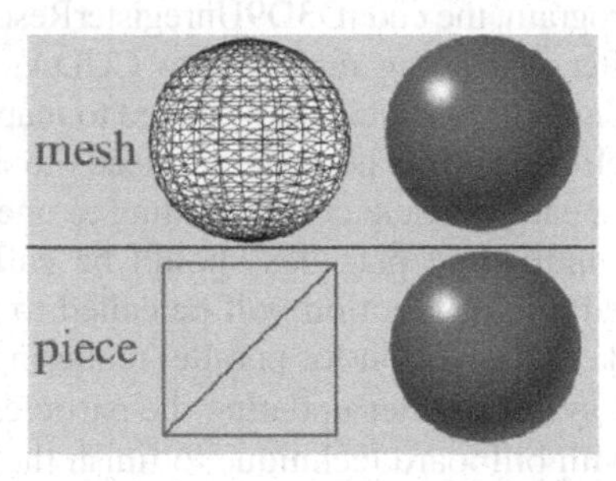

Fig. 3. Mesh compared to textured picture(the color is used to distinguish mesh and piece)

The principle of billboard is: use two triangle primitives[16] to form a rectangular picture, so that the rectangle's positive normal is always pointing to the viewer during the rendering process. In the Directx3D simulation program, the viewer refers to the virtual camera.

```
//set matrix
D3DXMATRIX mat World;
//set billboard matrix according to the camera's view matrix
D3DXMATRIX mat Billboard;
D3DXMatrixIdentity(& mat Billboard);
matBillboard._11 = matView._11;
matBillboard._13 = matView._13;
matBillboard._31 = matView._31;
matBillboard._33 = matView._33;
matBillboard._12 = matView._12;
matBillboard._21 = matView._21;
matBillboard._22 = matView._22;
matBillboard._23 = matView._23;
matBillboard._32 = matView._32;
D3DXMatrixInverse(& mat Billboard, NULL, & mat Billboard);
```

As the code shows above, every time before a single rectangular particle's rendering, the particle's world matrix is adjusted according to the virtual camera's view matrix, so that the rectangular particle's yaw angle, pitch angle, flip angle[17] on X,Y,Z directions can change as the camera's angle and position change, which is shown in **Fig.4**.

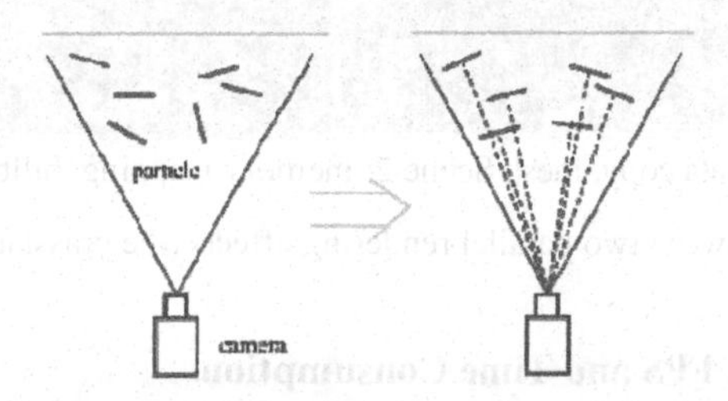

Fig. 4. Billboard principle(only yaw angle is displayed)

Here the process of thousands of billboard matrixes calculation is encapsulated into kernel functions which is more efficient than traditional serial billboard calculation. This process can be completed by CUDA parallel programming, resulting in great rendering speed improvement.

4 Simulation Results

4.1 Rendering Scene

According to the smoke diffusion model(depicted above), the CUDA programming concept and the billboard technique combined with particle's movement attributes, a C++ test project is constructed in this paper to realize the whole rendering process. All the simulation configurations are as follows:

- CPU: Intel X5650, 2.67GHz
- RAM: 8G
- GPU: NVIDIA Quadro2000
- Operating system:windows7-32bit
- IDE: VS2010
- Graphic API: Directx3D 9.0c
- CUDA version:v5.5

This paper simulates two scenarios both based on CUDA parallel rendering. As can be seen from the pictures (**Fig.5**) these two methods can almost achieve the same visual effect. However, the method which adopts the memory mapping and billboard has a relatively higher FPS(frame per second) value than the method which uses data copy and mesh entities.

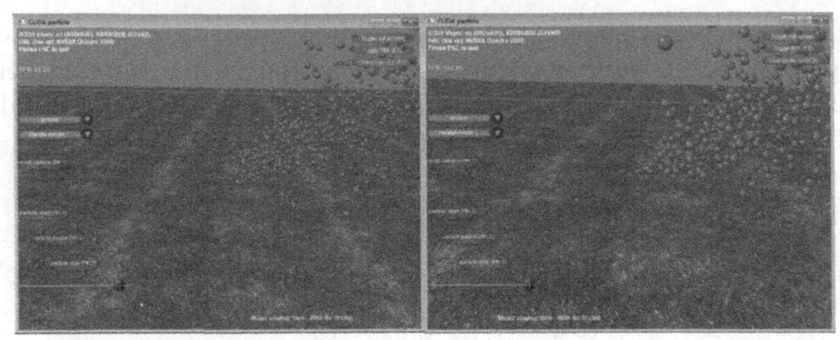

Scene 1: data copy, mesh Scene 2: memory mapping, billboard

Fig. 5. Comparison between two parallel rendering effects (the grassland is just decoration)

4.2 Comparison of FPS and Time Consumption

Table1 shows that the FPS of the method adopted in this paper is averagely one times larger than the old way, so this method can optimize the rendering efficiency greatly and make the rendering scene smoother.

Table 1. FPS values of two methods

Mode \ Number	5000	10000	30000	100000	700000
Parallel1(copy)	373.5	351.4	238.8	178.3	59.2
Parallel2(map)	1156.0	1087.3	719.1	302.2	112.1

Table 2. Time consumption of the two methods (Particle number:30000, Unit:ms)

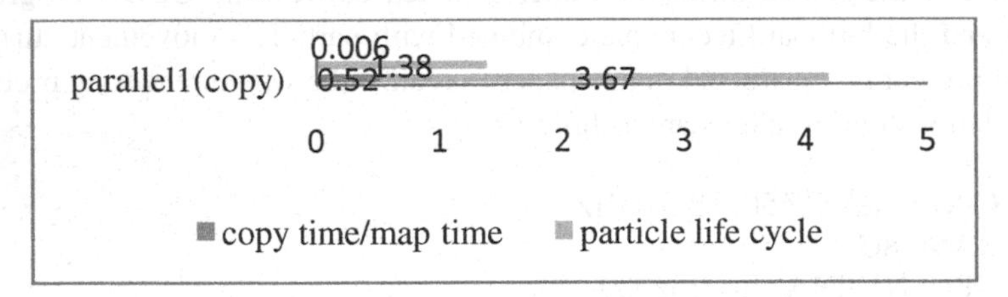

As can be seen from Table2, the memory copy will consume a lot of time, accounting for about 7% of the entire particle life cycle(generation, updating, rendering, dying),but the memory mapping occupies little time, it accounts for less than 1%. Therefore using memory mapping method can save the rendering time, reach conducive performance of simulation.

5 Conclusion

This paper establishes and implements a parallel rendering method of particle system based on CUDA, using memory mapping and billboard technique, optimizes the efficiency of the simulation. The further work will continue to improve the rendering effect and explore the application of parallel computing with CUDA in other visual fields.

Acknowledgements. This work was financially supported by the National Key Technology Research and Development Program (No. 2011BAK16B03) in China.

References

1. Chen, X., Liang, Y., Guo, F.: Fireworks simulation based on CUDA particle system. Journal of Computer Applications 33(7), 2059–2062 (2013)
2. Goldsworthy, M.J.: A GPU-CUDA based direct simulation Monte Carlo algorithm for real gas flows. Computers & Fluids 94, 58–68 (2014)
3. Yilmaz, E., Molla, E., Yildiz, C., Isler, V.: Realistic modeling of spectator behavior for soccer videogames with CUDA. Computers & Graphics 35, 1063–1069 (2011)
4. Molnar Jr, F., Szakaly, T., Meszaros, R., Lagzi, I.: Air pollution modeling using a Graphics Processing Unit with CUDA. Computer Physics Communications 181, 105–112 (2010)
5. Reeves, W.T.: Particle systems: a technique for modeling a class of fuzzy objects. In: Proceeding of the 10th Annual Conference on Computer Graphics and Interactive Techniques. ACM, New York (1983)
6. Matsumoto, M., Nishimura, T.: Mersenne twister: A 623-dimensionally equidistributed uniform pseudo-random number generator. ACM Transactions on Modeling and Computer Simulation 8, 3–30 (1998)
7. NVIDIA corporation, NVIDIA CUDA Programming guide, http://developer.download.nvidia.com/compute/cuda/2_1/toolkit/docs/NVIDIA_CUDAProgrammingGuide_2.1.pdf
8. Longmore, J.-P., Masrais, P., Kuttel, M.M.: Towards realistic and interactive sand simulation: A GPU-based framework. Powder Technology 235, 983–1000 (2013)
9. Li, B., Wang, C., Li, Z., Chen, Y.: A Practical Method for Real-time Ocean Simulation. In: Proceedings of 2009 4th International Conference on Computer Science & Education, Wuhan, China (2014)
10. Cha, M., Han, S., Lee, J., Choi, B.: A virtual reality based fire training simulator integrated with fire dynamics data. Fire Safety Journal 50, 12–24 (2012)
11. Amrasinghe, D., Parberry, I.: Real-time Rendering of Melting Objects in Video Games. In: The 18th International Conference on Computer Games
12. Westphal, E., Singh, S.P., Huang, C.C., Gompper, G., Winkler, R.G.: Multiparticle-coll-isiondynamics: GPU accelerated particle-based mesoscale hydrodynamic sim-ulations. Computer Physics Communications 185, 495–503 (2014)
13. Stein, A., Geva, E., El-Sana, J., Hull, C.: Fast parallel 3D convex hull on the GPU. Computer & Graphics 36, 265–271 (2012)
14. Creus, C., Patow, G.A.: Realistic rain rendering in realtime. Computers & Graphics 37, 33–40 (2013)
15. Zhao, L., Qin, K.: Method of generation and rapid showing for 3D rough models. Application Research of Computers 28(6)
16. Gumbau, J., Chover, M., Remolar, I., Rebollo, C.: View-dependent pruning for real-time rendering of trees. Computers & Graphics 35, 364–374 (2011)
17. Liu, X., Xiong, H., Jiang, L., Luo, Y.: Group Virtual Plants' Growth Using BillBoard. Computer Engineering 29(13)

Research on the Fast Parallel Recomputing for Parallel Digital Terrain Analysis

Shoushuai Miao[1,*], Wanfeng Dou[1,2,**], and Yan Li[1]

[1] School of Computer Science and Technology, Nanjing Normal University, Nanjing, 210023

[2] Jiangsu Research Center of Information Security and Privacy Technology, Nanjing, 210097

ms_shuai@126.com, bdouwf-fly@163.com, c467094983@qq.com

Abstract. With the rapid increasing of spatial data resolution, the high volume datasets make the geocomputation become more time-consuming especially operating some complex algorithms, i.e. viewshed analysis and drainage network extraction. Parallel computing is regarded as an efficient solution by utilizing more computing resource, which has also been proved its availability in digital terrain analysis according to many published literatures. Among them, the stable and credible services play an irreplaceable role in the high performance computing, especially when a failure occurs in the large-scale processing. However, litter research focuses on this issue. In this paper, a new approach for the parallel digital terrain analysis considering the performance of fault tolerance, named the Fast Parallel Recomputing (FPR) was proposed. FPR owns fast self-recovery ability. Once some failures are detected, all the surviving processes repartition the data block and recompute the sub-blocks in parallel to improve the efficiency of failure recovery. The experiments show that the proposed the FPR method achieves better performance than the traditional checkpointing method.

Keywords: digital terrain analysis, fault tolerance, parallel recomputing, fast self-recovery.

1 Introduction

Digital Terrain Analysis (DTA) is a digital information processing technology of computation of the terrain attributes and feature extraction on the basis of the Digital Elevation Mode (DEM) [3]. With increased precision and accuracy, DEMs have gone from 1,000 meter resolutions 5-10 years ago to 1-5 meter resolutions today in many areas. As a result of the increased precision and file sizes, many land surface parameters,

* Corresponding author.

** This work has been substantially supported by the National Natural Science Foundation of China (NO. 41171298).

Author's addresses: Wanfeng Dou, School of Computer Science and Technology, Nanjing Normal University, Nanjing,210023; douwf-fly@163.com.

F. Bian and Y. Xie (Eds.): GRMSE 2014, CCIS 482, pp. 244–251, 2015.

such as slope, profile curvature and hydrologic land-surface parameters for lower resolutions and smaller DEMs become prohibitively time-consuming when being applied to high-resolution and large-scale data. Hence, the parallel computing has becomes a fundamental tool for geographic information science [1] [2].

High performance Geo-computations take much time to finish their computation on distributed processors. At the same time, the reliability of the system becomes a foremost key while the stability of cluster with tens of thousands of processors is threatened constantly by a larger number of hardware and software failures, such as network, memory, processors and operating systems. Therefore, the fault-tolerance has become a necessary component of the reliable computations. The fault tolerant technique is mainly classified into hardware fault tolerance and software fault tolerance. Hardware fault tolerance usually adopts the hardware redundant technology to achieve the fault tolerance. However, with the updating and developing of hardware technology, the cost of hardware redundant technology is higher and will bring larger hardware overhead and performance overhead. Software fault tolerance not only deals with the hardware failure, but also finds and handles some design errors during the runtime of system [4]. Meanwhile, software fault tolerance avoids the overhead of hardware redundant and some design errors [5], so it can improves the overall reliability of the system. Therefore, software fault tolerant technique has been widely applied to the digital terrain analysis.

2 Related Works

Large-scale scientific computing brings more computation tasks so that the running time is longer than the applications in the small scale. Meanwhile, the failure risk become higher and it could cause a lot of resource waste including the time and storage. The checkpointing technique [6] is widely used for fault tolerance. The basic idea is that the current state and data should be saved to stable storage when processing the key position of a process. A large number of the checkpoints will be stored in the parallel system periodically. Whenever a process fails, all processes related to this process have to be rolled back to the last checkpoint to restart the computation. The operation of the checkpoints will lead to the huge volume of data transmission through I/O, which becomes a performance bottleneck. There are two improved methods for checkpointing, diskless checkpointing [7] and application-level checkpointing [8]. However, these two improvements both require the surviving processes, together with the failed one, to roll back upon recovery, and redo the task between the last checkpoint and the failure. This is not an efficient. Meanwhile, Theses methods introduce extra computation of encoding/decoding checkpoints and require a large memory space for large-scale scientific applications.

However, as the number of processes increases, the performance of traditional fault tolerance will be worse. The recomputing was first used for the online fault detection in an arithmetic logic unit. Patel and Fung [9] proposed the shift operation of time redundant in 1982. The principle is that comparing the computing results is required before and after the shift. The fault detection ability depends on the shifts of the digits.

To speed up the recovery procedure, Yang et al. put forward a new approach for fault tolerance of high performance computing, namely parallel recomputing [10]. The main features of parallel recomputing include the partition of task and data and parallel scheduling. According to the length of a parallel task, a range of the variables and constraints, task is divided into a series of sub-tasks. The partition is based on the user's guide of initial correction. The partition requires the experiences of the developers and the complexity of the specific procedures. The scheduling bookkeeping records the loop blocks that executed by every thread in the program. Each thread has been out of circulation block. When a thread failure occurs, it can make sure the collection of loop blocks in accordance with the scheduling bookkeeping of the thread. Then, it builds parallel recomputing scheduling code. However, the approach complicates the coding of the parallel program and it adds burden of the programmer.

To speed up the recovery procedure, Yang et al. [11] put forward a new application-level fault-tolerant scheme based on parallel recomputing, called Fault-Tolerant Parallel Algorithm (FTPA). When a failure occurs, all surviving processes recompute the subtask of the failed process in parallel. This method achieves fast fault recovery by utilizing surviving processes to recompute the workload of the failed process. However, the method is mainly applied to instruction level of a program that will affect whether or not the parallel recomputing is done. When the failure part of the program cannot be repartitioned again, then parallel recomputing could not be finished, so the performance of recomputing will not be provided. Moreover, Yang et al also proposed an improved approach that can solve program division and workload redistribution [12]. From the above, these approaches discussed the key issues about the program partitioning rather than data partitioning.

This paper presents a new approach based on parallel recomputing in the digital terrain analysis-Fast Parallel Recomputing (FPR). The FPR approach provides an effective fast recovery. When a process failure occurs, FPR redistributes the data block of the failed process to all the surviving processes. The data block will be partitioned into small data blocks, which are re-executed by some surviving processes in parallel. The number of data block is related to the number of surviving processes. The FPR can achieve minor overhead and improve the reliability of the system.

3 Fast Parallel Recomputing

In the field of the digital terrain analysis, DEM has a large amount of data in general. Processing the DEM data will take more time and increases the I/O cost. When a process fails, the work is needed to restart and all the computations will be rolled back. To avoid the failure, some measures are needed to guarantee the right service of the system. When facing a large number of the communication and computation among processes, multi-threaded or multi-process are adopted to process the partitioned data blocks, which can effectively improve the efficiency of data processing. Processing the sub-blocks has an advantage of fewer computing and storage. Fast Parallel Recomputing (FPR) is proposed in this paper. Upon a failure, the FPR does not require the roll-back of all surviving processes. Instead, it utilizes the surviving processes to compute the blocks of the failed process in parallel. This paper does not consider the master node failure. If the master node fails, all the data blocks will be recomputed again. In order to avoid the master node failure, the backup mechanism can be adopted. In the process of data distribution, a fast factor is adopted to regulate the number of the partitioned data block. If any other slave node fails, the data block will be repartitioned and recomputed again.

3.1 Designing Fast Parallel Recomputing

Designing a FPR is to incorporate the parallel fault recovery scheme in the parallel digital terrain analysis. The major characteristic of the FPR is that the partitioned data blocks of a failed process are recomputed and repartitioned by the surviving processes to gain the right result in the failed process. In order to simplify the relationship, there isn't any dependent relationship between data blocks. A collection of n processes is defined as P= $\{P_1, P_2, \ldots, P_n\}$, P_i denotes the current process. A set of data blocks is denoted by B= $\{B_1, B_2, \ldots, B_n\}$, B_i denotes the partitioned block that will be handled. A set of data results is denoted by R= $\{R_1, R_2, \ldots, R_n\}$, R_i represents the results of the computation. The whole process of FPR is shown in Fig. 4. The procedure of the FPR has four steps listed as follows:

Step1: Data partition. Before the recovery work, the main work is to partition the DEM data. Then, the partitioned data blocks will be read in main memory from the disk. According to the characteristics of the DEM data, the partition strategy of the data includes the row partition, column partition, and block partition. Hence, the whole data block is partitioned into a series of sub-blocks. The number of partitioned data blocks is 4. The partitioned strategy can be illustrated by Fig.1. The row partition, column partition, and block partition is shown in Fig. 1a, Fig. 1b, and Fig. 1c, respectively.

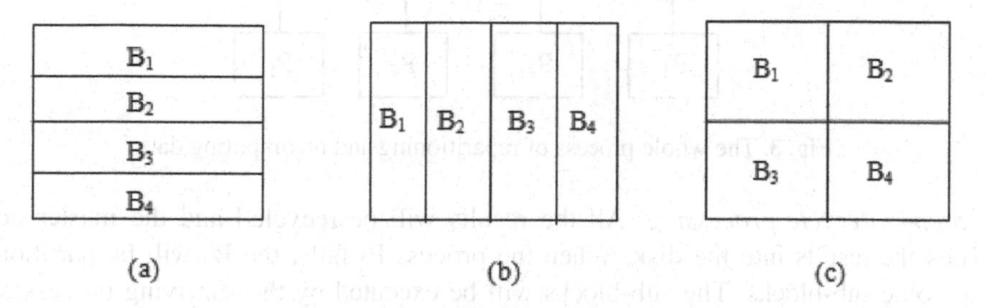

Fig. 1. The difference partition strategies of the DEM

Step2: Data computing. According to the digital terrain analysis algorithms, the processes deal with the partitioned data blocks. Some algorithms are dependent on data relationships which may become a tangled mess if it is improperly handled. In the paper, the algorithms have no data dependent relationships. Thus, the process can easily handle these data blocks.

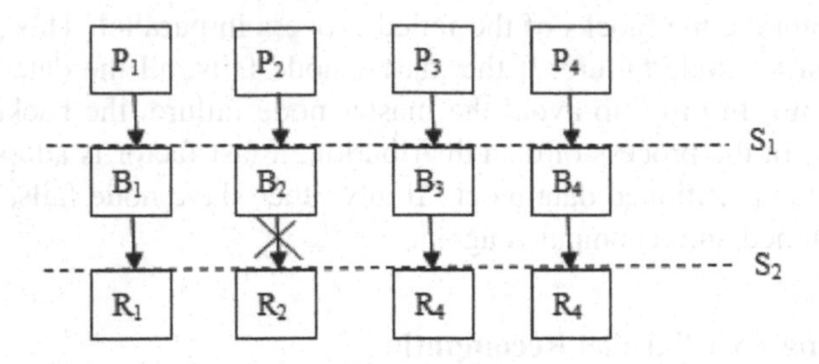

Fig. 2. The single failure case

Step3: Failure processing. At the entry of every block, one process needs to save the data block and the variables. In this paper, the process failed before the recomputing and the single failure case is shown in Fig. 2. S_i denotes the status. S_1 is the status of initial data and S_2 is the status of detecting failure. When a failure occurs in P_2, the recovery procedure firstly obtains the initial B_2. Then, the data block B_2 will be partitioned into four small data blocks. Other processes or leisure processes will compute the repartitioned data blocks. The overall process of the repartition is shown in Fig. 3.

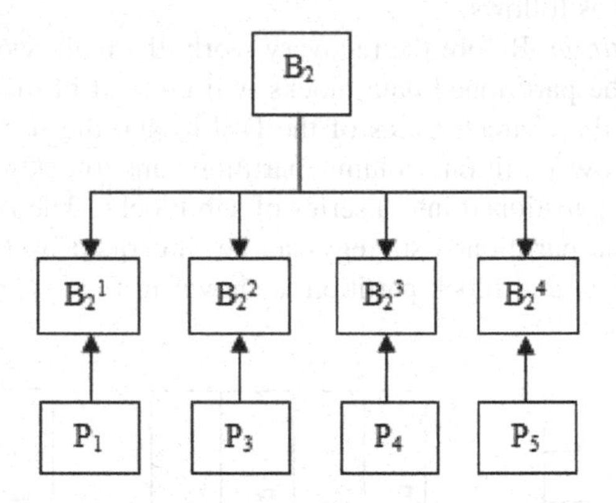

Fig. 3. The whole process of reparitioning and recomputing data

Step4: Recycle processing. All the results will be recycled and the master node writes the results into the disk. When the process P_2 fails, the B_2 will be partitioned into some sub-blocks. The sub-blocks will be executed by the surviving processes in parallel. The generated results set is denoted by $R_2 = \{R_2^1, R_2^2, \ldots, R_2^m\}$.

3.2 Fast Factor Settings

The number of partitioned blocks is a key issue. If the number of the data block is too big, every process will take a little time to deal with the data block. However, it will bring some problems in the communication of inter-processes and storage space. Hence, the symbol $\mathcal{E}$ is used to adjust the number of data block partitioned. The symbol $\mathcal{E}$ is named as fast factor. The number of the data blocks partitioned is related

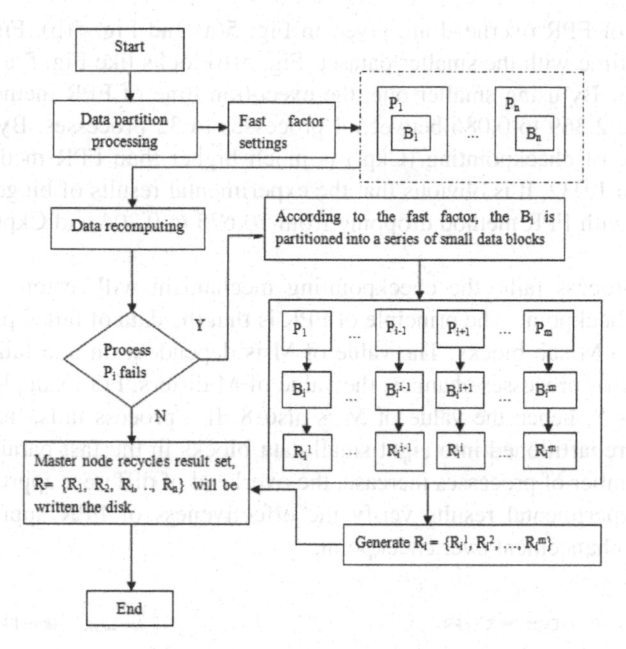

Fig. 4. The whole process of FPR

to the fast factor ε. N is the number of processes. Every process will handle a data block. The number of data blocks must be less than N. The value can be adjusted in accordance with data size of DEM.

Firstly, the value of ε is set to 4 and the number of data blocks partitioned is 4. When a process fails, ε is set to 4. In order to improve the efficiency of processing data , ε is set to 16. The value depends on the number of leisure processes and the size of a data block. So, it is more favorable for recomputing and will cost less time. However, as the number of the partitioned blocks grows, the number of the processes also increases. Then, every data block will be executed in parallel. The runtime of process and time of detecting error will be reduced greatly. It will cause the load imblance and the inter-processes communication will produce more overhead. To enhance the performance of fault tolerance, ε must be set to the appropriate value .

4 Experiments and Analysis

The experiments were performed on a small scale cluster system. Each node is equipped with an Intel XeonE5645 2.8 GHz quad-core processors and 8GB memory. The nodes adopt the Gigabit Ethernet connectivity. The master-slave parallel computing model is adopted. A primary node is responsible for distributing data and recycling results. The software environments have the GDAL 1.6.1, OpenMP 1.5.4, GCC 4.4.7, MPICH2. Two kinds of datasets are employed as the test data, the smaller one is 1.61GB and the bigger on is 3.29GB. The data type is floating-point and the type of image is TIFF format.

The charts of FPR overhead are given in Fig. 5(a) and Fig. 5(b). Fig. 5(a) shows the execution time with the smaller dataset. Fig. 5(b) looks like Fig. 5(a) and the data size is 3.29GB. By using smaller one, the execution time of FPR method drops dramatically from 2.369 to 0.086 between 4 processes to 32 processes. By contrast, the execution time of checkpointing (Ckpt) is much higher than FPR method, reducing from 11.358 to 1.032. It is obvious that the experimental results of bigger one show a similar trend, with FPR method dropping from 10.675 to 0.304 and Ckpt from 59.329 to 5.036.

When a process fails, the checkpointing mechanism will restore to recompute from the last checkpoint. The principle of FPR is that the data of failed process will be partitioned into M sub-blocks. The value of M is dependent on non-failed processes. As the number of processes changes, the value of M differs. For example, the number of processes is 8, hence the value of M is also 8. If a process fails, the related data block will be repartitioned into eight small data blocks in the fast parallel recomputing. As the number of processes increase, the overhead of different approaches will go down. The experimental results verify the effectiveness of FPR approach and the performance enhancement over checkpoint.

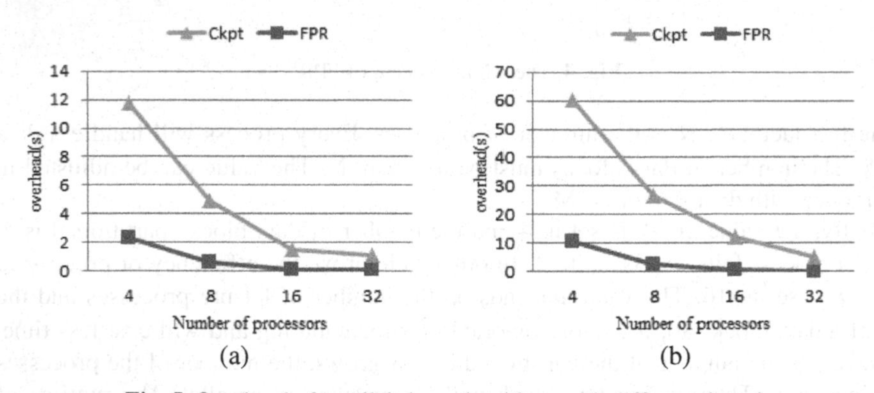

Fig. 5. Overhead of parallel slope algorithms with different data size

5 Conclusions

The digital terrain analysis has the characteristics of data intensive and computation intensive. The fault tolerance mechanism is necessary, especially the big data processing. In this paper, FPR achieves fast self-recovery. The approach can make full use of the non-fail processes to perform the failure recovery in parallel. It is difficult to deal with the operation of the data for the whole of DEM data. Hence, the new approach is used to solve the problems. The computing time will be shortened and the overall performance will be optimized by partitioning the data. Moreover, the performances of FPR are evaluated in different data size with the slope algorithm on a cluster system. The experimental results show that the overhead of FPR is less than the overhead of checkpointing when a failure occurs.

In the future, our efforts will mainly focus on two aspects. Firstly, the performance of the FPR will be evaluated in different digital terrain analysis algorithms. Secondly, further work will be focused on the fault tolerance with the dependent data blocks.

References

1. Chen, G.L., Sun, G.Z., Xu, Y.: Integrated research of parallel computing: Status and future. Chinese Sci. Bull. 54(11), 1845–1853 (2009)
2. Lecca, G., Pdtitdidier, M., Hluchy, L.: Grid computing technology for hydrological applications. Hydrology 403(1-2), 186–199 (2011)
3. Zhou, Q.M., Liu, X.J.: Digital terrain analysis. Science Press, Beijing (2006)
4. Du, Y.F., Tang, Y.H.: Classification and design of fault-tolerant parallel. Huazhong University of Science and Technology (Natural Science Edition) 39(4), 49–52 (2011)
5. Hanmer, R.S.: Patterns for Fault Tolerant Software. John Wiley & Sons Ltd (2007)
6. Chiueh, T., Deng, D.P.: Evaluation of checkpoint mechanisms for massively parallel machines. In: Proceedings of the Twenty-Sixth Annual International Symposium on Fault Tolerant Computing, FTCS 1996, pp. 370–379 (1996)
7. Plank, J.S., Plank, L.K., Puening, M.A.: Diskless checkpointing. IEEE Trans. Parallel Distrib. Syst. 9(10), 972–986 (1988)
8. Bronevetsky, G., Marques, D., Pingali, K., Stodghill, P.: Automated application-level checkpoint of MPI program. In: ACM SIGPLAN Symposium on Principles and Practice of Parallel Programming (PPoPP 2003), pp. 84–94. San Diego, CA (June 2003)
9. Patel, J.H., Fung, L.Y.: Concurrent Error Detection in ALUs by Recomputing with Shifted Operands. IEEE Transactions on Computer C.31(7), 589–595 (1982)
10. Wang, P.F., Du, Y.F., Fu, H.Y., Yang, X.J.: Parallel Recomputing: A New Approach for Fault-Tolerant Hight Performance Computing. Chinese Journal of Computer 36(3), 21–25 (2009)
11. Yang, X.J., Du, Y.F., Wang, P.F., Fu, H.Y., Jia, J., Wang, Z.Y., Suo, G.: The Fault tolerant parallel algorithm: the parallel recomputing based failure recovery. In: 16th International Conference on Parallel Architecture and Compilation Techniques, Brasov, pp. 199–209 (2007)
12. Fu, H.Y., Ding, Y., Song, W.: An application level checkpointing based on extended data flow analysis for OpenMP programs. Chinese Journal of Computers 32(10), 38–53 (2010)

Multi-source Remote Sensing Image Fusion Method Based on Sparse Representation

Xianchuan Yu, Yinggang Zhang, and Guanyin Gao

College of Information Science and Technology, Beijing Normal University, 100875, China
chuan.yu@ieee.org, yuxianchuan@163.com

Abstract. To improve the quality of the fused image, we propose a remote sensing image fusion method based on sparse representation. In the method, first, the source images are divided into patches and each patch is represented with sparse coefficients using an overcomplete dictionary. Second, the larger value of sparse coefficients of panchromatic (Pan) image is set to 0. Third, Then the coefficients of panchromatic (Pan) and multispectral (MS) image are combined with the linear weighted averaging fusion rule. Finally, the fused image is reconstructed from the combined sparse coefficients and the dictionary. The proposed method is compared with intensity-hue-saturation (IHS), Brovey transform (Brovey), discrete wavelet transform (DWT), principal component analysis (PCA) and fast discrete curvelet transform (FDCT) methods on several pairs of multifocus images. The experimental results demonstrate that the proposed approach performs better in both subjective and objective qualities.

Keywords: Image fusion, sparse representation, remote sensing, dictionary learning.

1 Introduction

As the limitation of remote sensing satellites onboard storage and bandwidth, the images with both high spatial resolution and high spectral resolution cannot be obtained directly. As a result, remote sensing satellites often provide panchromatic (PAN) image with high spatial resolution and multispectral (MS) image with high spectral resolution. Panchromatic (Pan) image whose spatial resolution is high contains rich spatial detail information. So it is able to express the detail feature clearly on surface. However, spectral information of Pan image is lacking. Multi-spectral (MS) image is rich in spectral information so that it is help to the feature identification, but its spatial resolution is low. Generally, remote sensing images with high spectral and high spatial resolutions are essential for complete and accurate description of the observed scene. Remote sensing image fusion is an effective technique to integrate spatial and spectral information of the PAN and MS images.

In the past decades, there are various methods available to implement image fusion have been proposed. Basically, the existing methods can be categorized into two classes: color space component replacement based methods [1] and ARSIS model

F. Bian and Y. Xie (Eds.): GRMSE 2014, CCIS 482, pp. 252–265, 2015.

based methods [2]. Color space component replacement based methods fuse images on the pixel-level. The representative component-substitution-based method is the intensity–hue–saturation (IHS) [3]. This category of methods can improve the spatial resolution of the image effectively, while they will produce serious spectral distortion. ARSIS model based methods improve spatial resolution of multispectral images by extracting the high-frequency component of panchromatic image and injected into the multispectral images. The wavelet transform based methods are the representative methods [4]. This category of methods solves a serious spectral distortion problem which is caused by color space component replacement based methods, but the fused image is prone to produce details injection over or offset phenomenon.

Obviously, in remote sensing image fusion, the extracting accurately for image information is very essential. Sparse representation can portray internal structure and characteristic of the signal effectively. As a result, sparse representation theory is gaining more and more attention in image fusion field. In [5], Hu proposed Remote sensing image fusion based on HIS transform and sparse representation. The method is applicable only when MS image has three bands, and it has certain shortcomings compared with traditional methods in terms of spatial resolution. Li et.al [6] proposed the image degradation model from high-resolution MS image to low-resolution MS image and Pan image. The degradation model approximately gets the optimal solution of high-resolution MS image sparse representation in a sparse domain. This method can achieve better fusion effect, but the fusion process is more complicated. In order to overcome the shortcomings of the existing methods, we propose a remote sensing image fusion method based on sparse representation. This method can preferably maintain spectral information of MS image, while improve the spatial resolution information as much as possible.

2 Basic Theory of Signal Sparse Representation

As a new theory born in field of signal processing, sparse representation theory has drawn more and more attention of researchers in related fields. Sparse representation is based on the idea that a signal can be constructed as a linear combination of atoms from a dictionary$D \in R^{n\times T}$, for example, $x \approx D\alpha$. Generally, $T > n$, this means that the dictionary is overcomplete. So there are numerous ways to represent the signal, among which sparse representation refers to the one with the fewest atoms. More formally, the sparse vector can be obtained by solving the following optimization problem:

$$\min_{\alpha}\|\alpha\|_0 \quad s.t. \quad \|x - D\alpha\|_2^2 < \varepsilon \tag{1}$$

Where $\|\cdot\|_0$ is the l_0 norm counting the number of nonzero entries in vector. $x\epsilon R^n$ is original signal. α is sparse representation coefficient. $\varepsilon \geq 0$ is the error tolerance.

The above optimization is an NP-hard problem and can only be solved by systematically testing all the potential combinations of columns [7]. Currently, there are more and more approximate solutions for this problem. As the simple structure and a small amount of computation, greedy iterative class algorithms attract lots of attention. Orthogonal Matching Pursuit (OMP) [8] algorithm is a typical representative of such algorithms. The algorithm has been widely used for its simplicity and effectiveness. In this paper, the OMP algorithm is used for image sparse decomposition.

OMP algorithm aims to find a group of sparse coefficients α which subjects to $x \approx D\alpha$ as follows:

1) Initialize the residual:

$$r_0 = x \tag{2}$$

2) Calculate the inner product value and find the index λ_t that solves the easy optimization problem

$$\lambda_t = \arg\max_j \langle r_t, d_j \rangle \tag{3}$$

3) Augment the index set and the matrix of chosen atoms

$$\Lambda_t = \Lambda_{t-1} \cup \{\lambda_t\} \tag{4}$$

$$\Phi_t = \left[\Phi_{t-1}\ D_{\lambda_t}\right] \tag{5}$$

4) Solve a least squares problem to obtain a new signal estimate

$$\alpha = \arg\min_x \|x - \Phi_t \alpha\|_2 \tag{6}$$

5) Update the residual

$$r_t = x - \Phi_t \alpha \tag{7}$$

6) Increment t, and return to Step 2) if $t < K$. K is the sparsity level.

As the above, OMP algorithm selects the atom which is the most relevant with the residual to represent the original signal, so the sparse coefficient represents the main components of the residual in each iteration. At the first iteration, the residual is equal to original signal. The sparse coefficient obtained in this iteration represents the main components of the original signal and its absolute value is larger. With the increase of the number of iterations, the residual is less and less and the absolute values of sparse coefficient are continuously declining overall.

3 Image Fusion Method

A. Sparse Representation for Image Fusion

Generally, nature image contains complicated and non-stationary information as a whole, while local small image patch has a simple and consistent structure. For this reason, the local image patch is more easy to sparse decompose than the whole image. In out method, as shown in fig. 1, we divide the source images into small patches by the sliding window technique. Then each patch is lexicographically ordered as a vector. All the vectors constitute one matrix V, in which each column corresponds to one patch in the source image. Assume that the sliding window interval is k, the size of patch is $\sqrt{n} \times \sqrt{n}$ and the size of image is $M \times N$, the size of V is $(\sqrt{n} \cdot \sqrt{n}) \times ((M - \sqrt{n} + k) \cdot (N - \sqrt{n} + k))$.

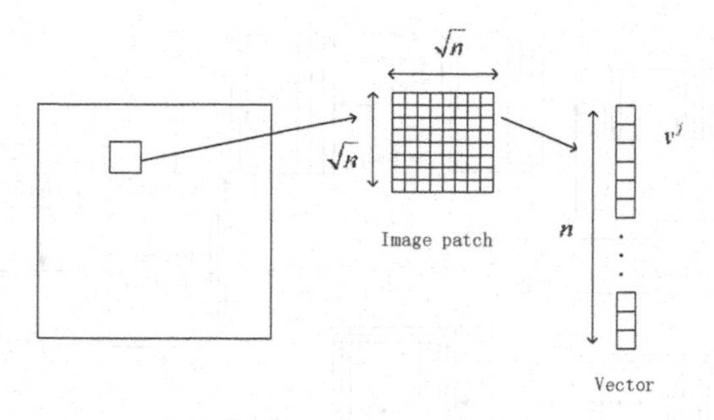

Fig. 1. Image patch selected and its lexicographic ordering vector.

The goal of remote sensing image fusion is to inject the high spatial resolution information of Pan image into the MS image, in order to obtain the fusion image including the high spatial resolution information and high spectral resolution information. The sparse representation describes the original signal accurately by a small amount of sparse coefficients and the dictionary. Generally, the larger value of sparse coefficients indicates the main component of image which is the low frequency component, while high frequency component mainly indicated by the smaller value of sparse coefficients. In the process of Pan image and MS image fusion, the low frequency component of Pan image as redundant information should be removed, while the high frequency information as the useful information should be injected into MS image. As shown in fig 2, fig 2(a) is Pan image. Fig. 2(b) represents the low frequency component which obtained by the larger sparse coefficients. Fig. 2(c) represents high frequency component which obtained by the smaller sparse coefficients.

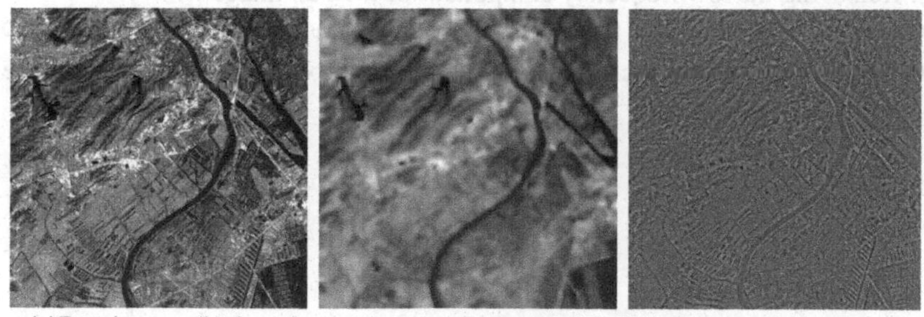

(a)Pan image (b) low frequency component (c) high frequency component

Fig. 2. Pan image decomposition

B. Proposed Fusion Scheme

The remote sensing image fusion method based on sparse representation is proposed, according to the theories above. The schematic diagram of this method is shown in fig 2.

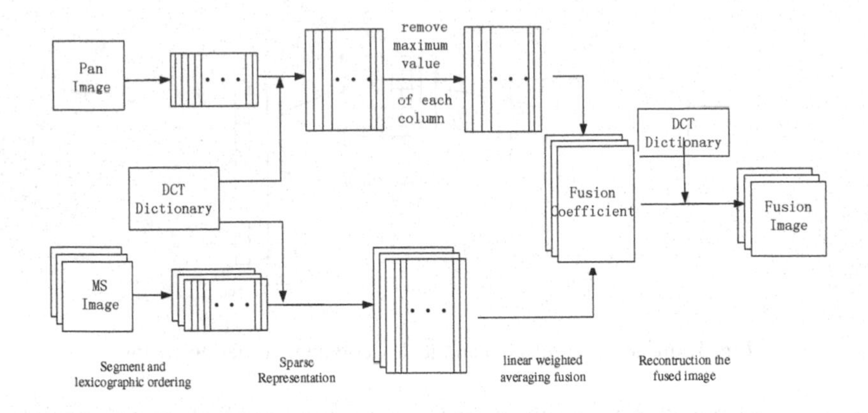

Fig. 2. The schematic diagram of the proposed fusion method

The steps are as follows:

1). Dividing image

The source images are first divided into small image patches. Then each patch is lexicographically ordered as a vector. All the vectors corresponding to Pan image constitute the matrix V_{Pan}. The vectors corresponding to three bands of MS image respectively constitute the matrix V_{Red}, V_{Blue}, V_{Green}.

2). Sparse representation

Each column vector v_j of the matrix V_{Pan} is sparse decomposed by orthogonal matching pursuit (OMP) algorithm. The sparse coefficients matrix is S_P. Similarly, the matrix V_{Red}, V_{Blue}, V_{Green} is sparse decomposed by OMP algorithm and the sparse coefficients matrixes are S_R,S_B,S_G.

3). Removing the low frequency component of the Pan image

Setting the weight T. For the jth column vector S_{P_j} in Pan image sparse coefficients matrix, the values which are greater than $T \cdot \max(S_{P_j})$ are set to 0, as the formula (8) and (9) are shown.

$$\mathrm{Max} = \max(S_{P_j}) \tag{8}$$

$$S_{P_j}(i) = 0, \quad \text{If: } S_{P_j}(i) > T \cdot Max \tag{9}$$

The range of T is 0~1.

4). Injecting details

The sparse coefficient S_P is fused with S_R, S_B, and S_G respectively by the linear weighted average rule. We get the fusion coefficients such as S_{FR}, S_{FB}, and S_{FG}. The fusion rule is show as formula (10) (11) (12) .

$$w_1 = \frac{S_P(i,j)}{S_P(i,j)+S_R(i,j)} \quad w_2 = \frac{S_R(i,j)}{S_P(i,j)+S_R(i,j)} \quad S_{FR} = w_1 * S_P(i,j) + w_2 * S_R(i,j) \tag{10}$$

$$w_1 = \frac{S_P(i,j)}{S_P(i,j)+S_B(i,j)} \quad w_2 = \frac{S_B(i,j)}{S_P(i,j)+S_B(i,j)} \quad S_{FB} = w_1 * S_P(i,j) + w_2 * S_B(i,j) \tag{11}$$

$$w_1 = \frac{S_P(i,j)}{S_P(i,j)+S_G(i,j)} \quad w_2 = \frac{S_G(i,j)}{S_P(i,j)+S_G(i,j)} \quad S_{FG} = w_1 * S_P(i,j) + w_2 * S_G(i,j) \tag{12}$$

5). Restoration of fusion coefficients

Fusion coefficients On the 4) such as S_{FR}, S_{FB}, S_{FG} were reconstructed, to get each band of fusion images, with different band combination image merging multiple spectrum fusion images.

The fusion image of each band is reconstructed from S_{FR}, S_{FB}, S_{FG}. All the fusion images are combined for multispectral fusion image.

The proposed method can remove the redundancy low frequency information in Pan image and extract the high frequency information as much as possible, so that the fusion images can preserve the spectral information better, at the same time, improve the spatial resolution information as much as possible.

4 Experiment Results and Analysis

The purpose of image fusion is to get the fusion images through some fusion methods, so that the fusion images can inherit both the spectral information of the multi spectral image and the detail information of high spatial resolution image. Meanwhile we must maintain a balance between the spectral information and detail information and prevent the supersaturation[9]. For the purpose of image fusion, both visual effects and quantitative analysis were used to evaluate the fusion results.

We select remote sensing data from the three satellites captured in different regions for the fusion experiments. The first set of data come from multispectral image composed of IKONOS multi band and high resolution images composed of panchromatic band, in which multispectral image is the false color image synthesized by B4, B3, B23 bands, and the experimental area is located in the main campus of Beijing Normal University. The second set of data sources involve multi-spectral image provided by China Brazil Earth Resources Satellite (CBERS) and panchromatic image provided by Landsat ETM+, in which multispectral image is the false color images synthesized by B3, B2, B13 bands, and the experimental area is located in Doumen District of Zhuhai City, Guangdong Province, China. The third set of data come from multispectral image provided by SPOT5 satellite and radar images provided by Terra SAR-X satellite, in which multispectral image is the false color image synthesized by B3, B2, B1 bands, and the experimental area is located in the Pearl River Delta region. Three sets of experimental images are sub images of 512 × 512 pixels intercepted from panchromatic or radar images, and the sub images are intercepted from multi-spectral images in corresponding region, as the fusion image data.

Classic intensity–hue–saturation (IHS)[2], Brovey transform[10], principal component analysis (PCA) [11], discrete wavelet transform (DWT) [4], and fast discrete curvelet transform (FDCT) [12] are selected for comparison. The over-complete DCT dictionary is selected as the sparse representation dictionary. The size of sliding window is 8 × 8 pixels and the interval sliding window is 1. The value of T is larger, the spatial detail information of fusion images is richer, and spectral information remains worse. In contrast, the value of T is smaller, the spectral information of fusion images is better, and the spatial detail information is less. In this paper, the value of T is set to 2/3, taking into account both spectral information and spatial detail information of fusion images.

As Fig. 3, Fig. 4 and Fig. 5 shown that the fusion images of these six algorithms are improved obviously in the visual effect and clarity compared with the original images. However, for different methods, the fusion effect is a big difference. The parts which are in the circle indicate the differences in maintaining spectral information between different fusion images. What can be seen from the Fig.s is that, the proposed method is the best in maintaining spectral information. The parts which are in the box show the differences in spatial detail information. The fusion images of DWT, FDCT and the proposed method are the clearest ones.

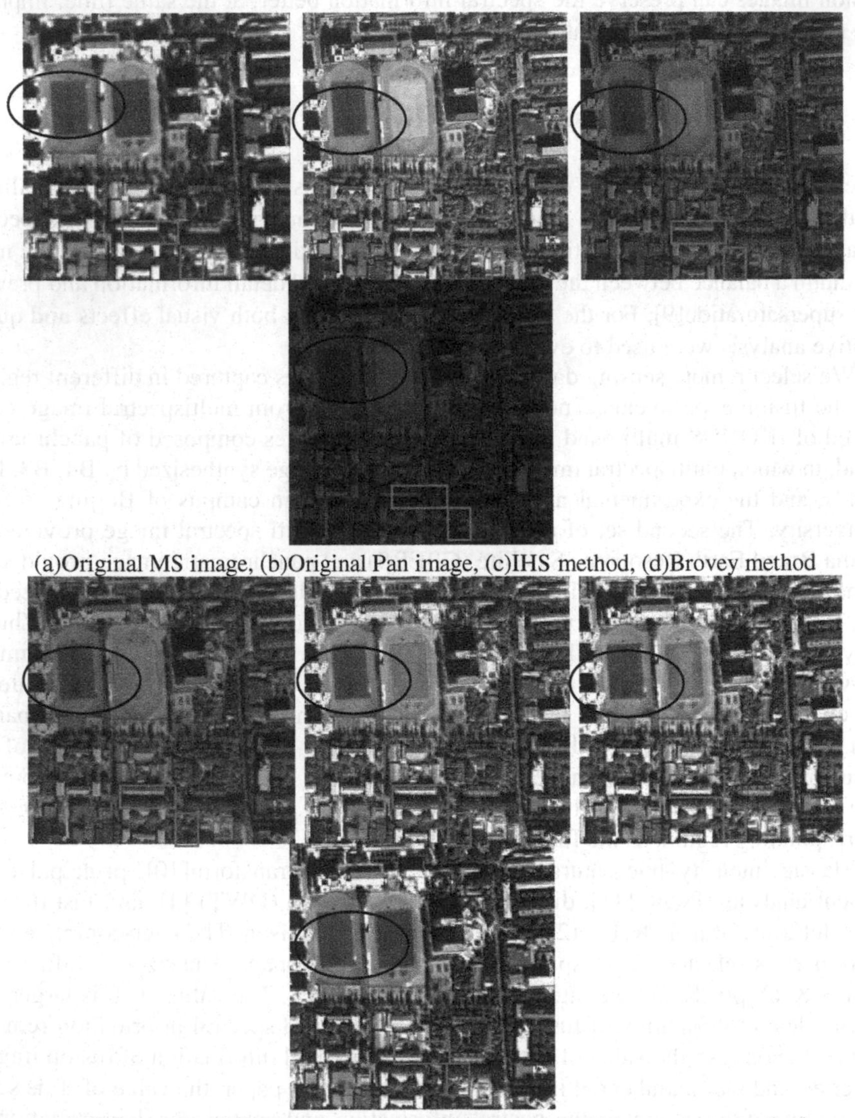

(a)Original MS image, (b)Original Pan image, (c)IHS method, (d)Brovey method

(e)PCA method, (f)DWT method, (g)FDCT method, (h)the proposed method

Fig. 3. High resolution fusion images based on different fusion methods (IKONOS datas)

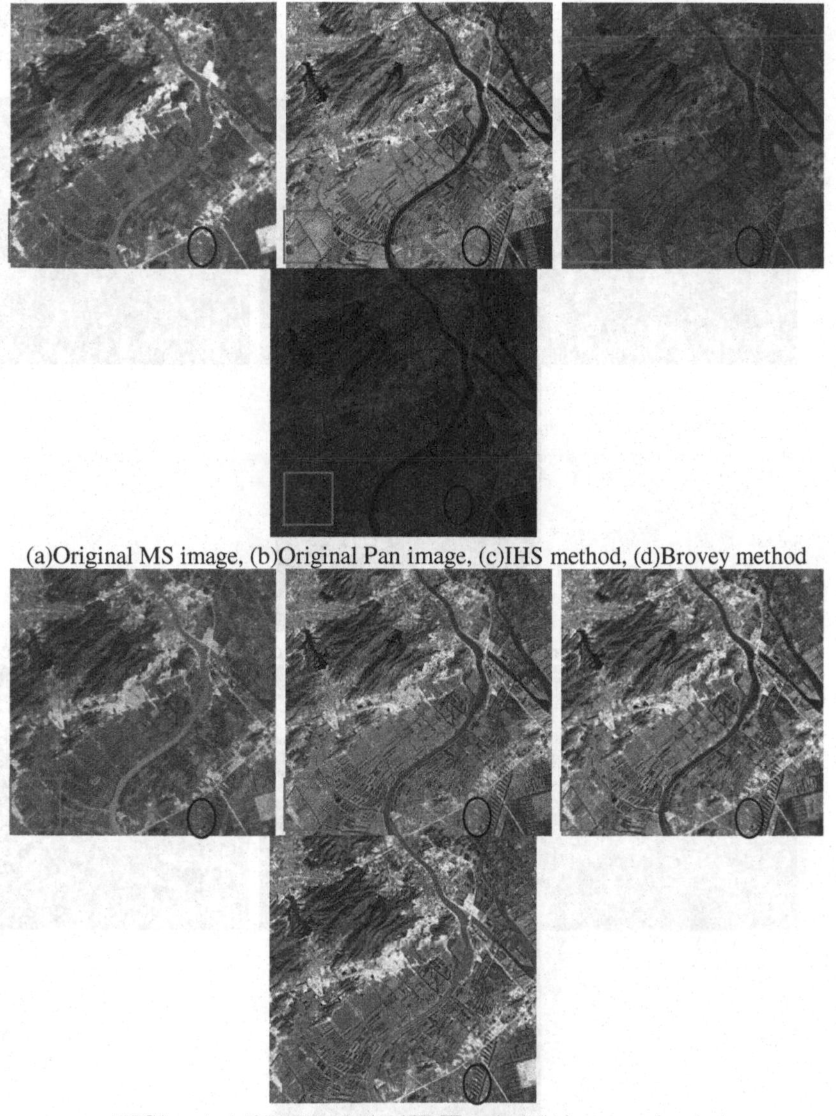

(a)Original MS image, (b)Original Pan image, (c)IHS method, (d)Brovey method

(e)PCA method, (f)DWT method, (g)FDCT method, (h)the proposed method

Fig. 4. High resolution fusion images based on different fusion methods (CBERS and Landsat ETM+ datas)

In order to further verify the fusion effect of the proposed method, six of the most commonly evaluation indexes are used to test the fusion images, such as correlation coefficient (CC)[13], standard deviation (STD)[14], spatial frequency (SF)[15], peak signal to noise ratio (PSNR) [16], information entropy (IE) [17], and relative global dimensional synthesis error (ERGAS) [18]. The fusion images should meet that CC, PSNR, STD, SF, and IE is great and ERGAS is low as soon as possible.

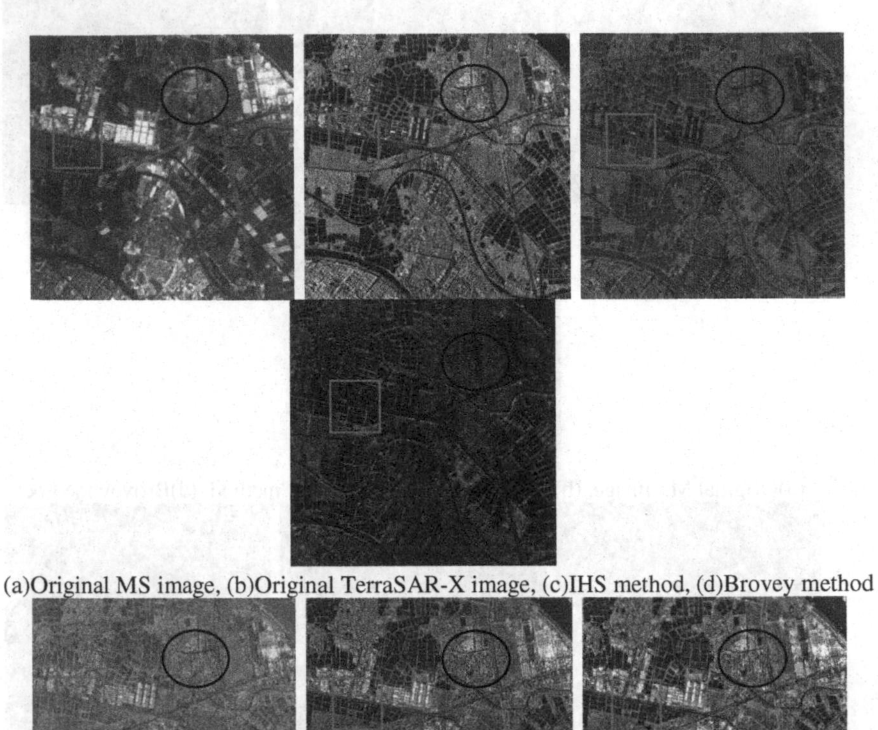

(a)Original MS image, (b)Original TerraSAR-X image, (c)IHS method, (d)Brovey method

(e)PCA method, (f)DWT method, (g)FDCT method, (h)the proposed method

Fig. 5. High resolution fusion images based on different fusion methods (SPOT and TerraSAR datas)

Table 1. The quantitative evaluation result of fusion images in the IKONOS datas

Evaluation index	IHS	Brovey	PCA	DWT	FDCT	Proposed method
CC	0.6184	0.6243	0.7858	0.8561	0.8521	**0.9166**
ERGAS	24.9308	61.0720	16.2894	9.3484	9.7734	**8.1209**
PSNR	28.9316	22.2654	34.5255	39.5859	38.7887	**44.3780**
STD	36.4523	20.5302	44.3223	61.2928	65.9582	**69.3641**
SF	25.7562	14.8732	25.6042	**46.7459**	45.3998	39.1260
IE	4.8891	4.3088	5.0615	5.3383	**5.3575**	5.3146

Table 2. The quantitative evaluation result of fusion images in the CBERS and Landsat ETM+ datas

Evaluation index	IHS	Brovey	PCA	DWT	FDCT	Proposed method
CC	0.6527	0.6109	**0.9180**	0.6735	0.6869	0.8082
ERGAS	26.0479	67.6681	9.8329	10.9126	11.2694	**9.2940**
PSNR	27.7790	20.1635	**41.6073**	34.7612	34.1570	38.1711
STD	43.1353	23.6950	57.5623	58.7570	63.2197	**65.3569**
SF	45.1778	27.3001	33.3351	**76.1133**	76.0836	71.3004
IE	4.9153	4.4517	5.218	5.3592	**5.3634**	5.3046

Table 3. The quantitative evaluation result of fusion images in the SPOT and TerraSAR datas

Evaluation index	IHS	Brovey	PCA	DWT	FDCT	Proposed method
CC	0.2041	0.3189	0.4193	0.6538	0.6812	**0.8358**
ERGAS	32.5021	85.9404	29.3061	16.2313	15.9199	**12.3254**
PSNR	28.3193	24.4214	28.3264	33.7217	33.8427	**41.4587**
STD	38.8342	27.9937	58.9631	53.0289	60.0785	**66.1889**
SF	28.9254	20.2429	36.6896	**50.8634**	49.4849	42.4588
IE	4.7621	4.2542	4.8296	**5.2406**	5.2354	4.8746

Fig. 6, 7 and 8 are corresponding with table 1, 2 and 3. The numbers of ERGAS take reciprocal. Red color represents the proposed method.

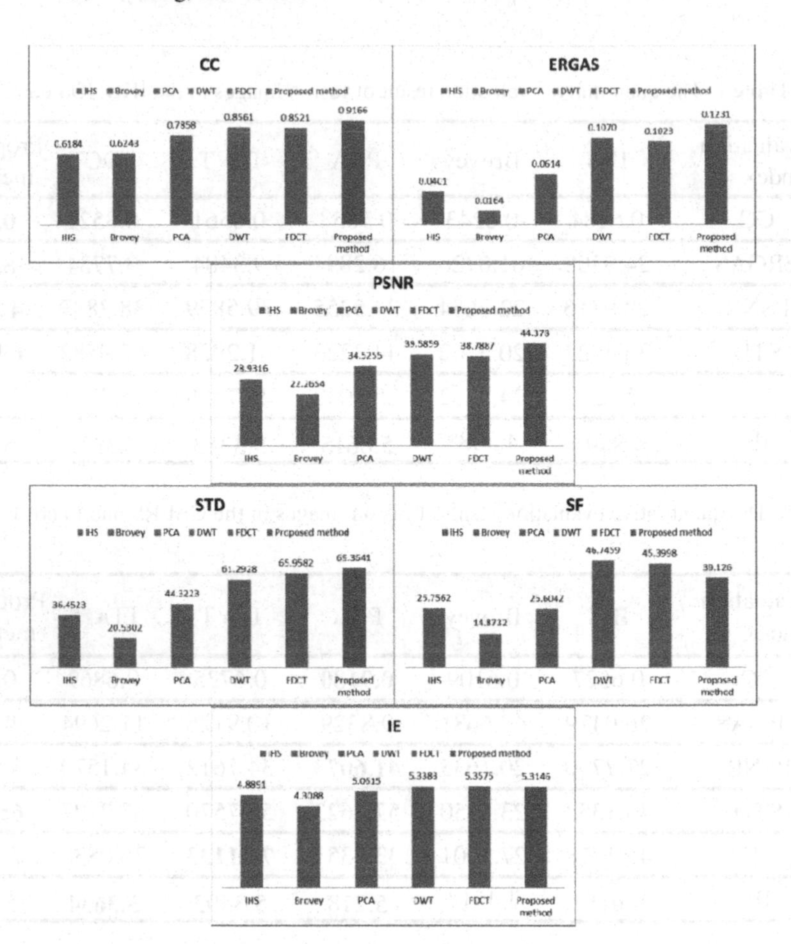

Fig. 6. The evaluation index histograms of different methods (the IKONOS datas)

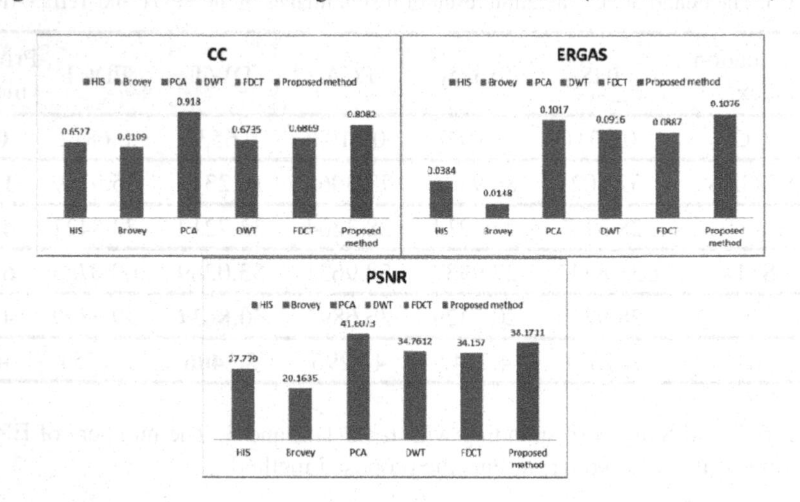

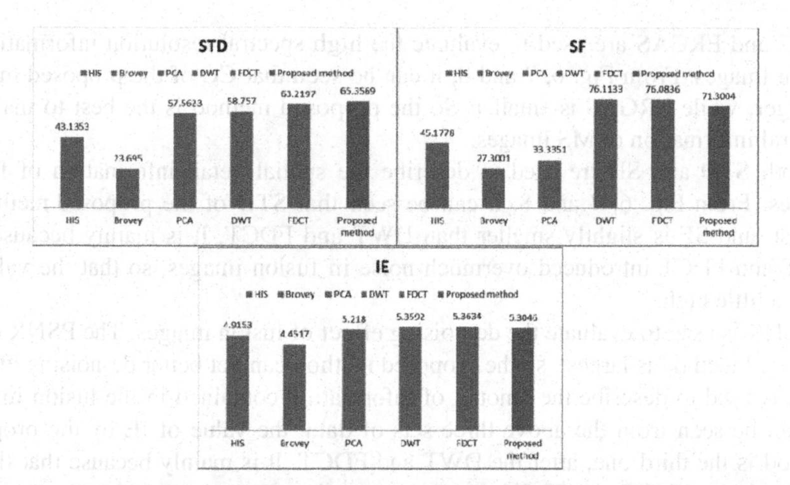

Fig. 7. The evaluation index histograms of different methods (the CBERS and Landsat ETM+ datas)

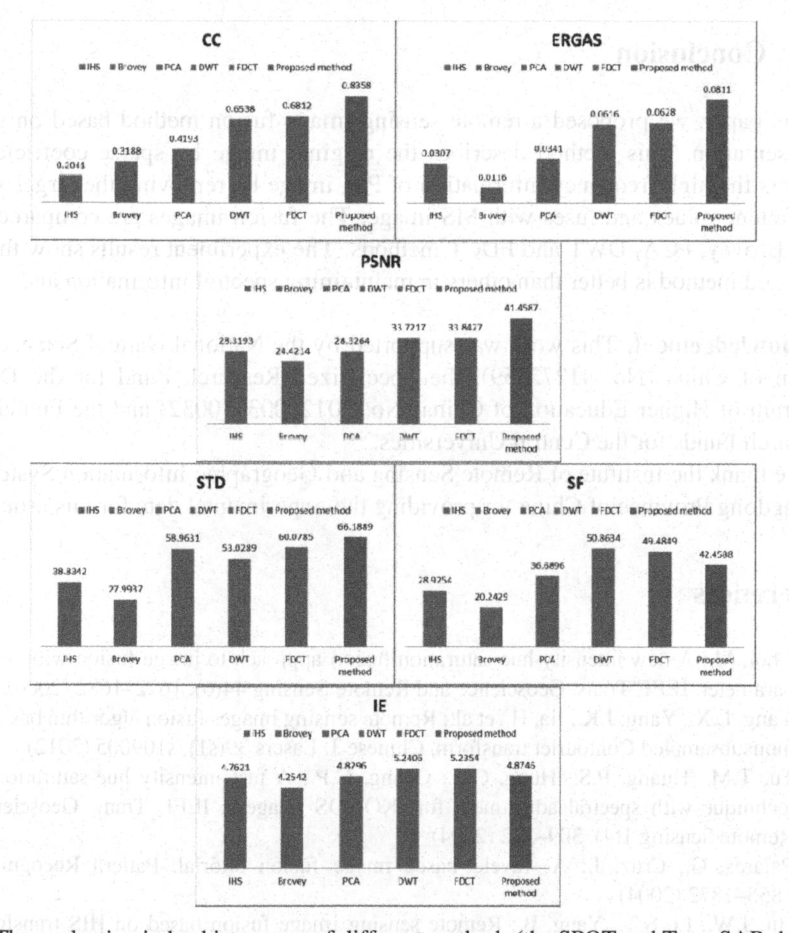

Fig. 8. The evaluation index histograms of different methods (the SPOT and Terra SAR datas)

CC and ERGAS are used to evaluate the high spectral resolution information of fusion images. From Fig. 6, 7 and 8, it can be seen that CC of the proposed method is larger, while ERGAS is smaller. So the proposed method is the best to maintain spectral information of MS images,

Both STD and SF are used to describe the spatial detail information of fusion images. From Fig. 6, 7 and 8, it can be seen that STD of the proposed method is largest, and SF is slightly smaller than DWT and FDCT. It is mainly because that DWT and FDCT introduced overmuch noise in fusion images, so that the value of SF is a little high.

PSNR is used to evaluate the de-noising effect of fusion images. The PSNR of the proposed method is largest, so the proposed method can get better de-noising effect.

IE is used to describe the amount of information contained in the fusion images. As can be seen from the above three sets of data, the value of IE by the proposed method is the third one, after the DWT and FDCT. It is mainly because that the fusion images of DWT and FDCT contain more noise.

5 Conclusion

In this paper, we proposed a remote sensing image fusion method based on sparse representation. This method describes the original image by sparse coefficient. It extracts the high frequency information of Pan image by removing the larger sparse coefficient values and fuses with MS image. The fusion images are compared with HIS, Brovey, PCA, DWT and FDCT methods. The experiment results show that the proposed method is better than others in maintaining spectral information and clarity.

Acknowledgement. This work was supported by the National Natural Science Foundation of China (No. 41272359), the Specialized Research Fund for the Doctoral Program of Higher Education of China (No. 20120003110032) and the Fundamental Research Funds for the Central Universities.

We thank the Institute of Remote Sensing and Geographic Information Systems in Guangdong Province of China for providing the experimental data for this article.

References

1. Choi, M.: A new intensity-hue-saturation fusion approach to image fusion with a tradeoff parameter. IEEE Trans. Geoscience and Remote Sensing 44(6), 1672–1682 (2006)
2. Yang, L.X., Yang, J.K., Jia, H., et al.: Remote sensing images fusion algorithm based on the monsubsampled Contourlet transform. Chinese J. Lasers 39(s1), s109005 (2012)
3. Tu, T.M., Huang, P.S., Hung, C.L., Chang, C.P.: A fast intensity hue-saturation fusion technique with spectral adjustment for IKONOS imagery. IEEE Trans. Geoscience and Remote Sensing 1(4), 309–312 (2004)
4. Pajares, G., Cruz, J.: A wavelet-based image fusion tutorial. Pattern Recognit. 37(9), 1855–1872 (2004)
5. Hu, J.W., Li, S.T., Yang, B.: Remote sensing image fusion based on HIS transform and sparse representation. In: CCPR, pp. 221–224 (2010)

6. Li, S.T., Yang, B.: A new pan-sharpening method using a compressed sensing technique. IEEE Trans. Geoscience and Remote Sensing 49(2), 738–746 (2011)
7. Davis, G., Mallat, S., Avellaneda, M.: Adaptive greedy approximations. Constr. Approx. 13(1), 57–98 (1997)
8. Aharon, M., Elad, M., Bruckstein, A.: K-SVD: An algorithm for designing overcomplete dictionaries for sparse representation. IEEE Trans. Signal Process. 54(11), 4311–4322 (2006)
9. Yu, X.C., Hu, D.: Blind Source Separation: Theory and Application. Wiley (2013)
10. Zhou, H.Z., Wu, S., Mao, D.F., et al.: Improved Brovey method for multi-sensor image fusion. Journal of Remote Sensing 16(2), 343–360 (2012)
11. Sun, Y., Zhao, C.H., Li, J.: Remote sensing image fusion algorithm based on NSCT and PCA transform domain. Journal of Shenyang University of Technology 33(3), 308–314 (2011)
12. Choi, M., Kim, R.Y., Nam, M.Y., Kim, H.O.: Fusion of multispectral and panchromatic satellite images using the curvelet transform. IEEE Geosci. Remote Sens. Lett. 2(2), 136–140 (2005)
13. Shi, W., Zhu, C.Q., Tian, Y., Nichol, J.: Wavelet based image fusion and quality assessment. International Journal of Applied Earth Observation and Geoinformation 6(3-4), 241–251 (2005)
14. Wu, W.B., Yao, J., Kang, T.J.: Study of Remote Sensing Image Fusion and Its Application in Image Classification. In: The International Archives of the Photogrammetry, Remote Sensing and Spatial Information Sciences, Part B7, Beijing, vol. XXXVII (2008)
15. Xydcas, C., Petrovi, V.: Objective Image Fusion Performance Measure. Electronics Letters 36(4), 308–309 (2000)
16. Karathanassi, V., Kolokousis, P., Ioannidou, S.: A comparison study on fusion methods using evaluation indicators. International Journal of Remote Sensing 28(10), 2309–2341 (2007)
17. Tsai, V.J.D.: Evaluation of multiresolution image fusion algorithms. Geoscience and Remote Sensing Symposium 9, 20–24 (2004)
18. Lillo-Saavedra, M., Gonzalo, C., Arquero, A., Martinez, E.: Fusion of multispectral and panchromatic satellite imagery based on tailored filtering in the Fourier domain[J]. International Journal of Remote Sensing 26(6), 1263–1268 (2005)

Research on Full-polarimetric Radar Interference Image Classification

Xiange Cao, Jinling Yang, Jianguo Hou, Haiyan Si, Jiang Liu, and Xianglai Meng

Shool of Surveying and Mapping Engineering, Heilongjiang Institute of Technology, Harbin, 150050, China
{caoxiange,yangjinlingkm,houjg2006}@126.com, {56199436,121281531,563827838}@qq.com

Abstract. The correlation coefficient of two interference image pairs is the basis of the polarimetric radar interference image classification; the different feature category can be obtained through the analysis of the correlation coefficient. The coherence matrix of polarimetric interference after dealing with multi-look obeys the complex Wishart distribution, with a judgment criterion which similar to the Bayesian maximum likelihood classifier, the polarimetric interference radar image can be classified; this method is called polarimetric interference Wishart ML unsupervised classification. This paper summarizes the classification step of this method, and based on full-polarimetric PALSAR data in Tahe area, analyzed the method of Wishart ML unsupervised classification, the classification results show that this classification method has strong adaptability, and has obvious boundary between classes, the classification results belongs to the type of information with the same scattering mechanism, these information provide references for the development and utilization of forest resources.

Keywords: Full-polarimetric radar, Scattering matrix, Correlation matrix, Polarimetric interference, Unsupervised classification.

1 Introduction

The polarimetric scattering matrix gained by full-polarimetric radar (PolSAR) measurement mode provides us access to more extensive feature information, the classification based on full-polarimetric radar images can provide a richer and accurate result compared to single polarimetric radar images. During polarimetric radar forestry applications: before biomass estimates often need to classify forest cover types, the development and utilization of forest resources, forest type identification, forest mapping, forest survey, determine the deforestation area and fire area are also need the classification[1,2]; therefore the research on polarimetric radar image classification of forest cover areas is necessary.

2 Principle of Polarimetric Interference Classification

At present, many polarimetric radar image classification method are proposed, and different to the conventional optical remote sensing classification which depend on

F. Bian and Y. Xie (Eds.): GRMSE 2014, CCIS 482, pp. 266–273, 2015.

the scattering intensity of object, the polarimetric radar classification methods are based on the parameters closely related to the target scattering mechanism which extracted from Sinclair scattering matrix and polarimetric Stokes matrix, and then by combining with other methods achieving the classification of ground object target.

The two interference image's correlation coefficient is the basis of polarimetric radar interference image classification, through the analysis of the correlation coefficient can obtain the different feature categories. Because in the time interval of two observations, the change of scattering mechanism physical characteristics of resolution cell and the scattering geometry relations and propagation characteristics of media can cause time decorrelation of the two images, thus the size of the correlation coefficient would also be affected. In addition, the two images correlation coefficient which used to deal with the interference is associated with the polarimetric mode of electromagnetic wave, the optimal coherent polarimetric radar interference decomposition results will contribute to understanding of the mechanism of ground object target scattering[3]. Thus the ground objects can be classified according to the interference correlation coefficient, and is expected to get a good classification results.

3 Polarimetric Interference Classification Process

Multi-look processing will be need before the polarimetric interference, it means the multi-point average calculation of range and azimuth respectively to single view of complex data; single view complex data can be multi-look display via multi-look processing. The (6×6) interference correlation matrix can get through multi-look processing of two radar images, the coherent interference matrix can be expressed as equation 1:

$$\langle T_6 \rangle = \frac{1}{n}\sum_{i=1}^{n} \vec{w}_i \vec{w}_i^{*T} = \begin{bmatrix} \langle T_{11} \rangle & \langle \Omega_{12} \rangle \\ \langle \Omega_{12} \rangle^{*T} & \langle T_{22} \rangle \end{bmatrix} \text{,and the } \vec{w} = \begin{bmatrix} \vec{k}_1 \\ \vec{k}_2 \end{bmatrix} \tag{1}$$

In equation 1, $\vec{k}_1$ is the scattering vector of first image, $\vec{k}_2$ is the scattering vector of second image vector, n is the number of multi-look.

The polarimetric interference coherence matrix obeys the complex Wishart distribution, the probability density function can be defined as follows[4,5], see equation 2:

$$p(\langle T_6 \rangle) = \frac{|\langle T_6 \rangle|^{n-6} \exp\left(-n \cdot Tr\left(\sum_w^{-1} \langle T_6 \rangle\right)\right)}{K(n,6)\left|\sum_w\right|^n} \tag{2}$$

In equation 2, $K(n,6)$ is a constant, $\sum_w = E(\vec{w}_i \vec{w}_i^{*T})$ is a (6×6) correlation matrix.

By define a Bayesian maximum judgment criterion to the correlation matrix $\langle T_6 \rangle$ which get by polarimetric radar interference, then the polarimetric interference radar image can be classified:

If for all of the polarimetric interference in the radar image pixels $j \neq m$, and $d(\langle T_6 \rangle, X_m) \leq d(\langle T_6 \rangle, X_j)$, then the pixels belongs to the category of X_m.

The distance define of above judgment standards see equation 3:

$$d\left(\langle T_6 \rangle, X_m\right) = \ln\left|\sum\nolimits_m\right| + Tr\left(\sum\nolimits_m^{-1} \langle T_6 \rangle\right) \tag{3}$$

In equation 3:

$$\sum\nolimits_m = \frac{1}{n_m} \sum_{i=1}^{n_m} \langle T_6 \rangle_i, \ \forall \langle T_6 \rangle_i \in X_m \tag{4}$$

$\sum_m$ is the average value of all the pixel's polarimetric interference matrix which belong to the category X_m, n_m is the number of pixels in class X_m.

In the process of polarimetric interference unsupervised Wishart ML classification also requires class aggregation according to the discrete degree W_i within the class and average distance B_{ij}. The definition of the discrete degree W_i within the class and average distance B_{ij} respectively see equation 5 and 6:

$$W_i = \frac{1}{n_i} \sum_{k=1}^{n_i} d\left[\left(\langle T_6 \rangle_k \in X_i\right), X_i\right] = \ln\left|\sum\nolimits_i\right| + 6 \tag{5}$$

$$\begin{aligned} B_{ij} &= \frac{\frac{1}{n_i} \sum_{k=1}^{n_i} d\left[\left(\langle T_6 \rangle_k \in X_i\right), X_i\right] + \frac{1}{n_j} \sum_{k=1}^{n_j} d\left[\left(\langle T_6 \rangle_k \in X_j\right), X_j\right]}{2} \\ &= \frac{W_i + W_j + Tr\left(\sum_i^{-1} \sum_j + \sum_j^{-1} \sum_i\right)}{2} \end{aligned} \tag{6}$$

In conclusion, the main steps of polarimetric interference unsupervised Wishart ML classification are as follows: first, utilize one of the polarimetric images which used for interference conduct Wishart unsupervised classification, then conduct optimal coherent decomposition to the two interference images, and conduct classification according to the distribution of the points on the $H_{\ln t} - A_{\ln t}$ plane; then class $m1$ and class $m2$ classification results are obtained respectively.

Step2: conduct the merger of class and get $N(\ N = m1 \times m2\ )$class as a result of classification, the specific practice is to assign the pixel belongs to the class X_{1i} in the class $m1$ and also belongs to the class X_{2j} in the class $m2$ to the class $X_{i+m1\times(j-1)}$.

Step3: utilize the equation $\sum_k = \frac{1}{n_k}\sum_{i=1}^{n_k}\langle T_6\rangle_i, \forall\langle T_6\rangle_i \in X_k$ to calculate the 6 order characteristic covariance matrix of each X_k class.

Step 4: each pixel is classified by minimum $d(\langle T_6\rangle, X_m)$ distance method.

Step5: according to the discrete degree within the class and average distance between class for class aggregation; Otherwise go to step 3 to continue calculate 6 order characteristic covariance matrix of each kind X_k.

The flow chart of this method shown in figure 1:

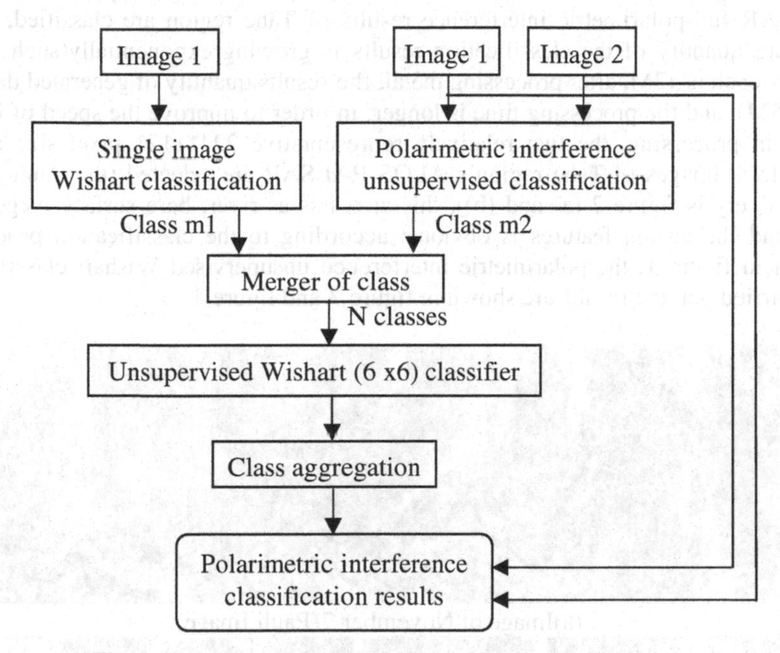

Fig. 1. Flow chart of polarimetric interference classification

4 Introduction of the Study Area and Data

The study area of Tahe is located in north of China and belongs to Heilongjiang Province, the specific geographical position is: east longitude 123.4°~124.9°, north latitude 52.3°~53.4°. This area is among the mountains with full forest cover and rich vegetation types, and a river goes through the zone of study, all the above provide convenient conditions for follow-up study.

ALOS satellite is the first full-polarimetric SAR satellite in the world, it can obtain the L-band full-polarimetric data (HH, HV, VH, VV), and can be used for repeat track interferometry[6]. The data of this study is the full-polarimetric PALSAR data of Tahe region obtained under full-polarimetric mode, and the data acquisition time is on May 7, 2007 and November 7, 2007, details are shown in table 1.

Table 1. Track and time of PALSAR data of Tahe region

Area	File name	Track	Time
China (Tahe region)	A0907325-033	421-1050	20071107
China (Tahe region)	A0907325-034	421-1050	20070507

5 Classification and the Results Analysis Based on ALOS PALSAR Polarimetric Interference

Based on polarimetric interference classification process, in this paper, the ALOS PALSAR full-polarimetric interference results of Tahe region are classified, due to the data quantity of the classification results is growing exponentially(such as two images each is 42M, after processing the all the results quantity of generated data will be 895M), and the processing time is longer, in order to improve the speed of interference in processing, the two relatively representative 224*1121 pixel size area in multi-look images of Tahe region's ALOS PALSAR are selected (the Pauli images respectively is figure 2 (a) and (b)), this area has as river, bare surface, vegetation, etc., and the terrain features is obvious; according to the classification process be shown in figure 1, the polarimetric interference unsupervised Wishart classification was carried out, the results are shown in figure 3 and figure 4.

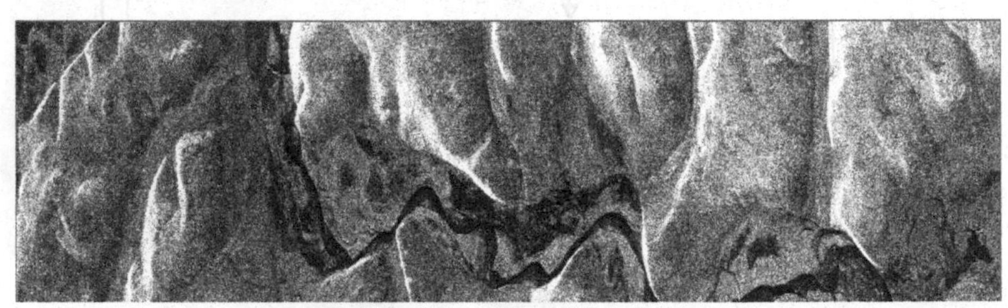

(a)Image of November 7(Pauli Image)

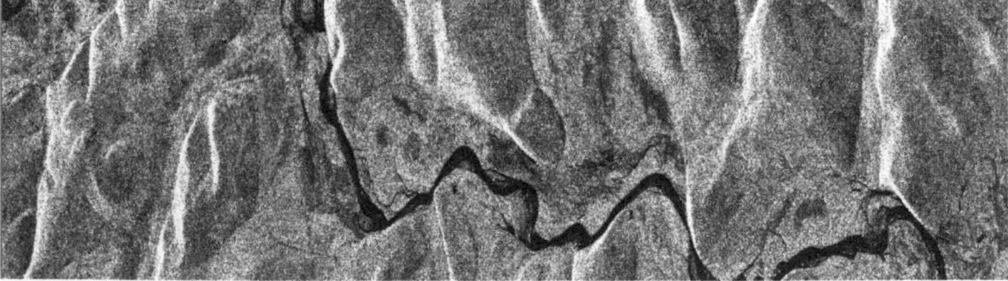

(b)Image of May 7(Pauli Image)

Fig. 2. Image of Study area

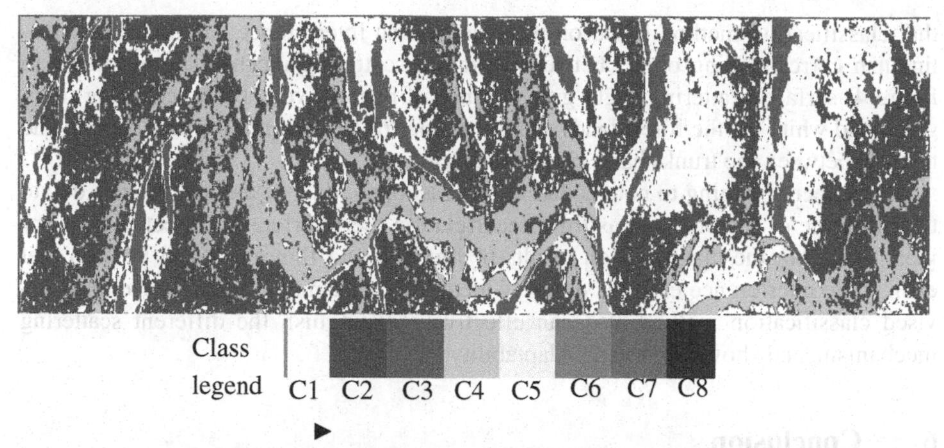

Fig. 3. Polarimetric interference unsupervised classification result of Wishart H/a (eight classes)

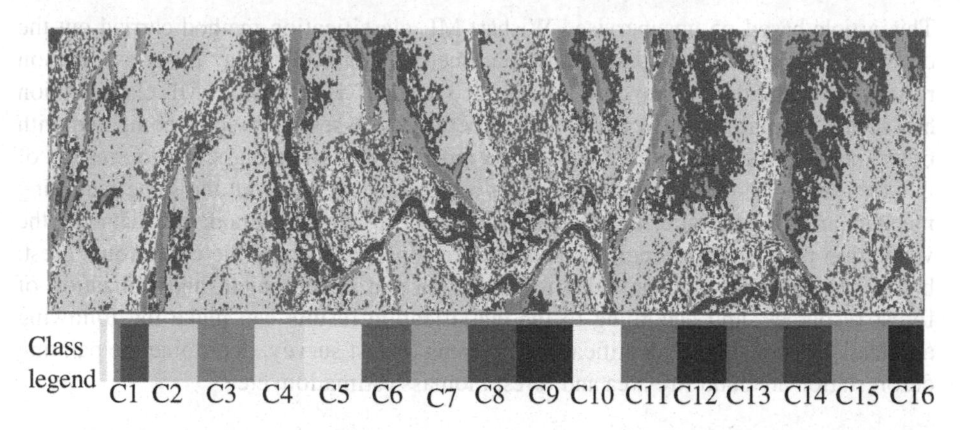

Fig. 4. Unsupervised Wishart H/a classification (8 classes) and unsupervised Wishart H/ A /a classification (16 classes)

The following conclusions can be concluded from the classification results of polarimetric interference unsupervised Wishart H/a classification (8 classes) and unsupervised Wishart H/ A /a classification (16 classes):

1) For ALOS PALSAR polarimetric images, the mountainous area which coverage with high vegetation, the above two kinds of methods can effectively classified out the classes with the same slope and slope direction; at the same time easy to distinguish the bare area and river valley.

2) The above figures also show that these two kinds of classification methods have strong comparison of clustering (such as in H/a classification results the bare surface along the river and river is divided into the same category), clear boundaries between classes, etc. For ALOS PALSAR full-polarimetric images application, polarimetric interference unsupervised classification results can meet the needs of different users.

3) It can be seen from the classification results of polarimetric interference unsupervised Wishart H/a classification and unsupervised Wishart H/A/a classification

that classification method based on the polarimetric interference can effectively distinguish corresponding object between different scattering mechanism, as shown in figure 4: surface scattering (light green on behalf of the river), body scattering (dark green and white for the forest land) and even order reflection (red for bare surface and the role between the trunk) has a clear boundary.

4) Another one thing to note is that in processing the data of the interference due to the influence of factors such as time decorrelation the interference effect is not very ideal, but from the classification results of figure 3 and figure 4 it can be seen that even if the interference effect is not very ideal, the polarimetric interference unsupervised classification method also can effectively distinguish the different scattering mechanism, and shows the strong adaptability.

6 Conclusion

This article based on unsupervised Wishart ML classification method carried out the classification of Tahe region's full-polarimetric PALSAR data, the classification results show that the polarimetric interference unsupervised Wishart ML classification has strong adaptability, and the boundary between classes is obvious, compared with other methods this method belongs to the most optimal. The classification results of this test belong to the type information, is a broad categories with the same scattering mechanism characteristic, the type information provide the foundation data for the vegetation parameters of region advantage tree species such as the coniferous forest, broad-leaved forest, and provides reference for the development and utilization of forest resources, and according to the classification results can make the following applications: forest type identification, mapping, forest survey, determine the painting deforestation area and fire area and forest biomass estimation, etc.

Acknowledgments. This work was financially supported by Heilongjiang Province Ordinary University key Laboratory of Spatial Information Comprehensive Laboratory Open Projects (KJKF-12-06), Dr. Fund project of Heilongjiang Institute of Technology (2012BJ03, 2012BJ06) and Reserve Talents of Universities Overseas Research Program of Heilongjiang([2013]350).

References

1. Mermoz, S., Toan, T.L., Villard, L., Réjou-Méchain, M., Seifert-Granzin, J.: Biomass assessment in the Cameroon savanna using ALOS PALSAR data. Remote Sensing of Environment, 1–11 (2014)
2. Cartus, O., Santoro, M., Kellndorfer, J.: Mapping forest aboveground biomass in the Northeastern United States with ALOS PALSAR dual-polarization L-band. Remote Sensing of Environment 124, 466–478 (2012)
3. Yang, Z.: Synthetic aperture radar interference and polarimetric interference technology research. The Chinese academy of sciences doctoral thesis (June 2003)

4. Ferro-Fami, L., Pottier, E., Lee, J.S.: Unsupervised Classification and Analysis of Natural Scenes from polarimetric Inteferometric SAR Data. In: IEEE Proceedings of IGARSS 2001, pp. 2715–2717 (2001)
5. Ferro-Famil, L., Pottier, E., Lee, J.S.: Classification and Interpretation of Polarimetric Interferometric SAR Data. In: IEEE Proceedings of IGARSS 2002, pp. 635–637 (2002)
6. Gan, T.H., Min, L.: The advanced land observing satellite-ALOS. Surveying and Mapping of Jiangxi, pp.11–15 (2007)

A Segmentation Method for Point Cloud Based on Local Sample and Statistic Inference

Yanmin Wang[1,2,3] and Hongbin Shi[1]

[1] Wuhan University, State Key Laboratory for Information Engineering in Surveying, Mapping and Remote Sensing, 129 Luoyu Road, Wuhan, 430079, China
[2] Beijing University of Civil Engineering and Architecture, Key Laboratory for Urban Geomatics of National Administration of Surveying, Mapping and Geoinformation, 1 Zhanlanguan Road, Beijing, 100044, China
[3] Engineering Research Center of Representative Building and Architectural Heritage Database, Ministry of Education, Beijing, 100044, China

Abstract. Terrestrial Laser Scanning has been established as a leading tools to collect dense point cloud over object surface. The collected point cloud does not provide semantic information about the scanned object. Therefore, different methods have been developed to deal with this problem, it may be the most effective one to segment point cloud into basic primitives. This paper intrudes a modified method based on RANSAC to identify planar, cylindrical and spherical surfaces in point cloud. The method firstly construct space division by 3D grid, draw a random sample to determine a sub-cell and carry local-RANSAC method to detect multi-primitive models in it, and get the candidate model(s) by local score, then the best model in the candidate modelscan be obtained by statistic inference, and its consensus set determined by distance and normal vector constrains in the global range. Finally the experimental results show that our method can segment man-made objects with regular geometry shape efficiently, and deal with the over and under segmentation properly.

Keywords: segmentation, multi-primitives, local sample, statistic inference, consensus set.

1 Introduction

Over the past few years, Terrestrial Laser Scanning (TLS) has been adopted as a popular tool to acquire spatial data. It can document objects with dense point clouds in a short time. However there are no interpretations and scene classification performed during data acquisition. Accordingly, it is very necessary for the collected point cloud to be processed to extract the required information for different applications, such as Building Information Model (BIM) construction[1], industrial site modeling[2],cultural heritage documentation[3].Segmentation and identification of geometric primitives are considered as the fundamental steps in data processing. This step is crucial since the accuracy of further laser scanning data processing activities depends on the validity of the segmentation results[4].

F. Bian and Y. Xie (Eds.): GRMSE 2014, CCIS 482, pp. 274–282, 2015.

Segmentation is the process to divide point cloud into several disconnected subsets according to some properties.These properties mainly include geometric information (curvature[5], Gaussian sphere[6], etc.), spectral information binding with geometry[7],[8]. The method based on geometric information can be divided into: edge-based segmentation, surface-based segmentation and other methods[9]. The edge-based approaches divide point cloud by first identifying the surface extents such as boundaries and intersections. Then each surface is isolated from each other by the identified extents. The surface-based approaches identify points with similar surface attributes, which are then grouped together. Other methods mainly includethe method based-on the scanline [10], the method based-on Level Set segmentation[11], the method based on Reeb graph partitioning[12], and so on.

In the artificial environment, most of the objects are made up of regular shapes, such as plane, cylinder, sphere. The method based-on surface can provide people with abstract expression of point cloud, which is very necessary for further analysis and expression, so this method has beenwidely recognized.In the method based-on surface, Random Sample Consensus(RANSAC) and HOUGH transform are the two most famous methods.However the original HOUGH transform has the deficiency of low efficiency and high memory consumption[13], RANSAC has the characteristics of low memory consumption, simple conceptionand easy to extend, so it has been widely used in the point cloud segmentation. For example: BOLLES extractedcylindersfrom range image using RANSAC method[14]; CHAPERON foundcylindersfrom point cloud by RANSAC method and Gauss image[15]; Wang segmented the airborne laser point cloud into planes to obtain the contour information[16]; Huangpuextractedrotating surface by RANSAC[17]; Tarek detected the planar datasets from scattered point cloud by Seq-RANSAC method[18]. All of these literatures only detect somekind of shape and lack the mechanism of multiple shapes detection. But in reality, artificial objects are generally made up of different kinds of regular shapes, so it is very necessary to design a method to detect multiple kinds of geometric primitives in the massive point cloud.

This paper proposes an automatic approach to deal with the shape detection in TLS point clouds. The algorithm proposed in our approach is able to rapidly and automatically detect planes, cylinders, spheres at the same time. The novelty of the approach lies in two aspects: (1) It adopts the local sample strategy to avoid most of the fake models in RANSAC and speed up the shape detection; (2) It uses statistic inference to distinguish the shape type in a local area and its neighbors.

Our proposed method comprises three major stages: First, the original dataset is divided by a 3d Grid structure, and the normal vector is computed by Principal Component Analysis in a local neighborof a given point.Then,the optimal models corresponding to each shape types are obtained by a local-RANSAC method in a local grid unit.At last the surface type which is most consistent to the real dataset is determined by a statistic inference.

The remainder of this paper is organized as follows. The segmentation method is presented in Sec. 2. Section 3 outlines the experiments conducted and analyzes the results obtained. Finally, we conclude and outline plans for future work in Sec. 4.

2 Methodology

RANSAC method was first put forward by Fischler and Bolles in 1981[19], it estimates the preset mathematical model from the observed datasets containing noise by an iterative manner. Each iterationconsists of two steps: hypothesis and scoring. When more observed data is involved in the process, more hypothetical models are generated with global sample and more time is needed to distinguish the optimal model. However, Myatt proved that the local sampling strategy can greatly increase the probability that the selected necessary sampling points belong to the same model in two-dimensional space, and argued that this conclusion can be extended to higher dimension space[20]. So in this section, we first divide point cloud by three-dimension gird and limit the sample into a local range, and then distinguish the shape type inthis local range. The algorithm flowchart is shown in Fig.1.

2.1 Data Preprocessing

Space partition. It is very difficult to query the neighbors and compute the properties in point cloud for its large data size and its scattered feature. So we build 3D Grid structure, which has the characteristic of easy implementation, higher data retrieval efficiency, for the point cloud in the global range and KDTree in the local range to establish the spatial index mechanism of point cloud.

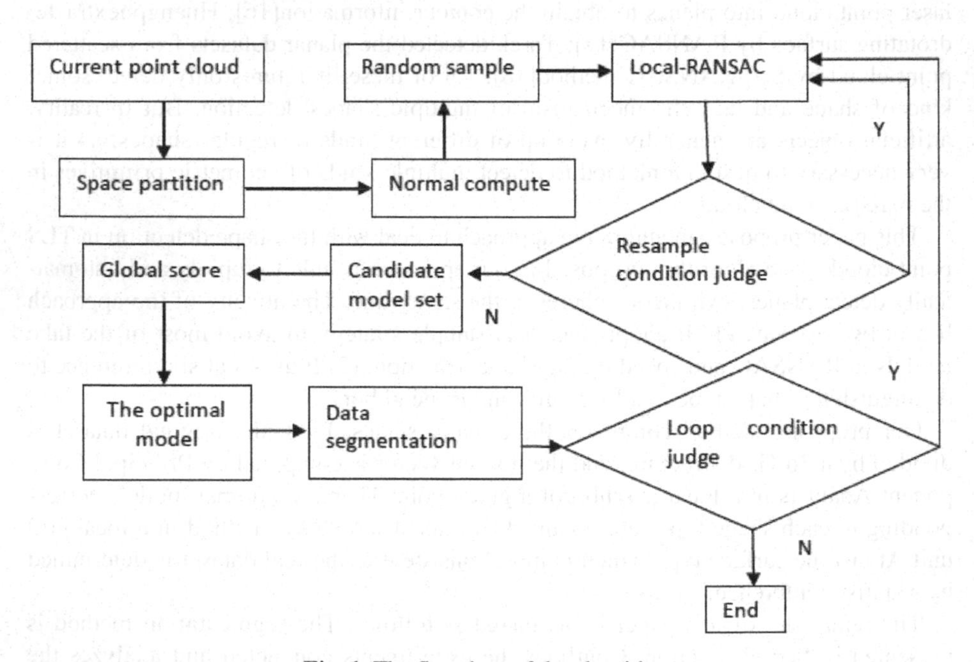

Fig. 1. The flowchart of the algorithm

Normal vector estimation. For a given point p, its neighbors $Ngh=\{p_1, p_2, p_3, \ldots\ldots p_k\}$ can be determined by $Grid + KdTree$, and the covariance matrix constructed by Eq(1) is decomposed into eigenvalue and eigenvector(see Eq(2)). The eigenvector corresponding to the least eigenvalue is considered as its normal vector.

$$C = \frac{1}{k}\sum_{i=0}^{k-1}(p_i - p)\cdot(p_i - p)^T \tag{1}$$

$$C \cdot \vec{V_i} = \lambda_i \cdot \vec{V_i}, i \in \{0,1,2\} \tag{2}$$

2.2 Local RANSAC

In the current point cloud dataset, a point is sampled randomly, and the grid unit can be determined according to its spatial location, then RANSAC method is implemented within the grid unit, in which the optimal models of plane, cylinder and sphere are detected in turn. The algorithm can bedescribed by the following steps:

(1) Model Construction. The first sample point is selected randomly in the current dataset, and the grid unit can be determined by its location, the remaining necessary samples are selected randomly within this local grid unit, then the random model M_i is constructed by the necessary samples. Plane: a point with norm vector can construct plane model; Sphere: two points with normal (p_1, nm_1), (p_2, nm_2) can determine two lines l_1, l_2 respectively, the midpoint O of the shortest line segment between l_1, l_2 is taken as the sphere center, and define $r = \frac{\| p_1 - o \| + \| p_2 - o \|}{2}$ as the sphere radius. Cylinder: we select two points with normal (p_1, nm_1), (p_2, nm_2) randomly and define the direction of the cylinder axis with $a = nm_1 \times nm_2$, the midpoint O of the shortest line segment between l_1, l_2 is taken as the center, and $\frac{op_1 \cdot \sin(a\cos(\frac{op_1}{\|op_1\|}\cdot a)) + op_2 \cdot \sin(a\cos(\frac{op_2}{\|op_2\|}\cdot a))}{2}$ is taken as the radius.

(2) Model Score. The point p satisfying Eq-(1) and Eq-(2)is selected as the consensus sets and its size is selected as the score of M_i.

$$\theta = \mathrm{acos}(n, n') \le th_\theta \tag{3}$$

Where n' is the normal vector at the position corresponding to the projected point of p on the model M_i.

$$D_M \leq th_d . \tag{4}$$

Where D_m is the vertical distance from P to the model M_i .

(3) Iteration. The optimal model M_i corresponding to each shape type in the local grid unit can be determined by an iterative manner. The onewhose score passesthrough the given threshold is added into the candidate model set M, $M = M \cup M_i$.

2.3 Optimal Model Determination

If there is only one model in M in the local grid unit, the one is the optimal model, the next thing is to determine its consensus set in the global range. If there are more than one model in M, the next thing is to determine the optimal model in the grid unit.In theory, the optimal model is the one that has the most consistent set in the global scope.However,it still takesmuch time to score the model even for plane model in the massive point cloud.

In the field of Mathematical Statistics, it is very common to inference the proportion of some characteristics in the overall sample space by limited sample. When the sample space is very large and the sample is simple random sampling without replacement, it is very feasible to simulate the distribution characteristics of random variables with normal distribution. In the current grid unit and its neighbors, Eq-(5) can describe the process whether a point belongs to the given model. If m points

$$pt_i = \begin{cases} 1, & Y \\ 0, & N \end{cases} \quad i = 1,2, \ldots\ldots, n \quad . \tag{5}$$

out of n satisfy Eq-(5), the expectation $E(p)$ equals $\frac{m}{n}$ in the whole data set, the variance $D(p)$ equals $\frac{p(1-p)(N-n)}{n(N-1)}$, and the confidence interval can be expressed as $E(p) \pm \sqrt{D(P)}$ under a given confidence. The model corresponding to the optimal confidence interval is selected as the optimal model. If there exist more models whose confidence intervals overlap on each other, the further distinction should be made in the random selected neighbor units.

3 Experiment and Result

3.1 Data Environment and Parameters Setting

In order to verify the validity of the algorithm in this paper, the dataset rich in plane, cylinder and sphere objects is chosen as the experimental data(shown in Figure 2). Its size is 3734518, and the average spacing is 5mm. The algorithm is implemented on a 64-bit workstation with a 2.40 GHz processor (quad-core) and 16 GB of system memory (RAM).

Fig. 2. Building Point Cloud

In data preprocessing,the size of grid cell is set to 40 times of the average point density to ensure there are enough sample space in the local grid unit. In local RANSAC, the maximum iterative times is set to 10000, the threshold of normal deviation set to $0.9rad$, the distance threshold set to 2cm uniformly, and the minimum set to 50.

3.2 Shape Detection

Suppose that a random sampling point falls in blue grid unitnumbered as E (as shown in Figure 3), in which the local RANSAC method is implemented, the size of consensus set of the optimal parameters of planar, cylindrical and spherical modelare1343, 1366, 1366 respectively(shown in Table 1) and their confidence intervals are shown in Table 2, so we cannot decide which shape type the points in grid unit E belongs to. Then some unit in E's neighbors is randomly selected to judge it further,suppose D is the random selected one, its three scores and confidence intervals are shown in Table 1 and Table 2 respectively, we can see that the confidence intervals of plane is very higher than others, so we can get the optimal model this time. If not so, other units remaining in the neighbors can be selected randomly to judge it further.

Fig. 3. Multi-primitives detection

Table 1. Local score statistic

Grid No		A	B	C	D	E	F	H	I	J
Sample Num		1373	1250	1435	1483	1366	1516	1812	1423	1618
	Plane	6	1243	61	49	1343	81	63	1422	124
	Cylinder	1230	1250	1435	1349	1366	1516	1451	1423	1618
	Sphere	538	292	606	305	1366	349	611	273	653

Table 2. Global inference statistic

	Total Sample	Total inlier	E(p)	D(p)	Conf Interval(Conf :0.95)
E	1366	Plane:1343 Cylinder:1366 Sphere:1366	0.983163 1 1	0.003482 0 0	[0.976338165,0.989987] [1,1] [1,1]
D	2849	Plane:1392 Cylinder:2715 Sphere:1671	0.488592 0.952966 0.586522	0.009363 0.003966 0.009224	[0.470240746,0.506944] [0.945193373,0.945193] [0.568441988,0.604601]

3.3 Over and Under Segmentation

After obtaining the optimal model in the local unit, the next thing is to determine the consensus set in the global range and segment its consensus set from the current point cloud dataset. But there exists over and under problems in segmentation.

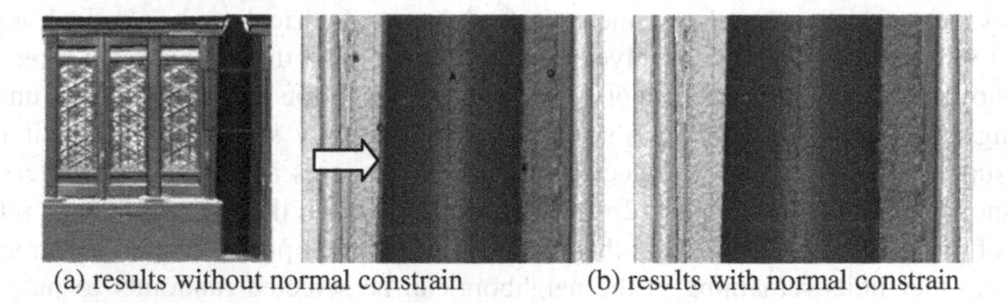

(a) results without normal constrain (b) results with normal constrain

Fig. 4. Segmentation results with normal constraint

In the process of consensus set determination, suppose the points corresponding to surface A is first segmented from the current point cloud and the points in the adjacent surfaces(see C,E in Figure4. (a)), which are segmented in a subsequent step.Considering the existence of distance tolerance, C and E are considered as a part of A and segmented from the point cloud, so it is oversegmentation for B and D, undersegmentation for A. Although the distance from the points in area C and E to the cylinder surface determined by the points in area A is less than the distance tolerance, the normal difference between them is very large, so the normal constrain can prevent C and E joining into A effectively(see Figure 4. (b)).

Another condition, which also belongs to the problem of under segmentation, is that the points belonging to the same parameter surface lie in different area (see Figure5. (a)). To deal with this condition, we use grid to part the points and spatial clustering to get the isolated objects. Coplanar segment and the final segmentation effect are shown in Figure 5. (b).

(a) Results without coplanar segmentation (b) results with coplanar segmentation

Fig. 5. Co-planar segmentation result

4 Conclusion

This paper proposed a segmentation method for TLS point cloud based on modified RANSAC method. In the proposed approach, the samplesare constrained into local range for the closer, the higher probability that the samples belong to the same surface, and theRANSAC method is carried out to detect the candidate optimal models in the local range. When there are more than one candidate optimal models after local-RANSAC, statistic inference is implemented in the local grid unit and its neighbors to distinguish which one is the most optimal. Normal vectors are pre-computed for every point in the dataset, it can decrease the number of necessary samples, which are used toconstruct hypothesis models in local-RANSAC procession, to improve the efficiency of local-RANSAC and deal with the over and under segmentation condition in the procedure of determining the consensus sets corresponding to some optimal model in the global range.The experiment results show that the method can effectively detect objects that made up of plane, cylinder and sphere. We plan to include automatic registration for point cloud based on surfaceprimitives, modeling and semantic classification based on point clouds in future work.

References

1. Xiong, X.H., Adan, A., Akinci, B., et al.: Automatic creation of semantically rich 3D building models from laser scanner data. Automation in Construction 31, 325–337 (2013)
2. Rabbani, T.: Automatic Reconstruction of Industrial Installations Using Point Clouds and Images. D. Delft. Delft University of Technology (2006)
3. Guarnieri, A., Vettore, A., Camarda, M., Domenica, C.: Automatic registration of large range datasets with spin-images. Journal of Cultural Heritage 12(4), 476–484 (2011)
4. Lari, Z., Ayman, H.: An adaptive approach for the segmentation and extraction of planar and linear/cylindrical features from laser scanning data. ISPRS Journal of Photogrammetry and Remote Sensing 93, 192–212 (2014)
5. Jiang, J.J., Zhang, Z.X., Ming, Y.: Data Segmentation for Geometric Feature Extraction from Lidar Point Clouds. In: IEEE IGRSS 2005, pp. 3277–3280 (2005)
6. Liu, Y., Xiong, Y.L.: Automatic Segmentation of Unorganized Noisy Point Clouds Based on the Gaussian Map. Computer-Aided Design 40, 576–594 (2008)
7. Shahar, B., Sagi, F.: Segmentation of Terrestrial Laser Scanning Data Using Geometry and Image Information. ISPRS Journal of Photogrammetry and Remote Sensing (2012)

8. Sapkota, P.P.: Segmentation of Coloured Point Cloud Data. Enschede, The Netherlands:International Institute for Geo-Information Science and Earth Observation (2008)
9. Belton, D.: Classification and Segmentation of 3D Terrestrial Laser Scanner Point Clouds. Bentley, Curtin University of Technology (2008)
10. Han, S.H., Lee, J.H., Yu, K.Y.: An Approach for Segmentation of Airborne Laser Point Clouds Utilizing Scan-line Characteristics. ETRI Journal 29(5), 642–648 (2007)
11. Xiao, C.X., Feng, J.Q., Miao, Y.W., et al.: Geodesic Path Computation and Region Decomposition of Point-Based Surface Based on Level Set Method. Chinese Journal of Computers 28(2), 250–258 (2005)
12. Yamazaki, I., Natarajan, V., Bai, Z.J., Hamann, B.: Segmenting Point Sets. In: Proc. IEEE Intl. Conf. Shape Modeling and Applications (SMI), vol. 6, pp. 4–13 (2006)
13. Iiiingworth, J., Kittler, J.: A Survey of the Hough Transform. Computer Vision,Graphics,and Image Processing 44(1), 87–116 (1988)
14. Bolles, R.C., Fischler, M.A.: A RANSAC-BASED Approach to Model Fitting and Its Application to Finding Cylinders in Range Data. In: Proceedings of the 7th International Joint Conference on Artificial Intelligence, pp. 637–643 (1981)
15. Chaperon, T., Goulette, F.: Extracting Cylinders in Full 3-d Data Using a Random Sampling Method and The Gaussian Image. In: VMV 2001, pp. 35–42 (2001)
16. Wang, Z., Li, H.Y., Wu, L.X., He, Z.X.: Building Outline Extraction from Airborne LiDAR Data Based on RANSAC Model. Journal of Northeastern University (Natural Science) 33(2), 271–275 (2009)
17. Huangfu, Z.M., Yan, L.H., Liu, X.M.: Parameters Extraction of Rotational Surface Based on RANSAC Algorithm. Computer Engineer and Design 30(5), 1295–1298 (2009)
18. Tarek, M.A.: Planar Surface Extraction for Complex Facades from Unstructured TLS Point Clouds. Wuhan University, Wuhan(2010)
19. Fischler, M.A., Bolles, R.C.: Random Sample Consensus:A Paradigm for Model Fitting with Applications to Image Analysis and Automated Cartography. Comm. of the ACM 24, 381–395 (1981)
20. Myatt, D.R., Torr, P.H.S., Nasuto, S.J., Bishop, J.M., Craddock, R.: Napsac:High Noise,High Dimensional robust Estimation-it's in the bag. In: BMVC 2002, pp. 458–467 (2002)

Declassified Historical Satellite Imagery from 1960s and Geometric Positioning Evaluation in Shanghai, China

Huan Mi[1], Gang Qiao[1], Tan Li[1], and Shujie Qiao[2]

[1] Center for Spatial Information Science and Sustainable Development Applications, College of Surveying and Geo-Informatics, Tongji Univeristy, Shanghai, China
[2] Software Engineering Institute, East China Normal University, Shanghai, China
{373302640,1038581087}@qq.com, qiaogang@tongji.edu.cn, qiaoshujie555@163.com

Abstract. The historical satellite imagery obtained by US since 1960s is of great significance for monitoring ecological environment changes and its sustainable development. Since 1995, US has declassified the following three batches of satellite imagery (Declassified Intelligence Satellite Photographs, DISP): DISP-1 in 1995, including the first reconnaissance satellite imagery CORONA, ARGON and LANYARD; DISP-2 in 2002, mainly GAMBIT satellite imagery and HEXAGON cartographic satellite imagery; DISP-3 from 2011, metadata (satellite imagery excluded) of GAMBIT, GAMBIT 3 and HEXAGON project information. This paper first presents in detail the declassified data, related plans, parameters and their coverages in China. To make full use of the imagery, we conducted geometric positioning analysis. With the satellite orbit parameters and other necessary parameters unavailable, we discussed geometric correction methods with KH-4B imagery covering Shanghai in 1970s as an example and came to the conclusion that the three order polynomial correction method is the most optimal. The mosaic map has also been generated, providing a basis for further application of the imagery.

Keywords: DISP, CORONA, ARGON, LANYARD, GAMBIT, HEXAGON, Geometric Correction.

1 Introduction

Historical remotely sensed imagery has always been of significant research and application value in many fields, such as ecological environment change, sustainable development, agricultural development, city expansion, shoreline change, archaeological exploration, ice dynamic monitoring, and so on. Landsat satellite images originated from 1970s is the earliest civil satellite imagery with a low resolution of 30-78 m. From 1960s to 1980s, US launched several reconnaissance satellite programs, obtaining a large number of historical surface imageries. For the concern of military and national defense security, all these satellite imageries have been kept highly confidential for a long time. Declassification of these imageries in recent years has pushed the history of studying surface change with remote sensing back to 1960s, which will greatly promote our knowledge of the earth surface half a century ago.

F. Bian and Y. Xie (Eds.): GRMSE 2014, CCIS 482, pp. 283–292, 2015.

When gathered, the images were film data and were then scanned as digital format. As a result, there was no corresponding geographic positioning information. Therefore, geographic positioning is a prerequisite for their application. Since the satellite orbit and sensor parameters are unknown, the majority of geographic poisoning study currently used approximate geometric correction models in different geographical regions [1, 2].

This paper first introduces the three batches of declassified data mentioned above and the related coverages in China. Since the images are of various distortions, geometric correction is needed before their application. With the camera parameter and exterior orientation elements unavailable, this paper then evaluates the geometric positioning accuracy of different correction methods with the KH-4B imagery of Shanghai in DISP-1 as an example, supporting the application in information management and sustainable development of historical resources.

2 Declassified Historical Satellite Imagery

Since the launch of the first reconnaissance satellite in 1960, US obtained a great number of high resolution satellite images of the earth surface. CORONA, the first batch of satellite imagery and declassified by President Clinton in 1995, is of great importance for related historical research. The second batch, including KH-7 and KH-9, was declassified in 2002, but it was not until 2011 that the system structure and metadata were declassified as the third batch. The following is a detailed introduction of the three batches of satellite images.

2.1 Declassified Satellite Imagery - 1

Table 1. DISP-1 specifications [3, 4]

	KH-1	KH-2	KH-3	KH-4	KH-4A	KH-4B	KH-5	KH-6
Function	Recon.	Recon.	Recon.	Recon.	Recon.	Recon.	Mapping	Surv.
Missions (successes)	10 (1)	7(3)	9(5)	26(20)	52(49)	17(16)	12(6)	3(1)
Acquisition Periods	8/1960	12/1960 - 7/1961	8/1961-12/1961	2/1962-12/1963	8/1963 - 9/1969	9/1967 - 5/1972	5/1962-8/1964	7/1963 - 8/1963
Type	Mono Pan.	Mono Pan.	Mono Pan.	Stereo Pan.	Stereo Pan.	Stereo Pan.	Mono Frame	Stereo Pan.
Scan(deg)	70	70	70	70	70	70	Unava.	22
Stereo(deg)				30	30	30		
Focal Length(in)	24	24	24	24	24	24	3	66
Ground Res. (ft)	40	30	25	25	9	6	460	6
Film Width(in)	2.1	2.1	2.25	2.25	2.25	2.25	5	5
Image Format(in)	2.1	2.1	2.25 ×29.8	2.18 ×29.8	2.18 ×29.8	2.18 ×29.8	4.5 ×4.5	4.5 ×25

Recon.: Reconnaissance. Surv.: Surveillance. Pan.: Panoramic. Unava.: Unavailable.

From 1960 to 1972, the first generation of American reconnaissance satellites obtained more than 860,000 images. The images declassified by President Clinton were CORONA, ARGON and LANYARD whose features were described by KeyHole (KH). CORONE systems include KH-1, KH-2, KH-3, KH-4, KH-4A and KH-4B. ARGON and LANYARD were KH-5 and KH-6, respectively.

Table 1 shows the parameters of DISP-1 mission, including launching parameter, time when the film was got and the corresponding resolution.

CORONA Satellite. The CORONA satellites were equipped with visible band panoramic cameras. The early systems (KH-1, KH-2, and KH-3) carried one panoramic camera and the latter systems (KH-4, KH-4A, and KH-4B) carried two panoramic cameras. The formation of 30° convergence angle by front and rear lens made it possible to obtain stereo image pairs as shown in Figure 1 (left).

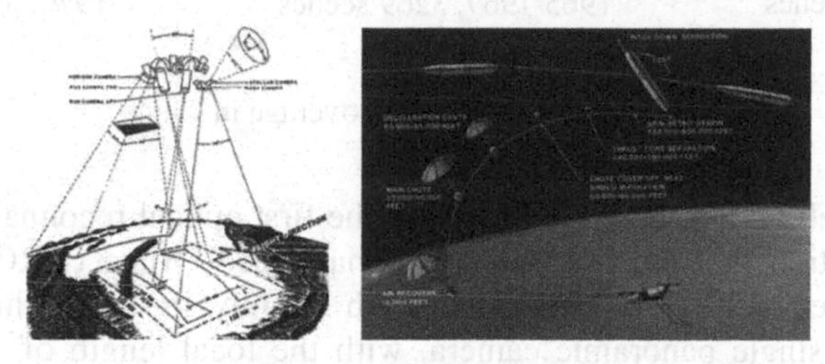

Fig. 1. Imaging geometry of CORONA (KH-4B) [5] (left) and CORONA film recovery maneuver (right) [6]

During the 1960s, digital communication systems was not developed enough to allow real time data transmission. CORONA had an interesting method of returning the pictures to earth. CORONA placed exposed film in a canister, then "de-orbited" the containers by sending them to earth through an orbit whose altitude decayed quickly. Airplanes then retrieved the canisters in mid-air according to these guidelines, and sent to the organizers for scanning. If mid-air recovery failed, a back-up water recovery with helicopter and ship would secure the capsule [7], shown in Figure 1 (right).

Figure 2 shows the coverage of CORONA satellites in China including all DISP images of KH-1, KH-2 and KH-3. Since the images obtained by KH-4, KH-4A and KH-4B were too many, only the KH-4 mission 9050, KH-4A mission 1024 and KH-4B mission 1112 were shown below.

ARGON Satellite. ARGON (KH-5) satellites were a series of reconnaissance satellites launched by US from 1961.2 to 1964.8. KH-5 tried 12 missions, among which three missions (9034A, 9058A, and 9059A) successfully delivered images of the Antarctic. KH-5 was equipped with a single frame camera and had a swath length of 556 kilometers [8]. Since the images were collected for mapping, the surface resolution was lower than that of other reconnaissance satellites. Figure 3(a) shows the coverage of KH-5 9059A in China, which almost covered the whole land of China.

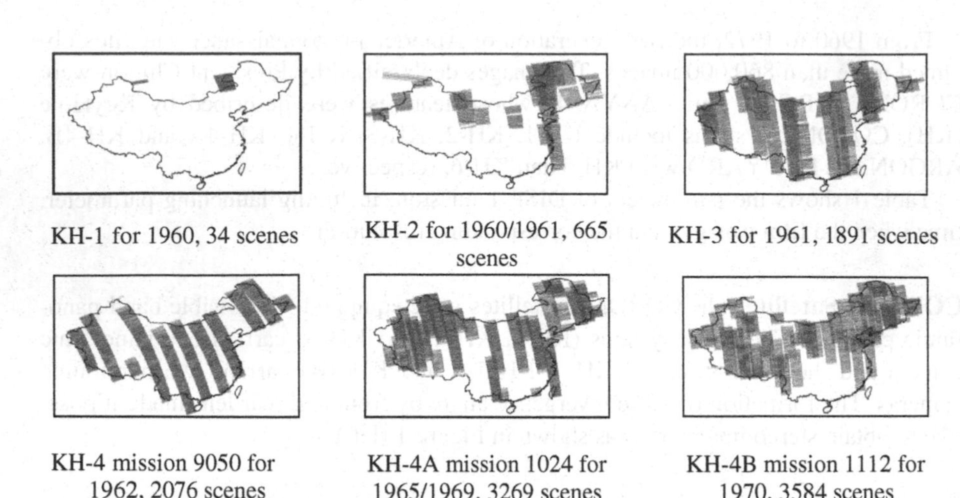

Fig. 2. CORONA images coverage in China

LANYARD Satellite. LANYARD (KH-6) is the first optical reconnaissance satellite with high resolution that the National Reconnaissance Office (NRO) attempted to develop. Launched several times from March to July in 1963, this satellite was equipped with a single panoramic camera, with the focal length of 1.67 m, ground resolution of 1.8 m and band length of about 14-74 kilometers. KH-6 tried three missions but only one was successful [9]. Figure 3(b) shows the distribution of KH-6 8003 mission in China, from which we can see that the data coverage is very small.

2.2 Declassified Satellite Imagery - 2

In 2002, US declassified DISP-2,i.e. GAMBIT (KH-7) and HEXAGON (KH-9). The principle of obtaining surface images by KH-7 and KH-9 is basically the same as that of CORONA. Both employed a telescope camera system and used the method of recovery tank transmission film. Table 2 shows the parameters of DISP-2.

Table 2. DISP-2 specifications [10]

	KH-7	KH-9
Function	Surveillance	Mapping
Missions(successes)	38 (28)	12 (12)
Acquisition Periods	7/1963-6/1967	3/1973-10/1980
Camera Type	Strip	Mapping
Ground Res.(ft)	2-4	30-35
Image Format(in)	9×4-9×500	9×18

GAMBIT Satellite. GAMBIT (KH-7) observation system is a high resolution imaging system in motion from 1963.7 to 1967.6. It flew 38 missions and obtained about

19000 pieces of panchromatic image and 230 pieces of color image. These images are of a width of 9 inches, a length of about 4-500 feet and a resolution of 2-4 feet [11]. Figure 3(c) shows the coverage of KH-7 in China.

HEXAGON Satellite. HEXAGON (KH-9) satellite mapping system was in motion from 1973.3 to 1980.10 and was used for mapping purposes and accurate geographic positioning. KH-9 frame image resolution is of 30-35 feet and the size of 9 by 18 inches. In its 12 missions, about 29,000 pieces of image were obtained. With each piece of image covering about 3400 km^2, the 29,000 pieces basically covered the whole earth [12]. Figure 3(d) shows the coverage of 684 pieces of image of mission 1211.

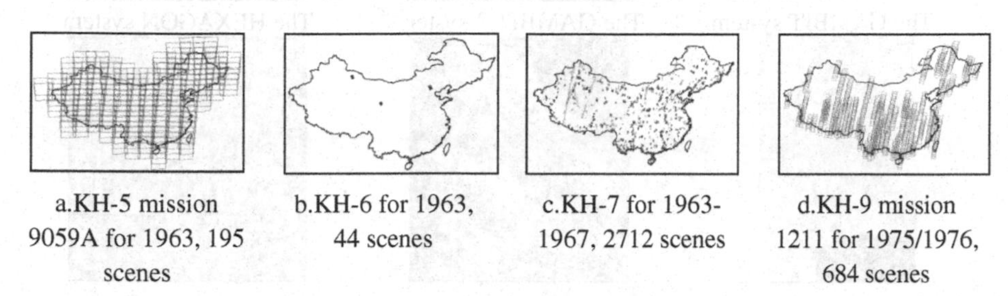

a.KH-5 mission 9059A for 1963, 195 scenes | b.KH-6 for 1963, 44 scenes | c.KH-7 for 1963-1967, 2712 scenes | d.KH-9 mission 1211 for 1975/1976, 684 scenes

Fig. 3. KH-5, KH-6, KH-7 and KH-9 coverages in China

2.3 Declassified Satellite Imagery - 3

Since 2011, US started to declassify DISP-3, i.e. the GAMBIT (KH-7), the GAMBIT 3 (KH-8) and the HEXAGON (KH-9) reconnaissance satellite. Though most of the images from the KH-7 and KH-9 mapping satellites were declassified in 2002, details of the satellite program and the satellite's construction were not declassified until 2011.

On September 17, 2011, The NRO published for the first time more than 90 records,consisting of documents, photographs, diagrams, 7 program histories, and 2 videos. The second release provided the public with a new collection of documents focused on the HEXAGON program on August16, 2012. On January 10, 2013 the NRO declassified the GAMBIT Dual Mode that was a project the NRO experimented on mission 4352 to provide the public with a new collection of documents [10]. Table 3 shows the related parameters of KH-7, KH-8, and KH-9.

Table 3. DISP-3 specifications [10]

	KH-7	KH-8	KH-9
Function	Surveillance	Surveillance	Surveillance
Missions (success)	38(28)	54(50)	20, 12 with the MCS(19)
Acquisition Periods	7/1963-6/1967	7/1966-4/1984	6/1971-4/1986
Camera Type	Strip	Strip	Panoramic
Lens	f/4.0	f/4.09	f/3.0
Aperture(in)	19.5	43.5	20
Focal Length(in)	77	175	60
Ground Res.(ft)	2-4	Better than 2	2-7
Film Length(ft)	3,000	Up to 12,241	320,000
Film Width(in)	9.46	5, 9	6.6

DISP-3 has so far been released 3 times. In addition to the 32 sample data, no other imagery was declassified. NRO has not announced later decryption time or plan. Figure 4 shows the system structure of KH-7, KH-8, KH-9 and the corresponding sample data.

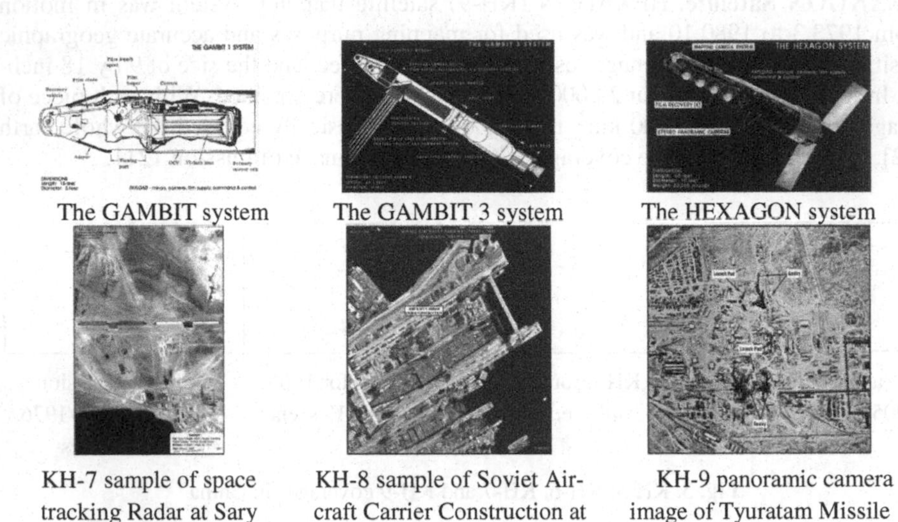

The GAMBIT system | The GAMBIT 3 system | The HEXAGON system

KH-7 sample of space tracking Radar at Sary Shagan, 5/28/1967 | KH-8 sample of Soviet Aircraft Carrier Construction at Mykolayiv, 7/4/1984 | KH-9 panoramic camera image of Tyuratam Missile Test Range, 8/1984

Fig. 4. The system structures and sample data of KH-7, KH-8, KH-9 [10]

GAMBIT 3 Satellite. GAMBIT includes two different systems: one is the GAMBIT (KH-7) and the other is GAMBIT 3 (KH-8). The KH-8 system was developed out of the KH-7 system and achieved a resolution better than 2 feet. Since the first launch in 1966.7, GAMBIT worked for nearly 20 years and was the reconnaissance satellite with the longest time in motion in the KH series [10,13].

HEXAGON Satellite. The HEXAGON (KH-9) mapping camera was flown on 12 of the 20 missions (1205 through 1216), from 1971.6 to 1986.4, devoted solely to mapping, charting, and geodesy, and was declassified in 2002. Total ground coverage for all missions was about 230,000,000 square nm [10].

3 Geometric Positioning Evaluation of Declassified KH-4B Imagery

DISP images were collected with panoramic cameras through slit scanning, so the distortion including panoramic distortion, scan positional distortion, image motion compensation distortion, and tipped panoramic distortion near the two ends of the photograph is extensive. Due to this effect, it is difficult to utilize the traditional photogrammetric method since they do not contain fiducial marks and no detail ephemeris information such as the position and velocity of satellite is available [14].

With the KH-4B CORONA data covering Shanghai as an example, this paper studies the reasonable correction model for geometric correction according to the image deformation characteristics. The method of polynomial correction is used with the sensor orbit ephemeris parameters and imaging parameters unavailable.

3.1 Study Area and Data

The study area is Shanghai, China, located between 120° 51' E - 122° 12' E and 30° 40' N - 31° 53' N, as shown in Figure 5. The research data consist of 11 scenes KH-4B images collected on December 6, 1970, of the mission 1112, with a ground resolution of 1.8 m.

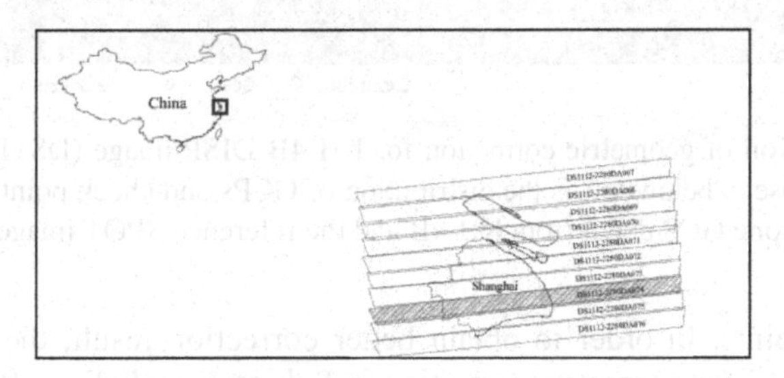

Fig. 5. Study area and KH-4B 1112 mission coverages in Shanghai, China

3.2 Distortion Characteristics and Geometric Correction of CORONA KH-4B Images

Based on the deformation characteristics of panoramic camera, geometric correction of KH-4B image was made using the partition polynomial and the Rubber Sheeting method.

The coordinates of the Ground Control Points (GCPs) were used to calculate the polynomial coefficients. Then the coordinates of any pixel of original image were calculated and interpolation was made using the image intensity to obtain geometric corrected image. The basic idea of the correct method is to simulate the image while assuming that image deformation is the result of translation, scaling, rotation, affine, bending and deformation of higher order. In Rubber Sheeting, uniform distribution of GCPs in the images forms a triangle. Then the coefficients of each triangle region were calculated one by one. Finally, conversion was made by the polynomial coefficients.

Figure 6 shows original data of a certain strip of the KH-4B mission 1112. In this research the strip is divided into 8 sub regions (Subsets), with two adjacent Subsets having a certain degree of overlap. Because great changes have taken place in Shanghai since 1960s, the selection of feature points with no changes as the control points is the main challenge.

GCPs were selected in the 5km grid with the 5m resolution SPOT orthophoto in Shanghai local coordinate system for reference. The left figure below in Figure 6 shows the distribution of GCPs and checking points in Subset 4, including 50 GCPs

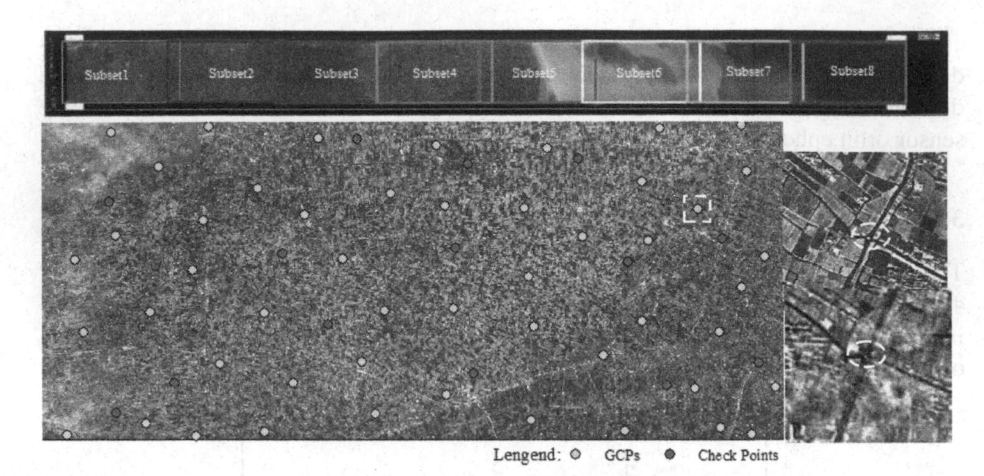

Fig. 6. Configuration of geometric correction for KH-4B DISP image (DS1112-2280DA074). Above is the 8 subsets, below left is the distribution of GCPs and check points in subset 4, and below right shows one GCP detail from KH-4B and the reference SPOT images.

and 14 check points. In order to obtain better correction result, the following four methods were used for geometric correction in Subset 4, including affine transformation, two order polynomial, three order polynomial and Rubber Sheeting. Correction results were assessed according to the accuracy of check points. Table 4 shows the accuracy of the check point in Subset 4.

Table 4. Accuracy of geometric correction results from four methods

	X Residual (m)	Y Residual (m)	RMS Error (m)
Affine	83.85	25.79	87.73
Second Order Polynomial	7.57	8.04	11.04
Third Order Polynomial	0.89	1.10	1.42
Rubber Sheeting	1.26	1.14	1.70

3.3 Result Analysis

The check point residuals of affine geometric correction in X direction and Y direction are 83.85 m and 25.79 m, respectively, with the Root Mean Square Error (RMSE) of 87.73 m. The larger error shows that the affine transformation method is not suitable for the geometric correction of panoramic camera image. When the two order polynomial geometric correction is used the RMSE of the check point is 11.04 m, far larger than the image resolution of 1.8 m.

The more complex the image distortion, the higher the polynomial order adopted in geometric correction should be. The check point residuals obtained through three order polynomial in X and Y directions are 0.89 m and 1.10 m, respectively, with the RMSE of 1.42 m. Since the check point residual is less than 1.8 m, which is 1 pixel,

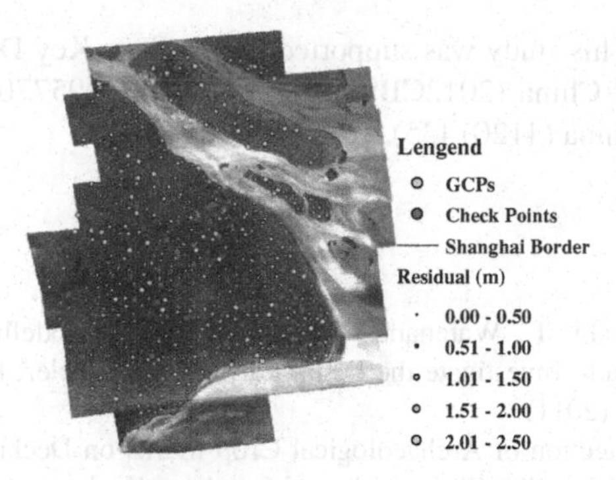

Fig. 7. DISP KH-4B mosaic of Shanghai in 1970 (698 GCPs and 215 check points)

it meets the requirement of geometric correction accuracy. The geometric correction result of Rubber Sheeting is also within 1 pixel, satisfying the accuracy requirement. However, its accuracy is worse than that obtained through three order polynomial and the imagery outside of the polygon area determined by the GCPs is cut. Thus, it is not suitable for KH-4B panoramic image geometric correction.

With the three order polynomial geometric correction method, geometric correction was made of KH-4B images covering Shanghai. Resampling was made using the nearest neighbor method with a sampling interval of 2 m. Thus obtained the corrected historical image mosaic, shown in Figure 7.

4 Conclusions

Remote sensing technology can be used to monitor the spatial variation and the natural and human factors impacting the ecological environment so as to achieve sustainable development. DISP imagery, which is of great significance, was declassified in 1995, 2002, and from 2011 to 2013. This paper first gives a detailed introduction of DISP imagery, including the related plans, detailed parameters and the coverage of some imagery in China. Since the imagery is of various distortions, geometric correction is needed. With the camera parameters and exterior orientation elements unavailable, this paper then evaluates the geometric positioning accuracy of different geometric correction methods with the KH-4B 1112 mission of DISP-1 in Shanghai, China as an example. The large error in the affine transformation and the two order polynomial method calculation shows that the correction process does not meet the panoramic camera distortion characteristics. Rubber Sheeting, although of satisfying correction precision, is not suitable for the entire image correction because the scope of the corrected image is the polygon area determined by the GCPs. The check points of RMSE obtained by using three order polynomial correction were within 1pixel, meeting the requirements of the geometric correction. DISP imagery, as historical high resolution satellite images, is not only applied in research in city expansion[15,16] and land use changes[17], but also is of high value in ecological environment change analysis and resource management.

Acknowledgments. This study was supported by the State Key Development Program for Basic Research of China (2012CB957704 and 2012CB957701) and National Science Foundation of China (41201425).

References

1. Lamsal, D., Sawagaki, T., Watanabe, T.: Digital Terrain Modelling Using Corona and ALOS PRISM Data to Investigate the Distal Part of Imja Glacier, Khumbu Himal, Nepal. JMS 8(3), 390–402 (2011)
2. Fowler, M.J.F.: Detection of Archaeological Crop Marks on Declassified CORONA KH-4B Reconnaissance Satellite Photography of Southern England. Archaeological Prospection 12(4), 257–264 (2005)
3. U.S. Geological Survey, http://www.usgs.gov
4. Kim, K.T.: Satellite Mapping and Automated Feature Extraction: Geographic Information System-Based Change Detection of the Antarctic Coast. The Ohio State University (2004)
5. KH-4B Camera System, http://fas.org/irp/imint/docs/kh-4_camera_system.htm
6. Wikipedia, http://en.wikipedia.org/wiki/Corona_(satellite)
7. Bayram, B., Bayraktar, H., Helvaci, C., Acar, U.: Coast Line Change Detection Using CORONA, SPOT and IRS 1D Images. In: XXth Congress of the International Society for Photogrammetry and Remote Sensing, Istanbul, pp. 437–441 (2004)
8. KH-5 Argon, http://en.wikipedia.org/wiki/KH-5
9. KH-6 Lanyard, http://en.wikipedia.org/wiki/KH-6
10. National Reconnaissance Office, http://www.nro.gov
11. KH-7 Gambit, http://en.wikipedia.org/wiki/KH-7
12. KH-9 Hexagon, http://en.wikipedia.org/wiki/KH-9
13. KH-8 Gambit 3, http://en.wikipedia.org/wiki/KH-8
14. Sohn, H.G., Kim, G.H., Yun, K.H.: Rigorous Sensor Modeling of Early Reconnaissance CORONA Imagery for Monitoring Urban Growth. In: Geoscience and Remote Sensing Symposium, pp. 1929–1931. IEEE Press (2002)
15. Thapa, R.B.: Monitoring landscape change in Kathmandu metropolitan region using multi-temporal satellite imagery. In: SPIE Asia-Pacific Remote Sensing, pp. 85281–85281 (2012)
16. Di Giacomo, G., Scardozzi, G.: Multitemporal High-Resolution Satellite Images for the Study and Monitoring of an Ancient Mesopotamian City and its Surrounding Landscape: The Case of Ur. International Journal of Geophysics (2012)
17. Dong, J., Chen, N., Ma, Y., Chen, J.: Land use change and information extraction of rural residential land based on Corona KH-4B imagery. In: 2012 2nd International Conference on Remote Sensing, Environment and Transportation Engineering, pp. 1–3 (2012)

Spatiotemporal Assessment of Water Quality of River Ndakotsu in Lapai, Nigeria

Naomi John Dadi-Mamud[1,*], Sonnie Joshua Oniye[2], Jehu Auta[2], and Victor Olatunji Ajibola[3,1]

[1] Biology Department, IBB University Lapai, Nigeria
[2] Biological Sciences Department, ABU Zaria, Nigeria
[3] Chemistry Department, ABU Zaria, Nigeria
dadimamud@yahoo.com

Abstract. The spatiotemporal assessment of Surface Water Quality of River Ndakotsu (latitude 9°.34N and longitude 6°.30E), Lapai-Nigeria was conducted from November 2010 to October 2012, using standard methods. The ranges obtained for the Physico-chemical parameters were air temperatures (23°c- 35 °C) water temperature (20 °C -31 °C), depth (8cm-63cm), Velocity (0.83cm/s-6.67cm/s), conductivity (10.28µScm/s-770.00µScm/s, pH (5.20-8.75), DO (0.28mg/l-6.70mg/l), BOD (1.10mg/l-31.35mg/l), water hardness (10.00mg/l-315.00mg/l), and total alkalinity (8.00mg/l-317.00mg/l). Water quality changes indicated significant differences ($p<0.05$) in water and air temperatures, depth, turbidity, velocity, conductivity, pH, water hardness, total alkalinity; Nitrate-Nitrogen and phosphate-phosphorus, but insignificant ($p>0.05$) in BOD_5 and COD between the five sampled stations within the months. Higher values of these parameters were observed at the impacted station 3.

Keywords: Anthropogenic, activities, physico- chemical parameters.

1 Introduction

Water is one of the most precious resources and as the human population grows the demands on water supply increase. In addition to the increasing demand on water, humans impact ecosystems directly through land use change and indirectly by generating non-point source pollution that is introduced into streams and rivers via urban runoff [1]. The need for clean water environment has aroused great interest and concern by many nations of the world. Contamination of streams, lakes, underground water, bays, or oceans by substances is harmful to living things. Studies have shown that biota differ greatly with changes in physico-chemical conditions in aquatic ecosystems [2]. The worldwide deterioration of surface water quality has become a growing threat to human society and natural ecosystems, hence the need to understand the spatial and temporal variability's of limnological parameters [3]. Water quality is determined by the physical chemical and biological state of the water body that influences the beneficial use of the water [4]. The physical environment of Rivers

* Corresponding author.

F. Bian and Y. Xie (Eds.): GRMSE 2014, CCIS 482, pp. 293–307, 2015.

places many constraints on organisms as well as on type and form of food that is available. Many water- dwelling organisms exploit the physical characteristics of river to obtain their foods. Generally, rivers are characterised by many interacting physical factors that produce spatial and temporal heterogeneity which may exert a major influence on benthic communities [5]. River Ndakotsu is potentially vulnerable to a variety of polluting influences. All through its course, there is a steady input of large quantities of detergents from laundry activities. At several points, the river receives large quantities of sewage and solid wastes. Further, when it rains, large volumes of run-off carrying agricultural and human wastes are discharged directly into the river. It serves as source of drinking water, domestic needs, artisanal fishing as well as dumping site of effluents. This work evaluates the water quality of the river to ascertain its suitability for aquatic biota, productive potentials and the possible disturbance of the natural ecosystem due to the effect of the influx of both agricultural and municipal effluents.

2 Materials and Methods

2.1 Description of the Study Area

The study area has the characteristic tropical climate of two distinct seasons; a dry season (November–April) and a wet season (May–October). The mean annual temperature is approx. 28°C (range 22–34°C) with mean annual relative humidity 85%. It is fed principally by precipitation, municipal effluent and surface run off from the riparian communities. It flows through the out sketch main town of Lapai from Makkara and then being join with river Etswan forming a confluence which empties into River Baro and finally empties into River Niger.

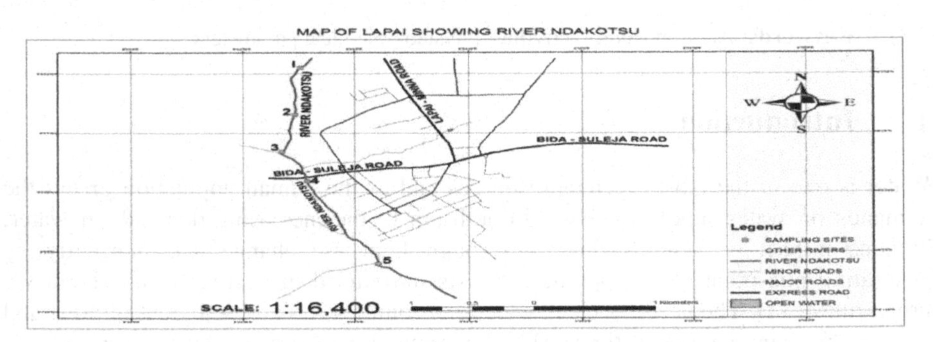

Fig. 1. Map of Lapai showing the five sampling stations

2.2 Sampling Stations

Five locations were chosen based on the different ecological factors and anthropogenic activities taking place around the river and confluence created by other tributaries, with 3km river distance between each station (Fig 1). Station 1 is located about 50m upstream from the source of the river and each. Human activities here include subsistence farming and bathing. Station 2 is located behind the Lapai general hospital and it is 3km distance from station 1. Human activities here include fishing, bathing and washing of clothes and vehicle, dumping of refuse. Station 3 is a tributary located off stream. Domestic waste and sewage are the major sources of pollutants being discharge. Station 4 is located at the discharge point of a reservoir by the Water Board with various activities like bathing, washing of clothes and vehicles, farming activities, and fishing it is characterized by tree canopy given shade to the water body. Station 5 is located downstream just below Lapai abbatoir. The abattoir effluent is composed of faecal matters, blood and ashes from burning and roasting of animals. It is being joined by river Dangana forming a confluence. Farming and fishing activities, washing and bathing also takes place at this station. The major plant species in these sampling stations are *Nymphaea sp* (water lilies), Utricularia sp. *Elaeis guineensis, Cocos nucifera* L. *Ceratophyllum submersum* L and Salvinia sp.

3 Water Quality Analysis

3.1 Field Determinations

Air and water temperature, pH, Conductivity and Total dissolved solids were measured in-situ using a HANNA portable combo waterproof pH//TDS/Temperature, Tester model HI 98130.Transparency was evaluated using Secchi disc as described by [6]. Water velocity was determined using the Pin-pong floatation technique and Water Depth was measured with the use of a rope and a metre rule. Dissolved Oxygen (DO) was determined using Digital D.O meter model 6 11-R Labtech.

3.2 Laboratory Analysis

Water samples were collected in sample bottles, labeled appropriately, stored in a portable cool box and transported to Chemical Laboratory, Water Board (Gidan Ruwa), Chanchaga, Minna Niger State for analysis. BOD was determined by modified Winkler Azide method [7], while COD and Alkalinity were determined as described by [8]. Nitrate- Nitrogen and Phosphate-Phosphorus were determined using HACH DR /2010.

3.3 Statistical Analyses

Paleontological Statistical Software Package (PAST) for windows (2007) statistical software was used for the data analyses. Individual and combined MANOVA was used to determine the level of significance among the parameters measured, and if significant, Tukey's honest significance difference test was employed to separate the means.

3.4 Results

Table 1 summarizes the mean values of the various parameters monitored at the five (5) selected stations over a 24-month time span along the river as revealed by multivariate analysis of variance (MANOVA).

Turkey's honest significant difference test (HSD) shows the significant difference between the parameters at all stations throughout the months within the four seasons.

The spatio-temporal air temp of the study area varied throughout the sampling period with station 3 having the highest air temp (35.00 °C) in April, while the lowest 23.00 °C was recorded in station 1 in December during the harmattan period (Fig. 2).Water Temperature highest value was recorded in station 3 during the month of April (31.00 °C), while station 4 had the lowest water temperature (20.00 °C) in the month of December (Fig. 3). Multivariate analysis of variance (MANOVA), shows significant variation at ($P<0.05$) for the stations, months and seasonal means (table 1). Also there was significant difference between the water temperatures at all stations and months, with station 3 the source of variation according to (HSD). Figure 4 shows the monthly water depth range variation from 63.00cm to 8.00cm, with the shallow depth at station 3 in the month of December and the deepest at station 4 in September the wet season. MANOVA shows that there were significant differences in the months, stations and seasons. Honest significant difference tests, revealed that station 1 had no significant difference with station 2, but had significant differences with stations 3, 4 and 5 (table 1). The river had the swiftest flow in the month of September (6.67cm/s) at station 4 and the slowest (0.83cm/s) in the dry season at station 2 (Fig. 5). Two ways analysis of variance (table 1), showed that there were significant differences ($P<0.05$) during the both wet and dry seasons throughout the 24months. The mean separation showed significant differences between mean water velocity at station 5.

The electrical conductivity (EC) of the river fluctuate all through the season and stations, with the highest value of 770.00μScm^{-1} in the month of May at station 3 and then decline to the value of 10.28μScm^{-1} in the same month of May at station 1(fig. 6)

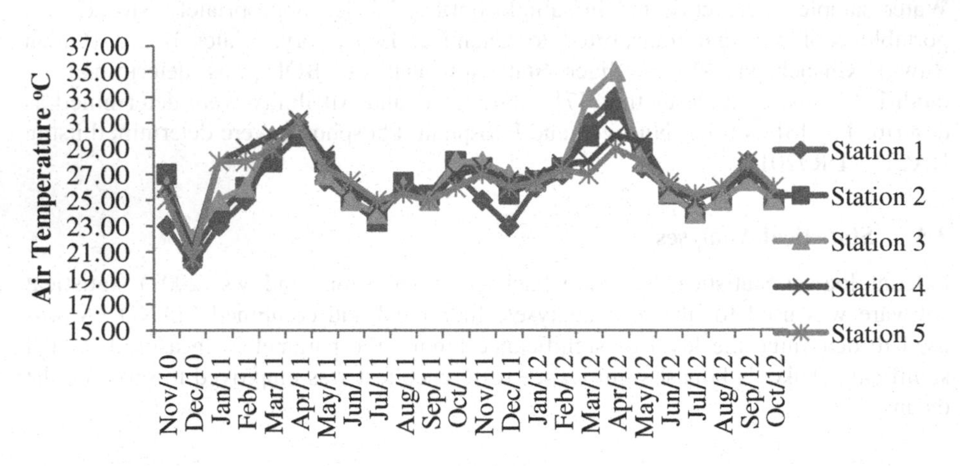

Fig. 2. Monthly variations in water and air temp of River Ndakotsu from November 2010 to October 2012

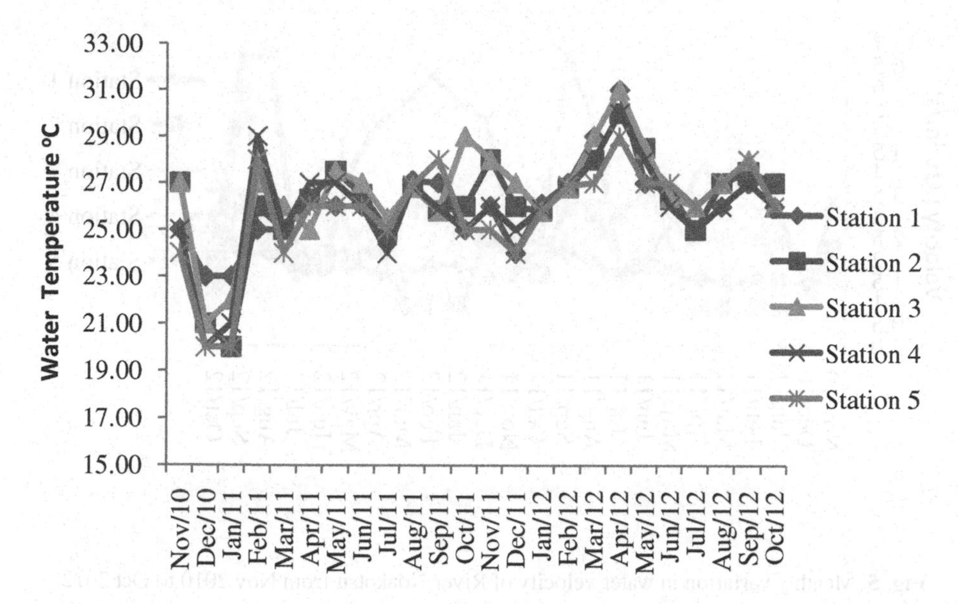

Fig. 3. Monthly variations in water and air temp of River Ndakotsu from November 2010 to October 2012

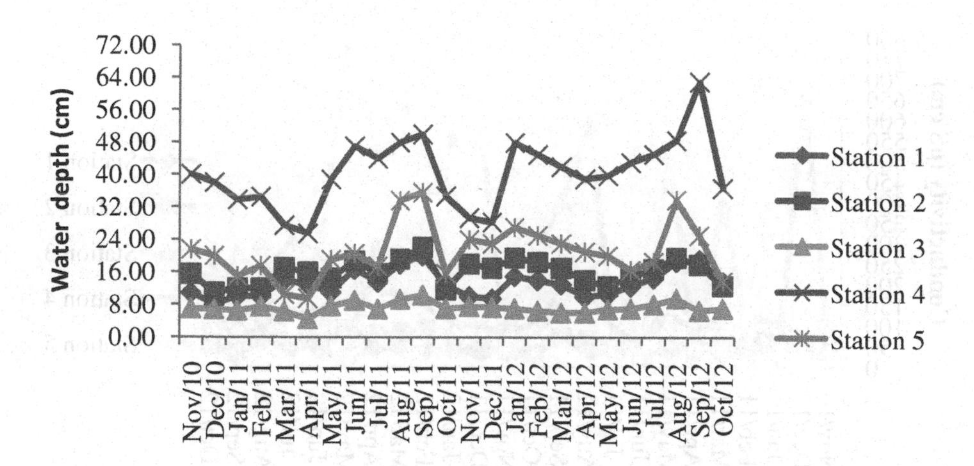

Fig. 4. Monthly variations in water depth of River Ndakotsu from November 2010 to October 2012

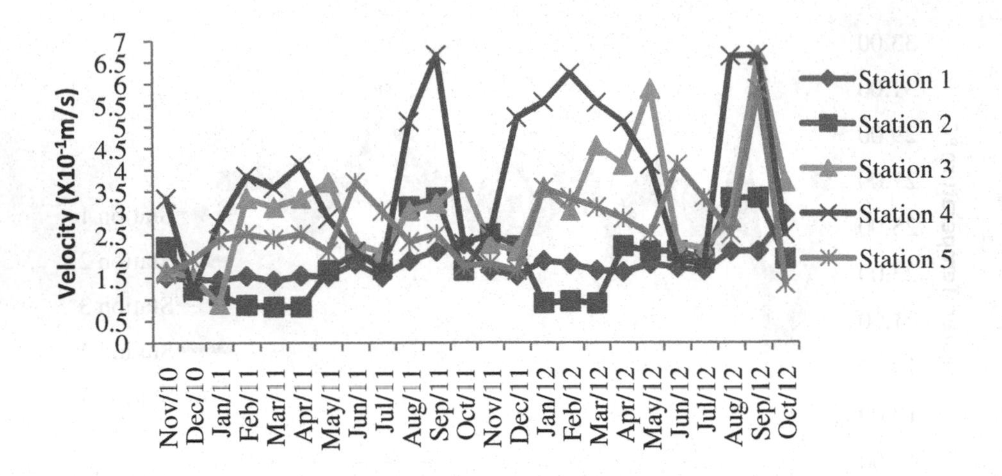

Fig. 5. Monthly variation in water velocity of River Ndakotsu from Nov 2010 to Oct 2012

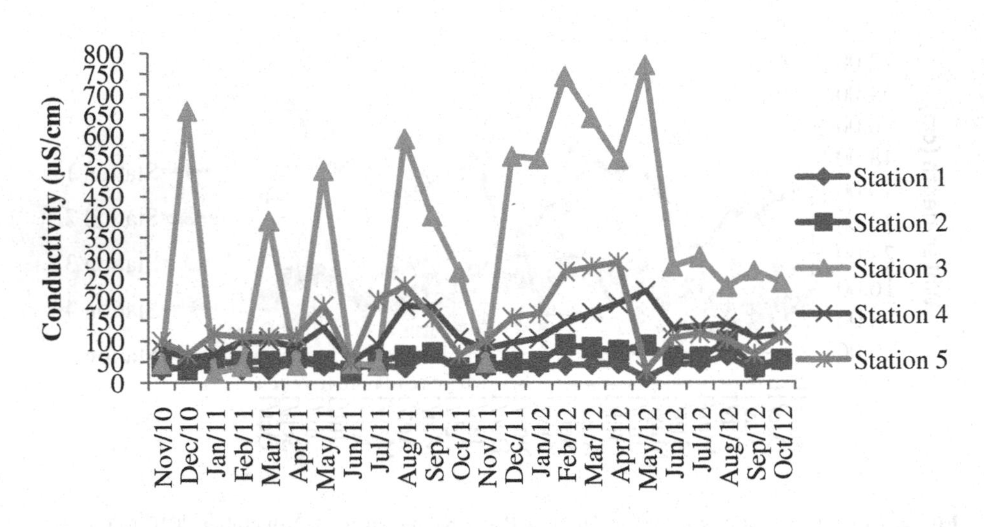

Fig. 6. Monthly variation in EC of River Ndakotsu from Nov 2010 to Oct 2012

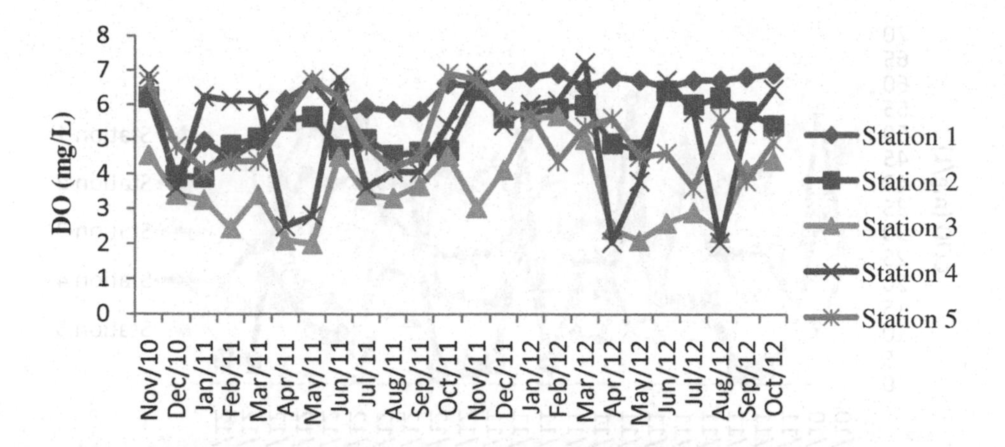

Fig. 7. Monthly variation in Dissolved Oxygen of River Ndakotsu from Nov 2010 to Oct 2012

The lowest dissolved oxygen (DO) concentration (0.28mg/l) was recorded for station 4 in May, while the highest value (6.70mg/l) was observed in November (Fig. 7). Dissolved oxygen had no significant differences within the months, but had significant difference among the stations. Stations 1 and 3 are the sources of the differences observed in these parameters as revealed by turkey's honest significant differences test (table 1). The highest Biological oxygen demands (BOD_5) concentration (31.35mg/l was recorded in December, at station 1 and the lowest concentration (1.1omg/l) was recorded in the dry season also at station 2 (Fig.8). There was no significant differences ($P>0.05$) between the mean water BOD_5 at all the stations and within the months (table 1).

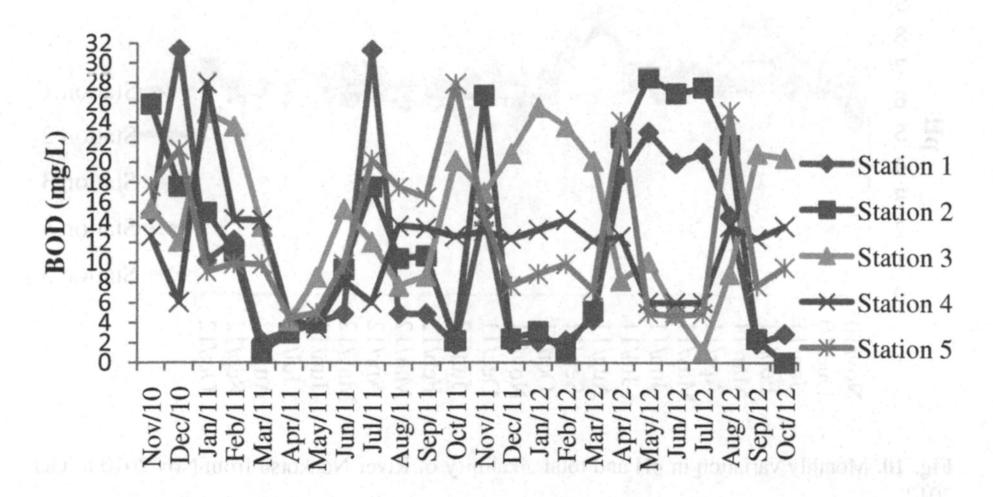

Fig. 8. Monthly variation in BOD5 and COD of River Ndakotsu from Nov 2010 to Oct 2012

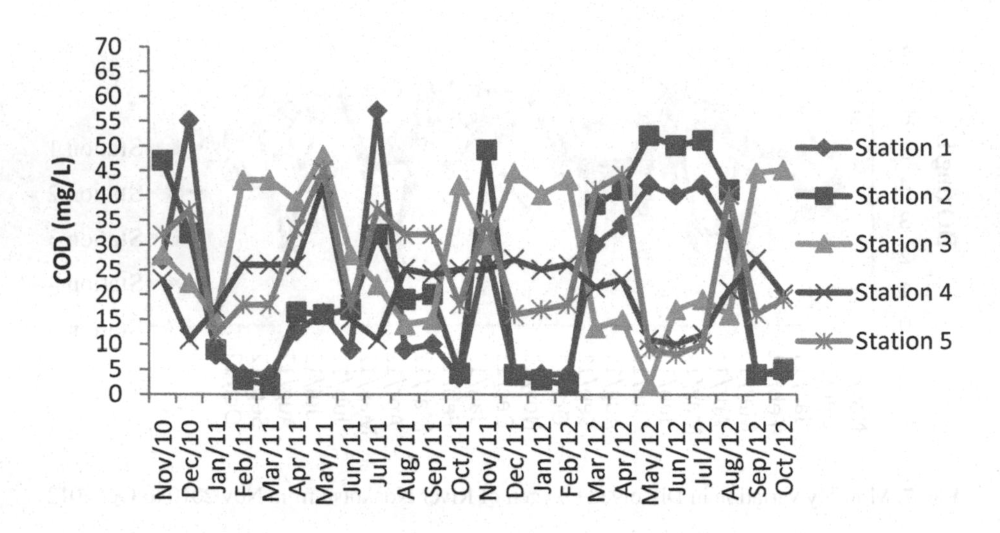

Fig. 9. Monthly variation in BOD5 and COD of River Ndakotsu from Nov 2010 to Oct 2012

Chemical oxygen demand (COD) concentrations were higher in stations 1 and 2 during the sampling period (Fig. 9). Multivariate analysis of variance (MANOVA), revealed that there were significant differences observed among the stations and within the months during both wet and dry seasons (table 1).

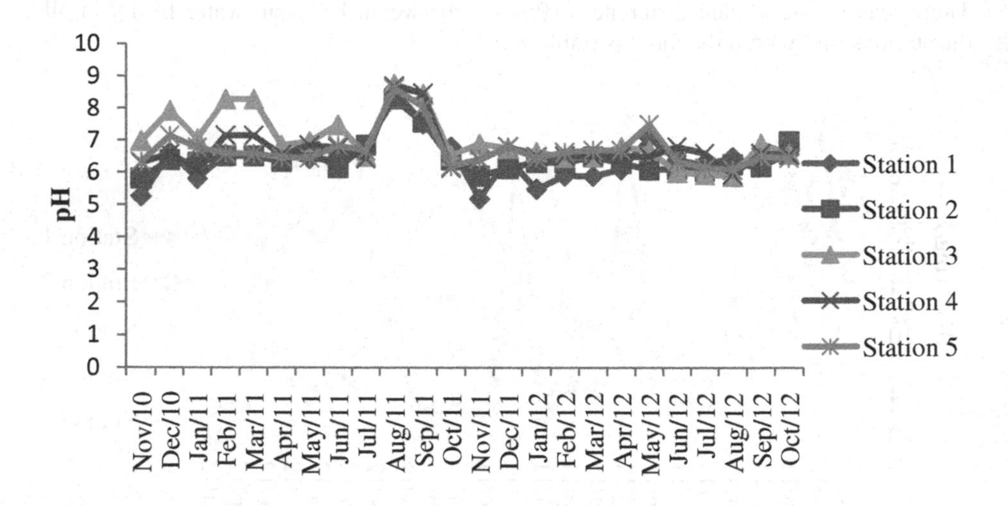

Fig. 10. Monthly variation in pH and total alkalinity of River Ndakotsu from Nov 2010 to Oct 2012

The pH of the water was found to be fairly stable during both seasons with the lowest and highest values of 5.20 and 8.75 for November and August in stations 1 and 3 respectively (Fig 10). The difference in pH values in the various stations as well as between the wet and dry season months did test statistically different (P<0.05). Station 3 had the highest value (317.00mg/l) of total alkalinity in the wet season, while station 2 had the lowest (8.00mg/l) in the month of January, the dry season (Fig. 11). Total alkalinity had significant differences among all the stations sampled within the months during the both seasons as revealed by MANOVA (table 1).

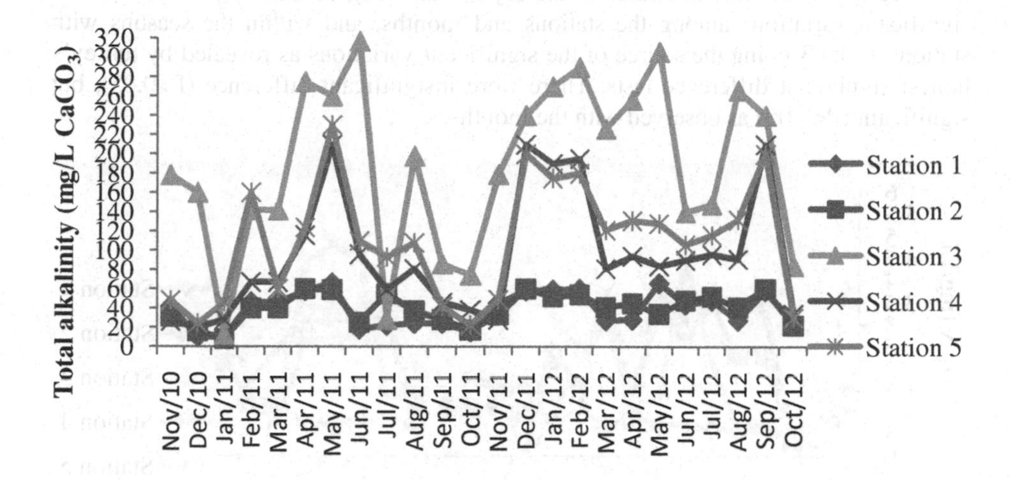

Fig. 11. Monthly variation in pH and total alkalinity of River Ndakotsu from Nov 2010 to Oct 2012

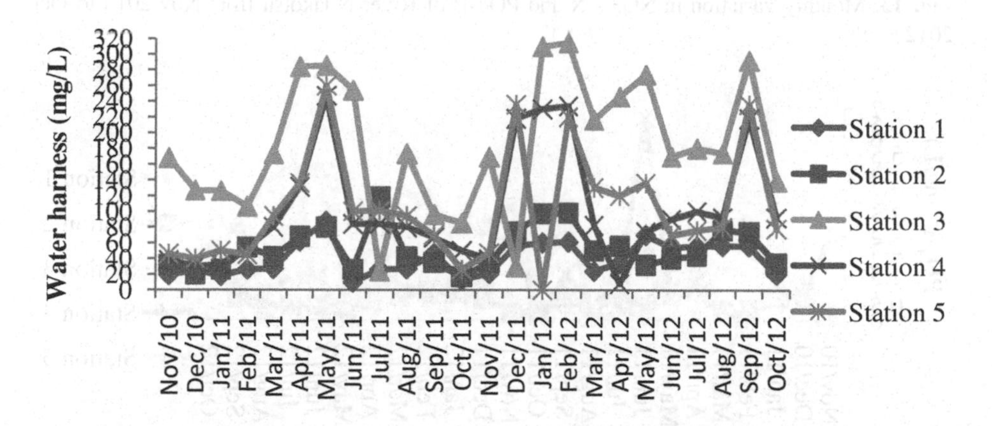

Fig. 12. Monthly variation in water hardness of River Ndakotsu from Nov 2010 to Oct 2012

Stations 1 and 2 had no significant difference, while significant differences were obtained for stations 3, 4 and 5. Water hardness had higher values during the dry months than the wet months. The highest value of 315.00mg/l was recorded in the month of February at station 3, while the lowest (10.00mg/l), at station 1 in June (Fig. 12). The hardness of the river varies within the month and among the stations, with station 3 being different from stations 1 and 2 which are not significantly different as well stations 4 and 5 which are also not significantly different as revealed by HSD (table 1).

Nitrate-Nitrogen (NO_3 - N) and Phosphate-phosphorus highest concentrations were recorded in wet months compared to the dry months (Fig 13 and 14). Table 1 shows significant variations among the stations and months, and within the seasons with stations 1 and 3 being the source of the significant variations as revealed by turkey's honest significant difference tests. There were insignificant difference (P>0.05), but significant (P<0.05) as observed with the months.

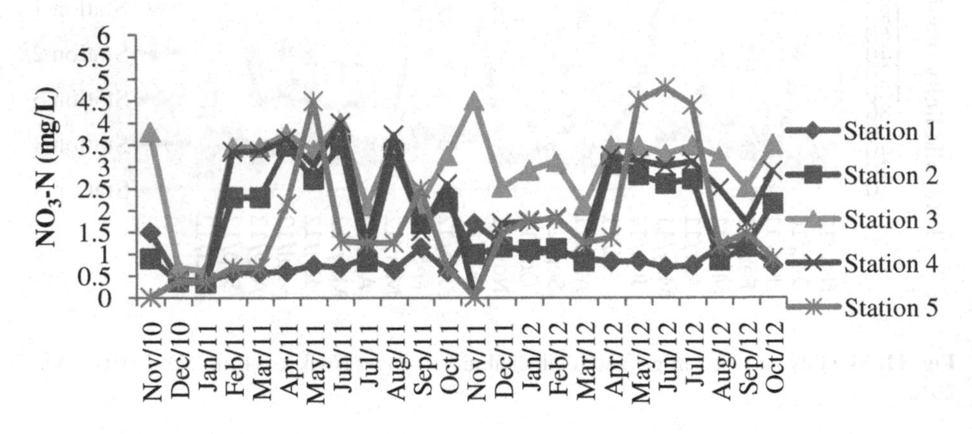

Fig. 13. Monthly variation in NO3 - N and PO4-P of River Ndakotsu from Nov 2010 to Oct 2012

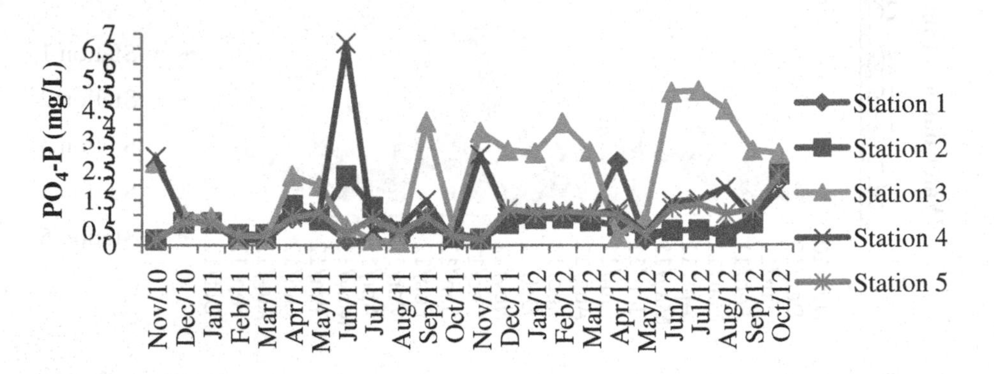

Fig. 14. Monthly variation in NO3 - N and PO4-P of River Ndakotsu from Nov 2010 to Oct 2012

Table 1. Multivariate Analysis of Variance for Spatiotemporal Variations of Physicochemical Parameters of the Study Stations, River Ndakotsu from Nov. 2010 to Oct. 2012

S/N	Parameter	Station 1	Station 2	Station 3	Station 4	Station 5	F-value ANOVA	
							Months	Stations
1.	Air temperature (°C)	26.03 ± 0.60^{a} (22.00-33.00)	26.51 ± 0.49^{ab} (21.00-32.00)	26.92 ± 0.60^{ab} (22.00-33.00)	27.01 ± 0.49^{b} (21.00-31.20)	26.65 ± 0.40^{ab} (20.50-31.00)	21.40*	2.95*
2.	Water temperature (°C)	25.98 ± 0.35^{ac} (23.00-31.00)	26.20 ± 0.43^{ac} (20.00-30.00)	26.63 ± 0.42^{a} (21.00-31.00)	25.90 ± 0.44^{bc} (20.00-30.00)	25.73 ± 0.45^{bc} (20.00-29.00)	24.12*	3.79*
3.	Water depth (m)	13.25 ± 0.67^{a} (18.00-20.00)	15.96 ± 0.61^{a} (11.00-22.10)	7.24 ± 0.26^{b} (4.50-10.30)	40.39 ± 1.75^{c} (25.50-63.00)	21.10 ± 1.37^{d} (9.50-35.60)	4.53*	234.0*
4.	velocity (m/s)	0.18 ± 0.07^{a} (0.14-0.29)	0.19 ± 0.0^{a} (0.08-0.33)	0.31 ± 0.03^{bc} (0.09-0.67)	0.39 ± 0.04^{c} (0.12-0.67)	0.27 ± 0.02^{d} (0.13-0.59)	2.70*	18.93*
5.	Turbidity (NTU)	37.81 ± 6.56^{a} (4.00-97.00)	20.14 ± 2.78^{a} (6.00-57.00)	72.49 ± 13.90^{b} (7.00-239.86)	42.20 ± 5.10^{a} (7.00-90.00)	37.25 ± 5.02^{a} (8.00-78.05)	1.29	6.48*
6.	pH	6.4 (5.2-8.6)	6.5 (5.8-8.3)	7.0 (5.9-8.8)	6.7 (6.0-8.7)	6.7 (6.0-8.6)	12.23*	10.67*

Table 1. (*continued*)

7.	Total alkalinity (mg/lCaCO$_3$)	38.65±3.72^a (10.0-66.00)	39.92±3.11^a (8.0-60.0)	188.25±17.94^b (12.40-317.0)	94.05±12.85^c (120.0-210.0)	109.80±12.37^{cd} (22.25-230.0)	5.57*	54.51*
8.	Water Hardness (mg/L)	39.75±4.31^a (10.0-87.00)	52.92±5.51^a (17.0-118.0)	183.96±17.34^c (24.00-315.00)	102.25±14.7^d (8.00-247.00)	112.55±14.30^d (26.0-260.00)	5.16*	38.8*
9.	Dissolved oxygen (mg/l)	6.14±0.17^a (3.9-6.9)	5.32±0.51^a b (3.91-	3.54±0.22^c (2.00-5.68)	5.08±0.34^b (2.05-7.20)	5.15±0.20^b (3.6-6.9)	2.27	22.04*
10.	Biochemical oxygen demand (mg/l)	10.57±1.94^a (1.85-31.35)	12.48±2.04^a (1.1-28.6)	14.55±1.46^b (1.12-25.60)	11.39±1.04^a (3.5-28.00)	12.47±1.46^a (4.5-28.0)	1.39	0.90
11.	Chemical oxygen demand (mg/l)	20.25±3.60^a (3.5-57.00)	23.26±3.83^a (2.0-52.0)	28.69±2.81^c (2.00-48.00)	21.65±1.50^{ab} (10.0-42.00)	25.21±2.44^b (8.00-45.00)	0.94	1.23
12.	NO_3-N (mg/L)	0.87±0.07^a (0.39-1.70)	1.85±0.20^b (0.35-35.0)	3.01±0.19^c (0.62-0.45)	2.23±0.25^d (0.01-4.00)	1.70±0.29^b (0.02-4.8)	3.47*	19.65*
13.	PO_4-P (mg/L)	0.71±0.13 a (0.12-	0.77±0.11 ad (0.18-	2.26±0.34 b (0.1-5.12)	1.35±0.28 c (0.18-6.7)	0.85±0.10 d (0.19-	1.76	10.28*

Note: *values are mean ± S.E. (minimum and maximum values are in parentheses). Different superscript letters in a row show significant differences (P < 0.05) indicated by Tukey Honest significant difference tests. * indicates significantly calculated F-value.*

4 Discussion

The surface water temperature of River Ndakotsu which ranged from 23.1 °C to 29.4 °C was similar and compared well with the ranges reported for the surface water temperature of most Nigerian rivers [4]. The high Air and Water temperature in the month of April and lower in the month of December and some months in wet season have been reported by [9]. The lower temperatures in December and January may be attributed to the cool dry north east trade winds, while during the rains, the low temperature may be due to the river losing heat rapidly, also to the mixing of the water, the river been shallow and highly exposed to wind action. The higher air and water temperature recorded in the dry season may be due to intense heat from solar radiation which was absorbed by the water body.This is in agreement with the findings of [10]. on some seasonal ponds in Northern Nigeria, [11] in the Sombreiro River and [4] in New Calabar-Bonny river. The highest mean depth obtained at station 4 in the wet season and lowest at station 3(impacted station), in the dry season may be due to rainfall during the wet season, which probably contributed to the varied values of bio-physicochemical parameters of this stations. For instance, depth has a profound influence in mixing and resuspension from sediment; physical, chemical and biological elements of aquatic system change with time and depth [12]. The swiftest water velocity in the wet season and slowest in the dry season was probably due to the high amount of flood which made the water flow faster and became turbulent most especially at station 4 which was more riverine than the other stations. The high turbidity observed during the dry season could be due to high level of sunlight and evaporation causing reduction in water level containing organic and inorganic matters. The temporal variation in electrical conductivity at the different stations may has been due to the types of human activities in the catchment area (high in station 3) and to the extent of usage of the river for irrigation. Higher values in the month of May, is an indication of more ions in the wet season which could be due to influx of allochotonous and inorganic materials to one of the tributaries surroundings. This is in agreement to the findings of [13] and contrary to [10]. The highest Dissolved oxygen in the dry season and lowest in the wet season, possibly could be due to wind action, characteristic of the dry season which enhances mixing of atmospheric oxygen with water. Dissolved oxygen is essential to most aquatic organisms and is greatly affected by their metabolism [14]. The result of this study is in agreement with the findings of [13]. But contrary to the finding of [4] where high dissolved oxygen concentrations were recorded in the wet season than the dry season for Lower Sombreiro River and the New Calabar-Bonny river respectively. The highest biological oxygen demand recorded in the wet season, indicate that the surface run off from the catchment area was organic in nature. The high BOD_5 value recorded in all the stations reflect high burden of organic pollution in an ecosystem, which adversely affect the water quality [5]. The mean range of chemical oxygen demand (20.25 – 28.69mg/l) showed the river to be highly polluted compared to other water bodies in Nigeria. The high COD Values are indications of discharge of domestic and agricultural effluents into the river, [15] linked high COD to pollution. This finding is contrary to [16], where the COD was higher in the dry season. The pH values were mostly maintained between he optimum ranges (6.5-8.5) for most biological activities. The only shift from this trend was observed in November in station 1 where the River was slightly acidic, which may have harmful effect on non tolerant algal species and in August in station 3 where the river was slightly alkaline [16] and [4]. The highest total alkalinity in station 3 in the wet season and lowest in station 2 the dry season could be due to the alkali condition of the River and the human activities in the catchment area and domestic waste associated with spent sour (a combination of alkali agent and detergent). The result obtained here is contrary to [4] study in new Calabar-Bonny River, who reported higher concentrations of alkalinity during the dry season than the wet season and attributed it to increased evaporation and concentration of ions in water during the dry season. The high values of water hardness in the dry months than the wet months could be attributed to reduction of the water level in the dry season due to evaporation. Water hardness reported for station 1 and 2 <40. 00mg/l caco$_3$ were in the range of soft waters by [17] classification. Stations 3, 4 and 5 with >, 100.00mg/l caco$_3$ is classified as hard water. Higher river discharge which contains much inorganic nutrients into station 3 as well as inputs from tributaries may be the reason why the total hardness of the station was higher than the other four stations. As stated by [18], inorganic nutrient levels are typically maximal in riverine zone of river. The higher concentration of No_3-N and PO4-P in the wet months compared to dry months, could have originated from agricultural activities on the banks of the River, which was washed into the river during the wet season. [19] linked nitrate-nitrogen origin to improper disposal of human excreta and domestic and agricultural waste. Nitrate concentration of 45mg/l in water used for baby's food could result in "blue baby syndrome" which is caused by a reaction known as meta-hemoglobin infertile mortality [20].

5 Conclusion

Nigeria contains ocean, lagoon, rivers, swamps and wetlands which are valued globally for their biodiversity. This study provides information on the baseline survey and present status of the water quality for long-term assessment and management of the river.

References

1. Dike, N.I., Ezealor, A.U., Oniye, S.J., Ajibola, V.O.: Pollution studies of River Jakara in Kano Nigeriaa, using selected physic-chemical parameters. Internal Journal of Research in Environmental Science and Technology 3(4), 123–129 (2013)
2. Sa'ad, M.A.H., Amuzu, A.T., Biney, C., Calamari, D., Imevbore, A.M., Naeve, H.F., Ochumba, P.B.: Scientific Bases for Pollution Control in Africa Inland Waters. Domestic and Industrial Organic Loads. FAO Fisheries Rep. (437), 6–24 (1990)
3. Adeogun, A.O., Chukwuka, V., Ibor, O.R.: Impact of Abbatoir and Saw mill effluents on Water Quality of Upper Ogun River. Abeokuta. American Journal of Environmental Sciences 7(6), 525–530 (2011)
4. Agbugui, M.O., Deekae, S.N.: Assessment of the Physico-chemical Parameters and Quality of Water of the New Calabar-Bonny River. Porthacourt, Nigeria. Cancer Biolog. 4(1), 1–9 (2014), http://www.cancerbio.net (ISSN: 2150-1041), (ISSN:2150-105X)

5. Arimoro, F.O., Ikomi, R.B., Iwegbue, C.M.A.: Water quality changes in relation to Diptera community patterns and diversity measured at an organic effluent impacted stream in the Niger Delta, Nigeria. Ecol. Indicators 7, 541–552 (2007b)
6. Boyd, C.E.: Water Quality in warm water fish ponds. Craftmaster public., 3rd edn. Agricultural extension station. Auburn university Alabama (1981)
7. APHA, Standard methods for the examination of water and Wastewaters 21st edition American Public Health Association American Waterworks Association Water Environment Federation, 1056 p. American Public Health Association, Washington (2005)
8. ASTM, Biological method for the assessment of water quality. American Society for Testing and Materials. Special Technical Publication No. 528 Philadelphia. D1771, 888 and 859 (2001)
9. Davies, O.A., Ugwamba, A.A.A., Abolude, D.S.: Physico-chemistry quality of Trans-Amadi (Woji) creek, Porthacourt, Niger Delta Nigeria. Journal of Fishery International 3(3), 91–97 (2008)
10. Chia, A.M., Iortsuun, D.N., Stephen, B.J., Ayobamire, A.E., Ladan, Z.: Phytoplankton responses to changes in macrophyte density in a tropical artificial pond in Zaria, Nigeria. Afrcan Journal of Aquatic Science 36(1), 35–46 (2011a)
11. Amakiri, N.E.: Aspects of ecology and population dynamics of Chrysichthys furcatus (Gunther, 1964) from Upper Sombreiro River, Nigeria. Unpublished M. Sc Thesis, Rivers State University of Science and Technology, 269p. (2011)
12. Ali, N., Oniye, S.J., Balarabe, M.L., Auta, J.: Concentration of Fe, Cu, Cr, Zn, Cd and Pd in Makera Drain, Kaduna, Nigeria. Chemclass Journal 2, 69–73 (2005)
13. Davies, O.A.: Seasonal abundance and distribution of plankton of Michinda stream, Niger Delta, Nigeria. American Journal of Scientific Research (2), 20–30 (2009)
14. Touliabah, H., Safik, H.M., Gab-Allah, M.M., Taylor, W.D.: Phytoplankton and Some features of El-bardawi, Lake, Sinai. Egypt. African Journal of Aquatic Sciences 27, 97–105 (2002)
15. Tepe, Y., Mutlu, E.: Physico-chemical characteristics of Hatay Harbiye Spring water, Turkey. Journal of the Institute of Science and Technology of Dumlupinar University 6, 77–88 (2005)
16. Mustapha, M.K.: Influence of watershed activities on the water quality and fish assemblages of a tropical African reservoir. Turkish Journal of Fisheries and Aquatic Sciences 9(1), 1–8 (2009a)
17. Kalff, J.: Limnology: Inland water ecosystems, 592 p. Prentice Hall, Upper Saddle River (2003)
18. Lind, O.T.: A Handbook of Limnological Methods Pubication, 199 p. C.V. Mosby Co. St. Louis (1979)
19. Gelinas, Y., Randall, H., Robidoux, L., Schmit, J.: Well water survey in two Districts of Conakry (Republic of Guinea) and comparison with the piped city water. Water Resources 9, 2017–2026 (1996)
20. Dike, N.I., Ezealor, A.U., Oniye, I.S.: Concentration of Pb, Cu, Fe and Cd During the Dry Season in River Jakara, Kano, Nigeria. Chem Class Journal, 78–81 (2004)

Suspended Sediment Concentration Distribution of Xiamen Sea Areas in China Based on Remote Sensing Analysis

Shuhua Zuo*, Hua Yang, Hongbo Zhao, and Yin Cai

Key Laboratory of Engineering Sediment of Ministry of Communications, Tianjin Research Institute of Water Transport Engineering, M.O.T., Tianjin 300456, China
zsh0301@163.com

Abstract. Six Landsat TM imageries from 2001 to 2008 were used as the data source to identify the spatial and temporal variations of the suspended sediment concentration in surface waters of the Jiulongjiang Estuary and Xiamen Bay. The results showed: (1) The spatio-temporal distribution of suspended sediment concentration in Xiamen Bay are not only governed by river flow and tidal current, but also are affected by factors of wind and waves, which will change greatly in different seasons and regions; (2) There were two relative high concentration zones in different time in the whole Xiamen Bay, the first is in Haimen Island near and the second is in the shallow water area of the east sea of Xiamen; (3) In the Xiamen waters, SSC has strong seasonal variation; surface suspended sediment concentration distribution is different in different waters in a tidal cycle.

Keywords: Suspended sediment concentration (SSC), Remote sensing, Jiulongjiang Estuary, Xiamen Bay.

1 Introduction

The suspended sediment concentration (SSC) is an important indicator to reflect the sediment transport and resuspension processes, which are the important factors to study the condition of an estuary and its navigation environment (Zuo et al., 2006). The SSCs in the estuary are controlled by multiple factors including freshwater and sediment discharges of the river, tides and waves (Li et al., 2010). Change of SSC in the estuary can indicate the changes of sediment sources and transport processes or dynamic conditions. So it is very significant that the investigation of sediment concentration is related to the study of the quality of water, morphology, ecological environment in the estuary, and the construction and maintenance of coastal engineering, harbor projects and navigation channels etc.

In recent years, a new measurement technique based on the satellite remote sensing imagery is quickly developed. Remote sensing is as a useful technique to study the SSC in estuarine and coastal waters (Zuo et al., 2007; Li et al., 2010).

* Corresponding author.

F. Bian and Y. Xie (Eds.): GRMSE 2014, CCIS 482, pp. 308–315, 2015.

The spatiotemporal distribution of SSC in a region can be obtained by analyzing the data of remote sensing, which is a very economical and efficient method. By establishing the relation between suspended sediment concentration and reflectance of water and combining field observations, the data of remote sensing are used helpfully to understand the sediment environment in the estuary more completely and precisely. Using different remote sensing imageries (e.g. MODIS, LANDSAT, SPOT) and retrieval techniques, some researchers studied the suspended sediment concentration in the coastal areas (Yun et al., 1981; Han et al., 2006).

In the paper, according to the remote sensing images and sediment concentration data acquired quasi-synchronously, quadric curvilinear regression model is developed by means of the statistic analysis of correlation. Based on the model the distribution diagram of surface SSC at the research area is obtained by retrieving the Landsat TM images from 2001 to 2007 of the Jiulongjiang Estuary and Xiamen Bay. This paper describes the status of distribution of surface SCC, sediment transportation and sedimentation in the different time of tide and season.

2 Study Areas and Methodology

2.1 Study Areas

The Jiulongjiang Estuary and Xiamen Bay are located in the south of Fujian Province coastal with subtropical climate with abundant precipitation, and its flood season is from April to September. The Jiulongjiang River is the second largest river in Fujian Province, located between 24°13′-25°51′N, 116°47′-118°02′E with a total length of 285km and a drainage basin area of 14740km^2. The river consists of Xixi distributary, Beixi distributary and Nanxi distributary. The hydrological records show that the average annual water discharge is 1.21×10^{10} m3 from 1991 to 2009 and the average annual sediments discharge is 2.69×10^6 t from 1952 to 2009. The biggest annual sediments discharge is 6.29×10^6 t in 2006.

The area of the Jiulongjiang Estuary and Xiamen Bay is the interaction region between rivers runoff and ocean tide. Meanwhile, due to the influence of its unique coast, underwater topography and coastal islands, it's hydrological and sediment conditions are more complex. The Xiamen Bay is a region with normal semi-diurnal tides and macrotidal range. The average tidal range at Gulangyu station (24°26′54″N, 118°04′12″E) is 4.08m, with a maximum of 6.88m. Figure1 shows the sketch map of the Jiulongjiang estuary and Xiamen bay.

2.2 Data Sources

The Landsat series satellites which were designed to collect earth resources data and landmass imageries by NASA with seven satellites launched from 1972 were used to analyze the SSC in the Jiulongjiang estuary and Xiamen bay. The Landsat 5 satellite with the sensors of Multispectral Scanner (MSS) and Thematic Mapper (TM) was launched in 1984 which is still working now. The sensors of Thematic Mapper (TM) has a 30 m's spatial resolution in visible lights and near infrared lights band (1-5 and

7 bands) and 120 m's resolution in the thermal infrared light band (6 band) (seen Table1).

The Landsat 5 satellite is sun-synchronous satellite with 16 days' period. The spatial range of each Landsat 5 TM image is 185 km × 170 km. The TM data have a higher spatial resolution and relatively long time series, which are suitable for estuarine and coastal studies. The spatial resolution of the images is 30 m (Table 1). Nine Landsat TM images from 2001to 2008 are chosen to analyze the SSC change in the Jiulongjiang estuary and Xiamen bay in recent years. Sea conditions of the satellite images are shown in Table 2.

Fig. 1. Sketch map of the Jiulongjiang estuary and Xiamen bay

Table 1. Landsat 5 TM data format

Landsat TM band	Wavelength/nm	Spatial resolution
Band 1	450-520	30
Band 2	520-600	30
Band 3	630-690	30
Band 4	760-900	30
Band 5	1550-1750	30
Band 7	2080-2350	30

Table 2. Sea conditions of the satellite images in the Xiamen sea area

Date	Data format	Season	Tidal stencils	Flood or ebb tide	Tidal level/cm	Mean monthly discharge /$m^3.s^{-1}$
2001-05-15	TM	Flood	Neap tide	Ebb tide	279	548
2003-07-08	TM	Flood	Neap tide	Ebb tide	270	224.6
2004-02-17	TM	Dry	Medium tide	High flood tide	506	112.9
2004-09-28	TM	Flood	Spring tide	Flood tide	472	437
2005-10-17	TM	Dry	Spring tide	Flood tide	491	460
2008-02-28	TM	Dry	Neap tide	Low ebb tide	200	/

Notes: data of the tidal level is taken from Gulangyu station; Mean discharge is the sum of the discharge of Punan station and the discharge of Zhengdian station.

2.3 Inversion Model Equation of SSC

In order to find the ideal correlative equation of suspended sediment concentration with remote sensing, an observation were carried out in Xiamen Bay on July, 24, 2000 to get the observed SSC under the natural dynamic factors of wind with the remote sensing data synchronism.

Based on the results of reflectance spectra tests in the lab and the observation in the Jiulongjiang Estuary and Xiamen Bay, The inversion model equation of SSC in the Jiulongjiang Estuary and Xiamen Bay is expressed as (Lin *et al.*, 2008):

$$S(X)=214.48X_1^2-141.20X_1+148.08X_2^2-72.71X_2 \quad (1)$$

where S is the SSC value (mg.L^{-1}), X_1 and X_2 are the SSC indexes. X_1 and X_2 are expressed as respectively:

$$X_1=TM3/TM2 \quad (2)$$

$$X_2=TM3/TM1 \quad (3)$$

$TM1$, $TM2$ and TM 3 are the band 1, band 2 and band 3 of the remote sensing data respectively.

Based on the calibration by ENVI, atmospheric correction by FLAASH and geometric correction to the satellite data, the SSC value was calculated according to Eqs (1), (2) and (3) using the software ENVI 4.2. The outside land region was erased by building a mask in the processing software ENVI.

3 Hydrodynamic Characteristics of Xiamen Sea Area

3.1 Characteristics of Tides and Tidal Currents

Based on the measured tidal data at the Xiamen ocean station from 1986 to 2006, the tides of Xiamen sea area are predominantly regular semi-diurnal. The characteristic values of tide stages are as follows: the mean high tide level is 5.53 m, the mean low tide level is 1.46 m, and the mean tidal range is 4.06 m.

Tidal currents are characterized as a type of reversing flow. Current moves westward during flood tides and eastward during ebb tides in the Jiulongjiang estuary, and northwestward during flood tides and southeastward during ebb tides in the Xiamen bay mouth area. However, tidal currents are a type of rotating flow. According to the field data on currents simultaneously measured in July 2000, Sep. 2008 and June 2009, the mean velocity is about 0.5m/s, the maximum velocity is about 1.2 m/s during spring tide. The current velocity of an ebb tide is larger than that of a flood tide for the same tidal range in the Jiulongjiang estuary, and flood tidal current velocity is larger than ebb tide in the Xiamen bay mouth area and outside waters.

3.2 Characteristics of Wind Waves

Based on the wind data of the Xiamen meteorological station from August 1952 to July 1999, in this region, the prevailing wind direction is NE and SE. The strongest wind direction and the inferior strongest wind direction are ESE and NE, respectively. In one year, the frequency of a wind scale greater than six degrees is 2.3% and the windy days are 27.7d.Yearly-average wind velocity is 5.0-6.0 m/s, the yearly maximum wind velocity is 19-28 m/s from August 1994 to July 1996.

In the Xiamen sea area, the primary prevailing waves are wind waves. The prevailing wave direction is E with the occurrence frequency of 43.2%, the strongest wave direction is SE. Annual average $H_{1/10}$ ($H_{1/10}$ is the wave height of the average for fore 10% waves) is about 1.0m, the highest wave height is 4.3m in the Xiamen bay mouth area. Due to the sheltering by the islands of mouth area, the wave action is less strong in the Jiulongjiang estuary.

Xiamen sea area is located in the west bank of Taiwan straits. Coastal area of Xiamen sea area is one of regions with the most severe storm tides in China. Based on statistical data there are 4.7 storm tides every year. The storm tides mainly occur in July to October with the extreme wind speed of 46m/s.

4 Distribution of SSC

Figure 2 shows the spatial distribution of surface suspended sediment in the Jiulongjiang Estuary and its joint waters. The spatio-temporal distribution of the suspended sediment concentration in Xiamen Bay are not only governed by river flow and tidal current, but also are affected by factors of wind and waves, which will change greatly in different seasons and regions. In this paper, based on the data of quantitative sediment concentration of remote sensing given recently and of the corresponding hydrology and meteorology, the distribution of average sediment concentration in surface water is analyzed.

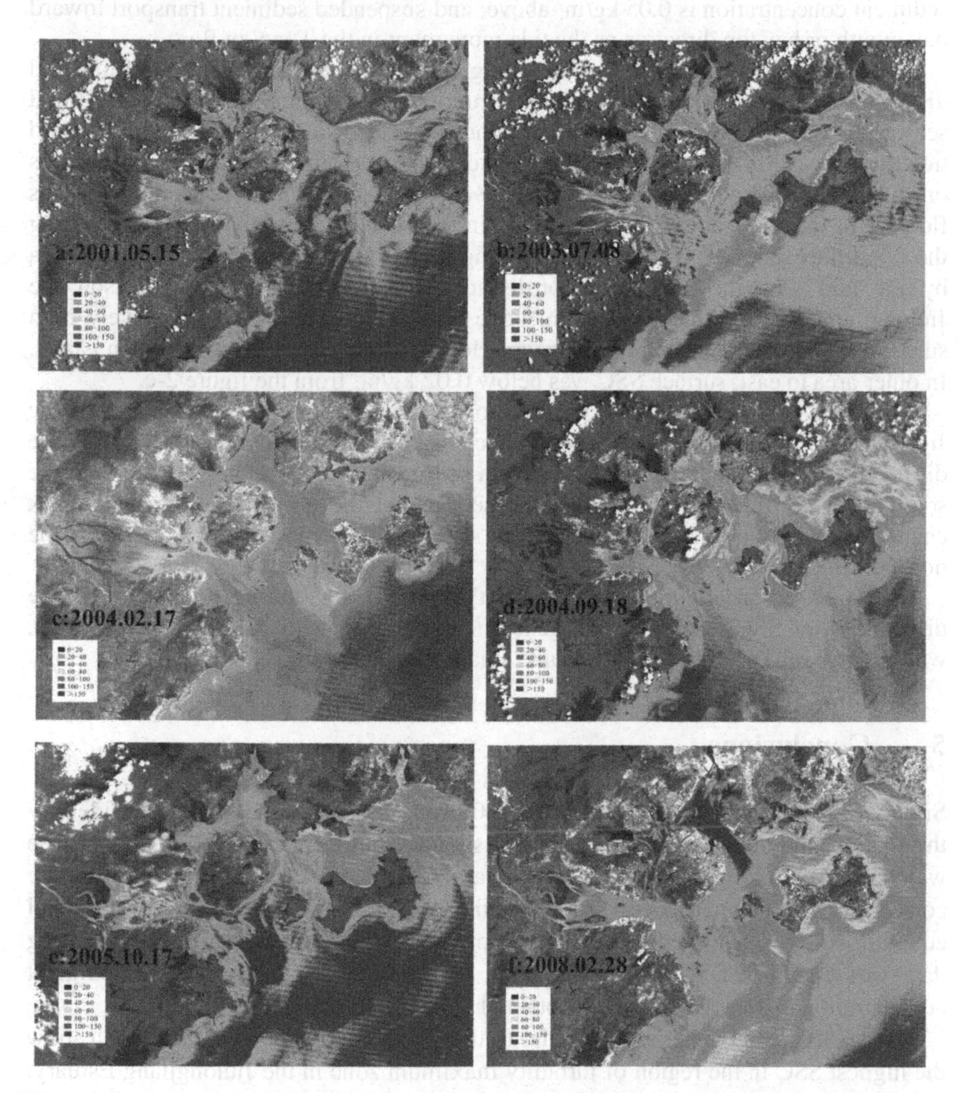

Fig. 2. Surface suspended sediment concentration (SSC) map at different time in the Jiulongjiang estuary and adjacent coastal waters

From the images, in general, it can be seen that there were two relative high concentration zones in different time in the whole Xiamen Bay. The first is in Haimen Island near, which is the maximum SSC area for estuary. The surface SSC generally is 0.06 kg/m^3 above, and is descending from the west to the east. The surface SSC of Dachengping shoals is highest with 0.08 kg/m^3 or above. The second is in the shallow water area of the east sea of Xiamen. The sediment concentration distributions present the strip along the shoreline of the Tong'an Bay with the surface SSC 0.05 kg/m^3 or so. Moreover, in the northern shoal region of the Tong'an Bay, the east coast shoals of Xiamen and the near shoals of the Aotou area, most of the sediment concentration is 0.05 kg/m^3 above, and suspended sediment transport toward bay mouth or bay top direction as the tide movement in the Tong'an Bay.

In the Xiamen waters, SSC has strong seasonal variation. The variation and distribution are largely controlled by the runoff of Jiulongjiang river. In the flood season, figure 2-b show the distribution trend of surface suspended sediment in flood tide during neap tide in the Jiulongjiang river estuary mouth and Xiamen bay waters on July 8th, 2003. The water with relatively higher sediment concentration was flowing in the channel to the Tajiao areas in the mouth. From the figure 2-d, during the flood tide, sediment from the Jiulongjiang river could be push to Xiamen west sea by tide, and affect the Dongdu harbor district of Xiamen port. In the dry season, the Jiulongjiang river sediment diffusion scope was obviously little, the relatively high suspended sediment concentration area was located in the west area of Haimen island. In other area to east, surface SSC was below 0.02 kg/m^3 from the figure 2-c.

In a tidal cycle, surface SSC distribution is different in different waters. In the Jiulongjiang river estuary, sediment concentration in south and north area are different obviously. During flood tide, north sediment concentration is higher than the south, it is the result of water dynamic action that the north is as the main back channel during flood tide. However, south sediment concentration is higher than the north during ebb tide.

Figure 2b, 2d and figure 2c, 2f present the surface SSC during flood season. The distribution of SSC is obvious. In general, SSC varied with the neap-spring cycle, with greater value during spring tide and smaller value during neap tide.

5 Conclusions

Six Landsat TM imageries from 2001 to 2008 were used as the data source to identify the spatial and temporal variations of the suspended sediment concentration in surface waters of the Jiulongjiang Estuary and Xiamen Bay. After calibration, atmospheric correction and geometric correction by the software ENVI, an inversion model equation between surveyed suspended sediment concentration and the remote sensing data was established. The spatio-temporal distributions of the suspended sediment concentration in Xiamen Bay were obtained. The results show:

(1) SSC spatial patterns are similar to the in situ observation results, which show the highest SSC in the region of turbidity maximum zone in the Jiulongjiang Estuary. The SSC pattern is controlled mainly by tidal dynamic conditions and wind speeds, rather than sediment discharges from the river.

(2) In general, there are two relative high concentration zones in different time in the whole Xiamen Bay. The first is in Haimen Island near, which is the maximum SSC area for estuary. The surface SSC generally is 0.06 kg/m^3 above, and the surface SSC of Dachengping shoals is highest with 0.08 kg/m^3 or above. The second is in the shallow water area of the east sea of Xiamen. The surface SSC 0.05 kg/m^3 or so.

(3) SSC has strong seasonal variation. The variation and distribution are largely controlled by the runoff of Jiulongjiang river. Surface SSC distribution is different in different waters. In the Jiulongjiang river estuary, sediment concentration in south and north area are different obviously. Surface SSC varied with the neap-spring cycle, with greater value during spring tide and smaller value during neap tide.

References

1. ENVI. ENVI Users Guide. U.S. RSI CO., Ltd., p. 613 (2005)
2. Han, Z., Jin, Y.-Q., Yun, C.-X.: Suspended sediment concentration in the Yangtze River estuary retrieved from the CMODIS data. International Journal of Remote Sensing 27, 4329–4336 (2006)
3. Li, J., Gao, S., Wang, Y.-P.: Delineating suspended sediment concentration patterns in surface waters of the Changjiang Estuary by remote sensing analysis. Acta Oceanol. Sin. 29(4), 38–47 (2010)
4. Lin, Q., Chen, Y.M., Huang, Y.-G.: An analysis of distribution of suspended sediment in Estuary by using remote sensing technology. Port & Waterway Engineering 2, 19–22 (2008) (in Chinese)
5. Yun, C.-X., Cai, M.-Y.: An analysis of the diffusion of suspended sediment discharged from the Changjiang River based on the satellite images. Oceanoligia et Limnoligia Sinica 12(5), 391–401 (1981) (in Chinese)
6. Zuo, S.-H., Li, J.-F., Wan, X.-N., Shen, H.-T.: Characteristics of temporal and spatial variation of suspended sediment concentration in the Changjiang Estuary. Journal of Sediment Research (3), 68–75 (2006) (in Chinese)
7. Zuo, S.-H., Yang, H., Zhao, Q., Zhao, H.-B.: Remote Sensing Analysis on Distribution and Movement of Surface Suspended Sediment in Coastal Waters of Wenzhou. Geography and Geo-Information Science 23(2), 47–50 (2007) (in Chinese)

The Spatial-temporal Distribution of Precipitation in Northwest China (1960-2008)

Chuancheng Zhao[1,2], Shuxia Yao[1,*], Jun Liu[1], Zhiguo Ren[1], and Jian Wang[2]

[1] Lanzhou City University, Lanzhou 730070, China
[2] State Key Laboratory of Cryospheric Sciences, Cold and Arid Regions Environmental and Engineering Research Institute, Chinese Academy of Sciences, Lanzhou 730000, China
{Zhao_chch1978,yaoshuxia}@163.com

Abstract. Northwest China, covering the provinces of Shaanxi, Gansu, Qinghai, Ningxia, Xinjiang, and a portion of Inner Mongolia, has an extremely complex climate due to the varied terrain where environment is sensitive to climate change. We employed the Mann-Kendal test to investigate trends of precipitation distributions using annual, seasonal and monthly data records from 1960 to 2008. On the whole, the trends of precipitation are more complex than those for temperature. The trends of annual, seasonal and monthly precipitation have shown remarkable difference between the east and west. In the west, such as in northern Xinjiang and western Qinghai, the trend has shown a significant increase, consistent with the temperature change. Whereas in the east such as in eastern Gansu, southern Shaanxi, the trend has shown a remarkable decline, opposite to the change observed in temperatures.

Keywords: Climate Change, Precipitation, Mann-Kendall Test, Northwest China.

1 Introduction

The climate warming trend has been well documented in many regions around the world during the last several decades, and the trend is projected to accelerate in the future[1]. The climate warming has been linked to serious ecological and environmental deterioration, extreme climate events such as flooding, drought, and other disasters, which has caused great social and economic losses[2]. Northwest China, as an example of mixed arid and semi-arid area, is sensitive to climate change, especially to precipitation and temperature change. Northwest China, affected by the Indian summer monsoons from the southwest, the East Asian summer monsoons from the southeast, also by westerly airflow in both summer and winter, and by northerly circulation in winter is climatically complex[3]. In recent years, several studies of these regions have shown that climate change on a regional and global scale has been detected by scientists. In the recent 50 years, the annual mean surface air temperature

* Author to whom all correspondence should be addressed: E-mail: yaoshuxia@163.com; Phone: +86 931 496 7160.

F. Bian and Y. Xie (Eds.): GRMSE 2014, CCIS 482, pp. 316–322, 2015.

has increased by1.1°C in China, i.e., a warming rate of 0.22°C/10a, significantly higher than the Northern Hemisphere's warming rate and the global average warming rate in the same period[4]. From the mid-1980s, the temperature in northwest regions began to rise rapidly, more than 0.6°C[5], slightly higher than the national average. Those observations above-mentioned show a climatic transition from a warm-dry pattern to a warm-wet pattern is only likely to happen in the western area of northwest China, while the climate is likely to remain very dry with no significant increase in precipitation in the central and east areas. Precipitation is predicted to increase in summer in the west region, but will decrease in the east part of northwest China with doubled CO_2[6].

The aim of this study is to examine long-term trends of precipitation using the Mann-Kendall trend test[7-8], which is widely used for the assessment of the significance of trends in hydro-meteorological time series. The Mann-Kendall trend test has the advantage of not being affected by the actual distribution of the meteorological data, e.g., precipitation; Therefore, is more suitable for detecting trends in hydrological time series, which are often skewed with outliers[9].

2 Materials and Methods

2.1 Study Area

The present study focuses on northwestern China, covering the provinces of Shaanxi, Gansu, Qinghai, and the autonomous regions of Ningxia, the Xinjiang Uygur Autonomous Region, and a portion of Inner Mongolia. The total area is 3. 09 million km^2, comprising approximately one-third of China's land area(Fig. 1). Precipitation, derived from the Asian summer monsoons, currently extends to the southern and eastern parts of our study area, although the summer monsoon penetrated markedly further north during the early to mid-Holocene. The annual precipitation has shifted from southeast to northwest, and the amount has declined from 800mm to 50mm. Arid regions occupy a vast area in northwest China, where the mean annual rainfall is less than 250 mm, but there are some parts of mountains such as Yili valley of Tianshan mountains, the precipitation amount is over 600 mm. Within the regions, annual mean precipitation in the west plain ranges from 50to 150 mm, and less than 25 mm in the Taklimakan Desert.

2.2 Data Collection

To investigate the general trends in climatic conditions for locations in northwest China, we obtained monthly precipitation data records from 204 stations during 1960-2008 from the China Meteorological Administration (CMA) observation archives (Fig. 1). These precipitation data sets are the longest consistent data series available and are the basis of analyses performed to assess monthly, seasonal and annual change. Several stations having more than 1year precipitation data missing were excluded from the analysis. The rest stations with data missing were processed in the following ways:

(1) If only one month's data is missing, the missing data is replaced by the average monthly precipitation value of its neighbor station.
(2) If data of consecutive months is missing, tit is estimated by simple linear correlation from its neighbor stations.

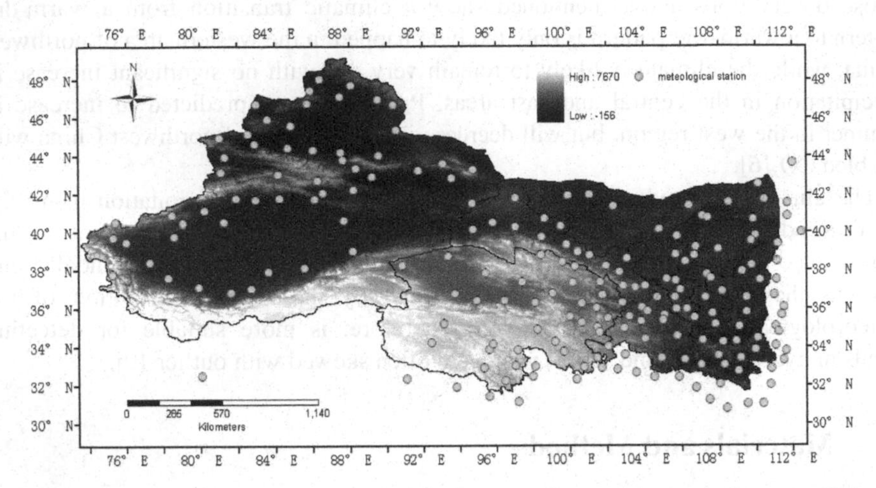

Fig. 1. Locations of temperature and precipitation recording stations in northwestern China

2.3 Trend Detection Test

The time series of precipitation were analyzed using the Mann-Kendall non-parametric trend test. This test enables us to investigate the long-term precipitation tendencies without assuming that a given data set follows a normal distribution.
The Mann-Kendall test was applied to annual, seasonal and monthly precipitation data. The test statistic, S, is calculated as:

$$S = \sum_{i=1}^{n-1} \sum_{j=i+1}^{n} sign(y_j - y_i) . \tag{1}$$

Where y_i and y_j are the data values at times i and j, n is the length of the data set, and

$$sign(y) = \begin{cases} 1 & if \;\; y > 0 \\ 0 & if \;\; y = 0 \\ -1 & if \;\; y < 0 \end{cases} . \tag{2}$$

With the null hypothesis that y_i are independent and randomly ordered, the statistic S is approximated to be normally distributed when $n \geq 8$. An increase or a decline in trend results in S being negative or positive, respectively. The expected value and variance of S are given by:

$$E[S]=0$$
$$V(S)=\frac{n(n-1)(2n+5)}{18}. \tag{3}$$

The standardized test statistic, Z, computed by

$$Z=\begin{cases}\frac{S-1}{\sqrt{V(S)}} & if\ S>0\\ 0 & if\ S=0\\ \frac{S+1}{\sqrt{V(S)}} & if\ S<0\end{cases}. \tag{4}$$

Follows a standard normal distribution. An increase or a decline in trend results in Z being negative or positive, respectively. In this study, 90%, 95% and 99% confidence levels were considered. When $/Z/$ is greater or equal to 1.28, 1.64, and 2.32, it indicates that confidence level has passed 90%, 95%, and 99%, respectively.
A non-parametric robust estimate of the magnitude of the slope of linear trend, β, determined by Hirsch et al.[10] , is given by

$$\beta=Median\left[\frac{(y_j-y_i)}{(j-i)}\right], \forall i<j. \tag{5}$$

in which $1<j<i<n$. The estimator β is the median of overall combination of record pairs for the whole data set, and is thereby resistant to the effect of extreme values in the observations. A positive value of β indicates that the time series has an upward trend. Otherwise, the time series has a downward trend.

3 Results and Discussion

To support the analysis of annual, seasonal and monthly precipitation trends, the precipitation concentration index is first calculated on a monthly basis for each station. The distribution of the precipitation concentration indices provides insight into regional heterogeneities in precipitation trends. On the whole, annual precipitation is affected greatly by variations in monthly precipitation in regions including most of Xinjiang, Qinghai, western Gansu and Inner Mongolia. But it is affected the most by variations of seasonal precipitation in regions such as Altay in Xinjiang, southeast of Gansu and Shaanxi.

3.1 Annual Trend

Due to topography, elevation and other factors, the change of annual precipitation mean is not consistent with the change of annual temperature mean. Fig. 2 has shown that the trend of annual mean precipitation has generally increased, yet a significant difference on the level of increment appears between the east and west during 1960-2008 in northwest China. The Altai Mountains and central Tianshan Mountains have experienced the greatest increases in annual mean precipitation, in excess of 10 mm per decade. In west central Xinjiang, southwest and central Qinghai, annual mean precipitation has increased above 5 mm per decade. The greatest decline has occurred in Gansu (east of the Yellow River) and central Shaanxi, where the change is more than -25 mm per decade. In Northeast of Gansu, Shaanxi, and south of Ningxia, the annual mean precipitation has decline by about -20 mm per decade. Thus, the trend of annual mean precipitation is evident.

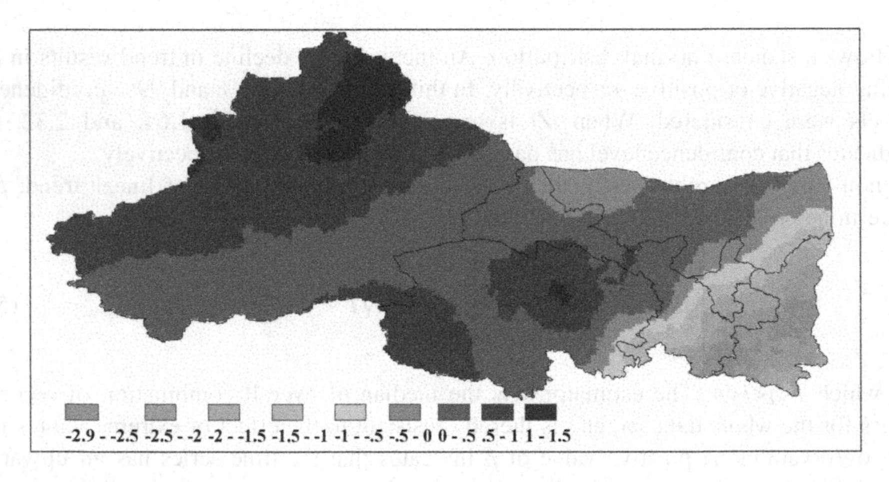

Fig. 2. Spatial distribution of β-values for annual mean precipitation during 1960-2008. Quantities show changes in annual precipitation in mm per decade.

3.2 Seasonal Trends

The seasonal precipitation is generally concentrated in spring, summer, and fall, accounting for 20%, 40%, and 35% respectively, whereas only 5% the annual precipitation arrives in winter. Thus, the trend of the seasonal precipitation in spring, summer and fall can be affected (be affected of affect?) the trend of annual precipitation in the study area significantly. Through *Z* value analysis, the trends in seasonal precipitation during 1960-2008 have not been found changed significantly for most of northwest China where trend analysis have shown weakly varying increase and decline. Based on the spatial analysis β of linear slopes (Fig. 3), the results have shown that precipitation in the whole study area has increased (or generally increases) in winter, mainly concentrated in northern Xinjiang, southern Qinghai, and central Shaanxi, with the highest rate of above 4 mm per decade.

The increase in spring precipitation is widespread in northwest China, the southern Qinghai and northern Xinjiang have seen the greatest increases of above 5 mm per decade. The decline has shown in Shaanxi, Ningxia, eastern Gansu and central Inner Mongolia, with the downtrend in southern Shaanxi having the greatest decline at rate of over 17 mm per decade. The increase of summer precipitation, as same as the spring, is widespread. However, the region of decline precipitation is greater for summer than for spring, and the same can be said for precipitation amount as well. The significant increase in fall precipitation is mainly concentrated in the Tianshan Mountains and southern Qinghai, while decline is most significant in the northeastern Gansu and southern Shaanxi.

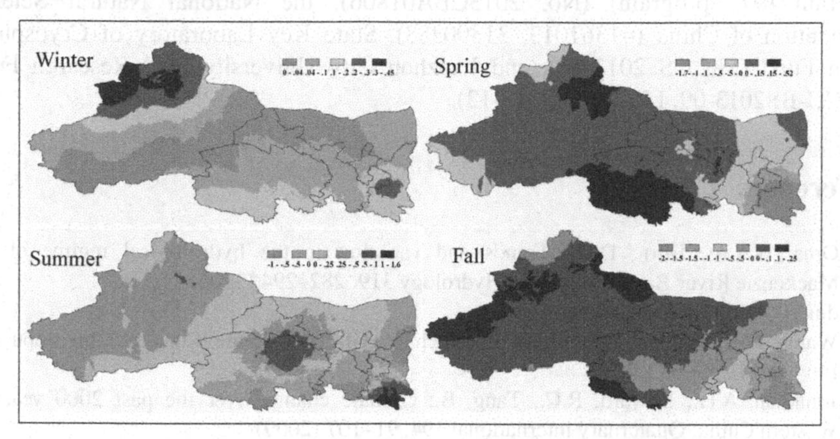

Fig. 3. Spatial distribution of β-values for seasonal precipitation during 1960-2008. Quantities show changes in annual precipitation in mm per decade.

3.3 Monthly Trends

Trends in monthly precipitation have obviously direct impact on seasonal and annual precipitation. So, the same spatial analysis has been carried out using monthly precipitation figures. It has shown that the two months have the most significant variation. The precipitation trend in most regions has increased in July, generally in excess of 5 mm per decade, concluded from observing most of Xinjiang area, north and central Qinghai. In contrast, precipitation trends have shown a decline in September in regions such as northern Xinjiang, southern Qinghai and Ningxia, eastern Gansu, and Shaanxi, where the largest decline is of 13 mm per decade, (for example, in northern Shaanxi.)

4 Conclusions

The object of this study is to catalog the spatial distribution of precipitation trends in northwest China during 1960-2008. The results show that the spatial distribution of annual mean precipitation has clearly changed over the recent 50 years. Elevation has a remarkable influence on temperature, whereas topography, hydraulic transport and geographic position have significantly affected by precipitation. The trends of annual, seasonal and monthly precipitation appear to be distinctly polarized by geography. In the west such as in northern Xinjiang, southern Qinghai, the trend shows significant increase, and in the east such as in Shaanxi, Ningxia, eastern Gansu, a remarkable decline is found. Thus, the trends of temperature and precipitation show a weak positive correlation in parts of northwest China such as Xinjiang and Qinghai, and a negative correlation in regions such as eastern Gansu, Shaanxi, and southern Ningxia.

Acknowledgments. This study was supported by major national science research program (973 program) (No. 2013CBA01806), the National Natural Science Foundation of China (41361013, 31300388), State Key Laboratory of Cryosphere Open Fund (SKLCS 2012-10), and Lanzhou City University PhD Research Fund (LZCU-BS2013-09, LZCU-BS2013-12).

References

1. Omar, I.A.A., Burn, D.H.: Trends and variation in the hydrological regime of the Mackenzie River Basin. Journal of Hydrology 319, 282–294 (2006), doi:10.1016/j.jhydrol.2005.06.039
2. Wang, J.M., Wei, D.: Cognition Research on Global Climate Change. Chn Popu Res Envi. 18(3), 58–63 (2008)
3. Jonathan, A.H., Edward, R.C., Tang, B.: Climate change over the past 2000 years in Western China. Quaternary lnternational 194, 91–107 (2009)
4. Ding, Y.H., Ren, G.Y., Shi, G.Y., Gong, P., Zheng, X.H., Zhai, P.M., Zhang, D.E., Zhao, Z.C., Wang, S.W., Wang, H.J., Luo, Y., Chen, D.L., Gao, X.J., Dai, X.S.: National assessment report of climate change (I): climate change in China and its future trend. Advances in Climate Change Research 2(1), 3–8 (2006)
5. Liu, Y.X., Li, X., Zhang, Q., Guo, Y.F., Gao, G., Wang, J.P.: Simulation of regional temperature and precipitation in the past 50 years and the next 30 years over China. Quaternary International 212(2010), 57–63 (2010)
6. Zhang, C.J., Gao, X.J., Zhao, H.Y.: Influences of fall precipitation in Northwest China with global warming. Journal of Glaciology and Geocryology 25(2), 157–164 (2003)
7. Mann, H.B.: Nonparametric tests against trend. Econometrica 13, 245–259 (1945)
8. Kendall, M.G.: Rank Correlation Methods, Griffin, London, 202 p. (1975)
9. Khaled, H.H.: Trend detection in hydrologic data: The Mann-Kendall trend test under the scaling hypothesis. Journal of Hydrology 349(2008), 350–363 (2008), doi:10.1016/j.jhydrol.2007.11.009
10. Hirsch, R.M., Slack, J.R., Smith, R.A.: Techniques of trend analysis for monthly water quality data. Water Resources Research 18, 107–121 (1982)

Assessment of Coastal Ecosystem Vulnerability of Dongshan Bay in Southeast China Sea

Jialin Ni, Jianping Hou, Jinkeng Wang, and Ling Cai[1,*]

Third Institution of Oceanography, State Ocean Administration, 361005 xiamen, China
cailing@tio.org.cn

Abstract. Coastal ecosystem vulnerability assessment was carried out for further better management of marine environment in southeast China. As a research sample area, the coastal of Dongshan Bay was divided into 138 evaluation units according to the geological and topographic features for anecosystem vulnerability assessment. The index system was established according to the "pressure-state-response" evaluation model. The calculated results were analyzed and evaluated through comprehensive assessment methods. The evaluation results illustrated that the fragile ecosystem in concentrated at two regions: one was the estuary of Zhangjiang river mangrove national nature reserve region, another one was the reclaim beach for aquaculture near the Shaxi town in north Dongshan Bay. In consideration of the evaluation results, more protection measures should be applied into these two areas.

Keywords: coastal ecosystem vulnerability, pressure-state-response, comprehensive assessment methods, Dongshan Bay.

1 Introduction

Ecological vulnerability refers to the sensitiveness and restorability of ecosystem within a certain spatial and temporal scale to the external interferences. It is not only the manifestation of its inherent attributes under external interferences, but also the consequence of natural properties and human activities[1],[2]. Ecological vulnerability is related with the type and strength of interference factors as well as the exposure degree and sensitiveness to interference factors of the ecosystem[3]. Abundant profound researches on the ecological vulnerability have been reported at both home and abroad, which have achieved various important research fruits[4],[5],[6]. As an important content of vulnerability study, the evaluation of ecological vulnerability not only supplements and improves the environmental evaluation theory, but also is of great theoretical and practical significance for the regional economic restructuring, eco-environment construction and protection.

Coastal regions where contribute the most global economic wealth are vital to the social and economic development [7]. However, under the background of climate change, human activities in these regions, such as tourism development, urban construction, embankment and dam construction, reclaiming land from the sea by

* Corresponding author.

F. Bian and Y. Xie (Eds.): GRMSE 2014, CCIS 482, pp. 323–331, 2015.

building dykes and aquaculture, are intensified continuously in recent years. These activities result in the continuous reduction of water and sediment fluxes from rivers to the sea, shrinkage of coastal wetland and deterioration of marine ecological environment, thus accelerating the ecosystem degradation in coastal regions. The evaluation of ecological vulnerability in coastal zone not only can provide references for the development layout of coastal economy, but also plays an important role in the harmonious development between coastal economy and marine environmental protection.

2 Brief Introduction to the Study Area

Dongshan Bay locates in the southeast of Fujian Province (23°43'~23°57'N,117°24'~117°37'E). The whole bay looks like an irregular pear. It is a semi-closed bay with narrow mouth and strong closure (Fig.1). Zhangjiang River flows in the bay from northwest. With a sea area of 247.89km2, Dongshan Bay is the largest bay in south of Fujian and one of the three good bays in Fujian Province. It enjoys a moderate climate, abundant precipitation, appropriate sea water temperature and salinity, good water quality environment and high productivity. It covers various ecological sensitive regions, such as Zhangjiangkou Mangrove Forestry National Nature Reserve and experimental plots of agriculture and herding. Additionally, it is surrounded by many scenic spots and historic sites, including Maluanwan Tourism Leisure Resort, Dongshan Coral Conservation Area and Jinluanwan Tourism Leisure Resort on the Dongshan Island.

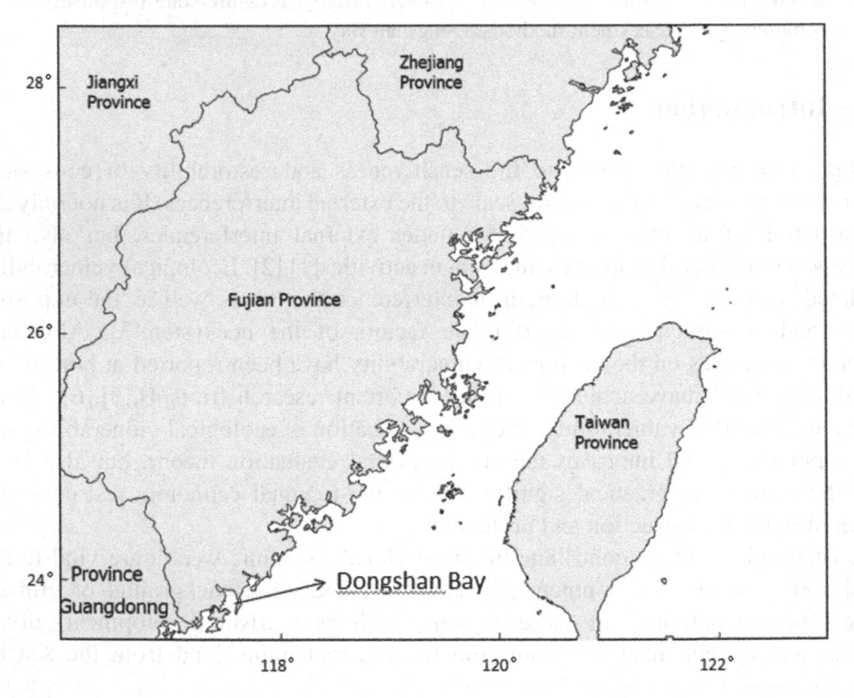

Fig. 1. The location of Dongshan Bay in China's Fujian Province

With the rapid economic development in surroundings of Dongshan Bay in recent years, various environmental problems emerged, such as increased discharge of domestic wastewater in offshore areas, backward aquaculture methods and pursuing production quantity and scale blindly, reclaiming coastal zone for buildings and dumping in the sea freely. The coastal protection forest is destroyed by exploitation and construction, human deforestation activities and coastal erosion, thus resulting in the worsening marine ecosystem in Dongshan Bay and influencing the sustainable development of surrounding societies and economies.

3 Research Method

3.1 Division of Evaluation Unit

There were many tidal wetland and reclaimed beaches for aquaculture in Dongshan Bay, and The project of land reclamation was increasing in recent yeas .Result in the area of bay keeping decreasing. In order to highlight the influence of transformation to the fragile ecosystem, The ecosystem in Dongshan Bay was divided into two subsystems: coastal area and littoral sea. coastal area covered the interval between high tide and low tide, which was submersed when the high tide is coming and revealed when the low tide is coming., littoral sea was covered by water throughout the year[8]. In this paper, only coastal area was evaluated According to the geological and topographic features, the coastal area of Dongshan Bay was divided into 138 evaluation units at an interval of 1km.

3.2 Establishment of Index System

The selection of evaluation index is very important for the ecological vulnerability evaluation. A scientific index system is the premise of accurate and real assessment of ecosystem vulnerability in the region. In this paper, the index system was established by using the widely applied "pressure-state-response" evaluation model. The index system was divided into objective level, criterion level and index level. The objective level is for estimating the ecological regime and structures in the study area. The criterion level includes the pressure (P), status(S) and response (R) of the system. The index level is established according to the principle of scientificity, representativeness, purpose and operability, which is determining the specific indexes for corresponding criterion level.

3.3 Index Selection

Through comprehensive analysis of the ecological environment, natural and human interference factors were included to compose the pressure index of the subsystem. As coastal area was the ecological transition belt between sea and continent, the pressure from both sea and continent should be considered. The pressure index of coastal area include alien species, total accumulative area of reclaimed land from sea, pollutant discharge in drain outlet and impact of land-based pollution on marine function zone respectively. Status criterion level reflected current situation of the ecosystem. So beach gradient and sedimentation, average wave height were chose to be index of criterion level of coastal area. Response level reflected the degree of ecosystem change within the pressure of natural and human interference. Response indexes of coastal area were biodiversity and biomass in intertidal zone and plankton and benthonic diversityrespectively. The background information of marine environment was collected from field investigation and published literatures, books, etc. The framework of index system for ecosystem vulnerability evaluation in Dongshan Bay was shown in Fig.3.

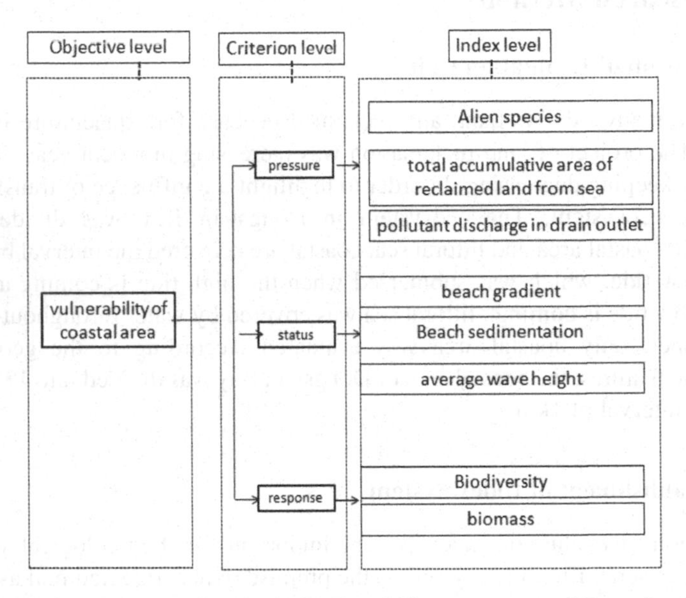

Fig. 2. The index system of ecological vulnerability assessment of Dongshan Bay

3.4 Index Classification

To solve the problem of incomparability between different dimensions, collected indexes were classified according to certain standard through two approaches: (1) cluster methodology according to statistical features of data; (2) literature review. Beach gradient, beach sedimentation and erosion velocity, average wave height, total accumulative area of reclaimed land from sea, biomass in intertidal zone, biodiversity in intertidal zone, plankton and benthonic diversity, are classified through cluster methodology. Geology and geomorphology, population density, alien species, pollutant discharge in drain outlet,, impact of land-based pollution on marine function zone, were classified through literature review. The relevant literatures of indexes of the second classification approach were presented as note in this paper. The specific classification standard of indexes was listed in the table of classification of ecological vulnerability evaluation indexes (Table 1).

Table 1. The classification of ecological vulnerability evaluation indexes in coastal area

Indexes	Score				
	0	1	3	5	7
Beach sedimentation*	Exposes rocky shore	Exposed wave-cut platforms in bedrock	Coarse-grai nd beaches	Mixes sand and gravel beaches	Tidal flats, marshes, wetland, mangroves
Beach gradient (%)	>0.115	0.115-0.055	0.055-0.035	0.035-0.022	<0.022
Average wave height (m)	<0.55	0.55-0.85	0.85-1.05	1.05-1.25	>1.25
Total accumulative area of reclaimed land from sea(ha^2)	0	<50	50-500	500-3000	>3000
Alien species	No alien species	No damage to ecosystem	Slight damage to ecosystem	Damage to ecosystem	Considerabl e damage to ecosystem
Pollutant discharge in drain outlet **	No drain outlet	D grade	C grade	B grade	A grade
Biomass (g/m^2)	<10	10-50	50-100	100-200	>200
Biodiversity,	<1	1-2	2-3	3-4	>4

*Environmental sensitivity index guidelines NOAA

**The Eco-environmental assessment guidance for terrestrial pollution source and near sea area (PRC ocean industrial standard. (.PRC National Standard HY/T086-2005)

3.5 Evaluation Method

Comprehensive Evaluation. In this paper, the ecological vulnerability in coastal were analyzed comprehensively. Equal weight method was applied to determine the weight of evaluation index. In other words, the weight of all indexes was determined 1 in order to decrease comprehensive evaluation error caused by the failure in distinguish relative significance of indexes accurately. The calculation formula was:

$$CVI = \sum_{k=1}^{n} C_k \quad (1)$$

whereCVI is the ecological vulnerability index of evaluation unit ;
Ck is the vulnerability of evaluation index k;
nis the number of indexes;

Vulnerability Evaluation. The threshold of ecological vulnerability index to different classifications is an important evaluation parameter of eco-environment vulnerability. Since the eco-environment vulnerability evaluation is still in exploration stage, no unified threshold of different classifications has been defined yet. This paper made a statistics on the frequency of ecological vulnerability in coastal area by using the total frequency division curve, which was used to draw the frequency histogram. The mutation point on the frequency curve was taken as the boundary between levels. The ecosystem of coastal area were divided into five levels, namely, non-vulnerable zone, slight vulnerable zone, moderate vulnerable zone, vulnerable zone and high-vulnerable zone. The division criterion was presented on the table of ecological vulnerability index classification (Table 2).

Table 2. The classification of ecological vulnerability value

-ecosystem	The value of vulnerability				
	non-vulnerable zone	Slight vulnerable zone	moderate vulnerable zone	vulnerable zone	high-vulnerable zone
coastal area	2-14	15-23	24-27	28-33	34-41

4 Evaluation Result

Based on the division criterion of regions with relative fragile ecosystem in Table2, evaluation units of coastal area were quantized by the software of mapinfo. The distribution of eco-environment vulnerability beach region was presented through a thematic map (Fig.5) and a statistical table of the evaluation results was made (Table3).

Table 3. Statistics of the vulnerability

Sub-ecosystem	non-vulnerable zone	Slight vulnerable zone	moderate vulnerable zone	vulnerable zone	high-vulnerable zone
Beach region	18.8%	20.3%	18.8%	35.5%	6.5%

According to the statistical table of the ecosystem vulnerability in Dongshan Bay, the whole coastal area is composed of 18.8% non-vulnerable zone, 20.3% slight vulnerable zone, 18.8% moderate vulnerable zone and 42% vulnerable zone and high-vulnerable zone. This indicated the relative fragile ecosystem in theDongshanbay.

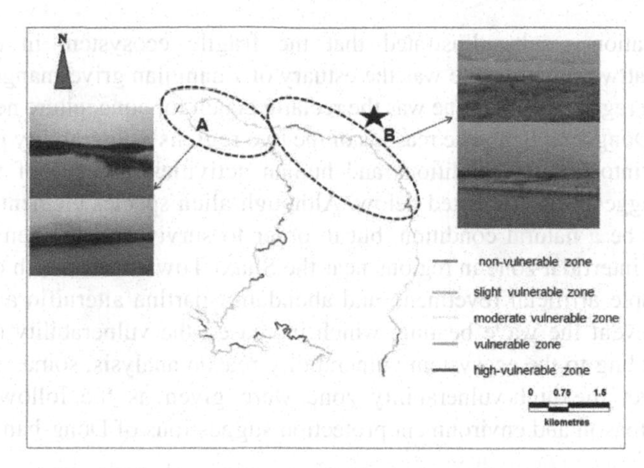

Fig. 3. Distributed of vulnerability value in beach region

The evaluation results also displayed the distribution of fragile beach region in Dongshan Bay. The vulnerable zone and high vulnerable zone are mainly distributed in Zone A along the bank of Zhangjiangkou Mangrove Forestry National Nature Reserve and Zone Bnear the Shaxi Town in north of the Dongshan Bay (Fig.5). These two regions have sludge substrate, flat terrain and high biodiversity as well as big population density and heavy pollution discharge in surroundings. Abundant Spartina alterniflora Loisel were also discovered in the Zhangjiangkou Mangrove Forestry National Nature Reserve.

5 Discussion

Presently, ecological vulnerability evaluation methods mainly include predictive evaluation based on scenario analysis and simulation[9],[10],[11]. as well as status evaluation based on natural and socio-economic indicators[12],[13],[14].The former achieves a predictive evaluation on the ecological vulnerability through subjective forecast or mathematical model simulation on the basis of various data collection, such as climate, hydrology, land utilization change, etc. The later one evaluates the ecological vulnerability with indicators based on the sensitiveness of natural conditions and adaptation of social economy. The major computational methods include comprehensive index method [15], analytic hierarchy process[16], fuzzy mathematics[17], and landscape ecology[18]

Existing evaluation researches of ecological vulnerability in coastal zone mainly focus on the climate change-induced sea level rise and predictive evaluation researches under interferences from natural disasters [19],[20],[21]. Few evaluations on the current ecological vulnerability in coastal regions caused by human activities have been reported yet. In this paper,we try to explores the current distribution of ecological vulnerability in Dongshan Bay in Fujian Province. The index system is established according to the "pressure-state-response" evaluation model. The current ecological vulnerability is evaluated comprehensively.

The evaluation results illustrated that the fragile ecosystem in coastal area concentrated attwo regions: one was the estuary of Zhangjian grivermangrovenational nature reserve region, another one was the reclaim beach for aquaculture near the Shaxi towninnorth Dongshan Bay. The reasons of the two regions' vulnerability mainly could be classified into natural conditions and human activities. A table of vulnerability reason and suggestions was listed below. Although alien species element was always considered to be a natural condition, but in order to survival, fishermen reclaimed a large scale of intertidal zone in regions near the Shaxi Town in the north of Dongshan Bay with simple artificial revetment, and abundant Spartina alterniflora Loisel were planted to prevent the wave beating, which increased the vulnerability of the beach region. According to the ecosystem vulnerability reason analysis, some advices about how to protect the high-vulnerability zone were given as the following Table4: vulnerability reason and environment protection suggestions of Dongshan Bay

Table 4. vulnerability reason and environment protection suggestions of Dongshan Bay

regions	reasons	Protect suggestions
Zone A mangrove reserve region	plenty of castoffs from industry, domestic wastewater and garbage were discharged in this zone	Through total Quantity Control, control pollution emissions, at the same time ecological restoration should be done
Zone B North reclaim beach	Reclamation of wetland and the invasion of alien species: SpartinaalternifloraLoisel	reduce human activities on the wet Land reclamation and damage to the forest, Control and governance of pollution.
Others zones	Prevention and control of ecological environment deterioration	Make evaluation and protection plan before development of ecological environment.

Acknowledgements. This work was supported by the research project of Third Institution of Oceanography, State Ocean Administration (Haisanke No. 2012010) .

References

1. Wang, R.H., Fan, Z.L.: Study on evaluation of ecological frangibility of Tarim River Basin. Arid environmental Monitoring 12(4), 39–44 (1998)
2. Zhang, X.N., Wang, K.L., et al.: The vulnerability of ecological environment in Beikeshi special zone Guilin. Acta. Ecologica. Sinica. 29(2), 749–757 (2009)
3. Newton, A., et al.: An overview of ecological status, vulnerability and future perspectivesof European large shallow, semi-enclosed coastal systems, lagoons and transitional waters. Estuarine, Coastal and Shelf Science 1-28 (2013)
4. Sun, L., Shi, Q.: Research on vulnerability of natural disaster in coastal city. Journal of Catastrophology 22(1) (2007)

5. Wang, N., Zhang, L.Q., et al.: Progress in research of coastal vulnerability assessment under the change of climate. Acta. Ecologica. Sinica., 32-7 (2012)
6. Tan, L.R., Chen, K., et al.: Assessment on storm surge vulnerability of coastal regions during the past twenty years. Science Geographical Sinica 31(9) (2011)
7. Turner, R.K., Adger, N., Doktor, P.: Assessing the economic cost of sea level rise. Environmental and Planing 27, 1777–1796 (1995)
8. Feng, S.Z., Li, F.Q., Li, S.Q.: An introduction to marine science. Beijing Higher Education Press (1999)
9. Wilhelmi, O.V., Wilhite, D.A.: Assessing vulnerability to agricultural drought: A nebraska case study. Natural Hazards (25), 37–58 (2002)
10. Downing, T.E.: Vulnerability to hunger in Africa: A climate change perspective. Global Environmental Change (1), 365–380 (1991)
11. Wang, J.L., Huang., J.X., et al.: Assessment to the vulnerability of ecosystem services to land use change – A case study of carbon stock of Taihu lake district in Jiangsu Province. Journal of Natural Resources 25(4) (2010)
12. Su, F., Zhang, P.Y., et al.: Assessment on the vulnerability of coal economy system in China. Geographical Research 27(4) (2008)
13. Santos, C.F., Carvalho, R., Andrade, F.: Quantitative assessment of the differential coastal vulnerability associated to oil spills. Journal of Coastal Conservation 17(1), 25–36 (2013)
14. Wang, X.D., Zhong, X.H., Liu, S.Z., et al.: Regional assessment of environmental vulnerability in the Tibeban Plateau: Development and application of a new method. Journal of Arid Environments (72), 1929–1939 (2008)
15. Qiao, Q.: Assessment on the vulnerability of ecological and landscape pattern in transition zone between cropping area and nomadic. Area in Sichuan Province. Beijing Forestry University (2005)
16. Lu, D.A., Tan, S.C., et al.: Establishment of ecological environment vulnerability index system in Jiangsu's coastal zone. Journal of Marine Science 27(1), 78–82 (2009)
17. Wu, J.Z., Li, B., et al.: Vulnerability of eco-economy in northern slope of Tianshan Mountains. Journal of Applied Ecology 19(4), 859–865 (2008)
18. Xie, H.L.: Ecology risk analysis of region by the method of landscape structure and spatial statistics. Acta. Ecologica. Sinica 28(10), 5020–5026 (2008)
19. Farhan, A.R., Lim, S.: Vulnerability assessment of ecological conditions in Seribu Islands, Indonesia. Ocean & Coastal Management (65), 1–14 (2012)
20. Palmer, B.J., Van der Elst., R., et al.: Preliminary coastal vulnerability assessment forKwaZulu-Natal, SouthAfrica. Journal of Coastal Research (64), 1390–1395 (2011)
21. Santos, M., del Río, L., Benavente, J.: GIS-based approach to the assessment of the coastal vulnerability to storms case study in Bay of cadiz (Andalusia, Spain). Journal of Coastal Research (2013)

The Chlorine Pollution Mechanism of Groundwater in Jiaozuo Area

XiuJin Xu[1,2], ZhenMin Ma[1,2], and Kai Man[1,2]

[1] School of Resources and Environment, University of Jinan, Jinan 250022, China
[2] Shandong Provincial Engineering Technology Research Center for Groundwater Numerical Simulation and Contamination Control
{867322749,939656278}@qq.com, stu_mazm@ujn.edu.cn

Abstract. It is found that by the survey on the industrial waste water emissions and unreasonable use of fertilizers and an analysis of the correlation between chloride pollutants and other chemical characteristics of shallow groundwater in Jiaozuo that the shallow groundwater has been seriously polluted with chlorides in this region. And this is attributed to the irrational emissions of industrial wastewater as well as incorrect utilization of fertilizers, which have caused an increase of chloride in the shallow ground water. In addition, we also found the sources polluting the shallow groundwater, the pollution mechanism and the transferring disciplinarian demonstrations.

Keywords: shallow groundwater, chloride, sources of pollution, pollution mechanism, migration patterns.

There is no life without water, water is the most critical resources of human production and living. It is also the necessity of human survival and development. Groundwater is an important part of water resources.It is very important to maintain a virtuous cycle of water resources. The nature of the goundwater is good and it is not polluted easily.[1] China is a country that is short of water. there many problems in our country, such as water shortages,water pollution, soil and water loss and so on.The water of south is more than the north, the west is short of water. According to the monitoring, at present the groundwater of the national most urban is polluted at a certain degree. The increasingly serious water pollution not only reduces the water functions, exacerbate the shortage of water resources, has brought the serious influence on our country's sustainable development strateg, but also make a serious threat to urban residents drinking water safety and people's health. Along with the urbanization and the development of economic and social, a large amount of land is occupied, non-agricultural irrigation water demand has increased dramatically, more and more industrial wastewater produce, a growing number of groundwater has been greatly polluted. Groundwater pollution is refers to the human activities cause the change of groundwater in the physical properties and chemical composition[2]. Groundwater pollution is different from the surface water pollution,

F. Bian and Y. Xie (Eds.): GRMSE 2014, CCIS 482, pp. 332–337, 2015.

Groundwater pollution occurs slowly and cannot be found easily. Once groundwater is polluted, it is hard to recovery it.

1 Jiaozuo Overview

1.1 Geographic Overview

Jiaozuo, a city in the northwestern Henan Province, bordering on the south of the Yellow River, right lying opposite to Zhengzhou City, connecting with Xinxiang in the east and neighboring Luoyang in the west, covers a total area of 4,072 sq. km and possesses 3.5203 million people. Not only located in the intersection of China and in the center of the New Eurasian Land Bridge, the city has gained a geographic advantage to serve as a hub of the four regions. Besides, the city is an owner of several transportation lines, including Zhengzhou-Jiaozuo, JiYang,etc. With this, Jiaozuo has grown more significant and was once even announced to be one of the major areas for energy base construction. There are many different kinds and quantities of superior mineral resources. A general survey found more than 40 kinds of mineral resources, accounting for 25% of all the sorts discovered in the province. In the city, the reserves of at least 20 kinds of minerals including coal, limestone, bauxite, refractory clay, pyrite, have been verified. Thereinto, the coal field, with a reserve of 3.24 billion tons of exclusive high-quality anthracite, starts from Xiuwu in the east, to Boai in the west, and bordering on Wuzhi in the south, extending as long as 65kilometres between the east and the west and as wide as 20 kilometres between the south and the north. However, the underground water in Jiaozuo, an important manufacturing base of chemicals, has become more severely polluted as a result of increasing emission of industrial waste water along with the rapid social and economic development.

1.2 Hydrogeological Review

There are five rivers whose drainage area exceeds 1000 sq. km, including Da Sha river, Bai Mamen river,Qu Ying river,etc. Jiaozuo has a very representative continental monsoon climate, with dry and windy Spring, rainy Summer, sunny Autumn and rarely snowy Winter. In the mountainous area, the mean annual precipitation is 594.4mm, with a maximum of 1101.7mm and a minimum of 260.3mm throughout the whole year, 416mm to the most in a month and 258mm in a day. Rainfalls differ a lot among the four seasons, the highest in July and August while June and September have a relatively lower amount. And the total amount of the four months has accounted for over 70% of the whole year. A majority of rivers in the city have been used as channels for pollution discharges. But as a result of the special climate and location, all rivers in the city are in accordance with seasonal changes, which means that no floods can be expected unless in certain years with abundant rainfalls. Da Sha river plays an essential role in our research on the shallow groundwater, since the wastewater from the chemical industries along its upriver flows directly into it. Consequently, wastewater and river water will replenish the shallow groundwater in this area by means of leakage.

2 Pollution Source and Pollution Way in Jiaozuo Area

Sources of pollution Mainly include the sewage ,Living garbage and the industrial waste water .Groundwater pollutants are affected by natural and anthropogenic factors. Household sewage and garbage can cause the content of the total dissolved solids hardness nitrate and rising levels of nitrate and chloride ,what is worse ,it can causes pollution of the pathogen .Industrial waste water and industrial waste make the concentrations of inorganic compound and inorganic compound of the groundwater high. the use of chemical fertilizer increase the level of the concentration of nitrate. Pesticide has a little effect on the groundwater, except in the case of shallow groundwater. Agricultural activities can oxidize organic contents in soil .For example it can turn organic nitrogen into inorganic nitrogen, going into the groundwater through infiltration. The natural salt water can pollute natural fresh water an so on. there are also chemical plants, textile mills and Mining factories in Jiaozuo area. The factories produce much complex waste. Solid waste is the main source of shallow groundwater pollution in Jiaozuo area. The solid waste produced by the factories mainly includes powder ash from stove, alkali dreg and red mud. Their major source is the power plants and chemical enterprise in Jiaozuo area. Unprocessed solid waste that is piling up haphazardly. Pollutants of the solid waste fall into the soil and shallow groundwater before washing away by rains.

3 The Pollution Mechanism of Shallow Groundwater in Jiaozuo Area

3.1 The Distribution of Chloride in Jiaozuo Area

When the amount of chloride exceeds 250mg/L, then the shallow groundwater is polluted. The chloride ion is typical of the conservative ion. After seeping into groundwater through the earth, due to chemical changes is not only barely retained in the vadose zone, and also its valence and quality remain unchanged .Based on that ,we can use the change of chloride ion in the groundwater to indicate how ground pollution may affect the purity of ground water[3].

The chloride in Jiaozuo area mainly is associated with Pollutant discharge from mining and other industrial enterprises. For example, the amount of alkali dreg is 225 thousands tons. The amount of sewage that is discharged by the factories is 2.25 million - 3.375 million tons, apart from 10% loss in evaporation, This sewage then infiltrates into the groundwater. The pollutants are located in fault, fracture belt. The alkali dreg contains a large amount of calcium chloride and magnesium salts. Due to atmospheric precipitation leaching effect and the hydrogeological conditions that are contributed to infiltration, the pollutants easily enter into the groundwater to cause pollution.

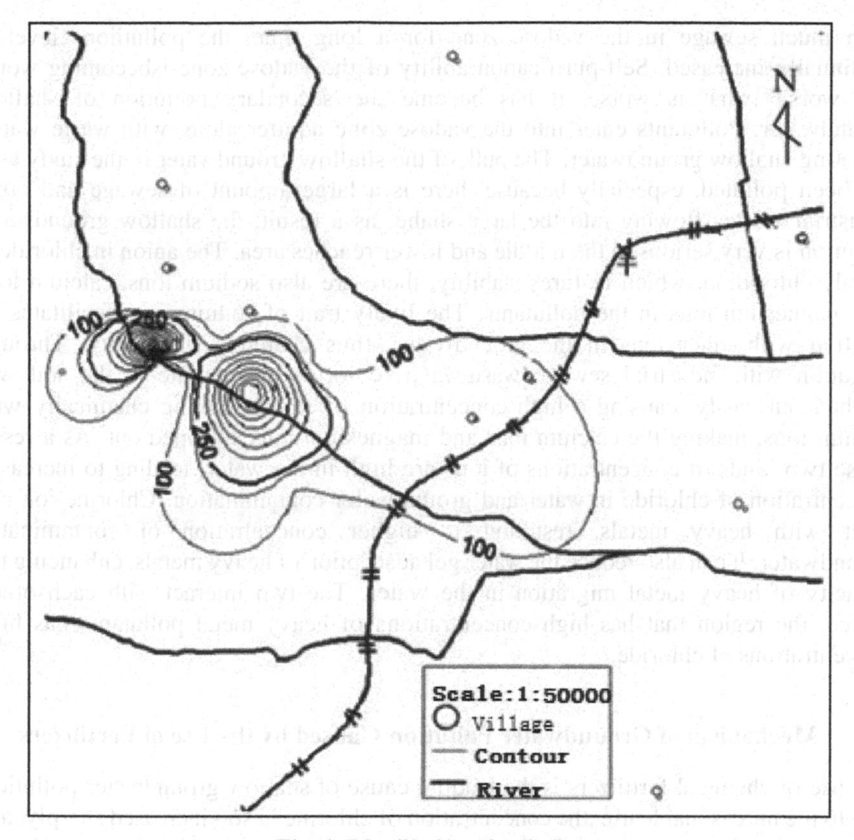

Fig. 1. Distribution of pollution

3.2 Mechanism of Groundwater Pollution Caused by Industrial Waste Water

The factories emit a large amount of waste water in Jinozuo area. Because of the polluted wastewater discharge pipes mostly mostly lead to rivers and channels, the groundwater depth is shallow in Jiaozuo area, most of the depth is less than 2 meters. Therefore, the pollutants easily enter into the groundwater to cause pollution. In addition, there are a lot of mining area inJiaozuo area. As a consequence of the long-term mining, groundwater drainage is occurred ,leading to acute water shortages in irrigation in this region. As a result, industrial waste water was used for agricultural irrigation, which in turn lead to groundwater pollution. All in all, the main sources of shallow groundwater pollution include: Mine drainage resulting in the formation of groundwater drawdown funnel ;Sewage irrigation ,river canal seepage and so on. The mine of the study area have discharged a lot of groundwater in a long time, which leads to groundwater drainage. In the study area ,the atmospheric precipitation fails to satisfy the needs of the agricultural irrigation, industrial waste water has long been used for irrigation. Most of the canals are damaged in the study area.The severe leakage has brought in pollutants of industrial sewage wastewater, mine water and sewage enter vadose zone. Rock and soil layer can adsorb some pollutions in water,to the extent that they can prevent pollutants from entering the groundwater. However, since there has been much sewage in the vadose zone for a long time, the pollutiions level is continually increased, Self-purification ability of the vadose zone isbecoming worse and worse ,what is worse, it has become the secondary pollution of shallow groundwater. Pollutants enter into the vadose zone aquifer along with waste water, polluting shallow groundwater. The bulk of the shallow groundwater in the study area has been polluted, especially because there is a large amount of sewage and other industrial wastes flowing into the large shahe, as a result, the shallow groundwater pollution is very serious in the middle and lower reaches area. The anion in chloride is mainly chloridion, which features stability, there are also sodium ions, calcium ions and magnesium ions in the pollutants. The lively trait of sodium ions facilitates its reaction with other ions in the groundwater, thus changing its quality. Through irrigation with industrial sewage wastewater, chloride and sulfate in the soil was leached out easily, causing a high concentration of salt, reactinng chemically with sodium ions, making the calcium ions and magnesium ions swapped out .As a result ,these two kinds of concentrations of ions are high in the water, leading to increased concentration of chloride in water and groundwater contamination. Chlorine ion can react with heavy metals, resulting in higher concentration of contaminated groundwater. It can also reduce the water gel adsorption of heavy metals, enhancing the capacity of heavy metal migration in the water. The two interact with each other. Hence, the region that has high concentrations of heavy metal pollutants has high concentrations of chloride.

3.3 Mechanism of Groundwater Pollution Caused by the Use of Fertilizers

The use of chemical fertilizers is the leading cause of shallow groundwater pollution. Due to the unreasonable use ,the concentration of chlorine in soil increased sharply, and pollutants are difficult to degrade in the soil, thus seeping into the groundwater with the sewage irrigation and atmospheric precipitation. The groundwater is polluted. The main source of organic chloride in the atmosphere is that the pesticide is lost in the application process, such as pesticide droplet drift; The volatilization of pesticides in application process; The volatilization of pesticide residue in plants and soil surface after the use of the pesticide; The loss during pesticide production and processing process, such as evaporation of pesticide products, emissions of waste gas, smoke and dust. The pollutants that leach through into the shallow groundwater pollute the groundwater[4].

4 Conclusion

To a conclusion, the serious chloride pollution of the shallow ground water has been caused by the unreasonable emissions of industrial wastewater and irrational use of fertilizers. It is due to improper production and development that the original elements, Ca2+ and Mg2+ are replaced by increasing pollutants. Consequently, the two ions expands, resulting in a critical increase of chloride in the water and groundwater pollution.

Acknowledgment. The author firstly would like to express their appreciation to the anonymous reviewers. This research was supported by National S&T Major Project No. 2012AA061705.

References

1. Baoxiang, Z.: Groundwater Vulnerability Assessment and Water Conservation Research in Yellow River Basin Groundwater. China university of geosciences institute of water resources and the environment, Beijing (2006)
2. Fried, J.J.: Groundwater Pollution. Elsevier, NewYrok (1975)
3. Shubin, Y., Zhenmin, M., Huishen, Z.: South-to-North Water Transfer Project of Jiaozuo Area of Shallow Groundwater Pollution Typical Characteristics 26(1), 91–95 (2012)
4. Hui, Z.: Present Situation and Prospect of the Migration Transformation Rule Research. Changchun university of science and technology, application of science, ChangChun13002

Calculations of the National Average Yield, Equivalence Factor and Yield Factor in Ten Years Based on National Hectares' Ecological Footprint Model -- A Case Study of Xiamen City

Juanjuan Dai*, Yaojian Wu, and Yurong Ouyang

Management and Development Strategy Research Center of Marine Environment, Third Institute of Oceanography, State Oceanic Administration, Xiamen, Fujian Province, China
{daijuanjuan,xmwyj,ouyangyurong}@tio.org.cn

Abstract. The concept of "national hectare" is introduced into the calculation of ecological footprint model in this article. That is, the national average yield is used instead of the global average yield. Based on this concept, the national average yields of plantation products, forest products, aquatic products and livestock products, the equivalence factors and the yield factors of Xiamen City from 2000 to 2009 are calculated. The main conclusions can be summarized as the following:(1) In the calculation of national average yield of livestock products, the types of productive land for beef, mutton and milk should be expanded from grassland to grassland and farmland according to the actual situation in China. The proportion of the productive land for livestock products attributed to grassland is lower, which affects the calculation result of equivalence factor of grassland. (2) The equivalence factor in the ecological footprint model based on the concept of national hectare is calculated by calorific value method. The results indicate that the equivalence factor of farmland is the maximum, and the grassland is the minimum. (3) It can be seen from the result that the yield factor of grassland in Xiamen City decreases continuously with the years. The biological productivity of grassland in Xiamen City declines continuously compared with the average biological productivity of the national grassland. However, there are large fluctuations for the yield factors across the years, but no obvious rules are identified. This study is conducive to the further calculation of ecological footprint and the comparison of ecological footprint across the provinces and cities. Thus, the utilization and sustainable development trends of natural capitals in China can be reflected.

Keywords: ecological footprint, national average yield, national hectare, equivalence factor, yield factor.

1 Introduction

Ecological footprint model is an assessment method for ecological sustainability proposed by a Canadian ecological economist William Reese in 1992 and improved

* Corrseponding author.

F. Bian and Y. Xie (Eds.): GRMSE 2014, CCIS 482, pp. 338–349, 2015.

by Dr. Mathis Wackernagel in 1996 [1-5]. The model studies the spatial measure of natural capital consumption from the perspective of biophysical quantities. It is a method used to measure the utilization degree of natural resource and life-supporting service function provided by the nature. Ever since the emergence of ecological footprint model, it attracts much attention and has a wide range of applications. In the early period of the application of the ecological footprint model, it was mostly used in the calculation and comparison on global and national scales. Later, the model was applied on the regional and urban scale. The concept of the ecological footprint was introduced into China for the first time by Zhang Zhiqiang [6], Xu Zhongmin [7] in 1999. The studies on the theory, method and calculation model of ecological footprint were extensively carried out since then. The empirical study extended from the nation [18, 19], provincial administrative region [10,11] to cities and counties [12,13], villages and towns [14] and even smaller scales (such as school) [15]. The global hectare method was adopted in most of these studies (global hectare, ghm2) to calculate the ecological footprints.

2 Concept of National Hectare

In the calculation of the ecological footprint model, various resources and energy consumption are converted into the following 6 types of biological production area: farmland, grassland, forest, water area, construction land and fossil energy land. In order to facilitate the international comparison, the global hectare method is generally used to calculate the ecological footprint. The resource or product consumption is converted into the area of ecological productive land according to the global average yield of various types of land. There are great differences in the natural and economic conditions between China and developed countries as well as in productivity structure and consumption structure. There exist significant differences in terms of the natural geographic conditions and social economic development level between the provinces and within one province. The global uniform equivalence factor and the national uniform yield factor cannot accurately reflect the characteristics of land productivity and social economic development[16]. Therefore, the ecological footprints in different provinces should be calculated and compared on the national scale, and the ecological footprint model should be established based on national hectare (nhm^2).

The natural resources consumed by human beings in various regions and services are converted into the area of biological productive land by the ecological footprint model based on national hectare and according to the national average productivity of various types of lands. After the adjustments of the equivalence factor and yield factor calculated by the national yield and data of the local actual land production, the comparative analysis of ecological footprint and ecological carrying capacity is carried out based on the concept of "national hectare".

3 National Average Yield of Biological Resources

The calculation formula of national average yield is as follows:

$$EP_{Ki} = P_{Ki} / A_{Ki} \quad (1)$$

here K is the type of consumption product, i is the year; EP_{Ki} is the national average yield of the k-th type of consumption product in the i-th year (kg•hm^{-2}); P_{Ki} is the national production yield of the k-th type of consumption product in the i-th year (kg); A_{Ki} is the national total production area of the k-th type of consumption product in the i-th year (hm^2). According to the above formula, the statistical data from "China Statistical Yearbook "(2000 -2009) and FAO were collected. The national average yields of plantation products, forest products, aquatic products and livestock products are calculated from 2000 to 2009 in China.

3.1 National Average Yield of Plantation Products

The plantation products in China mainly include grains, oil plants, sugar crops, tobaccos, vegetables, fruits and so on. The type of land occupation is farmland. The calculation results of national average yields are shown in Table 1.

3.2 National Average Yield of Forest Products

The main forest products in China include seven types: timber, rubber, resin, raw lacquer, tung oil seed, camellia seed and walnut. The type of the land occupation is forest land. Due to the lack of planting area for each type of forest product, the forestry land area for each year is taken as the output area of various forest products. The calculation results of national average yields are shown in Table 2.

3.3 National Average Yield of Aquatic Products

According to the division standard in the statistical yearbook, the aquatic products can be divided into two categories, freshwater products and seawater products. The freshwater products and seawater products can be also divided into natural products and cultured products. The type of ecological productive land is water area. Therefore, the national average yields of aquatic products are calculated by these four types. In the calculation, the area suitable for aquaculture in the Chinese inland waters is taken as the freshwater aquaculture area. The area of inland waters minus the area suitable for aquaculture is taken as the freshwater fishing area. The area suitable for aquaculture is taken as the seawater culture area. However, the area of fishing ground located on the continental shelf is taken as the seawater fishing area.. The calculation results of national average yields of aquatic products are shown in Table 3.

3.4 National Average Yield of Livestock Products

In the traditional ecological footprint model, the ecological productive lands occupied by livestock husbandry are all classified in grassland. This is because the large-scale breeding and industrialized operation mode are usually adopted for the livestock husbandry in developed countries. But China has a very different mode. The livestock breeding in China is mostly a sideline. It is a self-sufficient productive activity subordinate to the planting industry. The free-ranging and captive breeding are the main modes for livestock husbandry. Some abandoned agricultural resources and feed are mostly used for livestock. Therefore, the national average yield of livestock products in China cannot be calculated simply by using the traditional model of the global average yield.

In the calculation of ecological footprint for livestock products by Xie Hongyu [19], the ecological footprint can be calculated on the basis of the productive land occupied by the feed consumption by livestock. The areas of different productive lands required by the production of various types of products are calculated, and the result shows that 14% beef, 43% mutton and 28% milk products are from the grassland, and the rest from the farmland. The calculation results of national average yields are shown in Table 4.

According to the calculation results above, the national average yields of livestock products from 2000 to 2009 can be calculated, based on the national average yields of wheat, corn, barley and sorghum in Table 1. The calculation results are shown in Table 5 and Table 6.

Table 1. National average yields of the main planting products from 2000 to 2009 in China (10^3kg/hm^2)

Types of products	2000	2001	2002	2003	2004	2005	2006	2007	2008	2009	Types of lands
Grain crops	4.26	4.27	4.40	4.33	4.62	4.64	4.72	4.75	4.95	4.87	Farmland
Grain	4.75	4.80	4.89	4.87	5.19	5.22	5.31	5.32	5.55	5.45	Farmland
Rice	6.27	6.16	6.19	6.06	6.31	6.26	6.28	6.43	6.56	6.59	Farmland
Wheat	3.74	3.81	3.78	3.93	4.25	4.28	4.59	4.61	4.76	4.74	Farmland
Corn	4.60	4.70	4.92	4.81	5.12	5.29	5.33	5.17	5.56	5.26	Farmland
Sorghum	2.91	3.45	3.95	3.97	4.11	4.47	4.84	3.85	3.75	3.00	Farmland
Barley	2.47	3.76	3.64	3.51	4.10	4.14	3.91	3.61	3.56	3.70	Farmland
Beans	1.59	1.55	1.79	1.65	1.74	1.67	1.65	1.46	1.69	1.62	Farmland
Potatoes	3.50	3.49	3.71	3.62	3.76	3.65	3.43	3.47	3.54	3.47	Farmland
Oil plants	1.92	1.96	1.96	1.88	2.12	2.15	2.25	2.27	2.30	2.31	Farmland
Peanut	2.97	2.89	3.01	2.65	3.02	3.08	3.25	3.30	3.36	3.36	Farmland
Rapeseed	1.52	1.60	1.48	1.58	1.81	1.79	1.83	1.87	1.84	1.88	Farmland
Cotton	1.09	1.11	1.17	0.95	1.11	1.13	1.30	1.29	1.30	1.29	Farmland
Fiber plants	2.02	2.11	2.85	2.53	3.23	3.30	3.15	2.77	2.82	2.43	Farmland
Sugar crop	50.42	52.32	55.00	58.17	61.03	60.42	66.75	67.65	67.44	65.17	Farmland
Sugarcane	57.63	60.63	64.66	64.02	65.20	63.97	70.45	71.23	71.21	68.09	Farmland
Beet	24.52	26.81	30.23	24.93	30.83	37.52	39.77	41.36	40.75	38.52	Farmland
Tobacco	1.78	1.75	1.84	1.79	1.90	1.97	2.07	2.06	2.14	2.20	Farmland
Vegetables	29.52*	29.52	30.46	30.10	31.36	31.86	35.05	35.05*	35.05*	35.05*	Farmland
Tea	0.63	0.62	0.66	0.64	0.66	0.69	0.72	0.72	0.73	0.73	Farmland
Fruit	6.97	7.36	7.64	15.38	15.70	16.06	16.89	17.32	17.91	18.31	Farmland

Note: Data with * are the data of the adjacent year due to the lack of data on vegetables in the current year.

Table 2. National average yields of the main forest products from 2000 to 2009(kg/hm^2)

Types of products	2000	2001	2002	2003	2004	2005	2006	2007	2008	2009	Types of lands
Timber(m^3/hm^2)	0.18	0.17	0.17	0.18	0.20	0.21	0.25	0.27	0.31	0.27	Forest
Rubber	1.82	1.81	2.00	2.15	2.02	1.80	1.89	2.07	1.92	2.02	Forest
Rosin	2.09	2.14	2.14	2.38	2.36	2.69	3.19	3.39	2.98	3.42	Forest
Raw lacquer	0.02	0.02	0.02	0.03	0.03	0.05	0.07	0.05	0.05	0.07	Forest
Tung oil seed	1.72	1.54	1.48	1.42	1.34	1.29	1.34	1.27	1.30	1.20	Forest
Camellia seed	3.13	3.13	3.25	2.96	3.07	3.07	3.23	3.30	3.47	3.82	Forest
Walnut	1.18	0.96	1.29	1.49	1.53	1.75	1.67	2.21	2.91	3.20	forest

Table 3. National average yields of aquatic products from 2000 to 2009 in China ($10^3kg/hm^2$)

Types of products	2000	2001	2002	2003	2004	2005	2006	2007	2008	2009	Types of lands
Natural marine products	0.053	0.051	0.051	0.051	0.052	0.052	0.052	0.044	0.041	0.046	Waters
Marine products by artificial breeding	4.082	4.350	4.665	4.820	5.064	5.326	5.560	5.028	5.155	5.404	Waters
Natural freshwater products	0.211	0.200	0.210	0.230	0.226	0.241	0.237	0.210	0.210	0.204	Waters
Freshwater products by artificial breeding	2.242	2.363	2.509	2.626	2.803	2.979	3.183	2.920	3.071	3.284	Waters
Seawater products	1.737	1.942	2.166	2.277	2.436	2.625	2.809	2.599	2.677	2.854	Waters
Freshwater products	1.978	2.107	2.239	2.333	2.511	2.667	2.871	2.642	2.791	3.008	Waters
Aquatic products	1.835	2.010	2.196	2.301	2.469	2.644	2.837	2.619	2.731	2.927	Waters

Table 4. Required average grassland areas of the per kilogram livestock products from 2000 to 2009 in China($10^3m^2/kg$)

Items	2000	2001	2002	2003	2004	2005	2006	2007	2008	2009
beef	2.52	2.44	2.29	2.13	1.98	1.89	1.79	2.19	2.19	2.11
mutton	1.59	1.49	1.38	1.22	1.09	1.00	0.93	1.14	1.15	1.12
milk	0.81	0.65	0.52	0.38	0.30	0.24	0.21	0.19	0.19	0.19

Table 5. National average yields of the main livestock products in 2009 in China (kg/hm^2)

Types of products	Feeding way	Proportion of feeding way	Types of raw materials	Consumption of raw materials(kg)	Global average yield of raw material(kg/m^2)	Area of required biological productive land(m^2)		Average yields(kg/hm^2)	Types of lands
Pork	Feeding	1	Corn	1.31100	5258.49	2.4931	3.6199	2723.12	Farmland
			Barley	0.41700	3700.63	1.1268			
			Corn	1.42230	5258.49	2.7048	3.7246		
			Sorghum	0.30570	2997.66	1.0198			
Beef	Feeding	0.86	Corn	0.51256	5258.49	0.9747		8822.97	Farmland
	Grazing	0.14	Pasture	-	-	295.5254		4.74	Grassland
Mutton	Feeding	0.57	Corn	0.37164	5258.49	0.7067		8065.17	Farmland
	Grazing	0.43	Pasture			482.3072		8.92	Grassland
Poultry	Feeding	1	Corn	1.40840	5258.49	2.6783	2.9391	3402.35	Farmland
			Wheat	0.12360	4739.05	0.2608			
Eggs	Feeding	1	Corn	1.68540	5258.49	3.2051		3120.03	Farmland
Milk	Feeding	0.72	Corn	0.27310	5258.49	0.5194		13863.47	Farmland
	Grazing	0.28	Pasture	-	-	53.3754		52.46	Grassland

Table 6. National average yields of the main livestock products from 2000 to 2009 in China (kg/hm^2)

Items	Pork	Beef		Mutton		Poultry	Eggs	Milk	
2000	2304.1	7713.9	4	7051.3	6.3	2946.3	2727.8	12120.8	12.3
2001	2559.9	7883.3	4.1	7206.2	6.7	3009.9	2787.7	12386.9	15.3
2002	2677	8262.5	4.4	7552.8	7.2	3137.5	2921.8	12982.8	19.4
2003	2618	8074.8	4.7	7381.3	8.2	3085.6	2855.5	12687.9	26
2004	2816.9	8590.9	5	7853.1	9.1	3288	3038	13498.9	33.7
2005	2915.3	8871.4	5.3	8109.4	10	3386.6	3137.1	13939.5	41
2006	2928.3	8936.8	5.6	8169.2	10.8	3432.5	3160.3	14042.3	47.6
2007	2762.3	8668.9	4.6	7924.3	8.8	3339.8	3065.5	13621.4	52.6
2008	2895.5	9321.7	4.6	8521	8.7	3578.3	3296.4	14647	53
2009	2723.1	8823	4.7	8065.2	8.9	3402.3	3120	13863.5	52.5
Types of lands	Farmland	Farmland	Grassland	Farmland	Grassland	Farmland	Farmland	Farmland	Grassland

4 Calculation of Equivalence Factor

There are great differences in terms of average biological productivity for various types of land (farmland, forest land, grassland, water area). The average biological productivity should be multiplied by the equivalence factor for each one to convert into the standard area of biological production for direct comparison. In the calculation of the ecological footprint model based on national hectare, the equivalence factor of certain type of land is the ratio of the average biological productivity of this type of biological productive land to that of total national biological productive lands. The specific calculation formula is shown as follows[18]:

$$q_{ij} = \frac{\overline{p_{ij}}}{\overline{p_j}} = \frac{Q_{ij}}{S_{ij}} \Big/ \frac{\sum Q_{ij}}{\sum S_{ij}} = \frac{\sum_k p_{kj}^i \gamma_k^i}{S_{ij}} \Big/ \frac{\sum_i \sum_k p_{kj}^i \gamma_k^i}{\sum S_{ij}} \quad (2)$$

where q_{ij} is the equivalence factor of the i-th type of land in the j-th year (nhm^2/hm^2); p_{ij} is the average productivity of the i-th type of land in the j-th year ($10^9J/hm^2$); $\overline{p}_j$ is the average productivity of national lands in the j-th year ($10^9J/hm^2$); Q_{ij} is the total biomass of the i-th type of land in the j-th year (10^9J); S_{ij} is the biological productive area of the i-th type of land in the j-th year (hm^2). p_{kj}^i is the yield of the k-th type of biological products on the i-th type of land in the j-th year (kg). γ_k^i is the calorific value per unit of the k-th type of biological product on the i-th type of land ($\times 10^3 J/kg$).

The equivalence factors of various biological productive lands from 2000 to 2009 in China are calculated, based on the yield data of the main biological resources from 2000 to 2009 in "China Statistical Yearbook" and the unit of calorific value conversion of various biological products consulted in "Manual of Agricultural Technology and Economy (revised edition)" [19]. The calculation results are shown in Table 7. The calorific values of some of the products cannot be found in the "Manual". Therefore, according to the nature of the products, the calorific values of similar biological products or the average calorific values of multiple biological products are selected to replace them. It can be seen from Table 7 that the productivity of farmland per hectare is equivalent to that of 5.05 units of national hectare. It is 13 times larger than the water area, 30 times larger than the forest land and 63 times larger than the grassland. The biological productivity of farmland is the maximum, while that of grassland is the minimum.

5 Calculation of Yield Factor Taking Xiamen City as an Example

Based on the biomass conversion method in the calculation of equivalence factor and calculation result of national yields, the calculation formula [30] of the yield factor is shown as follows:

$$y_{ij}^{m} = \frac{\overline{p_{ij}^{m}}}{\overline{p_{ij}}} = \frac{Q_{ij}^{m}}{S_{ij}^{m}} \bigg/ \frac{Q_{ij}}{S_{ij}} = \frac{\sum_{k} p_{kj}^{im} \gamma_{k}^{i}}{S_{ij}^{m}} \bigg/ \frac{\sum_{k} p_{kj}^{i} \gamma_{k}^{i}}{S_{ij}} \tag{3}$$

where y_{ij}^{m} is yield factor of the i-th type of land in m region in the j-th year; $\overline{p_{ij}^{m}}$ is the average productivity of the i-th type of land in m region in the j-th year (10^9J/hm^2); $\overline{p_{ij}}$ is the average productivity of the national i-th type of land in the j-th year ($\times 10^9$J/hm^2). Q_{ij}^{m} is the total biomass of the i-th type of land in m region in the

Table 7. Caloric values per unit area and equivalence factors of various land types from 2000 to 2009 in China

Types of lands		2000	2001	2002	2003	2004	2005	2006	2007	2008	2009	Average
Farmland	Calorific value per unit area (10^9/hm^2)	76.44	75.40	76.82	73.80	79.95	82.35	84.33	84.62	89.68	0.69	81.41
	Equivalence factor (nhm^2/hm^2)	5.01	5	5.01	4.95	5.1	5.09	5.04	5	4.97	5.16	5.03
Forest	Calorific value per unit area (10^9/hm^2)	2.41	2.32	2.28	2.44	2.46	2.62	3.09	3.27	3.78	3.14	2.78
	Equivalence factor (nhm^2/hm^2)	0.16	0.15	0.15	0.16	0.16	0.16	0.18	0.19	0.21	0.18	0.17
Grassland	Calorific value per unit area (10^9/hm^2)	1.25	1.28	1.34	1.43	1.51	1.60	1.65	1.74	1.82	1.85	1.55
	Equivalence factor (nhm^2/hm^2)	0.08	0.08	0.09	0.1	0.1	0.1	0.1	0.1	0.1	0.11	0.1
Waters	Calorific value per unit area (10^9/hm^2)	5.23	5.44	5.75	6.05	6.38	6.78	7.17	7.57	7.92	8.38	6.67
	Equivalence factor (nhm^2/hm^2)	0.34	0.36	0.37	0.41	0.41	0.42	0.43	0.45	0.44	0.48	0.41
All	Calorific value per unit area (10^9/hm^2)	15.27	15.07	15.34	14.90	15.66	16.19	16.74	16.92	18.03	17.58	16.17
	Equivalence factor (nhm^2/hm^2)	1.00	1.00	1.00	1.00	1.00	1.00	1.00	1.00	1.00	1.00	1.00

Table 8. Yield factors of various land types in Xiamen from 2000 to 2009

Types of land			Farm land	Forest land	Grass land	Water area	Construction land
2000	Productivity (10^9J/hm^2)	Xiamen	145.25	1.51	7.82	26.84	145.25
		China	76.44	2.41	1.25	5.23	76.44
	Yield factor		1.90	0.63	6.25	5.13	1.90
2001	Productivity (10^9J/hm^2)	Xiamen	146.89	2.84	6.59	29.02	146.89
		China	75.40	2.32	1.28	5.44	75.40
	Yield factor		1.95	1.22	5.15	5.34	1.95
2002	Productivity (10^9J/hm^2)	Xiamen	140.25	1.84	6.02	27.84	140.25
		China	76.82	2.28	1.34	5.75	76.82
	Yield factor		1.83	0.81	4.50	4.84	1.83
2003	Productivity (10^9J/hm^2)	Xiamen	142.00	2.95	6.03	33.18	142.00
		China	73.80	2.44	1.43	6.05	73.80
	Yield factor		1.92	1.21	4.23	5.49	1.92
2004	Productivity (10^9J/hm^2)	Xiamen	126.40	3.82	5.77	37.36	126.40
		China	79.95	2.46	1.51	6.38	79.95
	Yield factor		1.58	1.55	3.83	5.85	1.58

Types of land			Farmland	Forest land	Grassland	Water area	Construction land
2005	Productivity (10^9J/hm^2)	Xiamen	142.95	3.72	5.39	36.28	142.95
		China	82.35	2.62	1.60	6.78	82.35
	Yield factor		1.74	1.42	3.38	5.35	1.74
2006	Productivity (10^9J/hm^2)	Xiamen	156.80	3.75	4.50	33.91	156.80
		China	84.33	3.09	1.65	7.17	84.33
	Yield factor		1.86	1.22	2.73	4.73	1.86
2007	Productivity (10^9J/hm^2)	Xiamen	117.71	3.23	3.88	39.29	117.71
		China	84.62	3.27	1.74	7.57	84.62
	Yield factor		1.39	0.99	2.22	5.19	1.39
2008	Productivity (10^9J/hm^2)	Xiamen	118.29	2.83	3.53	46.95	118.29
		China	89.68	3.78	1.82	7.92	89.68
	Yield factor		1.32	0.75	1.94	5.93	1.32
2009	Productivity (10^9J/hm^2)	Xiamen	139.61	3.21	3.65	46.79	139.61
		China	90.69	3.14	1.85	8.38	90.69
	Yield factor		1.54	1.02	1.97	5.58	1.54

j-th year ($\times 10^9$J); S_{ij}^{m} is the biological productive area of the i-th type of land in m region in the j-th year (hm^2); Q_{ij} is the total biomass of the i-th type of land in the j-th year ($\times 10^9$J); S_{ij} is the biological productive area of the i-th type of land in the j-th year (hm^2); p_{kj}^{im} is the yield of the k-th type of biological product on the i-th type of land in m region in the j-th year (kg); γ_k^i is the unit calorific value of the k-th type of biological product on the i-th type of land ($\times 10^3$J/kg); p_{kj}^{i} is the yield of the k-th type of biological product on the i-th type of land in the j-th year (kg).

The yield factors of various lands from 2000 to 2009 in Xiamen City are calculated based on the data of Xiamen City. The calculation results are shown in Table 8. The yield factor of construction land is equal to the corresponding value of the farmland. However, that part of construction land is not included in the calculation of yield factor. The yield factor of the productive land for fossil energy is 0. It can be seen from the calculation result that the yield factor of grassland in Xiamen City decreases continuously with years. It is indicated that the biological productivity of grassland in Xiamen City decreases continuously compared with the average biological productivity of national grassland. However, there are large fluctuations for the yield factors of the farmland, forest land and water area with the years, but there are no obvious rules.

6 Conclusion and Discussion

The concept of "national hectare" is introduced into the calculation of ecological footprint model in this article. That is, the national average yield is used instead of the global average yield. The main conclusions can be summarized as the following:

(1) The national average yields of plantation products, forest products, aquatic products and livestock products from 2000 to 2009 are calculated in this article. The national average yield of livestock products is calculated not by using the traditional model of the global average yield. However, according to the actual situation in China, the types of productive land for beef, mutton and milk should be expanded from grassland to grassland and farmland. The yield corresponding to feed consumed by livestock products is converted to the global average yield of livestock products. The proportion of the productive land for livestock products attributed to grassland is lower, which affects the calculation result of equivalence factor of grassland.

(2) The equivalence factor in the ecological footprint model based on the concept of national hectare is calculated by calorific value method. The equivalence factors of various biological productive lands from 2000 to 2009 are obtained. The results indicate that the equivalence factor of farmland is the maximum, and the grassland is the minimum.

(3) The yield factors of various types of lands from 2000 to 2009 in Xiamen City are calculated. It can be seen from the result that the yield factor of grassland in Xiamen City decreases continuously with the years. The biological productivity of grassland in Xiamen City declines continuously compared with the average biological

productivity of the national grassland. However, there are large fluctuations for the yield factors across the years, but no obvious rules are identified.

(4) The data about the yield of biological resources and area in this article were mainly from the statistical yearbook. Other statistical data needed but not found in statistical yearbook were referred from the data of FAO. There is a certain difference in the statistical methods between them. The adoption of both data sources has a certain impact on the calculation result. How to eliminate this effect is required to be further studied.

(5) The calculations of national average yield, equivalence factor and yield factor by the ecological footprint model based on national hectare lay a foundation for the calculation of ecological footprint on the province, city and county scale of China. Meanwhile, they can provide important parameters. This is conducive to the further calculation of ecological footprint and the comparison of ecological footprint across the provinces and cities. Thus, the utilization and sustainable development trends of natural capitals in China can be reflected.

References

1. Rees, W.E.: Ecological Footprints and Appropriated Carrying Capacity: What Urban Economics Leaves Out. Environment and Urbanization 4(2), 121–130 (1992)
2. Rees, W.E., Wackernagel, M.: Urban Ecological Footprints: Why Cities Cannot Be Sustainable and Why They are a Key To Sustainability. Environmental Impact Assessment Review 16, 223–248 (1996)
3. Rees, W.E., Wackernagel, M.: Our Ecological Footprint: Reducing Human Impact On The Earth. New Society Pubilishers (1996)
4. Wackernagel, M., Rees, W.E.: Perceptual And Structural Barriers To Investing In Natural Capital: Economics From An Ecological Footprint Perspective. Ecological Economic 20, 3–4 (1997)
5. Rees, W.E., Wackernagel, M.: Monetary Analysis: Turning A Blind Eye On Sustainability. Ecological Economic, 47–52 (1998)
6. Zhiqiang, Z., Zhongmin, X., et al.: Concept of Ecological Footprint and Evaluation of Calculation Model. Ecological Economy 10, 8–10 (2000)
7. Zhongmin, X., Guodong, C.: A Study on the Water Resources Carrying Capacity by Using the Method of Multi-objective Optimization Model -Taking the Heihe River as an Example. Journal of Lanzhou University(Natural Science Edition) 36(2), 122–132 (2000)
8. China Council for International Cooperation on Environment and Development, World Wide Fund for Nature: Report of Ecological Footprint in China (First). World Environment 5, 52–57 (2008)
9. China Council for International Cooperation on Environment and Development, World Wide Fund for Nature: Report of Ecological Footprint in China (Second). World Environment. 5, 63–69 (2008)
10. Zhongmin, X., Zhiqiang, Z.: Calculation and Analysis of Ecological Footprint in Gansu Province in 1998. Journal of Geographical Science 5, 607–616 (2000)
11. Xibang, H., Xinrui, Z., Zegen, Z.: Empirical Analysis of the Ecological Footprint and Development Capacity in Chongqing City During 1997-2005. Environmental Science and Management 33(2), 135–138 (2008)

12. Huihui, Z., Yuan, W., Xueming, G., et al.: Construction of the Sustainable Development Index System Based on the Material Flow and Ecological Footprint – By Taking Tongling City in Anhui Province as the Example. Journal of Ecology 32(7), 2025–2032 (2012)
13. Jianping, S., Xin, L., Yuanlei, Y.: Analysis of the Dynamic Changes of Ecological Footprint and Ecological Carrying Capacity in Mianyang City. Bulletin of Soil and Water Conservation 32(4), 276–280 (2012)
14. Jian, C., Ming, Z.: Evaluation of Sustainable Levels in Township Based on the Theory of Ecological Footprint. China Education Innovation Herald 8, 180 (2008)
15. Zheng, Y., Changchun, F., Junjie, K.: A Study of Sustainable Campuses Based on The Ecological Footprint Model: A Case Study Of Peking University. Resources Science 33(6), 1163–1170 (2011)
16. Hengyi, Z., Weidong, L., Shizhong, W., et al.: Calculations of Equivalence Factor and Yield Factor in the "Provincial Hectare" Ecological Footprint Model - A Case Study of Zhejiang Province. Journal of Natural Resources 24(1), 82–92 (2009)
17. National Bureau of Statistics of China: China Statistical Yearbook (2000-2009). China Statistics Press, Beijing (2000-2009)
18. FAO. FAOSTAT: http://faostat.fao.org/:
19. Hongyu, X.: Improvement and Application of the Ecological Footprint Model. Chemical Industry Press, Beijing (2008)
20. Zhongliang, G., Qinyu, Z., Xiujuan, T., et al.: Calculation and Application of Equivalence Factor and Yield Factor in "National Hectare" Ecological Footprint Model – A Case Study of Chongqing City. Journal of Anhui Agricultural Sciences 38(15), 7868–7871 (2010)
21. Ruofeng, N., Tianfu, L.: Manual of Agricultural Technology and Economy. Agriculture Press, Beijing (1984)

A Case Study on Environment Risk Assessment Methods of Coastal Petrochemical Project

Yuting Wu[1,2], Ling Cai[1], Honggang Yang[2,*], and Yaojian Wu[1,*]

[1] Third Institution of Oceanography, State Ocean Administration, Xiamen, Fujian Province, China
[2] School of Resources and Environmental Engineering, Wuhan University of Technology, Wuhan, Hubei Province, China

Abstract. With the advantages of good condition of port water, large environmental capacity, abundant transport capacity, low freight and modifiable tense situation of land use, the coastal region becomes the biggest petrochemical project region in the world. It is necessary to establish suitable environment risk assessment methods for coastal petrochemical projects so that we can keep away the environment from harm as well as minimize pollution to the environment in operation and after accident. In this research, a scientific nature and accuracy the coastal petrochemical projects' environmental risk system of Quanzhou petrochemical project is constructed, as well as the assessment method through the instance analysis.This system contains 3 citerion layer indexes, 2 expanding layer indexes and 22 factor layer indexes, forming a multi-level and multi-factor model of structure. After building the index system, the combination of empowerment method of analytic hierarchy process and entropy weight method have been carried out, the indexs are modified and improved using expert consultation method. A linear weighted group legal is appled to valuate the factor weight of the index system and calculate the environmental risk level of the coastal petrochemical projects, as a result, the risk of the project could be determined. It could provide a reference for the coastal petrochemical projects' environment protections: scientific location decision,environment management, risk prevention and control. It may reduce the impact of the accidents and prevent environmental risks by strengthening and perfecting the environmental risk management system combining with the sensitive degree of environmental factors.

Keywords: coastal petrochemical projects, environmental risk assessment, analytic hierarchy process, entropy weight method.

1 Introduction

Abundant water resources and energy sources are the prerequisite of basic petrochemical production[1]. With extensive water areas appropriate for port construction, big water environmental capacity, big water transportation capacity, low

* Corresponding authors.

F. Bian and Y. Xie (Eds.): GRMSE 2014, CCIS 482, pp. 350–360, 2015.

freight[2] and great contributions to relieving the intensive land utilization, coastal regions become the base of large petrochemical projects of developed countries. However, petrochemical production will discharge a lot of pollutants[3] which implies greater environmental risks compared to general industrial wastes.

Environmental risk assessment, an effective risk communication basis[4], can provide users and publics an effective risk communication platform. It is simple and feasible. However, the environmental risk assessment method and technology may cause excessive reliance on results with respect to uncertain risk areas involving various traditional methods and safety factors[5]. Sometimes, they may focus on a certain dangerous chemical rather than the global assessment[6]. Therefore, environmental risk assessment method determines the accuracy and scientificity of the result[7]. China's environmental risk assessment is still in the developing stage and has various shortcomings. No objective quantitative criterion and effective method for risk assessment have been found yet. Moreover, there are only few case studies and associated experiences available. To make scientific decision on coastal environmental protection, this paper established a method and index system for assessing marine environmental risks caused by local petrochemical project according to current risk characteristics.

2 Method

Although there's no unified standard on the selection of environmental risk assessment method have been established in China, appropriate analytical method still shall be selected according to project characteristics, technical data of project, risk assessment guidelines, feasibility, sensitivity, scientificity, transparency and globality in considering of characteristics and requirements of environmental risk assessment. In this paper, the environmental risks in coastal regions caused by petrochemical projects were evaluated through index method.

2.1 Establishment of Index System for Environmental Risk Assessment

To reflect the target requirements of the problem as much comprehensively as possible, the established index system for environmental risk assessment covers various aspects, such as social situation, economic policies, local environmental status, resources and research demands. Significant indicators which are influenced directly or indirectly by the management system were selected. Some influential index systems with more practical experiences were explored through literature review and analysis of environmental risk assessment theory. The environmental risk assessment system appropriate for petrochemical projects in coastal regions was determined according to these influential index systems and local situations.

2.2 Expert Consultation Method

Firstly, typical representative and significant assessment indexes were selected according to the influencing factors of environmental risk and practical situations of petrochemical projects in coastal regions. Secondly, the index system for assessing environmental risks caused by petrochemical projects in coastal regions was established based on system analysis and expert consultation. The index system includes three levels: 1) target layer: reflect the level of environmental risks; 2) criterion layer: measure the level of environmental risks from environmental risk source, control management mechanism of environmental risk source and environmental risk suffers; 3) factor layer: specific assessment indexes. Expert consultation tables include assessment system and weight data acquired through 25 questionnaires survey.

2.3 Index System

Generally speaking, a petrochemical project has a system with the highest environmental risk compared to rest systems. Dividing the environmental risk sources of a whole project into several systems can increase the determination accuracy of risk source carriers. A common index system for assessing environmental risks caused by petrochemical projects in coastal regions (Table 1) was established based on theoretical analysis and expert consultation. Indexes were screened by theoretical exploration.

Table 1. Index system of environmental risk assessment

Criterion layer	Extension layer	Factor layer
Environmental Sensitivity	Land Environmental Sensitivity	Proportion of exposure staff
		Enterprise layout
		Amount of surrounding high-risk enterprises
		Risks of working environment
		Number of surrounding environmental sensitive targets
		Effect of explosion and leakage on land environment
	Ocean Environmental Sensitivity	Ocean environmental functions
		Distance to water source protection zone
		Reach of prevailing wind
		Contamination risk capacity of marine environment
		Number of surrounding environmental sensitive targets
		Effect of explosion and leakage on marine environment
Environmental Risk Source	/	Danger reserve
		Danger categories
		Danger reactivity
		Danger inflammability
		Danger toxicity
		Area of project park
Environmental Risk Management	/	Environmental management system
		Green infrastructure construction
		Economic loss risks
		Emergency system

The criterion layer of the established index system includes Environmental Sensitivity, Environmental Risk Source and Environmental Risk Management[8]. Environmental Sensitivity refers to environment's responses to human beings and human activities. It is a comprehensive environmental risk assessment index and includes land environmental sensitivity and ocean environmental sensitivity. Higher Environmental Sensitivity represents higher environmental risk. Environmental Risk Source refers to substances that may damage the environments and is the prerequisite of environmental risk events. It includes not only the dangerous influence of contamination accidents on surrounding sensitive suffers, but also liquefiable original substances of contamination accidents or leakage. Environmental Risk Management includes environmental management system, green infrastructure construction, economic loss risks and emergency system. Poorer Environmental Risk Management means higher environmental risk.

2.4 Program Code

When determining the index weight, it is important to use appropriate method to evaluate the accuracy and objectivity of results. Weights of indexes of the established index system were determined by combining analytic hierarchy process and entropy weight method (Fig.1).

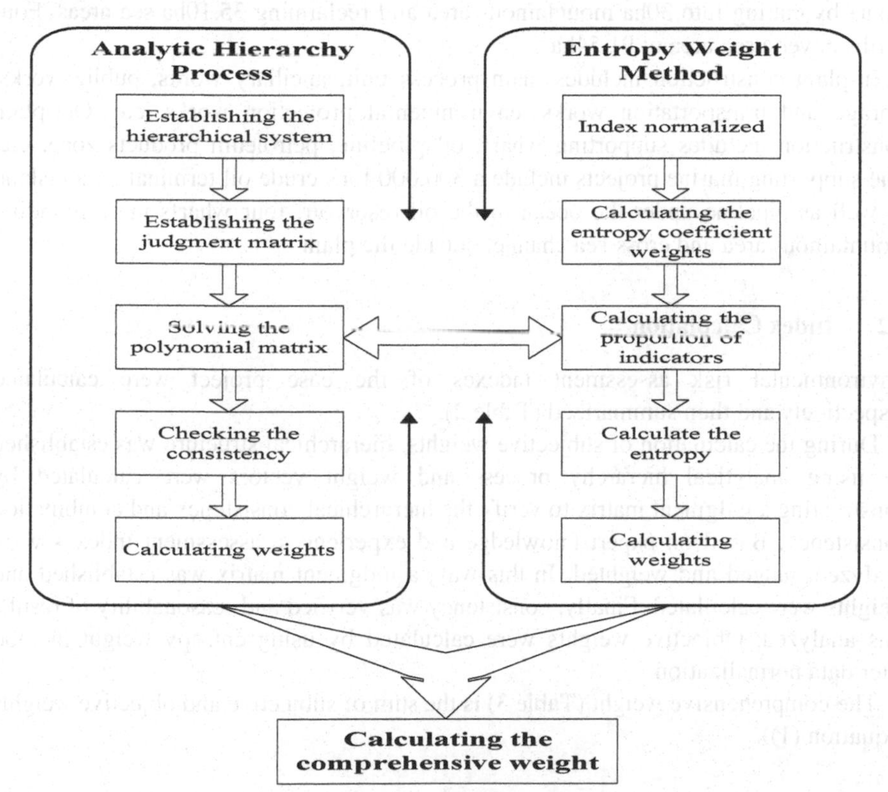

Fig. 1. Weight determination procedure

Comprehensive weight was determined by linear weighted method. It is sum of subjective weight calculated from analytic hierarchy process and objective weight calculated from entropy weight method:

$$W_0 = u\,W + (1-u)W_i\,. \quad (1)$$

where $0 \le u \le 1$.

W_0 varies with u. u was determined 0.5 in this paper according reference [9,10] and practical application.

3 Case Study

3.1 Introduction to Project

The case study was conducted in a petrochemical industrial area with an annual petroleum refining output of 12 million tons in Southeast China. The main plant covers a floor area of 234.8148ha (3,522mu) intertidal zone. The heavy oil deep processing project (5 million tons per year), the first-stage construction, includes four berths for oil products in a mountainous area. An oil reservoir is built behind the wharf by cutting into 30ha mountainous area and reclaiming 35.10ha sea areas. Four berths cover a sea area of 91.54ha.

In-plant construction includes main process unit, ancillary works, public works, storage and transportation works, environmental protection works, etc. Out-plant construction includes supporting wharf, oil pipeline, petroleum products zone, etc. The supporting marine projects include a 300,000 tons crude oil terminal on an island as well as pipeline under the ocean to the oil reservoir, four wharfs in surrounding mountainous area and cross-sea channel outside the plant.

3.2 Index Calculation

Environmental risk assessment indexes of the case project were calculated respectively and then summarized (Table 2).

During the calculation of subjective weights, hierarchical structure was established by using analytical hierarchy process and weight vectors were calculated by constructing a judgment matrix to verify the hierarchical consistency and combination consistency. Based on expert knowledge and experiences, assessment indexes were analyzed, judged and weighted. In this way, a judgment matrix was established and weights were calculated. Finally, consistency was verified and reasonability of results was analyzed. Objective weights were calculated by using entropy weight method after data normalization.

The comprehensive weight (Table 3) is the sum of subjective and objective weights (equation (1)).

Table 2. Summary of environmental risk assessment indexes of the case project

Target layer	Criterion layer	Extension layer	Factor layer	Weight
Environmental risk assessment system of a petrochemical project in coastal regions	Environmental Sensitivity (S)	Land Environmental Sensitivity	Proportion of exposure staff (Sl_1)	3
			Enterprise layout (Sl_2)	1
			Amount of surrounding high-risk enterprises (Sl_3)	3
			Risks of working environment (Sl_4)	3
			Number of surrounding environmental sensitive targets (Sl_5)	4
			Effect of explosion and leakage on land environment (Sl_6)	2
		Ocean Environmental Sensitivity	Marine environmental functions (So_1)	4
			Distance to water source protection zone (So_2)	2
			Reach of prevailing wind (So_3)	1
			Contamination risk capacity of marine environment (So_4)	2
			Number of surrounding environmental sensitive targets (So_5)	5
			Effect of explosion and leakage on marine environment (So_6)	4
	Environmental Risk Source (O)	/	Danger reserve (O_1)	5
			Danger categories (O_2)	5
			Danger reactivity (O_3)	4
			Danger inflammability (O_4)	3
			Danger toxicity(O_5)	2
			Area of project park (O_6)	2
	Environmental Risk Management (M)	/	Environmental management system (M_1)	1
			Green infrastructure construction (M_2)	1
			Economic loss risks(M_3)	5
			Emergency system (M_4)	1

Table 3. Comprehensive weight of environmental risk assessment indexes

Target layer	Criterion layer	W_0	Extension layer	W_0	Factor layer	W_0
Environmental risk assessment system of a petrochemical project in coastal regions	S	0.2262	Sl	0.42	Sl_1	0.1366
					Sl_2	0.0927
					Sl_3	0.1152
					Sl_4	0.1641
					Sl_5	0.2191
					Sl_6	0.2725
			So	0.58	So_1	0.1213
					So_2	0.1587
					So_3	0.0935
					So_4	0.1291
					So_5	0.2180
					So_6	0.2795
	O	0.3291	/		O_1	0.1699
					O_2	0.1402
					O_3	0.1185
					O_4	0.2076
					O_5	0.2665
					O_6	0.0975
	M	0.4448	/		M_1	0.2012
					M_2	0.2590
					M_3	0.1874
					M_4	0.3525

3.3 Result Analysis

The comprehensive environmental risk indexes were divided into low risk ((0-1]), relative low risk ((1-2]), general risk ((2-3]), relative high risk ((3-4]) and high risk ((4-5]). Meanwhile, the comprehensive risk was calculated 2.59 (general risk) by multiplying weight by the risk grade. This indicates that the case project has certain environmental risks and has to adopt some countermeasures, such as enhancing environmental risk inspection of enterprise departments, making perfect risk management contingency plans and strengthening staff safety education and trains.

Moreover, indexes weights and risk grades calculated by using combination weighting method and analytical hierarchy process were compared (Table 4).

Table 4. The comparing chart of synthetical weighted value and risk value between two different weighting methods

Criterion layer	W_0	wi	Extension layer	W_0	w_i	Factor layer	W_{0C}	Comprehensive risk value	wi_C	Comprehensive risk value
S	0.2262	0.1220	Sl	0.42	0.33	Sl_1	0.0130	0.0389	0.0041	0.0124
						Sl_2	0.0088	0.0088	0.0016	0.0016
						Sl_3	0.0109	0.0328	0.0024	0.0072
						Sl_4	0.0156	0.0468	0.0064	0.0191
						Sl_5	0.0208	0.0833	0.0105	0.0419
						Sl_6	0.0259	0.0518	0.0157	0.0314
			So	0.58	0.67	So_1	0.0159	0.0637	0.0052	0.0209
						So_2	0.0208	0.0416	0.0130	0.0261
						So_3	0.0123	0.0123	0.0035	0.0035
						So_4	0.0169	0.0339	0.0082	0.0164
						So_5	0.0286	0.1430	0.0205	0.1024
						So_6	0.0367	0.1467	0.0310	0.1239
O	0.3291	0.3196	/			O_1	0.0559	0.2796	0.0512	0.2560
						O_2	0.0461	0.2307	0.0323	0.1613
						O_3	0.0390	0.1560	0.0205	0.0821
						O_4	0.0683	0.2050	0.0804	0.2412
						O_5	0.0877	0.1754	0.1216	0.2433
						O_6	0.0321	0.0642	0.0136	0.0271
M	0.4448	0.5584	/			M_1	0.0895	0.0895	0.0948	0.0948
						M_2	0.1152	0.1152	0.1594	0.1594
						M_3	0.0834	0.4168	0.0405	0.2025
						M_4	0.1568	0.1568	0.2638	0.2638
Total							1	2.59	1	2.14

In Table 4, risk calculated 2.59 by the combination weighting method, but 2.14 by the analytical hierarchy process. Such difference is caused by the lower weight of Environmental Sensitivity (0.122) and Land Environmental Sensitivity (0.041) in analytical hierarchy process due to the influence from subjective factors. Consequently, indexes of Environmental Sensitivity can only have low weights and some (e.g. Sl_1, Sl_2, Sl_3 and Sl_4) even valued only ‰, thus lowering the assessment accuracy.

Moreover, combination weighting method and analytical hierarchy process calculated greatly different weights of indexes of Environmental Sensitivity. For example, difference of So_1, Sl_5 and Sl_6 reaches more than 0.01 and minimum difference is contributed by So_6 (0.0057), still higher than the weight of Sl_2 (0.0016), Sl_3 (0.0024), So_3 (0.0035) and Sl_1 (0.0041). As a result, effect of environmental factors on environmental risk is weakened, thus lowering the comprehensive risk of Environmental Sensitivity and finally influencing the results.

Based on above analysis, the combination of entropy weight method and analytical hierarchy process takes both objective and subjective factors into consideration and can offset such errors to a certain extent, thus increasing the reasonability and accuracy of results. According to the assessment result, the case project is suggested to focusing on preventive measures at the moment, such as enhancing protection to environmental sensitive targets and countermeasures to sudden environmental contamination accidents and perfecting the comprehensive risk management.

4 Discussion

Research on environmental risk assessment began in 1930s and achieved rapid development after 1970s, especially in United States[11-13]. Existing researches on environmental risk assessment are relative mature. They involve multiple influencing factors instead of single influencing factor and explore effects of different forms of pollutants on human health[14]. The new environmental risk assessment breaks the local application limit of risk assessment and is characteristic of global, regional and optimal assessment. China's environmental risk assessment achieved a diversified development after 2000, manifested by the good development in regional planning, farmland soil and petrochemical industry[15] as well as diversified and perfected assessment methods. The frequent petrochemical accidents in China in recent years highlight the importance of environmental risk assessment. The *Notice on Further Enhancing Environmental Risk Prevention of Environmental Impact Assessment Management* (UNCED [2012]77) reflects China's technical progress in comprehensive environmental risk assessment.

Frequent marine pollution accidents in recent years cause serious damages to marine environment and marine ecosystem[16-18]. Maine pollution is mainly caused by petrochemical accidents. Environmental risks caused by petrochemical project are characteristic of various sources, uncertainty, serious consequence, great chain effect, difficult quantification and complex influence[19]. With various pollutants, high environmental risks and uncertain protection goal, petrochemical accidents in coastal regions are easy to cause explosion and fire disaster. They develop quickly and will induce large-scale leakage of toxic chemicals, causing great property loss, causalities and serious environmental pollution[20]. For example, the oil spill accident in Penglai on June, 2011 discharged abundant crude oil into ocean, which polluted 870km2 surrounding sea and destroyed marine habitat seriously. The oil pipeline explosion accident in China Qingdao in 2013 polluted the beach seriously by spilled crude oil, which caused massive deaths of local crops and raising animals. Residual crude oil on stones and beach are difficult to be self-purified in a short period. To reduce petrochemical accidents in coastal regions and enhance countermeasures, China shall learn research methods from other countries, establish a standard environmental risk assessment system with Chinese characteristics, perfect related laws and policies and strengthen comprehensive risk management in future.

The case study also provides good references of comprehensive risk management to other petrochemical projects in coastal regions. Generally, China's petrochemical projects in coastal regions shall:

(1) Implement appropriate environment protection measures according to characteristics of specific petrochemical projects, analyze effect of project construction on environment according to environmental monitoring data, follow the principle of "simultaneous design, construction and service of safety facilities and the main construction", prevent environmental deterioration and ensure the sustainable economic development.

(2) Establish an environmental management administration and equip with environmental management staff during project construction and operation to enhance protections to environmental sensitive targets, countermeasures to sudden environmental contamination accidents, feedback to departments concerned and rehabilitation.

(3) Adopt fault liability system of total pollutant control, find causes of excessive pollutant emission and propose corresponding solutions.

(4) Implement pollution control measures, strengthen environmental protection management and make various environmental-friendly rules and regulations[21], mainly including operating instructions of environmental protection devices, pollution control parameters, in-plant and out-plant environmental monitoring system, environmental protection plan and in-plant environmental protection regulation.

(5) Provide regular or irregular trainings to environmental protection staff according to practical and job demands[22] to enhance environmental protection knowledge and environmental problem solving ability of environmental management staff, operation skills of pollution device operators, chemical analysis ability of environmental monitoring staff.

References

1. Hong Hua, Y., Li Bo, Z., Da Zheng, Z.: Pollution Characteristics of Groundwater from Petrochemical Enterprise in Coastal Area of Zhejiang Province. Environmental Science and Management 3, 011 (2012)
2. Peng, S.T., Wang, X.L., Dai, M.X., et al.: Discussion on identification and risk assessment of major hazard installations in petrochemical wharf. Journal of Waterway and Harbor 33(3), 241–244 (2012)
3. Zheng, L.N., Zhang, L., Zhang, L., et al.: Case Analysis of Environmental Risk Assessment for Petrochemical Industry. Advanced Materials Research 726, 1101–1104 (2013)

4. Bridges, J.: Human health and environmental risk assessment: the need for a more harmonised and integrated approach. Chemosphere 52(9), 1347–1351 (2003)
5. Pillay, A., Jin, W.: Formal Safety assessment. Technology and Safety of Marine Systems (2004)
6. Suter II, G.W., Vermeire, T., Munns Jr, W.R., et al.: An integrated framework for health and ecological risk assessment. Toxicology and Applied Pharmacology 207(2), 611–616 (2005)
7. Ya, Y., Ma, J.W.: Study on risk zoning technology of major environmental risk sources in urban scale and its application in Shanghai. China 12, 1050–1062 (2010)
8. Shao, C., Yang, J., Tian, X., et al.: Integrated Environmental Risk Assessment and Whole-Process Management System in Chemical Industry Parks. International Journal of Environmental Research and Public Health 10(4), 1609–1630 (2013)
9. Shao, L., Chen, Y., Zhang, S.S.: Comprehensive evaluation of cross–border sudden atmospheric environment based on AHP and fuzzy entropy source of risk. China Population, Resources and Environment. 27–31 (2010)
10. Li, X.H., Li, Y.M., Gu, Z.H., et al.: Based on analytic hierarchy process (ahp) and entropy weight method to the regional logistics development of competition analysis. Journal of southeast university: natural science edition 5, 398–401 (2004)
11. Du, S.J.: Environmental risk assessment research progress. Environmental Science and Management 31(5), 193–194 (2006)
12. National Research Council (US).: Committee on Risk Assessment of Hazardous Air Pollutants. Science and judgment in risk assessment. National Academy Press (1994)
13. Power, M., Mc Carty, L.S.: Trends in the development of ecological risk assessment and management frameworks. Human and Ecological Risk Assessment 8(1), 7–18 (2002)
14. Munoz, I., Tomas, N., Mas, J., et al.: Potential chemical and microbiological risks on human health from urban wastewater reuse in agriculture. Case study of wastewater effluents in Spain. Journal of Environmental Science and Health Part B 45(4), 300–309 (2010)
15. Chou, F., Wang, X.: Risk Assessment on Health Effects of Viruses in Reused Wastewater in City. Journal of Environment and Health 20(4), 197–199 (2003)
16. Castanedo, S., Juanes, J.A., Medina, R., et al.: Oil spill vulnerability assessment integrating physical, biological and socio–economical aspects: Application to the Cantabrian coast (Bay of Biscay, Spain). Journal of Environmental Management 91(1), 149–159 (2009)
17. Adler, E., Inbar, M.: Shoreline sensitivity to oil spills, the Mediterranean coast of Israel: assessment and analysis. Ocean & Coastal Management 50(1), 24–34 (2007)
18. Santos, C.F., Carvalho, R., Andrade, F.: Quantitative assessment of the differential coastal vulnerability associated to oil spills. Journal of Coastal Conservation 17(1), 25–36 (2013)
19. Power, M., McCarty, L.S.: Trends in the development of ecological risk assessment and management frameworks. Human and Ecological Risk Assessment 8(1), 7–18 (2002)
20. Aps, R., Fetissov, M., Herkül, K., et al.: Bayesian inference for predicting potential oil spill related ecological risk. WIT Transactions on the Built Environment 108, 149–159 (2009)
21. Liu, X.Y.: The Petrochemical Park Fire Safety Planning Study Based on Fire Risk Analysis. Advanced Materials Research 518, 1045–1051 (2012)
22. Gaudart, C., Garrigou, A., Chassaing, K.: Analysis of organizational conditions for risk management: the case study of a petrochemical site. Work: A Journal of Prevention, Assessment and Rehabilitation 41, 2661–2667 (2012)

Searching for Spatio-temporal Similar Trajectories on Road Networks Using Network Voronoi Diagram

Wenqiang Sha[1], Yingyuan Xiao[1,*], Hongya Wang[2], Yukun Li[1], and Xiaoye Wang[1]

[1] Tianjin Key Laboratory of Intelligence Computing and Novel Software Technology, Tianjin University of Technology, 300384, Tianjin, China
[2] Donghua University, 201620, Shanghai, China
yingyuanxiao@gmail.com

Abstract. Trajectory similarity search has been an attractive and challenging topic which spawns various applications. In this work, we study the problem that finds trajectories close to some query locations with timestamps designated by a user. We firstly define the similarity between the query locations and trajectory on road networks. Then, we propose a method to retrieve similar trajectories based on Network Voronoi Diagram. By taking advantage of Network Voronoi Diagram, the proposed method can accelerate the query processing through some pre-computation. Finally, we verify the efficiency of the proposed method by extensive experiments.

Keywords: trajectory similarity search, road networks, Network Voronoi Diagram, pre-computation.

1 Introduction

Due to the development of location-acquisition devices such as GPS and some social networks that provides the service of route sharing, a large amount of data recording the positions of moving objects, called trajectory, are generated and collected. The efficient utilization of spatio-temporal trajectory data is not only a problem that we must solve in the development of spatial databases, but also has significance in the LBS (Location-based Services) application. This enables many researchers to apply themselves to analyzing mass trajectory data. Trajectory similarity search (i.e., identifying similar trajectories with respect to a given query trajectory from the trajectory database) has long been an attractive and challenging topic due to its wide range of applications, such as trip planning, traffic analysis and carpooling. For example, travelers can improve their own travel experience by referring to the historical motion of others; the police may find the potential wrecker and witnesses by analyzing the historical spatio-temporal trajectories around the scene of the accident.

The majority of existing methods for trajectory similarity assume that objects are allowed to move freely in 2-D or 3-D space, without any motion restrictions. However, in a large number of applications, objects moving in a spatial network follow specific paths determined by the graph topology, and therefore arbitrary motion is prohibited. For example, vehicles in a city can only move on road segments. In such a

F. Bian and Y. Xie (Eds.): GRMSE 2014, CCIS 482, pp. 361–371, 2015.

case, the Euclidean distance between two trajectories does not reflect their real distance, that is, two trajectories which are similar regarding the Euclidean distance may be dissimilar when the network distance is considered. In the existing works, many similarity measures are based on Euclidean distance, like DTW [1], LCSS [2], ERP [3], EDR [4], which is not always meaningful in the case of spatial networks. To attack this problem, the spatial network is modeled as a graph, and the actual distance (i.e., network distance) between two objects should be defined by their shortest path distance on the graph, rather than their Euclidean distance.

Our research is motivated by two requirements. First, our method should be based on the characteristics of moving objects on road network space. Second, we should simultaneously consider a spatiotemporal similarity as well as spatial similarity. Based on these ideas, we propose a search method for finding trajectories on road networks from the trajectory database close to a set of query locations with timestamps.

In this work, our main contributions can be summarized as follows: (1) we propose a new storage scheme to speed up searching for similar trajectories; (2) we first use the Network Voronoi Diagram [5] to handle with the problem of trajectory similarity search. To the best of our knowledge, it is the first time to introduce the Network Voronoi Diagram to process the trajectory similarity search.

The rest of this paper is organized as follows. Section 2 reviews related work. Section 3 defines formally the problem studied in this work and related concepts used in this paper. Section 4 presents the processing technique for trajectory similarity search. Section 5 evaluates our method through extensive experiments, and Section 6 concludes this paper.

2 Related Work

In the last two decades, the problem of trajectory similarity search has been extensively studied. Some similarity measurements based on Euclidean distance were put forward, such as DTW (Dynamic Time Warping Distance) [1], LCSS (Longest Common Subsequences) [2], ERP (Edit Distance with Real Penalty) [3] and EDR (Edit Distance in Real Sequence) [4]. However, these Euclidean-based distance functions are not directly applied to road networks. On the other hand, these measures cannot fully explore the spatio-temporal information from the trajectory database. At the same time, some researchers focus on shape-based similarity search [6, 7], in which every trajectory contains a series of equally important sample points, while this is also not our research emphasis.

Hwang et al. [8] propose a simple similarity measure based on POI (Points of Interest) and TOI (Time of Interest) in road network. They search for similar trajectories with two steps: the filtering step based on spatial similarity and the refinement step based on temporal distance. Tiakas et al. [9] improve Hwang's method. They define new similarity measures based on spatial and temporal features of trajectories. They also introduce index structures to accelerate the query processing. Chen et al. [10] use a set of locations as the query source to search the Best-Connected Trajectories from a database, as is similar to our work. The main difference between [10] and our work is that authors only focus on spatial domain in [10], while this paper simultaneously considers spatial similarity and temporal similarity.

Kolahdouzan et al. [11] first use Network Voronoi Diagram to process NN (Nearest Neighbor) query in network metric space. They propose Voronoi-based Network Nearest Neighbor (VN^3) method. VN^3 method partitions a large road network space to many small Voronoi regions and then pre-computes distances both within and across the regions. By performing the pre-computations, VN^3 method avoids global computation for every node-pair, so it reduces the storage and enhances the query response time. Motivated by the properties of NVD, we introduce the NVD to the problem of trajectory similarity search in this paper. Through the Network Voronoi expansion, we can compute the distance between query point and nearest neighbor quickly in a relatively small region and thus avoid unnecessary searching area.

3 Preliminary

In this section, we first define the related concepts used in this paper, and then propose a similarity measure based on spatial and temporal features of trajectories. Finally, we formally define the query problem studied in this work.

A road network can be modeled as a graph $G = (V, E)$, where V is a set of vertexes corresponding to road junctions and E is a set of edges between two vertexes in V corresponding to road segments. If there is an edge in E connecting two vertexes in V, these two vertexes are adjacent to each other. Let $d_N(v_i, v_j)$ denote the network distance between the two vertexes v_i and v_j, and $d_E(v_i, v_j)$ denote their Euclidean distance. $d_N(v_i, v_j)$ is evaluated by using the existing algorithms for shortest paths between the two vertexes of the graph. Obviously, $d_E(v_i, v_j) \leq d_N(v_i, v_j)$, i.e., the corresponding Euclidean distance $d_E(v_i, v_j)$ lower bounds $d_N(v_i, v_j)$.

For simplicity, we assumed that all sample points in a trajectory have already been aligned to the vertexes in the corresponding road network.

Definition 1 (Trajectory): A trajectory in a road network is defined as a point series $< p_1, p_2, \ldots, p_n >$, where for each $1 \leq i \leq n$, $p_i = (v_i, t_i)$ with v_i being a vertex in G and t_i being its timestamp.

For convenience, we sometimes use $p_i(v)$ to denote v_i and $p_i(t)$ to denote t_i.

Let $q_i = (q_i(v), q_i(t))$ be a query point where $q_i(v)$ denotes the corresponding vertex of q_i in the road network and $q_i(t)$ is the timestamp designated by a user.

Definition 2 (The shortest matching point): Given a trajectory $T = < (v_1, t_1), (v_2, t_2), \ldots, (v_n, t_n) >$ and a query point $q_i = (q_i(v), q_i(t))$, we call $p_k = (v_k, t_k)$ *the shortest matching point* of q_i with respect to T, if for any $(v_j, t_j) \neq (v_k, t_k)$ $(1 \leq j, k \leq n)$, $d_N(v_j, q_i(v)) \geq d_N(v_k, q_k(v))$.

Definition 3 (Matching pair): Given a trajectory $T = < (v_1, t_1), (v_2, t_2), \ldots, (v_n, t_n) >$ and a query point $q_i = (q_i(v), q_i(t))$, if $p_k = (v_k, t_k)$ is the shortest matching point of q_i with respect to T, we define $< p_k, q_i >$ as the matching pair of q_i with respect to T.

Definition 4 (Spatial distance between a trajectory and a query point): Given a trajectory $T = < (v_1, t_1), (v_2, t_2), \ldots, (v_n, t_n) >$ and a query point $q_i = (q_i(v), q_i(t))$, the *spatial distance between T and* q_i is defined as follows:

$$\boldsymbol{dist_s}(T, q_i) = d_N(v_i, q_i(v))$$

where (v_i, t_i) is the shortest matching point of q_i with respect to T.

Definition 5 (Spatial distance between a trajectory and a set of query points): Given a trajectory $T = <(v_1, t_1), (v_2, t_2), \ldots, (v_n, t_n)>$ and a query point set $Q = \{q_1, q_2, \ldots, q_m\}$, the spatial *distance between T and Q* is defined as follows:

$$dist_s(T, Q) = \sum_{i=1}^{m} dist_s(T, q_i)$$

Further, we define temporal distance to measure the temporal similarity. If the difference of two points' timestamps is within a small range, we call the two points are similar in time.

Definition 6 (Temporal distance between a trajectory and a query point): Given a trajectory $T = <p_1, p_2, \ldots, p_n>$ and a query point $q_i = (q_i(v), q_i(t))$, the *temporal distance between T and q_i* is defined as follows:

$$dist_t(T, q_i) = | p_k(t) - q_i(t) |$$

where $<p_k, q_i>$ is the matching pair of q_i with respect to T.

Definition 7 (Temporal distance between a trajectory and a set of query points): Given a trajectory $T = <p_1, p_2, \ldots, p_n>$ and a query point set $Q = \{q_1, q_2, \ldots, q_m\}$, the *temporal distance between T and Q* is defined as follows:

$$dist_t(T, Q) = \sum_{i=1}^{m} dist_t(T, q_i)$$

Based on the abovementioned spatial distance and temporal distance, we define two similarity functions S_s and S_t using the following two equations:

$$S_s(T, Q) = \frac{1}{1+e^{-dist_s(T,Q)}} = \frac{1}{1+e^{-\sum_{i=1}^{m} dist_s(T,q_i)}}$$

$$S_t(T, Q) = \frac{1}{m \times Max_{i=1}^{m} | p_k(t) - q_i(t) |} \times dist_t(T, Q)$$

From the above equations, we know that both similarity functions take the range from 0 to 1 and for each similarity function the smaller function value means the more similar between T and Q. In this paper, spatial and temporal characteristics are equally important in the measure of spatio-temporal similarity. So we use the product of S_s (***T***, ***Q***) and S_t (***T***, ***Q***) to evaluate the spatio-temporal similarity between T and Q.

Definition 8 (Spatio-temporal Similarity between a trajectory and a set of query points): The spatio-temporal similarity function S_{st} between T and Q is defined as follows:

$$S_{st}(T, Q) = S_s(T, Q) \times S_t(T, Q)$$

From the above definition, we know S_{st} takes the range from 0 to 1 and a smaller value of S_{st} means a higher spatio-temporal similarity between T and Q.

Now we formally define the problem studied in this work.

Definition 9 (Spatio-temporal similarity trajectory query): Given a trajectory dataset D and a query point set Q, the *spatio-temporal similarity trajectory query* finds the trajectory $T \in D$, such that $\forall T^1 \in D - \{T\}$, $S_{st}(T, Q) \leq S_{st}(T^1, Q)$.

4 Searching for Spatio-temporal Similar Trajectories

In this section, we propose an efficient spatio-temporal similarity trajectory query algorithm based on Network Voronoi Diagram. Motivated by the observation that the trajectory that contains the *K- Nearest Neighbor* (*KNN*) of query points is usually spatially close to the query point set, we can take advantage of K nearest neighbor based on Network Voronoi Diagram to improve the performance of processing spatio-temporal similarity trajectory query.

4.1 Basic Idea

Due to the peculiarity of road network, the traditional index structure based on Euclidean space, like kd-tree, R-tree, quad-tree cannot be applied to road network directly. In this paper, we choose the Network Voronoi Diagram as our data structure.

A Voronoi Diagram divides a space into disjoint polygons where the nearest neighbor of any point inside a polygon is the generator of the polygon. Consider a set of limited number of points, called generator points, in the Euclidean plane. We associate all locations in the plane to their closest generator(s). The set of locations assigned to each generator forms a region called Voronoi Polygon or Voronoi cell, of that generator. The set of Voronoi Polygons associated with all the generators is called the Voronoi Diagram with respect to the generators set. The Voronoi Polygon and Voronoi Diagram can be formally defined as: Assume a set of generators $P = \{p_1, p_2, \ldots, p_n\}$, the region given by $VP(p_i) = \{p \mid d_E(p, p_i) \leq d_E(p, p_j)$ for $i \neq j$, $1 \leq i, j \leq n\}$ is called the Voronoi Polygon associated with p_i, and the set given by $VD(P) = \{VP(p_1), VP(p_2), \ldots, VP(p_n)\}$ is called the Voronoi Diagram generated by P. In this paper, we employ the Network Voronoi Diagram where the distance between two objects in space is their shortest path in the network rather than their Euclidean distance and hence can be used on road networks. The Network Voronoi Polygon and Network Voronoi Diagram can be formally defined as $NVP(p_i) = \{p \mid d_N(p, p_i) \leq d_N(p, p_j)$ for $i \neq j$, $1 \leq i, j \leq n\}$ and $NVD(P) = \{NVP(p_1), NVP(p_2), \ldots, NVP(p_n)\}$, respectively. A thorough discussion on regular and network Voronoi Diagrams is presented in [5].

The basic idea is that first we use Network Voronoi Diagram to partition a large road network to small Voronoi regions, as shown in Fig. 1. Note that only the vertex and edges inside $NVP(P_1)$ are drawn in Fig. 1. In this way, we can establish a one-to-one mapping relationship between all the vertexes and the *NVP*s. For a single *NVP*, the network distance between the border points inside the *NVP* must be pre-computed and be stored in a pre-computed table. Thus the query process can be accelerated.

Due to the large scale of trajectory database, it is prohibitively expensive to compute the S_{st} value with every trajectory in the database and choose the smallest one. Because it has to load every data trajectory from disk into memory in order to compute the exact S_{st} value, which would introduce too much I/O and thus bring large amount of computation. Now we note an observation: The trajectory which is spatially close has a higher probability to be similar to the query location. So secondly, for each query point q, we locate the *NVP* that contains q and then expand from q in the user-specific area and find all the adjacent vertexes of q within the given range. The trajectories that pass at least one of the vertexes within the given area are the ones we process. Thus we can reduce the number of processed trajectories and reduce the computation.

Finally, we find the trajectory that has the shortest matching point with regard to every query point, respectively and its corresponding distance are determined during the expansion. So we compute the exact S_s and S_t for these trajectories and return the trajectory with the smallest S_{st}.

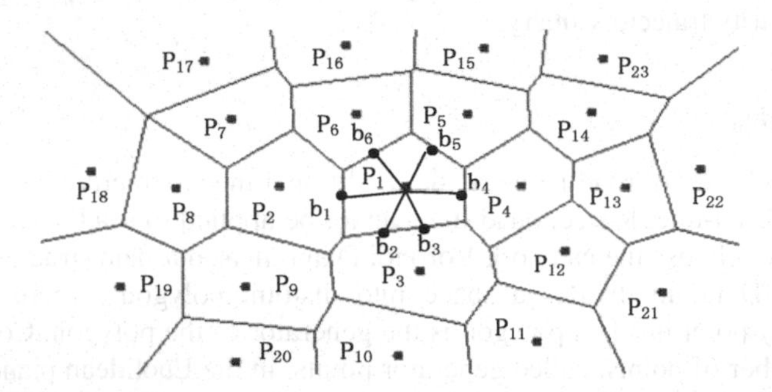

Fig. 1. Road network partitioning by using Network Voronoi Diagram

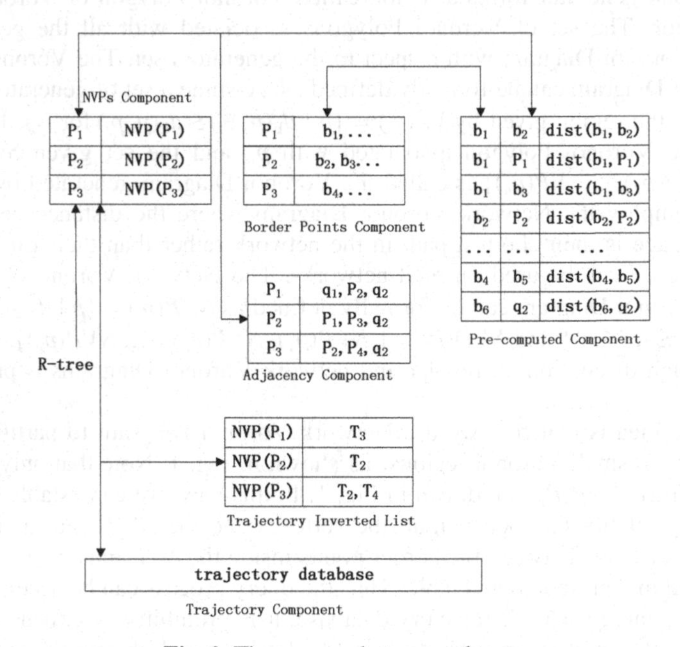

Fig. 2. The proposed storage scheme

To further speed up the query processing, we introduce the storage scheme similar to the one used in [11]. Fig. 2 shows the scheme intuitively. It mainly contains 6 parts: *NVP*s Component, Adjacency Component, Pre-computed Component, Border Points Component, Trajectory Inverted List and Trajectory Component, where the first four components are the same as the storage scheme mentioned in [11]. To locate its *NVP* for the query point quickly, we need use R-tree to pre-process *NVP*. As to Trajectory Inverted List, we build a trajectory inverted list for each *NVP*, which is a list of trajectory IDs that pass on the point of the generator of this *NVP*. This structure can be used in the inspection of trajectories for the candidate set *C*. Trajectory Component describes the spatio-temporal information of trajectories and is kept in disk.

4.2 The Proposed Search Algorithm

Based on the abovementioned idea, we formalize our search method in the following Algorithm 1.

Algorithm 1: *similarity query* (D, Q, Φ)

Input: trajectory database D, query point set Q and the distance threshold Φ
Output: spatio-temporal similar trajectory
1: construct a queue Qe_i for every query point q_i
2: **for** each q_i in the query point set Q **do**
3: form the voronoi polygon $NVP(q_i)$ according to q_i
4: **while** each neighbor P_j of q_i
5: **if** (P_j existed in Qe_i) then
6: neglect it
7: **else**
8: add P_j and $d_N(q_i,P_j)$ to Qe_i in the descending order of $d_N(q_i,P_j)$
9: **until** $d_N(q_i,P_j) > \Phi$
10: add those trajectories crossing through $NVP(q_i)$ and all $NVP(P_j)$ into candidate set C
11: **end for**
12: **for** each trajectory T in C
13: compute $S_{st}(T, Q)$
14: **end for**
15: **return** the trajectory with smallest S_{st} value

Further, we illustrate our algorithm by means of a concrete example. As shown in Fig. 3, T_1, T_2, T_3, T_4, T_5 and T_6 denote trajectories and $Q = \{q_1, q_2, q_3, q_4, q_5\}$ is a query point set designated by a user (the temporal attributes of Q are not shown in the Fig. 3). Taking q_2 for example, according to the Algorithm 1 we first locate the $NVP(q_2)$, and then expand from $NVP(q_2)$, that is to say, we add the neighboring vertexes P_1, P_2, P_3, P_4, P_5, P_6 into the expanding range and compute the distance between q_2 and the new inserted vertexes. During this expansion, we can use the Border Points Component and Pre-computed Component to accelerate the computation. By comparison, the current $d_N < \Phi$, then we continue adding the adjacent vertexes q_1, q_3, P_7, P_{11}, P_{12}, P_{17}, P_{18}, P_{19}, P_{20}, P_{21}, P_{22} into the Qe_i. The final searching area is shown within the area of red dashed line. It is also the same process for the other query points. Then T_1, T_2, T_3 and T_4 are added into the candidate set C. Finally, we compute S_{st} for T_1, T_2, T_3 and T_4, and T_1 with the smallest S_{st} is returned as result (line 12-15).

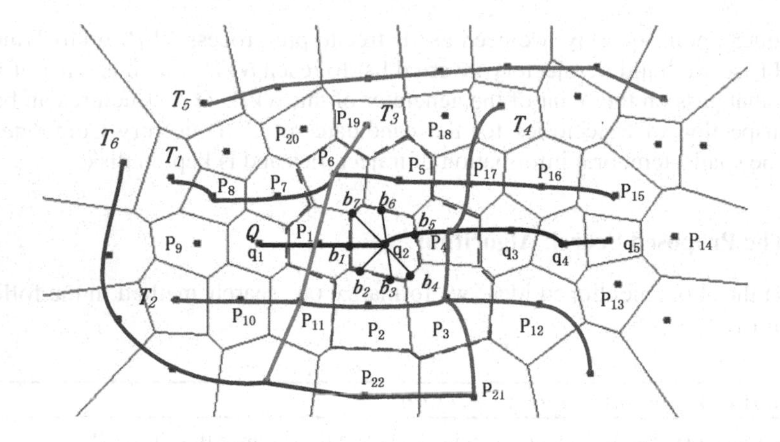

Fig. 3. An example of trajectory search

5 Performance Evaluation

In this section, we perform extensive experiments to demonstrate the performance of our proposed method. The road network datasets used in our experiments are Oldenburg Road Network and California Road Network [12], which contains 6,104 vertexes and 21,047 vertexes respectively. We implement our algorithm and simulation with Java 1.6 and Matlab 7.0 and test on a Windows platform with Intel(R) Core(TM) Duo CPU E7400 Processor (2.8GHz) and 2GB memory. The trajectories we used are generated on real road network by the generator developed by Brinkhoff [13]. We choose Hwang's method [8] as a baseline algorithm to evaluate the proposed method. The reason is that conceptually, Hwang's method and our method share some similarity (e.g., both of them simultaneously consider spatial similarity and temporal similarity). The main performance parameter is CPU time and the number of processed trajectories. We choose the number of processed trajectories as a metric because (i) it can describe the size of candidate set; (ii) it can reflect I/O to a certain degree. The main parameter setting is listed in table 1.

Table 1. Main parameters

	Oldenburg	California
Trajectory number	1000-10000(default 6000)	10000-20000(default 16000)
Query location number	2-12(default 8)	2-12(default 8)

Fig. 4(a) and 4(b) demonstrates the performance of the two methods in terms of CPU time, as a function of trajectory number with regard to two datasets. As shown in Fig. 4(a) and 4(b), CPU time grows as the trajectory number for each method. This is because the increment of trajectory number results in the increasing number of trajectories being analyzed in a given range. We can also see that our method gets a distinct advantage over Hwang's method in terms of CPU time.

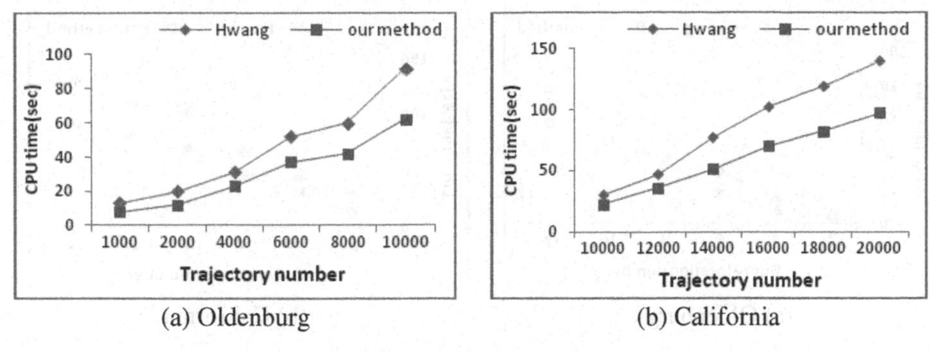

(a) Oldenburg (b) California

Fig. 4. CPU time vs. Trajectory number

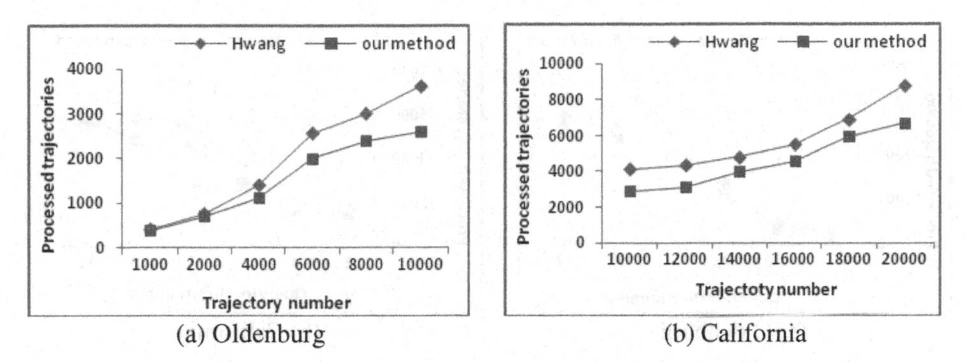

(a) Oldenburg (b) California

Fig. 5. Number of processed trajectories vs. Trajectory number

Fig. 5(a) and 5(b) demonstrates the performance of the two methods in terms of number of processed trajectories, as a function of trajectory number with regard to two datasets. As shown in Fig. 5(a) and 5(b), the number of processed trajectories grows as the trajectory number for each method. With the increasing number of trajectories, i.e. the trajectory data becomes denser, the number of the trajectory being analyzed in a given range increases. We can also see that our method is shown to be more efficient than Hwang's method in terms of the number of processed trajectories.

Fig. 6(a) and 6(b) presents the performance of two methods with the varying number of query location. The results of this experiment also show that our proposed method requires less CPU time than that of Hwang's. The CPU time increases linearly with an increase in the number of trajectory. This is because that the larger query locations cause larger query sources to expand and much more CPU time is consumed to process the query.

Fig. 7(a) and 7(b) shows the performance of two methods with the varying number of query location. The number of processed trajectories increases linearly with the increment in the number of query location. The larger query sources bring about more trajectories being added into the candidate set. The results of this experiment also show that the performance of our proposed method is better than that of Hwang.

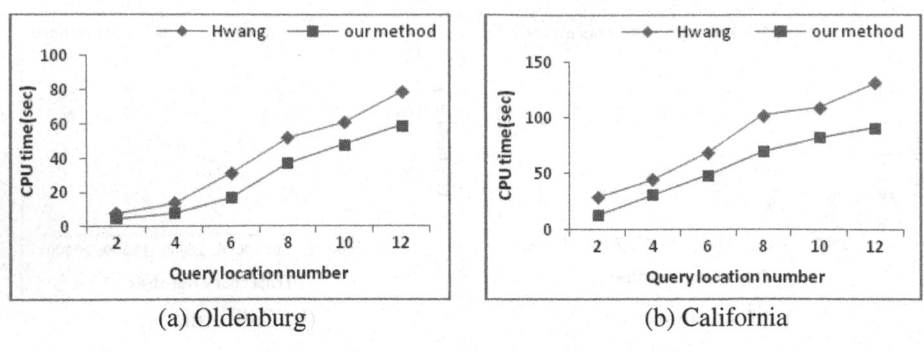

(a) Oldenburg (b) California

Fig. 6. CPU time vs. Query location number

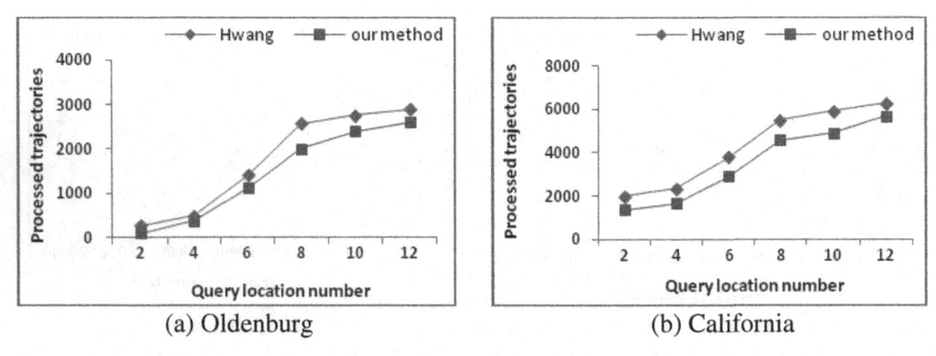

(a) Oldenburg (b) California

Fig. 7. Number of processed trajectories vs. Query location number

6 Conclusion

In this paper, we proposed a novel similar trajectory search. Different from previous work, in our study, we introduce Network Voronoi Diagram into trajectory similarity search and this is the first time, to the best of our knowledge. We believe that many real applications such as trip recommendation and traffic analysis will benefit from our work. To address the problem, a processing algorithm and the related storage scheme are proposed. We conduct extensive experiments along with comprehensive performance analysis to demonstrate the effectiveness of our method.

Acknowledgment. This work is supported by the Natural Science Foundation (NSF) of China under Grant No. 61170174, No.61370205 and No. 61170027, and the NSF of Tianjin under Grant No. 11JCYBJC26700.

References

1. Yi, B.-K., Jagadish, H., Faloutsos, C.: Efficient Retrieval of Similar Time Sequences under Time Warping. In: 14th IEEE International Conference on Data Engineering, pp. 23–27. IEEE Press, Orlando (1998)

2. Vlachos, M., Kollios, G., Gunopulos, D.: Discovering Similar Multidimensional Trajectories. In: 18th IEEE International Conference on Data Engineering, pp. 673–684. IEEE Press, San Jose (2002)
3. Chen, L., Ng, R.: On the Marriage of Lp-norms and Edit Distance. In: 30th International Conference on Very Large Data Bases, pp. 792–803. Morgan Kaufmann, Toronto (2004)
4. Chen, L., Özsu, M.T., Oria, V.: Robust and Fast Similarity Search for Moving Object Trajectories. In: ACM SIGMOD International Conference on Management of Data, pp. 491–502. ACM, Baltimore (2005)
5. Okabe, A., Boots, B., Sugihara, K., Chiu, S.N.: Spatial Tessellations, Concepts and Applications of Voronoi Diagrams. John Wiley and Sons Ltd, New York (2000)
6. Lin, B., Su, J.W.: Shapes Based Trajectory Queries for Moving Objects. In: 13th Annual ACM International Workshop on Geographic Information Systems, pp. 21–30. ACM, New York (2005)
7. Yanagisawa, Y., Akahani, J.-i., Satoh, T.: Shape-Based Similarity Query for Trajectory of Mobile Objects. In: Chen, M.-S., Chrysanthis, P.K., Sloman, M., Zaslavsky, A. (eds.) MDM 2003. LNCS, vol. 2574, pp. 63–77. Springer, Heidelberg (2003)
8. Hwang, J.-R., Kang, H.-Y., Li, K.-J.: Searching for Similar Trajectories on Road Networks Using Spatio-temporal Similarity. In: Manolopoulos, Y., Pokorný, J., Sellis, T.K. (eds.) ADBIS 2006. LNCS, vol. 4152, pp. 282–295. Springer, Heidelberg (2006)
9. Tiakas, E., Papadopoulos, A.N., Nanopoulos, A., Manolopoulos, Y., Stojanovic, D., Djordjevic-Kajan, S.: Searching for Similar Trajectories in Spatial Networks. Journal of Systems and Software 82, 772–788 (2009)
10. Chen, Z., Shen, H., Zhou, X., Zheng, Y., Xie, X.: Searching Trajectories by Locations: An Efficiency Study. In: ACM SIGMOD International Conference on Management of Data, pp. 255–266. ACM, Indianapolis (2010)
11. Kolahdouzan, M., Shahabi, C.: Voronoi-Based K Nearest Neighbor Search for Spatial Network Databases. In: 30th International Conference on Very Large Data Bases, pp. 840–851. Morgan Kaufmann, Toronto (2004)
12. Road Network Datasets, `http://www.cs.utah.edu/lifeifei/SpatialDataset.htm`
13. Brinkhoff, T.: Generating Network-Based Moving Objects. In: 12th International Conference on Scientific and Statistical Database Management, pp. 253–255. IEEE Press, Berlin (2000)

The Spatial Effect Identification of Regional Carbon Intensity and Energy Consumption Intensity of China

Liangxiong Huang[1] and Min Liu[2]

[1] Guangdong Research Institute for International Strategies, Guangdong University of Foreign Studies, Guangzhou, 510420, China
[2] Department of Ecnomics, School of Economics of Ji'nan University, Guangzhou, 510632, China
chickyliang@126.com, lium33@mail.sysu.edu.cn

Abstract. It is important and practical to search for an effective way to reduce the energy consumption and pollution emission. This paper structures a spatial econometric model of the panel for the identification of spatial effect on provincial carbon intensity and energy intensity, in turn to discuss the mutual positive influences between the areas. According to the empirical study, there exists significant positive mutual influence and convergence between provincial carbon intensity and energy intensity. The provincial level of the indicators also converges to country's average level in a picture of "boats rising with the tide". Additionally, provincial indicators only respond selectively to the domestic economic growth and sometimes even show a passive manner.

Keywords: carbon intensity, energy intensity, spatial effect, reduce the energy consumption and pollution emission, spatial econometric model of the panel.

1 Introduction

Global warming has been a global concern. To confront the environment problem, the necessary methods governments usually employ are energy saving and emission reduction. As for China, the second largest carbon dioxide and sulphur dioxide emission country, with more than 10 times higher level of emission per GDP than that of the developed countries (Li and Shen, 2008), the pressure and responsibility is heavy and urgent. China's energy efficiency is also unfavorable. The energy consumption per GDP is of triple that of US, quintuple that of Japan, and is also higher than those of Russia and India. Meanwhile, the loss caused by the extreme climate is significant in recent years.

Driven by both domestic and international factors, the Chinese government has been actively coping with the emission issue. In the eco-environment meeting held in Copenhagen, Chinese government guaranteed that by 2020 the carbon dioxide emission per GDP would be reduced by 40%~45% compared with that in 2005. Chinese government also promises to bring it as a restrictive indicator into the medium and long term prospect of domestic economic growth and society development. For example, the twelfth Five-Year-Plan passed in March, 2011 required the energy consumption per

F. Bian and Y. Xie (Eds.): GRMSE 2014, CCIS 482, pp. 372–381, 2015.

GDP to be reduced by 16% and CO2 emission per GDP to be reduced by 17%. It should be in both current and long-term consideration of China that how to achieve the goal and realize green development, build energy efficient and environment friendly society, balance the economic growth with population, resources and environment, and improve the harmonious relationship between nature and human being.

In fact, Chinese government's promise is to reduce the carbon intensity and energy consumption intensity. However, there will be a long way to go. On one hand, the primary difficulty of environment pollution treatment lies in the inherent public goods characteristic of environment and greenhouse gas emission. A lack in supervision and clear property right identification usually results in the tragedy of the common (Research team of Development Research Center of the State Council, 2009). On the other hand, the environment treatment issue is also a regional issue. Considering China's social background of "centralized politic regime and decentralized economic regime", the areas are actually proactively capable of protecting and improving the environment, meanwhile the performance evaluation carried out by central government and the competition between areas will affect the area environment policies deeply. These two reasons explain the mutual influence and spatial effect on carbon intensity and energy consumption intensity between areas (Case, et al, 1993). Therefore, to save energy and reduce emission, we should firstly catch the essential of China's current situation in terms of carbon intensity and energy consumption intensity. Based on this, we will conduct analysis on the mutual influence and make full discussion on the spatial effect.

Actually, if the correlation between the provincial carbon intensity and energy consumption intensity is positive, in which situation the drop of the two indicators in one area will be followed by neighbor area's declining indicators and vice versa. This brings convergence of the provincial indicators. If the correlation between the provincial carbon intensity and energy consumption intensity is negative, that is, the drop of one area's indicators will be accompanied by neighbor area's rising indicators. This brings divergence of the provincial performance on these two indicators. The provincial correlation on carbon intensity and energy consumption intensity should be of vital concern for it presents us a complete map of energy saving and emission reduction approaches. Empirically, we adopt spatial econometric method to identify the spatial effect of provincial carbon intensity and energy consumption intensity as well as to analyze the mutual influence between areas.

Previous researches mostly adopted spatial lag model (SLM, or spatial auto-regression model, SAR) or spatial error model (SEM), based on spatial econometric analysis. Unifying the two approaches and structure panel data MRSARAR (1, 1), the paper will go further to analyze the spatial effect on carbon intensity and energy consumption intensity in provinces and investigate which correlation exists between areas. Therefore this paper provides a more comprehensive analysis than previous ones.

2 Identification

2.1 Estimation Method

Spatial econometric models often include spatial lag model (SLM, or spatial autocorrelation model, SAR) and spatial error model (SEM). The recognition method in this paper is a combination of these two models. Considering the Kelejian's and Prucha's (1998) and Kapoor's (2007) methods, the model is as follows:

$$y_{it} = \rho_1 \sum_{j \neq i} w_{ij} y_{jt} + \theta y c_t + X_{it}{}' \beta + F_t' \delta + u_{it}$$

$$u_{it} = \rho_2 \sum_{j \neq i} w_{ij} u_{jt} + \varepsilon_{it} \qquad \varepsilon_{it} = \mu_i + \upsilon_{it}$$

$$\upsilon_{it} \sim i.i.d\left(0, \sigma_\upsilon^2\right)$$

$$\mu_i \sim i.i.d\left(0, \sigma_\mu^2\right)$$

$$E\left(\mu_i \upsilon_{it}\right) = 0$$

The examination of Kelejian and Prucha (1998) is the MRSARAR (1, 1) in the cross section data, which includes the explained variable spatial autocorrelation and common exogenous explanatory variable X_i, and its disturbance also has the spatial autocorrelation form; the examination of Kapoor (2007) is the MRSAR models with random effects in the panel data, which includes only ordinary exogenous explanatory variables without the explained variable spatial autocorrelation, and its disturbance also has the spatial autocorrelation form and individual random effect. Then, the model (1) combines the spatial lag and spatial error models organically.

In the last formula, y_{it} is the carbon dioxide emissions or energy consumption of unit GDP of province i in the No. t year, namely, carbon intensity and energy consumption intensity. y_{jt} is the carbon intensity and energy consumption intensity of other provinces j in the same period; w_{ij} is the relative importance of provinces j to province i with spatial weight matrix; $\sum_{j \neq i} w_{ij} y_{jt}$ is weighted variables province of i to provinces j; X_{it} is the representative of various factors influencing carbon intensity and energy consumption intensity; F_t is the control variables of national level that changes with time but not with individual, namely, common factor. In particular, when making decisions about carbon intensity and energy consumption intensity, provinces often need to find an evaluation benchmark to be the basis of decision-making, and the national average level is often the simplest decision-making basis of provinces. Especially in the case of cross section or having more provinces, the provinces will choose the carbon intensity and energy consumption intensity of national level to be a basis for decision-making. Further, under the background of Chinese governance model, especially with the hard constraints of the uncertain central goal of energy conservation and emission reduction, the regional competition is intense, and the safest way for provinces is to make a decision action that is higher than the national average level. In order to reflect this point, in the formula, yc_t

reflects the country's overall situation. When there are many cross sections or provinces, we can regard yc_t as one of thc decision-making basis of provinces. In other words, when the provincial government or individual enterprise in the province makes a decision, it depends on the overall situation yc_t. The overall situation characterizes with the corresponding national carbon intensity and energy consumption intensity, which changes with time but not with individual.

In the formula, the coefficient ρ_1 reflects the spatial effects between regions. If the coefficient ρ_1 is significantly positive, the provinces have positive influence with each other in carbon intensity or energy consumption intensity, which shows convergence; if the coefficient ρ_1 is significantly negative, the provinces have positive influence with each other in carbon intensity or energy consumption intensity, which shows divergence. Coefficient θ is the reflection of province to the national average level. If the coefficient is significantly positive, it shows that provinces follow or imitate the national average level; if the coefficient is significantly negative, it shows that provinces run in opposite directions with the national average level. If assuming that provinces make the national average level as a decision-making basis and θ is significantly positive, there will be a phenomenon of "a ship rising with the tide" in the regional competition.

The specific steps of estimation to (1) are as follows:

The first step: using the IV method to solve the endogenous problem which causes spatial lag and then get the consistent estimates of the parameters and the residual in the model (1). Wherein, the generalized residual, including fixed effects, proposed by Baltagi (2007) is used. Here, the selected instrumental variables set by one order and two order spatial changes in the explanatory variables are lagged.

The second step: according to the GM method proposed by Kapoor (2007), the consistent estimation of error spatial correlation coefficient ρ2 is obtained. Notes that $P=(i_T i_T'/T)\otimes I_N$, $Q=(I_T-i_T i_T'/T)\otimes I_N$, $\overline{\varepsilon}=(I_T\otimes W)\varepsilon$, $\sigma_1^2=\sigma_\upsilon^2+T\sigma_\mu^2$, the following 6 identities are taken as the overall moment conditions:

$$E\left(\varepsilon'Q\varepsilon/N(T-1)\right)=\sigma_\upsilon^2 \qquad E\left(\varepsilon'P\varepsilon/N\right)=\sigma_1^2$$

$$E\left(\overline{\varepsilon}'Q\overline{\varepsilon}/N(T-1)\right)=\sigma_\upsilon^2 tr\left(W'W\right)/N \qquad E\left(\overline{\varepsilon}'P\overline{\varepsilon}/N\right)=\sigma_1^2 tr\left(W'W\right)/N$$

$$E\left(\overline{\varepsilon}'Q\varepsilon/N(T-1)\right)=0 \qquad E\left(\overline{\varepsilon}'P\varepsilon/N\right)=0$$

Based on the generalized residual obtained in the first step, we can form the sample moments of the moment conditions and get the consistent estimates of ρ_2, σ_v^2, σ_μ^2 through NLS.

The third step: in order to solve the problems brought by non spherical perturbations, FGLS method is used to make Cochrane-Orcut-type transform for all variables (noted as z): $z^*(\hat{\rho}_2)=\left[I_T\otimes\left(I_N-\hat{\rho}_2 W\right)\right]z$. If there exist random effects, transform z^* as follows: $z_{it}^{**}(\hat{\rho}_2,\hat{\varphi})=z_{it}^*-\hat{\varphi}\overline{z}_{\cdot t}^*$, $\hat{\varphi}=1-\sqrt{\hat{\sigma}_\upsilon^2/\left(T\hat{\sigma}_\mu^2+\hat{\sigma}_\upsilon^2\right)}$. When there exist fixed effects, set $\hat{\varphi}=1$.

The fourth step: Use the IV estimation again for the transformed data to obtain the consistent estimates of ρ_l and θ. The instrumental variables set is the first-order and second-order spatial lag of transformed explanatory variables.

2.2 Spatial Weight Matrix

The spatial econometric model expresses the spatial interaction by introducing spatial weighted matrix. Spatial weighting matrix w is a symmetric matrix of $N \times N$. The diagonal element of w_{ii} is set to 0, and w_{ij} is the spatial relationship between province i and province j. Here, two kinds of weighted matrix are used: first is the more commonly used Rook adjacent spatial weight matrix. When the two areas have a common boundary, w_{ij}=1, while w_{ij}=0 when there is no common boundary. In particular, in order to avoid the "single island effect", it is set that Hainan province has common boundaries with Guangdong province and Guangxi Zhuang Autonomous Region. Second, in the robustness test, the K values of adjacent space matrix is used. Specifically, the weight of adjacent K regions selected in the given spatial unit is 1 and the other is 0. In addition, in order to reduce or eliminate the external influence between regions, the weight matrix is standardized $w_{ij}^* = w_{ij} / \sum_{j=1}^{N} w_{ij}$ to make the sum of elements is 1.

3 Data Description and Variable Declaration

The paper uses the provincial panel data from 1998 to 2008. Ignoring the Tibet and Chongqing sets which have too many missing data, we will use 29 provincial data sets in 11 years. The base year is set to be 2000. If the price inflation effect is to be eliminated, we will reduce the GDP deflation factor based on year 2000. All regression functions are given in the previous section (1).

Among the variables, y_{it} reflects the variables of provincial carbon intensity and energy consumption intensity, to be specific, carbon dioxide emission amount per GDP (*lgco2*), energy consumption amount per GDP (*lge*) and electronic power consumption (*lgel*). The measurement method for calculating carbon dioxide emission amount is referred to Du (2010). Correspondingly, yc_t represents the overall country level which is the average value of the respective variables. For example, if y_{it} represents the CO_2 emission amount per GDP of 29 provinces, yc_t represents the average CO_2 emission amount per GDP nationwide.

The factors influencing the provincial indicators are decided according to the analytical frame of Environment Kuznets Curve (EKC) (Grossman and Krueger, 1993). Besides, the method of collecting variables adopted by Peng and Bao (2006) is referred.

(1) The primary controlled variables include those reflecting economy size, industrial structure and technology advancement of the area. This paper employs the logarithm of real GDP per capita and its square (*lrgdp & lrgdp2*), the weight of

industrial production occupied in GDP (*wg2*), and the logarithm of amount of employee in R&D department (lrd2) to represent those variables respectively.

(2) Other provincial controlled variables include external dependence, i.e. the share of gross import and export amount in GDP (*ti*); the urbanization level, i.e. the proportion of urban population in the gross population, reflecting population distribution among urban and rural areas in the provinces; the logarithm of population intensity, reflecting the population loading strength of the environment; and last but not least, the quotient of gross emission expanse over unqualified industrial wastewater emission amount, similar to Effective Levy Rate (Zeng, 2008), reflecting the supervision effort of the local governments to protect the environment.

The common factors of national level (*Ft*) mainly include three observable kinds changing with time but not with individuals. The first reflects the country's economic situation with the national consumer price index (*cpi*) and the actual growth rate of per capita GDP (*gro*). The second kind of factor reflects the state of governance structure with the fiscal decentralization (*dec*) and political centralization (*cen*). The fiscal decentralization is the proportion of local government revenue and expenditure accounts in the national budget, reflecting the influence on the local governments' decision-making by the intergovernmental fiscal responsibility arrangement. And for the construction of political centralization variable, refer to the Edmark's and Agren's (2008) approach, charactering it with the dummy variable approach. The selection of Chinese officials adopts appointment system. The Chinese National Congress of the Communist Party held every five years and the 1st Session of National People's Congress held each year are important moments for personnel adjustment. We believe that in the year before these big meetings, provincial officials may change the pollution and energy consumption behavior, in order to pursue the promotion. So the value of 2001, 2002, and 2006 is 1, and the value of other years is 0. The third kind reflects the overall attitude of the nation to environmental governance represented by the ratio of the number of NPC bills delivered to the Environment and Resources Protection Committee to the total number of National People's Congress bills (*pcr*).

4 Empirical Analysis

This part identifies the provincial spatial effect of carbon intensity and energy consumption intensity and discusses the interaction of the two among provinces. The empirical results of the following part are based on the model specification (1), using the estimation method of the second part. Table 1 shows the basic results. The explained variables include unit GDP carbon dioxide emission and unit GDP power and energy consumption. Table 2 shows the robustness suggestion, replacing the spatial weight matrix with the spatial matrix most adjacent to *K* value .

Analyzing spatial effect coefficient is observing ρ_1. The results show that ρ_1 is significantly positive both in the carbon intensity and in the energy consumption (energy consumption intensity). This indicates that the regional carbon intensity and energy consumption intensity have positive effects on each other, which means that the improvement of the strength of a certain area results in the strength increase of the other areas. The positive mutual influence is shown in space as that the provinces with similar carbon intensity and energy consumption intensity are adjacent to each other and finally gets the convergence. The convergence can be explained by two reasons. (1) As the environment is public goods, when provinces consistently improve the carbon intensity or energy intensity under the decentralization system, "the tragedy of commons" occurs. (2) The local government has a certain decision-making power on the environment, which can affect the carbon intensity and energy consumption intensity, as Li and Shen's (2008) opinion that China's regional pollution control decision has obvious strategic characteristics. When a certain area gains competitive advantage by taking a strategic behavior, other regions will learn the experience and imitate it, resulting in a positive effect among regions and showing the convergence. It is the same as the example effect of Huang et al. (2011).

Table 1. The spatial effect of carbon intensity and energy consumption intensity

	lgco2		*lge*		*lgel*	
	(1)	(2)	(3)	(4)	(5)	(6)
ρ_1	0.539***	0.421**	0.536***	0.663***	0.532**	0.488**
	(0.187)	(0.207)	(0.116)	(0.127)	(0.208)	(0.226)
θ	0.167***	0.258***	0.259***	0.299***	0.454***	0.400***
	(0.063)	(0.069)	(0.038)	(0.041)	(0.058)	(0.058)
cpi	0.002	0.002	0.000	0.001	0.001	0.002
	(0.001)	(0.001)*	(0.001)	(0.001)	(0.001)	(0.001)
gro	0.029***	0.015***	0.022***	0.013*	0.001**	0.001**
	(0.010)	(0.002)	(0.007)	(0.008)	(0.000)	(0.000)
dec	0.013	0.004	0.000	-0.010	0.005	-0.002
	(0.011)	(0.014)	(0.006)	(0.009)	(0.010)	(0.011)
cen	-0.017	-0.030	0.008	0.015	0.030	0.000
	(0.027)	(0.031)	(0.017)	(0.018)	(0.021)	(0.022)
pcr	0.001	0.005	-0.012	-0.010	-0.017	-0.011
	(0.013)	(0.016)	(0.008)	(0.010)	(0.012)	(0.012)
cons	5.822***	1.956	8.657***	5.572***	-0.164	0.797
	(1.820)	(1.284)	(1.478)	(1.572)	(1.868)	(1.246)
ρ_2	0.760	0.746	0.755	0.729	0.856	0.852
N	319	290	319	290	319	290
R2	0.474	0.458	0.635	0.629	0.393	0.411
Moran I	0.000***	0.000***	0.000***	0.000***	0.000***	0.000***
Hausman	0.821	1.000	0.965	0.554	0.591	1.000
Anderson	0.000***	0.000***	0.000***	0.000***	0.000***	0.000***
Sargan	0.393	0.338	0.687	0.642	0.569	0.563

Notes: ①***, **, * respectively indicate the significance levels of 1%, 5% and 10%; ②it is standard error in the bracket; ③Hausman test is used to determine whether it is fixed effect or random effect. If it significant, it is fixed effect. ④Moran I tests the significance of ρ_2; ⑤Anderson and Sargan are the statistic magnitudes of Anderson LM and Sargan respectively. The above statistics all report the P value; ⑥Only concerned variables are listed.

Coefficient θ represents the reflection of provinces to the national average level. The results shows that θ is significantly positive. When the national average level is reduced, the provinces' also decrease, and vice versa, namely a rising tide raises all boats. That coefficient θ is significantly positive can be interpreted from two aspects: (1) the carbon intensity and energy consumption intensity of different provinces have positive effects on each other; the national average level changes with the provincial situation; (2) when provinces are in the change or make decision on the carbon intensity and energy consumption intensity, they often need to find an evaluation benchmark to be the basis of decision-making, particularly when the central government's objectives are not clear or hard enough. Since the national average level is the most simple decision basis for the provinces, the local governments (officials) have made an effort to set a higher level than the national average, leading to a phenomenon of rising tide raising all boats. In this paper, the national average level of carbon intensity and energy consumption intensity reflects the environmental status of the national macro level. The interpretation of the above two aspects can show that the provincial carbon intensity and energy consumption intensity may be endogenous. Is the provinces' decision-making based on national level, or is the national level the comprehensive result of provinces' decisions? In order to solve this problem, we make the y_{ct} lagged one period and put it into the original equation regression, as in equation (2), (4) and (6). Similarly, θ is significantly positive and coefficientρ_1 is still significant. Again, it verifies that the carbon intensity and energy consumption intensity of different provinces have positive effects on each other, showing the convergence; provinces also follow and imitate the national average level.

Among the common factors of national level, only the national economic growth (*gro*) is positive. The faster the national GDP growth, the greater the carbon intensity and energy consumption intensity are. This proves China's long-term development model of "high input, high output, high growth, and high pollution". The reflection of the national governance structure, namely, the fiscal decentralization (*dec*) and political centralization (*cen*), and the reflection of overall attitude of the state to environment governance, namely, the proportion of environmental protection bills of National People's Congress (*pcr*) were both not significant. It illustrates that Chinese provinces' competition is more reflected in the GDP, and provinces pay more attention to GDP growth environment and have less concern on other factors. From this perspective, provinces selectively react to the national level factors. They have a strong response to the growth of GDP, especially the factors on short-term growth. But provinces have less reaction to those factors not closely linked with GDP growth or even causing loss of GDP growth.

The conclusion of Table 1 is based on spatial weight matrix adjacent method. In order to show the robustness of the results, the K value of the adjacent space matrix is used instead of spatial weight matrix. The weight of the adjacent K regions chosen in the surrounding of given space is 1, and the weight of the remaining is 0. In general, K=4. The regression result is shown in table 2.

Table 2. Robustness test: K value the most adjacent spatial matrix

		lgco2	*lge*	*lgel*
The current period	ρ_l	0.416***	0.207**	0.395***
		(0.128)	(0.086)	(0.106)
	θ	0.149**	0.239***	0.365***
		(0.062)	(0.039)	(0.050)
A phase lag	ρ_l	0.559***	0.557***	0.405***
		(0.143)	(0.094)	(0.109)
	θ	0.284***	0.324***	0.300***
		(0.072)	(0.044)	(0.055)

Notes:①***,**,* respectively indicate the significance levels of 1%, 5% and 10%; ②it is standard error in the bracket; ③the above regression uses random effects.

The results of Table 2 verify the results of Table 1. Whether carbon intensity, or energy consumption intensity, the ρ_l which represents the provincial mutual influence is significantly positive. Moreover, the θ which represents the reaction of the province to the national average level is significantly positive. To avoid endogenous, a phase lag of y_{ct} is used, which shows that the provincial carbon dioxide intensity and energy consumption intensity have positive effect. Similar levels of provinces and autonomous regions are adjacent to each other, and eventually the provinces get convergence. Meanwhile, the provinces and autonomous regions in these two aspects will follow or imitate the effect of the national average.

5 Conclusion

This paper builds the MRSARAR model spatial panel (1, 1) to identify the spatial effect of carbon intensity and energy consumption intensity in Chinese provinces, in order to verify provinces' mutual influence on each other in the behavior of these two aspects. In particular, we also examined the reflection of the provinces to the national average level. The empirical results show that whether in the carbon intensity or energy intensity, Chinese provinces have positive influence on each other, making provinces of a similar level cluster together and showing convergence. The provinces will also present the effect of following and imitating the national average level, which results in a phenomenon of "boats rising with the tide". In addition, provinces only selectively react to the economic growth of the national level, without considering other factors of national level, even without enthusiasm. Chinese energy-saving and emission reduction will be in a dilemma without making energy-saving and emission reduction targets into "hard constraints".

The positive effect on carbon intensity and energy consumption intensity among Chinese provinces appears convergence, but provinces only react to the economic growth. These characteristics give a revelation that if we want to achieve energy-saving and emission reduction and green development, we must solve the problem of disordered competition of provinces fundamentally. And if we want to avoid the race to the bottom, we need to improve the evaluation mechanism and redress the competition between local governments.

The central government's evaluation system to local government officials guides their behaviors. The hard constraints of energy saving and emission reduction goals and the one-vote veto mechanism raise the attention of local governments. The lack of motivation for regional cooperation is a big obstacle for energy saving and emission reduction and green development. It is necessary to have local governments move from competition to competition and cooperation through institutional innovation.

Acknowledgement. This paper is sponsored by Youth project of philosophy and social sciences planning of Guangdong province (GD13YYJ04) and Guangdong University of Foreign Studies Youth Project (12Q01).

References

1. Du, L.: Influence Factors of China's Carbon Dioxide Emissions: a Study Based on Provincial Panel. The Economy of the South (11), 20–33 (2010)
2. Research group of the State Council Development Research Center: Global Greenhouse Gas Emission Reduction: Theoretical Framework and Solution. Economic Research (3), 4–13(2009)
3. Li, Y., Shen, K.: The Reduction Effect of China's Pollution Control Policy: An Empirical Analysis Based on Provincial Industrial Pollution Data. Management World (7), 7–17 (2008)
4. Peng, S., Bao, Q.: Economic growth and Environmental Pollution: the Environmental Kuznets Curve Hypothesis China Inspection. Research on Financial and Economic Issues (8), 3–17 (2006)
5. Zeng, W.: Transboundary Pollution Regulation: An Empirical Study on Inter Provincial Water Pollution Chinese. Economics 7(2), 447–464 (2008)
6. Baltagi, B.H., Song, S.H., Jung, B.C., Koh, W.: Testing for Serial Correlation. Spatial Autocorrelation and Random Effects Using Panel Data. Journal of Econometrics (140), 5–51 (2007)
7. Case, A.C., Rosen, H.S., Hines, J.R.: Budget Spillovers and Fiscal Policy Interdependence: Evidence from the States. Journal of Public Economics (52), 285–307 (1993)
8. Edmark, K., Agren, H.: Identifying Strategic Interactions in Swedish Local Income Tax Policies. Journal of Urban Economics (63), 849–857 (2008)
9. Grossman, G.M., Krueger, A.B.: Environmental Impacts of a North American Free Trade Agreement. In: Garber, P.M. (ed.) The Mexico-U.S. Free Trade Agreement, pp. 132–156. The MIT Press, Cambridge (1993)
10. Huang, L.X., Zhang, L., Shu, Y.: Pollution Spillover in Developed Regions in China-Based on the Analysis of the Industrial SO2 Emission. Energy Procedia (5), 1008–1013 (2011)
11. Kapoor, H., Kelejian, H., Prucha, I.R.: Panel Data Models with Spatially Correlated Error Components. Journal of Econometrics (140), 97–130 (2007)
12. Kelejian, H.H., Prucha, I.R.: A Generalized Spatial Two Stage Least Squares Procedures for Estimating a Spatial Autoregressive Model with Autoregressive Disturbances. Journal of Real Estate Finance and Economics (17), 99–121 (1998)

Effect of Expansion of Urban Region on Soil Animals in Farmland Ecosystem: A Case in Shenbei New District in Shenyang City, Liaoning Province

ZhenXing Bian[1,2], Miao Yu[3], JinHong Li[1], ZhenRong Yu[2], and Meng Kang[1]

[1] College of Land and Environment, Shenyang Agricultural University, Shenyang 110866, China
[2] College of Resources and Environmental Sciences, China Agricultural University, Beijing 100193, China
[3] College of Sciences, Shenyang Agricultural University, Shenyang 110866, China
Zhx-bian@263.net, 261180596@qq.com

Abstract. For mastering the effect of urban expansion on farmland ecosystem health, we adopted information entropy model, hand-picking and stereo microscope methods. Shenbei new district was divided into urban region, urban fringe and rural region. We took samples of large and medium size soil animals and analyzed their gradient variation of individual number, biomass and diversity, richness & evenness indices. The results show: (1) 620 soil animals belong to 2 phyla, 4 insectas and 5 orders. Dominant groups are Nematode, Collembolan and Coleopteran. (2) Individual number of soil animals from urban region to rural region show U-shape trend in both farmland and habitat scale. But species number doesn't change obviously. (3) There is a relationship of second order curve between the distance from city centre and individual number of soil animals. Using biomass of soil animals is a new try to reflect the effect of urban expansion on farmland ecosystem.

Keywords: urbanization, farmland ecosystem, soil animals, gradient variation, Shenbei new district.

1 Introduction

With the acceleration of urbanization process in China, urban expansion is approaching to rural regions rapidly. Some urban landscapes evolve from peri-urban agricultural landscapes and natural & half-natural landscapes gradually. Urban fringes are the transit zones from rural to urban landscapes. They are the most active areas in urbanization land development. By the impact of urban radiation and rural agglomeration effects, the problems of farmland landscapes and ecosystem health are particularly acute.[1] In recent years, research on urban fringes concentrate mainly on determining the borders of urban fringes, as well as planning, constructing and ordering ecological landscapes. While farmland biodiversity is mainly to investigate different lands in different areas and establish the databases of biodiversity, as well as reflect the ecological recovery of different habitats by sampling surveys on farmland biology, which is applied to pest

F. Bian and Y. Xie (Eds.): GRMSE 2014, CCIS 482, pp. 382–392, 2015.

management and landscape planning.[2] However, there are few studies to analyze whether urban sprawl impacts on farmland ecosystems with a parameter of biodiversity. In this study, we collect soil animals in the study area based on dividing the city by Geographic Information System. The purpose is to analyze the influence of urbanization on individual number and species of farmland soil animals, and then figure out the health of farmland ecosystems.

Farmland biodiversity reflects the stability of urban fringes and the influence on farmland ecology environment by pros and cons of farmland landscape pattern. Single structure of farmland landscapes will reduce the biodiversity in farmland landscapes.[3] There are a variety of claims about indicator organism in farmland biodiversity. Collins noted that an invertebrate was a good indicator to measure biodiversity after his study of termite [4]. Due to the heavy work of the investigation, it was suggested to take beetles, spiders and rove beetles as indicator organisms. Zhenrong Yu noted rove beetles played an important role in the ecosystem, which was also a kind of indicator organisms adopted and applied successfully abroad. Also arthropods could be used to evaluate biodiversity and environment in agroecosystem.[5] Zengping Yang pointed out that soil nematodes were particularly sensitive to the changes in farmland ecosystem, and it could be an indicator organism in farmland ecosystem health[6]. According to the real research area, arthropods and nematodes were selected as indicator organisms to study farmland biodiversity in this paper.

Shenbei New District (41 ° 54 '~ 42 ° 11'N, 123 ° 16' ~ 123 ° 48'E) is located in the north of Shenyang. It is the golden path connecting Jilin, Heilongjiang and Inner Mongolia. It is also a hub of northeast city corridor. Its total area is 10980 kilometers and average elevation is 58 meters. It is high in the east and low in the west. Most of the area is plain topography. It has continental monsoon climate and four distinct seasons in the north temperate zone. The annual average temperature is 7.5 degree centigrade and annual precipitation is 672.9 millimeters. With the acceleration of urbanization since 2006, the urban fringe in Shenbei constantly extend outward and enhance the disturbance by cities and humans. However, there is no deep understanding about the urban fringe impacted by urbanization and there is scare study about the composition, species and quantities of farmland animals. Therefore, it is representative and indicative to carry out this study in Shenbei new district.

2 Materials and Methods

2.1 Transaction Be Selected Based on Urban Fringe

The urban fringe of Shenbei New District in Sheyang city is divided by Cheng L.S.'s method for defining urban fringe and the information entropy model. By the calculation, the area of entropy value more than 0.8 was divided into urban fringe. According to land use types, the area of entropy value less than 0.8 was divided into urban region or rural region.

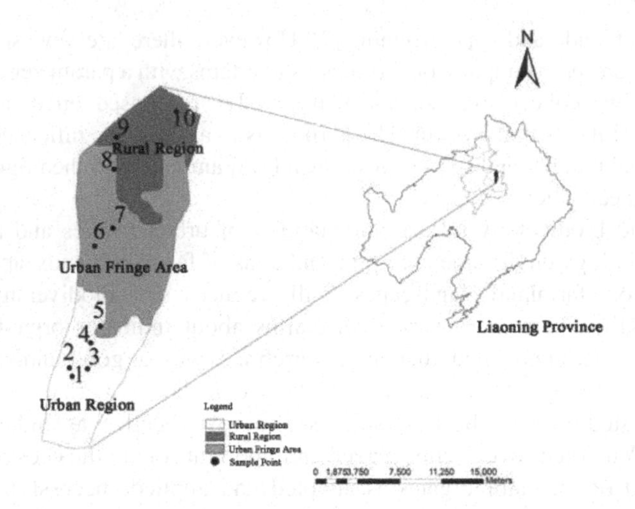

Fig. 1. The study area and location of the sample points

Based on the range of urban fringe in Shenbei New District, we selected the transaction in figure 1 which extends from the city center to the border of Shenbei New District. This transaction is located between Hada railway (from Harbin to Dalian) and G1 (from Beijing to Harbin), which has a relatively integrated farmland ecosystem. We have selected 10 sample points from urban region to rural region in accordance with the distance from the city center (Table 1).Among 10 sample points, samples 1 to 4 are located in urban regions 5-8 are located in urban fringe, 9-10 are located in rural region. Each sampling site includes three kinds of habitat types (arable land, grassland and woodland).Soil animals were collected in the arable land, grassland and woodland.

Table 1. The number of the sampling site and its distance to urban center Units:km

	Urban Region				Urban Fringe				Rural Region	
number	1	2	3	4	5	6	7	8	9	10
distance	9.4	10	10.7	12.5	14	21	23	28	31	34

2.2 Data Sources and Materials

Data Sources: The Landsat-TM video remote sensing data of Shenbei New District in 2012(downloaded from Science Data Services Platform); 1:50000 topographic map Shenyang (purchased from Liaoning Bureau of Surveying and Mapping).

Materials: Stereo microscope, environmental questionnaire of soil animals, 75% (v/v) alcohol, kraft paper, homemade boxes, shovels, tape, tweezers, markers, handheld GPS etc.

2.3 Sampling Method for Soil Animal

We collected soil nematodes and arthropods in 10 sample points (cultivated land, grassland and woodland) by the method of hand-picking in mid-to-late September 2013. Firstly, for each sampling site we selected three independent 25cm × 25cm quadrats as repeat[7] and used GPS to position. The interval between each quadrat is more than 10 meters. Sample points should not be potholes, mounds, slopes, rocks and roots, etc.[14] A complete flat site was selected to place 25 × 25cm box. Secondly, to facilitate to soil sampling, 25 × 25 × 25cm pits were dug outside the box in three directions. Finally, the visible soil animals were seized by hand picking method and saved into 75% (v/v) alcohol bottle and brought back to the laboratory for identification.

2.4 Laboratory Analysis

270 soil animal samples were identified by the stereo microscope. (Magnification is 20.) We observed soil animal morphology, structure, composition and other characters. Soil animals were classified and authenticated by referencing Chinese soil animal retrieving map[8]. Soil animals can be identified in Orders level and Families level.

2.5 Data Analysis

The sample data is the average of 3 replicates[12]. We selected Shannon-Wiener index, Pielou index and Menhunick index as biodiversity index for soil animal.

$$\text{Shannon-Wiener index: } H=\sum P_i \ln P_i \quad (1)$$

$$\text{Pielou index: } Jws = H / \ln S \quad (2)$$

$$\text{Menhunick index: } D = \ln S / \ln N \quad (3)$$

Pi is the proportion of soil animal individuals in Class i accounting for soil fauna total number. S is the number of groups. N is the number of soil animal individuals.

In analysis of variance, we use the least significant difference method (LSD). In linear analysis, we use SPSS software.

3 Results and Discussion

3.1 Distribution Characteristics of the Individuals and Species about Soil Animals in Farmland Ecosystem

On analysis of the individual quantity and species number about soil animals in the study area, it is concluded as follows (table 2). We collected 620 soil animals in 19 groups by hand picking from 10 samples in Shenbei new distinct. They belong to two phyla respectively, which are arthropoda and nematode. Arthropod as the main body in the study of farmland soil animals has 300 individuals and is 48.37% of the total. It is divided into five orders including coleoptera, collembolan, araneae, geophilomorpha and opiliones. It is respectively 27.41%, 8.71%, 2.90%, 4.19% and 5.16% of the total soil animals. Coleopteran is further divided into 14 families including staphylinidae, curculionidae, scaphidiidae, etc. Comparing farmland soil animals integrally, we conclude as follows. Nematode is the dominant species and 51.61% of the total. Common groups are 11 classes of arthropod including staphylinidae, curculionidae, scaphidiidae, scydmaenidae, histeridae, silphidae, pselaphidae, ollembola, araneae, geophilomorpha and opiliones, etc, which account for 45.32% of the total soil animals. Scarce groups are 7 classes of arthropod including erotylidae, carabidae and chrysomelidae, etc, which account for 3.06% of the total. There are 601 individuals in one dominant species and 11 common groups accounting for 96.94% of the total individuals. They constitute the main parts of large and medium-sized soil animals and play an important role during material circulation in ecological system.

According to the scope of urban fringe, sample 1-4 are located in urban region (U). Sample 5-8 are located in urban fringe (F). Sample 9-10 are located in rural region (R). With the increasing distance to city centers, there are different soil animal numbers and spatial distributions of species (Figure 2). The quantities, including soil arthropod, nematode animals and total individuals, reach maximum in rural samples. Maximum individual numbers of soil animals reach 165 in a single sample. The mediated animals are the soil animals in urban region, which are only second to rural region. The minimum animals are the samples in urban fringe. There is a changeable tendency about soil animal numbers, which are high in cities and low in urban fringes and high in countries. On analysis of variance, there are significant differences between U and F ($p < 0.05$). There are also significant differences between U and R ($p < 0.05$). There are extremely significant differences between F and R ($p < 0.01$). But there are no significant differences among the three regions ($p > 0.05$). The results show that the increasing expansion in cities has no significant influence on species of farmland soil animals while it has significant influence on individuals. We find the following reasons. In urban fringe, strong human interference destroyed the original stability of farmland ecosystem and reduced its stability resulting in the decrease of soil animal numbers with the increasing development and expansion of industry. In urban region, farmland ecosystem is relatively stable because of the stability of industrial layout, which leads to the increase of soil animal numbers. In rural region, it is the maintenance of originally high stability in farmland ecosystem, which makes soil animal numbers in high value range. While we cannot illustrate the reasons that there is no obvious change about farmland soil animal species among U, F and R. Further researches should be done about whether unimpaired species in Shenbei is due to a short time of urban expansion.

Table 2. The individual number of soil animals and species

	Orders	Families	Numbers	Individual to total(%)	Abundance
Arthropoda	Coleoptera	Staphylinidae	58	9.35	++
		Curculionidae	33	5.32	++
		Scaphidiidae	14	2.26	++
		Scydmaenidae	17	2.74	++
		Histeridae	12	1.94	++
		Silphidae	10	1.61	++
		Pselaphidae	7	1.13	++
		Erotylidae	4	0.65	+
		Carabidae	3	0.48	+
		Chrysomelidae	6	0.97	+
		Lathridiidae	3	0.48	+
		Silvanidae	1	0.16	+
		Cucujidae	1	0.16	+
		Tctrigoidae	1	0.16	+
	ollembola		54	8.71	++
	Araneae		18	2.90	++
	Geophilomorpha		26	4.19	++
	Opiliones		32	5.16	++
Nemata			320	51.61	+++
The total number of soil animals			620	100	

Note: +++ represents the dominant species (individual number accounts for over 10% of the total yield), + + represents the common groups (individual number accounts for 1% ~ 10% of the total), + represents rare groups (individual number are less than 1%). Soil arthropod is as the main body in the study of farmland ecosystem health. Nematode animals are as the auxiliary part. So the category of nematode animals is only classified to the phylum.

3.2 The Spatial Distribution Characteristic of Soil Animals in Different Habitats of Farmland Ecosystem

Figure 2 shows that there are different distribution and obvious regional change about numbers of individual animals and species in different habitats of the farmland ecosystem such as U, F and R. According to the trend lines of the three kinds of habitats, we conclude there is a high-low-high trend with the order of city-urban fringe-country about soil animal numbers. The species quantities have changed in different habitats. On analysis of variance, we can conclude that there is consistent otherness of the soil animal numbers in the cultivated land, grassland and forest land habitats. It represents that there are no significant differences between U and F and also between U and R ($p > 0.05$). But there are significant differences between F and R ($p < 0.05$). Therefore, city expansion influence in urban fringes of farmland ecosystem is much stronger than cities' and countries' on the three habitats. Analyzing species quantity change of soil animals in three kinds of habitats, there are no significant differences between each two of U, F and R ($p > 0.05$) in grasslands and forest lands. There is no obvious change on the amount of species in three habitats with the increase of distance. While there are significant differences between U and F and also between U and R ($p < 0.05$) in cultivated species quantities. The number of soil animals in urban fringes is less than cities', but the species quantity is higher. We have four reasons. First, as the planning and construction of cities trend to stabilization, the ecosystem constructed by the small ecological units recovers gradually. Small ecological units include cultivated land in cities and the grassland & forest land surrounding cities. And then biomass increases. Second, construction land cuts off the species communication and migration between different areas. Species numbers have not been well recovered. Third, urban fringes are strongly affected by humans, which resulting in the decrease of soil biomass and even below city level. Finally, species numbers have no significant reduce because of short time expansion in Shenbei New District.

Table 3 shows forest land owns the highest biodiversity index in urban fringe. Cultivated land takes second place and grassland is the lowest. Richness and evenness index are similar to biodiversity index. On analysis of variance, we can conclude that the diversity index of cultivated land presents non-significant differences between each two in the three regions ($p > 0.05$). Richness index has significant differences between F and R ($p < 0.05$). Evenness index has significant differences between U and F ($p < 0.05$). Other indices have non-significant differences between each two in the three regions ($p > 0.05$). The three indices of grassland present non-significant differences between each two regions. Diversity index of forest land has significant differences between U and F ($p < 0.05$). Richness index has significant differences between F and R ($p < 0.05$). Others have non-significant differences ($p > 0.05$).

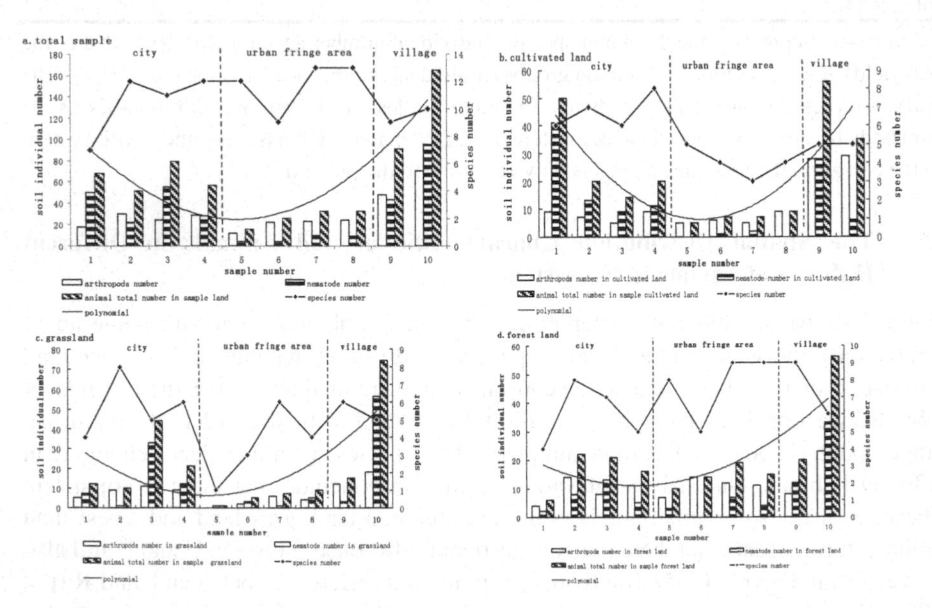

Fig. 2. Numbers of individual land species of soil animals in three habitats and sample points

3.3 The Variation analysis of Soil Animal Number and City Distance Change of Gradient

According to the scatter map of soil animal numbers from the sample, variation tendency corresponds to the polynomial and can be analyzed by dual conic model. The measure model is built by using SPSS software about the distance from city center and the amount of soil animal individuals. Regarding total individual number (N) as dependent variable to analyze the relationship between samples and the distances from city center and itself, we obtain the quadratic equation, which econometric model is N=258.786-0.026x+6.888E-7x2 .

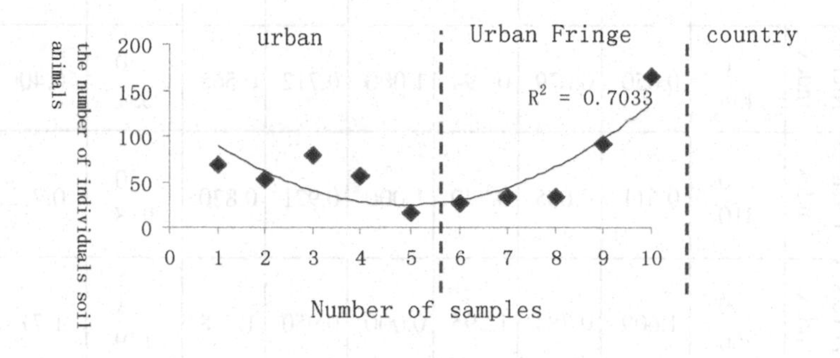

Fig. 3. Mathematical model analysis on individual numbers of soil animal in sample points

Figure 3 shows that the quadratic model derived from equations is better to the simulation of results. The fitting R^2 is 0.896 and approaches to 1. Variance is significant in the 1% confidence level. Analysis shows that individual number of soil animals reduces from a maintained high level with the increase of distance from city centers and then starts to increase rapidly in the nadir in the case of unchangeable other conditions. Figure 3 shows that the turning point is in conformity with the distance of both ends in urban fringe. It reflects the minimum individuals of soil animals are in urban fringes with the intense human disturbance. The reason is that stable ecosystem of soil animals is disrupted and destructed by high-intensity artificial exploitation and construction. It contributes number decrease caused by sensitivity of the soil instructions animals to the environmental change.

Table 3. Indices of biodiversity of soil animal in three habitats

Habitat type	Sample number	Urban Region				Urban Fringe				Rural Region	
		1	2	3	4	5	6	7	8	9	10
Arable land	Diversity index	0.735	1.190	1.227	1.538	1.609	1.277	0.956	1.273	1.162	1.283
	Richness index	0.458	0.650	0.679	0.694	1.000	0.712	0.565	0.631	0.400	0.453
	Evenness index	0.410	0.611	0.685	0.740	1.000	0.921	0.870	0.918	0.722	0.797
Grass land	Diversity index	0.960	1.609	0.784	1.295	0.000	0.950	1.748	1.149	1.714	0.763
	Richness index	0.442	0.778	0.425	0.589	0.000	0.683	0.921	0.631	0.662	0.374
	Evenness index	0.873	0.898	0.487	0.722	0.000	0.865	0.976	0.829	0.975	0.474
Forest land	Diversity index	1.099	1.783	1.234	1.424	1.973	1.438	1.908	2.061	1.435	1.196
	Richness index	0.613	0.673	0.589	0.580	0.903	0.610	0.746	0.811	0.694	0.445
	Evenness index	1.000	0.858	0.689	0.885	0.949	0.893	0.868	0.938	0.690	0.668

4 Conclusion and Discussion

According to this study, there are 620 soil animals in 10 sample points of the study area, and the dominant species are soil nematodes. Either the total individual number of soil animals or the individual number of soil animals in habitat, the minimum individual number is in urban fringe and the highest individual number is in rural region. This was consistent with Qin Zhong's research, which found that it was not conducive for soil animals to survive in rapid urbanization area[10].The reason was that some soil animals such as nematode and coleoptera, etc. could indicate the health of farmland ecology. They were more sensitive to the changes of soil and environment. Beginning in 2006, Shenbei New District has become one of the rapid economic development regions in Shenyang city. Urban fringe expands quickly with rapid urbanization. Continuous and intensive human activities have serious impact on the health and security of farmland ecosystems, which decline suitable ecological conditions for survive of soil animals and affect the survival of soil fauna. However, the amount of species of soil fauna in urban fringe is not the same as the individual number of soil animals. It does not decrease vastly, but remains at a relatively high level. Possible reason is the appearance of soil migration produced by urban construction and other niches, which promotes the exchange of species and habitats and then increases the number of soil animal species. It is short time of exploitation and construction in Shenbei New District's fringe not to result in the disappearance of all species.

The farmland in urban region has a larger individual number and analogous species of soil animals relative to urban fringe. The reasons are as follows. On one hand, urban construction tends to consummate and land use structure is relatively stable. Decreasing human interference gives more time to restore habitats. On thc other hand, city afforestation increases the number of niches, which creates relatively benign survival space for farmland soil animals. Establishing a statistical model to analyze sample points data, the result indicates that the individual number of soil animals shows a U-type trend from urban region to rural region at both farmland scale and habitats scale.

It is obvious that the results of this study have ecological significance on the interference of urbanization to urban fringes of farmland ecosystems. In the process of expanding urban fringes, ecosystem buffer zones should be built. We should fully utilize plant and animal resources to reduce the impact on farmland ecosystems and improve their health and stability in order to obtain better economic, social and ecological benefits.

This study also has shortcomings. Further research should be done on the internal relation between versatile information of soil animals and their habitats' ecological factors[13]. More samples and data should be added on this basis and the impact of soil environmental factors such as soil moisture, pH, etc[11] should be fully considered. At the same time, further research should also be done on the relationship among soil animals, human factors and environmental factors, which reveals the influence of urban fringes on farmland ecosystems.

References

1. Yu, L., Fu, Y., Yu, H.: Landscape pattern gradient dynamics and desakota features in rapid urbanization area: A case study in Panyu of Guangzhou. Chinese Journal of Applied Ecology 22(1), 171–180 (2011)

2. Wei, W., Zhang, Y., Zhao, B.: Impacts of urban expansion on spatio-temporal variation of landscape patterns during rapid urbanization: a case study of Kunshan city. Ecology and Environmental Sciences 20(1), 7–12 (2011)
3. Yang, F., Ge, F.: Effects of agricultural landscape patterns on insects. Chinese Journal of Applied Entomology 48(5), 1177–1183 (2011)
4. Qian, K., Liu, C., Li, J.: Biodiversity conservation and sustainable development of agriculture. Jiangsu Agricultural Sciences 41(12), 419–422 (2012)
5. Liu, Y., Chang, H., Yu, Z.: General Principles for Biodiversity Protection in Agro Landscape. Journal of Ecology and Rural Environment (6), 622–627 (2010)
6. Zhou, H., Chen, J., Cheng, R., et al.: Effects of ecological regulations of biodiversity on insects in ecosystems. Plant Protection (1), 6–10 (2012)
7. Wan, B., Xu, H., Ding, H., et al.: Methodology of comprehensive biodiversity assessment. Biodiversity Science (1), 97–106 (2007)
8. Zhou, H., Chen, J., Cheng, R., et al.: Effects of ecological regulations of biodiversity on insects in ecosystems. Plant Protection (1), 6–10 (2012)
9. Li, T., Li, C., Yu, D., et al.: Effects of heavy metals from road traffic on the community structure and spatial distribution of cropland soil animals. Acta Ecologica Sinica 30(18), 5001–5011 (2010)
10. Jang, G., Zhang, F., Kong, X., et al.: The different levels and the Protection of Multi-functions of Cultivated Land. China Land Science (8), 42–47 (2011)
11. Chen, X., Tang, J.: Advances in farmland field margin systems and biodiversity research. Chinese Journal of Eco-Agriculture 16(2), 506–510 (2008)
12. Liu, Y., Yu, Z., Wang, C., et al.: The diversity of ground-dwelling beetles at cultivated land and restored habitats on the Bashang plateau. Acta Ecologica Sinica 31(2), 0465–0473 (2011)
13. Qin, Z., Zhang, J., Li, Q., et al.: Community structure of soil and micro-fauna in different habitats of urbanized region. Chinese Journal of Applied Ecology 20(12), 0465–0473 (2009)
14. Chen, P.: Collection and survey methods of soil animals. Chinese Journal of Ecology 2(2), 46–51 (1983)
15. Liu, Y., Duan, M., Yu, Z.: Agricultural landscapes and biodiversity in China. Agriculture, Ecosystems and Environment 166, 46–54 (2013)
16. Hiron, M., Berg, Å., Eggers, S.: Bird diversity relates to agri-environment schemes at local and landscape level in intensive farmland. Agriculture, Ecosystems and Environment (176), 9–16 (2013)
17. Foley, J.A., Ramankutty, N., Brauman, K.A., et al.: Solutions for acultivated planet. Nature 478(7369), 337–342 (2012)

Review of NPS Pollution Research Using Hydrological Models Coupled with GIS and Remotely Sensed Data

Eblal Zakzok[1] and M. Faisal Rifai[2]

[1] Faculty of Civil Engineering, Aleppo University, Aleppo, Syria
[2] ETIC Executive Director
ezakzok@yahoo.co.uk, mfaisalriai@yahoo.com

Abstract. This paper reviews the accomplished research on water quality modelling with the use of Geographical Information Systems (GIS) and remotely sensed data. Basic definitions, theory and classifications of hydrological models usually used in the field of hydrological water quality modelling are highlighted. Examples of research in field of hydrology that benefited from modern technologies of GIS and remote sensing are discussed. Integration of hydrological models with topographic information known as digital elevation model (DEM) is investigated. Conclusions and possible research areas are also suggested.

Keywords: NonPoint Source pollution, Water Quality, Hydrological Modelling, Remote Sensing, GIS.

1 Introduction

Hydrologists have become more aware of water quality deterioration of surface waters. The increased concerns include both physical and chemical characteristics such as temperature, taste, suspended solids and dissolved materials [1]. Water pollution comes from either 'point' or 'non-point' sources (NPS). Point sources of pollution are defined as "pollutants that enter the water body from discrete, specified location and can usually be measured", whereas Non-point or diffuse sources cover "everything else" [2]. The most common chemicals transported by runoff in rural catchments are suspended sediment, Nitrogen and Phosphorus. Over-enrichment with the nutrients in aquatic systems leads to eutrophication [3] that results in a wide range of problems, including toxic algal blooms, loss of oxygen, fish deaths, loss of aquatic vegetation and loss of biodiversity. With the increase of human population and associated activities, it seems that the increase in NPS pollution of surface water is certain in the future. Therefore, NPS pollution needs to be understood and quantified at the watershed level in order to take some control measures to reduce the export of pollutants as Nitrogen and Phosphorus into watercourses.

The NPS pollution loadings, estimated as a function of the catchment land use data for different years, is the key variable reflecting the impact of human activities on the status of the river water quality [4]; [5]; and [6]. While identification and estimation

F. Bian and Y. Xie (Eds.): GRMSE 2014, CCIS 482, pp. 393–405, 2015.

of point source pollution is a relatively straightforward task, making management of point source pollution simpler, export estimation and management of non-point sources is, by contrast, more complicated [7]; [8]. Management of NPS pollution focuses on identifying sources of the pollution and cutting or reducing the pollution quantity released to the watercourses. Because of the diffuse nature of the pollution, it is difficult to identify the amount and source of pollution. Monitoring and modelling are the two general methods for identifying and quantifying pollution sources. Monitoring is often expensive and time consuming due to the equipment, the cost of samples analysis and the length of the record required for characterising the pollution. Modelling is the alternative option that can be used to simulate the NPS pollution, processes controlling NPS generation such as surface runoff, the effects of catchment characteristics and management practices on NPS pollution. Rainfall-runoff modelling is the first step towards modelling NPS pollution in many models [9]; [10]; [11]. Such modelling is usually used to evaluate the impact of other management practices enabling development of plans for controlling NPS pollution.

2 Literature Review

The impact of land use change on watershed hydrology can be substantial. Consequences of the land use change at the catchment level almost result in deterioration of water quality and increase of flooding risk. The major land use changes in many catchments are an increase in impervious/urban and agricultural areas, which has undesirable effects on water quantity and quality by increasing runoff and NPS pollution. In addition to the monitoring programs, hydrologic and water quality models are usually used for assessment of the impacts of land use changes on water quantity and quality at the catchment scale. Many hydrological models have been developed for a variety of scales and environments, and to provide a comprehensive review of literature of rainfall-runoff modelling would be impossible [12]. Therefore, this review will focus on some of the well-known models used in the field with emphasis on those including a water quality component and/or can be coupled with GIS and remotely sensed data.

2.1 Classification of Hydrological Models

A hydrological model is representation of a hydrological system, which allows transformation of some hydro-meteorological data into the desired output [13]. Hydrological models have been classified in several ways as deterministic, parametric, stochastic, empirical, physically based, lumped, distributed, event based, continuous [14]. Hydrological models in general are broadly classified as deterministic or stochastic (probabilistic) model. Stochastic models yield varied output due to randomly varied input. In contrast, deterministic models must produce a fixed output from a given input [15]; [16].

Distributed models have been distinguished from lumped models, in that the watershed is divided into elements and calculations are made for each element within the watershed. This is in contrast to the lumped models which calculate the output for a watershed based on average input values for the whole watershed [16]. In the lumped hydrological model the catchment system is considered as a Black Box where the areal diversity of the hydrological process can not be taken into account [13]. Such models have been called 'macroscopic' because they treat the entire hydrologic system as a single unit [15]. The distributed modelling system enables the spatial variation, characteristics and changes to be simulated and estimated inside a catchment [17]. Therefore the distributed models are preferred if remotely sensed data are used [13]. The distributed model subdivides the watershed domain into smaller computational elements. These elements are traditionally based on the concept of hydrologically homogenous areas in terms of land use/cover, soil and slope attributes and assigned homogeneous parameters [18]. However, the degree of spatial descretization or resolution used in the modelling defines whether the model is lumped, semi-distributed or fully distributed. Semi-distributed models attempt to calculate flow contributions from separate areas or sub-basins that are treated as similar units within themselves [15].

Watershed hydrology models may be classified in terms of scale as small, medium or large watersheds [19]. [13] indicated that some agreement was achieved in the hydrologic community about models of different scales:

1. Micro-scale: catchment areas up to 100 km^2 (scale in length 10 km).
2. Meso-scale: 100 to 10 000 km^2 (scale in length 100 km).
3. Macro-scale: order of magnitude about 1 Million km^2 (scale in length 1000 km).

Although he pointed that the orders of magnitude still under discussion, it is not clear what scale is a catchment of an area between 10 000 and 1 Million km^2. "Empirical" and "Physically Based" are other terms used with hydrological models. An empirical model is a model employing empirical relationships developed based on observation [14]. Unlike empirically based models, physically based models solve the differential equations used to describe the flow of water in the hydrologic system [18].

2.2 Remote Sensing and GIS in Hydrology

Remote sensing has been defined many times and a refined definition has been reported as "the practice of deriving information about the earth's land and water surfaces using images acquired from an overhead perspective, using electromagnetic radiation in one or more regions of the electromagnetic spectrum, reflected or emitted from the earth's surface" [20]. This definition may serves as a good expression to fit the nature of remote sensing practices in hydrology and water resources. The remote sensing ability of handling observational data at large scales makes this technology a valuable tool for the study of many features of the earth surface. Possibilities for the use of remote sensing in hydrology have been reviewed in several publications [21]; [22]; [23]; [24]). [22] highlighted examples of the research accomplishments and operational applications of remote sensing in hydrology and water resources. He cited many fields where remote sensing was used successfully including rainfall estimates, soil moisture measurements, snow cover extent and surface water inventory. However, until very recently, the use of remote sensing data in hydrological modelling was deemed low because of the lack of universally applicable operational methods and tools to convert remotely sensed data into information useful to water resources systems operators [25]; [26]. Although few remotely sensed data are directly applied in hydrology, some hydrological variables can be derived from these data and used in conjunction with other data from Geographic Information Systems (GIS) analysis as inputs into hydrological models [27].

Certain model parameters like evapotranspiration, interception and infiltration can be estimated based on land use classification and the normalized difference vegetation index (NDVI) derived from the Landsat data [13]. Information on the hydrologic system and the hydrologic processes with high resolution in space and time are required for modern hydrological modelling techniques, where conventional point measurements are not very suitable for this purpose. In contrast, remote sensing techniques have a great advantage in estimation of land use, vegetation canopy and soil moisture conditions. Remote sensing can be used to estimate input data, including topography, rainfall and evapotranspiration rates, and state variables, such as soil moisture, snow cover, snow water equivalent, and flooded areas. The model parameters values are mostly derived through the classification of soil and vegetation types from remotely sensed data [12].

In a study conducted by [28], a semi-distributed watershed model based on the Grouped Response Unit (GRU) approach using land cover classification derived from Landsat MSS (80 m resolution) gave better calibration and validation statistics than the lumped version of the same daily runoff model, Simple Lumped Reservoir Parametric (SLURP). In another study, a raster-based hydrologic model was developed consisting of several linked submodels which determine spatial and temporal distribution of precipitation, surface flow, infiltration and subsurface flow [29]. The satellite images of the Systeme Probatoire d'Observation de la Terre (SPOT) were used in conjunction with topographic maps, soil maps, literature and weather station data to directly or indirectly derive the model parameters and variables. It was concluded that satellite remote sensing and GIS tools for spatial analysis could improve understanding of spatially distributed hydrologic processes.

Spatially distributed physically based models require large amounts of spatial data and remote sensing fulfils part of this need. Several attempts have been devoted to introduce remote sensing data in distributed hydrologic modelling [30]; [29]; [31]; [32]; [33]. Using remotely sensed data in hydrologic and runoff modelling is of great benefit for modelling water quality. It can be helpful in determining watershed geometry, drainage network, and other map-type information for distributed hydrologic models and for empirical flood peak, annual runoff, or providing input data such as soil moisture or land use classes that can used to define runoff coefficients. One of the first applications of remote sensing data in hydrologic models was using Landsat data to determine both urban and rural land uses for estimating runoff coefficients [34]. Land use is a vital characteristic of the runoff process that affects infiltration, erosion, and evapotranspiration. Distributed models, in particular, need specific data on land cover and its location within the basin. The Soil Conservation Service (SCS) runoff Curve Number (CN) method [35] is probably the most suitable model capable of adapting remote sensing to hydrologic modelling. This is because the SCS-CN method allows remote sensing data to be used as a substitute for land cover maps obtained by conventional techniques [36]; [37]; [38]; [32].

In conjunction with an urban growth model, called SLEUTH, land use information obtained from the Landsat 5 Thematic Mapper (TM) sensor has been used to estimate future surface runoff and peak discharge as a result of an increase in urban development [39]. The urban growth model predicts urban sprawl on the basis of a variety of growth rules, the pattern of urban cells, and the interaction of urban cells and the surroundings (roads, terrain slope, and restricted areas). The study used the SCS-CN method over a watershed of 373 km^2 at Spring Creek Pennsylvania. The results suggest that increases in surface runoff and peak flow in the watershed will be rather small up to the year 2025 due to the prediction of reforestation in addition to urbanization. However, the main drawback of the study was using the seven bands of the Landsat images including the thermal band, which is normally used for locating geothermal activities [40] but not for land use classification. Furthermore, validation for land use classification was not conducted.

The potential contribution of remote sensing to runoff and erosion models was further examined by [41] on three examples: (1) an empirical erosion model at the regional scale, (2) a simple method of evaluating contributing surfaces, (3) runoff and erosion model (STREAM) working at a micro-scale catchment and on event-basis. The empirical erosion model used relies on four factors, topography, land use, intrinsic characteristics of the soil and characteristics of rainfall events. In such a model, the role of remote sensing data was updating the land use component that drives the processes of runoff and erosion. The maps resulting from such a model allow identification and ranking of the erosion risk at the catchment scale. Thus, control measures can be implemented on areas susceptible to high runoff and mudflow. Then, the STREAM model was modified to incorporate remote sensing data instead of field data. The modified version (STREAM-TED) requires slope aspect, land use classification from remote sensing data, surface roughness indices from Radarsat and antecedent rainfall amount. Although a good correlation between the runoff predictions of STREAM and STREAM-TED was obtained there was no indication to any validation against measured data.

The integration of the distributed hydrological models with GIS offers the advantage of using the full information content of the spatially distributed data to analyse the hydrologic processes [42]. GIS can overlay complex maps and perform spatial analysis to produce data that can be used as input data to the hydrological model [43]. The development in technologies such as GIS, Global Positioning Systems (GPS), and scientific visualization can facilitate the widespread use of hydrologic and water quality models [44].

There are other successful studies that have used remote sensing data in hydrological applications. For instance, [31] has used a spring and summer TM image to obtain land use data for the 1987 and 1993 in a study to analyse land use change effect upon river discharge for two catchments in Germany using the ANSWER model. Land use data of only three different years were used by [45] to estimate the impact of land use change on runoff depths in the Indian River Lagoon watershed (USA) using the L-THIA GIS model. The aerial photographs of the 1986 and 1996 were used to develop land use data for these years in the Taihu basin in China for a study of the impact of land use change on runoff [46]. Two satellite images of the 1995 and 2002 were used by [47] to estimate total nitrogen and total phosphorus loads based on the export coefficient model [48] in a large catchment in Australia.

2.3 Topography and Hydrological Modelling

Topographic data is a basic requirement to watershed hydrologic or geomorphologic studies [41]. Watershed topography may be represented by a series of point elevations, contour lines, triangular irregular network or a digital elevation model (DEM). These types of surface representation are extended to other attributes such as rainfall [49]; [50]. However, rainfall is mapped with a coarse resolution due to the spatial distribution of weather stations measuring rainfalls for a regional or national scale. In practice, landscape topography can be presented at a finer resolution than the rainfall can be presented. A DEM can be defined as "a digital representation of topography consisting of an ordered array of numbers representing the spatial distribution of elevation above a datum in a landscape" [18]. DEMs are the most common form of cartographic data from which elevation, slope and aspect can be computed using spatial analysis capabilities of GIS. The DEM is normally interpolated from line or points data [51]; [52] or by the use of digital photogrammetry [53].

Remote sensing can provide quantitative topographic information of suitable spatial resolution for model inputs. For instance, stereo SPOT imagery can be used to produce a Digital Elevation Model (DEM) with a horizontal resolution of 10m and vertical resolution approaching 5m in ideal cases [54]; [55]. A new technology using interferometric Synthetic Aperture Radar (SAR) has been used to illustrate similar horizontal resolutions with an approximately 2m vertical resolution [56]. A comprehensive review of using different satellite sensors and methods for extracting three-dimensional information about the earth's surface was given by [57].

DEMs are increasingly used in modelling distributed hydrological processes. Automated procedures of DEM analysis are commonly used to delineate watershed geometry and to derive subcatchment boundaries, stream drainage network, slope, aspects and flow directions [58]; [59]; [60]. However, some problems are commonly associated with the automatic delineation of DEMs such as presence of flat areas and depressions (sometimes called pits or sinks) [61]; [62]; [63]. Whether pits in the DEM are generated from artificial or natural origins they are considered erroneous and usually filled by raising the values of cells in depressions to the value of the depression's spill point [64]. Flat surfaces in DEMs can be attributed to: (1) the low vertical and/or horizontal DEM resolution to represent low relief landscape; (2) depressions filling; and (3) natural flat landscape, which seldom occur in nature. Methods and assumptions for treatment of these spurious features in the DEMs were discussed in detail by [18]; [65].

[64] used the automatic delineation of depressions, watershed and drainage networks from DEMs to compare the results of extracting slope and flow paths from three DEMs of varying resolution. Comparison of computed slope statistics for the three DEMs revealed that increases in cell size produce lower slope values. She concluded that the quality of derived information is a function of both the horizontal and vertical resolution of the DEM. However, only visual examination of the derived

drainage networks was carried out. Similarly, the watershed boundary derived from three DEMs were visually assessed by [66].

The topographically driven hydrologic model (TOPMODEL) has been developed to account for variability in the hydrological response of different areas of catchment using the topographic index (Equation 1) based on catchment topography [67].

$$Index = \ln(\alpha / \tan \beta) \tag{1}$$

Where ln is the natural logarithm, α is the area of the hillslope per unit contour length or the specified area; tan β is the local surface slope.

The topographic index represents a catchment wetness index where areas within the basin having the same values of the index are predicted to have similar hydrological responses, regardless of their position within the catchment. [66] mapped the topographic index spatial distribution for three DEMs. The valley network was much more defined in the fine resolution DEM than in the other two DEMs of coarser resolution. Then, they examined the effects of DEM accuracy on predictions of the Distributed Soil-Hydrology-Vegetation Model (DSHVM). The topographic effects on spatial distribution of short-wave radiation and air temperature were introduced into the model. The model was run using the three DEMs for a 4-year period and differences in runoff peaks, timing and volume were solely attributed to the vertical resolution of the DEM. The effect of the watershed extent derived from the three DEMs on the predictions was not considered as a possible reason. In another study, simulations based on automated digital terrain analysis to parameterize a semi-distributed hydrological model (SLURP) were very similar to those obtained with manual parameterization [68].

The quality of the DEM depends mainly on three major factors: the interpolation algorithm, the spatial distribution of the input data points, and the quality of the input data points (x, y, z) [17]. DEM quality assessment should be carried out particularly when important hydrological variables are estimated from a DEM and used in hydrological models.

2.4 NPS Pollution

In recent years, the concern about NPS pollution has broadly increased. Non-point sources of water pollution are now considered of greater importance than point sources in many locations [69]. Pollutants of main interest involve sediments, nutrients (Nitrogen and Phosphorus), pesticides, and pathogens. NPS pollution has been identified as the principal reason that U.S. waters do not meet water quality standards [70].

A dramatic increase in the development and application of hydrological and water quality models has occurred in the past two decades. These models evaluate complex environmental processes and assess the NPS pollution of soil and water resources. Monitoring water quality using remote sensing techniques depends on the ability to measure changes in the spectral signature backscattered from water, and then to relate such changes to water quality parameters. Whilst suspended sediments can be mapped with remotely sensed data, the limitation of the current spectral and spatial resolution of satellite data prevents the applications of monitoring water quality [24].

In the study of [71], three land cover sources for input into the Modified Universal Soil Loss Equation (MUSLE) were made available using GIS analysis to identify eroded sites as part of a program aimed to avoid costly filtration in New York 2000 square mile watershed. It was concluded that using Landsat Thematic Mapper (TM) imagery with a hybrid classification provided a rapid, objective means of producing land cover databases for use in the MUSLE over a large area. In another study, the GIS was used to provide input parameters to the Source Loading And Management Model (SLAMM); an empirical urban storm water quality model, for a small urban watershed located in Plymouth, Minnesota, USA [72].

The Agricultural NPS pollution model (AGNPS) [10], Geographic Resource Analysis Support System (GRASS), and GRASS WATERWORKS (a hydrologic modelling tool box being developed at the Michigan State University Centre for remote sensing) were integrated in a study to evaluate the impact of agricultural runoff on water quality in the Cass river; a sub-watershed of Saginaw Bay [73]. In their study, management scenarios including variations in crop cover, tillage methods, and other management practices were explored using the AGNPS model to minimize sedimentation and nutrient loading. [30] used the AGNPS as a test model for assessing NPS for two watersheds in Illinois. Input data were obtained from the GIS database including data on soil, land use, streams and water-bodies, farm boundaries, monitoring locations, land management practices. Digital Elevation Model (DEM) data were used to generate slope and drainage networks and other topographic properties. The land use information was identified from classification of remote sensing data (high-altitude scanned aerial photographs). There was no indication to method of classification used in their study.

The Water Erosion Prediction Project (WEPP) model was developed by the US-DA-ARS National Soil Erosion Research Laboratory (NSERL) [74]; [75]; [76]. WEPP can be run on a single storm or on continuous basis. Model outputs include both on-site and off-site erosion effects [2]. The on-site erosion effects include the average annual soil loss over the area on the hillslope. The output related to the off-site effects includes sediment loads and particle size information. The model has six components, which are climate generation (CLIGEN), hydrology, plant growth, soils, irrigation and erosion [76] The model makes adjustments to soil properties used in hydrology and erosion calculations on a daily time step as a result of tillage operations, freezing and thawing, compaction, weathering, or history of precipitation [2].

3 Conclusions

Although many models have been developed to simulate the NPS pollution at different scales, these models are generally complex and data demanding preventing wide application of these models. Therefore, there is a need for simplified methods to assess the environmental status of waters at a catchment scale. The revolutionary advances in computer technology, together with the rapid progress in geo-information technologies; remote sensing, GIS and GPS, over the last two decades have brought important developments to watershed modelling. However, this remains a dynamic and active area of research due to continuous development in computer and information technology. New studies are continuously brought into use as data is released from new remote sensing satellites with high spatial and temporal resolutions. Possible areas of future research in this field include:

1. Development and enhancement of modelling the effects of human activities on hydrologic and water quality.
2. Developing new algorithms for making use of the high-resolution remotely sensed and GIS digital spatial data.
3. Integration of Decision Support Systems (DSS) with GIS-based watershed models for NPS pollution assessment to provide regulatory information to modellers and decision makers to evaluate different NPS pollution control measures in which spatial decisions can be made more efficiently.

A literature review of research on NPS pollution modelling was presented. The roles of remote sensing and GIS in hydrological and water quality modelling were highlighted. The literature survey revealed the importance of these technologies in developing the current research into operational applications. However, quantification of NPS pollution loadings is still challenging due to the diffuse nature of the problem. These loadings are usually estimated indirectly using hydrological and water quality models. Among a large number of models, NPS pollution models coupled with GIS and remotely sensed data were reviewed with examples of few applications.

References

1. Ward, R.C., Robinson, M.: Principles of Hydrology. McGraw-Hill, Malta (2000)
2. Novotny, V., Olem, H.: Water Quality: Prevention, Identification, and management of Diffuse Pollution. Van Nostrand Reinhold, New York (1994)
3. Haygarth, P.M., Jarvis, S.C.: Soil derived phosphorus in surface runoff from grazed grassland lysimeters. Water Research 31(1), 140–148 (1997)
4. Mattikalli, N.M., Richards, K.S.: Estimation of Surface Water Quality Changes in Response to Land Use Change: Application of The Export Coefficient Model Using Remote Sensing and Geographical Information System. Journal of Environmental Management 48, 263–282 (1996)
5. Bhaduri, B.: A geographic information system-based model of the long-term impact of the land use change on nonpoint-source pollution at a watershed scale. Ph.D. Thesis. Purdue University. P: 189 (1998)
6. Tong, S.T.Y., Chen, W.: Modeling the relationship between land use and surface water quality. Journal of Environmental Management 66, 377–393 (2002)
7. Mitchell, G.: The quality of urban stormwater in Britain and Europe: database and recommended values for strategic planning models. School of Geography, University of Leeds, Leeds, UK. Technical Report (2001)
8. Schreier, H., Brown, S.: Multiscale approaches to watershed management: land use impact on nutrient and sediment dynamics. In: Tchiguirinskaia, I., Bonell, M., Hubert, P. (eds.) Scales in Hydrology and Water Management, pp. 61–75. IAHS Publication no. 287, Netherlands (2004)
9. HEC. Storage, Treatment, Overflow, Runoff, Model, STORM User's Manual, Generalized Computer Program, 723-S8-L7520. Corps of Engineers. Davis, CA (1977)

10. Young, R.A., Onstad, C.A., Bosch, D.D., Anderson, W.P.: Agricultural Nonpoint Source Pollution Model: A watershed Analysis Tool. Agriculture Research Service, U.S. Department of Agriculture, Morris, MN (1986)
11. Harbor, J.: A practical method for estimating the impact of land use change on surface runoff, groundwater recharge and wetland hydrology. Journal of the American Planning Association 60, 91–104 (1994)
12. Beven, K.J.: Rainfall-Runoff Modelling. John Wiley & Sons, Chichester (2000)
13. Schultz, G.A.: Hydrological Modeling Based on Remote Sensing Information. Advances in Space Research 13(5), 149–166 (1993)
14. Haan, C.T., Barfield, B.J., Hayes, J.C.: Design Hydrology and Sedimentology for Small Catchments. Academic Press, San Diego (1994)
15. Jones, J.A.A.: Global Hydrology: Processes, resources and environmental management. Longman, Singapore (1997)
16. Skidmore, A.K.: Taxonomy of environmental models in the spatial sciences. In: Skidmore, A.K. (ed.) Environmental Modelling with GIS and Remote Sensing, pp. 8–25. Taylor and Francis, London (2002)
17. Olsson, L., Pilesjo, P.: Approaches to spatially distributed hydrological modelling in a GIS environment. In: Skidmore, A.K. (ed.) Environmental Modelling with GIS and Remote Sensing, pp. 166–199. Taylor and Francis, London (2002)
18. Vieux, B.E.: Distributed Hydrological Modeling Using GIS. Kluwer Academic Publishers, Netherlands (2001)
19. Singh, V.P., Frevert, D.K. (eds.): Mathematical Models of Large Watershed Hydrology. Water Resources Publications, LLC, Michigan (2002b)
20. Campbell, J.B.: Introduction to Remote Sensing. The Guilford Press, New York (1996)
21. Engman, E.T., Gurney, R.: Remote Sensing in Hydrology. Chapman and Hall, Boca Raton (1991)
22. Rango, A.: Application of remote sensing methods to hydrology and water resources. Hydrological Sciences 39(4), 309–320 (1994)
23. Dubayah, R.O., Wood, E.F., Engman, E.T., Czajkowski, K.P., Zion, M., Rhoads, J.: Remote Sensing in Hydrological Modeling. In: Schultz, G.A., Engman, E.T. (eds.) Remote Sensing in Hydrology and Water Management, pp. 85–102. Springer, Berlin (1999)
24. Schmugge, T.J., Kustas, W.P., Ritchie, J.C., Jackson, T.J., Rango, A.: Remote sensing in hydrology. Advances in Water Resources 25, 1367–1385 (2002)
25. Kite, G.W., Pietroniro, A.: Remote sensing application in hydrological modelling. Hydrological Sciences 41(4), 563–591 (1996)
26. Pietroniro, A., Leconte, R.: A review of Canadian remote sensing applications in hydrology, 1995-1999. Hydrological Processes 14, 1641–1666 (2000)
27. Cruise, J.F., Miller, R.: Hydrologic Modeling Using Remotely Sensed Databases. In: Lyon, J.G. (ed.) GIS for Water Resources and Watershed Management, pp. 189–205. Taylor & Francis, London (2003)
28. Kite, G.W., Kouwen, N.: Watershed modelling using land classifications. Water Resources Research 28(12), 3193–3200 (1992)
29. Xiao, Q.F., Ustin, S.L., Wallender, W.: A spatial and temporal continuous surface-subsurface hydrologic model. Journal of Geophysical Research 101(D23), 29565–29584 (1996)
30. Lee, M.T., Kao, J., Ke, Y.: Integration of GIS, Remote Sensing, and Digital Elevation Data for a Hydrologic Model. In: Hydraulic Engineering: Proceedings of the 1990 National Conference, pp. 427–432. American Society of Civil Engineers, New York (1990)

31. Jurgens, C.: Application of a hydrological model with integration of remote sensing and GIS techniques for the analysis of land-use change effects upon river discharge. In: Owe, M., Brubaker, K., Ritchie, J., Rango, A. (eds.) Remote Sensing and Hydrology 2000, pp. 598–600. IAHS Publication No. 267, Wallingford (2001)
32. Melesse, A.M., Shih, S.F.: Spatially distributed storm runoff depth estimation using Landsat images and GIS. Computers and Electronics in Agriculture 37, 173–183 (2002)
33. Davenport, I.J., Silgram, M., Robinson, J.S., Lamb, A., Settle, J.J., Willig, A.: The use of earth observation techniques to improve catchment-scale pollution predictions. Physics and Chemistry of the Earth 28, 1365–1376 (2003)
34. Jackson, T.J., Ragan, R.M., Fitch, W.N.: Test of Landsat-Based Urban Hydrologic Modeling. ASCE J. Water Resources Planning and Management Div. 103(WR1, Proc. Papers 12950), 141–158 (1977)
35. Soil Conservation Service (SCS), Hydrology, National Engineering Handbook, Section 4, Chapter 10. Soil Conservation Service, USDA, Washington DC (1972)
36. Bondelid, T.R., Jackson, T.J., McCuen, R.H.: Estimating runoff curve numbers using remote sensing data. In: Proceeding of the International symposium on Rainfall-Runoff Modelling. Applied Modeling in Catchment Hydrology, pp. 519–528. Water Resources Publications, Littleton (1982)
37. Still, D.A., Shih, S.F.: Satellite data and geographic information system in runoff curve number prediction. In: Proceeding of the International Conference on Computer Application in Water Resources, Taipei, Taiwan, R.O.C., pp.1014–1021 (1991)
38. Sharma, T., Satya Kiran, P.V., Singh, T.P., Trivedi, A.V., Navalgund, R.R.: Hydrologic response of watershed to land use change: a remote sensing and GIS approach. International Journal of Remote Sensing 22(11), 2095–2108 (2001)
39. Carlson, T.N.: Analysis and Prediction of Surface Runoff in an Urbanizing Watershed Using Satellite Imagery. Journal of American Water Resources Association 40(4), 1087–1098 (2004)
40. Lillesand, T.M., Kieffer, R.W.: Remote sensing and image interpretation. John Wiley & Sons, USA (2000)
41. King, C., Baghdadi, N., Lecomte, V., Cerdan, O.: The application of remote-sensing data to monitoring and modelling of soil erosion. Catena 62, 79–93 (2005)
42. Vieux, B.E.: Geographic Information Systems and Non-point Source Water Quality and Quantity Modelling. Hydrological Processes 5, 101–113 (1991)
43. Chen, H., Liaw, S., Jan, J.: Applications of geographic information system and remote sensing in watershed hydrologic model. Quarterly Journal of the Experimental Forest of National Taiwan University 10(4), 77–93 (1996)
44. Tim, U.S.: Emerging technologies for hydrologic and water quality modeling research. Transactions of the ASAE 39(2), 465–476 (1996)
45. Kim, Y., Engel, B.A., Lim, K.J., Larson, V., Duncan, B.: Runoff Impacts of Land-Use Change in Indian River Lagoon Watershed. Journal of Hydrologic Engineering 7(3), 245–251 (2002)
46. Junfeng, G.: Impact of land use changes on runoff of the Taihu basin, China. In: Chen, Y., Takara, K., Cluckie, I., De Smedt, F.H. (eds.) GIS and Remote Sensing in Hydrology, Water Resources and Environment, pp. 219–226. IAHS Publication no. 289, Wallingford (2004)
47. Ierodiaconou, D., Laurenson, L., Leblanc, M., Stagnitti, F., Duff, G., Salzman, S.: Multitemporal land use mapping using remotely sensed techniques and the integration of a pollutant load model in a GIS. In: GIS and Remote Sensing in Hydrology, Water Resources and Environment, pp. 303–352. IAHS Publication no. 289, Wallingford (2004)

48. Jorgensen, S.E.: Lake Management. Pergamon, Oxford (1980)
49. Seed, A.: Modelling and forecasting rainfall in space and time. In: Tchiguirinskaia, I., Bonell, M., Hubert, P. (eds.) Scales in Hydrology and Water Management, pp. 137–152. IAHS Publication no. 287, Netherlands (2004)
50. Chaplot, V., Saleh, A., Jaynes, D.: Effect of the accuracy of spatial rainfall information on the modeling of water, sediment, and N03-N loads at the watershed level. Journal of Hydrology 312, 223–234 (2005b)
51. Desmet, P.J.J.: Effects of interpolation errors on the analysis of DEMs. Earth Surface Processes and Landforms 22(6), 563–580 (1997)
52. Wise, S.: The effects of GIS interpolation errors on the use of DEMs in geomorphology. In: Lane, S.N., Richards, K.S., Chandler, J.H. (eds.) Landform Monitoring, Modelling and Analysis, pp. 139–164. Wiley, Chichester (1998)
53. Konecny, G.: Geoinformation: Remote Sensing, Photogrammetry and Geographic Information Systems. Taylor and Francis, London (2003)
54. Case, J.B.: Report on the international Symposium on Topographic Applications of SPOT data. Photogrammetric Engineering and Remote Sensing 55(1), 94–98 (1989)
55. Endreny, T.A., Wood, E.F., Lettenmaier, D.P.: Satellite-derived digital elevation model accuracy: hydrogeomorphological analysis requirements. Hydrological Processes 14(1), 1–20 (2000)
56. Zebker, H.A., Madsen, S.N., Martin, J., Wheeler, K.B., Miller, T., Lou, Y., Alberti, G., Vetrella, S., Cucci, A.: The TOPSAR interferometric Radar Topographic Mapping Instrument. IEEE Transactions on Geoscience and Remote Sensing 30, 933–940 (1992)
57. Toutin, T.: Elevation modelling from satellite visible and infrared (VIR) data. International Journal of Remote Sensing 22(6), 1097–1125 (2001)
58. Jenson, S.K., Dominique, J.: Extracting topographic structure from digital elevation data for geographical information system analysis. Photogrammetric Engineering and Remote Sensing 54(11), 1593–1600 (1988)
59. Morris, D.G., Heerdegen, R.G.: Automatically derived catchment boundaries and channel networks and their hydrological applications. Geomorphology 1, 131–141 (1988)
60. Martz, L.W., Garbrecht, J.: Channel Network Delineation and Watershed Segmentation in the TOPAZ Digital Landscape Analysis System. In: Lyon, J.G. (ed.) GIS for Water Resources and Watershed Management, pp. 7–16. Taylor & Francis, London (2003)
61. Hutchinson, M.F.: A New Procedure for Gridding Elevation and Stream Line Data with Automated Removal of Spurious Pits. Journal of Hydrology 106, 211–232 (1989)
62. Martz, L.W., Garbrecht, J.: The treatment of flat areas and depressions in automated drainage analysis of raster digital elevation models. Hydrological Processes 12, 843–855 (1998)
63. Turcotte, R., Fortin, J.P., Rousseau, A.N., Massicotte, S., Villeneuve, J.P.: Determination of the drainage structure of a watershed using a digital elevation model and a digital river and lake network. Journal of Hydrology 240, 225–242 (2001)
64. Jenson, S.K.: Application of Hydrologic Information Automatically Extracted from Digital Elevation Models. Hydrological Processes 5(1), 35–45 (1991)
65. Garbrecht, J., Martz, L.W., Starks, P.J.: Technological Advances in Automated Land Surface Parameterization from Digital Elevation Models. In: Lyon, J.G. (ed.) GIS for Water Resources and Watershed Management, pp. 207–217. Taylor and Francis, London (2003)
66. Kenward, T., Lettenmaier, D.P., Wood, E.F., Fielding, E.: Effect of Digital Elevation Model Accuracy on Hydrological Predictions. Remote Sensing of Environment 74, 432–444 (2000)

67. Beven, K., Kirkby, M.J.: A physically based, variable contributing area model of basin hydrology. Hydrological Science Bulletin 24, 43–69 (1979)
68. Lacroix, M.P., Martz, L.W., Kite, G.W., Garbrecht, J.: Using digital elevation modeling techniques for the parameterization of a hydrological model. Environmental Modelling and Software 17, 127–136 (2002)
69. Miles, S.B., Ho, C.L.: Application and Issues of GIS as Tool for Civil Engineering Modelling. Journal of Computing in Civil Engineering 13(3), 144–152 (1999)
70. Olem, H.: U.S. Sharpens Focus on Non-Point Source of Water Pollution. Engineering News Records 230(24), E66–E68 (1993)
71. Fraser, R.H., Warren, M.V., Barten, P.K.: Comparative-evaluation of land cover data sources for erosion prediction. Water Resources Bulletin 31(6), 991–1000 (1995)
72. Haubner, S.M., Joeres, E.F.: Using a GIS for estimating input parameters in urban storm water quality modeling. Water Resources Bulletin 32(6), 1341–1351 (1996)
73. He, C.S., Riggs, J.F., Kang, Y.T.: Integration of Geographic Information-Systems and a Computer-Model to Evaluate Impacts of Agricultural Runoff on Water-Quality. Water Resources Bulletin 29(6), 891–900 (1993)
74. Foster, G.R., Lane, L.J.: User Requirements: USDA-Water Erosion Prediction Project (WEPP). USDA-ARS. NSERL Report No. 1. West Wafayette, IN (1987)
75. Lane, L.J., Nearing, M.A.: USDA-Water Erosion Prediction Project: Hillslope Profile Model Documentation. USDA-ARS. NSERL Report No. 2. West Wafayette, IN (1989)
76. Laflen, J.M., Lane, L.J., Foster, G.R.: WEPP-A new generation of erosion prediction technology. Journal of Soil and Water Conservation 46(1), 34–38 (1991)

Assessment of Terrestrial Ecosystem Health in the Yellow River Basin

Cuicui Wang[1], Xiangnan Liu[1], Weiguo Jiang[2], Wenjie Wang[3], Qiuling Li[4], Lihua Yuan[2], Ran Cao[2], and Yunfei Zhang[2]

[1] School of Information Engineering, China University of Geosciences, Beijing 100083, China
[2] Key Laboratory of EnvironmentalChange and Natural Disaster, Beijing Normal University, Beijing 100875, China
[3] Institute of Environmental Information,Chinese Research Academy of Environmental Sciences,Beijing 100012, China
[4] Foreign Language Department, China University of Geosciences, Beijing 100083, China
{icnwcc,liuxncugb}@163.com, jiangweiguo@bnu.edu.cn, 1262041863@qq.com

Abstract. Recently, research on ecosystem health assessment has become dominance in environmental and ecological science. In this paper, terrestrial ecosystem in the Yellow River Basin was chosen as the assessment subject and characteristics of the ecological pattern, functions and pressure were analyzed. 29 sub-watersheds were selected as assessment units and the terrestrial ecosystem health from 2000 to 2010 was assessed by using RS and GIS technologies with natural social economical features and influence factors. The results showed: (1) the overall health condition of the terrestrial ecosystem was moderate; (2) in the view of spatial distribution, ecosystem health condition in the upper reaches was better than that in the middle and lower reaches, and the condition in the lower reaches turned the worst; (3) in the view of time distribution, the terrestrial ecosystem health was degenerated from 2000 to 2010 and health conditions of only a few sub-watersheds were improved.

Keywords: The Yellow River Basin, terrestrial ecosystem health, assessment.

1 Introduction

Terrestrial ecosystem refers to ecosystem from the outside aquatic ecosystem to the basin watershed, providing habitats for various living beings and water conservation, soil conservation, biodiversity protection and product offerings, as well as other benefits for human beings. With the development of social economy, the degradation of ecological environment in basins and over-exploitation of resources by human beings have made deforestation, water and soil erosion, water pollution more serious, threatening the basin ecosystem health. Therefore, studies on basin ecosystem health have drawn more and more attentions. Authorities from different countries and regions have taken measures one after another, including assessing the health of watershed ecosystem, restoring the ecosystem [1-4] and complementing comprehensive treatments for the watershed environment in relation to the health of ecosystem. Yellow River

F. Bian and Y. Xie (Eds.): GRMSE 2014, CCIS 482, pp. 406–414, 2015.

Conservancy Commission (YRCC) put forward a new concept named "Maintaining the Healthy Life of the Yellow River" and constructed "1493" River Management System[5]. Changjiang Water Resources Commission (CWRC) raised "Maintaining the Health of Yangtze River for Enhancing the Harmonious Relationship between Humans and Water" slogan [6]. Hai River Water Resources Commission(HWRC) made investigation and study on water restoration and conservation in detail[7], and Songliao Commission conducted research on assessment index construction of watershed ecosystem health [8]. Based on the RS and GIS technologies, Zhang valued the watershed ecosystem health of Dongting Lake [9].Chen et al evaluated the ecological health of the Jiulong River Basin from its land, waters of the shore and tripartite with the RS and GIS technology and put forward corresponding management strategies[10].Shangguan made comprehensive assessments on the wetland ecosystem health of the Yellow River Basin Delta from the perspective of Pressure-State-Response based on PSR Model[11].By applying ecological factor scoring methods and GIS spatial analysis function, Fu et al made research on the sensitivity of the Maqu (upper stream of the Yellow River) Wetland ecosystem. The results show that the ecosystem of the Maqu (upper stream of the Yellow River) Wetland is very sensitive, and utilization of the ecosystem should be strictly controlled to protect the environment[12].

2 Study Area

The Yellow River Basin, as the study area, is vast in territory with most of the land situated northwest of China (ranging from 96°to 119°E and from 32°to 42°N), which stretches across four geomorphologic units including the Tibet Plateau, the Inner Mongolian Plateau, the Loess Plateau and the Huanghuaihai Plain from the west to the east with a drainage area of 794.5 thousand square kilometers. The climate there consists of arid climate, semi-arid climate and semi-humid climate and the annual precipitation

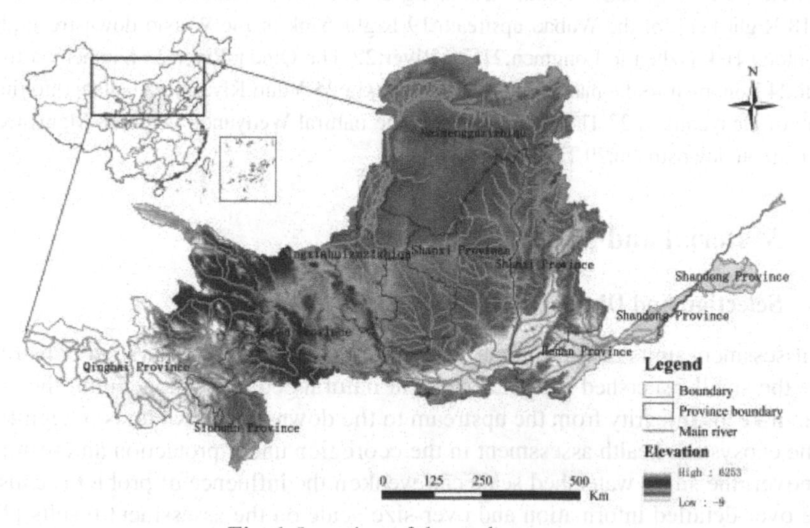

Fig. 1. Location of the study area

of most area is 300-800 mm. The annual mean temperature in the west ranks below 0°C-5°C with the midland 5°C-10°C and the southeast more than 10°C. The main vegetation are agricultural products, grasslands, woodlands and bushes, which account for 97% of the whole, while the rest 3% is covered by water body, tundra, bare land, etc.

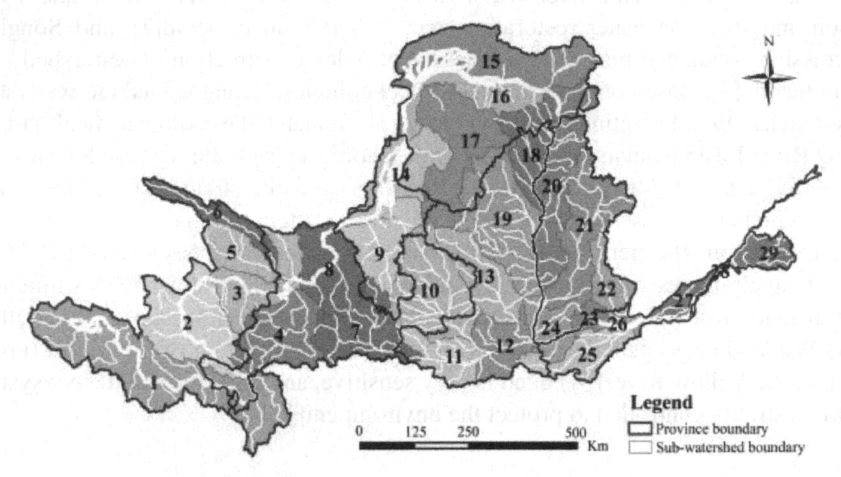

Fig. 2. Evaluation unit of the Yellow River Basin

Note: 1Heyuan to Maqu;2 Maqu to Longyangxia;3 Longyangxia to the main stream of Lanzhou;4 Daxia River and Tao River;5 Huangshui;6 Upstream of the gate of the Datong River;7 Upstream of Baojixia along the Wei River;8 Lanzhou to Xiaheyan;9 Qingshui River and Kushui River;10 Upstream of the Jing River in Zhangjiashan;11 Sub-watershed from the Wei River in Baojixia to Xianyang;12 Sub-watershed from the Wei River in Xianyang to Tongguan;13 Upstream of the Zhuangtou of the Beiluohe River;14 Xiaheyan to Shizuisha;15 North bank along Shizuishan to Hekouzhen;16 South bank along Shizuishan to Hekouzhen;17 Interior drainage area;18 Right bank of the Wubao upstream;19 Right bank of the Wubao downstream;20 Left bank along Hekouzhen to Longmen;21Fen River;22 The QindanRiver;23 Sanmenxia to Xiaolangdi;24 Longmen to the main stream of Sanmenxia;25 Yiluo River;26 Xiaolangdito the main stream of Huayuankou;27 The Jindi River and the natural Wenyanqu Basin;28 Main stream of Huayuankou downstream;29 The Dawen River

3 Material and Methods

3.1 Selection and Division of the Assessment Units

The assessment units directly affect the frame design and index system construction. Since the small watershed is an independent natural geomorphologic unit, the ecosystem in it keeps integrity from the upstream to the downstream, which is of significance for the ecosystem health assessment in the ecoregion under protection and restoration. Moreover, the small watershed scale can weaken the influence of problems caused by some over-detailed information and over-size scale on the assessment results [13], so the small watershed is chosen.

In this paper, the Yellow River Basin was divided into 29 small watersheds as assessment units based on the water system data of the Yellow River Basin and DEM data. The buffer zone is river-and reservoir-centered on the terrestrial areas and the range of them is limited by the belt on the bank (Fig.2).

3.2 Determination of the Index System

Watershed ecosystem is a complex combining natural, social and economic ecosystem. Terrestrial ecosystem health assessment is a significant factor referring to the influence study of man-made interference in social economy such as over-use of land and pollution, on the condition of the watershed ecosystem and water quality. Such human behaviors as over-grazing, over-mining and forest clearance for farmland expansion have turned the ecosystem of the Yellow River Basin more vulnerable [14]. Point source pollution was the major pollution there, with a low control rate of industrial effluents and treatment rate of domestic sewage [15]. The AHP and expert evaluation methods were used to ascertain the specific index and weight (Table 1). This paper also assessed the territorial ecosystem health of the Yellow River Basin between 2000 and 2010 from the perspective of ecological pattern, ecological function and ecological pressure.

Table 1. Index system of Terrestrial ecological health assessment in the Yellow River Basin

Evaluation objects	Indicators type	Evaluation indicators	Index weights
Terrestrial ecological	Ecological patterns (0.3)	Forest coverage(%)	0.6
		Landscape fragmentation	0.4
	Ecological functions (0.3)	Water Conservation Function Index	0.4
		Soil and Water Conservation Function Index	0.3
		Proportion of protected area in land area(%)	0.3
	Ecological pressure (0.4)	Percentage of construction land(%)	0.4
		Point source pollution load emission index	0.6

3.3 Standardization of Assessment Index Data

In order to eliminate the effects of differences in dimension and magnitude of each index on the assessment results, each index data was normalized to 0-100 by using standardization of data range.

Positive correlation:

$$Bi=100(X_i-X_{min})/(X_{max}-X_{min}) \quad (1)$$

Negative correlation:

$$Bi=100(X_{max} - X_i)/(X_{max}-X_{min}) \quad (2)$$

X_i, X_{max}, X_{min}, B_i respectively denote the original, maximum, minimum, and the normalized values of the i-th index.

3.4 Comprehensive Assessment Method

In this paper, weigh summation of land indicators and the obtained composite index (WHI) were used to indicate the terrestrial ecological health condition in each assessment unit. Among them, the composite index (WHI) is calculated as:

$$WHI = I_P W_P + I_F W_F + I_{Pr} W_{Pr} \quad (3)$$

Here I_P, I_F, I_{Pr} stands for terrestrial ecological pattern, ecological functions and ecological pressure health index, and W_P, W_F, W_{Pr} stand for the weights of them, respectively.

3.5 Assessment Criteria and Rank

According to composite index scores of terrestrial ecological health assessment, the rank of terrestrial ecological health was divided into excellent, good, moderate, poor and very poor(Table2).

Table 2. Evaluation criteria rank of terrestrial ecosystem health

Health condition	Excellent	Good	Moderate	Poor	Very poor
Composite index(WHI)	WHI≥80	60≤WHI<80	40≤WHI<60	20≤WHI<40	WHI<20

4 Results

4.1 Terrestrial Ecological Patterns Condition

The terrestrial ecological patterns index was about 30, 30, 35 points in 2000, 2005 and 2010 respectively in poor condition. The ecological health of the Yellow River source region was relatively good, and that of other regions were mainly poor or below. From 2000 to 2010, the terrestrial ecological health was improved, and the health condition of sub-watershed from Heyuan to Maqu was changed from moderate to good with 17 points increased in health scores and other 8 sub-watersheds were improved from very poor to poor. The JindiRiver and the natural Wenyanqu Basin, the main downstream of the Huayuankou had the worst ecological patterns health condition with the score less than 5 points (Fig.3). Terrestrial ecosystem of the Yellow River Basin had low forest coverage, and the structure and composition of its natural ecosystem had poor completeness.

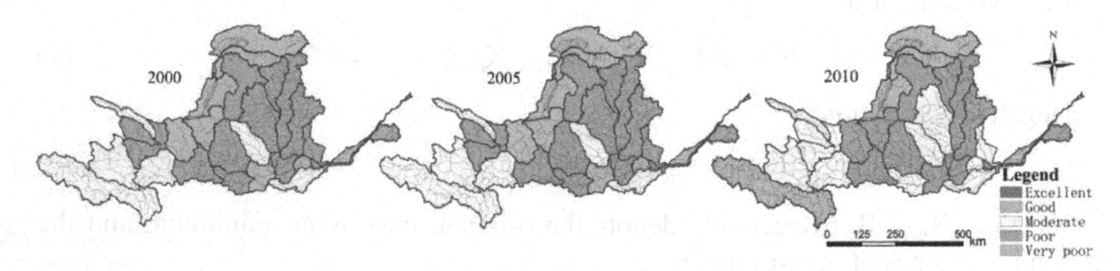

Fig. 3. Result of terrestrial ecological patterns assessment

4.2 Terrestrial Ecological Functions Condition

The terrestrial ecological functions index was about 48 points in moderate condition. The ecological functions had little improvements with 0.1 point increased in the health score from 2000 to 2010. Among those 29 sub-watersheds, only the health condition of the upstream of the Zhuangtou of the Beiluo River, was improved from moderate to good, while others' remained the same(Fig.4). The main factor threatening the terrestrial ecological functions health of the upstream and midstream of the Yellow River Basin was their poor soil conservation. Human activities, such as denudation, overgrazing and mining, destroyed excellent pastures and natural vegetation, which led to grassland degradation, desertification and soil erosion to different extents. The protection area in the downstream of the Yellow River Basin, which is the key to maintain the ecological functions health is relatively small.

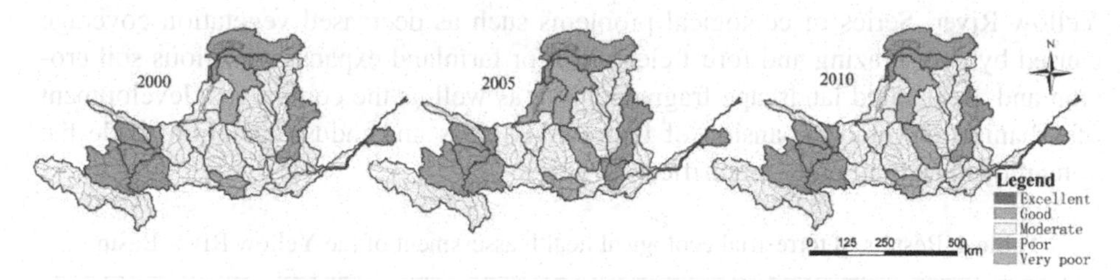

Fig. 4. Result of terrestrial ecological functions assessment

4.3 Terrestrial Ecological Pressure Condition

The terrestrial ecological pressure index was about 70 points in good condition in 2000, and 57 points in both 2005 and 2010 in moderate condition. From 2000 to 2010, terrestrial ecological health condition in the Yellow river source region and Inner Mongolia region degraded from good and excellent to moderate, while the condition of the FenRiver and the DawenRiver degraded from excellent and moderate to poor, respectively, with that of the Qindan River from good to moderate(Fig.5). 8 sub-watersheds got the health score of point source pollution load emission index less than 20, among which 2 had less than 3 scores. Point source pollution caused by urban expansion and industrial development was the main threat to terrestrial ecological health.

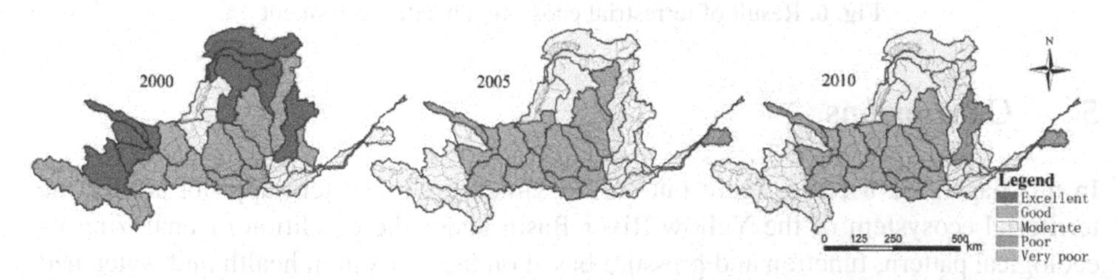

Fig. 5. Result of terrestrial ecological pressure assessment

4.4 Terrestrial Ecological Health Condition

The terrestrial ecological health composite index of the Yellow River was 51, 46 and 48 points in 2000, 2005 and 2010 respectively with a moderate health rank. From 2000 to 2010, the number of sub-watersheds in good health condition reduced by 24.67%, while that in poor health condition increased by 10.35% (Table 4). In the view of the spatial distribution, terrestrial ecological health condition of the downstream of the Yellow River was the worst. In 2000, the health condition of 6 sub-watersheds was good and they were the source region of the Yellow River, gate of the Datong River, interior drainage area, upstream of the Zhuangtou of the Beiluo River, Baojixiaalongthe Wei River and Xianyang. Until 2010, only the health condition of the upstream of the Zhuangtou of the Beiluo River, Baojixiaalong the Wei River and Xianyang remained good.(Fig.6)

Human activities form the greatest harm to the terrestrial ecological health of the Yellow River. Series of ecological problems such as decreased vegetation coverage caused by over-grazing and forest clearance for farmland expansion, serious soil erosion and accelerated landscape fragmentation, as well as the continuous development of urbanization, land expansion of impervious areas and industrialization made the contaminant loading more intensified.

Table 4. Results of terrestrial ecological health assessment of the Yellow River Basin

	Area proportion of different health level(%)		
Year	Good	moderate	poor
2000	30.18	63.56	6.26
2005	3.33	78.34	18.33
2010	5.51	77.88	16.61

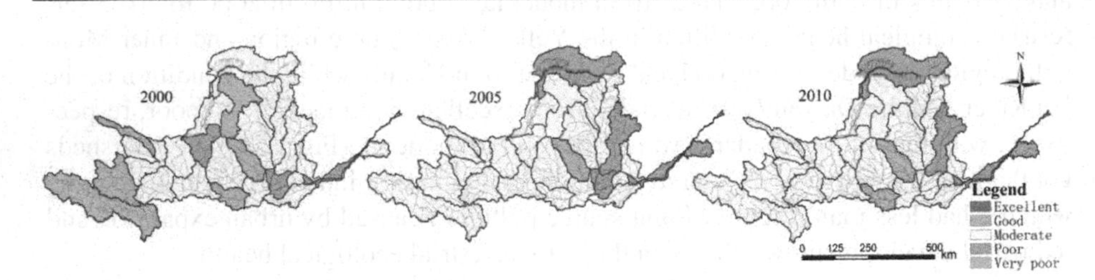

Fig. 6. Result of terrestrial ecosystem health assessment

5 Conclusions

In this paper, we tried to figure out an assessment index system appropriate for the territorial ecosystem of the Yellow River Basin under the condition of analyzing its ecological pattern, function and pressure based on the ecosystem health and watershed ecological theories. The time and spatial distribution were discussed as well as their driving force of the terrestrial ecosystem health condition from 2000 to 2010, providing some scientific basis for the conservation and sustainable development of the watershed ecosystem.

The results of the terrestrial ecosystem assessment showed that the terrestrial ecological health composite index of the Yellow River dropped from 51 to 48 points, and the health condition was moderate. In 2010, the health condition of 8 sub-watersheds showed the worst, which were sub-watersheds of the north and south bank from Shizuishan to Hekouzhen, the Fen River, sub-watershed from the Xiaolangdi reservoir to the main stream of the Huayuankou, the Jindi River, the natural Wenyanqu Basin, the main downstream of the Huayuankou and the Dawen River. Human activities such as deforestation and overgrazing, expansion of urbanization and industrialization were the main factors that threatened the terrestrial ecological health of the Yellow River.

Acknowledgements. This research is financially supported by National Natural Science Foundation of China (41171318), Tianjin Natural Science Foundation (13JCQNJC08600), National Key Technology Support Program (2012BAH32B03) and Remote Sensing Investigation and Assessment Project for Decade-change of National Ecological Environment (2000-2010).

References

1. Chen, X., Zhou, C.H.: Ecological security: Review of domestic and foreign. Progress in Geography 24(68-20) (2005)
2. Long, D., Zhang, S.C., Fan, C.Y.: The research of watershed ecosystem health assessment. Resources Science 28(4), 38–43 (2006)
3. Rapport, D.J.: Evolution of indicators of ecosystem health. In: Daniel, H. (ed.) Ecological Indicators, pp. 121–134. Elsevier Science Publishers Ltd., Barking (1992)
4. Ling, C.H., Cui, W., Pang, A.P., et al.: The progress of watershed ecological health assessment theory and methods. Progress in Geography 72(1), 9–17 (2008)
5. Hou, Q.L., Li, X.Q.: Discuss on healthy life of river. Yellow River Water Conservancy Press (2007)
6. Cai, Q.H.: Implementation opinions on maintain healthy Changjiang and promote harmony between human and water. Yangtze River (03), 1–3 (2005)
7. Wang, L.X.: Strengthen the fundamental role of hydrology, service watershed ecological protection. Hebei Water Resources (4), 22–41 (2005)
8. Ma, T.M.: Discussion on index system of health assessment in Liao River basin. Water Resources & Hydropower of Northeast China 26(293), 1–3+71 (2008)
9. Zhang, Z.: An assessment of watershed ecosystem health in Dongting Lake based on RS and GIS. Hebei Normal University (2010)
10. Chen, C., Wang, W.J., Wang, W., et al.: Assessment of ecosystem health in Jiulong river basin and measures for management. Journal of Hunan University of Science & Technology(Natural Science Edition) 28(3), 121–128 (2013)
11. Shangguan, X.M.: Research on the wetland ecosystem health assessment of the Yellow River Basin Delta. Shandong Normal University (2013)

12. Fu, Y.X., Zhao, J., Li, W.: Ecological Sensitivity Assessment of Maqu Wetland in Upper Yellow River. Yellow River 36(1), 65–66, 67 (2014)
13. Wu, B.F., Luo, Z.M.: Ecosystem health assessment of Daning river basin in the three gorges reservoir based on remote sensing. Resources and Environment in the Yangtze Basin 16(1), 102–106 (2007)
14. Zheng, M.H., Li, Z., Liu, L., et al.: Discussion on the Yellow River ecological protection measures. Water Resources Development Research (2012)
15. Zhang, S.K., Huang, J.H., Yang, C.Y., et al.: Survey and Analysis of Pollution Sources in Yellow River Basin. Yellow River 33(12), 45–47 (2011)

The Research of China's Urban Smart Environmental Protection Management Mode

Yan Zhang[1], Na Li[2], and Yun Zhang[3,*]

[1] College of Economic and Management, Wuhan University, No.129 Luoyu Road, Wuhan430072, Hubei

[2] College of Earth Science, China University of Geosciences, No.388 Lumo Road, Wuhan430074, Hubei

[3] International School of Software, Wuhan University, Luoyu Road, Wuhan 430079, Hubei

zy_yanzi000@126.com, yunzhang@whu.edu.cn

Abstract. The application of the Internet of Things (IOT) and next-generation information technology makes the digital environmental protection become smart environmental construction, and therefore makes our government's environmental management mode being changed. By studying some smart city application cases and literatures, this paper not only proposed the government management's operation mode and organization structure for the smart environmental protection, but also discussed an effective method to improve the working efficiency.

Keywords: Smart environmental protection, management mode, organization structure.

1 Introduction

The global urbanization develops fast in recent years, particularly in developing countries. Rapid growth and the continuous expansion of the city's population has brought a lot of problems, such as traffic congestion, environmental degradation, lack of basic resources, job stress, energy supply tension, increased crime, etc.. To solve these problems, it requires innovative urban operations and management mode, achieving scientific and rational planning, intelligent building and efficient management. In this context, the smart city construction become to a hot spot in the global field. The definition of smart city at present academia is uncertain. Professor Li Deren, Wuhan University, gave its definition: Digital City + Internet of Things (IOT) = Smart City. Therefore, I believe the city is to use the IOT, cloud computing, mobile networks, GIS and other next-generation computer technology to run the city scientifically and systemically, to achieve sustainable development of urban economy, society and the environment, and to improve the quality of urbanization, at the development pattern of smart construction, intelligence service, intelligence decision making and processing.

* Corresponding author.

F. Bian and Y. Xie (Eds.): GRMSE 2014, CCIS 482, pp. 415–423, 2015.

In the application field of smart city, the application of urban ecological environmental protection has caught more and more attention. China's urban pollution has been more serious, the city's development has brought a lot of waste water, gas and garbage which are difficult to deal with. Our government has realized the needs to intensify environmental protection, and adopt a series of policies and measures to control pollution and improve urban environmental quality. National Bureau of Statistics data show that environmental pollution control investments of gross domestic product (GDP) ratio, from 0.51 percent in the early 1980s, had increased to 1.59 percent in 2012[3]. As the emphasis on environmental protection continues to improve, in 2010 the Ministry of Environmental Protection, the Ministry of Industry and Information Technology jointly announced, in five years, the country will basically built up "smart green" system, establish environment information management system which fits the environmental protection work in a new era, form a reasonably and smoothly work mechanism, and make the environmental information network system covering the whole country. Smart environmental protection is to implement daily monitoring for the air pollution, water pollution, and solid waste by sensors and monitors. And through the cloud networking to integrate environmental protection IOT, conduct data collection, information processing, information sharing, and decision support, to achieve environmental management and decision-making dynamically, and ensure that people live in a healthy environment. The construction of smart environmental protection system helps the environmental protection departments monitor environment and pollution efficiently, helps to improve the efficiency of environmental management and scientific decision-making levels, helps to security the ecological environment, create a better living environment for people, improve people's lives quality, and build a harmonious social environment.

2 Development Situation at Home and Abroad

2.1 Development Situation Abroad

At the end of 2008, IBM proposed the "Smart Planet" concept for its own industrial transformation and commercial purposes of software and services market. With the rapid development of information technology in recent years, countries in the world generally accepted the "Smart Earth" ideas, and digital and intelligent is recognized to be the future trend of social development. United States took the lead out of the National Information Infrastructure (NII) and the Global Information Infrastructure (GII) plan, and in 2009 built America's first "smart city" in Iowa City, Dubuque. Connected all the resources of the city together for analysis and integration of all kinds of data by IBM's new technology, and respond intelligently, to serve the citizen' needs. Then the EU began promoting the "Information Society" program, set the information and communication technology as a European strategic development priorities, developed the "Europe 2020 strategy", and proposed three key tasks: smart growth, sustainable growth and inclusive growth. Japan developed the "i-Japan 2015 strategy" in July, 2009, aimed at social integration of digital information technology and promoting e-government reforms. In June 2006, Singapore launched the "smart country 2015" plan, intended to achieve national digitization, and make IT industry become a new economic growth point of Singapore.

2.2 Development Situation at Home

Our government attaches great importance to the development of smart city and IOT. In China, the smart city construction is government-led, research institutes involved in the construction business. When Premier Wen Jiabao inspected the CAS Wuxi R & D of wireless sensor network engineering center at August 7, 2009, he clearly required to establish Sensing Information Center as soon as possible. Science and Technology Minister Wan Gang's speech at the Shanghai World Expo, "Let the future development of science and technology to lead the city," pointed out to strengthen the application and popularization of information, intelligence and other technologies, so that "run the city with perception and adaptive capacity.

Under the encouragement of national policy, some provinces and cities have already put smart city on key research topic. Shanghai, Chongqing, Nanjing and other cities is based on the construction of information infrastructure to drive the construction of smart city. Wuxi, Hangzhou and other cities set the construction of IOT as the sally pot for the construction of smart city. Beijing, Wuhan, Suzhou and other cities are building demonstration projects to promote smart city construction.

Smart environmental protection is an important application area of "smart city", but the theory and practice about smart environmental protection at home and abroad are few. At present, China only has several limited smart environmental protection pilot, such as Weihai, Xiangtan, North Wharf District, Ningbo and other places, this is because our information infrastructure is imperfect, R & D level is weak, the sensor is difficult to promote the high cost of production. "The Second China Environment Information Technology Forum" was held in Nanning, Guangxi in June 2012, with "serve environmental protection, smart ahead " as the theme, further explore the new ideas of Chinese environmental protection industry around the e-wise and technology application, and promote the transformation of "digital environmental protection" to the "smart environmental protection". Xu Min , etc. discussed the concept of wisdom environmental protection in the " from digital environmental protection to smart", explained the IOT technology of environmental protection is the key to implement environmental protection transformation from digital to smart, and stated it should strengthen the construction of perception layer and smart layer; Professor Liu Rui, in Beijing Normal University, discussed the main task of China's environmental protection system construction content and suggestions for environmental construction in "the discussion of our smart environmental protection system".

2.3 The Overall Framework of Smart Environmental Protection

The construction of "smart green" mainly consists of the following application systems: on-line monitoring system of pollution sources, dust and noise, environmental information release systems, environmental data center, environmental emergency management systems, environmental comprehensive business processing systems, environmental geographic information systems[2]. The overall framework for the environmental smart is of four layers, they are the data collection layer, data communication layer, data management layer and data application layer from the bottom up. Data collection layer is to achieve real-time supervision of environmental factors such as environmental quality, pollution, ecology, and other radiation mainly through sensors monitoring equipment. Data communication layer is to use real-time monitoring data by the Internet technology, wired and wireless communications technology to realize interaction of environmental information and sharing. Data management layer means to take advantage of cloud computing, storage technology, to integrate and analyze vast amounts of cross-regional, cross-industry environmental information. Data application layer makes use of environmental information obtained at data management application layer, to establish object-oriented business application systems and information service portal, providing decision support for environmental quality, pollution control, ecological protection and other services[4]. The overall framework is shown in Figure 1.

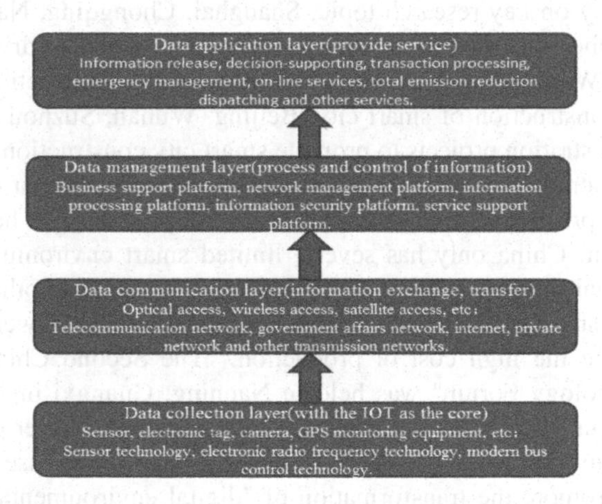

Fig. 1. Overall framework of smart environmental protection. From the bottom up are: data collection layer, data communication layer, data management layer and data application layer.

3 Smart Environmental Management Modes

Smart environmental protection requires not only strong environmental information technology as a support, but also environmental management pattern, organizational structure and business process innovation. This section focus on smart environmental management, organizational structure and mode of operation adapted to environmental concept.

3.1 Comparison with Conventional Smart Environmental Management

Government environmental management business can be summarized as the following three parts: environmental planning, integrated environmental business process, and environmental emergencies management. The main contents of this three-part business are as follows: Environmental Planning is mainly responsible for zoning, planning formulation, and foundation ability construction in terms of environmental protection, preparing comprehensive environmental function zoning and environmental protection planning, and reviewing specific environmental function division and environmental protection planning. Integrated environmental business process mainly includes the establishment of a basic system of laws, administrative regulations and economic policy in terms of environmental protection, environmental monitoring and information dissemination, and environmental pollution prevention. Environmental emergency management is managing against sudden environmental pollution problems. In traditional government environmental management, environmental information acquisition based primarily on intermittent detection and historical experience judgment, therefore the traditional means of environmental protection is primarily policy formulation and environmental governance. We can say that the traditional government environmental management is post-traceability and passive.

From the overall framework of smart environmental protection we can see, the smart environmental protection gave a new embedded to government environmental management. Smart environmental planning needed to support the planning and construction of the city. Through the collection of comprehensive information on real-time monitoring data, socio-economic, geographic, demographic, cultural, environmental, etc, and systematic structural analysis of massive data, help city managers and business operators understand the tendency of the environment, provide decision support for safety and risk prevention and environmental planning. At the same time, the integration of real-time monitoring of process data can be used to formulate the daily supervision and pollution emergency measures. And smart environment business process integrate different business information system resources through a real-time monitoring of sewage treatment, waste gas emissions, radioactive sources, water environment, atmospheric environment and acoustic environment, to implement administrative office, resource management, business process information into an organic whole, to achieve Daily supervision, process control and resource sharing of environmental quality monitoring information management, construction project approval, reporting and routine supervision of sewage charges, ecological environment management, environmental pollution and other census operations. Meanwhile through the powerful function of GIS, smart environmental integrated business process could manage, display dynamically in a 3-d way and predict its model for the combined results of monitoring network environment information and spatial information, in order to discovery in advance future environmental possible degradation problems and their causes. And help managers master the environment and its associated ecosystems potential threats and temporal trends, then support management decision-making for mitigation or prevention measures. Smart environmental emergencies management can receive timely report when environmental emergencies occur, which can buy time for timely processing. Simultaneously combined with surrounded space information, it can systematically understand the whole process and the possible impact of the incident and improve the timeliness of event processing and scientificity of decision-making for environment manager according to the simulation results provided by scientific analysis. Minimize the losses ultimately[5]. From the supervisor mode given by smart environmental protection, smart environmental protection management is based on cooperative management of the environmental monitoring data and multi-departments. Through forecasting, optimization, and simulation technology, it realizes initiative smart management. So to say it is a management mode emphasis on prospective.

In conclusion, the advantages of "smart green" application mainly in the following three points: First, it provides a comprehensive and accurate scientific basis for leadership decision-making. Types of "smart green" applications facilitate the inquiries for leadership and staff, and provide convenient services for leadership to keep abreast of the overall situation, the implementation of macro-management and scientific planning decisions. Second, it improves the work efficiency[1]. Application of a new generation of information technology and transmission technology enables staff to quickly take advantage of the LAN business processes, greatly improving the efficiency of environmental management, as well as quickly find sources of pollution for sudden environmental damage events, and make timely countermeasures, reducing emergency response time. Third, it improves the level of services. Through letting people understand the environmental situation and policies, sewage charges and other information timely online, and making it convenient for enterprises and the masses to complete the construction project approval, online reporting, and other services efficiently by the network, it has close the distance to the masses and improved service levels.

3.2 Smart Environmental Operation Mode and Organization Structure

The core of smart environment is not only the construction of environmental information systems, but also how to use it after completion, namely the formulation of operation mode. Traditional government management operation mode is a kind of "minding " approach, and the regulation content of the smart environment is not just the traditional environmental protection business, but also covers water-related affairs, population information, geographic information and other problems related to environment under the jurisdiction of other government departments. It is a cross-organizational operation mode. Our urban smart environmental operating mode is shown in Figure 2. The city should establish an environmental protection supervision center directly led by the municipal government. Urban environmental management supervision and command center can use the data acquired from the deployment of on-line monitoring system, environmental emergency response system, environmental quality management systems and other software, and from the city's resources cloud sharing platform, to conduct intelligent analysis, forecasting and early warning, analog simulation and other applications. Finally serve for environmental protection bureau and the city manager of environmental decision making, formulation of environmental policies and regulations, construction project management, environmental emergency management and so on. At the same time urban environmental management supervision and command center collect the masses of environmental problems from some channels of the website complaints, environmental protection hotline, and media reports, then distribute them to subordinate district of environmental management supervision and command centers. And district center then send law enforcement team to solve the hot environment problem timely.

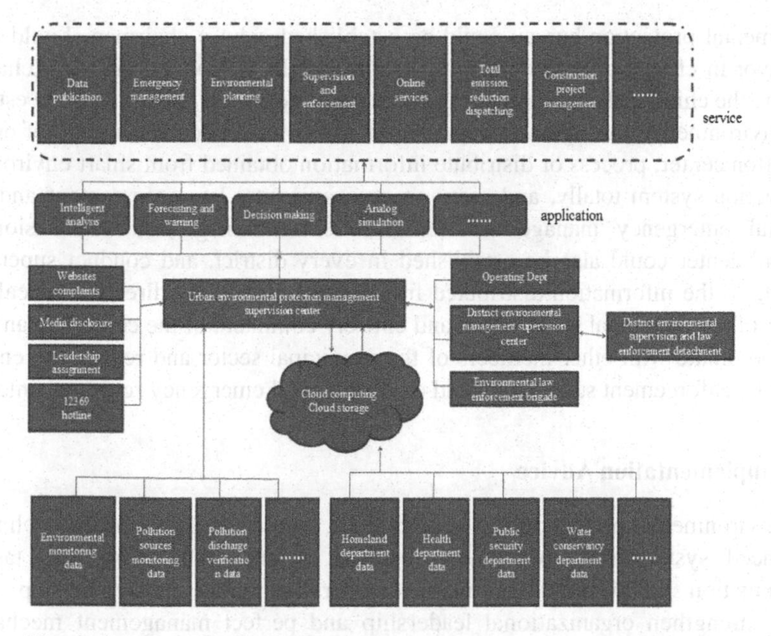

Fig. 2. Operation mode of smart environmental protection in Chinese urban

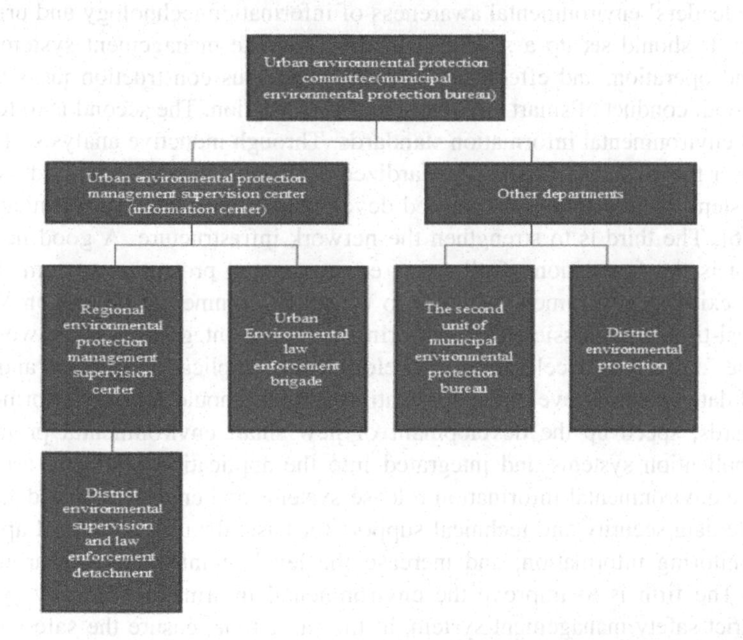

Fig. 3. Organization structure of smart environmental protection in Chinese urban

Therefore, the smart environmental protection management mode is the "two centers, overall command, hierarchical disposition" mode, shown in Figure 3. An office managed directly by the city environmental protection committee and the municipal environmental protection bureau could be established, whose chairman should be the vice mayor in charge of environment protection of the committee and vice chairman should be the chief of the municipal environmental protection bureau. Its next establish urban environmental management supervision and command center instead of original information center, process or distribute information obtained from smart environmental protection system totally, and carry on environmental law enforcement and environmental emergency management. Environmental management supervision and command center could also be established in every district, and conduct supervision according to the information distributed from urban central. By directly vertical management of the municipal government and entirely command of the center, it can effectively coordinate with other members of the municipal sector and reduce the environmental law enforcement supervision and environmental emergency response time.

3.3 Implementation Advice

Smart environmental protection is a huge project, whose covers are broad, technology is advanced, system is complex and investment is large. It is a long-time task. Its implementation should adopt the principle of overall planning and step by step.

First, strengthen organizational leadership and perfect management mechanism. Any project's carrying out can't do without the leadership's attention and needs enhancing the leaders' environmental awareness of information technology and urgency at all levels. It should set up a special team to formulate management system construction and operation, and effectively implement various construction tasks to ensure the smooth conduct of smart environmental construction. The second is to formulate unified environmental information standards. Through inductive analysis of business data over the years, establish a standardized environmental data standard. All the business systems should follow the unified development standards and technical specifications[6]. The third is to strengthen the network infrastructure. A good network environment is the foundation of all smart environmental protection system. It can modify the existing government network to create environmental protection WAN, ensuring real-time transmission of monitoring data and integration of network resources. The fourth is to accelerate the development of application systems and construction of data centers. Developing application systems should follow uniform technical standards, speed up the development of new smart environmental protection business application systems and integrated into the application platform, establish and improve environmental information release systems and environmental data centers, provide data security and technical support for basic data for advanced application of monitoring information, and increase the level of information sharing and utilization. The fifth is to improve the environmental information security system, establish strict safety management system, at the same time, ensure the safety of the system under the network environment with the application of data backup, duplicate supply, identity authentication and data encryption technology.

4 Conclusions

The new generation of information technology and IOT technologies impels the changes from digital environmental to a new smart environmental protection. This paper introduced the current development of domestic and foreign environmental protection, discussed the overall framework of smart environmental protection, analyzed the different environmental management between the content of traditional and smart environmental protections. We focused on the application of smart environmental management processes and organizational structure. Because fewer environmental studies at home and abroad are available, this paper has great significance in establishing an effective government management processes and innovative organizational structures.

Acknowledgments. This study was supported by grants from National Hubei Provincial Department of Education (No. B2013032), Fundamental Research Funds for the Central Universities (No.216274012).

References

1. Nam, T., Pardo, T.A.: Smart City as Urban Innovation: Focusing on Management Pohcy and Context. In: 5th International Conference on Theory and Practice of Electronic Governance. ACM, New York (2011)
2. Maiying, Y.U.E.: Smart City Practice Sharing Series. Electronic Industry Press, Beijing (2012)
3. Li, X., Deng, X.: Smart City Life in the Future. Posts and Telecom Press, Beijing (2012)
4. SOA Technical Committee of National Information Technology Standardization Technical Committee: Smart City Practice Guidelines. Science Press, Electronic Industry Press, Beijing (2013)
5. Wang, X., Cao, G.: Environmental Quality Assessment Based on Multivariate Statistics and GIS. Science Press, Beijing (2013)
6. Liu, R., Zhan, Z., Xie, T.: Discuss the System Construction of Our Smart Environmental Protection. J. Environmental Protection and Circular Economy 10 (2012)
7. Cheng, D.: Introduction of Top-level Design of Smart City. Science Press, Beijing (2012)

Spatial Autocorrelation Analysis of Regional Differences of Patent as Collateral Distribution in China

Yuanyuan Hu[1], Xin Gu[1,2], and Tao Wang[1,2]

[1] Business school of Sichuan University, Chengdu, 610064, P.R. China
[2] Innovation and Entrepreneurship Research Institute of Sichuan University. Chengdu, 610064, P.R. China
huyuanyuan0616@163.com, {gx6664,springer_wt}@sina.com

Abstract. The purpose of this paper is to explore the distribution of patent as collateral in the different region in China from 2008 to 2012, and analyze the spatial correlation of patent as collateral with GDP and enterprise patent application. Using the spatial autocorrelation analysis method, we analyze the Global and Local spatial autocorrelation of patent as collateral. Taking the GDP per capita and the number of enterprise patent application as indexes of measure, discusses the spatial autocorrelation of the patent as collateral. Some conclusions are drawn as follow: there is a great increase in the number of patent as collateral from 2008 to 2012, and the upward trend is more significant in east areas compared with middle and west areas in China; The Global and Local Moran's I of patent as collateral reveals that there is evolution from a discretization pattern to an agglomeration pattern, and the agglomeration area of patent as collateral increasingly concentrated in eastern China; the influences of regional economic development and enterprise patent application are not spatial autocorrelation with patent as collateral before 2012.

Keywords: spatial autocorrelation, patent as collateral, differences distribution, Moran's I.

1 Introduction

Patent right is not only a kind of invention protecting by laws and regulations, but also can get profits through some ways, for example pledging, transferring, selling and licensing (Merges, et al,2003; Grimpe and Fier,2010;Tour and Glachant,2011; Huang, et al,2011) [5,8,10,15]. And patent as collateral is an important way to the patent right transformation. Nowadays, this intangible assets have become an essential assets for small and medium-sized enterprises (SMEs). In the process of China's economic development, SMEs have played a significant role. However, the problem of financing has become the primary bottleneck restricting the development of SMEs (Serrasqueiro and Nunes, 2012; Rosenbusch, et al, 2011; Beck, 2006) [4, 18, 19]. Patent as collateral can provide a new financing channel for these enterprises. This kind of intangible asset

F. Bian and Y. Xie (Eds.): GRMSE 2014, CCIS 482, pp. 424–436, 2015.

can help resource-restricted ventures to access debt financing (Kaul, 2012; Hall and Jaffe, 2005; Jensen and Showalter, 2004) [11, 13, 14].Thus, this is a realistic way to alleviate the financial pressure for SMEs and optimize the bank guarantee structure.

Several studies have been performed on patent collateral.Amable, Chatelain and Ralf (2010) [1] studied how the assignment of patents as collateral determines the saving of firms and magnifies the effect of innovative rents on investment in research and development (R&D). The research found that high growth rates of innovations may be achieved despite financial constraints. And patent as collateral can maximize the growth rate of innovations. Meanwhile, in the process of patent pledging, banks play a very important role. According to the patents as collateral influences, the bank leaders need to determine its features and make patents more acceptable as collateral (Dang and Motohashi, 2012) [6]. Furthermore, most importantly, patents are an important form of collateral supporting the financing. If firms can more credibly pledge their patents as collateral, the patenting companies raise more debt financing, Thus, the R&D investment and patenting output also will increase (Mann,2013)[16]. However, what patents are invoked as collateral. An empirical analysis, using patent reassignment data, shows that technology-related characteristic is an important factor for patent as collateral, and lenders usually use high-quality technology patents to collateralize (Fischer and Ringler,2014)[7]. In recent years, a number of scholars' research on patent as collateral pay more attention on these aspects in China, as follows [12, 20]: relevant legal system, government policy, the model of financing loan and the current situation. However, little attention has been focused on the patent as collateral regional distribution and the interactions between patent as collateral and regional economic development and science and technology innovation.

This paper will use spatial autocorrelation analysis to explore the distribution of patent collateral from 2008 to 2012, and analyze the inherent relationship between patent as collateral distribution and the regional economic development, and the capability of regional science and technology. This study can help us to understand the patent as collateral distribution and the trend of patent collateral development from 2008 to 2012 in China, as well as know which factor will impact on patent collateral development. What is more, it provides a beneficial reference for making relevant patent collateral policies in some provinces of China.

2 Materials and Methods

2.1 Data Sources and Processing

In order to analyze the patent as collateral. regional distribution differences, this study involves the patent as collateral number(PP), GDP per capita and the enterprise patent application number (EPA)data in China ,which are respectively obtained from State Intellectual Property Office of China and China Statistical Yearbook (2008-2012). In fact, enterprise patent as collateral implementation is based on the development of regional innovation environment. GDP per capita can be used to measure regional economic differences, and the enterprise patent application number reflects the technological level of the regional enterprises. In addition, not all provinces have the enterprise patent as collateral from 2008 to 2012, as well as data of Hong Kong, Macao and Taiwan was not available either.

2.2 Analytic Methods

A four-step analysis process is followed: firstly, we will introduce the regional difference of the patent as collateral distribution in China; secondly, our analysis focuses on global spatial autocorrelation and local spatial autocorrelation of regional paten as collateral from 2008 to 2012. Following that, we will respectively analysis patent as collateral and GDP, patent as collateral and enterprise patent application with local spatial autocorrelation.

Global Spatial Autocorrelation. Spatial autocorrelation is an assessment of the correlation of a variable with reference to its spatial location and it deals with the attributes and the locations of the spatial features. Moran's I is a popular test statistic for spatial autocorrelation. It is a global test statistic for spatial autocorrelation, which is based on cross-products for measuring an attribute association. It is calculated for $n$observation on a variable x at location of i and$j$, and follows (Equation (1)) [9]:

$$I = \frac{n\sum_{i=1}^{n}\sum_{j=1}^{n} W_{ij}(x_i - x)(x_j - x)}{\sum_{i=1}^{n}\sum_{j=1}^{n} W_{ij}\sum_{i=1}^{n}(x_i - \bar{x})^2} \quad (1)$$

Where n is the number of observations of the whole region, x_i and $x_j$are the observations at locations of i and j; $\bar{x}$ is the average over all spatial units of the variable. w_{ij} is the spatial weight matrix that measures the strength of the relationship between two spatial units. The value of Moran's I will vary (-1, 1). A higher positive Moran's I implies that values in neighboring positions tend to cluster, while a low negative Moran's I indicates that high and low values are interspersed. When Moran's I is near zero, there is no spatial autocorrelation, meaning that the data are randomly distributed [17, 21].

Local Spatial Autocorrelation. Local Moran's I is a local test statistic for spatial autocorrelation, and identifies the autocorrelation between a single location and its neighbors. It is computed as follows (Equation (2)):

$$I_i = \frac{n(x_i - \bar{x})\sum_{j=1}^{n} W_{ij}(x_i - \bar{x})}{\sum_{i=1}^{n}(x_i - \bar{x})^2} \quad (2)$$

The notations in Equation (2) are as described for Equation (1), but the corresponding values are from the local neighboring region.

Before calculating Moran's I, standardization processing shall be carried out to observe data. To analyze and observe the spatial pattern of variations, test ofsignificance is required to ensure the correctness of inferred conclusion based on a certain probability. Test of significance adopts a Z test (Equation (3)):

$$Z(I) = \frac{I - E(I)}{\sqrt{VAR(I)}} \tag{3}$$

Where $Z(I)$ represents the significance level of Moran's I, $E(I)$ is the mathematical expectation of Moran's I, and $VAR(I)$ is variation.

Global spatial autocorrelation enables us to judge the existence of spatial discretization or agglomeration phenomena, but fails to detect the location of agglomeration or discretization as well as the relation between spatial units. Local spatial autocorrelation makes up for this deficiency [22].

It can reveal the similarity or correlation of attribution values between a spatial unit and its adjacent units, identify spatial agglomeration and spatial isolation, detect spatial heterogeneity, etc. Therefore,

The local indicator of spatial autocorrelation (LISA) was applied as an indicator of local spatial association. The LISA significance map was created incorporating information about the significance of the local spatial patterns. Especially, the map results in a spatial pattern consisting of five categories[1,2]: (1)"Low-Low" represents lower values surrounded by neighboring units with lower values, and it means positive spatial autocorrelation;(2)"High-High" shows higher values surrounded by neighboring units with higher values, and it means positive spatial autocorrelation; (3) "Low-High" indicates low values adjacent to neighboring units with higher values, and it means negative spatial autocorrelation; (4)"High-Low" represents higher values adjacent to neighboring units with lower values, and it means negative spatial autocorrelation; (5)"Not Significant" indicates that there is no spatial autocorrelation.

3 Results and Discussion

3.1 The Differences of Patent as Collateral Distribution

Since October 30, 2006, the first patent as collateral loan of China was birthed in Beijing. After that, enterprises which are situated on Guangzhou, Shanghai, Zhejiang and other coastal cities have begun to use patents as collateral to obtain loans, and then this trend quickly extended to the central and western provinces. Since 2008, State Intellectual Property Office of China began to publish the relevant data of patent as collateral. According to these data, there are several characteristics on the distribution of patent as collateral.

As is shown in the Fig 1, we can see the number of patents as collateral showed upward trends in the majority of provinces in China during this 5 years. It is noticeable that there was a rapid increase of patent as collateral in Beijing, Tianjin, Jiangsu, Shanghai, Zhenjiang and Guangzhou from 2008 to 2012. In contrast, Xinjiang, Xizang, Qinghai,

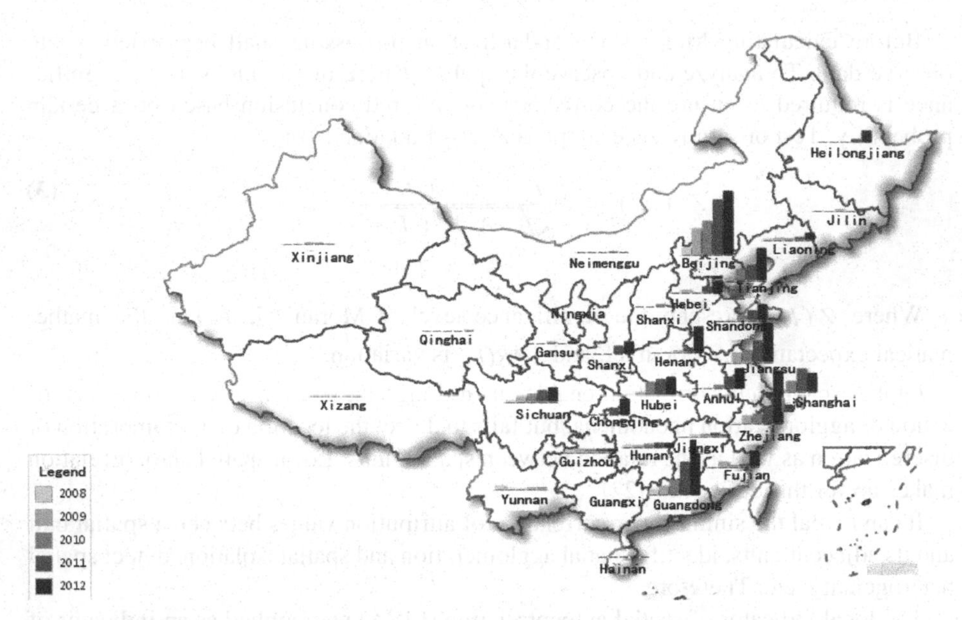

Fig. 1. The map of Paten As collateral Distribution in China from 2008-2012

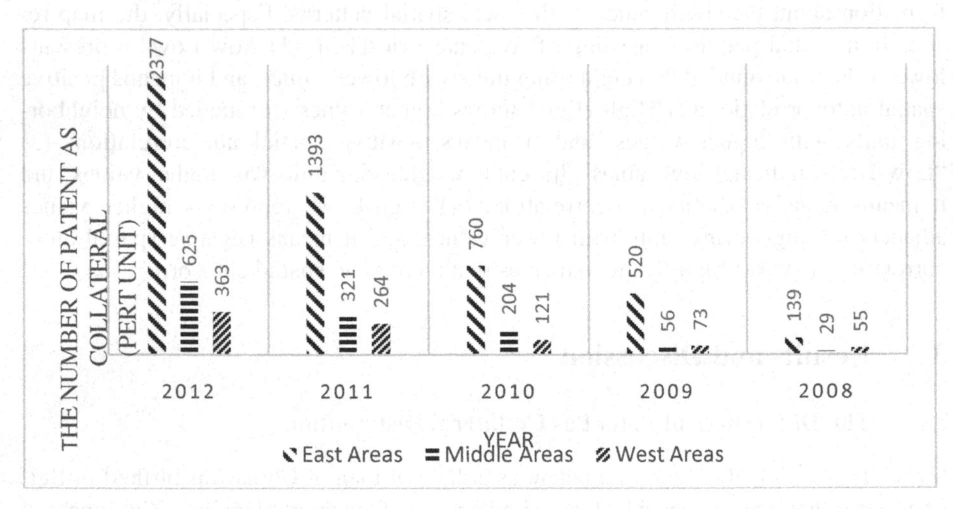

Fig. 2. Paten As collateral Distribution in Different Regional of China from 2008-2012

Shanxi and Guangxi, etc., which is located in west of China, had a gradual increase and the number of patent as collateral was far behind that of areas in eastern China.

Fig 2 is patent as collateral distribution in eastern, middle and western areas of China in the year between 2008 and 2012. We can see clearly that the number of patent as collateral in this three areas had noticeable rising trend. Especially in eastern areas, the number of patent as collateral is 2377 per unit in 2012, which is about 17 times higher than that part in 2008. Although there was a steady increase in the number of patent as collateral in middle and west areas, the rising speed was far behind the east areas. Patent as collateral number was 2377 per unit in eastern China in 2012 was respectively 3.8 and 6.55 times of counterparts in middle and west areas. Meanwhile, the difference of patent as collateral distribution number between middle areas and west areas had become wider. As is shown by the graph, the western areas number of patent as collateral was 55 and 73 in 2008 and 2009, which higher than the data of middles areas. However, the period from 2010 to 2012 saw a greatly increased in the number of paten as collateral of middle areas. Also the middle areas patent as collateral number began to outnumber the counterpart of western areas.

The above analysis can only reflect the changing trend and quantity of patent as collateral in China in the 5 years spanning from 2008 through 2012, but fail to reveal that the changes trends of patent as collateral distribution in a spatial unit.

3.2 The Analysis of Global Spatial Autocorrelation for the Patent as Collateral

Fig 3 and Table 1 is the Global Moran' I values of the number of patent as collateral distribution in China from 2008 to 2012. As we can see from these Figs, Global Moran's I of Chinese patent as collateral has an upward trend and Moran's I values arc positive. It was 0.039098 in 2008, increased to 0.311384 in 2012. And from 2010 to2012, the level of Moran' I confidence is greater than 90%. As we can see from Table 1, under the hypothesis 0.01 significance level, the Moran's I score of patent as collateral is higher than zero. It indicates that there is a positive correlation spatial characteristics, but the correlation is not obvious. What is more, the data reveals that the number of patents as collateral distribution difference is increasing in this 5 years.

3.3 The Analysis of Local Spatial Autocorrelation for the Patent as collateral

Moran's I reflects the overall situation of the patent as collateral in China spatial autocorrelation, but it masks the dynamic characteristics of the internal spatial pattern and fails to reflect the situation of a single spatial unit. Therefore, it is necessary to use the local coefficient Moran's I for a deep study of the correlation among different provinces of patent as collateral. The earlier analysis of global spatial autocorrelation shows that the trends of spatial autocorrelation of separate province are consistent and increasing during 5 years. Therefore, the study of the local spatial characteristic of the patent as collateral from 2008 to 2010 means the evolution of the entire local space. In order to reflect these changed in spatial patterns more directly, the Lisa-aggregation Figof 0.01 significance was made (Fig.4). In addition, the phoneme of patent as collateral does not happen in every province in China during this 5 years, but more and more provinces begin to have the data, which mean more and more enterprises use patent as collateral to get loans.

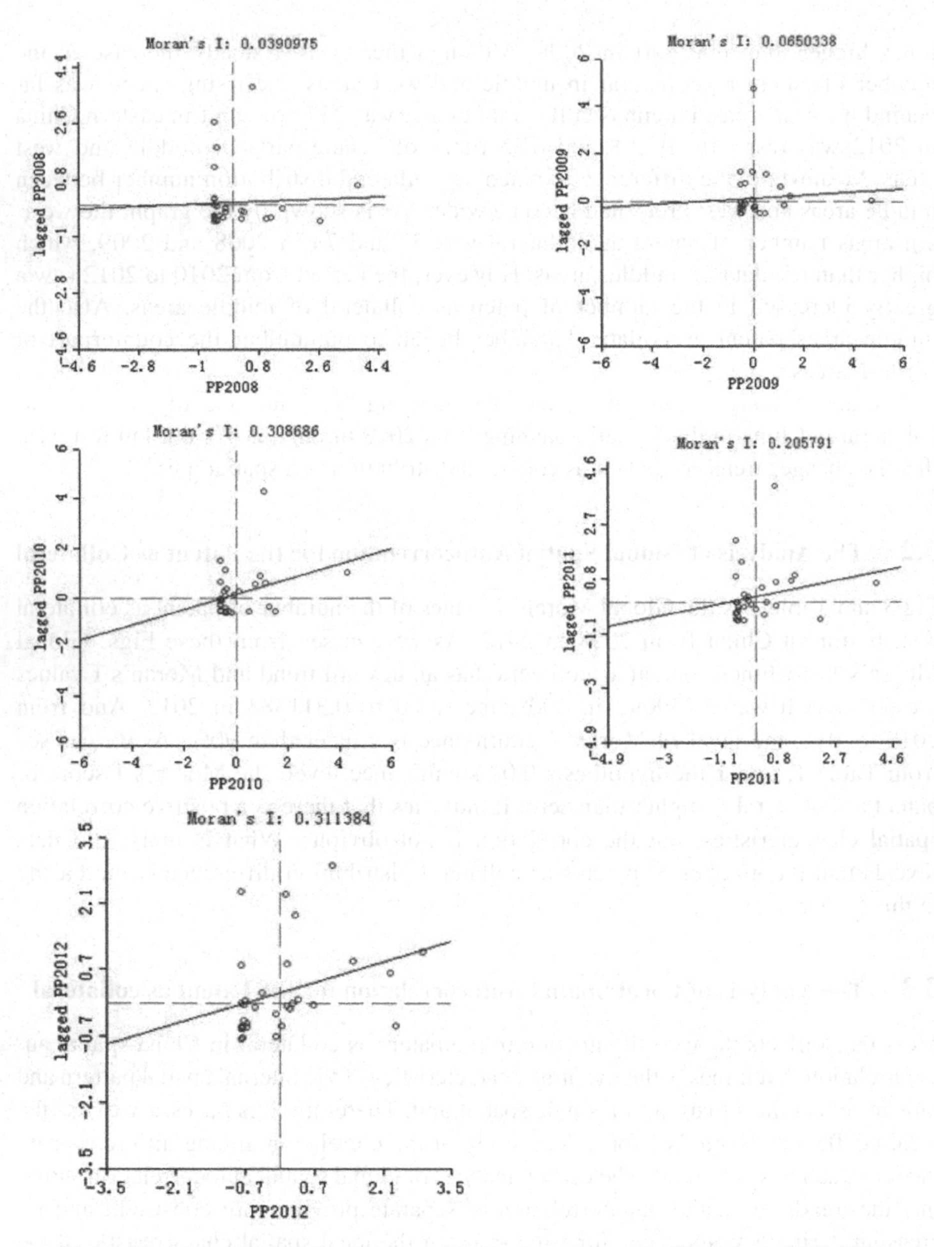

Fig. 3. The global Moran Scatter map of Paten As collateral in China from 2008-2012

Table 1. The Global Moran's I and Confidence of Patent As collateral from 2008-2012

	2008	2009	2010	2011	2012
Global Moran's I	0.039098	0.065034	0.308686	0.205791	0.311384
Confidence (%)	0.765	0.833	0.981	0.957	0.979

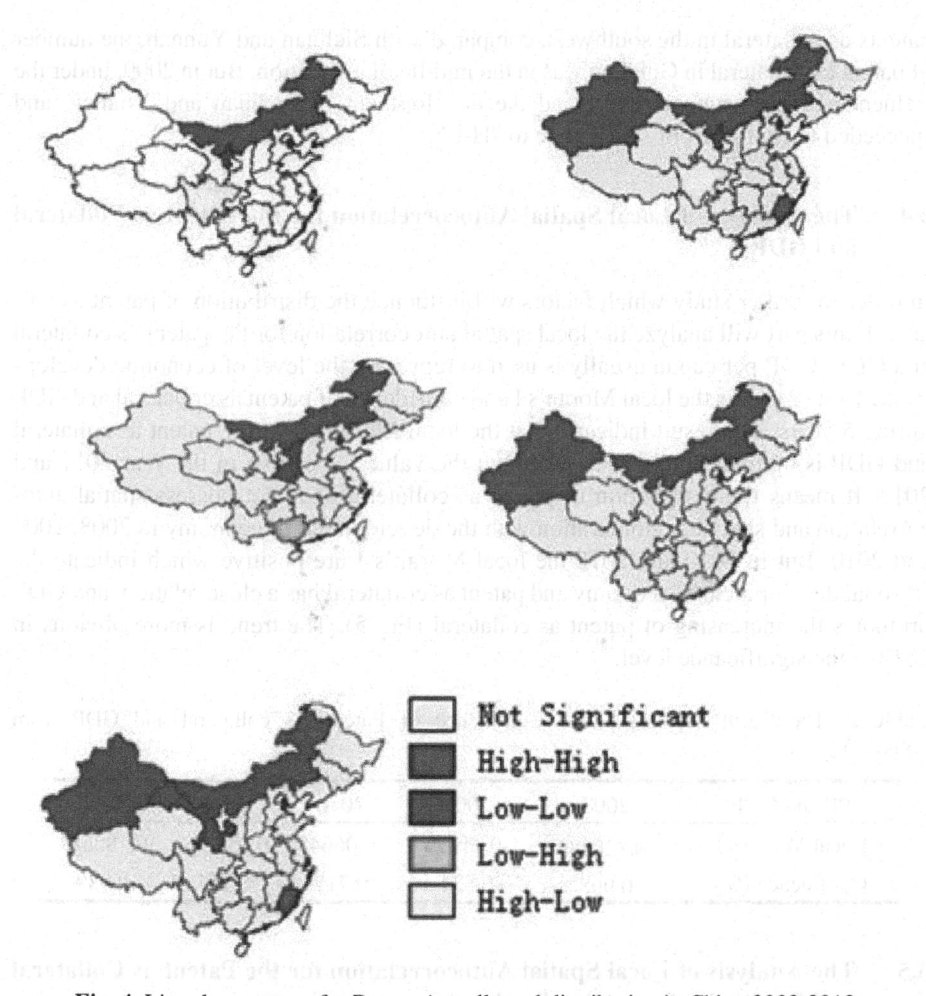

Fig. 4. Lisa cluster maps for Patent As collateral distribution in China 2008-2012

From the analysis of Fig 4, firstly, regions which form a significant "H-H" spatial association with spatial units are mainly located in eastern China during the study period, for example Beijing, Shanghai, Guangzhou . Obviously, this area is the most advanced area in China, especially for east coastal area, which obtain a large number technology enterprises, convenient transportation, better economic foundation etc. On the contrary, XinJiang, Neimenggu, Gansu are classified as "L-L" type significant from 2008 to 2012, due to weak economic foundation and a relatively backward scientific and technological level compared with eastern of China. Although Sichuan, Guizhou and Guangxi are in the western region of China, due to the local government policies and the level of economic development was relatively higher than those of other surrounding provinces. Therefore, its pattern of spatial association with the surrounding regions belongs to the significant "H-L" type. Besides, in 2008, Guizhou was classified as "L-H" type significant. Because there were only about 3 provinces having the number of patents as collateral in the southwest, compared with Sichuan and Yunnan, the number of patent as collateral in Guizhou was in the middle of a situation. But in 2009, under the influence of government policy and the development of Sichuan and Yunnan, and succeeded to change from "L-H" type to "H-L".

3.4 The Analysis of Local Spatial Autocorrelation for the Patent as Collateral and GDP

In order to further study which factors will influence the distribution of patent as collateral, this part will analyze the local spatial autocorrelation for the patent as collateral and GDP. GDP per capita usually is used to represent the level of economic development. Table 2 shows the local Moran's I and confidence of patent as collateral and GDP during 5 years. The result indicates that the local Moran's I of the patent as collateral and GDP is minus in the first 3 years, but the value is positive in the year 2011 and 2012. It means the distribution of patent as collateral does not possess spatial autocorrelation and spatial agglomeration with the development of economy in 2008, 2009 and 2010. But in 2010 and 2012 the local Moran's I are positive which indicate the regional development of economy and patent as collateral has a close relation, and GDP promotes the increasing of patent as collateral (Fig 5). The trend is more obvious in 2012 at the significance level.

Table 2. The Local Moran's I and Confidence of Patent As collateral and GDP from 2008-2012

PP and GDP	2008	2009	2010	2011	2012
Local Moran's I	-0.322289	-0.11075	-0.06648	0.054249	0.182681
Confidence (%)	0.998	0.834	0.712	0.757	0.939

3.5 The Analysis of Local Spatial Autocorrelation for the Patent as Collateral and Enterprise Patent Application

If SMEs want to acquire loans through patent as collateral, they must have the patent right. So in order to obtain the patent right, enterprises should apply for this patent firstly. Then we will talk about are there any relations between the patent as collateral and the number of enterprise patent application? According to Table3 and Fig 6, it can be seen that the local Moran's I values is positive only in 2009 and 2012. The data mean there is no significant spatial autocorrelation and spatial agglomeration between patent as collateral and the number enterprise patent application. Actually, the increasing number of enterprise patent application does not have great influence on the number of patent as collateral in region. What is more, from table 2 and table 3, we can see clearly that GDP and the enterprise patent application have significant spatial autocorrelation with patent as collateral in 2012. Therefore, with the development of social economy and the progress of science and technology, the number of patents as collateral in the region will have obviously increased.

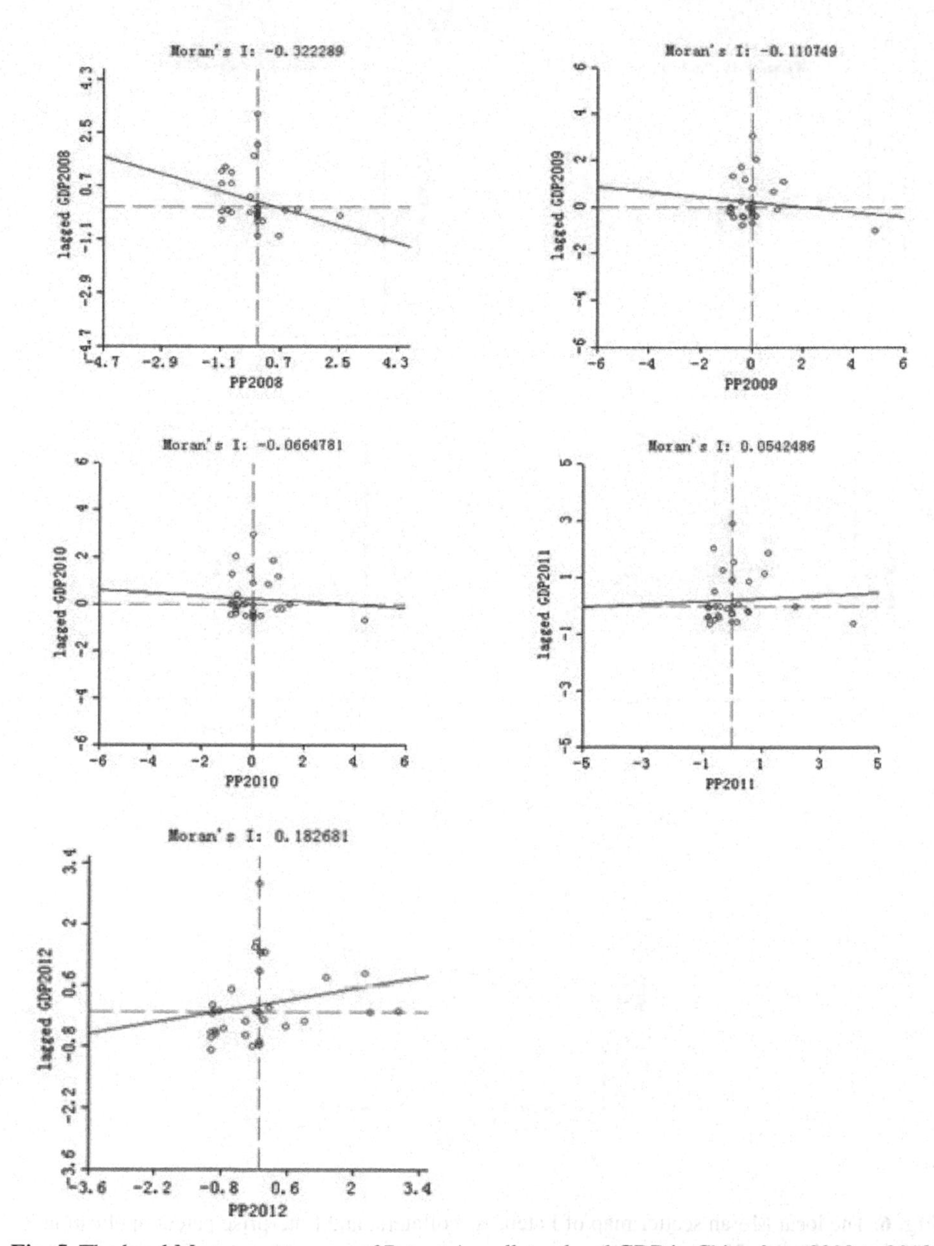

Fig. 5. The local Moran scatter map of Patent As collateral and GDP in China from 2008 to 2012

Table 3. The Local Moran's I and Confidence of Patent As collateral and Enterprise Patent Application from 2008 to 2012 in China

PP and EPA	2008	2009	2010	2011	2012
Global Moran's I	-0.178738	0.055168	-0.12547	-0.19031	0.198758
Confidence (%)	0.969	0.777	0.906	0.952	0.947

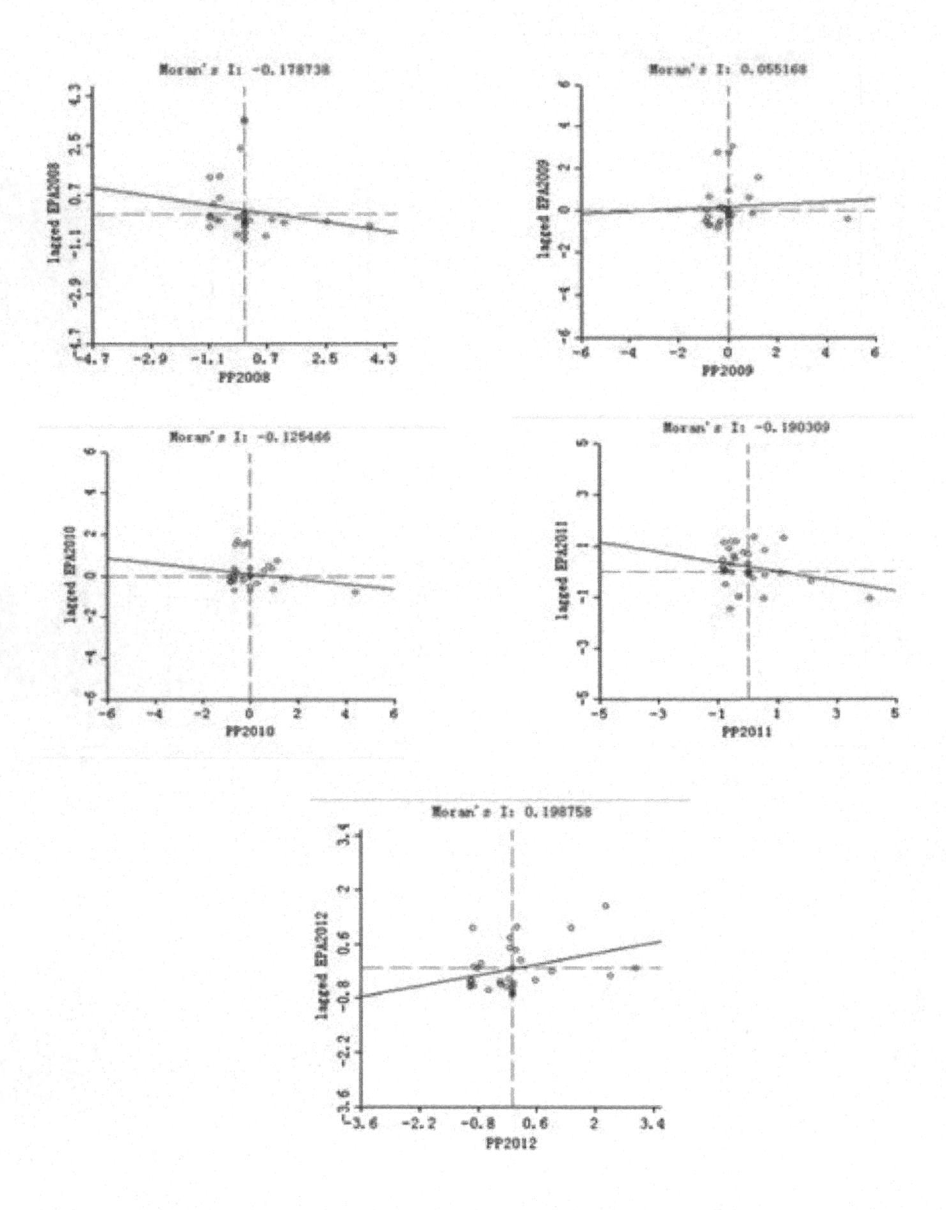

Fig. 6. The local Moran scatter map of Patent As collateral and Enterprise patent application in China from 2008 to 2012

4 Conclusion

Taking patent as collateral distribution as the research object of this paper, using the method of spatial autocorrelation analysis, we have studied the global and local spatial autocorrelation of the patent as collateral in the province of China, and further analyzed the local spatial of GDP, enterprise patent application and patent as collateral. We have drawn some conclusions as follows:

(1) Global spatial autocorrelation analysis of patent as collateral distribution in China shows strong spatial autocorrelation from 2008 to 2012, and the trend has gradually marked. The Global Moran's I is generally on the increase.

(2) Local spatial autocorrelation analysis indicates that the "H-H" areas of patent as collateral distribute mainly in eastern China; "L-L" areas have always been stabled in some provinces located in the west of China. In this 5 period, "L-H" and "H-L" areas mainly concentrate in Sichuan, Guizhou and Guangxi without specific regularity.

(3) The analysis of local spatial autocorrelation for GDP and enterprise patent application with the patent as collateral does not show obvious spatial autocorrelation.

The local Moran's I value of GDP and patent as collateral was negative number before 2011, it means at the beginning of the development of patent as collateral, regional economic development had not great influences on the number of patent as collateral, even had negative correlation. However, the situation is completely different from 2011. It reveals that enterprise and government begin to pay more attention to this new financing method of patent as collateral, and more and more enterprises use this way to obtain loans, which will promote regional economic growth. Meanwhile, economic development will accelerate stimulate enterprise loan.

The analysis of local spatial autocorrelation for patent as collateral and enterprise patent application shows a similar trend. The value of local Moran's I was negative in 2008, 2010 and 2011.

In other words, although patent as collateral bases on the amount of enterprise patent application, the most enterprise still does not use this kind of financing method to obtain loans. So the amount of enterprise patent application does not have greater impact on regional patent as collateral. But this phenomenon has some changes in improvement of the technology environment.

References

1. Amable, B., Chatelain, J.B., Ralf, K.: Patents as collaterall. Journal of Economic Dynamics and Control 34(6), 1092–1104 (2010)
2. Anselin, L.: The Moran scatterplot as an ESDA tool to assess local instability in spatial association. Spatial Analytical Perspectives on GIS 5, 111–125 (1996)
3. Anselin, L.: Local indicators of spatial association—LISA. Geographical Analysis 27(2), 93–115 (1995)
4. Beck, T., Demirguc-Kunt, A.: Small and medium-size enterprises: Access to finance as a growth constraint. Journal of Banking & Finance 30(11), 2931–2943 (2006)
5. De la Tour, A., Glachant, M., Ménière, Y.: Innovation and international technology transfer: The case of the Chinese photovoltaic industry. Energy Policy 39(2), 761–770 (2011)
6. Dang, J., Motohashi, K.: Patent value and liquidity: evidence from patent-collateralized loans in China (2012)
7. Fischer, T., Ringler, P.: What patents are used as collateral?—an empirical analysis of patent reassignment data. Journal of Business Venturing (2014)
8. Grimpe, C., Fier, H.: Informal university technology transfer: a comparison between the United States and Germany. The Journal of Technology Transfer 35(6), 637–650 (2010)

9. Goodchild, M.F.: Spatial autocorrelation. Geo Books, Norwich (1986)
10. Huang, C., Notten, A., Rasters, N.: Nanoscience and technology publications and patents: a review of social science studies and search strategies. The Journal of Technology Transfer 36(2), 145–172 (2011)
11. Hall, B.H., Jaffe, A., Trajtenberg, M.: Market value and patent citations. RAND Journal of Economics, 16–38 (2005)
12. Jir, T., Ding, Z.T.: Study on Countermeasures of Boosting the Development of Small and Medium-sized Enterprises by Patent Pledge Loan (2011)
13. Jensen, R., Showalter, D.: Strategic debt and patent races. International Journal of Industrial Organization 22(7), 887–915 (2004)
14. Kaul, A.: Technology and corporate scope: Firm and rival innovation as antecedents of corporate transactions. Strategic Management Journal 33(4), 347–367 (2012)
15. Merges, R.P., Menell, P.S., Lemley, M.A.: Intellectual Property in the New Technological Age: 2004 Case and Statutory Supplement. Aspen Publishers (2003)
16. Mann, W.: Creditor rights and innovation: Evidence from patent collateral. Available at SSRN 23(5), 6–15 (2013)
17. Overmars, K., De Koning, G., Veldkamp, P.A.: Spatial autocorrelation in multi-scale land use models. Ecological Modelling 164(2), 257–270 (2003)
18. Rosenbusch, N., Brinckmann, J., Bausch, A.: Is innovation always beneficial? A meta-analysis of the relationship between innovation and performance in SMEs. Journal of Business Venturing 26(4), 441–457 (2011)
19. Serrasqueiro, Z., Nunes, P.M.: Is age a determinant of SMEs' financing decisions? Empirical evidence using panel data models. Entrepreneurship Theory and Practice 36(4), 627–654 (2012)
20. Wang, X.-Y., Li, S.-Y.: On the Role of Government in Resolving the Risk of Patent Pledge. Tianjin Legal Science, 2–5 (2010)
21. Zhang, C., Mcgrath, D.: Geostatistical and GIS analyses on soil organic carbon concentrations in grassland of southeastern Ireland from two different periods. Geoderma 119(3), 261–275 (2004)
22. Zhao, X., Hung, X., Liu, Y.: Spatial Autocorrelation Analysis of Chinese Inter-Provincial Industrial Chemical Oxygen Demand Discharge. International Journal of Environmental Research and Public Health 9(6), 2031–2044 (2012)

The Design and Construction of WLAN-Based Indoor Navigation System

Rudong Xu[1,2], Jin Liu[1], Jiashong Zhu[2], and Xiaofan Jiang[2]

[1] State Key Laboratory for Information Engineering in Surveying, Mapping and Remote Sensing, Wuhan University, Wuhan 430079, China
[2] Shenzhen Research and Development Center of State Key Laboratory for Information Engineering in Surveying, Mapping and Remote Sensing, Shenzhen, 518057, China
xurudong@139.com, {41038331,david.jiang.gis}@qq.com, zhujiasong@163.com

Abstract. One of the fundamental technologies for smart city construction is location-aware technology. This paper discusses a number of key issues related to the implementation of indoor navigation systems. The paper proposes an innovative airport indoor navigation solution that combines Wi-Fi-based multi-mode positioning technology, flexible indoor map and simple route-planning algorithm, with an aim to improve positioning accuracy, stability and enhance user experience. It provide sgreat reference value to the indoor navigation system design and implementation of similar kind.

Keywords: smart city, indoor positioning, route-planning, indoor map.

1 Introduction

With the emergence and rapid development of smart city construction, in particular, the wide spread of Wireless Local Area Networks (WLAN) and smartphone, location-based services (LBS) applications are gradually evolving from the outdoor to the indoor. Indoor localization and navigation service is to become a very important research directions for LBS. Indoor localization and navigation service in a number of large shopping malls, airports, convention centers, hospitals and other complex indoor scenario have great applicable prospects, for example, it can provide users with localization information, planning the best path to indoor destination, pushing its around and other interesting information and so on.

The theoretical research, technical framework and application of GPS-based navigation have been well developed. Comparatively speaking, the key technologies for indoor localization: indoor localization algorithm [1-9], indoor map representation [10-14] and route planning algorithm [15-16] are still looking for significant breakthrough. In general, there do not exist a low-cost integrated indoor navigation solution that can provide a high resolution of accuracy as well as deep-going rout-planning for universal applicability.

In this paper, we propose an innovative airport indoor navigation solution that combines Wi-Fi-based multi-mode localization, flexible indoor map and simple route-planning

F. Bian and Y. Xie (Eds.): GRMSE 2014, CCIS 482, pp. 437–446, 2015.

with an aim to improve positioning accuracy, stability and enhance the user experience. The remarkableness of the solution lies in that it integrates core navigation components of localization and path-planning for airport indoor environments navigation at relatively low costs.

2 WLAN-Based Indoor Localization

Indoor localization technologies mainly include: radio frequency identification technology (RFID), Ultra-wideband (UWB) technology, wireless local area network (WLAN) technology, Bluetooth, infrared technology, etc. These positioning technologies have their own advantages and disadvantages. WLAN-Based indoor localization using RSS signal for positioning can make full use of existing wireless LAN infrastructure. It does not need to modify on existing infrastructure or add any additional hardware facilities. It uses a pure software approach to achieve positioning service for ordinary indoor environment with very low cost. WLAN-Based indoor localization algorithms are mainly classified into Wi-Fi geometric positioning with trilateral positioning method, positioning based on signal propagation model, positioning based on location fingerprint, as well as various improved algorithms. The location fingerprint positioning method has strong applicability in environment where the AP locations cannot be determined.

Wi-Fi location fingerprint algorithm(WLFA) is to characterize environment in the positioning space by abstract and formal description. The location in a positioning space is described by RSSI sequence of AP access points inside the positioning space. These RSSI sequences are gathered to form the Radio Map. Finally, the matching between the real-time measured RSSI sequence and radio map is performed to select the location with the maximum similarity in the radio map as the estimated location. This method consists of two stages: off-line training stage and online positioning stage.

Positioning relay solely on Wi-Fi is unstable and could contain large errors due to the instability of the Wi-Fi signal strength. Meanwhile, the mid-range and high-end smart phones usually contain inertial sensors, which consist of accelerometers, gyroscopes and magnetometers. The pedestrian walking steps can be derived by accelerometer data. Combined with ordinary pedestrian step length data, the pedestrian walking distance can be estimated. Further combining walking directions obtained by magnetometer, pedestrian dead reckoning (PDR) can be performed to derive pedestrian's relative displacement.

There are two key factors in PDR system: walking displacement S and course angle θ. S is obtained according to the principle of pedometer and step length estimation. Course Angle θ is the angle between the walking direction and direction of magnetic north is obtained from magnetometer or gyroscope or the combination of them.

The advantages of WLFA include ease of use, versatility, no requirement for additional positioning hardware, good long-term positioning result, and relatively accurate absolute positioning result. But large fluctuation of AP's transmit power , instability of AP signal, and other features limit the positioning accuracy, resulting in high frequency fluctuation in positioning accuracy. PDR has strong real-time and high frequency characteristic. But due to its inherent character of error accumulation, it exhibits large drift from accuracy location in long-term. The real-time combination of WLFA and PDR positioning technologies can complement each other, effectively suppressing high-frequency interference of WLFA positioning and correcting the accumulated error in PDR position. Thus stable long-term positioning result is obtained. The combined system performs well in both real-time and long-term. Wi-Fi

3 SVG-Based Indoor Map and Route Planning

SVG (Scalable Vector Graphics) is based on the XML, it is a kind of two-dimensional Vector Graphics format formulated by the W3C(World Wide Web Consortium). It is also the network vector graphics standard in network specification. SVG as vector graphic made up of text, is composed of point, line and filling. It has the advantages of enlarging and shrinking the vector graph size without losing image quality, and small file size. These characteristics make the SVG image independent of the underlying hardware and very suitable for network distribution. In addition, it provides a certain amount of semantic information for graphics object, thus convenient for graphic searching. SVG fully supports DOM, interoperating easily with JavaScript scripting language.

In geospatial hierarchical model, a map can be decomposed into several layers. Each layer is a collection of feature (Feature Set), each feature corresponds to a geographical entity that contains geometry properties and non-geometry attribute. Geometric properties include point, line, polygon and complex geometric objects. The point, line, polygon and complex geometric objects are described with < text >, < path >, < polygon > tag. The feature can be represented by a tag of SVG, layer by < g > tag description. Root element as layer node contains spatial reference information in its attribute. In summary, complex model of spatial information can be represented and rendered by SVG, which is especially applicable to the spatial visualization on mobile devices. In particular, as an open industrial standard, SVG data format can be converted directly from popular GIS packages such as ArcGIS, CAD and thus trivial to implement the data exchange and indoor/outdoor map unification under common geographical reference system.

Navigation function is driven by certain algorithms. The most commonly used path algorithm are Dijkstra algorithm, A* algorithm, Johnson algorithm, Floyd-Warshall algorithm, Bellman-Ford algorithm and SPFA algorithm. Among them, Dijkstra algorithm is one of the most typical and wider used algorithm in the shortest path routing. Dijkstra algorithm traverses all the vertices of the network from the starting point. After the execution of the algorithm, each of the results not only contain the shortest path from the starting point to the end point but also from the starting point to each of the network end point. Thus, when user changes destination, it is not required to rerun again. The shortest path to the new destination can be derived directly from the previous search results. Dijkstra algorithm used a total of two loops, therefore it has an order of O (n^2) complexity. Since the algorithm search for each of the network end node indiscriminately. It performs equal-probability-search. Therefore, the algorithm can guarantee finding the optimal solution, but with relatively low- efficiency in computation.

4 The Design and Application of Indoor Navigation System

Airport indoor navigation system (AINS) is developed on Android4.1 operating system, using Eclipse integrated development environment. The system is mainly divided into two parts: server and mobile terminal. The server is mainly responsible for the editing of indoor map data, conversion of spatial reference system, managing points of interest (POI), path-planning and map publishing service. The mobile terminal is responsible for loading and rendering of indoor map data, positioning and users' interactions. Taking into account the needs of cross-platform, indoor map framework bases on the webviewer components, using JavaScript language and the JQuery [17] open source library. Figure.1 shows the overall design architecture:

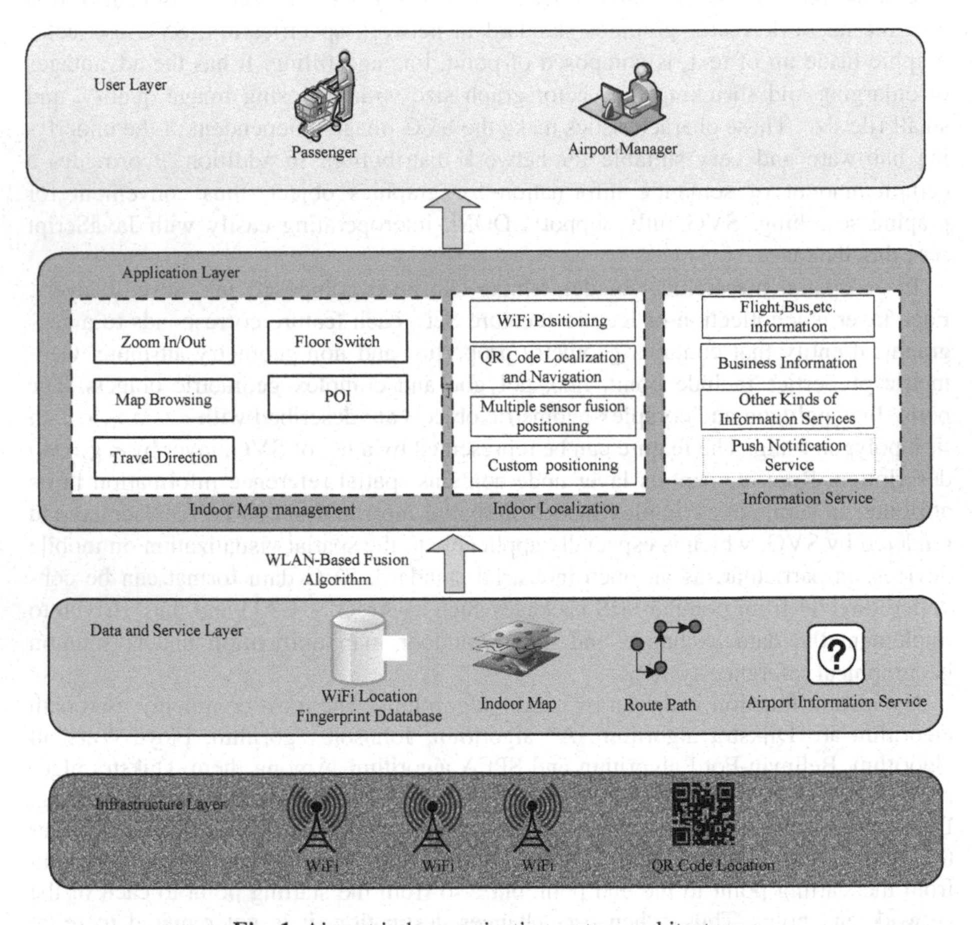

Fig. 1. Airport indoor navigation system architecture

4.1 Wi-Fi Location Fingerprint Data Collection

Wi-Fi fingerprint acquisition tool collect Wi-Fi location fingerprints, then filtered the location of the fingerprint according to the stability of each AP signal strength, keeping the stable AP information(Fig.2), and construct the airport Wi-Fi location fingerprinting database.

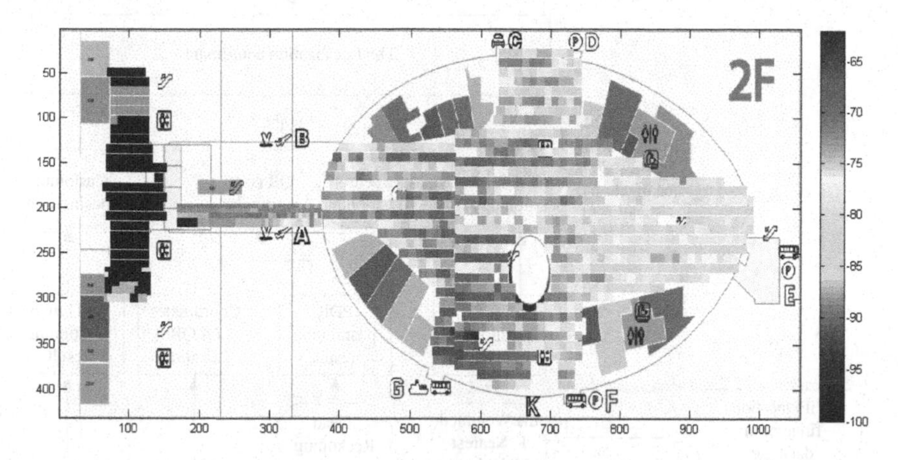

Fig. 2. The AP signal intensity distribution

4.2 Indoor Localization

The system provides four positioning modes: Wi-Fi positioning, multiple sensor fusion positioning, QR code positioning, customize positioning (Fig.3). Wi-Fi positioning: Delete the AP information of small signal strength (less than -95 dbm), and derive Wi-Fi positioning results with real-time fingerprint database matching, using k-nearest neighbors algorithm. The multiple sensor fusion positioning is using extended Kalman filter (EKF) to fuse inertial measurement data and RSSI data to improve the positioning accuracy; QR code positioning can also provide location information with few AP or weak Wi-Fi signal. that is positioned directly implemented by scanning the two-dimensional code icon; Customize positioning: when a user know he/she is around a significant landmarks (eg, KFC, information desk, etc.), he/she can input the location to the system by touch-screen operation, or when an positioning error occurs, user can correct the positioning errors.

4.3 Indoor Map

SVG map compilation: first, exclude the non- geospatial data from the existing building CAD floor plan data. Second, the data is converted into SVG format and input into the system. The SVG indoor map editor will provide all sorts of elements of the map symbol, elements classification and attribute information. Set scales, and transform the coordinate, especially, it can registration with OpenStreetMap to realize the integration of indoor and outdoor applications.

SVG interactions: SVG indoor map can be published as service, Mobile devices can access online map or perform off-line download. In the smartphone, all kinds of map management functions are provided, such as map zoom in or zoom out and map pan, map query, select and browse the attributes of POIs, set the navigation path, find location in the map, etc.

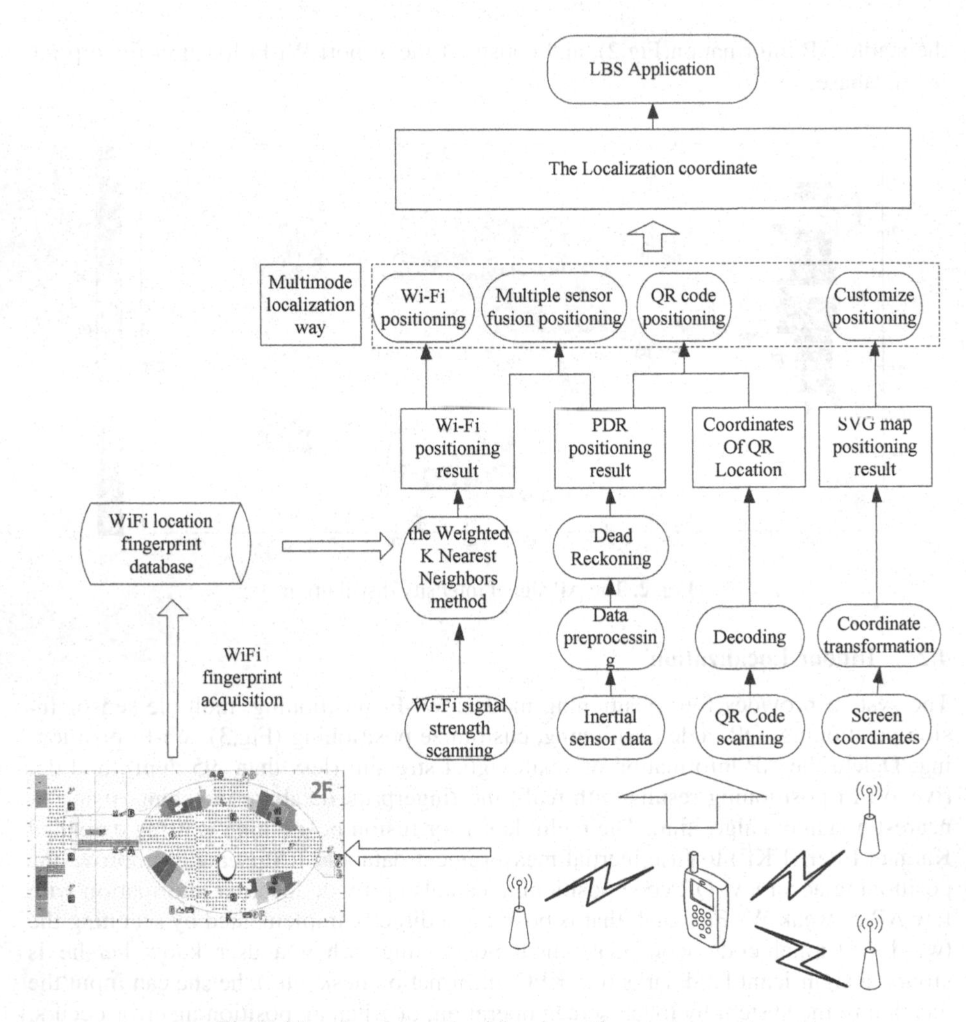

Fig. 3. The planning-path of the airport lounge to export

4.4 Route-Planning

Navigation path generation: navigation path consists of point and line, which can have attribute information. Together they form a navigation path with combination of each logically irrelevant path. Numbers are stored at points, according to the sequence of the point Each segment stores the coordinates of the beginning and end of the line segment, therefore the relationship between two adjacent segments is established. In two adjacent line segments, the end of a segment is the beginning of the next segment. Thereby all the segments are connected into a line, which is the navigation path.

Navigation implementation: obtain the current position information from the mobile devices, parse XML navigation path, real-time navigation path finding based on Dijkstra algorithm, redraw the map and refreshes the path information. The diagram (Fig.4) is drawn by the shortest path.

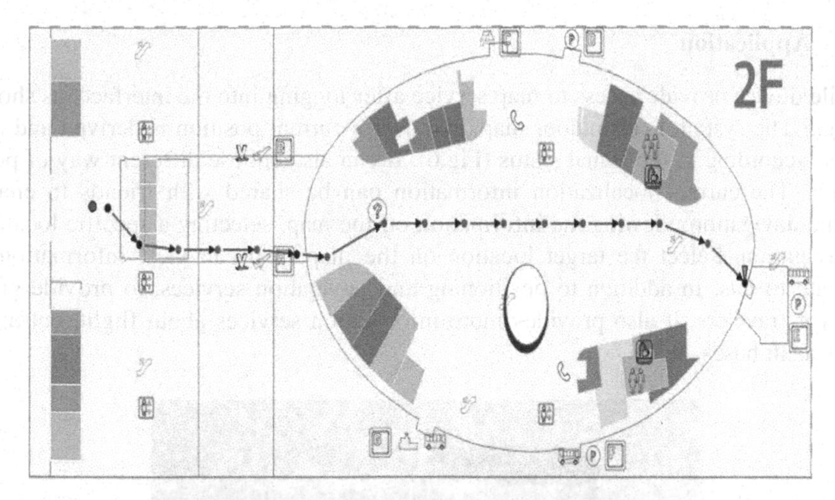

Fig. 4. The planning-path of the airport lounge to export

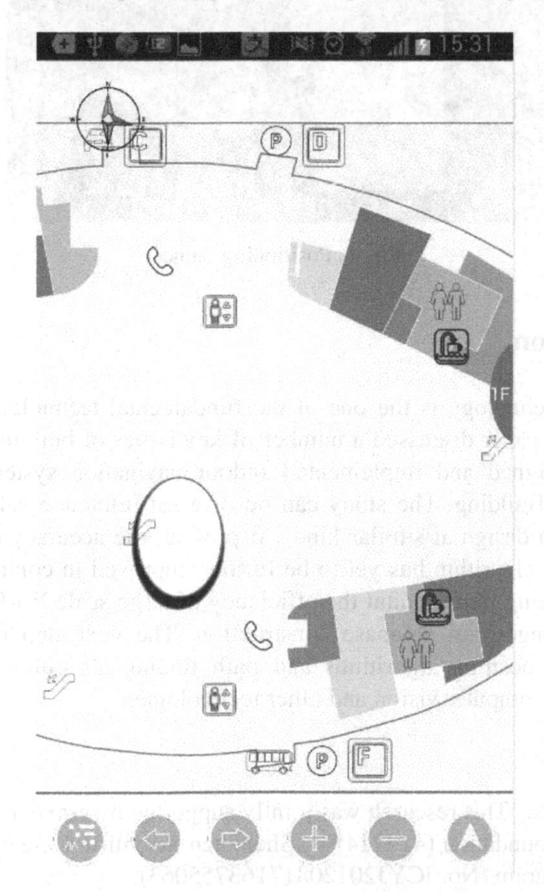

Fig. 5. The airport indoor navigation system

4.5 Application

Mobile device provide access to map service after logging into the interface, as shown in Fig.5.The system loads indoor map, and user's current position is derived and displayed according to the signal status (Fig.6). It can also choose different way of positioning. The current localization information can be shared with friends to enable linkage navigation, viewing the information on the map, selecting a specific location, or navigation. Select the target location on the map, you can view information or navigate to this. In addition to positioning and navigation services, to provide guidance for travelers, it also provides more information services about flight, entrance, exit, transit buses, etc.

Fig. 6. Positioning demo

5 Conclusion

Location-aware technology is the one of the fundamental technology for smart city construction. This paper discussed a number of key issues of building indoor navigation systems, designed and implemented indoor navigation system for Shenzhen Airport Terminal Building. The study can be of great reference value to the indoor positioning system design at similar kind. At present, the accuracy of WLAN-based indoor positioning algorithm has yet to be further improved in complex spatial environment, while taking into account the efficiency of large-scale Wi-Fi signal acquisition and Wi-Fi fingerprint database construction. The next step is to explore the fusion of various position algorithms and path finding algorithm, integrated with machine learning, computer vision and other technologies.

Acknowledgments. This research was jointly supported by grants from the National Natural Science Foundation (41271454), Shenzhen Scientific Research and Development Funding Program (No. JCYJ20120817163755063).

References

1. Alexis, K., Papachristos, C., Nikolakopoulos, G., Tzes, A.: Model predictive quadrotor indoor position control. In: 2011 19th Mediterranean Conference on Control & Automation (MED), pp. 1247–1252. IEEE (2011)
2. Medina, A.V., Gómez, I.M., Romera, M., Gómez, J.A., Dorrozoro, E.: Indoor Position System based on BitCloud Stack for Ambient Living and Smart Buildings. In: Liñán Reyes, M., Flores Arias, J.M., González de la Rosa, J.J., Langer, J., Bellido Outeiriño, F.J., Moreno-Munñoz, A. (eds.) IT Revolutions. LNICST, vol. 82, pp. 127–136. Springer, Heidelberg (2012)
3. Uchitomi, N., Inada, A., Fujimoto, M., Wada, T., Mutsuura, K., Okada, H.: Accurate indoor position estimation by Swift-Communication Range Recognition (S-CRR) method in passive RFID systems. In: 2010 International Conference on on Indoor Positioning and Indoor Navigation (IPIN), pp. 1–7. IEEE (2010)
4. Lin, Y.S., Chen, R.C., Lin, Y.C.: An indoor location identification system based on neural network and genetic algorithm. In: 3rd International Conference on Awareness Science and Technology (iCAST), pp. 193–198. IEEE (2011)
5. Nurminen, H., Ristimaki, A., Ali-Loytty, S., Piché, R.: Particle filter and smoother for indoor localization. In: 2013 International Conference on Indoor Positioning and Indoor Navigation (IPIN), pp. 1–10. IEEE (2013)
6. Werner, M., Kessel, M., Marouane, C.: Indoor positioning using smartphone camera. In: 2011 International Conference on Indoor Positioning and Indoor Navigation (IPIN), pp. 1–6. IEEE (2011)
7. Kohoutek, T.K., Droeschel, D., Mautz, R., Behnke, S.: Indoor Positioning and Navigation Using Time-of-Flight Cameras. In: TOF Range-Imaging Cameras, pp. 165–176. Springer, Heidelberg (2013)
8. Cheng, L., Wu, C.D., Zhang, Y.Z.: Indoor robot localization based on wireless sensor networks. IEEE Transactions on Consumer Electronics 57(3), 1099–1104 (2011)
9. Baniukevic, A., Sabonis, D., Jensen, C.S., Lu, H.: Improving wi-fi based indoor positioning using bluetooth add-ons. In: 12th IEEE International Conference on Mobile Data Management (MDM), vol. 1, pp. 246–255. IEEE (2011)
10. Nossum, A.S.: IndoorTubes a novel design for indoor maps. Cartography and Geographic Information Science 38(2), 192–200 (2011)
11. Puikkonen, A., Sarjanoja, A.H., Haveri, M., Huhtala, J., Häkkilä, J.: Towards designing better maps for indoor navigation: experiences from a case study. In: Proceedings of the 8th International Conference on Mobile and Ubiquitous Multimedia, p. 16. ACM (2009)
12. Han, D., Lee, M., Chang, L., Yang, H.: Open radio map based indoor navigation system. In: 8th IEEE International Conference on Pervasive Computing and Communications Workshops (PERCOM Workshops), pp. 844–846. IEEE (2010)
13. Becker, T., Nagel, C., Kolbe, T.H.: Supporting contexts for indoor navigation using a multilayered space model. In: 10th International Conference on Mobile Data Management: Systems, Services and Middleware (MDM 2009), pp. 680–685. IEEE (2009)
14. Worboys, M.: Modeling indoor space. In: Proceedings of the 3rd ACM SIGSPATIAL International Workshop on Indoor Spatial Awareness, pp. 1–6. ACM (2011)
15. Nakajima, M.: Path planning using indoor map data generated by the plan view of each floor. Pictogram 50, 50 (2011)

16. Goetz, M., Zipf, A.: Indoor Route Planning with Volunteered Geographic Information on a (Mobile) Web-based Platform. In: Progress in Location-Based Services, pp. 211–231. Springer, Heidelberg (2013)
17. Jquery Library, http://jquery.com/

A Regression Forecasting Model of Carbon Dioxide Concentrations Based-on Principal Component Analysis-Support Vector Machine

Yiou Wang, Gangyi Ding, and Laiyang Liu

Digital Performance and Simulation Technology Lab., School of Software,
Beijing Institute of Technology, 100081 Beijing, P.R. China
{Yiou Wang,Gangyi Ding,Laiyang Liu,wangyiou90}@163.com

Abstract. We propose Principal Component Analysis-Support Vector Machine (PCA-SVM) to forecast the changes of regional carbon dioxide concentrations. Firstly, we get the most valuable principal components of the influencing factors (IF) of carbon dioxide concentrations by PCA. Then we use the output of PCA as the input of non-linear SVM to learn a regression forecasting model with radial basis function. Due to the introducing of PCA, we successfully eliminate the redundant and correlate information in IFs and reduce the computation cost of SVM. The results of the comparative experiment demonstrate that our PCA-SVM model is more effective and more efficient than the standard SVM. Moreover, we have tested different kernel functions in our PCA-SVM model, and the experimental results show that PCA-SVM model with radial basis function performs best respecting to the learning ability and generalization capability.

Keywords: Principal component analysis, Support vector machine, Regression and forecasting model, Carbon dioxide concentrations.

1 Introduction

Greenhouse gases, as the most important factor of global warming, have attracted great attention of many governments and researchers. Carbon dioxide accounts for the largest volume of greenhouse gases in the atmosphere, and carbon dioxide's chemical properties are relatively stable, which is not easy to lift away once let off. Therefore correct researching, assessing and forecasting the changing trend of carbon dioxide concentrations are highly significant in many fields, such as the government decision-making, the people's living environment and the low-carbon economy. At present, some researchers have already got some achievements on monitoring the carbon dioxide concentrations. For example, Idso S. B., Idso C. D. and Balling C. R. published their research results about carbon dioxide concentrations of cities in Phoenix, Arizona, US in 2002 [1] and the distribution map of China's average carbon dioxide emissions appeared in the journal Nature in 2012 [2]. However, China started late on the research and had few monitoring data except the data got from some modern cities, e.g. Beijing [3] and Hangzhou [4], and Waliguan in the northwest. In order to improve the situation

F. Bian and Y. Xie (Eds.): GRMSE 2014, CCIS 482, pp. 447–457, 2015.

mentioned above, we established several experimental base stations to monitor carbon dioxide concentrations of Genhe city in Inner Mongolia from 2011. After three years of effort, we realized the real-time monitoring at any time and in any weather.

However, carbon dioxide does not belong to hazardous air pollutants or toxic air contaminants, so the research on the changing trend of regional carbon dioxide concentrations is relatively limited though some people begun to monitor carbon dioxide concentrations. At present, the main regression forecasting analysis is focused on pollutants [5-7]. Ballav S. and Patra P.K. proposed a WRF-CO2 model for regional transport simulations by using five different carbon dioxide fluxes [8], but they still did not give a forecasting method of regional carbon dioxide concentrations. The research on the forecasting of regional carbon dioxide concentrations usually employs traditional regression analysis methods [9], which are not suitable for many practical situations for assuming that the sample size is infinite and the distribution is prior known. Moreover, regional carbon dioxide concentration is affected by many influencing factors and these factors also interact with each other, which makes the forecasting process by traditional regression analysis very difficult. In order to overcome the demerit of the traditional regression analysis, Vapnik V. et al. proposed a SVM-based method which can achieve the global optimum because of the rigorous mathematical derivation and Structural Risk Minimization [10]. Vapnik V. firstly proposed the SVM algorithm using in classification in 1963 [11]. Moreover, Cortes and Vapnik also adopted the kernel function into SVM, which can change problems from nonlinearity space to linearity space and effectively solve the problem of "dimension disaster" existing in nonlinear classification process. Vapnik V., Golowich S. and Smola A. proposed the SVM which can be used to solve the regression problem and signal processing problem in 1997 [12]. After 2010, SVM method got a comprehensive and in-depth development which has become the main tool in the fields of machine learning and data mining, and achieved good effects in practice [13-15]. For the excellent performance, we utilize SVM to forecast carbon dioxide concentrations.

Carbon dioxide concentrations in the atmosphere have certain regional characteristics and are affected by various factors, e.g. climate change, geomorphic features, animals' and plants' breathing, human's living ways. Utilizing all influencing factors directly as the input of SVM will lead to the curse of dimensionality. In addition, the redundancy and correlation of the influencing factors will bring about a consequence that the model learned by SVM cannot reveal the intrinsic relationship between the influencing factors and the carbon dioxide concentrations. Therefore, we propose a regression and forecasting method based on the PCA-SVM algorithm, which is an improvement of SVM by introducing the PCA into the model.

2 Regression Forecasting Model Based on PCA-SVM

The main steps of establishing the regression and forecasting model based on PCA-SVM are as follows: firstly, we normalize the original data. Then, we use PCA to obtain the principal components which are responsible to the change of carbon dioxide concentrations, resulting in the reduction of the redundancy of the influencing factors. Finally, we use both the principal components obtained from the previous step and the historical carbon dioxide concentrations as the input of the SVM to learn a regression forecasting model with radial basis function.

2.1 Theory of PCA

PCA is a statistic analysis method based on the K-L decomposition, which is widely used in the pattern recognition and data dimension reduction. At the very beginning, PCA formats the original data into vectors and then computes the principal components of its covariance matrix. Essentially, PCA aims at seeking a projection subspace which can represent the original data with least mean square error criterion. We assume that the original data $X_1, X_2, \ldots, X_N$ are vectors of dimension N, where $X_i \in R^n$ and then the covariance matrix can be calculated as Equation (1).

$$C = \frac{1}{N}\sum_{i=1}^{N}\left(X_i - \overline{X}\right)\left(X_i - \overline{X}\right)^T . \quad (1)$$

where $\overline{X}$ is the mean vector of all training samples.

After the covariance matrix is gained, we can get the projection matrix V by the covariance matrix as shown in Equation (2).

$$J_c(V) = V^T C V . \quad (2)$$

where V is composed of vectors $V_1, V_2, \ldots, V_d$, which are orthogonal to each other and can make Equation (2) to obtain the maximum value. Once the eigenvectors are gained, we can get the main components Y of the original data X by Equation (3).

$$Y = \left(V_1, V_2, \ldots, V_d\right)^T X \quad (3)$$

Through these steps, PCA method can obtain the most useful information of the original data and achieve the purpose of dimensionality reduction.

2.2 Theory of SVM

In this paper, we use SVM for forecasting the carbon dioxide concentrations, thus we will give an introduction of support vector regression machine.

Assuming the training set is $S = \left\{\left(x_i, y_i\right) \mid x_i \in R^n, y_i \in R, i = 1, 2, \ldots, l\right\}$, where $x_i \in R^n$ and $y_i \in R$ are the ith input sample and the corresponding output respectively. If the hyperplane is $f\left(x\right) = \left\langle w \cdot x\right\rangle + b$, where $w \in R^n$, $b \in R$, then we can obtain Equation (4).

$$\left|y_i - f\left(x_i\right)\right| \le \varepsilon, i = 1, 2, \ldots, l . \quad (4)$$

From Equation (4), we can obtain the distance d_i between (x_i, y_i) and $f(x)$ by Equation (5).

$$d_i = \frac{\left|\langle w \cdot x \rangle + b - y_i\right|}{\sqrt{1 + \|w\|^2}} \le \frac{\varepsilon}{\sqrt{1 + \|w\|^2}}, i = 1, 2, \ldots, l . \tag{5}$$

that is, the optimal hyperplane can be achieved through minimizing $\|w\|^2$. Using the Lagrange function, the optimal problem can be replaced by Equation (6).

$$\min \frac{1}{2}\sum_{i,j=1}^{l} (a_i^* - a_i)(a_j^* - a_j)\varphi(x_i, x_j) + \varepsilon\sum_{i=1}^{l}(a_i^* + a_i) - \sum_{i-1}^{l} y_i (a_i^* - a_i),$$

$$s.t. \sum_{i=1}^{l}(a_i - a_i^*) = 0 . \tag{6}$$

where $0 \le a_i, a_i^* \le C, i = 1, 2, \ldots, l$, and a_i, a_i^* are Lagrange multipliers. Then the approximate regression forecasting function can be obtained by Equation (7).

$$f(x) = \langle w \cdot x_i \rangle + b = \sum_{i=1}^{l}(\bar{a}_i^* - \bar{a}_i)(x_i \cdot x) + b . \tag{7}$$

2.3 Regression Forecasting Model of Carbon Dioxide Concentrations Based on PCA-SVM

PCA can reduce the dimension of high-dimensional initial data and eliminate the correlation among varieties of influencing factors that account for the changes of regional carbon dioxide concentrations, but this method cannot model the non-linear relationship between regional atmospheric carbon dioxide concentrations and their influencing factors. In contrast, SVM can dig nonlinear characteristics of data tho-roughly and model the nonlinear relationship, however, due to the high dimension of training data, the training time is very long. In addition, SVM also does not consider the correlation among the data, which makes that the forecasting model learned by SVM does not forecast the tendency of carbon dioxide concentration changes well. Combining the PCA and SVM, not only can improve the calculation speed of SVM, but also can increase the accuracy of the forecasting model. Consequently, in order to take best advantage of both algorithms, we propose a regression forecasting model of regional carbon dioxide concentrations based on PCA-SVM. The algorithm flow chat is shown in Fig.1.

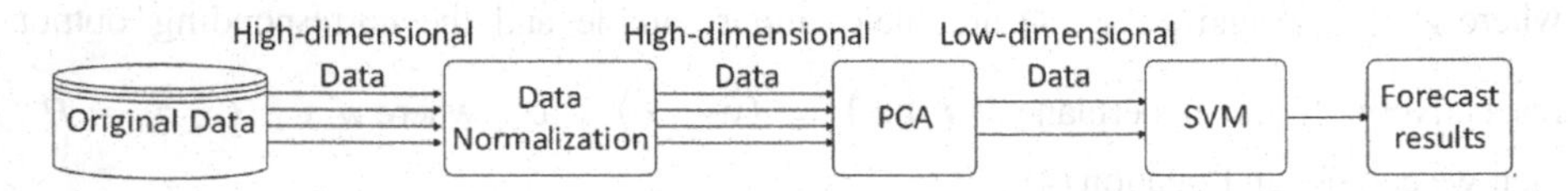

Fig. 1. The flow chat of forecasting model based on PCA – SVM

In this paper, we have considered eight main influencing factors that result in the changes of regional carbon dioxide concentration. There are various kinds of influencing factors and the sources of the factors are very different, which causes that the initial data are complex and heterogeneous. Thus we need to normalize these heterogeneous data to the range of $(0, 1)$ in advance by Equation (8).

$$X_i^* = \frac{X_i - X_{min}}{X_{max} - X_{min}} . \tag{8}$$

where X_i^* is the new value after normalization, X_i is the initial value, X_{min} and X_{max} are the minimum and maximum in the corresponding influencing factor set respectively. Because the initial values of temperature may be negative, we should convert them into positive ones at first when normalizing temperature data. The normalization equation of temperature data is shown in Equation (9).

$$X_i^* = \frac{X_i + X_{max} - X_{min}}{X_{max} - X_{min}} . \tag{9}$$

Since the contribution of each influencing factor to carbon dioxide concentration changes is different, in order to obtain main influencing factors that result in the changes of carbon dioxide concentration, we utilize PCA to reduce dimensions of eight factors after the initial data normalization. Then, we can obtain the principal component $V_1, V_2, \ldots, V_d$ extracted from raw data $X_1, X_2, \ldots, X_N$, where N is the number of data before dimension reduction, d is the number of data after dimension reduction and $d < N$. Through extensive experiments, we have found that, when $d/N \approx 0.6$, using the data whose dimension have been reduced as the input of SVM will get the optimal regression forecasting effect. Therefore, five influencing factors are selected.

We successfully eliminate the correlation and the redundancy of raw data by using PCA and get new data which can maximize the characterization of the raw data. Now we use the principal components extracted by PCA and the historical carbon dioxide concentration data as the input of SVM to learn a forecasting model. And PCA has projected the data from high-dimensional space to low-dimensional space, which improves the computing speed of SVM. Besides, because SVM uses the principal components as training samples, the accuracy of the algorithm has been greatly increased. By contrast, due to large amounts of the input data and not considering the correlation among these data, the standard SVM model performs not very well in the training time and in the regression forecasting process. Consequently, compared with the standard SVM, PCA-SVM achieves greater accuracy, better utility and higher robust in dealing with the problem of regional carbon dioxide concentration prediction.

In addition, when we solve regression problems by SVM, both the selection of kernel function and the strategy of parameters optimization have a great influence on the performance of the forecasting model because different kernel functions and different parameters optimization methods create different SVMs. Since radial basis function can determine the central point automatically, we select it as the kernel function of SVM. Moreover, we choose ε insensitive loss function as SVM's loss function. The prime advantage of ε insensitive loss function is the less number of support vectors extracted by SVM, which significantly reduces the complexity and saves a lot of computing time. ε insensitive loss function is shown in Equation (10).

$$l_\varepsilon(\delta) = \begin{cases} 0 & |\delta \le \varepsilon| \\ |\delta| - \varepsilon & others \end{cases}. \tag{10}$$

3 Experimental Results and Analysis

Since August 2011, we have carried out a field research and investigation in Genhe city of Inner Mongolia, and realized continuous, real-time and all-weather monitoring of regional carbon dioxide concentrations in the atmosphere of Genhe city by using wireless sensor network. Now eleven monitoring sites have already been set up in Genhe city for the project. The data-collection interval of each monitoring site is 1 second, thus as of April 30th of 2014, we have obtained 870 million historical carbon dioxide concentration data. Fig. 2 shows the distribution of wireless sensor network in

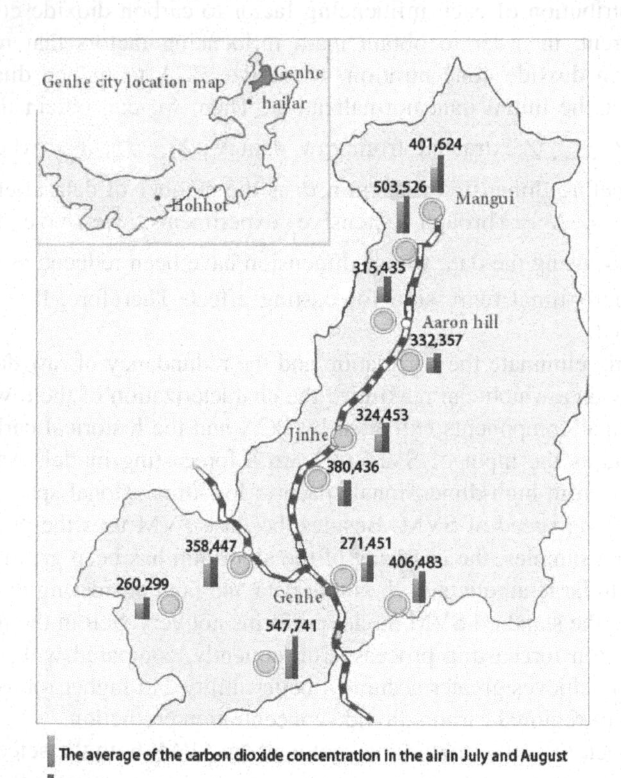

Fig. 2. The distribution of wireless sensor network in Genhe city (For interpretation of the references to color in this figure legend, the reader is referred to the web version of this article)

Genhe city. Left bars in Fig. 2 represent the average of carbon dioxide concentration of July and August in 2012. Right bars in Fig. 2 represent the average of carbon dioxide concentration of November and December in 2012. The numbers above these bars represent the corresponding average value.

There are various influencing factors of regional carbon dioxide concentration. Nevertheless, considering the contribution of these factors and thinking of whether these factor data can be measured or not, we study it from eight aspects, which are air temperature, humidity, lighting time, wind power, enterprise total coal combustion, traffic flow, residents total coal combustion and effective vegetation coverage. Effective vegetation coverage is defined as the vegetation coverage with the capacity of photosynthesis. All these factor data are eventually organized into units of days on average. Due to the instability of wireless network communication, data cleaning [16] should be done first to remove duplicate values, wrong values, null values and abnormal values. Then, we convert all these cleaned and heterogeneous factor data into a uniform format by normalization equations. For example, Table 1 shows some results of these eight influencing factors of regional carbon dioxide concentration after normalization from July 11, 2013 to July 15, 2013, where AT, H, LT, WP, ETCC, TF, RTCC and EVC are short for air temperature, humidity, lighting time, wind power, enterprise total coal combustion, traffic flow, residents total coal combustion and effective vegetation coverage, respectively.

Table 1. Examples of data normalization

Date	AT	H	LT	WP	ETCC	TF	RTCC	EVC
July 11th	0.9230	0.3107	0.9372	0.3333	0.3203	0.6322	0.0001	0.9999
July 12th	0.9316	0.3013	0.9372	0.0001	0.3319	0.5437	0.0001	0.9999
July 13th	0.9829	0.2745	0.8209	0.0001	0.3183	0.4387	0.0001	0.9999
July 14th	0.9401	0.3223	0.8209	0.0001	0.2902	0.3922	0.0001	0.9999
July 15th	0.8632	0.6783	0.8209	0.0001	0.3451	0.5911	0.0001	0.9999

For sake of selecting principal factors, we use PCA to analyze the normalized data and order the influencing factors according to their contribution to the carbon dioxide concentration changes. Air temperature takes the top spot, followed by humidity, residents total coal combustion, enterprise total coal combustion, wind power, effective vegetation coverage, lighting time and traffic flow. We pick five most important factors as the input vectors of SVM for further analysis.

We take Jinhezhengfu monitoring site of Genhe city as an example to compare the performance of proposed PCA-SVM with standard SVM. Now some details of the comparative experiments are given. In order to achieve a fair comparison, both of the algorithms choose the same kernel function for their support vector machines and experiment on the same data set. The uniform kernel is radial basis kernel function. Besides, both algorithms utilize a year of historical data from August 2011 to August 2012 as training samples and use historical data from August 2012 to April 2014 as test samples. All data are resized into units of days on average. Fig. 3. shows the comparison results of these two algorithms.

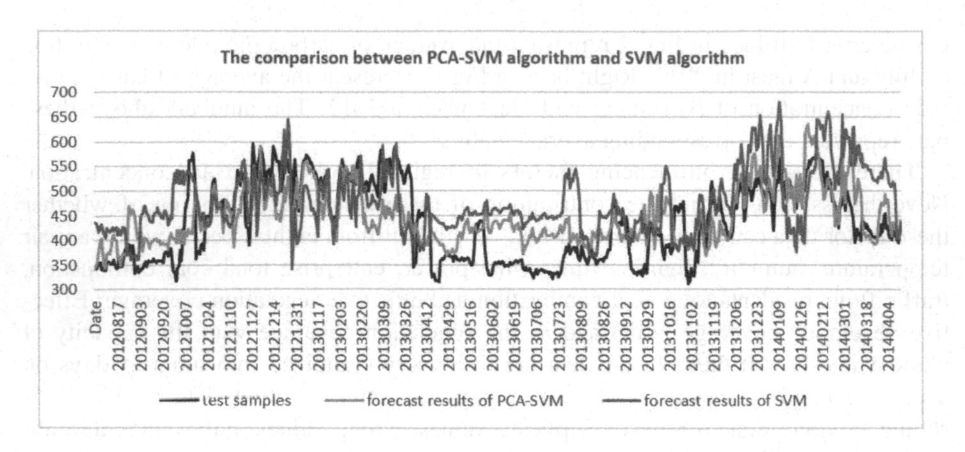

Fig. 3. The comparison of the forecasting results between PCA-SVM and SVM (For interpretation of the references to color in this figure legend, the reader is referred to the web version of this article)

As shown in Fig.3, both PCA-SVM and standard SVM can forecast regional carbon dioxide concentrations to some extent. However, standard SVM is sensitive to noise data on account of the interference of unessential influencing factors. Compared with PCA-SVM, the jitter of standard SVM is larger and the robustness of standard SVM is lower when data fluctuate significantly. PCA-SVM not only can eliminate negative impact of unessential influencing factors, but also can bring down the sample input dimensions of support vector machine because its input samples have been filtered by PCA. As a result, the curve produced by PCA-SVM is smoother and the data forecasted by PCA-SVM are more approximate to the actual data. Moreover, the training time of standard SVM is 5.6 minutes while the PCA-SVM is 4.3 minutes. PCA-SVM has greatly improved the training time of constructing a model. Monthly average of atmospheric carbon dioxide concentrations and relative error of these two algorithms are shown in Table 2, where relative error is calculated by Equation (11).

$$RelativeError = \frac{|ForecastingValue - TrueValue|}{TrueValue} \times 100\% . \tag{11}$$

From Table 2, we can calculate that the maximum relative error and the average relative error based on PCA-SVM are 20.48 and 7.24 respectively, while the values based on standard SVM are 25.43 and 11.7. All these data verify that PCA-SVM algorithm proposed by this paper is greatly effective and highly correct when solve the problem of forecasting regional carbon dioxide concentrations.

In addition, in order to demonstrate the effectiveness of radial basis function adopted by this paper, various experiments have been performed on different kernel functions, including polynomial kernel function, sigmoid function and radial basis function. The comparison regression forecasting results of using different kernel functions are shown in Table 3.

Table 2. Monthly average and relative error of PCA-SVM and SVM

Date	Test samples	PCA-SVM	SVM	Relative error of PCA-SVM	Relative error of SVM
Aug. 2012	357	390	409	8.46	12.71
Sep. 2012	355	398	446	10.80	2.04
Oct. 2012	441	436	491	1.14	10.18
Nov. 2012	485	473	479	2.53	1.25
Dec. 2012	543	484	510	12.19	6.47
Jan. 2013	511	495	467	3.23	9.42
Feb. 2013	525	497	493	5.63	6.49
Mar. 2013	541	449	490	20.48	1.04
Apr. 2013	394	401	424	1.74	7.07
May 2013	378	414	448	8.69	15.62
Jun. 2013	349	424	454	17.68	23.12
Jul. 2013	340	413	456	17.67	25.43
Aug. 2013	368	404	443	8.91	16.93
Sep. 2013	388	419	437	7.39	11.21
Oct. 2013	412	436	433	5.50	4.84
Nov. 2013	418	408	454	2.45	7.92
Dec. 2013	480	497	554	3.42	13.35
Jan. 2014	500	523	498	4.39	0.04
Feb. 2014	522	532	574	1.87	9.05
Mar. 2014	437	469	566	6.82	22.79
Apr. 2014	417	442	496	5.65	15.92

Table 3. The comparison results of using different kernel functions

Kernel function	Average relative error
Polynomial kernel function	18.42
Sigmoid function	12.79
Radial basis function	7.24

According to the results from Table 3, we find that PCA-SVM used radial basis function obtains the minimum relative error. And the corresponding model has the best learning ability and the strongest generalization capability. While polynomial kernel function has more parameters and the number of parameters determines the model complexity. And the model established by sigmoid function also produces higher average relative error. In addition, radial basis function can determine the central point automatically, so radial basis function is the best option for PCA-SVM.

Now we conclude that PCA-SVM has the advantage of higher forecasting accuracy, faster computing speed and greater running stability compared with standard SVM. And radial basis function is more suitable for PCA-SVM than polynomial kernel function or sigmoid function. Consequently, PCA-SVM established by radial basis function can be used for forecasting regional carbon dioxide concentrations.

4 Conclusion

We present a Principal Component Analysis-Support Vector Machine (PCA-SVM) algorithm to forecast regional carbon dioxide concentrations by combining PCA with SVM. Indeed, the practicality of PCA-SVM is well evidenced in the experimental results. There, compared with standard SVM, PCA-SVM algorithm reduces the values of maximum relative error and average relative error by 4.95 and 4.46 respectively, and saves 1.3 minutes to establish a regression forecasting model. Moreover, different kernel functions have been tested, and we find that radial basis function is the most suitable for PCA-SVM algorithm.

Acknowledgments. The authors would like to thank Forestry Bureau of Genhe city for providing their exterior grounds for our experiments.

References

1. Idso, S.B., Idso, C.D., Balling, C.R.: Seasonal and Diurnal Variations of Near-surface Atmospheric CO2 Concentration within a Residential Sector of the Urban CO2 Dome of Phoenix, AZ, USA. Atmospheric Environment 36, 1655–1660 (2002)
2. Guan, D., et al.: Nature Climate Change 2, 672–675 (2012)
3. Changsi, W., Yuesi, W., Guangren, L.: Characteristics of Atmospheric CO2 Variations and Some Affecting Factors in Urban Area of Beijing. Environmental Science 24, 13–17 (2003)
4. Leizhi, Z., Aiguang, Z.: Influence of CO2 Content in Atmosphere on Air Temperature in Hangzhou. Journal of Zhejiang Forestry College 17, 301–304 (2000)
5. Kit, Y.C., Jian, L.: Identification of Significant Factors for Air Pollution Levels using a Neural Network based Knowledge Discovery System. Neurocomputing 99, 564–569 (2013)
6. Sutapa, C., Debashree, D.: Mann-Kendall Trend of Pollutants, Temperature and Humidity over an Urban Station of India with Forecast Verification using Different ARIMA Models. Environmental Monitoring and Assessment 186, 4719–4742 (2014)
7. García, P.J., Álvarez, J.C.: Nonlinear Air Quality Modeling using Multivariate Adaptive Regression Splines in Gijón Urban Area (Northern Spain) at Local Scale. Applied Mathematics and Computation 235, 50–65 (2014)
8. Ballav, S., Patra, P.K., Takigawa, M., Ghosh, S.: Simulation of CO2 Concentration over East Asia Using the Regional Transport Model WRF-CO2. Journal of the Meteorological Society of Japan 90, 959–976 (2012)
9. Youmin, T., Dake, C., Dejian, Y., Tao, L.: Methods of Estimating Uncertainty of Climate Prediction and Climate Change Projection. Climate Change. InTech, Croatia (2013)
10. Vapnik, V.: Estimation of Dependencies Based on Empirical Data. Springer (1982)
11. Vapnik, V., Lerner, A.: Pattern Recognition using Generalized Portrait. Automation and Remote Control 24, 709–715 (1963)
12. Vapnik, V., Golowich, S., Smola, A.: Support Vector Method for Function Approximation, Regression Estimation, and Signal Processing. In: Mozer, M., Jordan, M., Petsche, T. (eds.) Advances in Neural Information Processing Systems, pp. 281–287. MIT Press, Cambridge (1997)

13. Zou, H.F., Xia, G.P., Yang, F.T., Wang, H.Y.: An Investigation and Comparison of Artificial Neural Network and Time Series Models for Chinese Food Grain Price Forecasting. NeuroComputing 70, 2913–2923 (2011)
14. Huali, D., Xiuquan, L.: A Prediction Method of Stock Price Inflection Point based on Chaotic Time Series Analysis. Statistics and Decision 5, 19–20 (2007)
15. Kyoung, K.: Financial Time Series Forecasting Using Support Vector Machines. NeuroComputing 55, 307–319 (2012)
16. Rahm, E., Do, H.H.: Data Cleaning: Problems and Current Approaches. IEEE Data Engineering Bulletin 23, 3–13 (2000)

Simulating Agricultural Land Use Changes in Uganda Using an Agent-Based Model

Jingjing Li[1] and Tonny J. Oyana[2,*]

[1] Department of Geography and Environmental Resources, Southern Illinois University, IL 62901, USA

[2] Research Center on Health Disparities, Equity, and the Exposome, University of Tennessee at Memphis, 66 Pauline Suite 300, Memphis, TN 38105, USA

jingjingli@siu.edu, toyana@uthsc.edu

Abstract. Agent based modeling, a processed based approach, is advantageous in simulating the interaction between human's decision processes and environmental systems. In this study, we apply an agent-based model to simulate potential agricultural land use change scenarios in Uganda. The simulation model incorporates decision making processes at small holder and commercial farmers' level on the basis of biophysical and socioeconomic factors and use these as basis to analyze how farmers' decisions may affect agricultural land use changes. Geographic information system (GIS) tools are employed to build spatial relations between farmer agents and land cover system. Satellite imageries are used to represent the initial land cover condition and serve as observed land cover dataset to calibrate the simulated results. The results of the simulation model are promising and the model was successful at representing historical and future scenarios of agricultural land use patterns at national-level.

Keywords: Agent-based model, agricultural land use, land cover change, simulation.

1 Introduction

Current land use/ land cover changes (LULCC) are being driven principally by human activities, which are causing significant impacts on the environmental systems on the earth [1] [2]. LULCC prominently contribute to the climate change at regional and global scales and the deforestation in tropical areas, which thus affect hundreds of millions of people in the world [3][4][5].Modeling LULCC is an effective methodology to simulate the dynamics of land-use systems and unravel the interaction between LULCC and the environment [6].

LULCC processes consist of various human's decisions acting at different scales, which gradually alter the landscape overtime [7]. Agricultural land use is largely determined by farmers' decisions on how to cultivate land and what to plant in the land [8]. Farmers' decisions on land use are critically influenced by complicated external variables such as physical constraints, policy scenarios, and economic factors [7] [9].

* Corresponding author.

F. Bian and Y. Xie (Eds.): GRMSE 2014, CCIS 482, pp. 458–470, 2015.

Agent-based modeling (ABM) is a popular approach to simulate land use changes, which provides a spatially explicit way to understand the interaction between human's decisions and LULCC[10]. It is a processed-based method consists of multiple agents interacting with each other within an environment according to formulated rules[11]. It has the capability to integrate human's decision-making processes into the land use change context, of which agents represent the decision-making actors [12]. Various external factors such as socioeconomic attributes have played key roles in human's decisions on land use, which have been considered as input data in the ABM model [13][14][15]. Using the ABM model, numerous studies have simulated decision-making processes at the household level [16].

Despite the existence of many land use change studies, there remains a lack of process-based approaches for studying agricultural land change processes at a national scale. This study intends to bridge this existing knowledge gap. We have developed an agent-based model to systematically represent or reproduce historical and future agricultural land use dynamics in Uganda. The model evaluated how farmers' decisions influenced agricultural land use changes and explored the interaction between farmers and external factors. Biophysical and socioeconomic factors contribute to farmers' decisions, such as physical constraints, policy scenarios, distance to towns, ownerships of land, and land use type of neighboring land parcels. Two research questions guide this study: (1) what biophysical and socioeconomic factors drive agricultural land use changes in Uganda? (2)What model and sub-models best represent the past, present, and future agricultural land use changes?

2 Methodology

2.1 Study Area

Uganda is a landlocked country located in East Africa and shares a border with Kenya, Tanzania, Rwanda, South Sudan, and Congo (Fig. 1)[17]. Uganda is rich in natural resources, and is particularly suitable for agriculture because of fertile soils, regular rainfall, and less exposure to pests [17]. Uganda has a total area of 24,155,058 hectares (ha).The temperature ranges from 16 to 30℃ , rainfall averages at 850–1700 mm/year [18].The population at a growth rate that stands at 3.2 percent increased continuously over time, from 24.2 million in 2002 to 35.4 million in 2013 mid-year, among which 82 percent lived in rural areas [18]. The economy in Uganda experienced steady growth over the last decade, with the gross domestic product (GDP) of US $3982.75 million and US $22167.5 million registered in the year of 2001 and 2013 respectively [18] [19]. Export of goods and services amounted to US $ 2.8 billion during 2012[18].

Agriculture is the most important economic sector in Uganda as it contributes to a large proportion of national GDP and exports [21]. It made contributions of approximately 24.1% to GDP and 51.9% to exports in fiscal year of 2010/11[17].Agriculture is a major source of employment, in which 66% of the working population is engaged [18] [22] [23]. This sector also plays a major role in poverty eradication, especially among the poor people who are primarily living in rural areas [24]. Research by IFPRI (2007) demonstrated that if agriculture in Uganda grew at 6 percent per annum,

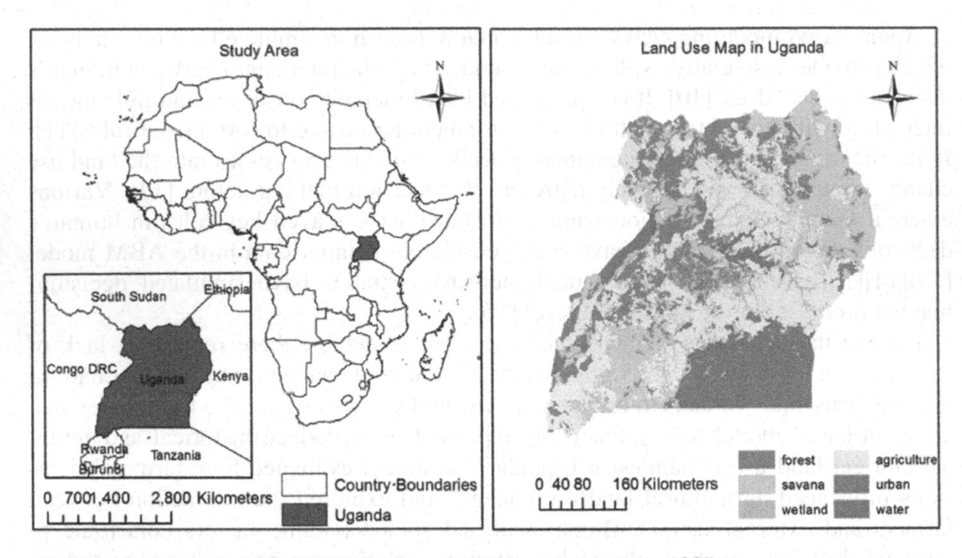

Fig. 1. Study Area (left figure data source: [20])

the national poverty headcount level would fall from 31.1 percent in 2005 to 17.9 percent by 2015 [25].

2.2 Datasets

Images and land usedatasets were acquired for the following time periods: 1996, 2001, and 2013. Land use datasets for 1996 and 2001 were derived from Landsat TM imageries[26], the 2013 land usedata were derived from Landsat 8 OLI imageries[26] (Table 1). For Landsat TM imageries in 1996 and 2001, we selected six spectral bands (band 1, band 2, band 3, band 4, band 5, and band7) with spatial resolution of 30-m. Band 6was not included as it is thermal infrared band. For Landsat 8 OLI imageries in 2013, we selected six spectral bands (band 2, band 3, band 4, band 5, band 6, band 7), which are corresponding to each of the six bands in TM imageries with the same spatial resolution and similar spectral ranges.Besides, we selected a panchromatic band (band 8) in Landsat 8 imageries with spatial resolution of 15-m for enhancement.The Digital Terrain Elevation data [27] were used to generate slope information.

Image Pre-processing. We rectified the images of 1996 to topographic maps, and then we co-registered the images of 2001 and 2013 to the corrected images of 1996. The images were re-projected to Universal Transverse Mercator (UTM) projected coordinate system (GS_1984_UTM_Zone_36N). Besides, we conducted radiometric correction. The data have been converted to Top of Atmosphere (TOA) reflectance with a correction for sun angle. In order to improve the classification accuracy, we enhanced the images using principle component analysis approach. Studies indicate a data fusion approach–the integration of principle component analysis and panchromatic image significantly improve the classification accuracy [28]. In this study, we employed this data fusion method to enhance the 2013 Landsat 8 images. We conducted

Table 1. Attributes of Landsat 8 imageries in 2013

Satellite	Sensor	Path & Row	Acquisition date	Resolution
Landsat 8	Operational Land Imager	P170R057	2013-07-24	30×30m
Landsat 8	Operational Land Imager	P170R058	2013-08-09	30×30m
Landsat 8	Operational Land Imager	P170R059	2013-06-22	30×30m
Landsat 8	Operational Land Imager	P170R060	2013-07-24	30×30m
Landsat 8	Operational Land Imager	P170R061	2013-07-08	30×30m
Landsat 8	Operational Land Imager	P171R057	2013-05-28	30×30m
Landsat 8	Operational Land Imager	P171R058	2013-05-28	30×30m
Landsat 8	Operational Land Imager	P171R059	2013-07-15	30×30m
Landsat 8	Operational Land Imager	P171R060	2013-07-15	30×30m
Landsat 8	Operational Land Imager	P171R061	2013-07-15	30×30m
Landsat 8	Operational Land Imager	P172R057	2013-08-07	30×30m
Landsat 8	Operational Land Imager	P172R058	2013-08-07	30×30m
Landsat 8	Operational Land Imager	P172R059	2013-07-06	30×30m
Landsat 8	Operational Land Imager	P172R060	2013-07-06	30×30m
Landsat 8	Operational Land Imager	P172R061	2013-08-07	30×30m
Landsat 8	Operational Land Imager	P173R057	2013-07-13	30×30m
Landsat 8	Operational Land Imager	P173R058	2013-07-13	30×30m
Landsat 8	Operational Land Imager	P173R059	2013-06-11	30×30m
Landsat 8	Operational Land Imager	P173R060	2013-07-13	30×30m
Landsat 8	Operational Land Imager	P173R061	2013-06-11	30×30m

principle component analysis to transform the images into six principle components. The first three principle components (PC1, PC2, and PC3) were selected as they best represent the most variances of the image, which account for 99.17 percent of total variances. We substituted the first principle component (PC1) with the panchromatic image. After that we applied the inverse principle component analysis to the images.

Image Classification. We applied the United States Geological Survey (USGS) land use/land cover classification system [29], which was developed in 1976. Given that Landsat imageries havea medium spatial resolution (30-m), we classified the images at level I classification system, so we ended up with six land use/land cover classes for this study area, which include water, wetland, urban, agriculture, forest, and grassland. We employed hybrid classification method to classify the satellite images. The hybrid classification method combines supervised and unsupervised methods, which involves training samples and ISODATA algorithm [30].We selected five training samples for each class. Then we generated a cluster image at ERDAS Imaging [31] platform by inputting the signature file and setting iteration number and convergence tolerance as 30 and 0.95 respectively. We categorized the 30 types of clusters into six land use classes and recoded the image to get the final classification image.

Accuracy Assessment. After classification, we conducted accuracy assessment using Google aerial photos [32] as reference. We employed an equalized random sampling algorithm to select reference points. We selected 75random points for each class in the classified image, and the software automatically selected the corresponding points in the reference maps. An error matrix was generated and the overall accuracy, kappa statistics, commission error, and omission error were applied to evaluate the accuracy of the classification.

2.3 Model Description

Conceptual Framework. There are two sets of agents in this model: generic and vector agents. Generic agents consist of agricultural and non-agricultural developers. Vector agents consist of land parcels with land use attributes. Fig. 2 shows the conceptual framework of this model.We considered several external factors that would affect the parcel agents' decisions about converting their status. Physical constraints such as steep slope (> 18 degree), water bodies, and wetland buffer areas largely limit the development of agriculture (Fig. 2). Other factors include policy scenarios such as conservation/ reserve land use constraints area, suitability such as land tenure ownership (customary, mailo, and native), distance to towns, and adjacent parcels' development status.

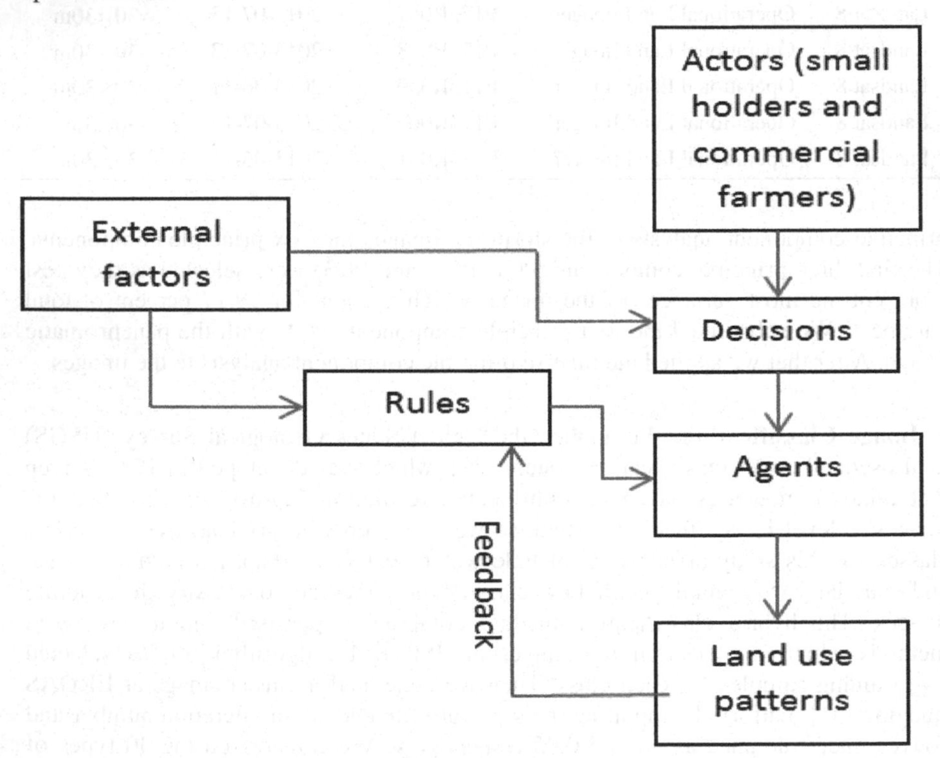

Fig. 2. Conceptual framework

Computational Framework. The model was createduringAgent Analyst—an Analyst Agent extension toolkit for ArcGIS 10.2 [33]. The baseline map of this model was the land use/land cover data of Uganda in 1996.The generic agents—agricultural and non-agricultural developers were derived from the baseline map. The actors—smallholders and commercial farmers were represented by land parcels. As input data of the model, the external variables were constructed and reclassified with the same resolution of the baseline map (30-m), which affected and regulatedactors' decisions on land parcel agents. The input GIS data layers include physical constraints, conservation constraints, land tenure, and distance to towns (Fig.3). Neighboring land use status list were generated and imported in a text file format. The time step was once a year. Land parcel agents evaluated the status of land use at each time step to determine their development—keeping their status or converting to the other status.

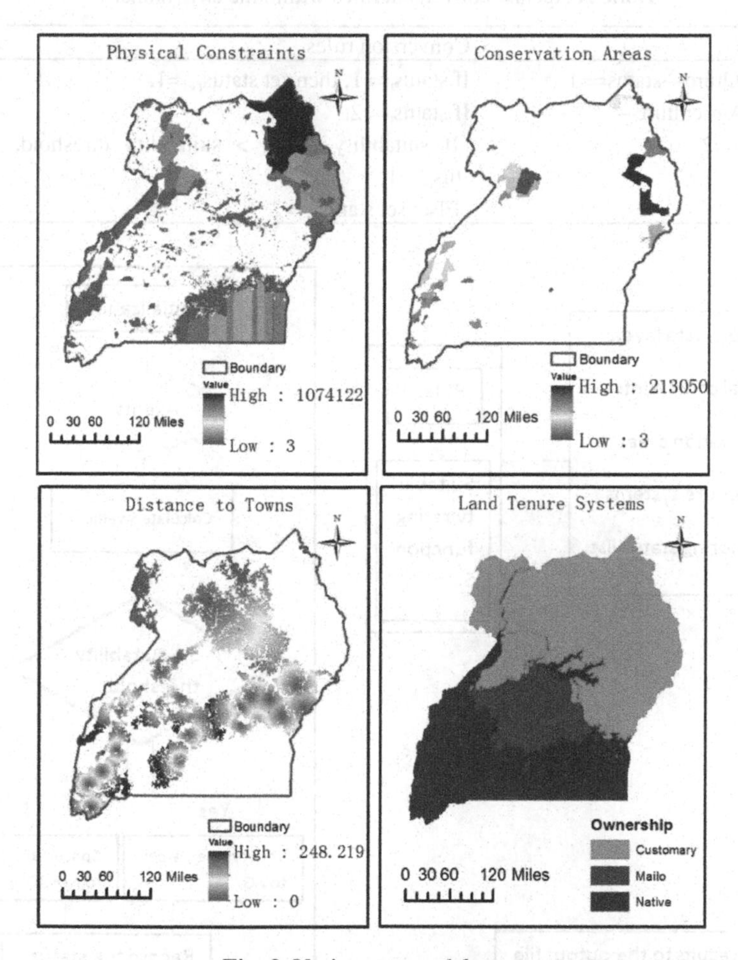

Fig. 3. Various external factors

Agents' Rules. In this study, we computed a suitability value to assess the agents' behaviors. Suitability rating value S is a function of physical constraints, customary ownership probability, conservation constraints, distance to towns, and neighborhood status.

$$S=f\,(w_1*p,\ w_2*l,\ w_3*c,\ w_4*d,\ w_5*n). \tag{1}$$

Where p is physical constraint percent, l is customary ownership probability, $c$refers to conservation constraints percent, d refers to distance to towns for each cell, $n$represents neighboring status,w_1, w_2, w_3, w_4, *and* $w_5$are weights assigned to these factors to measure the degree of their influence. If the suitability rating value is more than suitability threshold (preset), the agent parcels converted to agricultural land.Agents' transition rules are shown in Table 2, and computational framework is shown in Fig. 4.

Table 2. Agents' conversion rules within the environment

Status	Conversion rules
Agriculture—status==1	If $status_t$==1, then set $status_{t+1}$=1.
Non-Agriculture—status==2	If $status_t$==2: If suitability rating > suitability threshold, set $status_{t+1}$=1; Else: set $status_{t+1}$=2.

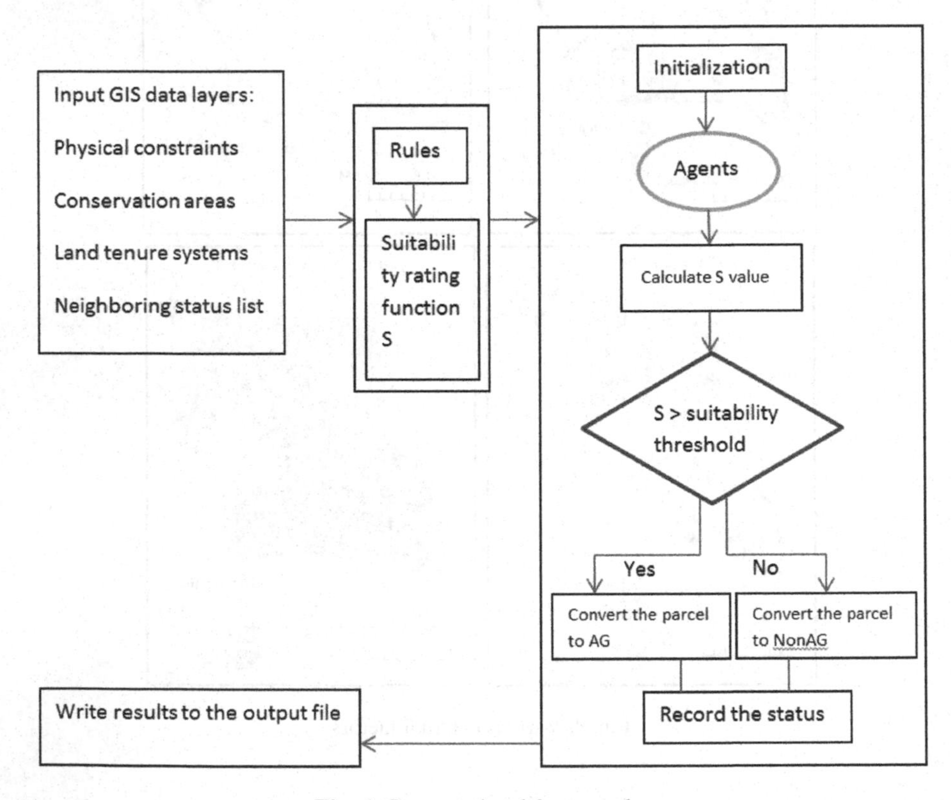

Fig. 4. Computational framework

3 Results

3.1 Land Use Changes

Agricultural land increased significantly during the period of 1996–2013. It increased from 8,983,409 ha (37.19%) in 1996 to 9,177,742 ha (38%) in 2001, and to 10,311,565.07ha (42.69%) in 2013 (Table 3). Agricultural land expanded in central, southwestern, and northeastern part of the country (Fig. 5). The growing demand for food and cash crops stimulated the expansion of agricultural land. Urban areas increased from 27421.16 ha in 1996 to 41,802.13 ha in 2013 with a growth rate of 52.44 percent.

Forest land cover shrank dramatically during 1996–2013 period. Forest experienceda slight increase during 1996–2001;however, it decreased sharply duringthe period of 2001–2013. Forest occupied 25.10 percent of the total area in 1996, whereas it dropped drastically to 12.60 percent in 2013 (Table3). On the other hand, grassland has the reversing trend as forest. Grassland decreased during 1996–2001 and then increased during 2001–2013. Grassland increased by 28.18 percentduring 1996–2013.

Table 3. Land use changes in Uganda (1996–2013)

Land class		water	wetland	urban	AG	forest	grass	total
1996	Area (ha)	3,738,815	400,587.30	27,421.16	8,983,409.00	6,063,652.00	4,941,173.00	24155058.00
	Percent	15.48%	1.66%	0.11%	37.19%	25.10%	20.46%	100.00%
2001	Area (ha)	3,699,472.00	1,172,552.00	18,479.47	9,177,742.00	7,227,859.00	2,858,954.00	24155058.00
	Percent	15.32%	4.85%	0.08%	38.00%	29.92%	11.84%	100.00%
2013	Area (ha)	3,796,711.19	628,532.57	41,802.13	10311565.07	3042996.86	6333450.18	24155058.00
	percent	15.72%	2.60%	0.17%	42.69%	12.60%	26.22%	100.00%
1996–2013 changes	Area (ha)	57,896.19	227,945.27	14,380.97	1,328,156.07	-3,020,655.14	1,392,277.18	
	Percent	1.55%	56.90%	52.44%	14.78%	-49.82%	28.18%	

3.2 Accuracy Assessment

The kappa statistics measures the agreement the classified image to the reference image, and kappa coefficient κ > 0.8 represents strong agreement and good accuracy [30]. The accuracy assessment of the observed land use data of Uganda in 2013 is shown in Table 4. The overall accuracy is 88%, and overall kappa statistics is 0.856 (Table 4), which presents good agreement between classified image and the reference map.

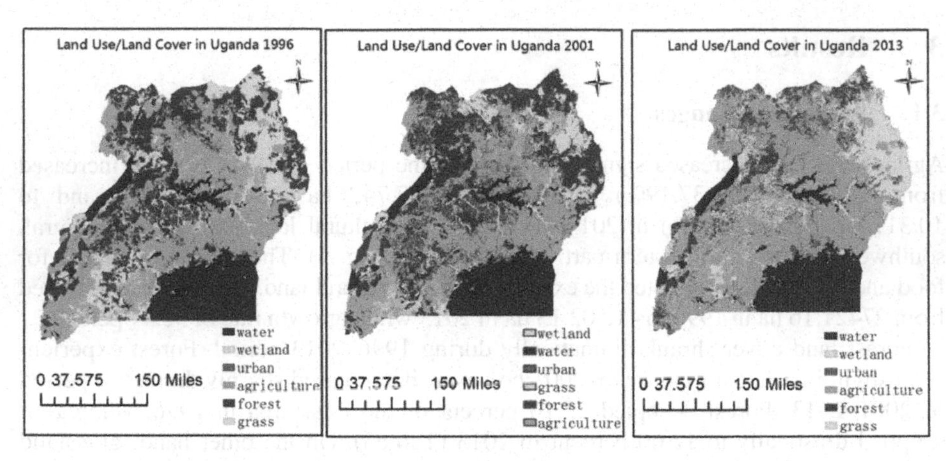

Fig. 5. Land Use/Land Cover in Uganda (1996, 2001, and 2013)

Table 4. Accuracy assessment of Landsat 8 imagery data in 2013

		reference							
		water	wetland	urban	agriculture	forest	grass	total	User's A%
classified	water	73	2					75	97.33%
	wetland		70				5	75	93.33%
	urban			69	5		1	75	92.00%
	agricul-ture				62	3	10	75	82.67%
	forest				8	59	8	75	78.67%
	grass				10	2	63	75	84.00%
	total	73	72	69	85	64	87	450	
	Pro-duc-er'sA %	100.00%	97.22%	100.00%	72.94%	92.19%	72.41%		
overall classification accuracy =(73+70+69+62+59+63)/450=88%, overall kappa statistics =0.856									

3.3 Simulation Results

The simulation results of the year 2001 and 2013 are shown in Fig.6, which exhibit some similarities to the observed land use/land cover data derived from satellite imageries (Fig. 6).The simulated land use patterns in the year 2013 were compared with the observed land use data derived from Landsat 8 imageries. We combined the simulated and observed land use data and developed a cross-tabulation image (Fig. 7). The combined image contained four categories: correctly simulated agriculture; correctly simulated non-agriculture; omission error of agriculture; commission error of agriculture.

Apart from the spatial comparisons for the simulated and observed land use data, we conducted accuracy assessment for the simulation data in 2013. We employed the observed land use data in 2013 as reference and generated an error matrix (Table 5) to calibrate the accuracy of simulated data. The overall accuracy score is82.98%, and the kappa coefficient is 0.66, which suggests medium agreement to the observed data.

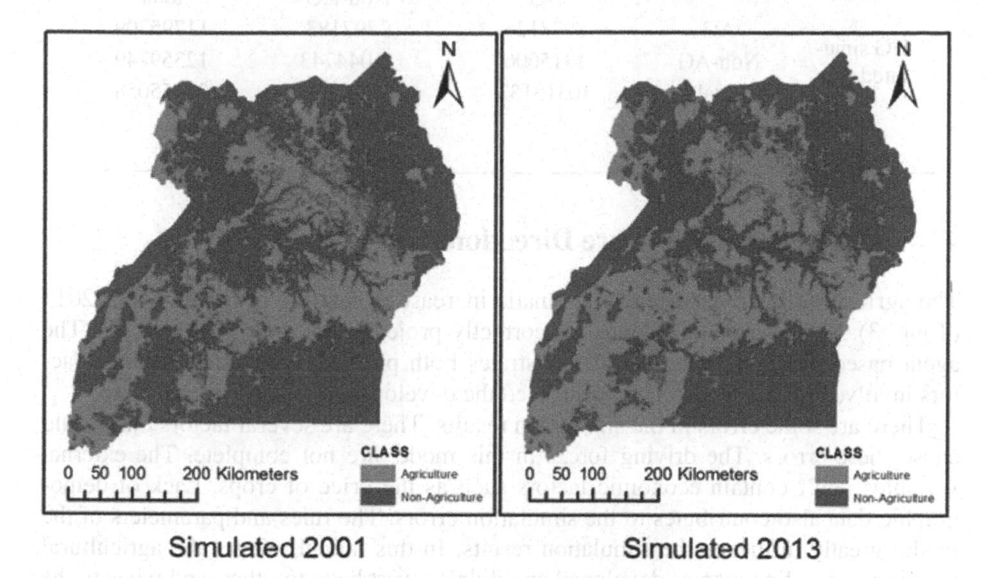

Fig. 6. Simulated land use/landcover data in Uganda

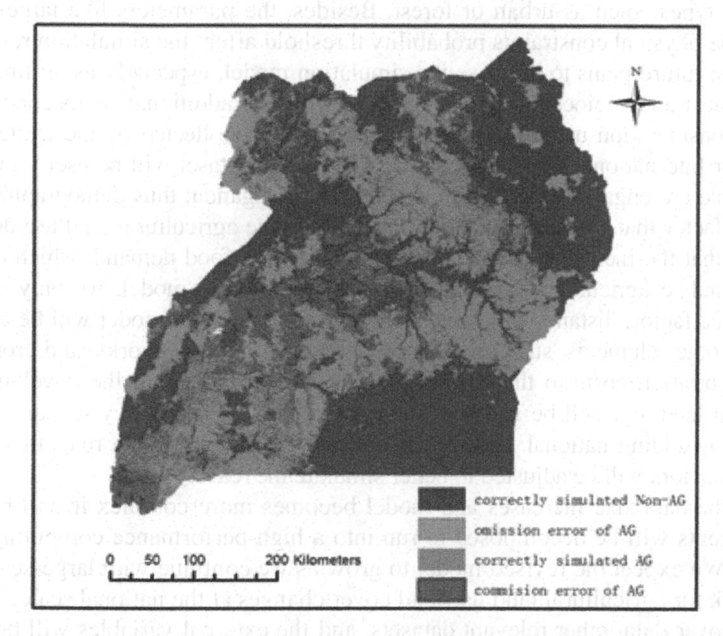

Fig. 7. Spatial comparison for 2013 simulated and observed land use data in Uganda

Table 5. Accuracy assessment of simulated land use data in 2013

		UG observed data		
		AG	Non-AG	total
UG simulated data	AG	8998126	2797183	11795309
	Non-AG	1315006	11044743	12359749
	total	10313132	13841926	24155058
overall accuracy			82.98%	
Kappa Statistics			0.66	

4 Discussion and Future Directions

The agricultural land experienced dramatic increase during the period of 1996–2013 (Table 3). Most agricultural areas are correctly projected by the model (Fig. 7). The agent-based modeling approach demonstrates both physical and socioeconomic factors involved in this model have influenced the development of agricultural land.

There are some errors in the simulation results. There are several factors that would cause these errors. The driving forces in this model are not complete. The external variables don't contain economic factors such as the price of crops. Lack of demographic data also contributes to the simulation errors. The rules and parameters of the model greatly influence the simulation results. In this model, we set the agricultural land in the baseline map as developed and didn't revert back to other land types in the iteration process. However, in real world, agricultural land could probably change to other land types, such as urban or forest. Besides, the parameters like largest parcel size and the physical constraints probability threshold affect the simulation results.

We have future plans to improve the simulation model, especially using finer scales of biophysical and socioeconomic factors. We will take additional socioeconomic factors into consideration using the household survey data collected by the United States Agency for International Development (USAID). This dataset will be useful given that many people are engaged in agricultural activities in Uganda; thus demographic data is a weighty factor that has a significant influence over the agricultural land use decisions. We know that the increasing population spurs increased food demand, which does primarily stimulate agricultural land expansion. In the current model, we only used one market force factor, distance to towns,sothe revised simulation model will be enhanced by adding other elements, such as distance to transportation networks, and crop prices. The transportation costs to the market play an important role in the development of agricultural land and will be added to the revised model. The policy scenarios will be improved by adding national agricultural policies. Additionally, the rules that regulate agents' behaviors will be adjusted to better simulate the real world.

When the data size increases and model becomes more complex it will be scaled and the agents will be decomposed to run into a high-performance computing (HPC) platform. We expect the revised model to grow as we continue with large-scale simulation work for agricultural land use/land cover changes at the national scale. The land use land cover data, other relevant datasets, and the external variables will be too big to be dealt with under the Desktop computing environment. The HPC environment will overcome the big data challenge,thus enabling us to develop several historical and future agricultural land use scenarios in Uganda.

References

1. Turner, B.L., William, B.M., David, L.S.: Global land-use/land-cover change: towards an integrated study. Ambio-Stockholm 23, 91–91 (1994)
2. Turner II, B.L., Lambin, E.F., Reenberg, A.: Land change science special feature: the emergence of land change science for global environmental change and sustainability. Proc. Natl Acad. Sci. 104, 20666–20671 (2007)
3. Bonan, G.B.: Effects of land use on the climate of the United States. Climatic Change 37(3), 449–486 (1997)
4. Pielke, R.A., Marland, S.G., Bets, R.A., Chase, T.N., Eastman, J.L., Neils, J.O., Niyogi, D.D.S., Running, S.: The influence of land-use change and landscape dynamics on the climate system: relevance to climate-change policy beyond the radiative effect of greenhouse gases. Philosophical Transactions of the Royal Society A 360, 1705–1719 (2002)
5. Manson, S.M.: Agent-based modeling and genetic programming for modeling land change in the Southern Yucatan Peninsular Region of Mexico. Agriculture, Ecosystems & Environment 111(1), 47–62 (2005)
6. Verburg, P.H., Schulp, C.J.E., Witte, N., Veldkamp, A.: Downscaling of land use change scenarios to assess the dynamics of European landscapes. Agriculture, Ecosystems & Environment 114(1), 39–56 (2006)
7. Valbuena, D., Verburg, P.H., Bregt, A.K., Ligtenberg, A.: An agent-based approach to model land-use change at a regional scale. Landscape Ecology 25(2), 185–199 (2010)
8. Rindfuss, R.R., Walsh, S.J., Turner II, B.L., Fox, J., Mishra, V.: Developing a science of land change: challenges and methodological issues. PNAS 101, 13976–13981 (2004)
9. Beratan, K.K.: A cognition-based view of decision processes in complex social–ecological systems. Ecol. Soc. 12(1), 27 (2007)
10. Matthews, R.B., Gilbert, N.G., Roach, A., Polhill, J.G., Gotts, N.M.: Agent-based land-use models: a review of applications. Landscape Ecology 22(10), 1447–1459 (2007)
11. Macal, C.M., North, M.J.: Agent-based modeling and simulation. In: Winter Simulation Conference, pp. 86–98 (2009)
12. Valbuena, D., Verburg, P.H., Bregt, A.K.: A method to define a typology for agent-based analysis in regional land-use research. Agriculture, Ecosystems & Environment 128(1), 27–36 (2008)
13. Naivinit, W., Page, C.L., Trébuil, G., Gajaseni, N.: Participatory Agent-Based Modeling and Simulation of Rice Production and Labor Migrations in Northeast Thailand. Environmental Modelling & Software 25, 1345–1358 (2010)
14. Saqalli, M., Gérard, B., Bielders, C., Defourny, P.: Testing the Impact of Social Forces on the Evolution of Sahelian Farming Systems: A Combined Agent-Based Modeling and Anthropological Approach. Ecological Modelling 221, 2714–2727 (2010)
15. Acosta, L.A., Rounsevell, M.D.A., Bakker, M., Doorn, A.V., Gomez-Delgado, M., Delgado, M.: An Agent-Based Assessment of Land Use and Ecosystem Changes in Traditional Agricultural Landscape of Portugal. Intelligent Information Management 6, 55–80 (2014)
16. Mena, C.F., Walsh, S.J., Frizzelle, B.G., Yao, X.Z., Malanson, G.P.: Land Use Change on Household Farms in the Ecuadorian Amazon: Design and Implementation of an Agent-Based Model. Applied Geography 31, 210–222 (2011)
17. MAFAP (MONITORING AFRICAN FOOD AND AGRICULTURAL POLICIES).: Review of food and agricultural policies in Uganda. MAFAP Country Report Series, FAO, Rome, Italy (2013)
18. UBOS (Uganda Bureau of Statistics).: Statistical Abstract 2013. UBOS, Uganda (2013)
19. UBOS (Uganda Bureau of Statistics).: Statistical Abstract 2002. UBOS, Uganda (2002)

20. World borders dataset, http://thematicmapping.org/downloads/world_borders.php (accessed on June 24, 2014)
21. Benin, S., Thurlow, J., Diao, X., Kebba, A., Ofwono, N.: Agricultural growth and investment options for poverty reduction in Uganda. Intl. Food Policy Res. Inst. (2008)
22. UBOS (Uganda Bureau of Statistics).: Statistical Abstract 2002. UBOS, Uganda (2010)
23. MAAIF (Ministry of Agriculture, Animal Industry and Fisheries).: National Agriculture Policy 2011. Kampala, Uganda (2011)
24. FAO.: The State of Food Insecurity in World 2012 (2012)
25. IFPRI.: Agricultural growth and investment options for poverty reduction in Uganda. International Food Policy Research Institute (2007)
26. Landsat Satellite imageries for Uganda, http://earthexplorer.usgs.gov/USGS (accessed on February 9, 2014)
27. Digital Terrain Elevation Data, http://data.geocomm.com/catalog/UG/group121.html (accessed on March 5, 2014)
28. Lu, D., Weng, Q.: Urban classification using full spectral information of Landsat ETM+ imagery in Marion County, Indiana. Photogrammetric Engineering & Remote Sensing 71(11), 1275–1284 (2005)
29. Anderson, J.R.: A land use and land cover classification system for use with remote sensor data, vol. 964. US Government Printing Office (1976)
30. Jensen, J.R.: Introductory digital image processing: a remote sensing perspective, 3rd edn. Prentice-Hall Inc. (2005)
31. ERDAS Imagine. ERDAS Inc. Norcross, Geogia
32. Aerial Photos for Uganda, Imagery @ 2014 CNES/AstriumDigitalGlobe, Google Maps and Google Earth, http://maps.google.com (accessed on June 10, 2014)
33. Agent Analyst. ESRI Inc. Redlands, California

Correlation Analysis of Extraction Mechanism of Remote Sensing Anomaly with Mineralization and Ore-Controlling-Illustrated by the Case of Qimantage Area, Qinghai Province*

Mingchao Zhang [1,2,**], Renyi Chen[2], Chao Tang[1], Yongsheng Li[2], Ke Li[1,2], Zhiying Lun[1,2], Lei Yao[1,2], and Xiaodong Gong[1,2]

[1]Faculty of Earth Sciences and Resources, China University of Geosciences, Beijing 100083, China
[2] Technical Guidance for Mineral Resources Exploration, MLR, Beijing 100120, China
Correspondence: Mingchao Zhang, Faculty of Earth Sciences and Resources, China University of Geosciences, Beijing, NO.29 Xueyuan Road, Haidian District, Beijing, China
chris.zhang0210@gmail.com

Abstract. Based on geological background of the workspace, this study determined possible mineralization and alteration types and alteration combinations after analysis; extracted targeted remote sensing anomalies related to the mineralization in the study area; obtained the alteration types corresponding to the mineralization by sorting out the achieved mineralization clues and categories of altered minerals according to the basic geological information; selected appropriate data and reliable algorithm to extract remote sensing alteration information by analyzing the discernible diagnostic spectral characteristics through the mineral and alteration identification physical basics. It also made extensive significant analysis through the combination of remote sensing information about anomalies and deposit type, stratum, tectonic feature etc., as well as previous geophysical and geochemical data and delineated anomalies. Taking into consideration the basic geological background and the distribution conditions of contact metsomatic deposits, hydrothermal deposits and porphyry deposits, the delineation of anomalies is of certain significance for the regional two-dimensional metallogenic prediction and the determination of prospecting targets.

Keywords: Qimantage, Qinghai Province, Remote sensing anomaly, Anomaly analysis, Correlation analysis of mineralization and ore-controlling.

1 Introduction

As an important part of the metallogenic prediction research, remote sensing has decades of history with rich experience in application. In addition, the establishment of

* This work was supported by the Geological Survey of China Geological Survey Project Dynamic (12120113090000).

** Corresponding author.

F. Bian and Y. Xie (Eds.): GRMSE 2014, CCIS 482, pp. 471–485, 2015.

remote sensing geology discipline has made great contributions to the development of geological research investigation and led to a number of fruitful achievements [1]. With the promotion of remote sensing geological survey, the application of remote sensing is constantly expanding, from interpretation of the tectonic framework to extraction of the qualitative micro-structure, and from lithology determination to quantitative inversion of the mineral [2-4].

Mineralization and alteration phenomena generated by metasomatic alteration during the mineralization geological processes are reflected as anomalies in the remote sensing images. At present, the majority of scholars conducting research related to the process of mineralization forecast or remote sensing geological surveys will consider the extraction of remote sensing anomalies as an important factor, highlighting the convenience, speed and effectiveness of remote sensing on geological mineralization prediction [3, 4]. Nevertheless, according to the previous studies some problems become prominent, mainly in the following aspects: instead of analyzing geological background of the study area, many studies blindly extract some remote sensing abnormal areas which may be unrelated to the study area or of little correspondence with mineralization when identify alteration types or alteration combinations of potential mineralization. Some regions in which mineralization has little or no relation with hydrothermal deposit are not suitable for using the remote sensing extraction method as part of mineralization prediction. Different minerogenetic series (minerals) have different characteristics depending on various alteration geneses. And a considerable number of minerals do not report alteration phenomenon. Therefore, the "OHA (Hydroxyl Anomaly) + FCA (Ferric Contamination Anomaly)" remote sensing anomalies are not universal representative for mineralization, which means that mineralization and alteration is not simply equal to "OHA+FCA". Metasomatic alteration information that can be identified by means of remote sensing is limited by now. And whether the remote sensing anomalies showed in the mineralized alteration images can be extracted and identified are depended on the spectroscopic model of the altered eigen-minerals [2-5].

How to deepen the understanding of the existing problems, and on the basis, to further improve the geology and physical models achieved from remote sensing data in both theory and practice is the main content of this paper. Illustrated on the analysis of remote sensing anomaly extraction and ore-controlling in Qimantage, Qinghai Province, this paper is trying to sort out the correlation between information extraction of remote sensing anomalies and mineralization factors in the study area from the bottom area of remote sensing and alteration phenomena, to provide necessary data and theory support for the metallogenic prediction.

2 Overview of Geological Background

2.1 Geology

Located in the southwest margin of Qaidam Basin, Qimantage metallogenic sub-belt is in the north of Qingzang Plateau, East Kunlun offshoot-Qimantage between the East

Kunlun Mountains and Altun Mountains. The tectonic structure belongs to secondary tectonic unit of the East Kunlun orogenic belt, which situated in the southwestern margin of the ancient Qaidam rift site constituting three basins-two ridges. The unique basin-ridge structure supplies not only a good space environment for mineralization but also good conditions for preservation of the formed deposits. The faults in the central and northern Kunlun Mountains go across the study area. The geographic structure experienced at least multi-cycle and multi-system orogeny from Luliang to Indosinian. The complex crustal structure and historical tectonic evolution formed complex tectonic background. Ancient plate edges are the most active areas with positive and frequent energy exchanges and extensively metallogenic fluids communication as the favorable places for the activation, migration and enrichment of deposits. Qimantage is one of the areas with the most obvious metallogenesis in the East Kunlun.

2.2 Analysis of Mineralization Types

The number of clarified deposits (mines) is more than a dozen, to be specific, thirty-forty, after research, including Galinge polymetallic iron-cobalt deposit, Kendekeke polymetallic iron-cobalt deposit, Yemaquan iron polymetallic deposit, Ka'erqueka copper polymetallic deposit, Hutouya copper polymetallic deposit, Sijiaoyang-Zhukutou iron-copper-lead-zinc polymetallic deposit, Jingrendong copper deposit, Wulanwuzhu'er copper-tin deposit and deposits of Fe, Cu, Co, Pb, Zn, Sn, Ag, W and so on. It is said deposits are mainly high-temperature hydrothermal and superimposed ones, followed by skarn or hydrothermal-exhalative sedimentary ones. Porphyry copper deposit also has a good prospect. And Ka'erqueka porphyry-skarn copper polymetallic deposit has made a great breakthrough.

Through the search and summary of extensive literature, the minerals and the corresponding alteration types in the study area as well as ore-controlling factors of the deposit have been summed up, as shown in Table 1. Remote sensing data obtained through the study area were targeted extraction of anomalies. It is said preliminary abnormal areas will be screened according to ore-controlling factors and then determine final anomalies [5].

Table 1. Minerals and corresponding alterations and ore-controlling factors in Qimantage, Qinghai

Deposits	Minerals	Alterations	Major ore-controlling factors
Yemaquan (skarn deposit)	Fe, Cu, Pb, Zn, Ag Polymetallic	Silicification, albitization, epidotization, chloritization, pyrite, taconite, tremolite, actinolitization, garnet, diopside and phlogopite	Magma, rock, tectonic feature (skarn)

Table 1. (*Continued*)

Kendekeke (skarn deposit)	Fe, Co Polymetallic	Skarnization (garnet, chloritization, carbonatization, silicification), silicification, chloritization etc.	Stratum, surrounding rock
Galinge (skarn deposit)	Non-ferrous metals (Co, Au, Bi, Mo, Ni)	Developed Skarnization, seen chloritization, epidotization and silicification	Tectonic structure, stratum, hydrothermal liquid
	Iron Polymetallic	Dominated by skarnization and chloritization, followed by serpentinization, carbonization, sodication, epidotization, secondary amphibolization, tremolite etc.	Tectonic structure
Hutouya (sedimentary deposit)	Iron Polymetallic	Dominated by Skarnization and serpentinization, often seen chloritization, potassium feldspathization, sericitization, carbonization, epidotization, tremolite etc.	Stratum, Syngeneic homogenous breccia and effusive rock-secondary effusive rock (volcanic eruption)
	Pb, Zn	Skarnization, silicification, marbleization, chloritization	Stratum, tectonic structure (stratabound deposit)
Ka'erqueka (porphyry, medium-low temperature hydrothermal combined with skarn deposit)	Pb-Zn polymetallic	Potassium, garnet, diopside, epidotization, chloritization, carbonization	Magmatic body, surrounding rock, tectonic structure (skarn)
	Polymetallic	Silicification, chloritization, epidotization, tremolite, garnet, diopside	Magmatic hydrothermal liquid
	Porphyry copper	Skarnization (developed diopside, chlorite, epidote, actinolite, garnet, calcite etc.)	Tectonic structure, magmatic hydrothermal liquid
	Cu, Mo	Kaolinization, sericitization, potassium, silicification etc.	Tectonic structure, magmatism
	Copper polymetallic	Beresitization, Skarnization	Rock body, fault-fracture zone

According to the foregoing analysis, the main mineralization and alteration types in the study area are chloritization, epidotization, carbonization of quartz, calcium, iron and magnesium, iron oxidation, Kaolinization and so on.

3 Mapping and Analysis of Altered Minerals

3.1 Extraction Algorithm of Remote Sensing Anomaly Spectrum

Based on the main alteration types in the following study area, hydrothermal deposit, skarn deposit and porphyry copper deposit, the remote sensing anomalies can be divided into four types: Al-OH (e.g. sericitization and chloritization corresponding to the absorption characteristics in Band 5 and Band 6 according to Aster), Fe (e.g. limonitization, corresponding to the absorption characteristics in Band 1 and Band 3 according to Aster), silica in thermal infrared band and carbonate minerals.

According to the diagnostic spectral characteristics of alteration, the extraction of remote sensing abnormal information of minerals with hydroxyl groups is mainly by adopting principal component analysis (PCA) in Band 1, Band 3, Band 4 and Band n of Aster. The abovementioned n is determined by the spectral characteristics of altered minerals in the study area [6, 7]. The extraction method of ferric contamination anomaly makes mask principal component analysis for Band 1, Band 2, Band 3 and Band 4, thereby attracting components relevant to the ferric contamination anomaly. Band ratio calculation method of identification of alteration minerals is also very helpful. Utilizing the Band 4, Band 6 and Band 1 of Aster as RGB false color combination, then the absorption of Al-OH at Band 6 is purple while the absorption of Fe of Band 1is yellow. Adopting Band 4/5, Band 4/6, Band 4/7 as RGB false color combination and the absorption of Al-OH at Band 5 and Band 6 and absorption of Fe-OH at Band 7 both shows white [8, 9].

According to the corresponding analysis, determined by the diagnostic spectral characteristics of the remote sensing data, the extraction method of the anomalies of the study adopted band ratio method supplemented by PCA to deal with the anomaly data achieved by Aster. For minerals with combination of Fe^{3+}, kaolinite, silicification, epidote, calcite, chlorite, etc., alteration extraction method was used, as shown in Table 2.

Table 2. Main extraction methods of alteration minerals

Minerals	Extraction methods	Minerals	Extraction methods
Fe^{3+}	Band2/Band1	Iron oxide	Band 4/Band3
Fe^{2+}	(Band5/ Band3) + (Band2/ Band1)	Carbonate minerals	Band 4/Band7
Kaolinite	Band4/ Band5	Epidote/calcite/chlorite	(Band7+Band9)/Band8

In the thermal infrared range of atmospheric window (8-12μm), silica, silicate and carbonate minerals corresponding to Si-O and C-O shows strong spectral characteristics. Studies have shown that, the content of SiO_2 is positively correlated with Band 13 and Band 14 in the thermal infrared band of Aster while negatively correlated to Band

10 to Band 12 three bands. Previous experimental results validated that 13×14/ (10×12) is the optimal band combination. On the basis of analysis of emissivity spectral characteristics, carbonate index, CI, is D13 /D14 and emissivity of Band I is DI. The higher the value of CI, the greater the possibility of calcite or dolomite. Quartz index, QI is (D11×D11)/ (D10×D12). Higher value of QI corresponds to quartzite; lower value corresponds to sulfate rocks (such as gypsum) and alkali feldspar (e.g. K-feldspar). Mafic index, MI is D12/ (D13×CI3). MI is negatively correlated with SiO_2 content in silicate rocks. And ultrabasic rocks correspond to higher MI value. While from felsic rocks to quartz rocks, the corresponding MI value decreases (below Fig.1).

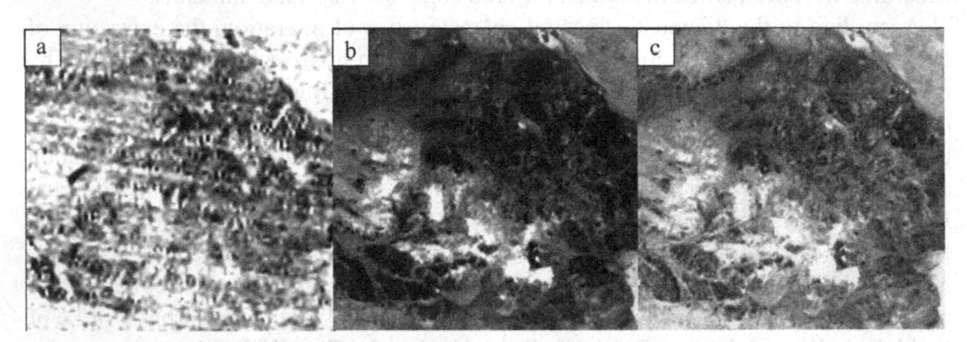

Fig. 1. The relationship between the content of SiO_2 and the band combination of ASTER (a) Aster Quartz Index, higher value corresponds to quartzite; lower value corresponds to sulfate rocks (such as gypsum) and alkali feldspar (e.g. K-feldspar). (b) In Aster RGB 4/5, 4/6 and 4/7, the white area indicates developed siallitization (higher values of Band 4/5 and Band 4/6 reflecting the strong absorption of Al-OH) and limonitization (higher value of Band 4/7). (c) In Aster RGB 4/5, 4/6 and 3/1, the white area indicates developed siallitization (higher values of Band 4/5 and Band 4/6 reflecting the strong absorption of Al-OH) and rich hematite and magnetite iron oxides (Fe^{2+}=> Fe^{3+}) (higher value of Band3/1)

3.2 Analysis of Remote Sensing Anomalies

After treatment of band ratio and principal component analysis, the grayscale reflectance of processed remote sensing images were inversed and then stretched to 0-255. DN value statistics and density separation to the said images were conducted. Proper threshold to determine the alteration area on the basis of the advanced luminance distribution histogram was selected. Median screening and other processing methods to enhance the extracted remote sensing anomaly data were utilized [10]. After the screening of the obtained alteration types and remote sensing abnormal areas, the overall silica is located in Quaternary except that partial anomalies are in Quaternary River or alluvial fan. The Quaternary object spectrum and silica spectra in the visible and near infrared bands are relatively similar. In light of the imprecision of algorithm may cause the erroneous provision of silica in the process of extraction, silica shall be excluded as well. Finally, the distribution of remote sensing anomalies was screened out shown in below Fig.2.

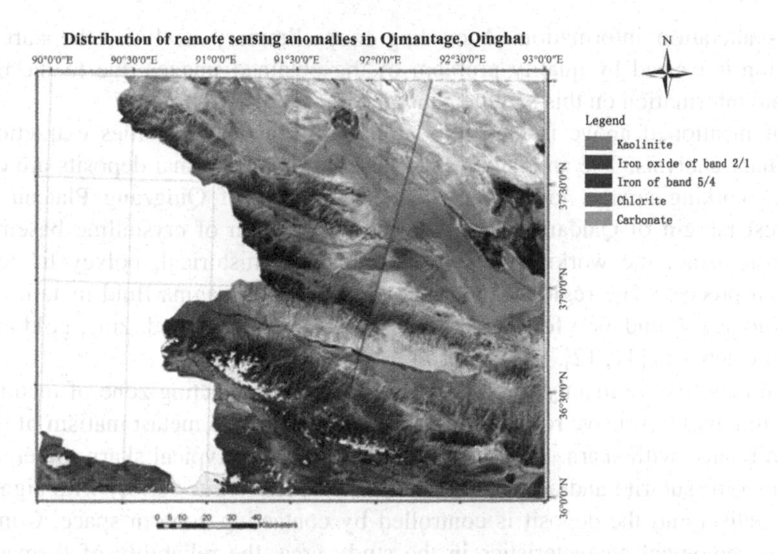

Fig. 2. Distribution of remote sensing anomalies in Qimantage, Qinghai

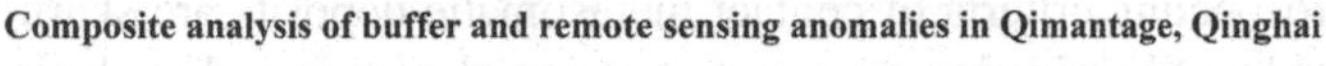

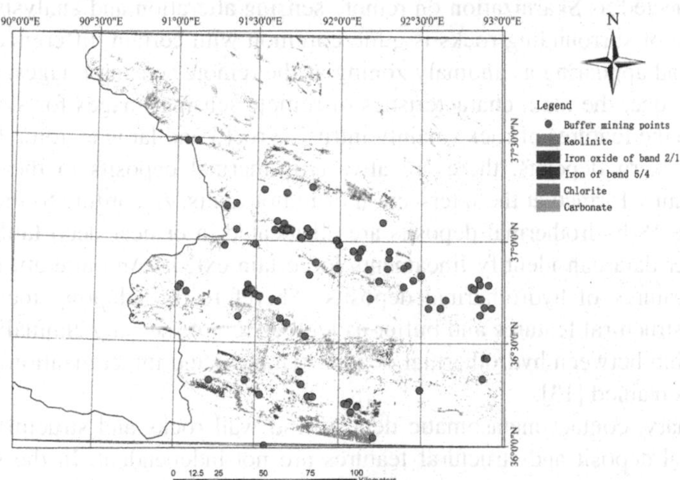

Fig. 3. Composite analysis of buffer and remote sensing anomalies in Qimantage, Qinghai

After screening, the extracted study area becomes 5441.35 km^2 consisting of five minerals corresponding to deposit types in the study area and alteration minerals combination. And ferric contamination anomalies accounts for over 70% of all the statistical area, coinciding with the fact that this area is mainly skarn-iron minerals.

A total of 1500 buffer analyses of the 99 deposits (mines) in the study area were conducted and composite analysis was combined with extracted alteration information. There are 87 deposits (mines) within the scope of extracted remote sensing anomalies. As shown in the bottom right part of Fig.3, some deposits (mines) have no obvious alteration information. According to preliminary analysis, the part under discussion is caused by quality problem of the obtained images due to the unclear abnormal information on this sample band.

What mentioned above is the regional remote sensing anomalies extraction and preliminary analysis. The anomalies and causes of hydrothermal deposits are closely related. Working area is located in the northern part of Qingzang Plateau in the southwest margin of Qaidam Basin. Since the formation of crystalline basement in Paleoproterozoic, the working area has experienced historical, polycyclic tectonic evolution process. The regional revolution of structure-magma-fluid-metamorphism laid geological foundation for the formation of copper, iron, lead, zinc, gold and polymetallic deposits [11, 12].

Skarn deposits are mainly formed at or around the contacting zone of medium acid-medium mafic intrusive rocks and carbonate rocks due to metasomatism of ore-gas solution related with skarn. Such deposits generally have typical skarn mineral combinations (grossularite-andradite series, diopside-hedenbergite series) with significant metasomatism and the deposit is controlled by contacting zone in space. Compared with the geological characteristics in the study area, the reliability of formation of contact metasomatic deposits can be mutually confirmed.

The main prospecting criteria of contact metasomatic deposits are skarn and Skarnization reflected as Skarnization on remote sensing alteration and analysis of buffer. Skarnization of surrounding rocks is quite common with certain differences for various bands and appearing as anomaly zoning at the remote sensing images. Controlled by tectonic zone, the main characteristics of remote sensing images for skarn deposit include also the features of rock (mainly intrusive rock) and large tectonic features. In addition to skarn deposits, there are also hydrothermal deposits in the study area which is mainly formed at the intersection of major faults. According to the statistical analysis, the 18 hydrothermal deposits are all located in or near deep faults. Utilization of Aster data can identify lineaments to certain extent. And alteration is one of the basic features of hydrothermal deposits. Therefore, by adopting the composite analysis of structural features and buffer as well as remote sensing anomalies, a sense of relationship between hydrothermal deposits (mines) and mineralization and alteration can be obtained [13].

In summary, contact metasomatic deposit and wall rocks and structural features, hydrothermal deposit and structural features are not independent. In the study area, the formation of deposits mentioned-above and the geological features are closely associated. The formation of different deposits is mutually effected as well.

Composite Analysis of Remote Sensing Alteration Information and Ore-Controlling Factors for Hydrothermal Deposit

Based on the 1:250,000 mapping made by previous scholars, there are 3191 faults in total in the study area mainly trending NW, NE, and near EW. Structural features are not only the ore-controlling factors, but also dominating mineralization. Therefore, the correlation analysis of structures and alteration is important in theoretically guiding the study of hydrothermal deposits [14].

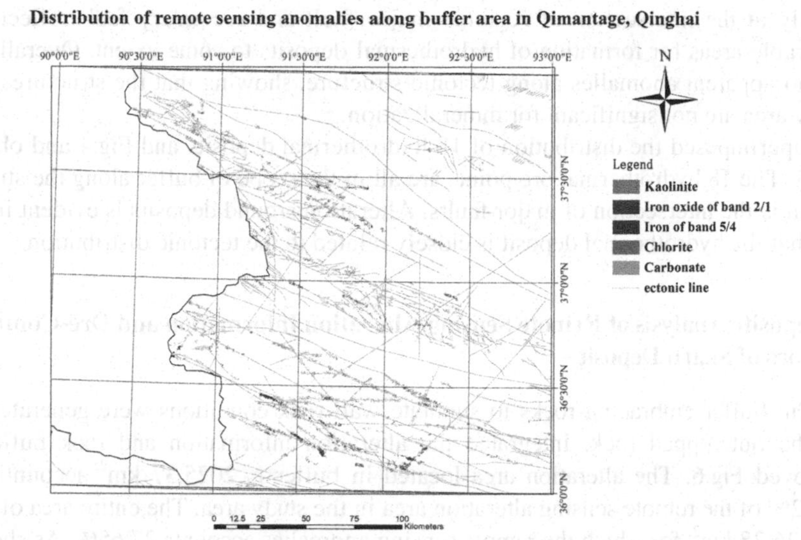

Fig. 4. Distribution of remote sensing anomalies along buffer area in Qimantage, Qinghai

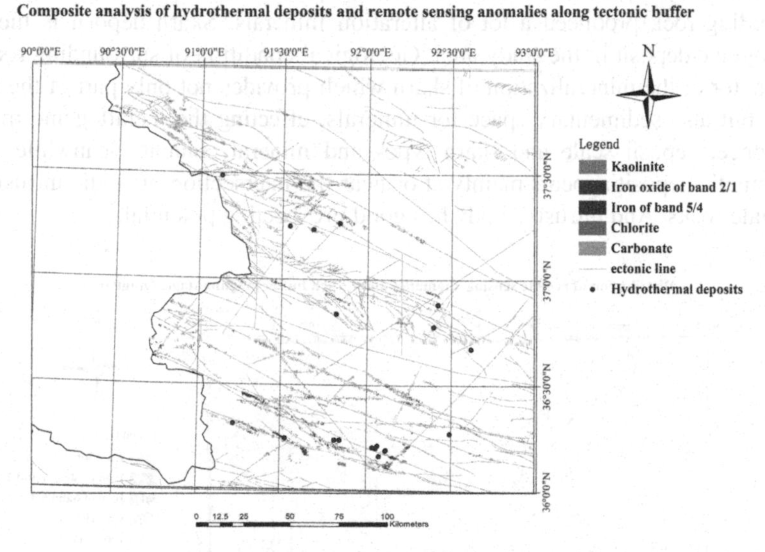

Fig. 5. Composite analysis of hydrothermal deposits and remote sensing anomalies along tectonic buffer area in Qimantage, Qinghai

According to mineralization and alteration abnormal images (see Fig.4) most of the anomalies distribute as bands trending NNW, near EW or along the great fault trending near NS. Major fault in the study area generated 1500m buffer area and the statistical alteration area within the buffer is 1752.635 km^2, accounting for 32.21% of the total alteration area in the workspace, indicating that the alteration in that area is controlled by fault but not particularly evident. Remote sensing anomalies appearing mainly at the intersection of regional major fault and secondary fault reflected the favorable areas for formation of hydrothermal deposits to some extent. Overall, there are no apparent anomalies along tectonic structure, showing that the structures in the study area are not significant for mineralization.

Superimposed the distribution of 18 hydrothermal deposits and Fig.4 and obtained Fig.5. The 18 hydrothermal ore points are all in the scope of buffer along the structure which is the intersection of major faults. Alteration around deposits is evident indicating that the hydrothermal deposit is closely related to the tectonic distribution.

Composite Analysis of Remote Sensing Alteration Information and Ore-Controlling Factors of Skarn Deposit

1500m buffer embracing rocks to simulate wall rock conditions were generate based on the outcropped rock. Integrated the alteration information and rock buffer and achieved Fig.6. The alteration area located in buffer is 2025.37 $km^{2,}$ accounting for 37.22% of the remote sensing alteration area in the study area. The entire area of buffer is 7326.28 $km^{2,}$ for which the remote sensing anomalies accounts 27.65%. As shown in Fig.7, 40 out of 41 are located in the rock margin or within the 1500m buffer. The above data fully explained that the metasomatic process of magmatic hydrothermal and surrounding rock produced a lot of alteration minerals. Skarn deposit is the mainly metallogenic deposit in the study area. Geological condition of surrounding rock is the main factor of the mineralization of skarn which provides not only part of the material source but also sedimentary space for minerals, effecting the metallogenic mode, ore occurrence, deposit scale and skarn types and mineral content meanwhile. Contact metasomatic deposit appears mainly at or near the intersection of acidic intrusions and carbonate rocks. And intrusive body has good prospecting potential.

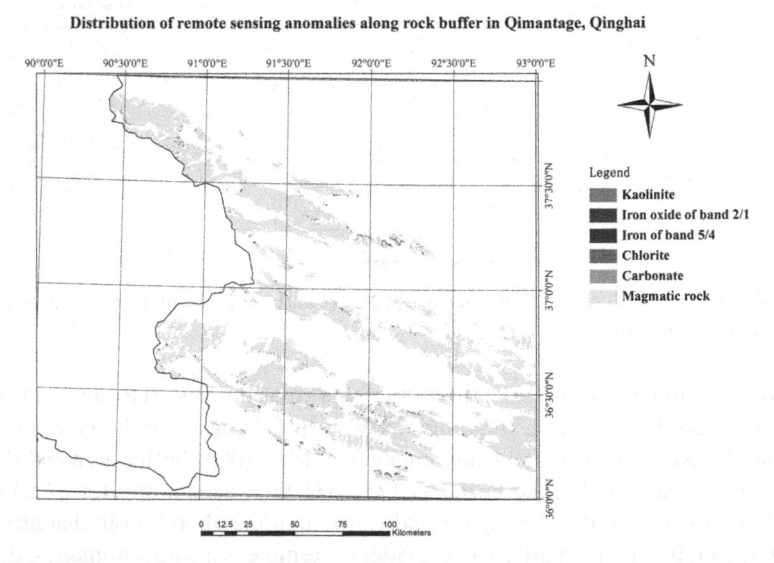

Fig. 6. Distribution of remote sensing anomalies along rock buffer in Qimantage, Qinghai

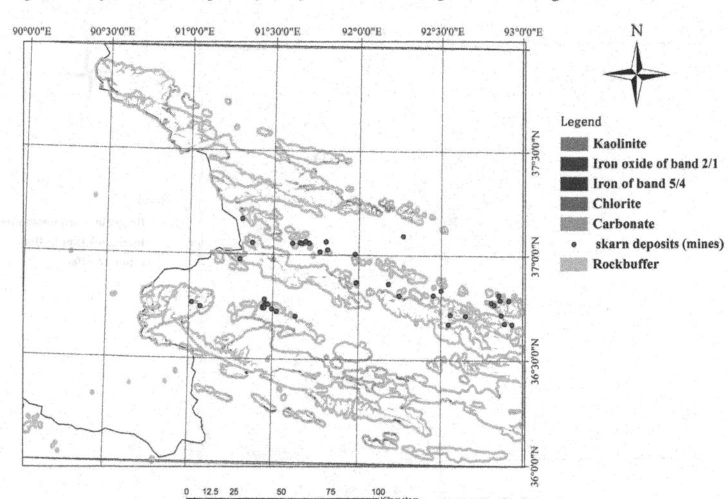

Fig. 7. Composite analysis of Skarn deposits and remote sensing anomalies along tectonic buffer area in Qimantage, Qinghai

Selection of Recommended Anomalies with Priority

Remote sensing anomaly map of the study area was screened to get prior anomaly coordinates based on the principle that the preferred anomalies are associated with mineralization [15, 16]. Such anomalies include abnormalities associated with structures, which does not belong to ore-controlling factors while plays a dominating role in mineralization. Therefore, correlation analysis of structure and alteration is of important significance for prospecting study. Anomalies of Level A mainly occur at the intersection of regional great fracture and secondary fracture, the area under discussion is a favorable generation place for hydrothermal deposits. Nevertheless, when screening anomalies associated with structure, do note that many abnormalities only associated with structure in spatial instead of tectonic alterations. There is no better way to distinguish each other except field study despite that it's particularly confusing.

Volcanic and intrusive rocks as well as ophiolite are obvious signs of hydrothermal deposits. Take full account of the mineralization and alteration based on structure and wall rocks during the selection of anomaly coordinates.

Firstly, take the surface center coordinates of alteration anomaly areas and generate 1000m buffer, then take the center coordinates of the intersection part of intrusive rocks and ophiolite located in the 1500m buffer. The point closest to tectonic features should be selected as priority. According to the above method, the study area was recommended to focus on verification of 145 statistical anomalies (see Fig.8).

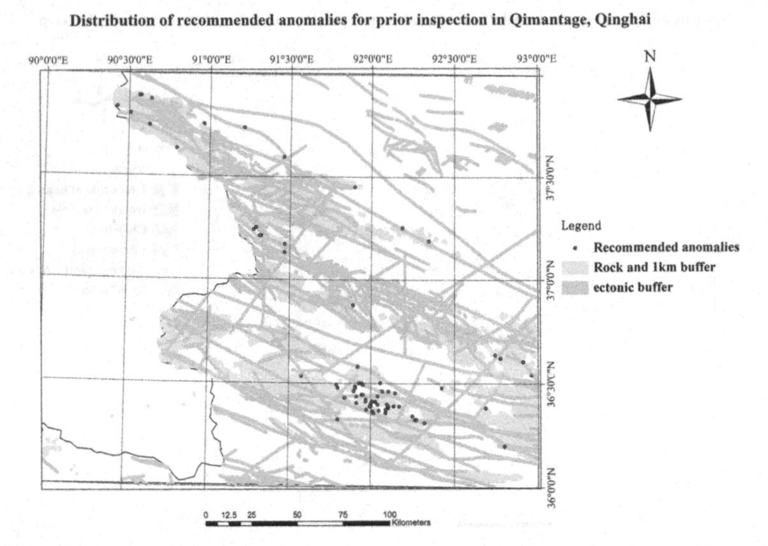

Fig. 8. Distribution of recommended anomalies for prior inspection in Qimantage, Qinghai

3.3 Delineation of Anomalies

Make comprehensive utilization of all kinds of data include geological, mineral, geophysical, geochemical and remote sensing etc., especially information that is conductive to obtain and realistic [17]. With the platform of GIS, make projection transformation and coordinate registration for vectors of remote sensing anomaly layer, geophysical combination anomaly layer, geochemical combination anomaly layer and geological layer. Through correlation analysis, discriminant analysis and weighted overlay analysis to study the morphology, strength and distribution rules of anomalies as well as the significance of prospecting. Then optimize the inspection sequence for all anomalies to guide targeted prospecting.

The purpose of remote sensing is prospecting while anomaly information extraction is one processing step. Optimize the conduction sequence of anomalies through comprehensive analysis of various data and then verify abnormal areas to test the optimal anomalies after field research on detailed geological conditions. Make further improvements and adjustment for pre-processed remote sensing images timely.

Comprehensively analyze the superimposed results of various types of anomalies information, coupled with various geological information including the regional geological setting and distribution of rock formations, magma emplacement, geological tectonic characteristics, regional geochemical anomalies, wire and loop imaging features of remote sensing and alteration information to extract prospecting information. Then combine the above information with ductile shear zone mineralization analysis to determine the extent and location of anomalies [18, 19].

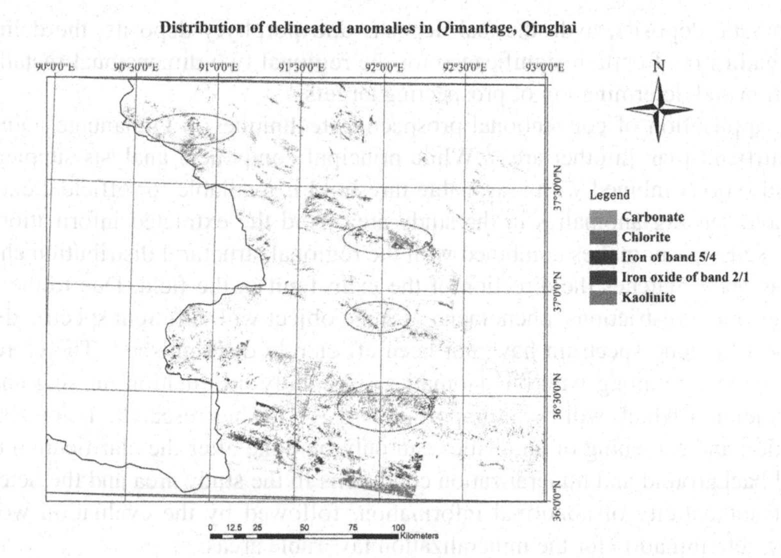

Fig. 9. Distribution of delineated anomalies in Qimantage, Qinghai

Four anomalies were determined through the entire workspace with an area of 2,774.3097km^2 and 286.85km^2 for remote sensing abnormal area. And the second anomaly mentioned above is the study area of this paper. Through the comparison of Fig.8 and Fig.9, all of the four anomalies are located in the intersection of structure and magmatic rocks and zones, with a large part of outcropped surrounding rocks. Such zones are favorable area for the wall rock alteration of hydrothermal deposits, contact metasomatic deposits. In light of the basic geological conditions and the distribution of contact metasomatic deposits, hydrothermal deposits and porphyry deposits, the delineation of anomalies is of certain significance for the regional two-dimensional metallogenic prediction and determination of prospecting targets [20].

4 Conclusions

Based on geological background of the workspace, the study determined possible mineralization and alteration types and alteration combinations after analysis; extracted targeted remote sensing anomalies related to the mineralization in the study area; obtained the alteration types corresponding to the mineralization by sorting out the achieved mineralization clues and categories of altered minerals according to the basic geological information; selected appropriate data and reliable remote sensing alteration extraction algorithm to extract abnormal information by analyzing the discernible diagnostic spectral characteristics through the mineral and altering physical basics. Extensive significant analysis was made through the combination of remote sensing anomalies and deposit type, stratum, structure etc., as well as previous geophysical and geochemical data and cuts off an area of anomalies. Taking consideration of the basic geological background and the distribution conditions of contact metasomatic deposits, hydrothermal deposits and porphyry deposits, the delineation of anomalies is of certain significance for the regional two-dimensional metallogenic prediction and determination of prospecting targets.

The application of conventional prospecting techniques in Qimantage, Qinghai is more difficult than in other areas. While principal component analysis supplemented by band ratio combined with reasonable threshold is available for efficient extraction of remote sensing anomalies in the study area. And the extracted information about remote sensing anomalies combined with the regional structural distribution characteristics exactly matches the direction of the main fault in the field. Due to the remote sensing source restrictions, phenomena of same object with different spectra, different objects with same spectrum have not been efficiently distinguished. Therefore some errors may come along with the anomalies, especially information missing and erroneous mention which will be an important part for further research. To be short, the extraction and screening of anomalies can only be done after the clarification of geological background and mineralization conditions in the study area and the determination of authenticity of abnormal information, followed by the evaluation work and priority determination for the mineralization favorable areas.

Interpreting the remote sensing images and combining the structural characteristics and mineralization alteration make the extraction of alteration more accurate.

Compared with ETM, the spectral and spatial resolution remote sensing data of Aster has been greatly improved: more short-wave infrared bands; a stronger advantage in terms of information extraction of remote sensing anomalies; higher spectral resolution and greater ability to identify alteration minerals and mineral combinations; availability to make wider range of alteration mineral mapping. The improvements and developments mentioned-above are of great significance for remote sensing prospecting.

References

1. Ninomiya, Y., Fu, B., Cudahy, T.J.: Detecting Lithology with Radiance at the Sensor Data of ASTER Multispectral TIR. Remote Sensing of Environment 99, 127–139 (2005)
2. Fei, Y., Li, Y.: Extraction of Remote Sensing Alteration Information and Analysis of Metallogenic Prognosis based on ETM-A Case Study of Danghe-Nanshan Area in Gansu. Remote Sensing Technology and Application 26(4), 482–488 (2011)
3. Zhang, Y., Yang, J., Chen, W.: A Study of the Method for Extraction of Alteration Anomalies from the ETM+(TM) Data and Its Application: Geologic Basis and Spectral Precondition. Remote Sensing for Land & Resources (4), 30–36 (2002)
4. Tian, F., Dong, L., Yang, S., Wang, R.: Application of Combined Spectra of Mixed Minerals to Mapping Altered Minerals: A Case Study in the Yunnan Region Based on Hyperion Data. Geology and Exploration 46(2), 331–337 (2010)
5. Chen, J., Wang, Q., Dong, Q., Yuan, C.: Extraction of Remote Sensing Alteration Information in Tuotuohe, Qinghai Province. Earth Science-Journal of China University of Geosciences 34(2), 314–318 (2009)
6. Jing, F., Chen, J.: The Review of the Alteration Information Extraction with Remote Sensing. Remote Information 2, 62–66 (2005)

7. Chen, J., Wang, A.: The Pilot Study on Petro-Chemistry Components Mapping with ASTER Thermal Infrared Remote Sensing Data. Journal of Remote Sensing 11(4), 601–608 (2007)
8. Liu, C., Wang, D., Li, X.: Extracting Clay Alteration Information of Medium Vegetation Covered Areas Based on Linear Model of Spectral Mixture Analysis. Remote Sensing Technology and Application 18(2), 95–98 (2003)
9. Yang, C., Zhu, Q., Jiang, Q., Liu, B.: Silicon Dioxide Content of Surface Rock Quantify Inversion and Retrieval by thermal Infrared ASTER Data. Geology and Exploration 45(6), 692–696 (2009)
10. Chen, J., Sun, W., Yan, B., Yu, H.: The Application and Research of the Anomaly Extraction Process Base on the ASTER Multi-Spectral Remote Sensing in Tianhu Iron Ore Mine. Xinjiang Geology 27(4), 268–272 (2009)
11. Chen, X., Hu, G.: The Extraction of Metallogenesis and Alteration Information of West Gejiu, Yunnan with ASTER. Yunnan Geology 27(4), 502–508 (2010)
12. Wang, Q., Chen, J.: Extraction and Grading of Remote Sensing Alteration Anomaly Based on the Fractal Theory. Geological Bulletin of China 28(2/3), 285–288 (2009)
13. Costantiti, M., Farina, A., Zirilli, F.: The Fusion of Different Resolution SAR Images. Proceedings of the IEEE 85(1), 139–146 (1997)
14. Coutis, B.A.: Remote Sensing of Cold Deserts: Spectral Reflectance Properties of Weathered Rock Surfaces. A. In: Proceedings of the Seventh Thematic Conference on Remote Sensing for Exploration Geology, ERIM, Calgary, Alberta, Canada, pp. 478–500 (1989)
15. Crosta, A.P., Sabine, C., Taranik, J.M.: Hydrothermal Alteration Mapping at Bodies, California, Using AVIRIS Hyper-spectral Data. Remote Senses Environ. 65(3), 309–319 (1998)
16. Feldman, S.C., Taranik, J.V.: Comparison of Techniques for Discriminating Hydrothermal Alteration Minerals with Airborne Imaging Spectrometer Data. Remote Sensing Environ. 24(1), 67–83 (1988)
17. Liu, J., Hu, Z., Qian, Z., Li, H., Sun, J., Yan, Z.: Characteristics of NW linear Structure and Its Significance for Prospecting in East Kunlun Mountains. Journal of Xi'an Engineering University 22(2), 18–21 (2006)
18. Lu, F., Xing, L., Fan, J., Pan, J., Meng, T., He, Q.: Extracting Alteration Information of Remote Sensing Based on Alteration Information Field. Geology and Prospecting 42(2), 65–68 (2006)
19. Tang, H., Dun, P., Fang, T., Shi, P.: The Analysis of Error Sources for SAM and Its Improvement Algorithms. Spectroscopy and Spectral Analysis 25(8), 1180–1183 (2005)
20. Zhao, H., Yao, F., Sun, J.: Metallogenic Regularity and Mineralization Conditions of the Porphyry Type Polymetallic Ore Deposit in Eastern Jilin Province-Taking Xiaoxinancha Gold-Copper Deposit for Example. Gold 26(7), 12–14 (2005)

On Merging Business Process Management and Geographic Information Systems: Modeling and Execution of Ecological Concerns in Processes

Xinwei Zhu[1], Guobin Zhu[1], Seppe vanden Broucke[2], and Jan Recker[3]

[1] International School of Software, Wuhan University
Luoyu Road 37, Hongshan, Wuhan, Hubei, China
[2] Department of Decision Sciences and Information Management, KU Leuven
Naamsestraat 69, B-3000 Leuven, Belgium
[3] Information Systems School, Queensland University of Technology
2 George St, Brisbane QLD 4000, Australia
{xinwei.zhu,gbzhu}@whu.edu, seppe.vandenbroucke@kuleuven.be,
j.recker@qut.edu.au

Abstract. Business Process Management describes a holistic management approach for the systematic design, modeling, execution, validation, monitoring and improvement of organizational business processes. Traditionally, most attention within this community has been given to control-flow aspects, i.e., the ordering and sequencing of business activities, oftentimes in isolation with regards to the context in which these activities occur. In this paper, we propose an approach that allows executable process models to be integrated with Geographic Information Systems. This approach enables process models to take geospatial and other geographic aspects into account in an explicit manner both during the modeling phase and the execution phase. We contribute a structured modeling methodology, based on the well-known Business Process Model and Notation standard, which is formalized by means of a mapping to executable Colored Petri nets. We illustrate the feasibility of our approach by means of a sustainability-focused case example of a process with important ecological concerns.

Keywords: geographic information systems, business process management, business process model and notation, sustainable processes, data-aware processes, process modeling, process execution, coloured petri nets.

1 Introduction

Business Process Management (BPM) has evolved as a holistic, systematic management practice for managing, documenting, modelling, analyzing, simulating, executing and improving end-to-end business processes [1]. BPM considers the use of information systems as a key driver for successful business processes [2] by allowing the monitoring, validation and execution of entire business processes as workflows.

F. Bian and Y. Xie (Eds.): GRMSE 2014, CCIS 482, pp. 486–496, 2015.

Much attention within the BPM community has traditionally been given to studying control-flow aspects of business processes, meaning the sequence and ordering in which activities can be performed. In recent years, however, other aspects, or "contexts" within this view, have received an increasing amount of attention [3]. As such, scholars have shifted towards studying various approaches that integrate additional contexts so that processes can be rapidly modified and adapted to (new) external data-governed inputs, such as social aspects, logistic information, or resource and inventory artifacts [4].

One particular context which has been gaining traction and we wish to emphasize in this work is that of "Green BPM" [5]. In managing operational processes, enterprises have traditionally been geared towards optimizing economic imperatives, such as time, cost, efficiency and quality. Whilst doing so, they have been a major contributor to environmental degradation caused by consumption of earthly resources, CO_2 emissions and waste. Consequently, organizations are increasingly encouraged to improve operations from an ecological perspective, working towards making their processes environmentally sustainable. From a BPM perspective, such ecological concerns are another context element that should be merged with traditional approaches towards the modeling, analysis or execution of business processes.

This paper contributes to the existing body of work by proposing a structured and formalized methodology towards enabling sustainable process management. Whilst a great deal of work has already been put towards investigating how this context can be taken into account during the planning, design and modeling of processes, the execution of business processes against this context has received less attention. Such an approach would help to govern and constrain control-flow and process behavior (making a decision to execute a particular activity, for instance) based on ecological aspects during the actual "running" of the process. Naturally, it self-evidently follows that such an approach cannot be constructed in isolation from other information systems that are highly relevant in this context. One type of information system that we are interested in specifically are Geographic Information Systems (GIS) [6]. This is because these systems provide a means to store, manipulate and manage various types of geographical data, and thus are also able to provide information regarding ecologic and resource-based aspects that could be used to describe the ecologically relevant context of a business process.

We proceed as follows. Section 2 provides an overview of related work and preliminaries. Section 3 introduces our new structured process modeling approach based on the Business Process Model and Notation (BPMN) standard in order to show how non-control-flow information can be modeled in an explicit manner when designing business processes. In Section 4, we discuss how the execution of such processes can be supported by introducing a formalized mapping to executable colored Petri nets. In Section 5, we describe a case example showing the feasibility of our approach in the context of sustainable processes, where the link with GIS also becomes evident. Section 6 concludes the paper.

2 Background

2.1 Related Work

We regard the environmental context as one of the key variables in the wider setting of a business process. In the layered process context model proposed by Rosemann et al. [3], this context is explicitly included as a separate layer influencing the business process—including factors stemming from ecological aspects and geospatial artifacts. Considering including this context towards making BPM sustainable, we refer to the seminal work of vom Brocke et al. [5], where the notion of "Green BPM" was put forward and a framework for sustainable information systems was presented. Other approaches on the topic of Green BPM exist, but it should be noted that the discussions on Green BPM methods are still in the early stages and so far only a few approaches exist [5, 7-10]. In particular, although works have been presented in order to annotate process models with relevant sustainability information or propose modelling extensions to explicitly include ecological information such as carbon footprint effects [11], we wish to take this a step further by also including such aspects during the execution of process models, as to constrain and govern the control-flow of processes. Houy et al. [8, 9] proposed a semi-automated process improvement approach based on environmental information, by constructing a library of process fragments based on which an optimal process model can be constructed. A downside of this approach is that modelers first need to construct such a collection of feasible process fragments and that the optimization is not performed during execution but rather as a post-hoc analysis. Note that we do not only consider this to be a relevant aspect for organizations aiming to optimize their processes in terms of minimalizing the environmental impact of their activities, but also to optimize their efficiency by taking such environmental information into account in the first place. Consider for example a farming process where the decisions regarding soil preparation, crop selecting, seeding/planting, pest prevention, and harvesting can be based on weather information or ground health indicators about the location of the farming processes.

2.2 Preliminaries

Business Process Model and Notation. A general introduction to BPM is available elsewhere [12]. We focus on the industry standard BPMN [13] to present a structured modeling approach to combine process control-flow aspects (which are supported by standard BPMN) with other contextual information such as geospatial or environmental concerns.

BPMN is a standard for business process modeling that provides a graphical notation for designing business processes, similar to a flowchart. Fig. 1 provides an overview of the BPMN graphical elements that will be used throughout the paper. The graphical constructs are categorized in four categories, these being:

- Flow objects: events, activities and gateways;
- Connecting objects: sequence flow, message flow and association;
- Swim lanes: pools and lanes (grouping other elements);
- Artifacts: data objects, groups, and textual annotations.

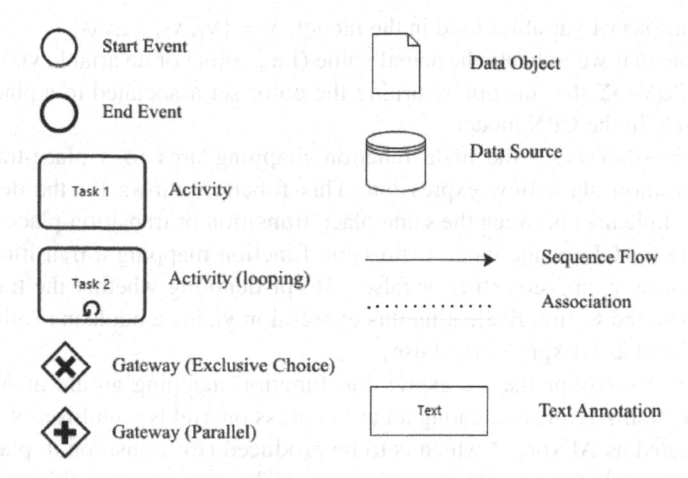

Fig. 1. Overview of common BPMN constructs used throughout the paper

Every BPMN model starts and ends with a Start Event and End Event element respectively. Sequence Flow arrows connect these elements with a series of Activities. Branching and merging points in the process are represented with Gateway elements, either as an exclusive choice or a parallel branch. Data Objects and Data Sources represent external data which can be associated to sequence elements by means of an Association arc.

In order to examine how structured BPMN can be extended to support geospatial and/or ecological variables, in the following, we base ourselves on earlier published work [14], in which we extended BPMN to support location-dependent patterns. We generalize this here to enable the inclusion of *any* contextual information using existing BPMN constructs. We discuss this approach in Section 3.

Coloured Petri Nets. To support the execution of our structured, ecological-aware process models and to support integration with GIS systems, we will provide a formalized mapping from our structured BPMN methodology to coloured Petri nets (CPN) [15, 16]. CPNs are an extension of Petri nets [17], a mathematical modeling language for the description of distributed systems CPN are the standard formalism to define executable process models and allow for their execution in typical workflow engines.

In the following we briefly review the most important CPN formalisms of relevance to this paper. The interested reader is referred to Jensen et al. [16] for more detailed background information.

Definition 1. A CPN is a tuple (P, T, A, Σ, C, V, N, G, E, M, I) with:

- P the set of places, $P = \{p_1, p_2, ..., p_{|P|}\}$
- T the set of transitions, $T = \{t_1, t_2, ..., t_{|T|}\}$ with $P \cap T = \emptyset$
- A the set of arcs, $A = \{a_1, a_2, ..., a_{|A|}\}$
- Σ the set of color sets defined within the model
  A color set is a grouping of colors. A color is an attached value to a token

- V the set of variables used in the model, $V = \{v_1, v_2, \ldots, v_{|V|}\}$
  Note that we indicate the actual value (i.e., color) of a variable $v \in V$ as v^*
- $C: P \cup V \rightarrow \Sigma$ the function returning the color set associated to a place or a variable in the CPN model
- $N: A \rightarrow P \times T \cup T \times P$ the node function mapping arcs to a place-transition or transition-place flow expression. This function allows for the definition of multiple arcs between the same place-transition or transition-place pair
- $G: t \in T \rightarrow GExpr$ the guard expression function mapping a transition $t \in T$ to a boolean expression (true or false) GExpr denoting whether the transitions is permitted to fire. Evaluating this expression yields a boolean result value, indicated as $GExpr^* \in \{true, false\}$
- $E: a \in A \rightarrow AExpr$ the arc expression function mapping an arc $a \in A$ to an expression AExpr. Evaluating an arc expression yields a multiset of tokens, indicated as $AExpr_{MS}^*$ which is to be produced (for transition to place arcs) or consumed (for place to transition arcs). The expression itself can use one or multiple variables in V
  The color sets of the input and outputs of the arc expression must correspond to the color sets of the places the arcs connects to: $a \in A: [\exists \sigma \in \Sigma: [\forall \tau \in E(A)_{MS}^*: [\tau \in \sigma] \wedge C(P \cap N(a)) = \sigma]]$
- $M: p \in P \rightarrow C(p)_{MS}$ the marking function, returning the multiset of tokens contained in a place with $\forall p \in P: [\forall \tau \in M(p): [\tau \in C(p)]]$
- $I: p \in P \rightarrow IExpr$ the initialization function, this function initializes places in the model with a state, expressed as colored tokens. The evaluation of an IExpr yields a token multiset, indicated as $IExpr_{MS}^*$ with $\forall p \in P: [\forall \tau \in IExpr_{MS}^*: [\tau \in C(p)]]$

We also define the following functions:

- Let $p: A \rightarrow P$ be a function returning the place attached to an arc, i.e. $p: a \in A \rightarrow P \cap N(a)$
- Let $t: A \rightarrow T$ be a function returning the transition attached to an arc, i.e. $t: a \in A \rightarrow T \cap N(a)$
- Let $Type: A \rightarrow \Sigma$ be a function returning the type (color set) of the associated place to an arc, $Type: a \in A \rightarrow C(p(a))$

The actual execution of CPN models involves "firing" enabled transitions, which move tokens from their input to output places according to the expression on the arcs connecting them.

Definition 2. For a transition $t \in T$ to be enabled, the following criteria need to hold:

- All expressions of the incoming arcs should be satisfied: $\forall a \in A, t(a) = t: [E(a)_{MS}^* \neq \emptyset]$
- The guard condition of the transition must evaluate to true, $G(t)^* = true$

Enabled transitions can be fired. Output and input places are updated accordingly given the input and output arc expressions. Firing an enabled transition brings a marking $M1 \rightarrow_t M_2$ as follows by updating the marking for each connected input and output place. Tokens are removed from input places according to the arc expressions on the arcs connecting those places to the transition which is fired, whereas output places receive tokens according to the arc expressions on the arcs connecting those places to the transition which is fired.

3 Modeling Contextual Processes

This section describes a structured BPMN modeling approach in order to define process models with contextual artifacts. These contextual artifacts are in the following represented as instances of data artifacts, whereby the type of data can be defined as any contextual information of relevance, including geospatial or ecological information.

Our approach is structured since we extend BPMN by means of making the approach stricter, following a clear set of guidelines. The following definition provides an overview of these guidelines.

Definition 3. A structured contextual BPMN model is defined as a PBMN model so that:

- Control flow is represented using Start Event, End Event, Activity, Exclusive Choice Gateway, Parallel Gateway and Sequence Flow constructs only (see Fig. 1).
- The Data Source element represents a collection of data elements belonging to the same type. For instance, a Data Source can be defined to represent resources, people, countries, and so on.
- The Data Object element represents a particular data element belonging to a Data Source. This implies that each Data Object has one type.
- Activity elements can involve a number of Data Objects which will be bound to that activity at the time of execution. These Data Objects are connected to the activity by means of an Association arc (from the Activity element to the Data Object element). All Data Source elements are also connected to the Activity Element to indicate the Activity is taking data from the types they represent.
- The execution of Activity elements can be constrained by free-form constraints which are annotated to the Activity element: a Text Annotation connected to the Activity with an Association arc. This allows to govern control-flow at execution time based on external data. Constraints may—but do not have to—involve other Data Objects bound to previous Activity elements, in which case an Association arc is added from the input Data Object to the constrained Activity element.

Note that the strictness of our approach is not limiting by any means, as any BPMN model can be converted to its structured counterpart (e.g., by merging lanes and pools). In our case example in Section 5, Fig. 2 provides an example of the structured modeling approach, in the form of a build-to-order process.

4 Executing Contextual Processes

To support the execution of our structured BPMN process models and to support integration with GIS systems, we now provide a formalized mapping from our structured BPMN methodology to CPN. To be more precise, the conversion from a BPMN model to an executable CPN model is performed as follows.

Definition 4. A structured BPMN model is converted to a CPN as follows:

- $\Sigma=\{U, D_1, \ldots, D_n\}$ with U = {unit} the color set containing on control-flow oriented color and $D_i = \{d_1, \ldots, d_{|Di|}\}$ a color set for each Data Source contained in the BPMN model with $d_1, \ldots, d_{|Di|}$ the Data Elements contained in this Data Source. I.e. Customer={customer1, customer2, …}
- Control flow elements from the BPMN model (Activities, Sequence Flow and Gateways) can be immediately mapped to CPN models as the semantics of the latter allow all control-flow constructs of the former (i.e. sequence, loop, exclusive choice and parallel split)
- For each Data Source D_i, an input place $p^iD_i \in P$ is added to the CPN model with $C(p^iD_i) = D_i$ and $I(p^iD_i) = \{d_1, \ldots, d_{|Di|}\}$
- For each Data Element de_i acting as an output for an Activity, an output place $p^o de_i \in P$ is added to the CPN model with $C(p^o de_i)$ equal to the data type corresponding with the Data Element (i.e. the Data Source the Data Element is contained in)
- Constraints in the BPMN model are added as guards to the transitions in the CPN model
- For each Data Element de_i acting as an output for an Activity, we add four arcs to the CPN model: two arcs to move a token from an input place p^iD_i to the transition and move it back to the input places, an arc to remove a currently bound Data Element from the output place $p^o de_i$ and an arc to assign the chosen Data Element to the output place.
- Whenever a Data Element acts as an input in a constraint of another Activity, we add two arcs from and to the output place containing this Data Element to the transition.

The following section describes a case study which applies our approach in a practical, sustainability-driven context.

5 Case Study

We illustrate the feasibility of our approach against the backdrop of sustainable process modeling by means of a case example. Fig. 2 depicts a structured BPMN model that represents a build-to-order process. The example process model follows a fairly standard and simple sequence of activities in terms of control-flow: whenever an order comes in, the availability for the ordered product is checked. When the product is not

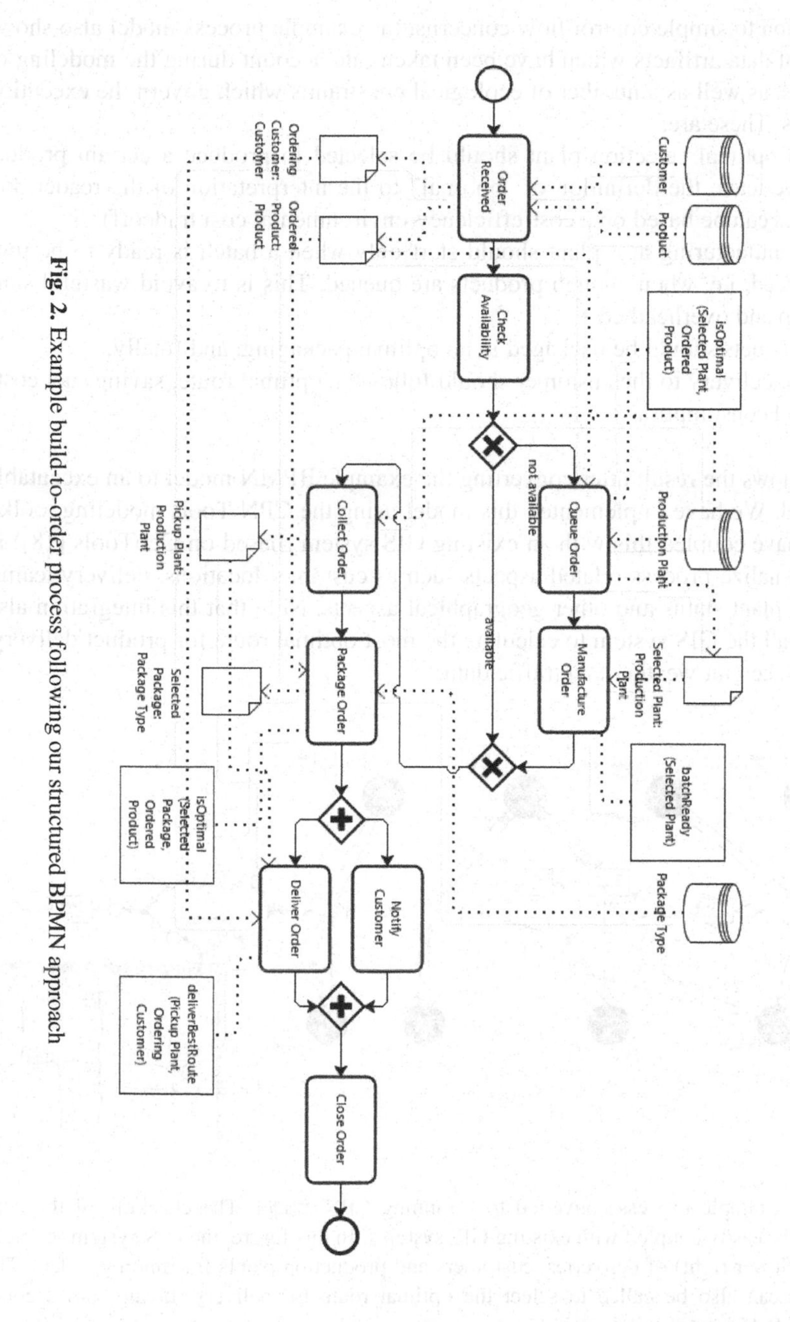

Fig. 2. Example build-to-order process following our structured BPMN approach

available, a job is queued in a manufacturing plant to start production. Once the product is finished (or available), it can be collected and prepared for packaging. Next, a notice is sent to the customer while the product is on the way to be delivered. Once these steps have been taken, the case can be closed.

In addition to simple control flow concerns, the example process model also shows a number of data artifacts which have been taken into account during the modeling of this process, as well as a number of ecological constraints which govern the execution of activities. These are:

(i) an optimal selection plant should be selected to produce a certain product (we leave the definition of "optimal" to the interpretation of the reader, but this can be based on a cost-efficiency-environmental cost tradeoff),
(ii) manufacturing at a plant should start only when a batch is ready to be produced, i.e. when enough products are queued. This is to avoid wasteful startup and overhead costs.
(iii) products should be packaged in an optimal packaging, and finally,
(iv) the delivery to the customer should follow an optimal route, saving fuel costs and consumption.

Fig. 3 shows the result after converting the example BPMN model to an executable CPN model. We have implemented this model using the CPN Tools modeling toolkit [16], and have coupled this with an existing GIS system (based on GeoTools [18]) in order to visualize process-related aspects such as customer locations, delivery teams, production plant status and other geographical aspects. Note that this integration also allows to call the GIS system to calculate the most optimal route for product delivery, taking into account weather and traffic data.

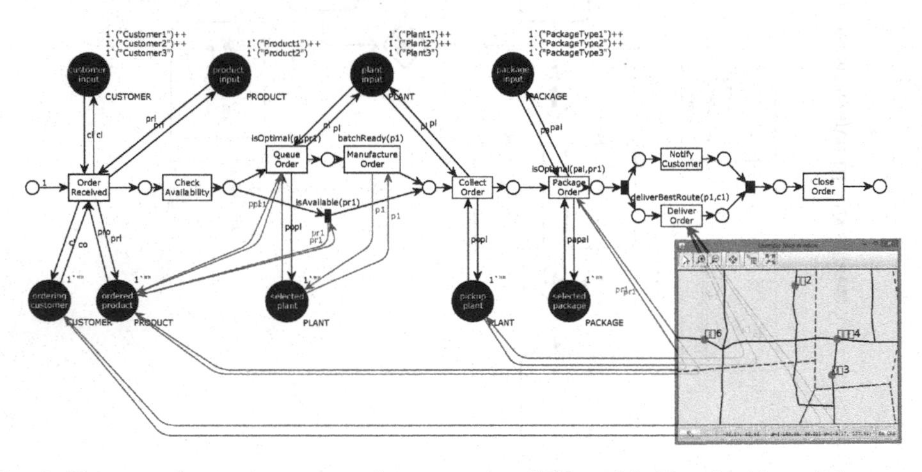

Fig. 3. The example process converted to a running CPN model. The checking of the constraints can be easily coupled with existing GIS systems. In this figure, the GIS system provides a live map (lower right) of deliveries, customers and production plants for running orders. The GIS system can also be called to select the optimal route for delivery, taking into account weather and traffic data.

During execution, the CPN model allows to track process instances as they are running, governed by the ecologic constraints. The system checks these constraints at execution-time and determines which transitions can be fired. In addition, the model approach allows for integration with existing GIS systems, for instance to enable detailed geographical monitoring functionality. This model thus allows analysis of ecological-related concerns and to derive possibilities for improvement.

6 Conclusions

We have presented a first approach towards enabling the modeling and execution of processes that can incorporate context information, in our case ecological and geographical concerns. We do so by first proposing a general approach towards constructing process models with any type of contextual data, and then show how to convert this model to a CPN model using a formalized mapping which allows the model to be executable, and to be integrated with GIS systems, for example. This integration with GIS can then perform supporting tasks to "feed" and constrain the execution of activities based on geographic data, or the running process can update the GIS system as activities are executed.

In future work, we plan to expand on this initial work by (i) providing mathematical proofs for our formalizations, (ii) designing and executing empirical evaluations using real-life cases, and (iii) exploring the various integration possibilities between GIS and BPM systems in the explicit context of sustainability concerns related to business processes.

Acknowledgments. This work is supported by the National Key Technology R&D Program, China (grant 2012BAH01F02) and by the KU Leuven research council (grant OT/10/010).

References

1. Vom Brocke, J., Rosemann, M.: Handbook on Business Process Management: Strategic Alignment, Governance, People and Culture. Springer (2010)
2. van der Aalst, W.M.P., ter Hofstede, A.H.M., Weske, M.: Business Process Management: A Survey. In: van der Aalst, W.M.P., ter Hofstede, A.H.M., Weske, M. (eds.) BPM 2003. LNCS, vol. 2678, pp. 1–12. Springer, Heidelberg (2003)
3. Rosemann, M., Recker, J.C., Flender, C.: Contextualisation of business processes. International Journal of Business Process Integration and Management 3(1), 47–60 (2008)
4. Rosemann, M., Recker, J.C.: Context-aware process design: Exploring the extrinsic drivers for process flexibility. In: The 18th International Conference on Advanced Information Systems Engineering. Proceedings of Workshops and Doctoral Consortium. Namur University Press (2006)
5. Vom Brocke, J., Seidel, S., Recker, J.: Green business process management: towards the sustainable enterprise. Springer (2012)
6. Chang, K.-T.: Introduction to geographic information systems. McGraw-Hill, New York (2010)
7. Ghose, A., et al.: Green business process management: A research agenda. Australasian Journal of Information Systems 16(2) (2010)

8. Houy, C., Reiter, M., Fettke, P., Loos, P.: Towards green BPM – sustainability and resource efficiency through business process management. In: Muehlen, M.z., Su, J. (eds.) BPM 2010 Workshops. LNBIP, vol. 66, pp. 501–510. Springer, Heidelberg (2011)
9. Houy, C., et al.: Advancing business process technology for humanity: Opportunities and challenges of green BPM for sustainable business activities. In: Green Business Process Management, pp. 75–92. Springer (2012)
10. Recker, J.C., et al.: Business process modeling: a comparative analysis. Journal of the Association for Information Systems 10(4), 333–363 (2009)
11. Recker, J., et al.: Modeling and Analyzing the Carbon Footprint of Business Processes. In: Green Business Process Management, pp. 93–109. Springer (2012)
12. Dumas, M., et al.: Fundamentals of Business Process Management. Springer, Berlin (2013)
13. Mendling, J., Weidlich, M.: Business Process Model and Notation (2012)
14. Zhu, X., et al.: Exploring Location-Dependency in Process Modeling. Business Process Management Journal 20(6) (2014)
15. Jensen, K.: Coloured petri nets. Springer (1987)
16. Jensen, K., Kristensen, L.M., Wells, L.: Coloured Petri Nets and CPN Tools for modelling and validation of concurrent systems. International Journal on Software Tools for Technology Transfer 9(3-4), 213–254 (2007)
17. Murata, T.: Petri nets: Properties, analysis and applications. In: Proceedings of the IEEE (1989)
18. Turton, I.: Geo tools. In: Open source approaches in spatial data handling, pp. 153–169. Springer (2008)

Geostatistical Modeling of Topography in the Land Rearrangement Project

Jiumao Zhou[1], Hongyi Li[2,*], and Chenglong Dan[2]

[1] Jiangxi Provincial Bureau of Coal Geology Surveying and Mapping Team, Nanchang, 330001 China

[2] Jiangxi University of Finance and Economics, Nanchang, 330013 China

zhoujiumao@163.com, lihongyi1981@zju.edu.cn, danchenglong@sina.com

Abstract. Precision topography map in the field scale is very important for land rearrangement planning. In this study, the elevation in the field scale from such measurements we encountered a strong trend in a land rearrangement project, of Yichun, China, which we had to take into account in addition to correlated random variation. We separated the two sets of effects and estimated their parameters by OLS. A spherical function seemed best to describe the random effect, and its parameter values were inserted into the equations for prediction by ordinary kriging. We compared the results by cross-validation and calculated the mean error (ME), mean squared error (MSE) and mean squared deviation ratio (MSDR). The fitted model gave small ME, as expected---kriging is unbiased. The MSE is also very small. The MSDR was 1.032, which ideally should equal 1. The goodness of the model encourages regression kriging is a good enough method for modeling topography in land rearrangement project.

Keywords: DEM, Interpolation technique, Regression kriging, Land rearrangement.

1 Introduction

Land rearrangement plays an important role in the enhancing of the quantity and quality of the cultivated land in China. From 2006 to 2012, there were 8 290 000 km^2 of the land was rearranged in China [1]. Digital Elevation Model (DEM) is a digital representation of the topography, the major input to quantitative analysis of topography, also known as geomorphometry [2]. Because the land unknotting work, which calculating the earthwork quantity depends on the DEM, is the core content of a land rearrangement project. Its budget is amount to 45.3% of the total budget [3]. So, a precision DEM at the field scale is very important for the land rearrangement projects planning.

There are two popular methods to obtaining a DEM, which are direct extraction of topographic information using well established photogrammetric techniques and

---

* Corresponding author.

F. Bian and Y. Xie (Eds.): GRMSE 2014, CCIS 482, pp. 497–504, 2015.

using laser and radar-based sensors to automatically derive elevation information [4]. However, it is always at a resolution of 30 m, and at 10 m resolution for specific areas. Interpolation techniques, such as Inverse Distance Weighting (IDW), local polynomial, Natural Neighbor (NN), Radial Basis Functions (RBFs), polynomial interpolation functions, and geostatistics are based on the principles of spatial autocorrelation, which use the available sample points to generate predictions for a particular point [2,5]. Kriging is a geostatistical interpolation method that utilizes variogram which depend on the spatial distribution of data rather than on actual values. It is quite feasible for modelling DEM [2, 6-9].

Ordinary kriging, the familiar `workhorse' of geostatistics, is based on the assumption that the variable of interest is intrinsically stationary. If there is trend, such as the topography in this study, however, the assumption is untenable, and a more elaborate model of the variation is needed to take into account the trend. Regression Kriging (RK) in which the interdependence of the trend and variation in the residuals is disregarded enjoys some popularity in soil science [10]. And it is the one we use here to explore the topography in a land rearrangement project

2 Site and Methods

2.1 Study Area and Sampling

The study area, totally 435.87 ha, is located in a land rearrangement project supported by the government's investment in the Yichun City, Jiangxi Province, China. There are 1106 elevation points were georeferenced by a Trimble Global Positioning System with the support of a local Continuous Operational Reference System (CORS) built by the Jiangxi Provincial Bureau of Coal Geology Surveying. Fig.1 is a `bubble plot' of the data with `bubbles', i.e. circles or discs, of diameter proportional to the measured elevation.

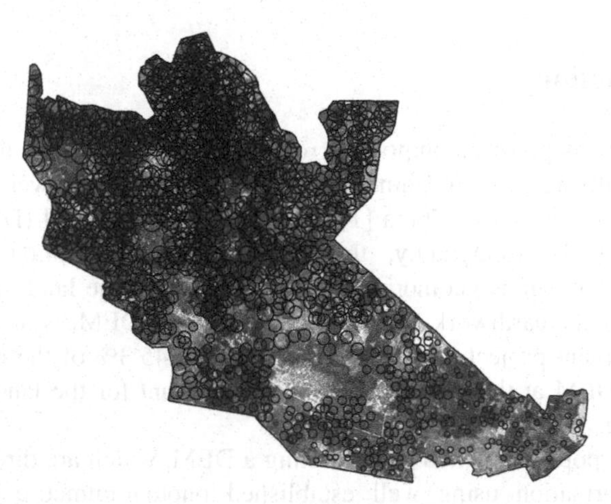

Fig. 1. Location of the region studied and bubble plot showing the positions where measurements were made with `bubbles' of diameter proportional to the elevation

2.2 Regression Kriging

The bubble plot (Fig. 1) shows an obvious trend of topography towards the southeastern to northwestern of the region. The pattern seems to have the general form of a cubic trend surface, and we therefore fitted such a surface by ordinary least squares (OLS) regression [11]:

$$z(x) = u(x) + \varepsilon(x) = \sum_{i=0}^{K} \beta_j f(x_j) + \varepsilon(x) \quad (1)$$

in which for K=9 the set f(x) is the expansion of a constant, b_0, plus cubic terms in f(x). Then, the experimental variogram of the residuals, $\varepsilon(x)$, was computed by the method of moments. We used GenStat [12] for the purpose. Finally, the residuals of elevation were predicted by ordinary kriging, and the spatial trend was added back for mapping the topography. We kriged in Matlab, version R2012b, and mapped the results in ArcGIS 10.

3 Result and Discussion

3.1 Summary Statistics

Table 1 summarizes the statistics of the measured elevation points; the histogram of the data with a normal distribution curve is shown in Fig.2. The elevation is range from 54.07 m to 77.67 m, with a variance of 21.46.

Table 1. Summary statistics of the observed elevation points (m)

Minimum	Maximum	Mean	Median	Std dev.	Variance	Skewness
54.07	77.67	67.78	68.16	4.63	21.46	-0.23

3.2 Variogram of the Elevation

We first computed the experimental variogram of the elevation by the method of moments:

$$\gamma(h) = \frac{1}{2m(h)} \sum_{j=1}^{m(h)} \left\{ z(x_j) - z(x_j + h) \right\}^2 \quad (2)$$

Where z(x) and z(x+h) are observed values transformed to logarithms at places x and x+h separated by the lag vector h and m is the number of paired comparisons at that lag. We treated the variation as isotropic, so that the lag is a scalar in distance only: h=|h|. The points plotted in Fig. 2(a) show the result. They follow a steady upward curve increasing without bound to which we could fit by weighted least squares a power function,

$$\gamma(h) = 0.3429 + 0.000448h^{1.45} \tag{3}$$

in which h is in metres. Although the exponent, 1.45, is well within the bounds for an intrinsic stationary process.

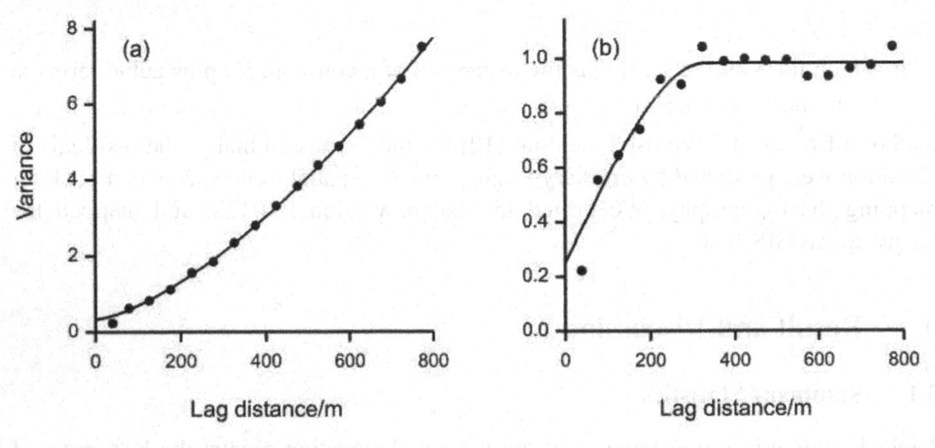

Fig. 2. Experimental variograms: (a) of the observed elevation computed by the method of moments, shown as points, and the power model fitted to it. (b) of the OLS residuals from a cubic trend with spherical model shown by the curve.

3.3 Variogram of the OLS Residuals

Figure 2(b) shows by the plotted points the experimental variogram of the OLS residuals from the cubic trend, again computed by the method of moments, Equation (2). These clearly follow a bounded curve, and the spherical model, the solid line, is the model well fitted to them. Its formula is

$$\gamma(h) = \begin{cases} c_0 + c_1\left\{\dfrac{3h}{2a} - \dfrac{1}{2}\left(\dfrac{h}{a}\right)^3\right\} & 0 \le h \le a \\ c_0 + c_1 & h > a \end{cases} \tag{4}$$

Where c0, c1, *a* is the nugget, partial sill and range of the spherical model, respectively.

Table 2 lists the coefficients of the cubic trend, and Table 3 lists the parameters of the fitted spherical model.

Table 2. Fixed effects of cubic OLS trend surface

Terms	OLS Spherical	t
Constant	30.08	14.55
x	3.56×10^{-2}	15.90
y	3.95×10^{-2}	15.26
xy	2.10×10^{-5}	-3.40
x^2	1.38×10^{-4}	-5.62
y^2	1.40×10^{-5}	-2.06
yx^2	2.96×10^{-9}	11.09
xy^2	3.57×10^{-9}	11.69
x^3	1.68×10^{-9}	13.61
y^3	2.18×10^{-9}	11.77

Table 3. Parameters of spherical models of the observed elevation fitted to the OLS. The symbols are c0 for the nugget variance, c1 for the sill of the autocorrelated variance, and r for the range.

Terms	OLS Spherical
c_0	0.25
c_1	0.99
r	334.86

3.4 The Predictions

Table 4 summarizes the results by cross-validation on punctual supports with the local predictions, which the nearest 20 data are used to interpolate. The table lists mean error (ME), mean squared error (MSE), mean kriging variance, σ^2, the median of the squared error, the mean squared deviation ratio (MSDR). The ME, MSE and MSDR are defined as follows.

Table 4. Cross-validation of spherical models for OLS residuals predictions*

		SDR		σ^2		
ME	MSR	Mean	Median	Mean	Median	Kurtosis
-0.002	0.446	1.032	0.097	0.4394	0.4279	112.5

**ME is mean error, MSE is mean squared error, σ^2 is the kriging variance, SDR is the squared deviation ratio and kurtosis is the kurtosis coefficient of the cross-validation errors.*

$$\text{ME}=\frac{1}{N}\sum_{i=1}^{N}\left\{z(x_i)-\hat{z}(x_i)\right\} \tag{5}$$

$$\text{MSE}=\frac{1}{N}\sum_{i=1}^{N}\left\{z(x_i)-\hat{z}(x_i)\right\}^2 \tag{6}$$

$$\text{MSDR}=\frac{1}{N}\sum_{i=1}^{N}\frac{\left\{z(x_i)-\hat{z}(x_i)\right\}^2}{\sigma^2(x_i)} \tag{7}$$

Kriging is unbiased, and so the ME should ideally be zero. The ME is small compared with the mean of the elevation, 67.78 m. In using kriging, we attempt to minimize the squared errors. We also acquired a small MSE. The model that most accurately describes the variation should produce squared errors equal to the kriging variances, so the MSDR is ideally 1. Table 4 reveals the goodness of the model in this study, with a MSDR of 1.032

A model for which the MSDR=1 is generally regarded as good. However, Lark [13] pointed out that a better diagnostic is the median of the SDR. If the cross-validation errors are normally distributed then the expected value of the median SDR is 0.455. In our study all of the median SDRs were much less, even though the mean SDR in the best combination was close to 1.0. The likely explanation is that the distributions, though symmetric, are leptokurtic. Figure 3 shows the distribution for global cross-validation with the OLS variogram model, for which the kurtosis coefficient is 112.5 (Table 4).

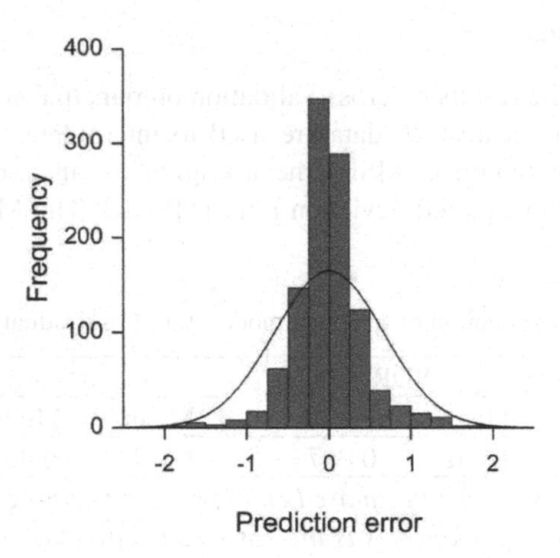

Fig. 3. Histograms of the cross-validation prediction errors by RK with OLS spherical variogram and local search on observed elevation points. The curve is of the normal distribution for the calculated means and variances of the errors.

The predictions are for blocks of size 5 m at 5 m intervals on a grid. Fig 4a shows the predictions. We calculated and mapped also the prediction variances, Fig.4b. The kriging variances are smallest close to the sampling points and small generally throughout the region away from the boundaries. The variances become large only near the field boundary and the village, beyond which there are no data.

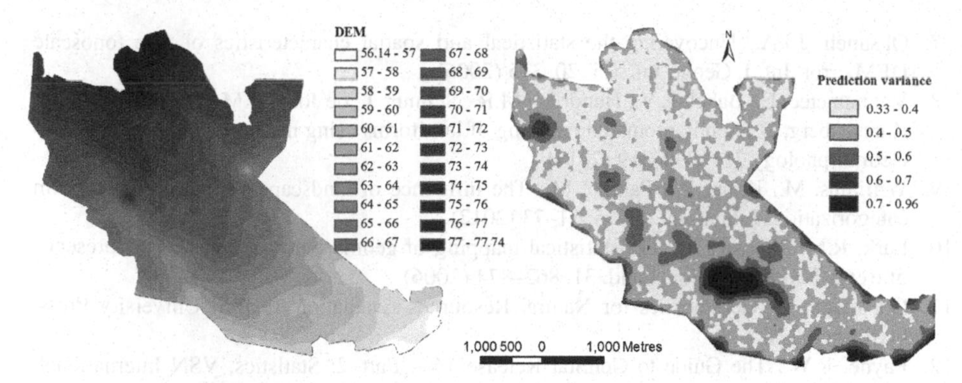

Fig. 4. (a) Map of DEM; (b) corresponding prediction variances

4 Conclusion

Land rearrangement planners need precision topography map for planning. In modelling the DEM in the field scale from such measurements we encountered a strong trend, which we had to take into account in addition to correlated random variation. We separated the two sets of effects and estimated their parameters by OLS. A spherical function seemed best to describe the random effect, and its parameter values were inserted into the equations for prediction by ordinary kriging. The parameters of the goodness of the model encourage regression kriging is a good enough method for modelling topography in land rearrangement project.

Acknowledgement. This research was financially supported by the National Science Foundation (NO. 41101197), Ministry of Education, Humanities and social science project (No 10YJC910002), and the Natural Science Foundation of Jiangxi Province (No. 20114BAB213017).

References

1. Yang, X., Jin, X., Guo, B., Xu, G., Pan, Q., Zhou, Y.: Spatial-temporal differentiation of land consolidation investment in China from 2006 to 2012. Trans. Chinese Soc. Agric. Eng. 8, 227–235 (2014) (in Chinese)
2. Hengl, T., Bajat, B., Blagojevic, D., Reuter, H.I.: Geostatistical modeling of topography using auxiliary maps. Comput. Geosci. 12, 1886–1899 (2008)
3. Fu, M., Hu, Z., Mi, J.: Contribution analyses and control counterm easures of land consolidation and reclamation project cost. Trans. CSAE. 4, 291–294 (2003) (in Chinese)
4. AeroMetric, `http://www.aerometric.com/services/digital-elevation-models/`
5. Peralvo, M.: Influence of DEM interpolation methods in Drainage Analysis. In: CE 394K GIS in Water Resources, pp. 301–312 (2009)
6. Fisher, P.: Improved Modeling of Elevation Error with Geostatistics. GeoInformatica 3, 215–233 (1998)

7. Oksanen, J.J.A.: Uncovering the statistical and spatial characteristics of fine toposcale DEM error. Int. J. Geogr. Inf. Sci. 20, 345 (2006)
8. Vannametee, E., Babel, L.V., Hendriks, M.R., Schuur, J., de Jong, S.M., Bierkens, M.F.P., Karssenberg, D.: Semi-automated mapping of landforms using multiple point geostatistics. Geomorphology 221, 298–319 (2014)
9. Williams, M., Kuhn, W., Painho, M.: The influence of landscape variation on landform categorization. J. Spat. Inf. Sci. 5, 51–73 (2012)
10. Lark, R.M., Webster, R.: Geostatistical mapping of geomorphic variables in the presence of trend. Earth Surf. Proc. Land. 31, 862–874 (2006)
11. Goovaerts, P.: Geostatistics for Natural Resources Evaluation. Oxford University Press, Oxford (1997)
12. Payne, R.W.: The Guide to GenStat Release 16 – Part~2: Statistics. VSN International, Hemel Hempstead (2013)
13. Lark, R.M.: Kriging a soil variable with a simple non-stationary variance model. J. Agr. Biol. Envir. St. 14, 301–321 (2009)

A Self-adjusting Node Deployment Algorithm for Wireless Sensor Network

Kaiguo Qian[1], Shikai Shen[2,3,4,*], and Zucheng Dai[1]

[1] Department of Physics Science and Technology, Kunming university, Kunming, China
[2] Kunming IOT & Ubiquitous Engineering Center, Kunming, China
[3] School of Information and Technology Kunming University, Kunming, China
[4] Future University, Hakodate, Japan
{qiankaiguo,958524088}@qq.com, kmssk2000@sina.com

Abstract. A self-adjusting node deployment algorithm is proposed to improve the coverage rate after nodes randomly deploying in wireless sensor network. The proposed algorithm sets the standard distance between sensor nodes that is calculated according to the required minimum nodes meet the full coverage task area. The randomly deployed Nodes are self-adjusted to its new position, which makes nodes move away from each other when mutual distance is less than the standard. On the contrary the nodes gathered themselves together when mutual distance is more than the standard. Simulation experiments show that the new algorithm does more quickly complete coverage then the virtual force deployment algorithm, and it is an efficient method to complete the deployment of sensor networks.

Keywords: Wireless sensor network, Self-adjusting, Node deployment, The standard coverage distance.

1 Introduction

Wireless sensor networks [1] is a monitoring networks composed by a large number of micro computer equipment which is randomly deployed to task area through wireless self-composition. Wireless sensor networks can be applied in military invasion, environmental monitoring, industry data collection,health monitoring, smart home and fine agriculture. Sensor nodes deploymentis a basic problem for the other networking technology such as routing design,data fusion and positioning technology for the sensor network applications [2].Deterministic method that sensor nodes are deployed in planned location in advancecan effectively control the cost of the networking for the overall optimization deployment, but is not suitable for changeable and inaccessible places, in where,administrator can randomly sow the sensor nodes. In order to meet the requirement of the network coverage, the number of sensor nodes ismore than the actual needin monitoring area, which causes redundant nodes, shapes of overlapping, and increases the cost of meshing.With the development of microelectronics technology, in

* Corresponding author.

F. Bian and Y. Xie (Eds.): GRMSE 2014, CCIS 482, pp. 505–512, 2015.

recent years, there appearsmobilityand low-cost sensors node [3-4].So, we can move sensor nodes to the ideal positionafter randomseeding sensors with the help of the node mobility to ensure the quality of network coverage.Virtual force algorithm [5] effectively controls the sensor nodes move to the idea locationthrough introducing a stress modelbetween the nodes andobstacles and among nodes.It effectively improves the network coverage, but there is no moving target position.The algorithm [6] spreads the mobility nodes to eliminate coverage blind area on the basis of the static random deployment of sensor nodes.Methods[7-8] are recent improvement mechanism of the virtual force algorithm which easily cause collisionin the process of node moving because there is no clear node moving targets. At the same time,they needtemporary global position informationto calculate the virtual force, which causes high communication price increases energy consumptionin the iteration process. AASR [9] reduces the node energy waste and improves the efficiency of coverage throughadjusting node's perception radius. The strategy [10] introduced thepartition that the task area is divided into several sub area,everyarea selects minimum cover sets with the energy efficiency standards.The literature [11] proposed node deployment scheme based on perceived probability model, using the evidence theory by calculating the probability of nodes in the surrounding area comprehensive perception to modify the virtual force algorithm to realize the biggest coverage of the monitoring area. These virtual force algorithm need global topology information of network nodes to calculate virtual force stressed of every node which is a challenge for resource-constrained node.We propose a new algorithm adjusts sensor node locationbased on the node distributionwhenmaximizingcoverage the task area that referencesgreedyideas, which realizes local optimal coverage for adjusting the local node position to achieve the overall optimal coverage. It rapidly improves the coverage performance, and the node does not need to maintain global network topology information.

2 System Model

The following network model and node covering model are applied to analyze the performance of the new algorithm.

2.1 Network Model

Generally, wireless sensor networks can be abstracted as an undirected graph $G=(V,E)$, $V=\{s_1,s_2,\ldots,s_{i,}\ldots,s_n\}$ is the set of nodes in wireless sensor networks. $E=\{e_1,e_2,\ldots,e_{j,}\ldots e_m\}$ is the set of edges that can communicate with each other between the nodes, $e_j=(u,v)$represents the link by which node u and v can transmit data directly. The sensing radius of sensor node is r, communication radius is R, when the network covers the task area, it requires the mutual communication between nodes, and there is no isolated node. Related research shows that task area is fully covered by the sensor nodes realizes completely communication each other if R is twice of r. Therefore, assuming that the previous conditions established, that is the network connection is not considered in the research process, and also assuming that all nodes in the network have the same sense, communication and mobility capability.

All nodes have positioning capability, and can always perceive their location information, furthermore, obstacles in the monitoring area are not considered in the research.

2.2 Coverage Model

Definition 1: The Euclidean Distance of sensor node $S(x_s,y_s)$and $P(x_p,y_p)$ is shown as formula 1.

$$disp(s, p) = \sqrt{(x_s - x_p)^2 + (y_s - y_p)^2} \tag{1}$$

Definition 2:if any point within a certain area is not covered by any sensor node, then this area is called blind zone.

Definition 3:the coordinate of sensor node S in two-dimensional plane R^2 is (x,y), then the coverage area of node S is a circular regioncalled as the coverage circlewith its center of point(x,y) and its sensor radius is r, which shows as formula 2.

$$\text{ca}(s) = \{p \mid p \in R^2, Dist(s, p) \le r\} \tag{2}$$

Definition 4: The coverage model is described asformula 3.

$$p(s,q) = \begin{cases} 1, & if \quad dist(s,q) \le r \\ 0, & otherwise \end{cases} \tag{3}$$

2.3 Evaluation Index

(1) The number of coverage node

The effective coverage area of sensor node is πr^2 ,and task area is M * N. The number of need node is calculated as equation 4.

$$n = \frac{2 \times M \times N}{3\sqrt{3}\pi r^2} \tag{4}$$

(2) Coveragerate

Coverage rate is defined aspercentage of the coverage sets of all sensor nodes and the area of the task area, as shown in equation 5.

$$\text{d}_{coverage} = \frac{\bigcup_{i=1}^{N} \text{cov}\, erarea(s_i)}{M \times N} \tag{5}$$

(3) The convergence time

The convergence time is the iteration rounds to cover the performance requirements. The faster deployment of sensor network, algorithm is more applicable to practical application.

(4) Moving distance

It is the average distance of the sensor node moving to cover the performance requirements. The shorter the distance, the lower the sensor network deployment cost.

3 Self-adjusting Node Deployment Algorithm

The sensor nodes adjust themselves to optimal location according to the relationship between each other after nodes random deploying. It makes nodes move away from each other that gather together and makes nodes close to each other that spread out to a certain distance. The sensor nodes randomly deployed are evenly distributed in the task area. It uses local node adjustment strategy to avoid long distance movement, which only adjust the neighbor node within the scope of the node communication to achieve global equilibrium distribution.

3.1 Local Adjustment Strategy

It sets the sensing radius of sensor nodes to r, communication radius R. Assuming that R is 2 times of r, which can guarantee communication with each other. In general, the local nodes distribution after random deployment is shown in Fig.1.

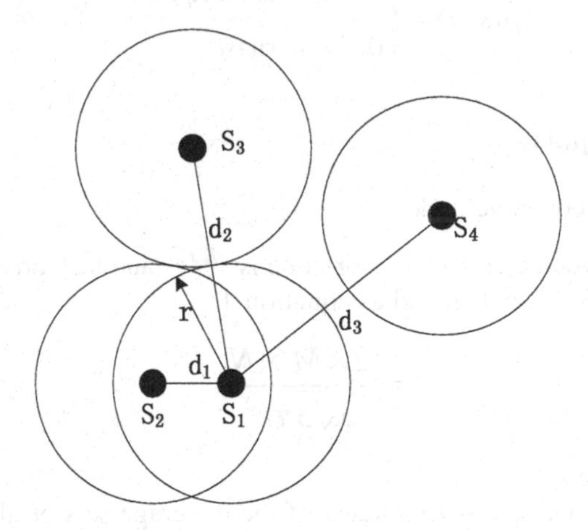

Fig. 1. Local Node Distribution

Node S1 is distributed around with S2, S3 and S4.The distance d1 between the S1 and S2is less than the coverage radius r of sensor nodes, and overlapping area of the two nodes is larger. The distance d2 between the nodes S1 and S3 d2 is equal to the sensor node coverage radius 2times, there is one intersection point between each other, and no overlapping area, which effective coverage area between two nodes is the largest. It is difficult to determine the position of the third node S4, if the distance d3 of S4 to S1 is greater than or equal to R, d4 is greater than or equal to the distance to the S3, which causes the coverage blind zone. The distribution of S4, S1 and S3 is shown asFig.2 (a).

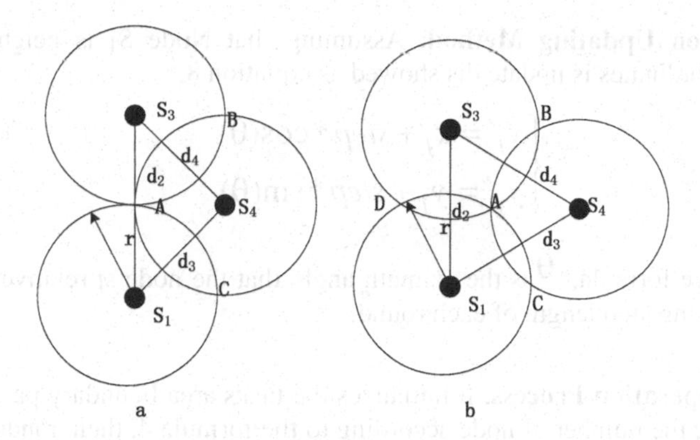

Fig. 2. Sensor Node Distribution in The Local Area

The area of the overlapping is calculated as equation 6.

$$S_a = 4(\frac{1}{4}\pi r^2 - \frac{1}{2}r^2) \tag{6}$$

Another way of eliminating cover hole that nodes are placed is shown in Fig 2 (b),and the area of the overlapping is calculated as equation 7.

$$S_b = 6(\frac{1}{6}\pi r^2 - \frac{\sqrt{3}}{4}r^2) \tag{7}$$

Obviously, $S_a > S_b$,So, the coverage performance in Fig.2 (b) is better than that of Fig.2 (a). In general, Fig.2 (b) is the best way of node deployment with local sensor node distribution, then, the three nodes are triangle and the distance between nodes is the side of the triangle that is equal to $d_0 = \sqrt{3}r$

3.2 Self-adjusting Judgement and Movement Rules

Self-adjusting Judgement Rule. According to local best coverage performance, sensor nodes self-adjusting strategy is described as follows:

When the distance between the two nodes is less than threshold value d0, they move away from each other to the location that the distance is equal to the position. It is shown in Fig.1 that node S2 move to the left from S1 that the distance is equal to the position.

When the distance between the nodes is greater than threshold value d0, they move close to the distance that is equal to threshold value d0.As shown in Fig1, node S3moves in the direction of S1 to the position that the distance is equal to d0.

In order to avoid the node long distance moving, the node does not move when the distance is greater than the communication distance.

Node Location Updating Method. Assuming that Node Sj is neighbor node of Si,node Sj coordinates is update dis showed as equation 8.

$$\begin{cases} x_j' = x_j + step * \cos(\theta) \\ y_j' = y_j + step * \sin(\theta) \end{cases} \tag{8}$$

In the above formula, θ is the azimuth angle that the node sj relatives to the node si.Step is moving step length of each round.

Algorithm Operation Process. It initializes the tasks area boundary parameter m * n and determine the number of node according to the formula 4, then, randomly deploys n nodes in the task area. Running the self-adjusting deployment algorithm is described as follows.

Step1:Calculating the coverage rate, if the coverage rate is greater than the covering quality threshold C0, it will quit, otherwise perform Step2.

Step2: Every node broadcast their location (xi, yi) and node number Sid with the scope of the radius R. The node received the radio signal records the information. Every node maintains a neighbor node information table through the exchange of information.

Step3: Every node calculates the distance between each other according to their own neighbor node list.

Step4: Every node determines oneself whether update their location meet the need of self-adjusting decision rules. If they need a new location, enter the Step5, otherwise enter the Step1.

Step5: Every node updates their new position and jumps to Step1.

4 Simulation Performance Analysis

In MATLAB R2012a, with task area A=1000m*1000m, 60 sensor nodes are deployed with radius of perception as 90 meter, iteration time is set to 25 rounds, and moving step is set to 10 m.

4.1 Convergence Time and Coverage Rate

The coverage rate of virtual force algorithm and self-adjusting deployment algorithm experimental result is shown in Fig.3.

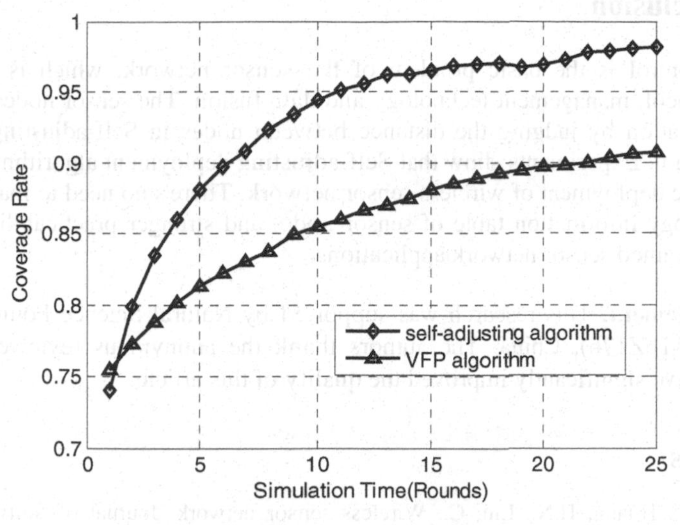

Fig. 3. Coverage rate

From the result of the experiment, the self-adjusting algorithm can quickly complete the deployment of wireless sensor network (WSN), which only need10 rounds to achieve the coverage requirement that the coverage rate reached 95% .It is a very fast and efficient method of the deployment.

4.2 Sensor Node Movement

Node movement trajectory is shown in Fig.4.It is can be seen from the diagram, the node after self-adjusting deployment, almost evenly distributed to the task area. The node moved average distance of 10.47 mper round, which is smaller relative to the node's perception radius.

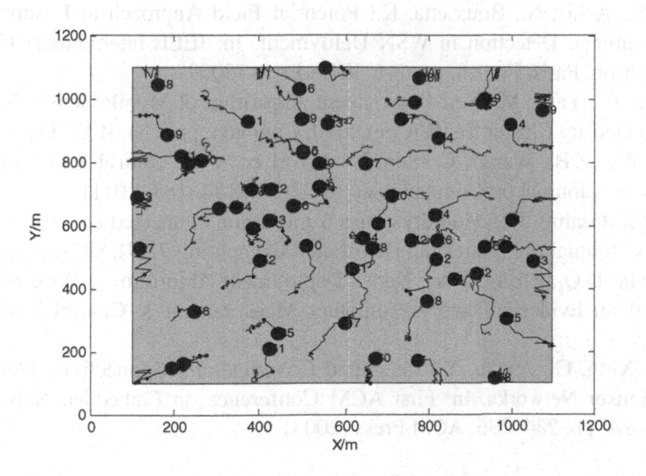

Fig. 4. Node movementtrajectory

5 Conclusion

Coverage control is the basic problem of the sensor network, which is basic for routing protocol, management technology and data fusion. The sensor nodes move to optimum location by judging the distance between nodes in Self-adjusting deployment algorithm. Experiments show that Self-adjusting deployment algorithm is a fast, efficient node deployment of wireless sensor network. There's no need to maintain the global topology information table of sensor node, and stronger practicability for resource-constrained sensor network applications.

Acknowledgement. This research was supported by Natural Science Foundation of Yunnan (2011FZ176), China. The authors thank the anonymous reviewers whose comments have significantly improved the quality of this article.

References

1. Ren, F.Y., Huang, H.N., Lin, C.: Wireless sensor network. Journal of Software 14(7), 1282–1291 (2003) (in Chinese)
2. He, Z.S., Zhu, Y.G., Zhuang, Y.B.: Survey on wireless sensor network Coverage Control technology. Computer Measurement & Control 20(7), 1737–1739 (2012) (in Chinese)
3. Bang, W., Hock, B.L., Di, M.: A survey of movement strategies for improving network coverage in wireless sensor networks. Computer Communications 32(13), 1427–1436 (2009)
4. Liu, B., Brass, P., Dousse, O., et al.: Mobility Improves Coverage of Sensor Networks. In: Proceedings of the 6th ACM International Symposium on Mobile Ad Hoc Networking and Computing, Urbana-Champaign, USA, pp. 300–308. ACM (2005)
5. Zou, Y., Chakrabarty, K.: Sensor Deployment and Target Localization Based on Virtual Forces. In: Proceedings of the IEEE INFOCOM, San Francisco, USA, pp. 1293–1303. IEEE (2003)
6. Zhou, T., Hong, B.G., Piao, S.H.: Hybrid Sensor Networks Deployment Based on Virtual Force. Journal of Computer Research and Development 44(6), 965–972 (2007)
7. Aitsaadi, N., Achir, N., Boussetta, K.: Potential Field Approach to Ensure Connectivity and Differentiated Detection in WSN Deloyment. In: IEEE international Conference on Communication, Paris, Franch, pp. 1–6. IEEE Press (2009)
8. Tan, L., Yu, C., Yang, M.: Self-Deployment Algorithm of Mobile Sensor Network Based on uniform Density Cluster. In: WiCom 2010, Chengdu, pp. 1–4. IEEE Press (2010)
9. Han, Z.J., Wu, Z.B., Wangr, C., Sun, L.: Novel coverage control algorithm for wireless sensor network. Journal on Communications 32(10), 174–183 (2011)
10. Khedr, A.M., Osamy, W.: Mobility-assisted minimum connected cover in a wireless sensor network. Journal of Parallel and Distributed Computing 72(7), 827–837 (2012)
11. Li, Q.Y., Ma, D.Q., Zhang, J.W.: Nodes Deployment Algorithm of Wireless Sensor Network Based on Evidence Theory. Computer Measurement & Control 21(6), 1715–1717 (2013)
12. Wang, X., Xing, G., Zhang, Y.: Integrated Coverage and Connectivity Con guration in-Wireless Sensor Networks. In: First ACM Conference on Embedded Networked Sensor Systems, USA, pp. 234–236. ACM Press (2003)

A New Index Model NDVI-MNDWI for Water Object Extraction in Hybrid Area

Weifeng Zhou*, Zhanqiang Li, Shijian Ji, Chengjun Hua, and Wei Fan

Key Laboratory of Fishery Resources Remote Sensing and Information Technology,East China Sea Fisheries Research Institute, Chinese Academy of Fishery Sciences, 200090 Shanghai, China

Abstract. Extracting accurate water boundary plays a critical role in oceanography research. In this study, we tested several water extraction methods using CCD and IRS images from satellites HJ-1A and HJ-1B in Liaodong Bay that has miscellaneous topography structure. These tested methods are based on variant water indices, including NDVI, NDWI, and MNDWI. Moreover, we proposed a new water index NDVI-MNDWI as the fittest model for water object extraction in hybrid area. Relative to these traditional indices mentioned above, the new NDVI-MNDWI model can enhance the differentiation between land and water, and can decrease the interference of artificial objects as well. The experimental results show that the NDVI-MNDWI model can get accurate borders of water objects in hybrid regions.

Keywords: water object extraction, water index model, NDVI-MNDWI, hybrid area, miscellaneous structure.

1 Introduction

Extracting water boundary to monitor water resource on a macro scale by means of remote sensing is an important research in water resource management. Therefore, more and more people focus on how to identify water boundary quickly and accurately from satellite remote sensing image. With the development of remote sensing technology, many traditional water extraction methods keeps improving and more methods that are new appear. However, most methods are aim at mountain water or water targets in urban area whose cover is generally smaller [1-4]. Therefore, the improvement of those methods used to extract large span and detailed water is still needed [5].

Liaodong Bay, located in the northeast of the Bohai Sea, is china's important economic development region. It's rich in fishery resources and its offshore belongs to intertidal zone where there are many aquaculture areas. A plurality of inland rivers input water source to Liaodong Bay. The water resource in Liaodong Bay is belongs to case-II water and very particular.

* Corresponding author. E-mail address: zhwfzhwf@163.com. Supported by Key Technologies R&D Program of China (No.2013BAD13B06).

F. Bian and Y. Xie (Eds.): GRMSE 2014, CCIS 482, pp. 513–519, 2015.

The study uses several traditional indices including NDVI, NDWI, and MNDWI as well as a new water index NDVI-MNDWI to extract water boundary in Liaodong Bay. And we have compared the 4 indices to find the fittest one.

2 Materials and Methods

2.1 Satellite Remote Sensing Data

The data we used was from HJ-1A/B satellite, which is to use China's own small satellite constellation for eco-environment and disaster remote sensing monitoring and to improve the collection, processing and application of China's environment and disaster RS information. HJ-1A and HJ-1B each comprise a satellite platform and payloads. Payloads include data transmission, a wide-coverage multispectral charge-coupled device (CCD) camera (HJ-1A and HJ-1B), hyper-spectral imager (HSI, HJ-1A only), and infrared camera (IRS, HJ-1B only). The CCD camera has four bands of blue, green, red and shortwave infrared spectral wavelengths. The design uses two CCD cameras for field splicing to realize 720 km width. The infrared camera has four bands of near infrared, shortwave infrared, middle infrared and thermal infrared. The infrared camera's nadir pixel resolution is 150 m and the width of view is 360 km. We chose both CCD and IRS images in July 5, 2010 and April 13, 2011 as research data.

The MNDWI method need the green band of CCD and the middle infrared band of IRS, so we integrated the four bands of CCD and the middle infrared band of IRS, then took some necessary pre-processing such as geometric correction and atmospheric correction. There is a slight cloud in the images of July 5, 2010, but they do not make sense because we just extract the water information. We need further processing after the extraction of water.

2.2 NDVI-MNDWI Index Model

Water has strong absorption of incident energy. In the wavelength range of sensors, the reflectance of water is always weak and decreasing with increasing wavelength. In RS model of clear water, its reflectance can be expressed as "blue > green > red > near infrared > middle infrared". When the wavelength is larger than 740 nm, water almost absorbed all of the incident energy so that the clear water in this wavelength range never reflects energy at all. Nevertheless, the reflectance of water would change with the increase of water turbidity (the concentration of organic and inorganic substances). For example, if water is containing more sediment, accordingly, its reflectance will increase and the peak of its spectral curve will shift to long wavelength direction. Therefore, the wavelength range is usually used to research water boundary or to identify the range of water[6-8].

Water Indices Models. Water indices method is one of water extraction methods that use two bands where water has strong reflectance and weak reflectance to calculate by ratio operation. There are some commonly used indices models such as normalized difference vegetation index (NDVI), normalized difference water index (NDWI)and modified normalized difference water index (MNDWI). The water indices methods are easy-to-use. Moreover, the ratio operation result can weaken background and enhance water features. At present, they are widely used in the research of water [9-13]. The following is the specific formula of each index model.

Table 1. Water Indices Models

Models	Formula	Description
NDVI	NDVI=(NIR-Red)/(NIR+Red)	NIR and R represent the reflectance of near infrared and red band respectively and correspond to the 4th and 3rd band of CCD camera.
NDWI	NDWI=(Green-NIR)/(Green+NIR)	Green represents the green band and corresponds to the 2ndband of CCD camera.
MNDWI	MNDWI=(Green-MIR)/(Green+MIR)	MIR represents the middle infrared band and corresponds to the 2nd band of IRS camera

NDVI-MNDWI. In terms of the research region, the traditional models cannot extract the water information of Liaodong Bay commendably because of the complex topography and vast range. Therefore, we use the form of manifold combination of indices models of NDVI-MNDWI as a new method to extract water information. Actually, the water information presents low value in NDVI images but high value in MNDWI images. Using NDVI-MNDWI to enhance the difference between water information and other land uses would be better in some extent.

3 Results

We use NDVI, NDWI, MNDWI and NDVI-MNDWI the four models to extract the water boundary in Liaodong Bay with HJ-1B images. The results are given in Figure 1.

Compared with figure 2, the NDVI model (Fig. 1a.) is less than ideal to extract this kind of detailed water in Liaodong Bay. The NDWI model (Fig. 1b) always extracts other objectives except water only because their reflectances are similar with water. The MNDWI model (Fig. 1c) shows that the extraction is not complete enough and the method could ignore some important water region such as coastal aquaculture areas and rivers. It is perfect to remove the noise of city buildings from small-scale water region but not good enough when it is applied in Liaodong Bay.

The NDVI-MNDWI model is the optimal one for its perfection to extract the two main rivers flowing into Liaodong Bay and reduce noises of other objectives. The threshold settings of each model are given in Table 2.

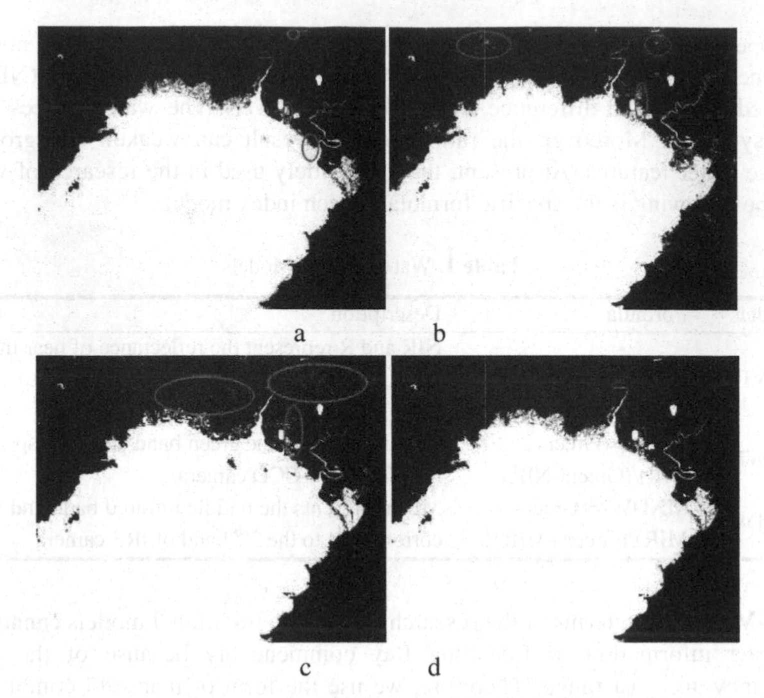

Fig. 1. The water extraction results. (a. NDVI, b. NDWI, c. MNDWI, d. NDVI-MNDWI)

Table 2. The threshold settings of each model

Models	Threshold settings
NDVI	0.13
NDWI	0.15
MNDWI	0.79
NDVI-MNDWI	1.08

We pick out 10 points (yellow points shown in Fig.2.in the research area to verify the accuracy for each model. These points represent different objectives including land and water. If a point was proved to be in accord with the actual, the point would be masked as Y. If not, it would be N. The result is shown in Table 3.

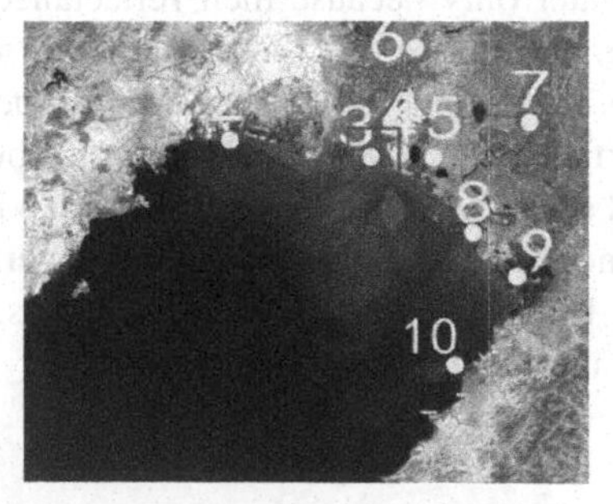

Fig. 2. CCD (342) multispectral image

For the random 10 points, the precision of each model is 60% for NDVI, 80% for NDWI, 60% for MNDVI, and 90% for NDVI-MNDWI. The best is the model of NDVI-MNDWI. Besides, the extraction result of NDVI-MNDWI (see Fig. 1d.) is more accurate and the points mistaken for water are relatively less.

Table 3. Accuracy verification of each model

Points \ Models	NDVI	NDWI	MNDVI	NDVI-MNDWI
1	N	Y	N	Y
2	Y	Y	Y	Y
3	Y	Y	Y	Y
4	Y	N	Y	Y
5	N	N	N	N
6	N	Y	N	Y
7	N	Y	N	Y
8	Y	Y	Y	Y
9	Y	Y	Y	Y
10	Y	Y	Y	Y
Precision	60%	80%	60%	90%

In addition, the estuary region (40°5′37.51″N-40°8′38.77″N, 121°2′9.02″E-121°2′28.08″E) in the research area (a red rectangle shown in Fig. 2.)is selected as the validation region to conduct a contrast test.

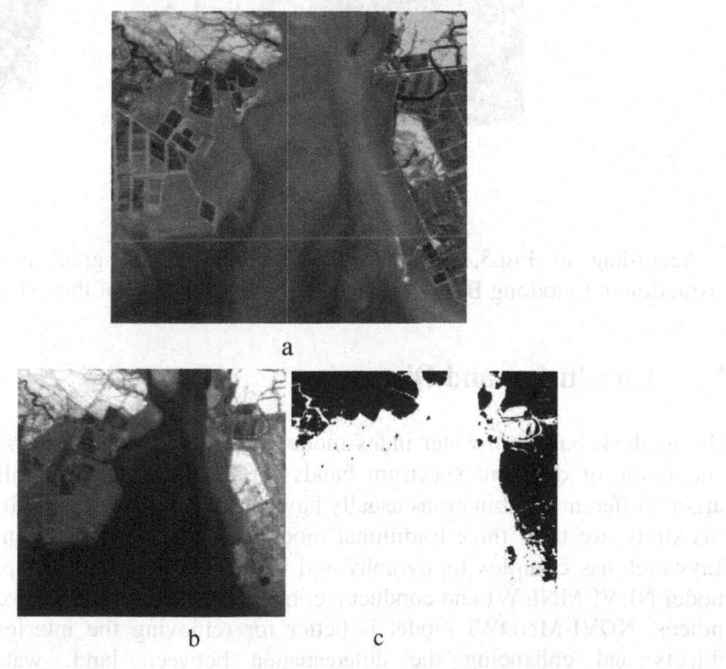

Fig. 3. The result of the five models and the corresponding water extraction. (a.CCD (342) multispectral image, b. NDVI, c. NDVI mask, d. NDWI, e. NDWI mask, f. MNDWI, g. MNDWI mask, h. NDVI-MNDWI, i. NDVI-MNDWI mask).

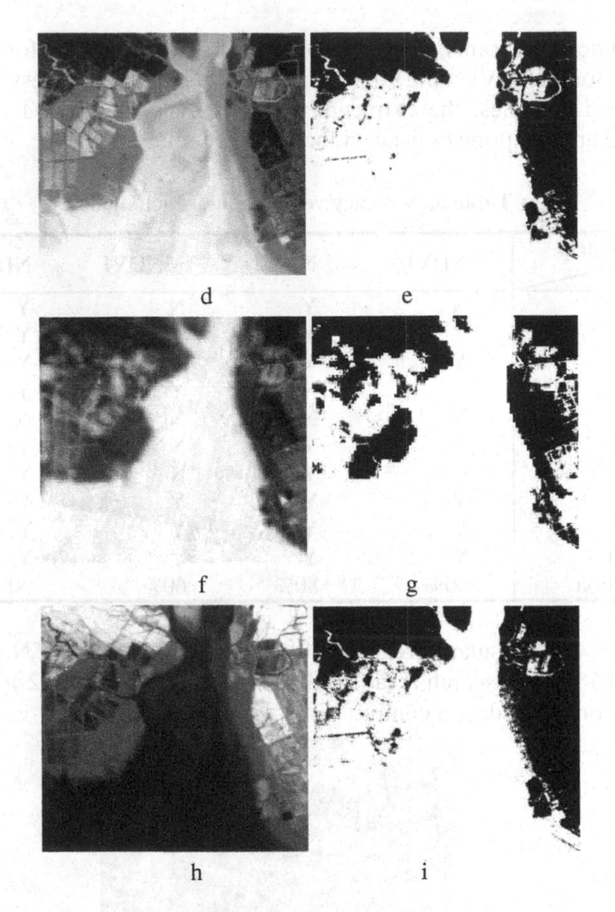

Fig. 3. (*Continued*)

According to Fig.3, NDVI-MNDWI model has a great advantage in water extraction in Liaodong Bay for the precision or the range of the extraction.

4 Conclusion and Discussion

The methods based on water index model have higher efficiency by the combination calculation of different spectrum bands for the purpose of highlighting the water target. Different combinations usually have the various effects for different regions. In this study, we used three traditional models to extract water boundary in Liaodong Baywhich has complex topography and vast range. As well, we put forward a new model NDVI-MNDWI and conduct a comparative analysis. Relative to the traditional indices, NDVI-MNDWI model is better for removing the interference of artificial objects and enhancing the differentiation between land, water and artificial structures.The experimental results show that NDVI-MNDWI model can get accurate water boundary in hybrid area that has miscellaneous topologhy structure efficiently.

in practice, the threshold plays an important role in models.. It depends on the specific situation such as regional conditions or weather conditions. Besides, the extraction of water boundary is affected by the resolution of RS images as well as the tidal variation, and so on. For different research region, we should choose the most appropriate model. For example, the NDWI model adapts to to those water areas that not affected by artificial structures, as well the MNDWI model has a good affect to extract water targets in urban area. However, to hybrid areas like Liaodong Bay with complex topography structure, vast range and estuarine distribution, it should employ the NDVI-MNDWI model or combine multiple indices models to get a better result.

References

1. Fan, D.K., Li, M., He, S.M.: Comparison and Research on Different Indices for Water Extraction Based on CCD Images from HJ Satellite. Geography and Geo-Information Science 28(2), 14–19 (2012)
2. Chen, S.H., Liu, Z.H., Liu, X., Ni, J.S.: Information Extraction of Water in Nanjing based on TM Data. Journal of Anhui Agri. 39(11), 6863–6865 (2011)
3. Jiang, C.Y., Li, M.C., Liu, Y.X.: Full-automatic Method for Coastal Water Information Extraction from Remote Sensing Image. Science of Surveying and Mapping 40(3), 332–337 (2011)
4. Han, D., Yang, X.M., Ji, K.: The research on the method of automatically extracting water body information from small satellite images. Science of Surveying and Mapping 33(1), 51–54 (2008)
5. Chen, J.B., Liu, S.X., Wang, C.Y.: Research on Urban Water Body Extraction Using Knowledge-based Decision Tree. Journal of Remote Sensing 28(1), 29–37 (2013)
6. Mei, A.X., Peng, W.L., Qin, Q.M.: An Introduction to Remote Sensing. Higher Education Press (HEP), Beijing (2001)
7. Ling, C.X., Zhang, H.Q., Lin, H.: Research on Extracting Coastal Wetlands Water Information Using Composition (CIWI) Water Index. Resources and Environment in the Yangtze Basin 9(2), 152–157 (2010)
8. Ding, F.: A New Method for Fast Information Extraction of Water Bodies Using Remotely Sensed Data. Remote Sensing Technology and Application. 24(2), 167–171 (2009)
9. Xu, H.Q.: A Study on Information Extraction of Water Body with the Modified Normalized Difference Water Index (MNDWI). Journal of Remote Sensing 9(5), 589–595 (2005)
10. El-Asmar, H.M., Hereher, M.E.: Change Detection of the Coastal Zone East of the Nile Delta Using Remote Sensing. Environmental Earth Sciences 62(4), 769–777 (2011)
11. Cheng, Q., Luo, J.C., Sheng, Y.W., Shen, Z.F., Zhu, Z.W., Ming, D.P.: An Adaptive Water Extraction Method from Remote Sensing Image Based on NDWI. Journal of the Indian Society of Remote Sensing 40(3), 421–433 (2012)
12. Lu, S.L., Wu, B.F., Yan, N.N., Wang, H.: Water Body Mapping Method with HJ-1A/B Satellite Imagery. International Journal of Applied Earth Observations and Geo-Information 13(3), 428–434 (2011)
13. Shanlong, L., Bingfang, W., Nana, Y., Hao, W.: Water body mapping method with HJ-1A/B satellite imagery. International Journal of Applied Earth Observation and Geo-Information, 428–434 (2011)

A Novel Combinational Forecasting Model of Dust Storms Based on Rare Classes Classification Algorithm

Zhenhua Zhang[1,3,*], Chao Ma[2], Jinhui Xu[2], Jiangnan Huang[1], and Longxin Li[1]

[1] School of International Trade and Economics, Guangdong University of Foreign Studies, 510006 Guangzhou, China

[2] Cisco School of Informatics, Guangdong University of Foreign Studies, 510006 Guangzhou, China

[3] Faculty of Business, Environment, Society, Coventry University, CV1 5FB Coventry, United Kingdom

zhangzhenhua@gdufs.edu.cn, statistics_zhang@aliyun.com, ab7364@coventry.ac.uk

Abstract. It is very important for people, who are facing the dust storm disaster, to forecast dust storm accurately. Traditional prediction algorithms tend to be suitable for discovering the rules of majority classes instead of that of minority classes or rare classes. In this paper, according to the monthly data of dust storm observation and the data of occurrence regularity of dust-storm provided by observation points in China, we have discovered that the dust storm occurrence data are merely the rare classes, while the data of non-occurrence of dust storm are the majority classes. Considering that current adopted methods are only suitable for excavating the time period of non-occurrence of dust storm rather than the regularity of dust storm occurrence, we find that the current algorithms are defective in studying rare classes, thus their accuracy is relatively low and is difficult to be improved. Taking this into account, according to the principles of rare classes algorithms, we combine SMOTE algorithm with adaboost algorithm as well as the random forest algorithm and propose a combination machine learning method applicable for the study of rare classes regularities. In this combination algorithm, we first balance the samples of different classes according to the idea of SMOTE algorithm, and then we make predictions utilizing random forest algorithm according to the adaboost system. This new combination algorithm possesses a total predictive accuracy which reaches 96.51%, a false alarm rate of zero, and a missing report rate of merely 0.28%. In general, this combination algorithm is one with practicability, effectiveness and simple feasibility, thus it can be applied and popularized to realistic dust storm forecasting.

Keywords: Sand-dust storm forecasting, rare classes, SMOTE algorithm, adaboost algorithm, random forest.

---

* Corresponding author.

F. Bian and Y. Xie (Eds.): GRMSE 2014, CCIS 482, pp. 520–537, 2015.

1 Introduction

Nowadays, dust storm has received extensive concern for its severe destruction of the vegetation and ecotype in occurring places. It greatly accelerates the process of land desertification in these regions and has impacts on atmospheric visibility and optical characteristics as well as earth-atmosphere system radiation balance, causing vast damages to natural environment. As the sand particles are sent into high altitude and continue their movement forced by wind, they will cause dust fall in a large range and augment of aerosol concentration and then produce certain influence on regional climate. Therefore, further study of the occurrence mechanism of dust storm and that of the prediction of dust storm are both scientific problems urgent to be solved.

At the early stage of dust storm and sand storm research, most of the researchers focused on the causes and the factors, and now they are also among the important and hot issues in this field. For example, Ye, D.Z., et al. (2000, [1]) analyzed some causes of dust weather in northern China and control measures, some scholars (Zhang R.J., et al., 2002, [2]) studied new characteristics and origins of dust storm weather in China, Zhang Q., et al. (2011, [3]) applied Copula function to analyze the spatial variability of probability distribution of extreme precipitation in Xinjiang, and Li N., et al. (2013, [4]) also used Copula function to study the return period based on joint probability distribution of three indexes of hazards in dust storm disasters. Lin Z.H., et al. (2012, [5]) discussed specific mechanisms by which land surface processes affect sand storm modeling and they made recommendations of further improvements on numerical sand storm simulations. Guan Q.Y., et al. (2013, [6]) studied the processes and mechanisms of severe sandstorm development in the eastern Hexi Corridor in China during the Last Glacial Period. Wang P., et al. (2011, [7]) applied network connectivity of wireless sensor networks to the study of sandstorm monitoring. Prezerakos N.G., et al. (2010, [8]) studied the synoptic scale atmospheric circulation systems associated with the rather frequent phenomenon of coloured rain and the very rare phenomenon of dust or sand deposits from a Saharan sandstorm triggered by a developing strong depression.

Over the last 20 years, researchers have been observing and predicting sandstorms using various methods. Zhang Y., et al. (2004, [9]) used space-borne optical sensors data over northern China to detect sandstorm according to its optical thickness data. Ali Mamtimin & Yang X.H., et al. (2011, [10]) utilized sand transport empirical formulas to estimate the sand flux in Taizhong and they found that the values calculated by Lettau's sediment discharge formula were close to those produced by instrument measurements. Some scholars made use of remote sensing and set up mathematical model for tracing dust storms (Kaskaoutis D.G., et al., 2012, [11]).

In recent years, applications of machine learning methods to sand-dust storm forecast have become more and more popular. Some scholars (Lu Z.Y., et al., 2006, 2008, [12, 13]; Chang T., et al., 2007, [14]; Fu Q.Q., et al., 2014, [15]) applied support variable machine (SVM) to sand-dust storm forecasting. There are also researchers using all kinds of neural network algorithms to forecast the sand-dust storm, such as the Back Propagation Neural Network (Rem, Z.H., 2007, [16]; Gou M.M., et al., 2009, [17]; Zuo H.J., et al., 2010, [18]), the Genetic Algorithms Neural Network (Lu Z.Y., et al., 2005, [19]), the Fuzzy Neural Network (Wang H.Z. et al., 2004, 2005, [20, 21]), the Radial Basis Function Neural Network (Wang J. et al., 2004, [22]; Lu Z.Y., et al.,

2008, [13]), and other Artificial Neural Network algorithms (Lu Z.Y., et al., 2007, [23]; Jamalizadeh M.R., et al., 2008, [24]). In the combination forecasting model research on sand-dust storm, Lu Z.Y., et al. (2008, [13]) first utilized single particle swarm optimization and improved particle swarm algorithm to optimize the parameters of RBF-SVM, and then they proposed SPSO-RBF-SVM, PSO-RBF-SVM, and WPSO-RBF-SVM, all of which possess higher recognition accuracy and efficiency than SVM does. Moreover, the accuracy of PSO-RBF-SVM and WPSO-RBF-SVM are respectively 71.2% and 84.6%, both of which are higher than the accuracy of SVM (58.2%) and BP neural networks (48.03%) respectively. And Zhang W.Y. et al. (2009, [25]) also designed a sand-dust storm warning system based on grey correlation analysis and particle swarm optimization SVM.

Ju H.B. (2004, [26]) studied sand and dust storms monitoring and proposed some problem on early warning. Li D.K., et al. have done some research on methods of dust storm monitoring and early-warning since 2006 ([27]), and then Wang H., et al. (2010, [28]) developed a new-generation sand and dust storm forecasting system to predict the sandstorm and do some early warnings. In 2012, Zhao, H.L. et al. ([29]) improved the simulation of soil moisture and applied an improved IAP dust storm numerical modeling and prediction system (IAPS 2.0) to the prediction of dust storm events over North China for two typical dust storm episodes with different characteristics with respect to dust sources and the dust affected regions. Some researchers, such as Al-Yahyai, S & Charabi, Y. et al. (2014, [30]), forecasted the real transport path of the dust storms by trajectory calculation.

Although many researchers have achieved considerably outstanding research achievements in sand-dust storm forecasting and early warning, most of them are suitable for finding the rule of major class but not the rule of minor classes. In the observational data sets, dust storm data and normal data are unbalanced distribution. Taking this into account, we present a novel combination prediction algorithm according to some machine learning algorithms on rare classes.

The issue of a short-term prediction of dust storm is a pattern classification problem. Most of the observational data from observation stations are normal, and the rest observational data indicate the occurrence of dust storm. In fact, among training samples, the dust storm class is a minority class while the normal class is the majority class. And the dust storm class often takes up much smaller proportion of total samples relative to the normal class. Thus, the short-term prediction of dust storm is a kind of imbalanced data classification. In general, the traditional methods mentioned above are not efficient enough to cope with the recognition and prediction of minority classes and rare classes. The classification performances in most traditional methods decline because of a large margin on imbalanced datasets. Great biases also exist in the traditional classifiers, one of which is that the recognition rate of minority classes and rare classes are much lower than that of majority classes.

From the analysis above, we should set up a novel combinational model suitable for the rare classes' classification to cope with the classification of dust storm.

Over the last 20 years, many scholars and researchers had found some algorithms in statistical machine learning study (Witten I.H. et al., 2011, [31]). In recent years, more and more researches focus in rare classes' classification algorithm in imbalanced data sets (Zhi W.M. et al., 2010, [32]). In 1995, Cortes C. and Vapnik V. launched support vector machine theory (SVM, [33]), which is a statistical machine learning method for small sample. After that, imbalanced data sets and their classification learning were presented and studied (Kubat M. & Matwin S., 1997, [34]). During this period, Dietterich T.G. proposed the ensemble Learning in machine learning (2000, 2002, [35, 36]), and Freund Y. et al. studied some boosting algorithms (1996, 1997, [37, 38]), which had been expanded into adaboost later (Lee Y. et al., 2013, [39]). Several years later, many scholars presented (Chawla N.V. et al., 2002, [40]) and studied (Hu F. et al., 2013, [41]) SMOTE algorithm. Meanwhile, random forests, derived from decision tree, was presented (Breiman L., 2001, [42]) and studied (Ließ M. et al, 2014, [43]) by some scholars. Some researchers (Meinshausen, N., 2006, [44]; Dong L.J. et al, 2014, [45]) analyzed the difference among these machine learning algorithms, such as random forest, SVM, Naïve Bayes method, etc.

To solve the problem of rare classes' classification on dust forecasting, we propose a new model with good performance on prediction of short-term dust storm by utilizing the theory of statistic ensemble learning in this paper. Focusing on this problem and according to the theory of statistic ensemble learning, we present a novel model which is the combination of SMOTE and adaboost random forest (ARF) and it contributes a solution for the problem of rare classes' classification. In this model, firstly SMOTE algorithm is used to preprocess the monthly data of dust storm observation and balance samples in different classes. And then, for preprocessed data, random forest will function as a weak classifier while adaboost algorithm will be adopted for statistic learning. This model (SARF) is a combination of SMOTE algorithm, adaboost algorithm, and random forest algorithm. SARF algorithm possesses superior performances, such as eliminating the necessity of feature selection, better generalization features, strong noise resistance, low requirement on the quality of the training data, etc. The simulation shows that the accuracy of SARF will be 96.51%, which is much higher than the accuracy of traditional methods, such as the accuracy of SVM (32.46%) and that of FNN (54.28%). In one word, SARF model possesses very good performances in dust storm prediction, thus it is an effectively practicable method.

2 Impact Factor and Feature Selecting

In 2000, a report, named "Causes and control measures on the dusty weather in North China" and proposed by the Department of Geology, Chinese Academy of Sciences (Ye, D.Z. et al., 2000, [1]), indicated that in the coming decades, precipitation would change little, temperatures would rise significantly, and that surface evaporation would increase, according to observational data at that span of several decades. All of these conditions would cause soil moisture to significantly decrease and form climates needed for the occurrence of dust storms. Thus in this report and further report (2006, [46]), the researchers concluded that there would be distinct drought trend in northern China.

Zhang R.J., et al. (2002, [2]) found the physical mechanism of dust storms occurrence quite complex and it must involve the following three conditions: sand sources, strong winds and atmospheric stratification, all of which include: Wind, strong wind days, precipitation, evaporation, humidity, temperature, etc. Zhang Q., et al. (2011, [3]) analyzed the impact of extreme precipitation on dust storm with Copula Function. Li N. et al. (2013, [4]) discussed the extension of Copula joint distribution model in 3D multiple hazards for disaster comprehensive analysis according to three basic conditions of formation of severe dust storms: wind speed, abundant sand source and unstable atmospheric stratification. According to the three basic characteristics of meridional circulation index, the daily average maximum wind speed and soil moisture, they established a Copula joint distribution by taking the case of dust storm events occurring in Xianghuanqi station in Inner Mongolia from 1990 to 2008. And the 3D Frank Copula fits better the occurrence probability in middle and high parts.

Based on the research above and derived from Chinese terrestrial climate data sets for monthly value data sets (1961-2005, Zhou Z.J. et al., 2008, [47]), indexes including monthly average gale speed, frequency of gale occurrence in the given month, monthly average temperature, precipitation, evaporation capacity, monthly average relative humidity, and frequency of dust storm occurrence in the given month are selected to be the conditional attributes of dust storm forecast.

3 Algorithm Introduction

In 1997, Kubat and Matwin launched the Synthetic Minority Over-sampling Technique (SMOTE) algorithm to construct classifiers from imbalanced datasets, combining over-sampling of minority class with under-sampling of majority class. The idea of this method is to synthetically generate new minority class samples using the nearest neighbors of these cases, which may avoid over-fitting to some extent. Furthermore, the majority class is also under-sampled, leading to a more balanced dataset.

Adaboost algorithm is an iterative algorithm proposed by Freund and Schapire (1996) on the basis of on-line allocation algorithm. The core idea of adaboost is to better maintain strong classifiers by gathering different weak classifiers based on the same training set. Two key points endowing this algorithm with strong applicability in short-term prediction of dust storm are concluded as follows: (1) modify the distribution of training samples and focus on wrongly classified samples. This is significant for the evaluation of dust storm rarity, since the weight of each sample is then defined according to the correctness of classification in each training set, and this process depends on the accuracy of provided overall classification; (2) combine multiple classifiers to create a strong classifier. Initially, all weights are set equally. Along with each iteration, the weights of incorrectly classified samples are increased so that the weak learner is forced to focus on hard examples in the training set, which is crucial to dust storm forecasting since the occurrence of dust storm is the case of a minority class. Furthermore, by linear weighting method, where greater weight is given to a weak classifier with lower error, adaboost combines many weak classifiers (or base classifiers) to generate a strong classifier with high accuracy. These characteristics make adaboost very adaptable to short-term prediction of dust storm.

Random forest is an ensemble learning algorithm developed by Breiman (2001). It is proved to be a preeminent algorithm for high-dimensional regression and classification in machine learning, data mining and pattern recognition. Random forests are combinations of decision tree classifiers where all the trees are valued independently according to random vector samples and with the same distribution. It has been proved by Breiman that as more trees are added, instead of overfitting, random forests produce limited value of the general error. He also found that the random selection of features is preferable to split the yielding error rates of each node compared to adaboost. Performance of adaboost will be deteriorated markedly if mixed with 5% noise, while the random forest procedures generally show small changes. Thus, a combination of random forest and adaboost algorithm can eliminate the negative effects of noise. Moreover, random forest algorithm has the following outstanding advantages: (1)free from feature selection; (2)performs well even in the case of missing data; (3)computes faster than bagging or boosting, and is very powerful to solve high-dimensional problems; (4)overcomes overfitting using out-of-bag techniques; (5)relatively robust to outliers and noise; (6)simple to be parallelized.

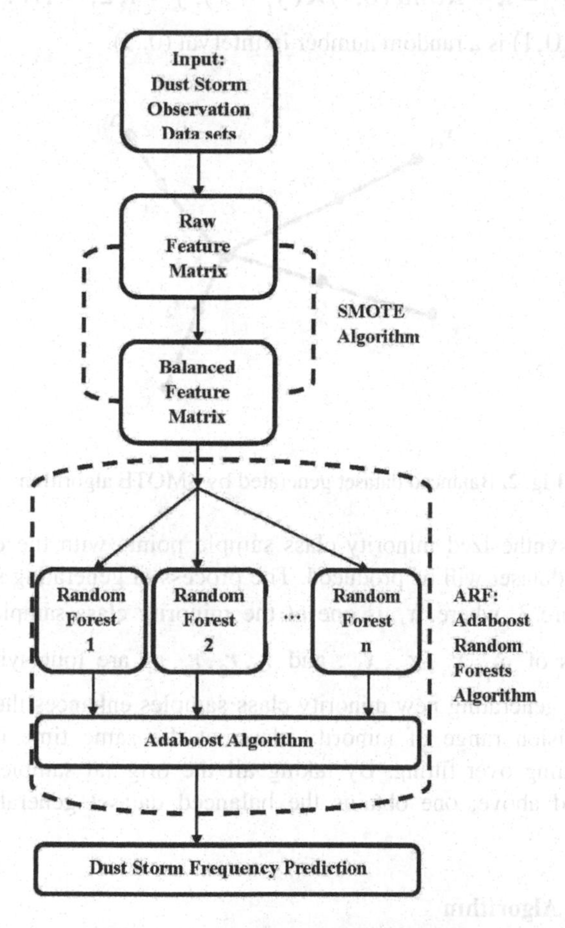

Fig. 1. The process of SARF algorithm

Considering that the forecast of dust storm is a classification problem of rare class, we propose a method combining SMOTE with adaboost random forest (ARF).

In order to balance the dataset, further process and better utilize classical classification methods, the most simple and intuitive approach to classify imbalanced dataset is to modify the distribution of the original ones. Thus, the combination forecast method involves two steps: SMOTE algorithm and ARF algorithm.

3.1 SMOTE Algorithm

Suppose a minority class sample is denoted by x. The algorithm randomly chooses N neighbors from k minority class nearest neighbor samples, denoted by $y_j (j = 1, 2, \ldots, N)$.. Then, proceed the randomly linear interpolations between x and $y_j (j = 1, 2, \ldots, N)$. and construct synthetic samples p_j as follows:

$$p_j = x + Rand(0,1) \times (y_i - x), \; j = 1, 2, \cdots, N.$$

Where $rand(0,1)$ is a random number in interval (0, 1).

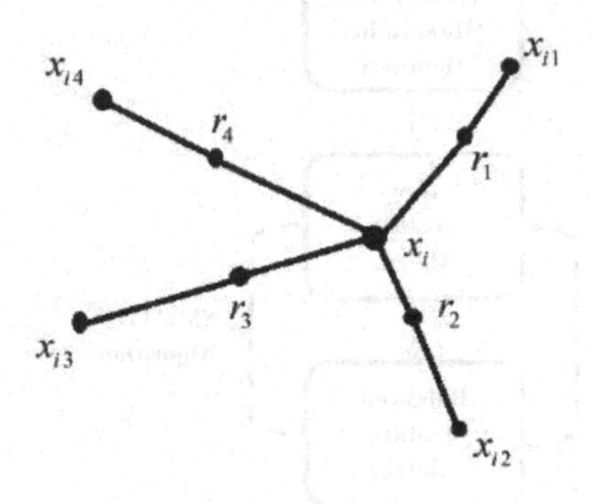

Fig. 2. Balanced dataset generated by SMOTE algorithm

Combine the synthesized minority class sample points with the original dataset, and new training dataset will be produced. The process of generating synthetic data is illustrated in figure 3, where x_i is one of the minority class samples, x_i are four nearest neighbors of $x_{i_1}, x_{i_2}, x_{i_3}, x_{i_4}$, and r_1, r_2, r_3, r_4 are four synthetic samples. This approach of generating new minority class samples enhances the generality and enlarges the decision range of minority class, at the same time it possesses the capacity of avoiding over-fitting. By taking all the original samples and synthetic samples produced above, one obtains the balanced dataset generated by SMOTE algorithm.

3.2 Adaboost Algorithm

The computing steps of adaboost algorithm are as follows.

Input: A set of m samples $\{(x_1, y_1), \ldots, (x_m, y_m)\}$ with their classifications $y_i \in Y = \{1, 2, \ldots, k\}$; A chosen weak classifier; Iteration number T.

Initialize for all i : $D_1(i) = 1/m$, i.e. uniform distribution.

Enter a t times computing cycle, where $t = 1, 2, \ldots, T$. Computing Steps will be:

Step1, Call weak classifier, endow it with weight D_t;

Step2, Train the weak classifier and then derive a corresponding hypothesis $h_t \in X \to Y$;

Step3, Calculate the error of h_t : $\varepsilon_t = \sum_{i:h_t(x_i) \neq y_i} D_t(i)$, if $\varepsilon_t > 1/2$, then *T=t-1*, break out of the loop;

Step4, $\beta_t = \varepsilon_t / (1 - \varepsilon_t)$;

Step5, Update distribution D_t : $D_{t+1}(i) = \frac{D_t(i)}{Z_t} \times \begin{cases} \beta_t & \text{if } h_t(x_i) = y_i \\ 1 & \text{otherwise} \end{cases}$,

where Z_t is a normalized constant (chosen so that $D_{t+1}(i)$ becomes a new weight.

Output the desired ultimate hypothesis : $h_{fin}(x) = \arg\max_{y \in Y} \sum_{t:h_t(x)=y} \log \frac{1}{\beta_t}$.

3.3 Random Forests Classifier

The operating steps of the Random Forests algorithm are as follows:

Step1, Select n bootstrap samples from the original dataset. In a typical bootstrap sample, approximately 63% of the original observations occur at least once.

Step2, Construct a fully-grown CART tree for each bootstrap sample respectively and a forest is produced. At each node, only a small number m of randomly selected variables ($m<M$) are available for the binary partitioning.

Step3, Adopt voting system to put together all the numerical values calculated by decisive trees, according to the predicted results of n CART trees, with ties splitting randomly.

Observations included in the original dataset but do not occur in a bootstrap sample are called out-of-bag observations. CART trees are tools for predicting the out-of-bag observations. The process of random forests using out-of-bag observations to estimate the generalization errors is known as out-of-bag estimate, which is similar to cross-validation to some extent. However, Breiman has proved that unlike cross-validation, the out-of-bag estimate is an unbiased estimate ensuring that random forests do not overfit as the number of combinations increases.

The three principle parameters to be set in random forest are number of CART trees, number of randomly selected variables, and the minimal number of examples at terminal nodes. Based on the feature of dust storm forecasting, these parameters are set by 10, 4 and 1 respectively in this paper.

4 Random Forests Classification Based on Adaboost Ensemble Learning: The SARF Model

As is illustrated in previous sections, random forest is a powerful ensemble learning method with outstanding performances such as fast calculating speed, strong robusticity to noise, and it does not overfit. However, it has been proved by Segal and Statnikov (2004) that random forest can overfit, especially in noisy data and small sample datasets. Moreover, it suffers from the difficulty of tree's growth without pruning, which indicates that there is still room for improvement.

In this paper where random forest is adopted as a weak classifier, we combine adaboost algorithm with random forest algorithm and generate a new forecasting model with better performance. Furthermore, since our dust storm dataset is imbalanced, we apply SMOTE algorithm to balance the training set before building model on datasets. Thus, the hybrid algorithm for new forecasting model can be operated as follows:

Step1, Input data: A set S of m training samples $\{(x_1, y_1), \ldots, (x_m, y_m)\}$ with labels $y_i \in Y = \{1, 2, \ldots, k\}$;Number of iterations *T*;Number of generated trees *G*; Number of neighbors in SMOTE *N*; Number of minority samples *C*; Number of randomly selected variables *r*.

Step2, Initialize $D_1(i) = 1/m$ for all i , i.e. uniform distribution.

Step3, Balance datasets using SMOTE algorithm:
$\hat{S} = SMOTE(\{(x_1, y_1), \ldots, (x_m, y_m)\}, C)$.

Step4, for $t = 1, 2, \ldots, T$, Do:

Step4.1, Call random forest, endow random forest with the distribution D_t;

For $g = 1, 2, \ldots, G, \hat{S}_g = BoostTrap(\hat{S}), CART_g = BuildCART(\hat{S}_g, r)$, each tree cast a vote: $y_g = CART_g(x)$, next g ;

Step4.2, Derive a weak hypothesis from the training weak classifier $h_t \in X \to Y$ by voting: $h_t(x) = \arg\max_{y \in Y} \sum_{g:CART_g(x)=y} y_g$, with the error $\varepsilon_t = \sum_{i:h_t(x_i) \neq y_i} D_t(i)$, .if $\varepsilon_t > 1/2$, then *T*=*t*-1, break out of the loop.

Step4.3, $\beta_t = \varepsilon_t / (1 - \varepsilon_t)$.

Step4.4, Update distribution D_t : $D_{t+1}(i) = \frac{D_t(i)}{Z_t} \times \begin{cases} \beta_t & \text{if } h_t(x_i) = y_i \\ 1 & \text{otherwise} \end{cases}$,

where Z_t is a normalized constant ((chosen so that $D_{t+1}(i)$ will be a distribution).

Step4.5, Output: the desired ultimate hypothesis will be:

$$h_{fin}(x) = \arg\max_{y \in Y} \sum_{t:h_t(x)=y} \log \frac{1}{\beta_t} \cdot$$

This thesis combines adaboost with random forest Algorithm, using random forest as a weak classifier and together generating a strong classifier with higher precision under the framework of adaboost, thus enhancing the classification performance of Random Forests. This new adaboost random forest algorithm (abbreviated ARF, the same as follows) combines the advantages of adaboost algorithm and random forest algorithm, which possesses the following superior performances: Good classification performance with low classification error rate; High prediction accuracy and good generalization performance, thus avoiding over-learning; Strong noise resistance and robusticity; Low requirement on the quality of the training data, maintaining good prediction performance even under high dimension or data missing; Eliminating the necessity of data preprocessing and feature selection; Realizing that random forest algorithm itself is a learning of equal weight integration of CART decision tree while the framework of adaboost provides weight-based decision methods, the combination of them can further improve the classification performance of the original random forest algorithm and better outputting classification results; Adopting random forests as weak classifiers makes the cross-validation for unbiased estimation on data unnecessary.

5 Dust Storm Prediction Based on SARF Model

5.1 SARF Algorithm

STEP 1: Dust storm is a kind of disastrous dusty wind weather phenomenon resulted from strong wind drawing much sand and dust near the ground surface into sky, leading to worsened air condition with a visibility within 1Km.

Three dominant factors causing dust storm are respectively: Gale factor; Dynamic-thermal factor; Sand-dust source factor.

The data for short-term prediction are as follows: (1) Numbered lists and place names of 193 observation points in 6 northwestern provinces; (2) Each observation point offers monthly meteorological data related to dust storm from year 1961 to 2005.

A=Year; B=Month; C= Monthly average gale speed; D=Frequency of gale occurrence in the given month; E=Monthly average temperature; F=Precipitation; G=Monthly average relative humidity; H= Evaporating capacity; I=Frequency of dust storm occurrence in the given month.

The first 8 items (i.e. year, month… monthly average relative humidity) are selected as sample features while frequency of dust storm occurrence is extracted as sample classification label and is given a value range from 0 to 9 defined as class 0 to 9 correspondingly. Note that the frequency of dust storm occurrence is also identified as categorical variable.

Now take 4 samples as instances from observation station 52323, and the sample format is as below:

Table 1. Schematic diagram of sample formats

No.	A	B	C	D	E	F	G	H	I
52323	1981	1	4.2	3.3	-12	1.4	50.2	52	0
52323	1981	3	4.5	4.6	-0.7	0.9	179.9	33	0
52323	1981	4	4.2	6.7	6.6	2.3	287.8	31	0
52323	1981	5	5.2	4.3	12	1.7	484	18	0

STEP 2: Extract 500 out of 6899 samples which have been randomly layered as training set, and extract another 500 samples as testing set following the same procedure.

STEP 3: Preprocess the training set by SMOTE algorithm, proceed synthesis operation on the minority classes (i.e. Class 5,6,7,8, and 9) by generating new samples of the minority classes. This method may balance the dataset without causing overfitting.

STEP 4: Apply combination prediction model by taking the 8 features of dataset as input while the frequency characteristics of dust storm as output, according to the new dataset obtained from STEP 3. There's no need to normalize the sample data because normalization will deteriorate the performance of ARF algorithm. Meanwhile, benefited from the character of ARF algorithm, pretreatment or screening is also unnecessary. In this condition, ARF algorithm still maintains good performance.

STEP 5: Evaluate the generalization ability and prediction performance of the model by utilizing the test set which contains 500 samples after obtaining ARF classifier.

STEP 6: Train the samples aiming at specific dust storm observation points via measuring the values of climatic factors, such as month, wind speed, time interval of gale occurrence, precipitation, temperature, etc. Corresponding prediction of days of dust storm is carried out after entering the data into ARF model, from which we can define risk degree of dust storm. In other words, the higher the output value is, the greater the occurrencing possibility of current dust storm will be. Hence actions of advanced warning and early prevention should be taken.

5.2 Classifier Evaluation Indexes

Avoiding false prediction and underreporting is the most fundamental requirement for disaster prediction model. Therefore, calculating false alarm rate and missing report rate of a predictive model is an important procedure of evaluating its performance. Based on the confusion matrix, a new evaluation method is proposed.

Confusion matrix is denoted by $(a_{ij})_{n \times n}$, where a_{ij} represents the number of samples, classified as class *i* with their actual class denoted by *j*. Assume that samples in class1 are negative samples, and then we define: False alarm rate as $\eta = \sum_{j=2}^{n} a_{1j} \Big/ \sum_{i=1}^{n} a_{1i}$; Missing report rate as $\mu = \sum_{i=2}^{n} a_{i1} \Big/ \sum_{i>1, j>1} a_{ij}$.

To evaluate the performance of the classifier put to use in this paper, we define a set of usual evaluation measurements according to confusion matrix (Table 2), including Precision, Recall rate, and ROC curve as follows:

Precision= $\frac{TP}{TP+FP}$; Recall rate= $\frac{TP}{TP+FN}$.

Table 2. Confusion matrix

		Predicted Class	
		Positive	Negative
Actual Class	Positive	True Positive(TP)	False Negative(FN)
	Negative	False Positive(FP)	True Negative(TN)

Since dust storm prediction is a multi-classification problem, we can adopt these measures to evaluate the performance of a multi-classifier in each individual class respectively.

6 Results and Analysis

In previous sections, we have proposed the short-term forecast algorithm for dust storm according to SMOTE algorithm and ARF algorithm. In this combination model, we first apply SMOTE algorithm to rebalance the dataset. And after pre-processing of data, we use the random forest as a weak learner and embed it into the adaboost algorithm to generate an ensemble learning model, which owns desirable characteristics, such as free from feature selection, excellent generalization performance, strong robust city, and low training set quality requirement.

Now we move on to the evaluation of the dust storm forecast capacity, using the proposed combination algorithm derived from the idea of forecast procedure constructed in the previous section. The dataset described earlier in the previous part will be used in the experiment under weka environment. For further examination of the combined algorithm's applicability in dust storm forecasting, weka version 3.6 is selected to evaluate its performance and effectiveness. We utilize weka and build the process presented above to arrive at the following conclusions.

Experiments are performed using a 10-fold cross-validation to reduce the bias associated with the random sampling strategy on the dust storm dataset from table 1. The dataset consists of a training set (9 folds) and a test set (remaining fold). Part of our experiment results are produced in figure 3 and table 3.

Table 3. Accuracy of proposed forecasting method

Algorithm	SMOTE +ARF	SMOTE + RF	SMOTE + Logistic	SMOTE+ Decision Stump	SMOTE + J48	SMOTE + Decision Table	SVM	Fuzzy ANN
Precision	96.51%	93.25%	26.58%	17.43%	72.55%	32.46%	32.46 %	54.28%
Class1 Precision/recall	0.989/1	0.947/0.989	0.552/0.711	0.216/0.844	0.837/0.856	0.604/0.644	0.585/0.69	0.739/0.669
Class2 Precision/recall	0.944/0.944	0.904/0.944	0.271/0.211	0/0	0.742/0.733	0.36/0.2	0.273/0.33	0.096/0.214
Class3 Precision/recall	0.976/0.9	0.949/0.822	0.167/0.067	0/0	0.7/0.7	0.325/0.3	0.246/0.14	0.093/0.103
Class4 Precision/recall	0.957/0.978	0.912/0.922	0.273/0.1	0/0	0.716/0.644	0.435/0.111	0.227/0.46	0/0
Class5 Precision/recall	0.98/1	0.961/0.98	0.278/0.2	0/0	0.756/0.62	0.292/0.38	0/0	0/0
Class6 Precision/recall	0.963/1	0.963/1	0.115/0.269	0/0	0.625/0.962	0.273/0.115	0/0	0/0
Class7 Precision/recall	0.857/1	0.917/0.917	0.065/0.167	0/0	0.5/0.583	0.096/0.583	0/0	0/0
Class8 Precision/recall	1/1	0.833/1	0.08/0.4	0/0	0.286/0.4	0.064/0.6	0/0	0/0
Class9 Precision/recall	1/1	1/1	0.063/0.5	0.037/1	1/0.5	0.25/0.5	0/0	0/0
Class10 Precision/recall	1/1	1/1	0.053/0.5	0/0	0.667/1	0.667/1	0/0	0/0
False Alarming Rate	0%	1.1%	28.89%	15.55%	14.44%	35.55%	31.28 %	26.04%
Miss Reporting Rate	0.28%	1.35%	14.09%	74.79%	4.06%	10.29%	12.28 %	73.05%

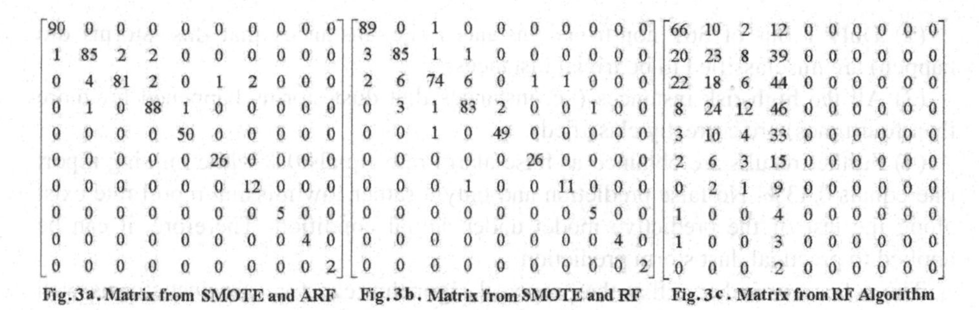

Fig. 3. Confusion matrix

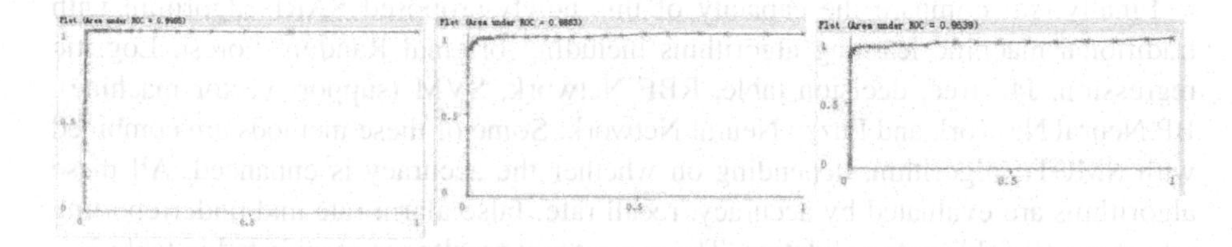

Fig. 4. Receiver Operating Characteristic Curve from SMOTE and ARF

Table 3 displays the accuracy of our forecasting method, which shows that the SARF method and SRF method have improved accuracy up to 96.51% (443 correctly classified instances among 459 instances) and 93.25% respectively. Table 3 illuminates further evaluation measures, including precision, recall rate, false alarm rate and missing report rate, which are shown in table 3 for details and figure 4 for the Receiver Operating Characteristic Curve. However, we place strong emphasis on the fact that when a data set is unbalanced (in our case, dust storm rarely happens), the error rate of a classifier does not represent its true performance. More importantly, we need confusion matrix as a tool to visualize the true performance of SARF method.

Figure 3 provides the confusion matrix of the proposed algorithm to examine whether the algorithm is confusing different classes or not. Each column of the matrix indicates the instances in the predicted class, while each row points out the instances in an actual class. The experiment result reveals that most of the non-vanishing elements lie along the diagonal of the confusion matrix derived from our method. It is manifested that the proposed algorithm is capable of recognizing the distinction of any two arbitrary classes. Hence, this new algorithm is highly effective. Moreover, figure 3b depicts the confusion matrix of the original random forest (RF) combined with SMOTE algorithm, and figure 3c the confusion matrix of RF without SMOTE. The comparison reveals that as SMOTE algorithm significantly boosts the performance of RF on imbalanced dataset, adaboost can indeed make improvement to RF, which yields a 100% accuracy on high-risk instances(i.e. instances that dust storms happened for more than four times).

Further excellent traits of the ARF algorithm are as follows:

(1) All of the negative instances (i.e. instances where no dust storms happened) are accurately predicted;

(2) Only 1 out of 369 non-trivial instances (i.e. instances that dust storms did happen) are misclassified to be trivial instances;

(3) All the high-risk instances (i.e. instances that dust storms happened for more than four times) are correctly classified;

(4) Further results are obtained as false alarm rate equals 0% while missing report rate equals 0.43%. No false prediction and only a rather low missing report rate exist along the test of the predictive model under actual condition. Therefore, it can be applied to practical dust storm prediction.

These features indicate that the proposed algorithm can forecast dust storms well with low false alarm rate and underreporting rate that are crucial to the practical application in dust storm forecast.

Finally, we compare the capacity of this newly-proposed SARF algorithm with traditional machine learning algorithms including original Random Forest, Logistic regression, J48 tree, decision table, RBF Network, SVM (support vector machine), BP Neural Network and Fuzzy Neural Network. Some of these methods are combined with SMOTE algorithm, depending on whether the accuracy is enhanced. All these algorithms are evaluated by accuracy, recall rate, false alarm rate and underreporting rate using 10-fold cross-validation. The experiment results are manifested in table 3.

Compared with traditional machine learning algorithms, the result generated from the proposed algorithm is much more preferable both in accuracy and recall rate, through 10-fold cross-validation test. Our algorithm also possesses better adaptability with low false alarm rate and underreporting rate. On the contrary, suffering from interference of unbalanced dataset, traditional algorithms cannot distinguish positive instances from negative ones, thus resulting in high false alarm rate and underreporting rate.

7 Conclusion and Future Improvement

In this paper, we propose a new combination algorithm for building a dust storm forecast model based on SMOTE algorithm and adaboost random forest algorithm. Results of multi-rare-class classification tasks are carried out. SMOTE algorithm is applied to rebalance the dataset and improve the performance of multi-rare-class classification while random forest is utilized as a weak learner of adaboost ensemble learning technique. Compared with traditional dust storm forecast models, the novel SARF algorithm produces better results in all evaluation measurements, such as accuracy, regression rate, false alarm rate and underreporting rate. As for future application, we need to optimize the parameters and make further improvement as some researches (Lu Z.Y. et al., 2008, [13]; Zhang Z.H. et al., 2013, [47]), and to use other hybrid ensemble learning algorithms (e.g. SMOTE boost methods) and cost-sensitive coefficient methods.

Acknowledgements. This paper is funded by the National Natural Science Foundation of China (Grant No. 71271061, Grant No. 61073147), the "Twelfth Five-Years" Philosophy and Social Sciences Planning Project of Guangdong Province

(Grant No. GD12XGL14), Student Innovation Training Program of China (Grant No. 201411846013), the Science and Technology Innovation Project of Department of Education of Guangdong Province (Grant No. 296-GK13201, No. 2013KJCX0072), the Business Intelligence Key Team of Guangdong University of Foreign Studies (Grant No. TD1202), the Teaching Quality and Teaching Reform Project for Higher Education of Guangdong Province (Grant No. 110-GK131021, No. 2013176), the "Twelfth Five- Years" Education Planning Project of Guangdong Province (Grant No. 2012JK129), and the Major Education Foundation of Guangdong University of Foreign Studies (Grant No. GYJYZDA12011), Philosophy and Social Sciences Planning Project of Guangzhou (Grant No. 2014GZZGJ0067).

References

1. Ye, D.Z., Chou, J.F., Liu, J.Y.: Causes of sand-stormy weather in northern China and control measures. Acta Geographica Sinica-Chinese Edition 55(5), 513–521 (2000) (in Chinese)
2. Zhang, R.J., Han, Z.W., Wang, M.X., Zhang, X.Y.: Dust storm weather in China: new characteristics and origins. Quaternary Sciences 22(4), 374–380 (2002) (in Chinese)
3. Zhang, Q., Li, J.F., Chen, X.H., Bai, Y.G.: Spatial variability of probability distribution of extreme precipitation in Xinjiang. Acta Geographica Sinica 66(1), 3–12 (2011)
4. Li, N., Gu, X.T., Liu, X.Q.: Return period analysis based on joint distribution of three hazards in dust storm disaster. Advances in Earth Science 28(4), 490–496 (2013) (in Chinese)
5. Lin, Z.H., Levy, J.K., Lei, H., Bell, M.L.: Advances in Disaster Modeling, Simulation and Visualization for Sandstorm Risk Management in North China. Remote Sensing 4, 1337–1354 (2012) (in Chinese)
6. Guan, Q.Y., Pan, B.T., Yang, J., Wang, L.J., Zhao, S.L., Gui, H.J.: The processes and mechanisms of severe sandstorm development in the eastern Hexi Corridor China, during the Last Glacial period. Journal of Asian Earth Sciences 62, 769–775 (2013)
7. Wang, P., Sun, Z., Vuran, M.C., Al-Rodhaan, M.A., Al-Dhelaan, A.M., Akyildiz, I.F.: On network connectivity of wireless sensor networks for sandstorm monitoring. Computer Networks 55, 1150–1157 (2011)
8. Prezerakos, N.G., Paliatsos, A.G., Koukouletsos, K.V.: Diagnosis of the relationship between dust storms over the Sahara desert and dust deposit or coloured rain in the South Balkans. Advances in Meteorology 2010(760546), 1–14 (2010)
9. Zhang, Y., Guan, Y., Guo, S., Nie, Y., Lin, Q.: Sandstorm detection using space-borne optical sensors data over the Northern China. Journal of Electromagnetic Waves and Applications 18(2), 153–160 (2004)
10. Mamtimin, A., Yang, X.H., Xu, X.L., He, Q., Yu, B., Tang, S.H.: Sand flux estimation during a sand-dust storm at Tazhong area of Taklimakan Desert, China. Journal of Arid Land 3(3), 199–205 (2011)
11. Kaskaoutis, D.G., Prasad, A.K., Kosmopoulos, P.G., Sinha, P.R., Kharol, S.K., Gupta, P., Elaskary, H.M., Kafatos, M.: Synergistic Use of Remote Sensing and Modeling for Tracing Dust Storms in the Mediterranean. Advances in Meteorology 2012(861026), 1–14 (2012)
12. Lu, Z.Y., Zhang, Q.M., Zhao, Z.C.: Sand-dust storm forecasting model based on SVM. Journal of Tianjin University 39(9), 1110–1114 (2006) (in Chinese)

13. Lu, Z.Y., Li, Y.Y., Lu, J., Zhao, Z.C.: Parameters optimization of RBF-SVM sand-dust storm forecasting model based on PSO. Journal of Tianjin University 41(4), 413–418 (2008) (in Chinese)
14. Chang, T., Fu, W.D., Qin, R.: Study on forecasting of sandstorm in Xinjiang based on LS-SVM. Journal of Shanxi Meteorology 5, 6–9 (2007) (in Chinese)
15. Fu, Q.Q., Xie, Y.H., Tang, B., Zhang, H.D.: Research on sand-dust storm warning based on SVM with combined kernel function. Computer Engineering and Design 35(2), 646–650 (2014) (in Chinese)
16. Rem, Z.H.: The use of BP neural net in meteorological phenomena data forecasting. Agriculture Network Information 11, 42–44 (2007) (in Chinese)
17. Gou, M.M.: Study on sandstorm forecast model by using BP neural network in the Xilin Gol area. Degree Paper. Inner Mongolia Agricultural University, Hohhot (2009) (in Chinese)
18. Zuo, H.J., Gou, M.M., Li, G.T., Li, X.: Study on sandstorm forecasting with BP neural network method. Journal of Desert Research 30(1), 193–197 (2010) (in Chinese)
19. Lu, Z.Y., Yang, Y.F., Zhao, Z.C., Pang, Y., Liu, H.Z.: The study of the sand-bust storm forecasting model based on GA-neural network. Computer Engineering and Applications 41(33), 220–222 (2005) (in Chinese)
20. Wang, H.Z.: The forecasting model of sand-dust storm based on FNN. Degree Paper. Tianjin University, Tianjin (2004) (in Chinese)
21. Wang, H.Z., Liu, Z.Q., Wang, P.: Apply of fuzzy neural networks with fuzzy weights to the forecasting of sand-dust storm. Journal of Tianjin University of Science & Technology 20(2), 64–67 (2005) (in Chinese)
22. Wang, J., Jiang, D.H.: Establishment of radial basis function neural network to predict the impact of sandstorm on Shanghai. Pollution Control Technology 17(1), 3–5 (2004) (in Chinese)
23. Lu, Z.Y., Yang, L., Zhao, Z.C., Yang, Y.F.: A field feature extraction method of sand-dust storm ensemble forecast system based on ANN. Computer Simulation 24(6), 341–344 (2007) (in Chinese)
24. Jamalizadeh, M.R., Moghaddamnia, A., Piri, J., Arbabi, V., Homayounifar, M., Shahryari, A.: Dust storm prediction using ANNs technique (A case study: Zabol City). Proceedings of World Academy of Science: Engineering and Technology 45, 529–537 (2008)
25. Zhang, W.Y., Liu, X., Xiao, W., Chi, D.Z.: Software design of sand-dust storm warning system based on grey correlation analysis and particle swarm optimization support vector machine. In: Proceedings of the 2nd IEEE International Conference on Power Electronics and Intelligent Transportation System (PEITS 2009), vol. 2, pp. 47–50 (2009)
26. Ju, H.B.: Study on sand and dust storms monitoring and early warning. Degree Paper. Chinese Academy of Sciences, Beijing (2004) (in Chinese)
27. Li, D.K., Du, J.W.: Research on methods of dust storm monitoring and early-warning. Journal of Catastrophology 21(1), 55–58 (2006) (in Chinese)
28. Wang, H., Gong, S.L., Zhang, H.L., Chen, Y., Shen, X.S., Chen, D.H., Xue, J.S., Shen, Y.F., Wu, X.J., Jin, Z.Y.: A new-generation sand and dust storm forecasting system GRAPES_CUACE/Dust: Model development, verification and numerical simulation. Chinese Science Bulletin 55(7), 635–649 (2010)
29. Zhao, H.L., Jason, K., Levy, H.L., Michelle, L.: Bell: Advances in Disaster Modeling, Simulation and Visualization for Sandstorm Risk Management in North China. Remote Sensing 4, 1337–1354 (2012)
30. Al-Yahyai, S., Charabi, Y.: Trajectory calculation as forecasting support tool for dust storms. Advances in Meteorology 2014(698359), 1–6 (2014)

31. Witten, I.H., Frank, E., Hall, M.A.: Data mining: practical machine learning tools and techniques. The Morgan Kaufmann Series in Data Management Systems. Elsevier, Amsterdam (2011)
32. Zhi, W.M., Fan, M.: Research on classification of rare classes. Computer Technology and Development 20(7), 250–253 (2010) (in Chinese)
33. Cortes, C., Vapnik, V.: Support vector machine. Machine Learning 20(3), 273–297 (1995)
34. Kubat, M., Matwin, S.: Addressing the curse of imbalanced training sets: one-sided selection. In: ICML1997, pp. 179–186 (1997)
35. Dietterich, T.G.: Ensemble methods in machine learning. In: Kittler, J., Roli, F. (eds.) MCS 2000. LNCS, vol. 1857, pp. 1–15. Springer, Heidelberg (2000)
36. Dietterich, T.G.: Ensemble learning. In: The Handbook of Brain Theory and Neural Networks, pp. 405–408. The MIT Press, Cambridge (2002)
37. Freund, Y., Schapire, R.E.: Experiments with a new boosting algorithm. In: Proceedings of the 3rd International Workshop and Conference on Machine Learning, pp. 148–156. Morgan Kaufmann Publishers, Inc. (1996)
38. Freund, Y., Schapire, R.E.: A decision-theoretic generalization of on-line learning and an application to boosting. Journal of Computer and System Sciences 55(1), 119–139 (1997)
39. Lee, Y., Han, D.K., Ko, H.: Reinforced adaBoost learning for object detection with local pattern representations. The Scientific World Journal 2013(153465), 1–14 (2013)
40. Chawla, N.V., Bowyer, K.W., Hall, L.O., Kegelmeyer, W.P.: SMOTE: Synthetic minority over-sampling technique. Journal of Artificial Intelligence Research 16, 341–378 (2002)
41. Hu, F., Li, H.: A novel boundary oversampling algorithm based on neighborhood rough set model: NRSBoundary-SMOTE. Mathematical Problems in Engineering 2013(694809), 1–10 (2013)
42. Breiman, L.: Random Forests. Machine Learning 45(1), 5–32 (2001)
43. Ließ, M., Hitziger, M., Huwe, B.: The sloping mire soil-landscape of southern Ecuador: Influence of predictor resolution and model tuning on random forest predictions. Applied and Environmental Soil Science 2014(603132), 1–10 (2014)
44. Meinshausen, N.: Quantile regression forests. The Journal of Machine Learning Research 7, 983–999 (2006)
45. Dong, L.J., Li, X.B., Xie, G.N.: Nonlinear methodologies for identifying seismic event and nuclear explosion using Random Forest, Support Vector Machine and Naive Bayes Classification. Abstract and Applied Analysis 2014(459137), 1–8 (2014)
46. Department of Geology: Chinese Academy of Sciences. Causes and control measures on the dusty weather in North China. Advance in Earth Sciences 15(4), 361–364 (2006) (in Chinese)
47. Zhou, Z.J., et al.: Reference Department, National Meteorological Center, China Meteorological Administration: Severe dust storms sequence and their supporting data Sets in China (1951-2007), `http://www.cma.gov.cn/` (2008) (in Chinese)
48. Zhang, Z.H., Wang, M., Hu, Y., Yang, J.Y., Ye, Y.P., Li, Y.F.: A Dynamic Interval-Valued Intuitionistic Fuzzy Sets Applied to Pattern Recognition. Mathematical Problems in Engineering 2014(408012), 1–16 (2014)

The Study on Spatial Distribution of Floor Area Ratio Based-on Kriging——The Case of Wuhan City

Xinyan Li[1] and Lei Miao[2]

[1] School of Architecture and Urban Planning, Huazhong University of Science and Technology, 1037 Luoyu Road, Wuhan, China
[2] School of Information and Safety Engineering, Zhongnan University of Economics and Law, 182 Nanhu Avenue, Wuhan, China
lixinyanemail@263.net, miao_lei01@163.com

Abstract. FAR (floor area ratio) is an important indicator for urban management. But now in China, FAR is basically determined by the planners with experience, which makes the authority and scientific of the urban planning been widely questioned and criticized. This article attempts to apply quantitative methods to determine the value of FAR. This research utilized the spatial analysis by Geographical Information System (GIS), combined it with geostatistics method to study the spatial distribution of FAR. We used ArcGIS software to carry out the experiment study. The case of Wuhan study shows that the result of Kriging interpolation is superior to other interpolation algorithms and it can produce the minimum forecast error. According to the cross validation table, it is considered that Kriging interpolation is the optimal method for estimation and projection of FAR.

Keywords: FAR, spatial interpolation, geostatistic, kriging.

1 Introduction

Nowadays, cities in China are developing rapidly. On the one hand the city expand outwards rapidly, on the other hand buildings are higher and higher as well as its' density increases gradually. Urban land development is necessary to emphasize the economic benefits, but also to ensure a comfortable life. Therefore, to determine a reasonable urban land development intensity is an important task of urban planning and management. In china, urban land development intensity is controlled primarily at regulatory plan, while the floor area ratio (FAR) is the most important control indicator. But now in China, FAR is basically determined by the planners with experience, which makes the authority and scientific of the urban planning been widely questioned and criticized. This article attempts to apply quantitative methods to determine the value of the FAR. This research utilized the spatial analysis by Geographical Information System (GIS), combined it with geostatistics method to study the spatial distribution of urban land development intensity.

The quantitative study is mainly based on spatial interpolation algorithms. At present, the major spatial interpolation methods include Inverse Distance Weighting

F. Bian and Y. Xie (Eds.): GRMSE 2014, CCIS 482, pp. 538–547, 2015.

(IDW), spline interpolation, polynomial interpolation, neighborhood interpolation and minimum curvature interpolation, etc. All the above interpolation methods belong to deterministic interpolation methods, and another type of interpolation method is geostatistical data interpolation, which is a branch of spatial statistical analysis. The major difference between geostatistical and deterministic interpolation methods lies in that, geostatistical interpolation has introduced probability model. It is considered by the geostatistical interpolation method that accurate predicted value cannot be obtained by one statistical model, so the deviations of predicted value are given to indicate the probability of rational predicted values[1][2].

Based on regionalized variables, geostatistics utilizes variant functions to explore natural phenomenon with both random and structural properties, or those with spatial correlation and dependency. The theory and approach of geostatistics can be applied to any studies on the random and structural properties of spatial data, or spatial correlation and dependency, or spatial pattern and variations, or optimal and unbiased interpolation estimates on such data, or simulations on the discreteness and volatility of such data. The difference between geostatistics and classical statistics lies in that, geostatistics not only considers the value of the sample, but also the spatial position of the sample and distance between samples, which remedies the defect of classical statistics in this regard. Geostatistical analysis has been widely applied to geological and mining fields, and now is gradually used for urban and regional studies[3][4][5][6][7].

This article takes Wuhan as research area, the results of different spatial interpolation algorithms are compared, and then select the most suitable one. Finally, the values of FAR are predicted of inner city of Wuhan.

2 Method

Kriging, also known as spatial local interpolation, is one of two major content geostatistics. It is built on the theory and structure variogram analysis, based on values in a limited area of regionalized variables were an unbiased optimal estimation method [2]. Specifically, it is limited number of sample points in the neighborhood of data points have been determined to be estimated in accordance with point (or to be estimated block segment), and taking into account the sample shape, size and spatial relationship of mutual position, they point the positional relationship between each other space to be estimated and after variograms provide structural information, a linear treat valuation point values unbiased optimal estimation.
Kriging can be formulated as follows:

$$Z(s) = \mu(s) + \varepsilon(s)$$

S represents the position of different points, the spatial coordinates may be planimetric rectangular coordinates or latitude and longitude coordinates. Z (s) is the attribute value of the position of the point. μ (s) is the trend value, ε (s) is the random errors of self-correlation.

Kriging weights not only considered the distance between the known points and the point to be estimated, but also the orientation of the spatial distribution, through semi-variogram the weight value is determined. Kriging method does not require the data were normally distributed, but when the data were normally distributed, Kriging would be the best kind of unbiased estimate interpolation method. For non-normally distributed data, data conversion can be performed, such as Log logarithmic transformation, Box-Cox transformation, etc. Normally distributed data can be interpolated.

Kriging can also be expressed as:

$$Z(x_0)=\sum\lambda_i Z(x_i)$$

Where, $Z(x_0)$ is the point value to be estimated. $Z(x_i)$ is the known point values around the point to be estimated. λ_i is the weight of the i-th point, the weight value not only considered the distance between the known points and the point to be estimated, but also the orientation of the spatial distribution. Through semi-variogram the weight value is determined.

Semi-variogram and covariance functions are quantitative expressions of spatial autocorrelation function. Semi-variogram is defined as follows:

$$R(h_{ij})=1/2\cdot var(Z(s_i)-Z(s_j))$$

Where, h_{ij} is the distance between the sample points of s_i and s_j, $Z(s_i)$ and $Z(s_j)$ representing the properties of points respectively. If s_i and s_j are seemed as the sample points as a pair, the R (h) represented the half variance of distance between the sample points. For a uniform distribution of sample data, the distance between any sample points is a multiple of h, while sample data which is generally randomly distributed, the distance between pairs of sampling points may be a unique value. In order to facilitate the calculation of R (h), the distance between pairs of samples divided into n segments of n, and those who located between the samples h_n and h_{n-1} are recorded as a set of requirements to find R (h) value. Wherein, h is called the step length, n is called the step size groups.

Sill refers the apex that the semi-variogram can reach, the maximum value of r(h) of the samples. Theoretically, if $h = 0$, R (h) should be 0, but the variation of the measurement error and the presence of microscopic, r (h) is nonzero when $h = 0$, then the value is the nugget value.

2.1 Data Source

The experimental zone of this study is the city of Wuhan, Hubei. A total of 1,715 land development cases are collected from the central area of Wuhan from 2001 to 2010. And such information has been extracted from each case, including spatial position, land use, land area, FAR and registration date, etc. But the information about FAR is missing in 516 cases, so the valid data about FAR only includes 1,199 figures. Different land uses can cause huge difference in FAR. All the land development cases have been classified into 4 categories according to the land uses, i.e. commercial land, residential land, industrial and warehouse land and other lands. The first categories are the focus of this study.

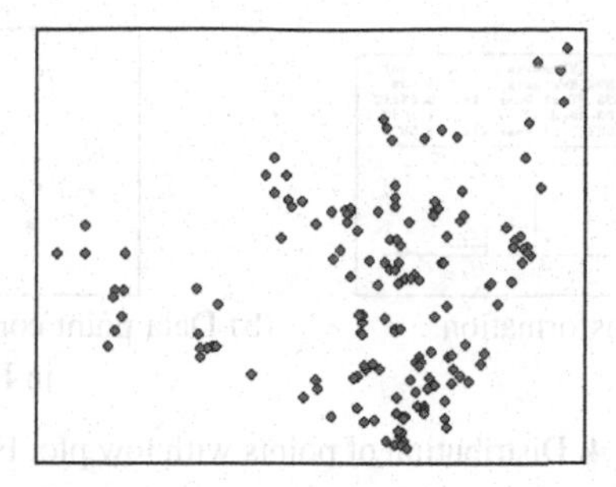

Fig. 1. Data about the commercial land in Hankou District

2.2 Data Distribution Test

If the data shows normal distribution, interpolation methods can generate the best surface. If the data shows skewed distribution, i.e. leaning to one side, normal distribution can be achieved through data transformation. So it is quite important to know about the data distribution before creating the surface. The variant (FAR) distribution of data set is detected according to the frequency histogram of data attribute, as shown in Fig. 2:

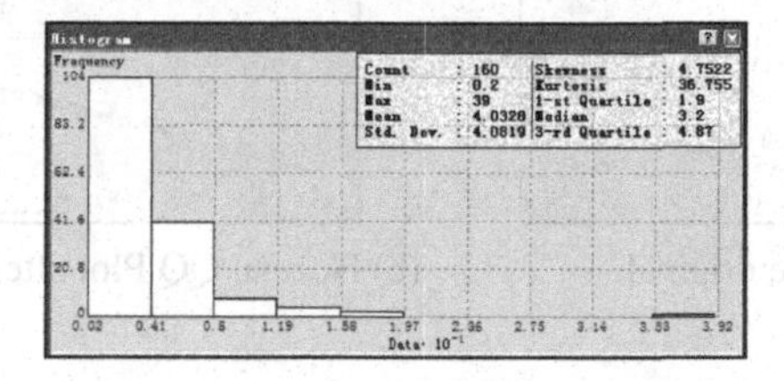

Fig. 2. Original data histogram

As shown in the histogram, the FAR does not shown normal distribution. Data transformation is required before trend surface analysis. Data distribution is re-inspected after logarithmic transformation, as shown in Fig. 3 and Fig. 4:

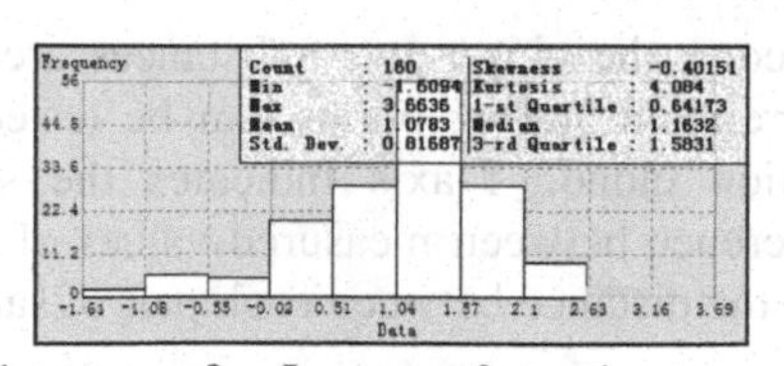

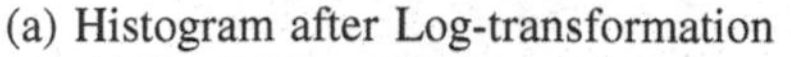

(a) Histogram after Log-transformation

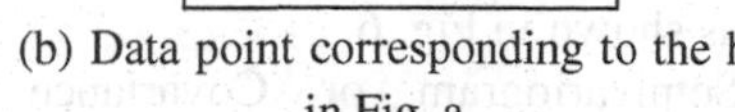

(b) Data point corresponding to the highlight in Fig. a

Fig. 3. Distribution of points with high FAR

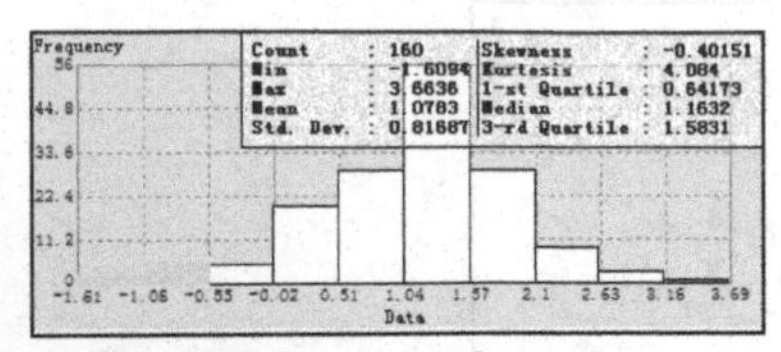

(a) Histogram after Log-transformation

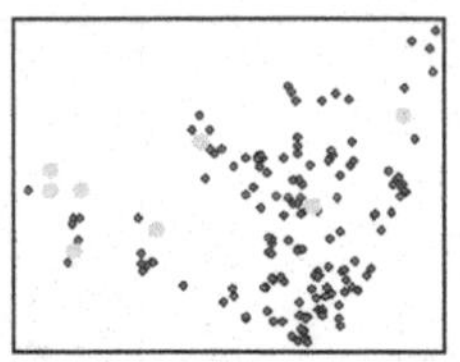

(b) Data point corresponding to the highlight in Fig. a

Fig. 4. Distribution of points with low plot FAR

After logarithmic transformation, FAR shows approximately normal distribution. The highlight part on the right of Fig. 3 indicates the data points with higher FAR, while the highlight part on the left of Fig. 5 shows the data points with lower FAR. The spatial position of data points can be traced according to the original data point map. Data points with higher FAR are generally along the Yangtze River, while data points with lower FAR are generally outside the central area of Hankou.

The normal distribution property of data can also be tested by Normal QQ-Plot. The Normal QQ Plot of the original value of FAR and its Normal QQ Plot after log-transformation are as follows:

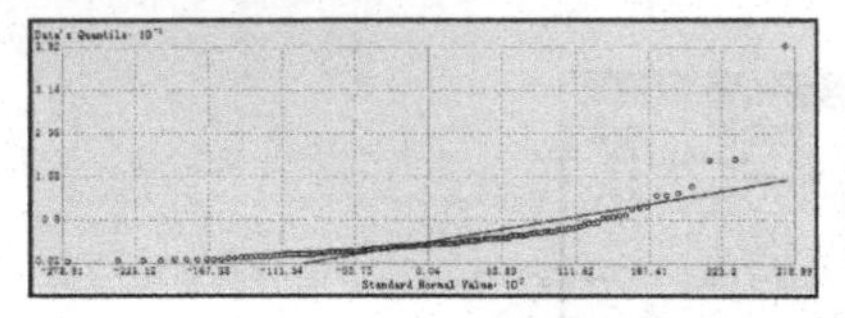

(a) Normal QQ Plot of the original value of FAR

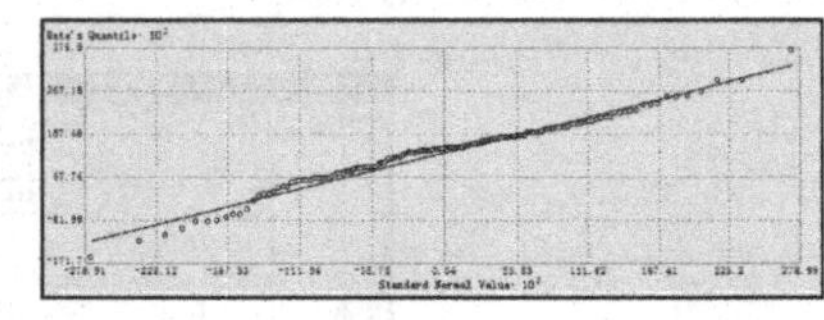

(b) Normal QQ Plot after log-transformation

Fig. 5. Normal QQ Plot

As shown in the QQ Plot, FAR after log-transformation shows approximately normal distribution, because the data points are almost in a straight line.

2.3 Variance Variation Test

According to the spatial autocorrelation theory, the closer two substances are, the more likely they are alike. Spatial autocorrelation among points can be tested by semi-variant function or covariance function cloud. Y-axis indicates the semi-variogram values, i.e. the square of the difference between measured values of each pair of sample points, while X-axis indicates the distance between each pair of sample points, as shown in Fig. 6.

In Semivariogram or Covariance Cloud, adjacent points have smaller semivariogram values. As the distance between the pair of sample points increases, the semivariogram values increase accordingly. This indicates strong spatial autocorrelation among variables. But if we carefully observe the semivariogram, it can be found out that some points in close proximity to each other (in highlight part) have several exceptionally higher semivariogram values. According to the inspection,

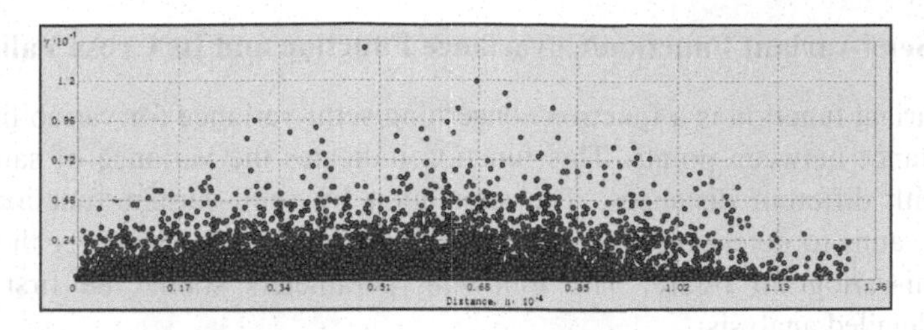

Fig. 6. Semivariogram or covariance cloud

all these exceptional values are related to one point, which is a commercial complex in central area of Hankou. Its FAR (floor area ratio) reaches as high as 39, but this value is not incorrect, so it cannot be deleted.

2.4 Trend Surface Analysis

The trend analyzers can transform the sampling points into 3-D graphics that take attribute values as the altitude axis. The overall trend of sampling data can be analyzed from different angles. Since the trend surface cannot accurately demonstrate the actual surface, residual error model is established after eliminating the trend. When establishing the residual error model, short-term variability in the surface should be analyzed.

Fig. 7. Trend analysis

As shown in the trend analysis figure, the green trend line projected in the southeast and northwest directions shows in the inverted U shape, and smoothly transits in cascades from south to north. We can draw a general conclusion from this figure that, the FAR(floor area ratio) of commercial land in Hankou gradually decreases from south to north, and east-west direction on both sides of the low, middle high. The analysis result is consistent with the actual situation. According to the overall trend of data distribution, a second-order polynomial can be used to model the data. This trend is first eliminated from the data, and statistical analysis is made on residual error or short-term variability of the surface. When creating the final surface, this trend is re-added to produce more meaningful results. Analysis will not be affected by eliminating the overall trend. Once the overall trend is added again, a more accurate surface will be generated.

2.5 Semi-variant Function/Covariance Function and Its Cross Validation

Semi-variant function is a function concerning semi-variance (or variability) of points and distance between points. This function indicates the variance of sampling point pairs with different distances. The simulation by semi-variant function/covariance function aims to determine an optimal fitted model, which can cover those points in the semi-variogram figure. The following parameters should be first determined before detailed analysis:

① Step value: determining an appropriate step length for dividing the values of semi-variant function. Pairs of points are divided into different distance levels to avoid a great amount of combinations. The size of this distance level is step length. The step length can be defined by the distance between each pair of points. It is calculated that the average distance between neighboring points in the data set is 263 m, which will be considered as step length.

② Directional effect value: directional effect is also called anisotropy, which means neighboring things are more similar in some directions than other directions. Directional effect has some influence on the points in semi-variant function and their fitting models. According to the calculation by direction distribution model provided by ArcGIS, the directional effect of this data set is Southeast-northwest 490.

The geostatistical analyst provided by ArcGIS is utilized to draw up the distribution map of FAR of commercial land in central area of Hankou. The FAR(floor area ratio) after log-transformation is fitted by spherical semi-variant function model (fitting very well in all directions). The cross validation result of the model is shown as follows:

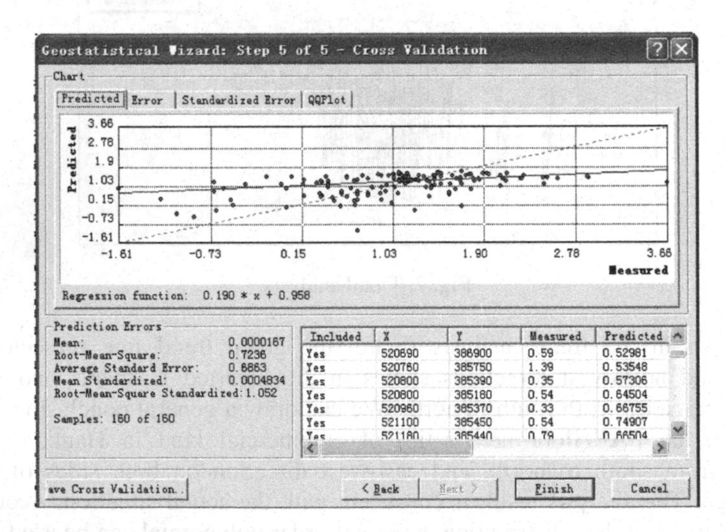

Fig. 8. Cross validation of the model

Table 1 shows the forecast error of the model. The average forecast error and divided difference standard deviation are close to 0, and root mean square standard deviation is approximately 1. Root mean square error and average standard error are quite small and close to each other. Judging from all the indicators, the model has finc fitting precision. This shows that it is quite appropriate to predict the distribution trend of FAR of commercial land in central area of Hankou by Kriging Interpolation method.

Table 1. Forecast error of the model

average forecast error	Divided difference standard deviation	Root mean square	average standard error	Rms standard deviation
0.0000167	0.0004834	0.7236	0.6863	1.052

2.6 Comparison of Forecast Models

There is no need to include page numbers. If your paper title is too long to serve as a running head, it will be shortened. Your suggestion as to how to shorten it would be most welcome.

ArcGIS geostatistical module has provided several spatial statistical algorithms. It is uncertain whether Kriging Interpolation can achieve the best result. In order to find out the optimal algorithm, we predicted the trend of the same data set by different algorithms. See Table 2 for forecast errors of all the algorithms (other algorithms have only two error indicators).

Table 2. Comparison of forecast errors of all the algorithms

Index	Kriging	IDW	Global Polynomial	Local Polynomial	Radial Basis Function
Mean	0.0000167	-0.005325	0.002198	-0.03514	0.03128
RMS	0.7236	0.773	0.7394	0.7214	0.7203

After comparing the forecast results of different algorithms, it is found that the result of Kriging Interpolation is superior to other interpolation algorithms and it can produce the minimum forecast error. According to the above cross validation table, it is considered that Kriging Interpolation is the optimal method for data analysis in the experimental zone.

3 Forecast the FAR of Hankou Commercial Land

Based on the above analysis, Kriging Interpolation method is utilized to forecast the FAR distribution of commercial land in the central area of Hankou District. See the result in Fig. 9:

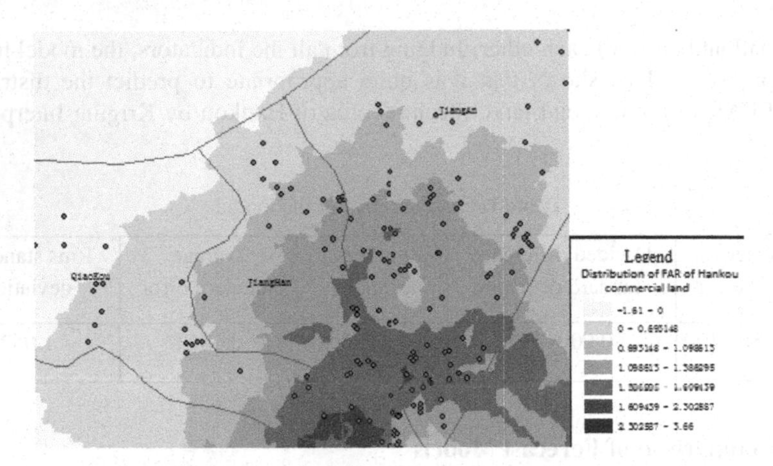

Fig. 9. Distribution map of FAR of Hankou commercial land

In Fig. 9, the red line represents the administrative boundary of Wuhan. As shown in the figure, the FAR of commercial land in Hankou District is the highest in the regions along the Yangtze River and Han River, especially along the Han River, and then gradually decreases towards the surrounding areas and shows a fan-shape distribution, which is consistent with geographic conditions of Hankou District.

4 Conclusion

In this paper, GIS and geostatistical methods are applied to quantitatively research the spatial distribution of urban land development intensity. Through the contrast analysis, Kriging interpolation result is better than other several interpolation methods, therefore Kriging method can be used to estimate and predict the value of FAR of area to be developed and other indicators. Quantitative prediction of FAR provide a necessary technical support, which will greatly improve the scientific and authority of urban planning. But this article's research is still based on the static prediction, that is, according to the known sample points to predict the FAR of unknown points over the same period. The spatial statistics for dynamic prediction method still needs further study. In addition, Kriging cannot eliminate the influence of terrain features, if there are obstacles in the study area, such as rivers and lakes, the estimation and prediction of FAR is more suitable for sub-blocks.

Acknowledgments. This paper is based in part on work supported by the National Natural Science Foundation of China (50808089), National Social Science Fund (09BZZ045), National Natural Science Foundation of China (51278211).

References

1. Hou, J., et al.: Practical geostatistics: spatial information statistically. Geological Publishing House (1998)

2. Wang, Z.: The geostatistics in ecology. Science Press (1999)
3. Du Daosheng, C.F.: Application of the Integration of Spatial Statistical Analysis with GIS to the Analysis of Regional Economy. Geomatics and Information Science of Wuhan University 27(4), 391–395 (2002)
4. Gong, J., Xia, B., Liu, Y.: Study on spatial-temporal heterogeneities of urban ecological security of Guangzhou based on spatial statistics. Acta Ecologica Sinica 30(20), 5626–5634 (2010)
5. Yang, Z., Cai, J.: Progress of Spatial Statistics and Its Application in Economic Geography. Progress in Geography 29(6), 757–768 (2010)
6. Calderón, G.F.-A.: Spatial Regression Analysis vs. Kriging Methods for Spatial Estimation. International Advances in Economic Research 15, 44–58 (2009)
7. Xie, H.: Regional eco-risk analysis of based on landscape structure and spatial statistics. Acta Ecologica Sinica 28(10), 5020–5026 (2008)

Open Source Point Process Modeling of Earthquake

Xinyue Ye[1], Jiefan Yu[2], Ling Wu[3,*], Shengwen Li[4], and Jingjing Li[5]

[1] Department of Geography, Kent State University, USA
[2] Department of Geology, Bowling Green State University, USA
[3,*] College of Criminal Justice, Zhongnan University of Economics and Law, China
[4] Department of Information and Engineering, China University of Geosciences, China
[5] Center for Yellow River Civilization and Sustainable Development, Henan University, China

xye5@kent.edu, yjiefan@bgsu.edu,
{wuxianhaoshen,lsw4000}@gmail.com, 450079492@qq.com

Abstract. As the most destructive natural disaster, earthquake may cause catastrophic demolition which leads to devastating outcomes. Earthquake's hypocenter is a point associated with spatial footprint, occurrence time, and magnitude. Open source point process modeling is adopted to detect the possible interactions among earthquakes with spatial features, temporal characteristics and effect of the other covariates. Integration of geologic variables in the point process models is suggested for the investigation of the magnitude 6.1 earthquake and the aftershocks occurred on Jun 30, 1975 in Yellowstone National Park. A Multitype Strauss Process Model is utilized, and it successfully captures the spatial pattern of epicenters under the effect of interaction radii among different types of earthquakes. In addition, the distance to faults and earthquake focal depths are implemented as the geologic variables in explaining the seismic pattern.

Keywords: point process modeling, Multitype Strauss Process Model, earthquakes, open source.

1 Introduction

An earthquake is the trembling of the earth due to a sudden release of strain built up over a long period of time [1]. In point processing modeling, earthquakes can be treated as events with spatial footprints, temporal stamps, and geologic/environmental variables [2]. An earthquake can cause catastrophic destruction and devastating consequences, making it one of the most damaging natural disasters. According to the report of United States Geological Survey (USGS), there are approximately 500,000 earthquakes detected in the world each year, and most of them are significant. For example, the magnitude-8.0 earthquake on May 12, 2008 in Sichuan, China caused 69,197 deaths, 374,176 injuries, and 18,222 missing population [3]. The magnitude-9.0 Tōhoku earthquake in 2011 in Japan has resulted in 10,804 deaths, 2,777 injuries

---

* Corresponding author.

F. Bian and Y. Xie (Eds.): GRMSE 2014, CCIS 482, pp. 548–557, 2015.

and 16,244 missing [4]. Throughout the history, earthquakes have caused hundreds of thousands of death, tremendous emotional suffering, massive environmental damage as well as financial losses. The long-term losses due to earthquakes reach approximately $4.4 billion annually in the United States. Hence, an in-depth understanding of earthquakes' nature and mechanism will help minimize the loss via the help of better predictions and preparation.

Mathematical methods have gradually been recognized in various seismology studies including hazard estimation, locations or magnitudes quantification, and prediction model assessment over the last two decades [5]. The Gutenberg-Richter law and Omori law are two representative examples. Gutenberg-Richter law describes the systematic relationship between number of earthquakes and the magnitude, arguing that larger earthquakes tend to be less frequent than smaller ones [6]. Omori's law claims that aftershock activity decays over time after the main shock [7].

Point process modeling is used to detect possible interaction among points with spatiotemporal characteristics and the effect of other covariates [8]. This approach has been widely applied to a variety of domains such as epidemiology, ecology, and environmental science. In the recent decades, the application of point process modeling has received increasing attention in seismic hazard estimation. A series of questions regarding earthquakes can be addressed by point process modeling: whether the distribution of earthquakes is clustered, whether it is possible to generate a hazard assessment map, etc. Ogata examined space-time cluster of earthquake pattern with goodness-of-fit of point process modeling [5]. Point process modeling is also used to show correlations between seismic patterns and geologic data such as surface heat flow data and the distance to the plate boundary in southern California area as well as that in northern Pakistan [8, 9].

Earthquake's hypocenter is a point with spatial information, time of occurrence and magnitude. Hence, a seismic pattern can be studied with point process modeling. Point process modeling fits the point data distribution pattern into different distributions such as Poisson Point Process for random distributed patterns and Strauss Point Process for correlated pair points [2]. The point pattern can also be classified by the type of point, and in this case it is the magnitude of earthquake [10]. Traditionally, point pattern simulation and prediction is purely based on the data distribution. Recently, covariates have been used in point process modeling. For instance, geologic information can be utilized as a covariate of hidden mechanism process and distribution pattern, including distance to plate boundary or fault, ground deformation, the migration of pore fluids, in order to provide a better estimation of hazard's location, time, frequency, and magnitude to increase the accuracy of model-fitting of earthquakes.

Point process modeling reveals the geologic relationships within statistical analysis, and it may also enhances the understanding of physical mechanism of earthquake. Open source statistical tools such as R packages have greatly facilitated the progress of point process modeling [10].

2 Data and Methodology

The main objective of this study is to implement a point process model to investigate the interaction between earthquakes of different magnitude and background volcanic or tectonic factors in Yellowstone National Park area. The magnitude 6.1 earthquake on Jun 30, 1975 was the largest earthquake happened within Yellowstone National Park in history. It creates two new hydrothermal systems near the Norris Geyser Basin, leading to speculation that this event was related with the hydrothermal fluid fluctuation associated with volcanic activity. To evaluate this hypothesis, earthquakes are categorized into two groups by magnitude and analyzed with a Multitype Strauss Model to analyze the correlation between the main shock and its aftershocks in the Yellowstone area (Fig. 1). The best model will be selected basing on maximum likelihood approach in the seismic pattern recognition. Furthermore, the simulation and prediction results could contribute to the understanding of hidden physical earthquake mechanisms, estimation of further potential seismic hazard, and exploration of the possible application of point process modeling in seismic studies.

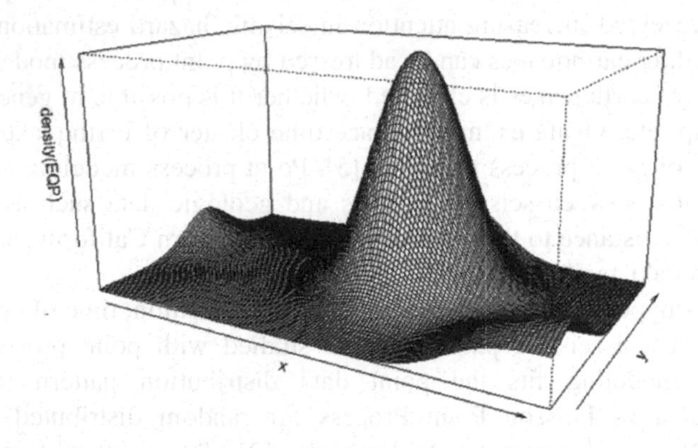

Fig. 1. The density of earthquakes on Jun 30, 1975 and its aftershocks

2.1 Study Area

As one of the most seismically active regions, the Yellowstone National Park area has long been the focus of earthquake studies [11]. From the earliest earthquake report in 1872, there have been over 32,000 earthquakes recorded in the Yellowstone National Park area [12]. Most of them were less than 3 magnitude, occurred as earthquake swarms in shallow crust and sometimes lasted for several days or more. Those swarms were located both within the Yellowstone calderas, between the calderas and from the rupture zone of the 1956 Hebgen Lake earthquake in the northwest. The largest earthquake swarm occurred in 1985, it consists of over 3,000 earthquakes and lasted over 3 months. Another large earthquake swarm in Yellowstone National Park occurred from December 27, 2008 to January 7, 2009, it includes more than 1,000 earthquakes. Although most earthquakes were not significant by magnitude, some large-magnitude quakes also occurred in this region, causing much damage. The largest one was the magnitude 7.5 Hebgen Lake earthquake on August 18, 1959, and it led to 28 deaths and over $11 billion in damage. The location was about 24 km northwest of the Yellowstone caldera. The most recent and the only large earthquake within the caldera happened on Jun 30, 1975 [12]. This magnitude 6.1 earthquake and its aftershocks generated two new hydrothermal areas near Norris Geyser Basin. In addition, there were a total of 17 earthquakes happened with magnitude greater than 3 in almost the same day due to the landslide hazard [14]. Based on the previous studies, all these large and small earthquakes appear to be highly correlated with Yellowstone volcanic activity and fault systems as described below.

As the most volcanically and seismically active region in the United States, Yellowstone National Park covers an area of 8,983 km^2 [11]. It is mostly located in Wyoming, with a small portion in Montana and Idaho. There are several unique geologic features in and around the national park area, including Yellowstone caldera, the Old Faithful Geyser and unique ecological systems. Yellowstone National Park is also a classic case for studying the relationships among seismic activity, volcanic activity, and fault system.

The Yellowstone National Park is one of the most geologically valued regions in United States. The park lies on the 2 km-high Yellowstone Plateau and intersects with a 1,300 km-long Intermountain Seismic Belt, which extends from northern Montana south to northern Arizona [13].

2.2 Point Process Model

Point process refers to how existing point pattern is distributed in space and/or time. As a spatial statistical method, point process modeling aims to explore, simulate, and predict point patterns by parametric model-fitting [10]. The point can be unmarked or marked with specific characteristics attached, such as the type of trees, the size of the area, the magnitude of the earthquakes, etc. This point pattern could be totally random or highly clustered when evaluated by certain statistical tests. The modeling process could be assisted with or without continuous explanatory covariates, which are assumed to explain the behavior of the point pattern [2]. Point process modeling has long been an essential strategy to explore the point patterns in areas including neuroscience, forestry and plant ecology, astronomy, and seismology [10].

Point process modeling consists of determining first and second order effects in order to understand their statistical significance. The first order effect shows the average point number in each unit, which can be measured by intensity and density [10]. The point pattern could be homogeneous or clustered in specific units to tell the distribution trend of point process modeling. The second order effect represents variation of point number in each unit, illustrating the interaction among different (or same) types of points by distance [10]. Point pair tests in second order effects include Ripley's K, F, G, and L-function, which are also called as nearest neighborhood distribution functions. Point process modeling was implemented to simulate the earthquake distribution and test the hypothesis of captured geologic variables for further seismic hazard assessment and prediction. In the Multitype Strauss Process, pair-wise interaction depends on not only the interaction among points, but also the type of the points [2]:

$$c(u,v)=\begin{cases}1 \; if \; \|u-v\|>r \\ \gamma \; if \; \|u-v\|\le r\end{cases} \quad (1)$$

3 Results

Four different Multitype Strauss Models were employed to investigate the distribution interaction among each type of earthquake mark and the possible geologic variables, including the occurrence of geothermal features, distance to the faults, and focal depth of earthquakes.

An object named “ppm” is built to fit the Multitype Strauss Model in R. X represents the dataset of the point pattern; trend represent the general distribution orientation of the point dataset, often shown as polynoms of different degrees depending on the complexity of the pattern [2]. The trend could be ~1 for stationary Strauss process, or ~x + y for a non-stationary Poisson process of a loglinear intensity, or ~ polynom(x, y, 2) under 2 order polynoms in the Cartesian coordinates in which the intensity is in a log-quadratic spatial trend [2]. The interaction reveals the correlation between the each data type, illustrating by the interaction radius among small and large earthquakes in this study.

The first model is purely based on the interpoint relationship among large and small magnitude earthquakes. The radii showing interpoint correlation was determined from the second order effect as discussed above. The matrix of radius cannot totally rely on the result of G-function, because the Multitype Strauss Model limits the radius matrix only to be a symmetric equation. However, the best radius applied in the study can be determined by observation of the real seismic pattern, result indicated by G-function, and numerous attempts for achieving a result most similar to reality [15]. The final radius applied in this study is:

```
r <- matrix (c (1500, 10000, 10000, 2000), nrow=2, ncol=2) .
```
(2)

The appropriate Multitype Strauss Model based on marked earthquakes and interaction radius among different marks is written as:

```
ppm (EQP, ~ polynom (x, y, 2), MultiStrauss (c ("small", "large"), r)) .
```
(3)

The above model achieves the best fit with AIC = 2536.31. The result of the Multitype Strauss Model returns estimated values of interaction parameters and fitted coefficients. The value of interaction parameters demonstrates the clustering relationship among earthquake marks. For instance, the interaction radii between large and large earthquakes and between small and small earthquakes are both larger than 1, meaning the seismic pattern is highly correlated among same type of earthquake. Since the interaction radius between different types of earthquakes is approximately 1, there is a possibility for those interactions to be clustered more than inhibition. The fitted trends of each type of earthquake are shown in Fig. 2 illustrating the possibility of earthquake events by graduated colors. The goodness-of-fit can be carried out by evaluating the fitness between observation data and simulated envelop of K -function (Fig. 3), in which the upper and lower boundaries are developed by randomly generating realistic data under the fitted model and calculating the distance from the nearest neighborhood [2]. If the observation pattern lies between the upper and lower boundaries of generated simulation envelop, the model is considered to reflect an adequate fit. As the Fig. 3 demonstrates, most of the observation data fall between the upper and lower boundaries of simulated envelopes indicating the Multitype Strauss Model and the interaction radius explain most of the variability of earthquake distribution. However, there is still some observation data that fall out of the simulated envelop, which means some variables related to the seismic pattern are still unexplained.

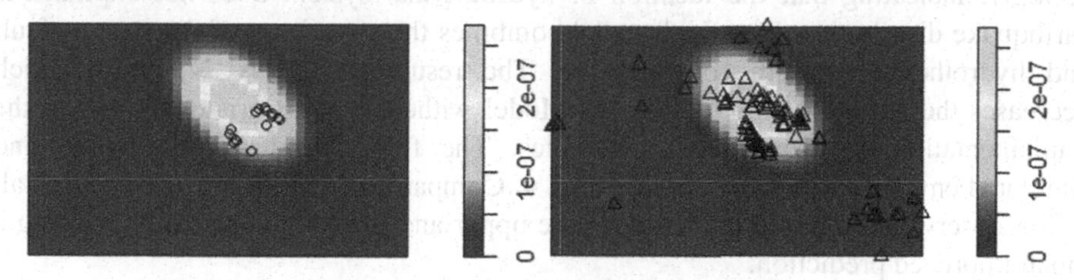

Fig. 2. Fitted trend from Multitype Strauss Modeling of the probability of occurrence among small (right) and large (left) magnitude earthquakes

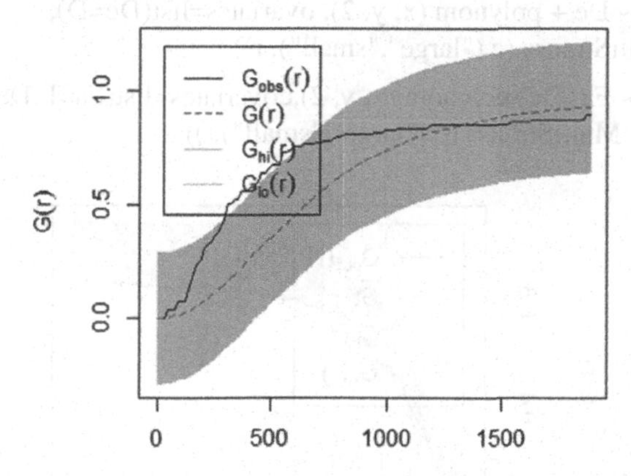

Fig. 3. The result of K test envelope function in meters. The envelope is created by 99 randomly generated points, and the black line is the observation data.

Multitype Strauss Model can also be utilized for the scenarios in which the seismic pattern depends on spatial covariates [10]. Based on the previous studies, the presence of faults, location of hydrothermal areas, and focal depth of earthquakes are suggested to control the earthquake distribution. In this research, those geologic factors are separately tested as covariates in Multitype Strauss Model to investigate and verify those geologic influences on interpoint interaction between earthquakes.

To prepare for the model fitting, the covariates are integrated into a format of pixel image or data list. The "distmap" function in R is used to calculate the distance to line segments by meter as the covariate of distance-to-fault in which the fault system contains the Intermountain Seismic Belt and caldera ring faults. The occurrence of hydrothermal system is converted from a shapefile in ARCGIS and clipped by the extent of earthquake patterns. The earthquake focal depth covariate is interpolated and fitted into a trend surface by Generalized Least-squares in "surf.gls" function. All the covariates are projected to 'NAD_1983_UTM_Zone_12N' which uses meter as the unit. Significantly smaller AIC values demonstrate the suitability of the model and effect of these geological covariate (The AIC of the distance-to-fault model is 2534.617 and earthquake focal depth is 2499.595, while the original model is 2536.31), as indicated by equation (4) and (5). However, the Multitype Strauss Model with covariate of hydrothermal system location reported an increasing value of AIC 2538.31, indicating that the location of hydrothermal system does not explain the earthquake distribution. The final model combines the covariates of distance-to-fault and hydrothermal system (Equation 6). The resulting AIC is 2491.954, which decreases the AIC of Multitype Strauss Model without any covariates by 45 via the implementation of two geologic variables. The fitted trend of this model and simulated envelope is demonstrated in Fig. 4. Compared to Fig. 3, Fig. 4 witnesses all of the observation data falling between the upper and lower boundaries, exhibiting a much improved prediction.

```
fit <- ppm (EQP, ~ Fa + polynom (x, y, 2), covariates=list(Fa=F),
            MultiStrauss (c ("large","small"),r)) .                        (4)
```

```
fit <- ppm (EQP, ~ De + polynom (x, y, 2), ovariates=list(De=D),
            MultiStrauss (c ("large","small"), r)) .                       (5)
```

```
fit <- ppm (EQP, ~ Fa+De+polynom (x, y, 2),covariates=list(Fa=F,De=D),
                MultiStrauss (c ("large","small"),r)) .                    (6)
```

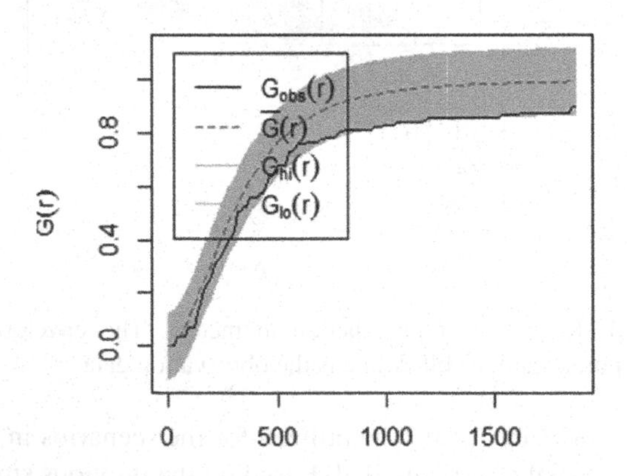

Fig. 4. The result of K test envelope function in meters. The envelope is created by 99 randomly generated points, and the black line is the observation data.

4 Discussions

Multitype Strauss Model is investigated for its capability of capturing the earthquake associated geologic factors such as the location of faults, the distribution of hydrothermal system, etc. The fitted trend and simulated envelop not only demonstrate the adequately prediction of interaction between pair-wise point, but also the fitness of associated geologic covariates. However, the interaction radius is limited as a symmetric matrix to fit in Multitype Strauss Model, which is not always suitable for the reality of an earthquake pattern. In this study, the applied symmetric interaction radius is kept close to the real value. However, the result is still insufficient in simulating the exact real distribution pattern such as the outliers discussed above. Similarly, the research extent and category of earthquake marks based on magnitude changes by different choices. However, the amplitude of influence on model fitting remains unchanged under the same interpoint interaction and geologic covariate.

The data format for the covariate should be continuous throughout the entire research area. Unfortunately, it is impractical to capture geologic value on every location for all the covariates. For instance, the focal depth of earthquakes was found to be highly associated with earthquake location and mechanism, but it is unrealistic to obtain the focal depth of every single location. As a result, the earthquake focal depth covariate is generated by interpolating and fitting into a trend surface to test the relationship between the focal depth value and the possibility of earthquake occurrence. The generated trend surface of focal depth is accepted in Multitype Strauss Model, because it is possible to obtain similar focal depth with surrounding earthquake under the same mechanism in the same major earthquake event.

The point process modeling of earthquake study is also limited by technical devices and computational software. Due to the size of the earthquake dataset and complexity of the Multitype Strauss Model, the computational device needs relatively more physical memory for the high intensity computational spatial seismology. A dataset with more than a hundred earthquake events requires at least a 64-bit computer with 4GB memory.

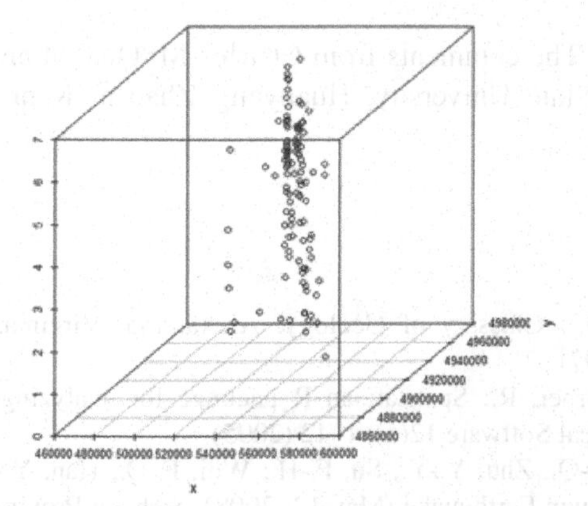

Fig. 5. 3D visualization of earthquake hypocenters. The x and y axes show the locations of earthquakes in meters, and the z axis shows the focal depth.

The R computation package for point process model in seismic study allows 3D space-time computation with visual and simulation model for 3D earthquake data. With the 3D space-time analysis of the earthquake distribution, it is possible to capture the potential factors in point process modeling in further study. 3D "SpatStat" computational package allows the Multitype Strauss Model to compute the correlations between 3D earthquake distribution and 3D fault location, with an expectation of a more accurate and close simulation of seismic pattern in a space-time context. Fig. 5 is an example of 3D visualization of the applied earthquake dataset.

In general, the fitted northwest-trending seismic pattern matches seismic background from the previous study of Yellowstone National Park. The goodness-of-fit and simulated envelope also shows that the pair-wise interaction radius is capable to capture most of the variables among different earthquake types. Moreover, three geologic covariates further explain the correlation between earthquake location and geophysical mechanism. The occurrence of faults and focal depth of earthquakes are considered to be adequately correlated with the seismic pattern. This result is also supported by previous geophysical earthquake analysis. The focal depths of the magnitude 6.1 main shock and its aftershocks are consistent with most of other earthquakes that occurred within the extent of caldera. The distance-to-fault was also found to be associated with earthquake pattern. On the other hand, the covariate of hydrothermal system does not further explain the seismic pattern in Multitype Strauss Model. Although the hydrothermal fluids fluctuation might be responsible for the occurrences of small magnitude earthquakes (Waite and Smith, 2002), it is not necessary for the earthquakes to cluster around hydrothermal systems. Since the transport of hot fluids between the ductile and brittle rocks does not have to be near the hydrothermal locations, the earthquakes triggered by fluctuations of hydrothermal fluids also do not have to cluster around hydrothermal locations. After all, unobserved geologic mechanism in the deep Earth are likely to be fully captured and explained in statistical model, but the result of this study still can contribute to further geophysical seismic study by verifying and indicating potential geologic features not accounted for by the mathematics-based model.

Acknowledgement. The comments from Charles M. Onasch and Peter V. Gorsevski at Bowling Green State University, Huanyang Zhao at Kent State University are appreciated.

References

1. Jackson, J.A. (ed.): Glossary of Geology: Alexandria, Virginia. American Geological Institute, 769 (1997)
2. Baddeley, A., Turner, R.: Spatstat: an R package for analyzing spatial point patterns. Journal of Statistical Software 12(6), 1–42 (2005)
3. Cui, P., Chen, X.-Q., Zhu, Y.-Y., Su, F.-H., Wei, F.-Q., Han, Y.-S., Liu, H.-J., Zhuang, J.-Q.: The Wenchuan Earthquake (May 12, 2008), Sichuan Province, China, and resulting geohazards. Natural Hazards 56(1), 19–36 (2011)

4. Hoshiba, M., Iwakiri, K.: Initial 30 seconds of the 2011 off the Pacific coast of Tohoku Earthquake (Mw 9.0)—amplitude and Tc for magnitude estimation for Earthquake Early Warning—. Earth Planets Space 63, 553–557 (2011)
5. Ogata, Y.: Space-time point-process models for earthquake occurrences. Annals of the Institute of Statistical Mathematics 50(2), 379–402 (1998)
6. Gutenberg, B., Richter, C.F.: Seismicity of the Earth and Associated Phenomena, 2nd edn., pp. 17–19. Princeton University Press, Princeton (1954)
7. Omori, F.: On the aftershocks of earthquakes. The Journal of the College of Science, Imperial University of Tokyo 7, 111–120 (1894)
8. Anwar, S.: Implementation of Strauss point process model to earthquake data. M.Sc. Thesis, University of Twenty, Enschede, p. 47 (2009)
9. Enescu, B., Hainzl, S., Ben-Zion, Y.: Correlations of seismicity patterns in southern California with surface heat flow data. The Bulletin of the Seismological Society of America 99(6), 3114–3123 (2009)
10. Baddeley, A.: Analysing spatial point patterns in R. Workshop notes. Version 4.1. CSIRO online technical publication (2010)
11. Rubeis, D.: Space-time combined correlation between earthquakes and a new, self-consistent definition of aftershocks. In: Modeling Critical and Catastrophic Phenomena in Geoscience Lecture Notes in Physics, vol. 705, pp. 259–279 (2006)
12. Chang, W.-L., Smith, R.B., Farrell, J., Puskas, C.M.: An extraordinary episode of Yellowstone caldera uplift, 2004–2010, from GPS and InSAR observations. Geophysical Research Letters 37(23), 6–11 (2010)
13. Smith, R.B., Arabasz, W.J.: Seismicity of the Intermountain Seismic Belt, Neotectonics of North America. Geological Society of America, Decade Map 1, 185–228 (1991)
14. Pitt, M.A., Weaver, C.S., Spence, W.: The Yellowstone Park earthquake of June 30, 1975. Bulletin of the Seismological Society of America 69(1), 187–205 (1979)
15. Isham, V.S.: Multitype Markov point processes: some approximations. Proceedings of the Royal Society of London, Series A 391, 39–53 (1984)

A Novel Approach for Computing Inverse Relations of Basic Cardinal Direction Relations

Yiqun Dong, Wenxing Xu, Jiandong Liu, Shuhong Wang, and Huina Jiang

College of Information Engineering, Beijing Institute of Petrochemical Technology, Beijing, China
hover3917@163.com

Abstract. Cardinal direction calculus (CDC) which contains 218 basic cardinal relations (BCDRs) is an expressive method for qualitative direction information of regions. The inverse relations of BCDRs, however, are still an important issue of spatial direction reasoning. In this paper, a quadruple model is proposed to represent BCDRs based on minimum bounding rectangle (MBR). Then a novel approach to computing the inverse relations of BCDRs is presented in terms of constraints in the quadruple model. Compared with mainly reasoning approaches which do not consider the inverse relations, the algorithm proposed are effective to reasoning performance for BCDRs.

Keywords: basic cardinal direction relation, minimum bounding rectangle, inverse relations.

1 Introduction

As an important issue of modeling spatial information, qualitative reasoning about direction relations between regions has attracted extensive attention in recent years, and its theories and methods have been widely applied to many fields such as geographical information, robot navigation, pattern recognition, image processing, etc. Most existing work approximates a region by a point (e.g., its centroid)[1~5], which is certainly imprecise in real-world applications such as describing the direction information between two countries, say, Portugal and Spain[6]. Goyal and Egenhofer [7, 8] proposed the direction relation matrix (DRM) for representing direction relations between connected plane regions. The original description of the DRM lacks formality and does not consider limit cases. Skiadopoulos fixed this problem by CDC in [6] where 218 basic cardinal direction relations were defined formally for regions which are closed, connected and have connected boundaries. Present researches are mainly based on CDC [9 ~13].

Inverse is a basic operator for spatial direction relation. To describe the relative direction of two regions *a* and *b* more precisely, we need to specify both the relation of *a* with respect to *b* and the relation of *b* with respect to *a*. Moreover, for the reasoning problem of a qualitative calculus, one frequently used approach is to devise local consistency algorithms to solve the decision problem over a subclass which

F. Bian and Y. Xie (Eds.): GRMSE 2014, CCIS 482, pp. 558–570, 2015.

contains basic relations. It is known that path-consistency implies the consistency of complete basic networks for the calculus based on IA[14] and RCC8[15]. However, unlike these calculus, CDC is not closed under the inverse operator, which means that the inverse relation of a BCDR may not be a BCDR[6,9,16,17]. In order to make use of constraint-based methods, in particular path-consistency, to reason about CDC, computing the inverse relations of BCDRs is an important and inevitable problem. Early works generally assume the inverse relations of BCDRs are single-tile relations or devise reasoning algorithms only for rectangular relations which is a subset of BCDRs[18~20]. All those restrictions make the reasoning insufficient. Therefore, efficient algorithms for computing the inverse relations of each BCDR and related formal proof are needed to be studied.

2 Basic Cardinal Direction Relations: Definitions and Basic Notations

In this paper, we focus on the direction relations for regions which are closed, connected and have connected boundaries. All these regions are denoted by $\mathcal{REG}$ [21]. For regions a and b in $\mathcal{REG}$, when representing the direction of a (named primary region) with respect to b (named reference region), b is approximated with its MBR, denoted by $MBR(b)$, as shown in Fig.1(1). $MBR(b)$ divides the plane into nine areas which we call tiles to express the nine directions of b respectively. The reference system of b consists of all these tiles, denoted by $\pi(b)$, as shown in Fig.1(2). The exact geometry of a is used to a certain extent for the representation of the direction in $\pi(b)$. Then basic cardinal direction relations (BCDRs) are defined as follows:

Definition 1. R_1: ...: R_k is a BCDR where (i) $1 \leq k \leq 9$, (ii) $R_1 \ldots R_k \in \{B, S, SW, W, NW, N, NE, E, SE\}$, (iii) $R_i \neq R_j$ for every i and j such that $1 \leq i, j \leq k$ and $i \neq j$, (iv) there exists regions $a_1,\ldots,a_k \in REG$, $a_1 \in R_1(b),\ldots,a_k \in R_k(b)$ and $a_1 \cup \ldots \cup a_k \in REG$ for any reference region $b \in REG$ [6].

For example, in Fig.1(3), the direction relation of region a with respect to b is $NE{:}E$, denoted by $a\ NE{:}E\ b$.

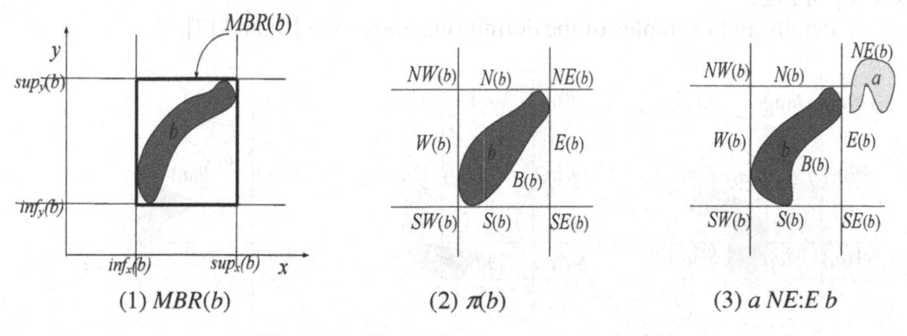

Fig. 1. Cardinal direction relation calculus

According to definition1, there are 218 BCDRs, denoted by set D. Relations in D are jointly exhaustive and pairwise disjoint.

Definition 2. Let $R\in D$, R is called rectangular relation, iff there exist two rectangles regions (with sides parallel to the x- and y-axes) a and b such that $a\ R\ b$ is satisfied, otherwise it is called non-rectangular [6].

There are 36 rectangular relations, denoted by set $Drec$ which is a subset of D.

Definition 3. Let $R_1=R_{11}:\ldots:R_{1n}$ and $R_2=R_{21}:\ldots:R_{2m}$ be two cardinal direction relations. R_1 includes R_2 iff $\{R_{21}, \ldots, R_{2m}\}\subseteq\{R_{11}, \ldots, R_{1n}\}$ holds [6].

Definition 4. Let $R\in D$, the bounding relation of R, denoted by $Br(R)$, is the smallest rectangular relation (with respect to the number of tiles) that includes R [6].

By rectangular relation, it is easy to see the following implications hold:

$(\forall\ a, b\in \mathcal{REG})\ a\ R\ b \rightarrow MBR(a)\ Br(R)\ MBR(b)$.

$(\forall\ a, b\in REG)\ MBR(a)\ Br(R)\ MBR(b) \rightarrow (\exists\ c, d\in REG)\ c\ R\ d \wedge (c\in MBR(a)) \wedge (d\in MBR(b))$.

Definition 5. Let $R\in D$, the rectangular relation formed by the westernmost tiles of $Br(R)$ is denoted by $Most(W,Br(R))$. Similarly, there are rectangular relations $Most(S, Br(R))$, $Most(N, Br(R))$ and $Most(E, Br(R))$. Moreover, the rectangular relation formed by the southwesternmost tiles of R is denoted by $Most(SW, Br(R))$. Similarly, there are single-tile relations $Most(SE, Br(R))$, $Most(NW, Br(R))$ and $Most(NE, Br(R))$. Finally, as a special case, let $Most(B, Br(R))=R$ [6].

We use function symbolδas a shortcut. For arbitrary single-tile cardinal direction relations $R1,\ldots,Rk$, the notation $\delta(R_1,\ldots,R_k)$ is a shortcut for the disjunctive relations in D that can be constructed by combining single-tile relations $R1,\ldots,Rk$. For instance,

$\delta(B, NW, W)=\{B, W, NW, B{:}W, W{:}NW, B{:}W{:}NW\}$.

Moreover,

$\delta(\delta(R_{11},\ldots,R_{1i}),\delta(R_{21},\ldots,R_{2j}),\ldots,\delta(R_{m1},\ldots,R_{mk}))=\delta(R_{11},\ldots, R_{1i}, R_{21},\ldots,R_{2j},\ldots, R_{m1},\ldots,R_{mk})$.

Definition 6. Let $R\in D$. The inverse of relation R, denoted by R^{-1}, is another cardinal direction relation such that: for arbitrary regions a, $b\in REG$, $a\ R^{-1}\ b$ holds, iff $b\ R\ a$ holds [6].

However, CDC is not closed under inverse, which means that the R^{-1} is not necessarily a BCDR but it may be a disjunction of basic relations. Therefore, we have $R^{-1}=\{R_i|(\exists a_i, b_i\in REG)(a_i\ R\ b_i \wedge b_i\ R_i\ a_i \wedge R_i\in D)\}$.

For instance, $W{:}NW{:}N^{-1}=\{B{:}S{:}E, S{:}E{:}SE, B{:}E{:}SE, B{:}S{:}SE, B{:}S{:}E{:}SE, B{:}E, B{:}S, B\}$, as shown in Fig.2.

More details and examples of the definitions above see [6] and [7].

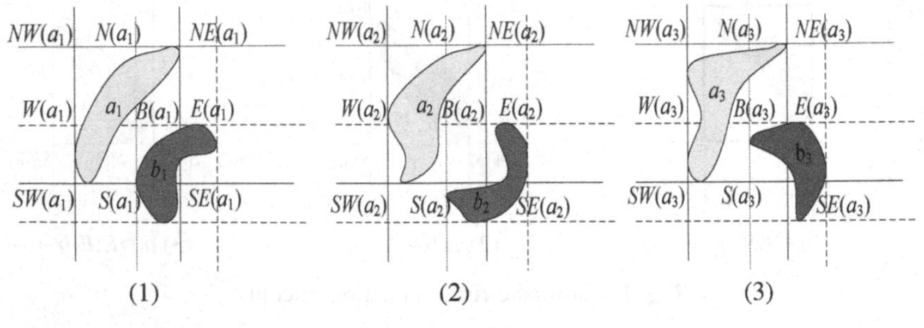

Fig. 2. W:NW:N-1

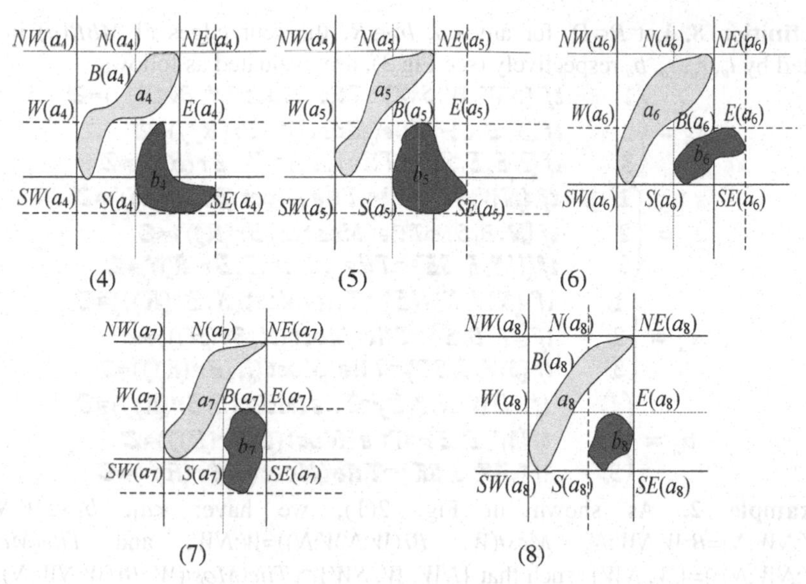

Fig. 2. (*Continued*)

3 Quadruple Model of CDC

Definition 7. Let $R \in D$ and $R = R_1:\ldots: R_k$, all the direction tiles of R is denoted by $Tile(R)$, i.e., $Tile(R)=\{R_1, \ldots, R_k\}$.

Example 1. $Tile(SE)=\{SE\}$, and $Tile(SE{:}E{:}NE)=\{SE, E, NE\}$.

BCDRs are jointly exhaustive and pairwise disjoint, then it is easy to know that for any $R\in D$, $Tile(R)$ can be determined uniquely, i.e., there exist a bijection from 218 BCDRs onto 218 different sets of tiles.

We have introduced BCDRs informally in Section2. Using set-theoretic notation, each R in D can also be defined as a *binary relation* which consists of pairs of regions satisfying R, i.e., $\{<a_i, b_i> \mid a_i\ R\ b_i\}$ where $a_i\in REG$ is the primary region and $b_i\in REG$ is the reference region. Then we use quadruple to represent and investigate the distribution of the edges of $MBR(a_i)$ and $MBR(b_i)$

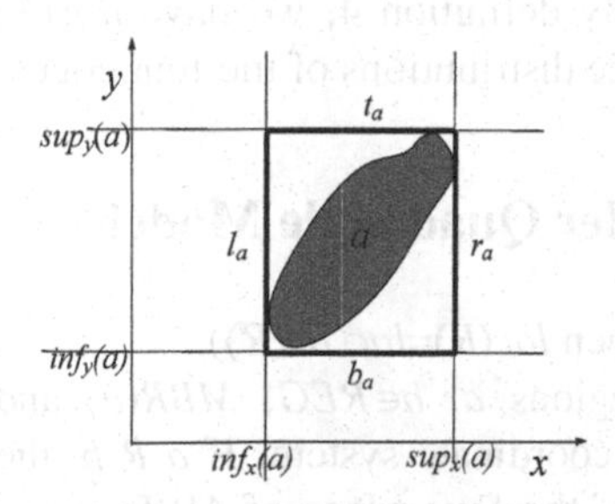

Fig. 3. The edges of $MBR(a)$

Definition 8. Let $R\in D$, for any $<a, b>\in R$, the four edges of $MBR(a)$ in $\pi(b)$, denoted by l_a, r_a, t_a, b_a respectively (see Fig.3), are evaluated as follows:

$$l_a = \begin{cases} 1, & if\ \{NW, W, SW\}\cap Tile(Most(W, Br(R)))\neq\emptyset \\ 2, & if\ \{N, B, S\}\cap Tile(Most(W, Br(R)))\neq\emptyset \\ 3, & if\ \{NE, E, SE\}\cap Tile(Most(W, Br(R)))\neq\emptyset \end{cases}$$

$$r_a = \begin{cases} 1, & if\ \{NW, W, SW\}\cap Tile(Most(E, Br(R)))\neq\emptyset \\ 2, & if\ \{N, B, S\}\cap Tile(Most(E, Br(R)))\neq\emptyset \\ 3, & if\ \{NE, E, SE\}\cap Tile(Most(E, Br(R)))\neq\emptyset \end{cases}$$

$$t_a = \begin{cases} 1, & if\ \{NW, N, NE\}\cap Tile(Most(N, Br(R)))\neq\emptyset \\ 2, & if\ \{W, B, E\}\cap Tile(Most(N, Br(R)))\neq\emptyset \\ 3, & if\ \{SW, S, SE\}\cap Tile(Most(N, Br(R)))\neq\emptyset \end{cases}$$

$$b_a = \begin{cases} 1, & if\ \{NW, N, NE\}\cap Tile(Most(S, Br(R)))\neq\emptyset \\ 2, & if\ \{W, B, E\}\cap Tile(Most(S, Br(R)))\neq\emptyset \\ 3, & if\ \{SW, S, SE\}\cap Tile(Most(S, Br(R)))\neq\emptyset \end{cases}$$

Example 2. As shown in Fig. 2(1), we have: $<a_1, b_1>\in W{:}NW{:}N$, $Br(W{:}NW{:}N)=B{:}W{:}NW{:}N$, $Most(W, Br(W{:}NW{:}N))=W{:}NW$, and $Tile(Most(W, Br(W{:}NW{:}N)))=\{W, NW\}$, such that $\{NW, W, SW\}\cap Tile(Most(W, Br(W{:}NW{:}N)))\neq\emptyset$. Then $l_{a_1}=1$. Similarly, $r_{a_1}=2$, $t_{a_1}=1$, $b_{a_1}=2$.

Definition 9. Let $<a, b>\in R$, the quadruple $<l_a, r_a, t_a, b_a>$ of $MBR(a)$ in $\pi(b)$ is called a *location* of R, and all the *locations* of R are denoted by $loc(R)$.

REG is an infinite set in two-dimensional plane space. For each R in D, we have $loc(R)=\{ <l_{a_i}, r_{a_i}, t_{a_i}, b_{a_i}> \mid i\geq 1, a_i\in REG \}$.We extend R to cardinal direction relations, i.e., $R\in 2^D$, if $R=\{R_1, R_2,\ldots,R_k\}$ where $R_1, R_2,\ldots, R_k$ are BCDRs, then we have $loc(R)=\{loc(R_i) \mid R_i\in R, 1\leq i\leq k\}$.

Example 3. In Fig. 2(1), $<l_{a_1},r_{a_1},t_{a_1},b_{a_1}>=<1, 2, 1, 2>$, and $<1, 2, 1, 2>\in loc(R)$.

Suppose $<a, b>\in R$, by definition 8, there are 81 possible quadruples. According to IA and point algebra[22], however, $inf_x(a)<sup_x(a)$ and $inf_y(a)<sup_y(a)$ hold, see Fig.1(1), such that $l_a\leq r_a$ and $t_a\leq b_a$. Therefore, only 36 quadruple are realizable for $MBR(a)$ in $\pi(b)$.

Definition 10. If $<a, b>\in R$, then the $<l_b, r_b, t_b, b_b>$ of $MBR(b)$ in $\pi(a)$ is called an *inverse location* of R, and all the *inverse locations* of R are denoted by $iloc(R)$.

For each $<a, b>\in R$, we have $MBR(a)\ Br(R)\ MBR(b)$ and $MBR(b)\ Br(R)^{-1}\ MBR(a)$ hold, where $Br(R)^{-1}\in 2^D$. By definition 4, we have $iloc(R)=loc(Br(R)^{-1})$. Therefore, $iloc(R)$ describes all possible distributions of the four edges of $MBR(b)$ in $\pi(a)$.

4 Constraints under Quadruple Model

Proposition 1. Let $R\in D$, then $loc(R)=loc(Br(R))$.

Proof. For any two regions $a, b\in REG$, $MBR(a)$ and $MBR(b)$ are rectangular regions in the same plane coordinate system. If $a\ R\ b$, then $MBR(a)\ Br(R)\ MBR(b)$. $loc(R)$ is the distribution of the four edges of $MBR(a)$ in $\pi(b)$ and $\pi(b)=\pi(MBR(b))$, then we have $loc(R)=loc(Br(R))$, proposition 1holds.

The surjection from 218 BCDRs onto 36 rectangular relations is investigated with IA in [20]. According to proposition1, there exist a bijection from 36 rectangular relations onto the 36 *locations*. Table 1 shows the mapping among the rectangular relations, *locations* and sets of direction tiles.

Proposition 2. For any $R \in D$, $loc(R)$ is a singleton set.

Proof. Suppose $<a_i, b_i>$, $<a_j, b_j> \in R$, and $i \neq j$, then $MBR(a_i)\ Br(R)\ MBR(b_i)$ and $MBR(a_j)\ Br(R)\ MBR(b_j)$ hold. Moreover, reference regions b_i and b_j divide the plane in the same way respectively, such that the tiles in $\pi(b_i)$ and the ones in $\pi(b_j)$ have the same spatial topological structure. For any $R \in D$, $Br(R)$ is unique [20], such that rectangular relation $Most(W, Br(R))$ can be determined uniquely. Thus, $Tile(Most(W, Br(R)))$ is an unique set of tiles. According to definition 8, we have: $l_{a_i}=l_{a_j}$. Similarly, $r_{a_i}=r_{a_j}$, $t_{a_i}=t_{a_j}$, $b_{a_i}=b_{a_j}$, such that: $<l_{a_i}, r_{a_i}, t_{a_i}, b_{a_i}>=<l_{a_j}, r_{a_j}, t_{a_j}, b_{a_j}>$. Therefore, Proposition 2 holds.

Table 1. The mapping among rectangle relations, *locations* and sets of tiles

Rectangular relation *R*	*loc*(*R*)	*Tile*(*R*)	Rectangular relation *R*	*loc*(*R*)	*Tile*(*R*)
B	{<2, 2, 2, 2>}	{*B*}	*B:S*	{<2, 2, 2, 3>}	{*B, S*}
S	{<2, 2, 3, 3>}	{*S*}	*NE:E*	{<3, 3, 1, 2>}	{*NE, E*}
SW	{<1, 1, 3, 3>}	{*SW*}	*E:SE*	{<3, 3, 2, 3>}	{*E, SE*}
W	{<1, 1, 2, 2>}	{*W*}	*NW:N:NE*	{<1, 3, 1, 1>}	{*NW, N, NE*}
NW	{<1, 1, 1, 1>}	{*NW*}	*W:B:E*	{<1, 3, 2, 2>}	{*W, B, E*}
N	{<2, 2, 1, 1>}	{*N*}	*SW:S:SE*	{<1, 3, 3, 3>}	{*SW, S, SE*}
NE	{<3, 3, 1, 1>}	{*NE*}	*NW:W:SW*	{<1, 1, 1, 3>}	{*NW, W, SW*}
E	{<3, 3, 2, 2>}	{*E*}	*N:B:S*	{<2, 2, 1, 3>}	{*N, B, S*}
SE	{<3, 3, 3, 3>}	{*SE*}	*NE:E:SE*	{<3, 3, 1, 3>}	{*NE, E, SE*}
NW:N	{<1, 2, 1, 1>}	{*NW, N*}	*NW:N:W:B*	{<1, 2, 1, 2>}	{*NW, N, W, B*}
N:NE	{<2, 3, 1, 1>}	{*N, NE*}	*N:NE:B:E*	{<2, 3, 1, 2>}	{*N, NE, B, E*}
W:B	{<1, 2, 2, 2>}	{*W, B*}	*W:B:SW:S*	{<1, 2, 2, 3>}	{*W, B, SW, S*}
B:E	{<2, 3, 2, 2>}	{*B, E*}	*B:E:S:SE*	{<2, 3, 2, 3>}	{*B, E, S, SE*}
SW:S	{<1, 2, 3, 3>}	{*SW, S*}	*NW:N:NE: W:B:E*	{<1, 3, 1, 2>}	{*NW, N, NE, W, B, E*}
S:SE	{<2, 3, 3, 3>}	{*S, SE*}	*W:B:E:SW: S:SE*	{<1, 3, 2, 3>}	{*W, B, E, SW, S, SE*}
NW:W	{<1, 1, 1, 2>}	{*NW, W*}	*NW:N:W:B: SW:S*	{<1, 2, 1, 3>}	{*NW, N, W, B, SW, S*}
W:SW	{<1, 1, 2, 3>}	{*W, SW*}	*N:NE:B:E: S:SE*	{<2, 3, 1, 3>}	{*N, NE, B, E, S, SE*}
N:B	{<2, 2, 1, 2>}	{*N, B*}	*NW:N:NE: W: B:E:SW:S:SE*	{<1, 3, 1, 3>}	{*NW, N, NE, W, B, E, SW, S, SE*}

Proposition 3. Let $R_i, R_j \in D$, if $loc(R_i)=loc(R_j)$, then $R_i^{-1}=R_j^{-1}$.

Proof. Let $A=\{R_{ik} | Br(R_{ik}) \in Br(R_i)^{-1}, k \geq 1\}$, for each $R_{ik} \in A$, we have

$(\exists a, b \in REG)\ MBR(a)\ Br(R_i)\ MBR(b) \wedge MBR(b)\ Br(R_{ik})\ MBR(a)$

According to the implications we mentioned in section 2.1, we have :

$(\exists c, d, e, f \in REG)\ c\ R_i\ d \wedge e\ R_{ik}\ f$

$(\exists c, d, e, f \in REG)\ c\ R_i\ MBR(d) \wedge e\ R_{ik}\ MBR(f)$

where $c, f \in MBR(a)$, $d, e \in MBR(b)$ and $MBR(c)=MBR(f)$, $MBR(d)=MBR(e)$, such that:

$(\exists c, d, e, f \in REG)\ c\ R_i\ MBR(e) \wedge e\ R_{ik}\ MBR(c)$

$(\exists c, d, e, f \in REG)\ c\ R_i\ e \wedge e\ R_{ik}\ c$

Thus we have $R_{ik} \in R_i^{-1}$.

Conversely, for each $R' \in R_i^{-1}$, we have $(\exists a', b' \in REG)\ a' R_i\ b' \wedge b' R' a'$, Then

$(\exists a', b' \in REG)\ MBR(a')\ Br(R_i)\ MBR(b') \wedge MBR(b')\ Br(R')\ MBR(a')$

i.e., $Br(R') \in Br(R_i)^{-1}$. Thus we have $R' \in A$. Therefore, $A=R_i^{-1}$.

Similarly, let $A'=\{R_{jt} | Br(R_{jt}) \in Br(R_j)^{-1}, t \geq 1\}$, then $A'=R_j^{-1}$. Since $loc(R_i)=loc(R_j)$, by proposition1, we have $loc(Br(R_i))=loc(Br(R_j))$. Then $Br(R_i)=Br(R_j)$ and $Br(R_i)^{-1}=Br(R_j)^{-1}$, such that $A=A'$. Therefore, Proposition 3 holds.

Proposition 4. Let $R_i, R_j \in D$, if $R_j \in R_i^{-1}$, iff $loc(R_j) \in iloc(R_i)$.

Proof. For each $R_j \in D$, $R_j \in R_i^{-1}$, we have $(\exists a, b \in REG)\ a\ R_i\ b \wedge b\ R_j\ a$. Then

$(\exists a, b \in REG)\ MBR(a)\ Br(R_i)\ MBR(b) \wedge MBR(b)\ Br(R_i)^{-1}\ MBR(a)$

$(\exists a, b \in REG)\ MBR(a)\ Br(R_i)\ MBR(b) \wedge MBR(b)\ Br(R_j)\ MBR(a)$

Thus $Br(R_j) \in Br(R_i)^{-1}$ and $loc(Br(R_j)) \in loc(Br(R_i)^{-1})$. By definition 10, $iloc(R_i)=loc(Br(R_i)^{-1})$ holds. By proposition1, $loc(R_j)=loc(Br(R_j))$ holds. Therefore, $loc(R_j) \in iloc(R_i)$.

Conversely, for each $R_j \in D$, $loc(R_j) \in iloc(R_i)$, we have $loc(Br(R_j)) \in loc(Br(R_i)^{-1})$ and $Br(R_j) \in Br(R_i)^{-1}$ hold.Then

$(\exists a, b \in REG)\ MBR(a)\ Br(R_i)\ MBR(b) \wedge MBR(b)\ Br(R_i)^{-1}\ MBR(a)$

$(\exists a, b \in REG)\ MBR(a)\ Br(R_i)\ MBR(b) \wedge MBR(b)\ Br(R_j)\ MBR(a)$

According to the implications we mentioned in section 2.1, then we have :

$(\exists c, d, e, f \in REG)\ c\ R_i\ d \wedge e\ R_j\ f$

$(\exists c, d, e, f \in REG)\ c\ R_i\ MBR(d) \wedge e\ R_j\ MBR(f)$

where $c, f \in MBR(a)$, $d, e \in MBR(b)$ and $MBR(c)=MBR(f)$, $MBR(d)=MBR(e)$, such that: $(\exists c, d, e, f \in REG)\ c\ R_i\ MBR(e) \wedge e\ R_j\ MBR(c)$ holds. Then $(\exists c, d, e, f \in REG)\ c\ R_i\ e \wedge e\ R_j\ c$. Thus $R_j \in R_i^{-1}$. Therefore, proposition 4 holds.

5 Determine the Inverse Relation of BCDRs

5.1 Relation *Con* and *Dcon*

Under the same plane coordinate system, for any two regions a, $b \in REG$, there are four pairs of corresponding parallel edges determined by $MBR(a)$ and $MBR(b)$ (t_a and t_b, b_a and b_b, r_a and r_b, l_a and l_b). Now we introduce relations *Con* and *Dcon* to describe the *connect* and *disconnect* topological structures between $MBR(a)$ and $MBR(b)$ respectively. The formally definition is as follow:

Definition 11. *Dcon* and *Con* are binary relations, for any two regions a, $b \in REG$, if $inf_x(a) \neq inf_x(b) \wedge inf_y(a) \neq inf_y(b) \wedge sup_x(a) \neq sup_x(b) \wedge sup_y(a) \neq sup_y(b)$ holds, then a *Dcon* b, otherwise a *Con* b.

For two regions, *Dcon* describs a simple position that no corresponding parallel edges of their MBRs are coincident. Obviously, it doesn't exist regions a, $b \in REG$, such that a *Dcon* $b \wedge a$ *Con* b. Thus, *Dcon* and *Con* are two are jointly exhaustive and pairwise disjoint relations. Moreover, if a *Dcon* b, then b *Dcon* a holds. Therefore, *Dcon* and *Con* are symmetric.

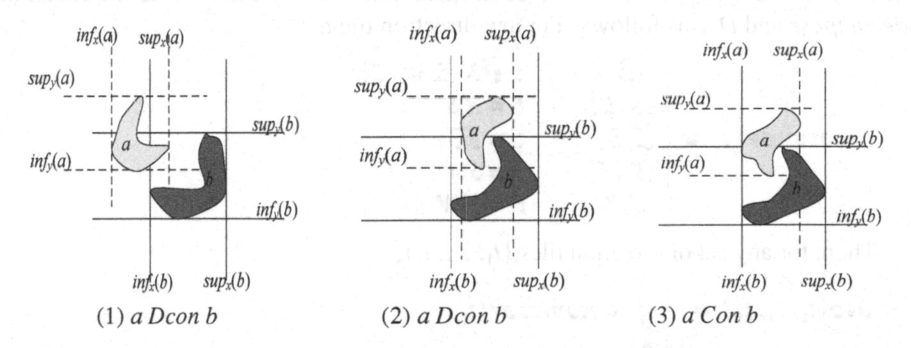

Fig. 4. Con and *Dcon*

Example 4. Regions in Fig.4(1) and Fig.4(2) satisfy relation *Dcon*, regions in Fig.4(3) satisfy relation *Con*.

5.2 Algorithm for Inverse Locations of BCDRs

For each R in D, it can be viewed as a set of ordered pairs of regions that satisfy R. Based on relations *Con* and *Dcon*, we divide R into two subsets, denoted by $Con(R)$ and $Dcon(R)$ respectively. For each $<a, b>$ in R, if a *Con* b, then it belongs to $Con(R)$, otherwise it belongs to $Dcon(R)$, namely, $R=Con(R) \cup Dcon(R)$. Similarly, we have $R^{-1}=Con(R^{-1}) \cup Dcon(R^{-1})$, where $Con(R^{-1})=\{<b_i, a_i> \mid <a_i, b_i> \in Con(R)\}$ and $Dcon(R^{-1})=\{<b_j, a_j> \mid <a_j, b_j> \in Dcon(R)\}$. Then we have

$$iloc(R)=loc(Con(R^{-1})) \cup loc(Dcon(R^{-1}))$$

By the proposition 4, R^{-1} can be determined with $iloc(R)$. Now, we turn to compute $iloc(R)$. First, we consider the *locations* of the ordered pairs in $Dcon(R^{-1})$

Algorithm 1. **Loc_Inv_{Dcon}**
Input: R
Output: $loc(Dcon(R^{-1}))$
Let $loc(R)= < l, r, t, b>$
$loc(Dcon(R^{-1}))\leftarrow\varnothing$
if (l=r=1 ∨ l=r=3) **then** $l'\leftarrow 4-l$. $r'\leftarrow 4-r$.
else $l'\leftarrow 4-l-1$. $r'\leftarrow 4-r+1$.
if (t=b=1∨ t=b=3) **then** $t'\leftarrow 4-t$. $b'\leftarrow 4-b$.
else $t'\leftarrow 4-t-1$. $b'\leftarrow 4-b+1$.
$loc(Dcon(R^{-1}))\leftarrow\{< l', r', t', b>\}$

We distinguish *Dcon* and *Con* with the coincident condition of corresponding parallel edges of two MBRs. The difference between *Dcon* and *Con* can also be described in a quantitative way by considering the numerical relations among the edges, i.e., $loc(Con(R^{-1}))$can be deduced from $loc(Dcon(R^{-1}))$. Next, we introduce algorithm Loc_Inv based on algorithm Loc_Inv_{Dcon} to compute $loc(Con(R^{-1}))$ and $iloc(R)$. To analyze the direction tiles in more fine-grained way, we define functions *decompose* and *Dec* as follow: For any direction tile t,

$$decompose(t) = \begin{cases} \{t\}, & t \in \{N, S, W, E\} \\ \{N, E\}, & t = NE \\ \{S, E\}, & t = SE \\ \{S, W\}, & t = SW \\ \{N, W\}, & t = NW \end{cases}$$

Then, for any set of direction tiles $\{t_1, \ldots, t_k\}$,

$$Dec(t_1, \ldots, t_k) = \bigcup_{1 \le i \le k} decompose(t_i)$$

For instance, $Dec(B, NE, N)=\{B, N, E\}$.
Suppose $loc(R)=\{<l, r, t, b>\}$, for each $<a_i, b_i>\in R$, all the possible values of the left, right ,up and down edges of $MBR(b_i)$ in $\pi(a_i)$ are denoted by set $\mathcal{L}$, $\mathcal{R}$, $\mathcal{T}$and $\mathcal{B}$. Let $\mathcal{M}_1$={N, E, W, S}, $\mathcal{M}_2$={NE, SE, SW, NW}, $\mathcal{M}_3$={N, S}, $\mathcal{M}_4$={W, E}.

Algorithm 2. Loc_Inv
Suppose $loc(R)=\{<l, r, t, b>\}$
Input: R
Output: $iloc(R)$
Loc_Inv_{Dcon} (R)
$iloc(R)\leftarrow\varnothing$
$\mathcal{L}\leftarrow\{l'\}$. $\mathcal{R}\leftarrow\{r'\}$. $\mathcal{B}\leftarrow\{b'\}$. $\mathcal{T}\leftarrow\{t'\}$.$\mathcal{L}\leftarrow\{l'\}$. $\mathcal{R}\leftarrow\{r'\}$. $\mathcal{B}\leftarrow\{b'\}$. $\mathcal{T}$ $\leftarrow\{t'\}$
if $B\in Tile(Br(R))$ **then**
if $S\in \mathcal{M}_3-Tile(Br(R))$ **then** $\mathcal{B}\leftarrow\mathcal{B}\cup\{b\}$.

if $N\in \mathcal{M}_3$–*Tile*(*Br*(*R*)) **then** $\mathcal{T}\leftarrow\mathcal{T}\cup\{t\}$.
if $W\in \mathcal{M}_4$–*Tile*(*Br*(*R*)) **then** $\mathcal{L}\leftarrow\mathcal{L}\cup\{l\}$.
if $E\in \mathcal{M}_4$–*Tile*(*Br*(*R*)) **then** $\mathcal{R}\leftarrow\mathcal{R}\cup\{r\}$.
else
if $Tile(Br(R))\cap\mathcal{M}_3\neq\varnothing$ **then**
if $W\in \mathcal{M}_4$–*Dec*(*Tile*(*R*)) **then** $\mathcal{L}\leftarrow\mathcal{L}\cup\{l\}$.
if $E\in \mathcal{M}_4$–*Dec*(*Tile*(*R*)) **then** $\mathcal{R}\leftarrow\mathcal{R}\cup\{r\}$.
if $Tile(Br(R))\cap\mathcal{M}_4\neq\varnothing$ **then**
if $S\in \mathcal{M}_3$–*Dec*(*Tile*(*R*)) **then** $\mathcal{B}\leftarrow\mathcal{B}\cup\{b\}$.
if $N\in \mathcal{M}_3$–*Dec*(*Tile*(*R*)) **then** $\mathcal{T}\leftarrow\mathcal{T}\cup\{t\}$.
$iloc(R)\leftarrow\{\mathcal{L}\times\mathcal{R}\times\mathcal{T}\times\mathcal{B}\}$

Example 5. In Fig.2, *R*=*W*:*NW*:*N*, it is known from Loc_Inv that $\mathcal{L}$={2}, $\mathcal{R}$={2, 3}, $\mathcal{T}$={2} and $\mathcal{B}$={2, 3}. Therefore, *iloc*(*B*:*W*:*NW*:*N*)={{2}×{2, 3}×{2}× {2, 3}}={<2, 2, 2, 2>, <2, 3, 2, 3>, <2, 3, 2, 2>, <2, 2, 2, 3>}.

5.3 Algorithm for Inverse Relations of BCDRs

Next, algorithm INV is given to compute R^{-1} with *iloc*(*R*).

Algorithm 3. INV

Suppose *iloc*(*R*)={ q_1, ... , q_i, ... , q_k } where $1\leq i\leq k$. The rectangular relation corresponding to q_i by table1 is denoted by R^{q_i}

Input: *R*
Output: R^{-1}
$R^{-1}\leftarrow\varnothing$.
iloc(*R*)← Loc_Inv(*R*).
foreach $q_i\in iloc(R)$ **do**
INV←∅.
INV←δ(*Tile*(R^{q_i})).
foreach $R'\in$ INV **do**
if $loc(R')\in iloc(R)$ **then** $R^{-1}\leftarrow R^{-1}\cup\{R'\}$.

Example 6. In Fig.2, *R*=*W*:*NW*:*N* and *loc*(*R*)={<1, 2, 1, 2>}. Algorithm INV takes *W*:*NW*:*N* as input. By algorithm Loc_Inv, *iloc*(*W*:*NW*:*N*)= {<2, 3, 2, 3>, <2, 3, 2, 2>, <2, 2, 2, 3>, <2, 2, 2, 2>}. For <2, 3, 2, 3>, it corresponds to an unique rectangular relagion *B*:*S*:*E*:*SE* (see table.1), then δ(*Tile*(*B*:*S*:*E*:*SE*))={*B*, *S*, *E*, *SE*, *B*:*S*, *B*:*E*, *S*:*SE*, *E*:*SE*, *B*:*S*:*E*, *B*:*S*:*SE*, *B*:*E*:*SE*, *S*:*E*:*SE*, *B*:*S*:*E*:*SE*} where *B*:*S*:*E*, *B*:*S*:*SE*, *B*:*E*:*SE*, *S*:*E*:*SE* and *B*:*S*:*E*:*SE*'s *locations* belong to *iloc*(*W*:*NW*:*N*), i.e., *B*:*S*:*E*, *B*:*S*:*SE*, *B*:*E*:*SE*, *S*:*E*:*SE* and *B*:*S*:*E*:*SE* are in R^{-1}. Similarly, repeating the process above for the other *locations* in *iloc*(*W*:*NW*:*N*), we have

$$R^{-1}=\{\ B{:}S{:}E,\ B{:}S{:}SE,\ B{:}E{:}SE,\ S{:}E{:}SE,\ B{:}S{:}E{:}SE,\ B{:}E,\ B{:}S,\ B\}.$$

Next, we integrate our algorithm with the widely used qualitative reasoning method proposed in [9], and compare it with the approaches that do not consider the inverse relations.

Example 7. Let constraint set $C=\{a_1\ N{:}NE\ a_2,\ a_1\ NE\ a_3,\ a_2\ B{:}N{:}NE\ a_3,\ a_2\ SW\ a_1\}$

When reasoning about cardinal direction relations with constraint satisfaction problem, such as [9, 10, 11, 20, 22], C is converted with the principles and properties of IA or RA. For $a_1\ N{:}NE\ a_2$ (see Fig.5), we have the following constraints hold:

$$(\exists e_1, e_2 \in REG)\ e_1\ N\ a_2 \wedge e_2\ NE\ a_2 \wedge (e_1 \cup e_1 = a_1)$$

Since $e_1\ N\ a_2$, then $inf_x(a_2) \leq inf_x(e_1) \wedge sup_x(e_1) \leq sup_x(a_2) \wedge sup_y(a_2) \leq inf_y(e_1)$. Similarly, by $e_2\ NE\ a_2$ we have $sup_x(a_2) \leq inf_x(e_2) \wedge sup_y(a_2) \leq inf_y(e_2)$.$e_1, e_2 \in REG$, such that

$$inf_x(e_1) < sup_x(e_1) \wedge inf_y(e_1) < sup_y(e_1) \wedge inf_x(e_2) < sup_x(e_2) \wedge inf_y(e_2) < sup_y(e_2) \wedge$$
$$inf_x(a_2) < sup_x(a_2) \wedge inf_y(a_2) < sup_y(a_2).$$

Since $e_1 \cup e_1 = a_1$, then we have $e_1 \subseteq a_1$and $e_1 \subseteq a_1$, such that

$$inf_x(a_1) \leq inf_x(e_1) \wedge sup_x(e_1) < sup_x(a_1) \wedge inf_y(a_1) \leq inf_y(e_1) \wedge sup_y(e_1) \leq sup_y(a_1),$$
$$inf_x(a_1) < inf_x(e_2) \wedge sup_x(e_2) \leq sup_x(a_1) \wedge inf_y(a_1) \leq inf_y(e_2) \wedge sup_y(e_2) \leq sup_y(a_1).$$

According to the topological structure of the $\pi(a_2)$, we also have $inf_x(a_2) \leq inf_x(a_1) < sup_x(a_2) < inf_y(a_1)$ and $sup_x(e_1) \leq inf_x(e_2)$. Finally, we obtain set O by converting all the BCDRs constraints in C.

$O=\{inf_x(a_2) \leq inf_x(e_1),\ sup_x(e_1) \leq sup_x(a_2),\ sup_y(a_2) \leq inf_y(e_1),\ sup_x(a_2) \leq inf_x(e_2),\ sup_y(a_2) \leq inf_y(e_2),\ inf_x(e_1) < sup_x(e_1),\ inf_y(e_1) < sup_y(e_1),\ inf_x(e_2) < sup_x(e_2),\ inf_y(e_2) < sup_y(e_2), \ldots\}$.

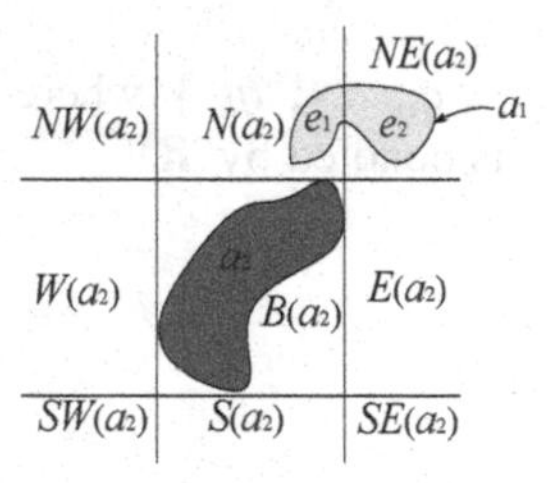

Fig. 5. a1 $N{:}NE\ a_2$

The reasoning about set C with BCDRs turns to the consistency decision problem of set O with orderd constraints. Then path-consistency methods are frequently applied to it. It is obvious that these approaches ignore the impact of inverse relations to the efficiency and correctness of reasoning. Next, we illustrate how our work improves the reasoning performance of Example 7. Given $a_2\ N{:}NE\ a_1$, by the quadruple model we have $loc(N{:}NE)=\{<2, 3, 1, 1>\}$. Algorithm INV takes $N{:}NE$ (i.e.,<2, 3, 1, 1>) as input, then we have $iloc(N{:}NE)=\{<1, 2, 3, 3>, <2, 2, 3, 3>\}$, therefore, $N{:}NE^{-1}=\{S{:}SW, S\}$, such that $a_2\ S{:}SW\ a_1 \wedge a_2\ S\ a_1$ holds. Then we obtain constraint set C' by propagating C with this new constraint.

$C'=\{a_1\ N{:}NE\ a_2,\ a_1\ NE\ a_3,\ a_2\ B{:}N{:}NE\ a_3,\ a_2\ SW\ a_1,\ a_2\ S{:}SW\ a_1 \wedge a_2\ S\ a_1\}$

Obviously, one of the necessary and sufficient conditions for the consistency of C is that : C' is consistent. While the confliction between $a_2\ SW\ a_1$ and $a_2\ S{:}SW\ a_1 \wedge a_2\ S\ a_1$ makes C' not consistent. Therefore, set C is not consistent.

6 Conclusion

In this paper, we solved the problem of the inverse relations of BCDR. We first define the *locations* to describe BCDRs quantitatively based on MBR. Then a quadruple model is proposed, by which the constraints and principles between a BCDR and its inverse relations are investigated. Two binary relations *Con* and *Dcon* are defined to describe the *connect* and *disconnect* topological structures between MBRs respectively. Then a novel algorithm is proposed for inverse relations of all BCDRs, including single-tile, multi-tile and rectangular relations. Integrating with widely used qualitative reasoning works, the algorithm is applied on checking consistency problem of BCDRs and yield promising results.

References

1. Haar, R.: Computational models of spatial relations. TR-478, MSC-72-03610. Computer Science, University of Maryland, College Park, MD (1976)
2. Frank, A.U.: Qualitative spatial reasoning about cardinal directions. In: The 7th Austrian Conference on Artificial Intelligence, pp. 157–167 (1991)
3. Moratz, R., Lücke, D., Mossakowski, T.: A condensed semantics for qualitative spatial reasoning about oriented straight line segments. Artificial Intelligence 175, 2099–2127 (2011)
4. Wolter, D., Lee, J.H.: Qualitative reasoning with directional relations. Artificial Intelligence 174, 1498–1507 (2010)
5. Mossakowski, T., Moratz, R.: Qualitative reasoning about relative direction of oriented points. Artificial Intelligence 180-181, 34–45 (2012)
6. Skiadopoulos, S., Koubarakis, M.: Composing cardinal direction relations. Artificial Intelligence 152(2), 143–171 (2004)
7. Goyal, R.K.: Similarity assessment for cardinal directions between extended spatial objects. PhD thesis, The University of Maine (2000)
8. Goyal, R., Egenhofer, M.J.: The direction-relation matrix: A representation for directions relations between extended spatial objects. The Annual Assembly and the Summer Retreat of University Consortium for Geographic Information Systems Science (1997)
9. Skiadopoulos, S., Koubarakis, M.: On the consistency of cardinal direction constraints. Artificial Intelligence 163(1), 91–135 (2005)
10. Liu, W., Zhang, X., Li, S., Ying, M.: Reasoning about cardinal directions between extended objects. Artificial Intelligence 174, 951–983 (2010)
11. Liu, W., Li, S.: Reasoning about cardinal directions between extended objects: The NP-hardness result. Artificial Intelligence 175, 2155–2169 (2011)
12. Dong, Y., Liu, D., Wang, F., Wang, S., Lv, S.: A MBR based Approach for Modeling Direction Relations between Uncertain Regions. Acta Electronica Sinica 39(2), 329–335 (2011)

13. Wang, S., Liu, D.: Knowledge representation and reasoning for qualitative spatial change. Knowledge-Based Systems 30, 161–171 (2012)
14. Allen, J.F.: Maintaining knowledge about temporal intervals. Communications of the ACM 26(11), 832–843 (1983)
15. Randell, D.A., Cui, Z., Cohn, A.G.: A spatial logic based on regions and connection. In: Proceedings of the Third International Conference on Principles of Knowledge Representation and Reasoning (KR 1992), pp. 165–176 (1992)
16. Cicerone, S., di Felice, P.: Cardinal directions between spatial objects: The pairwise-consistency problem. Information Sciences 164(1-4), 165–188 (2004)
17. Liu, W., Zhang, X., Li, S., Ying, M.: Reasoning about cardinal directions between extended objects. Artificial Intelligence 174(12-13), 951–983 (2010)
18. Wang, J., Jiang, G., Guo, R.: Reversal Operation of Spatial Direction Relationship. Surveying and Mapping 25(5), 324–328 (2008)
19. Wu, J., Cheng, P., Chen, F., Mao, J.: Qualitative Reasoning on Direction Relationship of Spatial Target. Surveying and Mapping 35(2), 160–165 (2006)
20. Navarrete, Sciavicco, G.: Spatial reasoning with rectangular cardinal direction relations. In: Proc. of the ECAI 2006 Workshop on Spatial and Temporal Reasoning, pp. 1–10 (2006)
21. Lipschutz, S.: Set Theory and Related Topics. McGraw Hill, New York (1998)
22. van Beek, P.: Reasoning about qualitative temporal information. Artificial Intelligence 58, 297–326 (1992)

Renewable Energy Challenges and Opportunities: Geospatial and Qualitative Analysis of Southern California

Monica Perry*, James Pick, and Jessica Rosales

School of Business, University of Redlands, 1200 E. Colton Ave. P.O. Box 3080, Redlands, California 92373-0999 USA
Monica_Perry@Redlands.Edu

Abstract. Renewable energy has become an increasingly important and viable approach for mitigating the negative impacts of generating electrical energy with fossil fuels, such as oil and gas. In our study of renewable energy we conducted geospatial analysis of an array of demographic characteristics and renewable energy facilities (manufacturers, installations) for wind and solar energy. The analysis focused on Southern California, but also included two benchmark urban regions in other states, Maryland and Texas with well-developed solar and wind sectors, respectively. Qualitative data on renewable energy was also collected through personal interviews to supplement our geospatial analysis. Implications for sustainability are drawn by identifying key aspects of renewable energy development in Southern California.

Keywords: Renewable Energy, Solar Energy, Wind Energy, Eco-innovation, geospatial analysis.

1 Introduction

Global climate change continues to be a fundamental and critical issue for both developing and developed countries. The United Nations' Convention Framework on Climate Change is a global collaboration which illustrates the significance of climate change, as well as the widespread commitment to reducing and minimizing the negative aspects of climate change [1]. Towards that end, greenhouse gases represent the primary cause of negative climate change, with fossil fuel activities associated with electricity generation being the primary cause of increased greenhouse gases in the United States as well as around the world [2].

A wide array of alternative energy sources exists for fossil and nuclear fuels, the most significant being renewable energy sources. Renewable energy represents an increasingly important and viable approach for reducing the negative environmental impacts of generating electrical energy with fossil fuels, such as oil and gas. The most important renewable energy sources in Southern California are solar, wind and geothermal energy sources. The relative abundance of wind, solar and geothermal energy sources varies across different geographic regions, thus laying the foundation for differential potential for these forms of renewable energy generation.

F. Bian and Y. Xie (Eds.): GRMSE 2014, CCIS 482, pp. 571–580, 2015.

As part of an in-depth research project on renewable energy development and manufacturing potential in Southern California, we conducted a multi-method study utilizing geospatial analysis. The geospatial analysis was supplemented primarily with qualitative analysis through personal interviews.

In the discussion that follows we address how geospatial and qualitative analysis were applied to the renewable energy sector to identify meaningful patterns and draw implications for the renewable energy sector in Southern California. We focused our efforts on geographic analysis of a variety of renewable energy, demographic, and economic characteristics to illustrate the opportunities and challenges for renewable energy development and manufacturing in Southern California. Characteristics included location of key renewable energy facilities and resources, the population and other key complementary attributes such as transportation infrastructure.

Renewable energy resources, markets and manufacturing represent important features of the Coachella Valley in Southern California. The Coachella Valley (the Valley) extends southeast approximately 45 miles from the southeastern part of the San Bernardino Mountains to the Salton Sea. The Salton Sea is a water basin filled by rainfall and irrigation runoff that flows northwards from the Imperial Valley draining into the Sea.

The sections that follow address three aspects of the research study. In Section 2, we present a detailed discussion of the methodology and data used for both the geospatial and qualitative analysis. The methodology sub-section addresses data sources as well as challenges with data availability. In Section 3 we provide an overview of the primary geographic area of interest in Southern California, the Coachella Valley, by focusing on demographic, economic and natural resource characteristics. In the last section we present preliminary implications and conclusions for renewable energy and sustainability in the Valley.

2 Methodology: Geospatial, Qualitative Data and Analysis

2.1 Geospatial Data and Analysis

The renewable energy project was designed with the goals of understanding the challenges and opportunities for renewable energy development in the Valley. The specific renewable energy activities of interest included potential production activities such as manufacturing and assembly, as well as aspects of current and future demand for renewable energy. To achieve those goals the project primarily focused on analysis of key aspects of the renewable energy supply chains for solar, wind and geothermal combined with in-depth analysis of the political, economic, socio-demographic and physical characteristics of the Valley.

Effort was spent identifying relevant data sources related to each of the three renewable energy sectors (solar, wind and geothermal) as well as data on the cities and urban area of the Coachella Valley. Extensive searches yielded secondary data sources which provided considerable information and data for analysis. The data sources included: federal, state and local government agencies, trade associations, as well as articles, reports and books authored by experts and participants in renewable energy, and governmental, business, and nonprofit websites.

Government resources included census and other data from the federal government, California state agencies and nonprofits concerned with renewable energy. These resources were used to describe and map relevant characteristics of the population and businesses of nine cities and two unincorporated areas or census designated places (CDPs) within Coachella Valley. The nine cities included Cathedral City, Coachella, Desert Hot Springs, Indian Wells, Indio, La Quinta, Palm Desert, Palm Springs, Rancho Mirage. The two CDPs were Thousand Palms and Mecca. U.S. Census social and economic data include population, educational attainment, income, net worth, occupation, employment, age, home ownership, internet use, and crime.

Data on businesses included business types from the North American Industry Classification System (NAICS), size (employment) and location. Given the project focus on the Valley, additional detailed data on manufacturers were identified, mapped and analyzed. Population estimates of the Valley were projections from Southern California Association of Governments (SCAG) and the California Department of Finance were utilized with occasional modified assumptions. We estimated population for 2020 based on continuation of the U.S. Census' average yearly city and non-city 2008-2013 growth rates.

Although our primary focus was on renewable energy in the Valley, some aspects of the renewable energy sector were relatively under-developed. As a result, we sought to identify some well-developed renewable energy sectors outside of the Valley as benchmarks or exemplars. We identified geographic regions in Maryland and Texas for solar and wind energy respectively, as well-developed renewable energy exemplars.. Similar population data for those regions were also collected, mapped and analyzed. Data on renewable energy facilities, policies and initiatives was also gathered from federal agencies, such as the US Department of Energy, U.S. Geological Survey, and U.S. Energy Information Administration, as well as state agencies for California, Maryland, andTexas . These sources were used primarily to collect information and data on renewable energy supply chain participants, policies, incentives, initiatives, and environmental issues.

In addition to government resources, trade and non-profit associations provided a wealth of specific data on various aspects of renewable energy in the Valley as well as in the United States. Data on renewable energy supply chain activities in specific geographic areas were collected from renewable energy trade associations. Individuals within select trade associations were contacted to investigate the availability of additional relevant data for the study. Data on manufacturers for major wind, solar and geothermal related products were collected, mapped and analyzed for California, solar for Maryland and wind for Texas. Similarly, data on the locations of major solar, wind and geothermal installations (existing and in development) were collected, mapped and analyzed. Market characteristics included the existing status and forecasts for major installations and jobs in renewable energy came from nonprofits especially the Solar Energy Industries Association (SEIA) and American Wind Energy Association (AWEA).

Additional data and information were provided from the following trade associations and non-profit organizations: Alternative Energy Institute, American Council on Energy Efficient Economy, As You Sow, National Renewable Energy Laboratory (NREL), The Solar Foundation, and the Utility Variable-Generation Integration Group.

2.2 Qualitative Data Analysis

Supply chains represent relatively complex ecosystems of interdependent organizations. The interdependent organizations provide various goods, information and services necessary to serve ultimate customers [3]. The process of serving the ultimate renewable energy customer consists of many prior stages where a variety of organizations engage in a wide assortment of activities to transform and distribute raw materials, manufacture, assemble and distribute components, as well as provide financing and information.

While broad commonalities exist in the respective supply chains for the three renewable energy industries, the nature of each industry, products, and organizations involved create particularities for each specific renewable industry. For example, significant differences in physical principles and technologies exist in each of the different renewable energy industries [4]. As such, the qualitative approach utilized in-depth personal interviews with open-ended questions to provide sufficient opportunities to identify specific information on supply chain structures, issues, requirements and activities for solar, wind and geothermal industries. A key benefit of personal interviewing with open-ended questions is that they allow the opportunity for interviewers to probe participants for clarification and additional depth.

We designed and conducted personal interviews with a total of twelve participants. Participants included renewable energy experts from non-profits, providers of renewable energy products and services, as well as officials and managers from the public sector. The information from the personal interviews with renewable energy experts was useful in providing additional context for the geospatial analysis generated with secondary data. Similarly, the personal interviews with renewable energy executives helped identify organizational, inter-organizational, and strategic factors not easily addressable with geospatial analysis. As a result, the combination of geospatial and qualitative data analysis provided a relatively rich and robust basis for understanding the challenges and opportunities of renewable energy and drawing meaningful conclusions.

3 Human, Economic and Natural Resources in the Valley

According to the 2010 US Census and Southern California Association of Governments, the Valley has almost half a million residents (490,000) with the overwhelming majority (347,000) residing in its 9 largest cities and Thousand Palms CDP. Approximately 100,000 are seasonal residents, predominantly present in the winter. Fig. 1 illustrates the total population by zip code for the Valley, and is indicative of the types of maps used for demographic analysis. Population projections for the Valley indicate more than 550,000 residents by 2020. At two percent annually for 2010 the population growth rate for the Valley is about double the growth rate for both the state of California and the United States as a whole.

The economy of the Valley is based on agriculture, tourism, and retirement. Its farming is typified by fruits and vegetables, including national prominence in dates, palm trees, and citrus. As a well-known winter tourist location, winter visitors are attracted by warm temperatures, dryness, and desert landscapes. Retired people comprise a meaningful portion of the residents. Approximately 20 percent of the population is 65 years of age or older, which compares to California's 11 percent [5]. There are high-end retirement communities and residential areas, as well as moderate priced retirement developments in the northeastern and southwestern parts of the Valley.

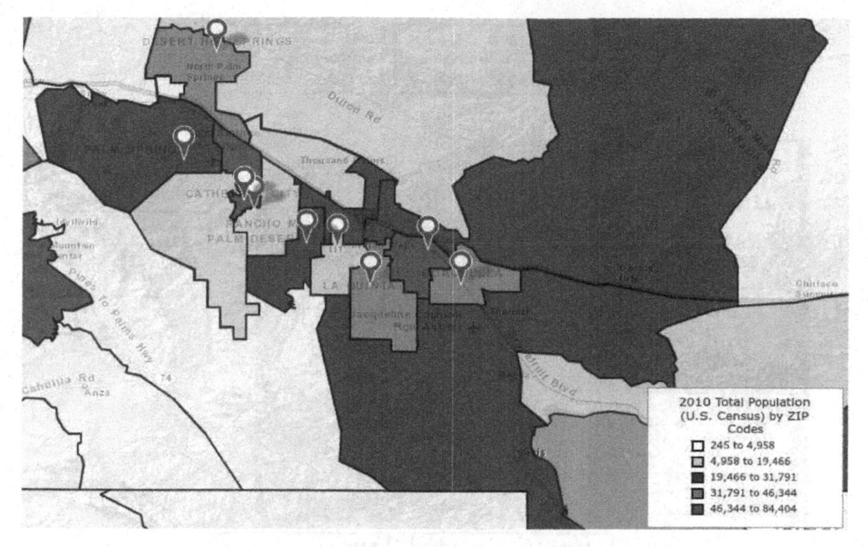

Fig. 1. Population by zip code for Coachella Valley cities (2010 Census)

The Valley is endowed with some of the most plentiful natural resources for renewable energy in the U.S. The western part of the Valley possesses areas with large and exposed areas of high winds. Winds have historically ranged from a high maximum of 59 mph and an average high maximum of 29 mph [6]. The Valley's desert location is linked to high sunlight intensity which favors solar energy production, both within the Valley and in the unpopulated areas of Riverside County to the West extending parallel to the Interstate 10 all the way to the Arizona border. The Valley had an average of 838 solar watts per square meter (w/m^2) and an average yearly total of 3,538 hours of sunshine between 2009 and 2014 [6]. In addition to plentiful sun and wind, the south end of the Valley borders the Salton Sea. The Salton Sea has one of the largest geothermal deposits in the U.S. However extraction of the geothermal deposits has been limited due to high salinity and the varied composition of its brines [7] [8].

The Valley's resources are presently utilized for renewable energy, with considerable growth potential, subject to natural barriers, community challenges and restrictions especially in the urban areas. In the West, in the area of San Gorgonio Pass and mostly visible from Interstate 10, and along State Highway 62, hundreds of wind turbines output considerable electrical energy that is purchased by utilities for the electrical grid of southern California. For example, in Fig. 2 one can see that the Valley has a number of small solar plants. The solar plants currently have a capacity totaling 55 MW [9], which equates to 5.0% of the Valley's electrical generating capacity. By contrast, along Interstate 10, between the western Coachella Valley and Arizona, a massive series of solar electrical generating plants are in development or construction with some already in operation. The total capacity in development/construction is 3,077 MW [9], which equates to one and a half nuclear plants.

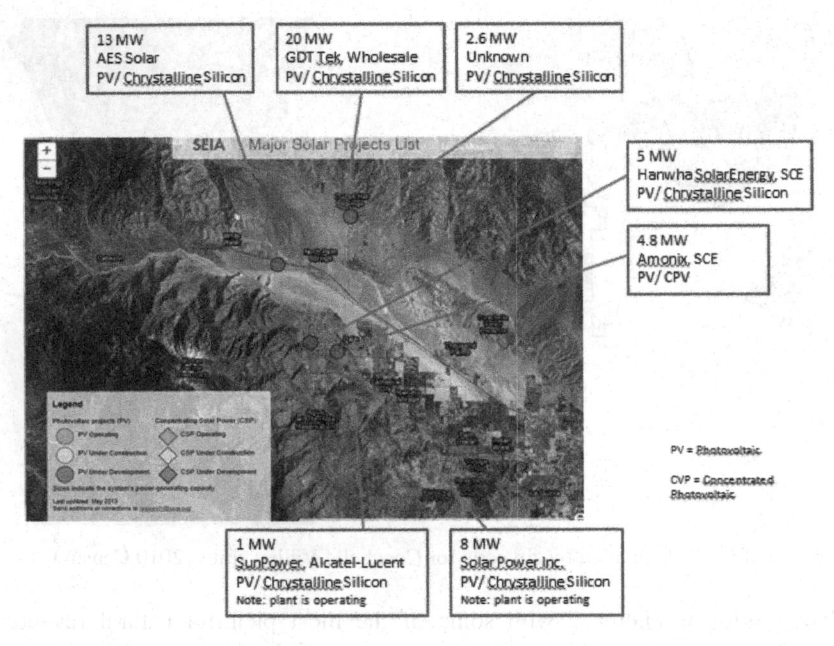

Fig. 2. Solar Installations in Coachella Valley (Adapted from SEIA)

In residential solar energy, based on Coachella having 1.5x the California prevalence, we estimate that the Valley had installed residential solar capacity in 2013 of 60.3 MW [1] [9], which constitutes 5.3% of estimated total Valley energy capacity. As a result of rapid growth in residential solar projected for 2014, we estimate the Valley will have 99.1 MW of installed residential solar capacity by the end of 2014. It is clear that the Valley produces a high level of solar energy, about an estimated 10.3%in its energy supply in 2013.

Another aspect is the challenge of evenness in solar production, daily and seasonally, coupled with seasonal fluctuations in demand in the Coachella Valley, which is accentuated by the seasonal flows of tourists. Unevenness can be mitigated through use of smart grid and battery storage technologies by utilities, organizations, and communities. These technologies are being actively researched and developed by several small enterprises in the Valley. It might advance to a larger scale if Lithium extraction were to expand as part of geothermal energy growth in the Salton Sea area.

Geothermal energy is a renewable form of energy that utilizes the heat flow of the earth to generate very hot steam and hot water (350-480 degrees F). Geothermal deposits in the earth can be drilled into from the ground and used for electrical generation, direct uses of hot water, or extraction of minerals from the hot brines being brought to the surface. Although geothermal sub-surface fields eventually lose their heat and expire, they are regarded as "renewable," because there is a continual flow and because the water itself is not lost, but is recycled at lower temperature [7]. In 2014, the U.S. had 5,410 MW of installed geothermal capacity, constituting 29 percent of the world capacity of 55,709 MW of installed geothermal capacity. This represented only 0.43 percent of U.S. electrical demand [7]. Worldwide, nearly all geothermal is concentrated along converging tectonic plate edges. The Salton Sea geothermal capacity is presently 437 MW, or a sixth of the state resource [10].

Since the Coachella Valley is already well endowed nearby with electricity generation from fossil, solar, and wind, geothermal capacity scale-up of electricity generation in the Salton Sea area does not have much economic impact on the Valley related to electricity. The number of new workers needed is limited and would likely be drawn from the experienced pool of geothermal workers located mostly in Imperial County, just to the south of Coachella Valley.

However, concomitant direct use of the Salton Sea brines for extraction of lithium and possibly other substances has a chance of providing greater economic stimulus for the Valley. One firm is in the process of completing a successful pilot test plant, receiving brines from the adjacent EnergySource's Hudson Ranch One Geothermal Plant in Calipatria [11] [12]. The intent is to scale up the pilot extraction plant to a full-size plant producing substantial amounts of lithium, which is a scarce substance that is principally produced in South America. Lithium is essential for lithium batteries, which constitute a primary battery source for modern electronics as well as electric automobiles. There is some opportunity for prospective industrialization in the Valley based on the lithium extraction. The lithium extraction could encourage the location of some battery manufacturers for renewables in the Valley, building on a base of some small start-up firms already present.

While the Valley has some advantages to support the renewable energy sector, challenges are also present. The major advantages include proximity to large renewable operating facilities; a fairly attractive, medium-sized urban area; support from nonprofits and somewhat from local governments, vast renewable energy resources in and adjacent to it and the presence of low-level manufacturing.. In spite of these advantages the manufacturing of renewable products and components is just beginning in the Valley with a handful of small, early-stage firms. The Valley faces considerable challenges in scaling up renewables manufacturing, including insufficient workforce size and skill levels, limited R&D environment in the Valley, a lack of renewable energy-related training and education, unfavorable transportation compared to coastal areas in Southern California, the risk of sun-setting of the federal subsidy for renewables, and a limited local financial investment environment.

4 Implications and Conclusions for Sustainability in the Valley

A multitude of factors have an impact on sustainability with respect to energy generation and use in general, and the Valley in particular. These factors represent the relative complexity of sustainability. Although complex, the factors can be classified in three broad categories, which Camarinha, Afsarmanesh, and Boucher [13] describe as the three pillars of sustainability. The three pillars are environmental, social and economic. What is central to sustainability with these three pillars is the importance of collaboration within and across the pillars to better support sustainability. Coordination and interaction of the various stakeholders within each pillar as well as across the three pillars is likely to lead to enhanced sustainability in the Valley

Some elements of each of the three pillars were present in our study of renewable energy in and around the Valley. The Valley possesses considerable natural resources which are elements of the environmental pillar. The abundance of natural resources that are relatively underutilized support an increased potential for renewable energy, especially with respect to solar, wind and geothermal energy. In conjunction with the abundant natural environment, our study identified several companies in the Valley which were relatively innovative in their development of renewable energy businesses. For example, one innovative company currently has developed a patented process for lithium extraction from Salton Sea brines [12], while another has developed special solar arrays/storage for a variety of applications, including limited-use vehicles [14].

While most of the renewable companies were relatively small and emerging, they represented a potentially critical component of the economic pillar. If the existing companies are able to expand and attract new companies to the Valley, renewable energy innovation may continue to support the economic pillar of sustainability. Although there were no major solar or wind energy related manufacturers in the Valley, clusters of manufacturers existed near the Valley in other parts of Southern California (See Fig. 3). These manufacturers can serve the Valley's need for solar or wind products.

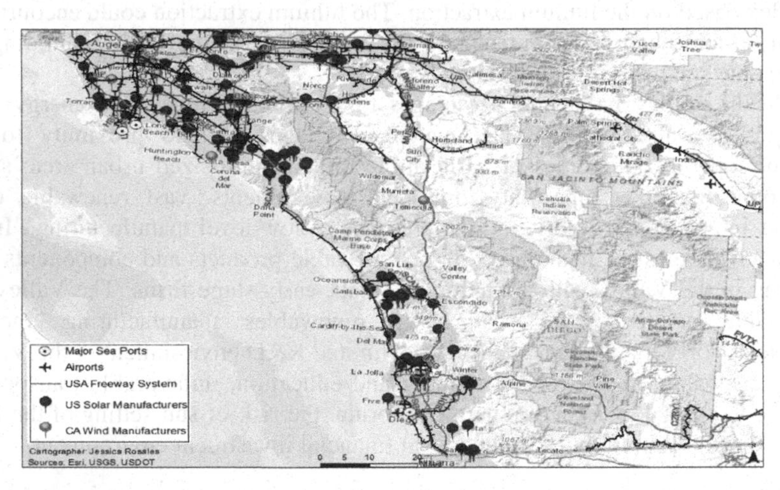

Fig. 3. Solar and Wind Renewable Energy Manufacturing near the Valley. Data Sources: AWEA 2013 and SEIA 2014.

Growth of renewables in the Valley also reflects the social pillar, which can potentially support and interact with either the environmental or economic pillar. In the Valley, the social pillar includes community participation and characteristics that are relatively favorable. The state of California's Renewable Portfolio Standard (RPS) addresses goals to reach a certain percentage of energy production from renewable sources of energy by a certain future date, and is 33% by 2020 [15]. At the local level, a number of renewable energy initiatives exist that are supported by the cities and constituents. San Bernardino County exhibits relatively positive support for renewable energy as evidenced by its inclusion in the general county plan, as well as the San Bernardino County Partnership for Renewable Energy and Conservation (SPARC) initiative [16]. Similarly, Riverside County exhibits considerable support for renewable energy though activities and programs. Riverside County is part of the Salton Sea Authority and provides an array of commercial and residential incentives and programs that support renewable energy [15], while cities in western Riverside County actively provide financing through the HERO program [17]. Various cities, such as Palm Springs, have participated actively in The Clean Cities Coachella Valley Region's Coalition which focuses on alternative-fuel transportation.

Overall, cities in Coachella Valley have displayed considerable interest in renewable energy, with some cross-city initiatives as well as particular cities taking the lead in a variety of renewable energy initiatives. The CV Link is a proposed multi-purpose pathway connecting cities in the Coachella Valley which would be accessible to low-speed (neighborhood) electric vehicles [18]. An independent study on electric vehicle readiness for The Coachella Valley Association of Governments sponsored by the California Energy Commission and the Coachella Valley illustrates the commitment of Palm Desert and Palm Springs to renewable energy [18]. Both cities have already invested in and developed public electrical vehicle charging stations to support electric vehicles. Such relatively abundant natural resources create considerable market opportunities for commercial or industrial utilization of renewable energy as well as with consumer utilization of renewable energy for households and personal transportation.

As discussed in Section 3, the population of the Valley is expected to grow at a rate greater than that of the state of California and the U.S. As a result, increased demand for energy generation would be expected as the population expands. Such increased demand for energy in the Valley could be addressed with requisite increases in the utilization of renewable energy such as solar, wind and geothermal energy. There is also the potential of increasing the relative proportion of renewable energy to the Valley's energy supply, beyond what might be required to address increased energy demand. Doing so would also further sustainability efforts in the region.

Further geospatial and qualitative analysis is warranted to draw definitive conclusions regarding sustainability related to renewable energy development in the Valley. Addressing some limitations of availability of detailed geographic data on renewable energy participants could improve the geospatial analysis, and provide more robust conclusions. In particular, U.S. Census data with detailed NAICS codes for renewable energy (currently not available but planned for inclusion) could add considerable depth to geospatial analysis. Such geospatial analysis would be useful to better address the current state of renewable energy and potential growth in specific geographic regions.

Nevertheless, our study was useful in helping to understand the current state of renewable energy in the Valley. The comparative geospatial analysis of exemplary renewable energy manufacturing sectors in other states illustrated how well developed and highly productive certain urban areas can become and suggested policies for

Coachella Valley governments to consider for advancing renewable energy. Lastly, supplementing our geospatial analysis with interview data provided the opportunity to both confirm conclusions drawn from the geospatial analysis as well as develop a more robust assessment of the challenges and opportunities for renewable energy.

Acknowledgement. We acknowledge support of grant 07-69-06995 from U.S. Department of Commerce.

References

1. United Nations Framework Convention on Climate Change, http://unfccc.int/2860.php
2. United States Environmental Protection Agency. Sources of Greenhouse Gas Emissions, http://www.epa.gov/climatechange
3. Mentzer, J.T., DeWitt, W., Keebler, J.S., Min, S., Nix, N.W., Smith, C.D., Zacharia, Z.: Defining Supply Chain Management. Journal of Business Logistics 22, 1–25 (2001)
4. Boyle, G.: Renewable energy: Power for a sustainable future. Oxford University Press in association with the Open University, Oxford (2012)
5. U.S. Bureau of the Census. American FactFinder. Washington, DC, http://www.census.gov
6. Desert Weather, http://www.desertweather.com
7. Garnish, J., Brown, G.: Geothermal Energy. Chapter 9. In: Boyle, G. (ed.) Renewable Energy, pp. 409–460. Oxford University Press, Oxford (2010)
8. Butler, E.W., Pick, J.B.: Geothermal Energy Development. Plenum Publishing Company, New York (1982)
9. Solar Energy Industries Association (SEIA), http://www.seia.org
10. Matek, B., Gawell, K.: Report on the State of Geothermal Energy in California. Geothermal Energy Association, Washington, DC (2014)
11. The Desert Review. Tesla Motors' Gigafactory, the 10-million-square-foot facility, May Have the Valley on their Short List of Where to Build, April 14. The Desert Review, Brawley (2014)
12. Simbol Inc. Interview with Tracey Sizemore, VP of Simbol Inc. (April 2014)
13. Camarinha-Matos, L.M., Afsarmanesh, H., Boucher, X.: The role of collaborative networks in sustainability. In: Camarinha-Matos, L.M., Boucher, X., Afsarmanesh, H. (eds.) PRO-VE 2010. IFIP AICT, vol. 336, pp. 1–16. Springer, Heidelberg (2010)
14. Solaris Power Cells. Interview with Lenny Caprino, President of Solaris (May 2014)
15. Database for State Incentives and Renewable Efficiency (DSIRE), http://www.dsireusa.org/
16. Land Use Services Department. Renewable Energy and Conservation Planning Grant Application, http://www.sbcounty.gov/Uploads/lus/SPARC_Initiative_Application.pdf
17. Western Riverside Council of Governments, HERO Financing Now Available for Energy/Water Conservation Retrofits, http://www.wrcog.cog.ca.us
18. Coachella Valley Association of Governments, CVLink FAQ, http://www.cvag.org/library/pdf_files/trans/CM%20RFP/CVLink_FAQ.pdf

Identification of Soil Erosion Types in Nyaba River Basin of Enugu State, Southeastern Nigeria Using Remote Sensing and Geographical Information System Techniques.

V.U. Okwu- Delunzu[1], K.E. Chukwu[1], W.O. Onyia[2], A.O. Nwagbara[3], and B.A. Osunmadewa[4]

[1] Dept of Geography and Meteorology, Enugu State University of Science and Technology, Nigeria
[2] Dept of Building, Enugu State University of Science and Technology, Nigeria
[3] Dept of Architecture, Enugu State University of Science and Technology, Nigeria
[4] Institute of Photogrammetry and Remote Sensing, Dresden University of Technology, Germany
dvirginiaugo@yahoo.co.uk

Abstract. This study presents possible way of soil erosion identification on a tropical watershed of Nyaba river basin in Enugu State combining remote sensing and GIS techniques. The aim of this study is to identify and map out soil erosion types in Nyaba river basin and its spatial distribution in 2011. This study identifies soil erosion types in Nyaba River Basin subcatchment units in Enugu Urban of Enugu State, Nigeria, which is a mountainous region with steep slope and an upland ecosystem. To achieve the aim of this study, data were collected from LandSat satellite image of 2011, rainfall data from NIMET for 2011, hydrological data and soil samples were collected in the entire subcatchment unit. The data collected were processed using different methods; Erdas Imagine 9.1 was used in the Land use land cover image classification while the soil type was determine using Soil Textural Triangle by David Whiting, 2011 U.D.S.A , the hydrological data generated was analysed using rational method formula and were categorized into three classes: high, medium and low. Spatial multi-criteria evaluation operation in Arc GIS 10 was used to generate the soil erosion map of the study area. The result showed that there were various erosion types in the study area. Erosion types identified by the research work are rill, sheet, and gully erosion. In conclusion, integrated Remote Sensing, Geographical Information System and Spatial Multi-criteria Evaluation Model tool used to identify soil erosion types in mountainous regions will assist decision makers on environmental management policy.

Keywords: Soil erosion types, Enugu Urban, identification, environmental problem, spatial multi-criteria evaluation model, GIS, remote sensing, land-use.

F. Bian and Y. Xie (Eds.): GRMSE 2014, CCIS 482, pp. 581–592, 2015.

1 Introduction

Soil erosion by water is a serious environmental problem affecting large areas of arable land in Enugu State, Southeastern Nigeria. Erosion is facilitated by numerous factors and processes such as land use type, topography, climate, and soil. Thus, the actions of man such as encroachment of agricultural activities on forest areas, deforestation for commercial and industrial purposes, urbanisation and general misuse of land, as well as the effect of climatic changes such as high rainfall regime, drought, and desertification, tend to exacerbate the impact of soil erosion on the environment UNESCO [6]. The erosion rate of an area at any given point in time is dependent mainly on climatic factors of precipitation, temperature, seasonality, wind speed and geological factors such as; sediment or rock type, porosity or permeability, and the slope (gradient) of land.

Various erosion-focused studies have been carried out by researchers in Nigeria Ofomata[3], Egboka and Okpoko[1], Okagbue and Uma[1], Akpokodje *et al.*[1], Ezezika and Adetona[1], Obiadi *et al.*[1], none of the research has try to identify the erosion types in comprehensively modelling erosion types and mapping it out within Nyaba Catchment in Enugu,Nigeria.

Therefore, for scientific research of the phenomenon function to establish proper measures for its control and management of erosion problem in the area, it is important to identify and map out erosion types in the area for effective planning. To do this, Methodical approaches used are remote sensing and GIS techniques which provided valuable tools for identification and mapping of soil erosion types in the Nyaba River basin subcatchment units in Enugu State, Nigeria.

1.1 Study Area

The study area is Enugu urban which lies within 7^0 26’ to $7^0$37’E longitudinal extent of the Greenwich Meridian and Latitudinal space of $6^0$21’ to $6^0$30’N of the equator (Fig. 1). It is the political and administrative headquarters of Enugu State, is located in the Nyaba Drainage Basin, a humid tropical watershed in the Southeastern Nigeria (Fig. 2). The study area occupies a land area of 8000 km^2 covering the spatial entity under Enugu Town Planning Authority Area, Chukwu [5]. The climate of the area is mainly dry and wet seasons; the wet season which is ushered in by the prevalence of the southwest tropical maritime (mT) air mass lasts from mid March to October with double maxima while the dry season lasts from November to early March inclusive. Nimet [5]. The surrounding areas are farmlands and vegetation zones. The rocks are highly porous and the soil is sandy, making it prone to erosion. The river is seasonal and therefore is frequently explored and drained of sand for building and construction. This also contributes to the expansion and increase in the gully erosion in the area. Chukwudelunzu [5].

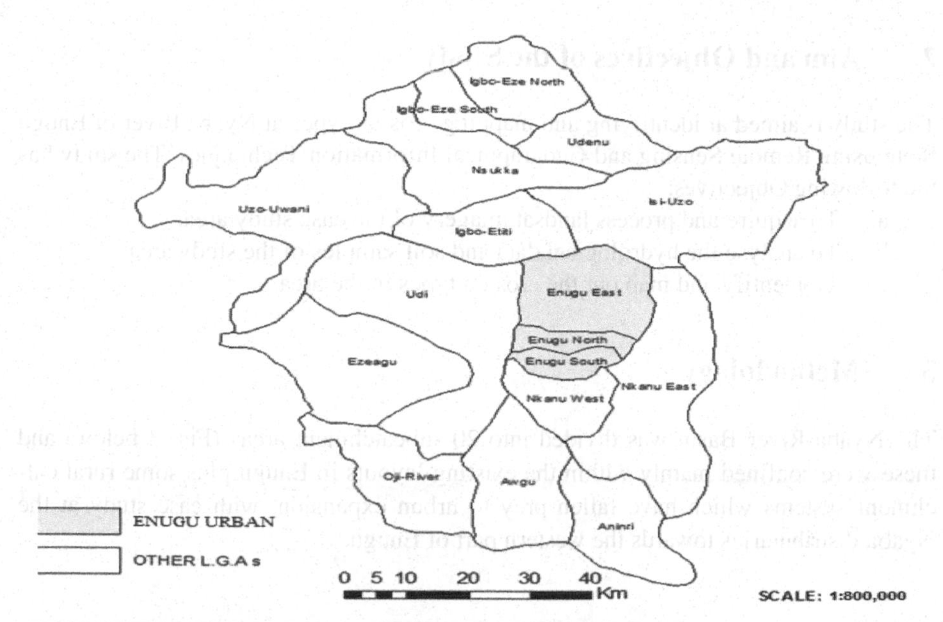

Fig. 1. Enugu State and the Location of Study Area, Enugu Urban
Source: Enugu State Ministry of Land and Surveying, Administrative Map of Enugu State, 2008.

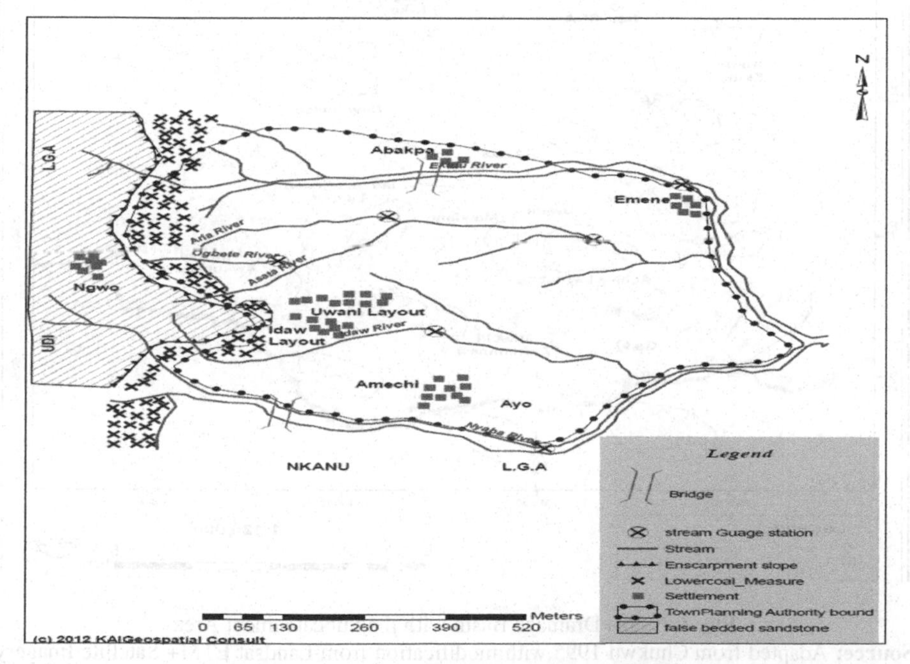

Fig. 2. Enugu Urban in Nyaba Drainage Basin
Source: Adapted from Chukwu 1995 with modification from Landsat ETM+ Satellite Imagery from U.S.G.S (United State Geological Survey) and Field Work 2011

2 Aim and Objectives of the Study

The study is aimed at identifying and mapping erosion types at Nyaba River of Enugu State using Remote Sensing and Geographical Information Techniques. The study has the following Objectives:

a. To acquire and process landsat imagery of the case study area.
b. To analyse the hydrological data and soil samples of the study area.
c. To identify and map out the erosion types in the area.

3 Methodology

The Nyaba River Basin was divided into 20 subcatchment areas (Fig. 3 below) and these were confined mainly within the existing layouts in Enugu plus some rural catchment systems which have fallen prey to urban expansion, with case study at the Nyaba distributaries towards the western part of Enugu.

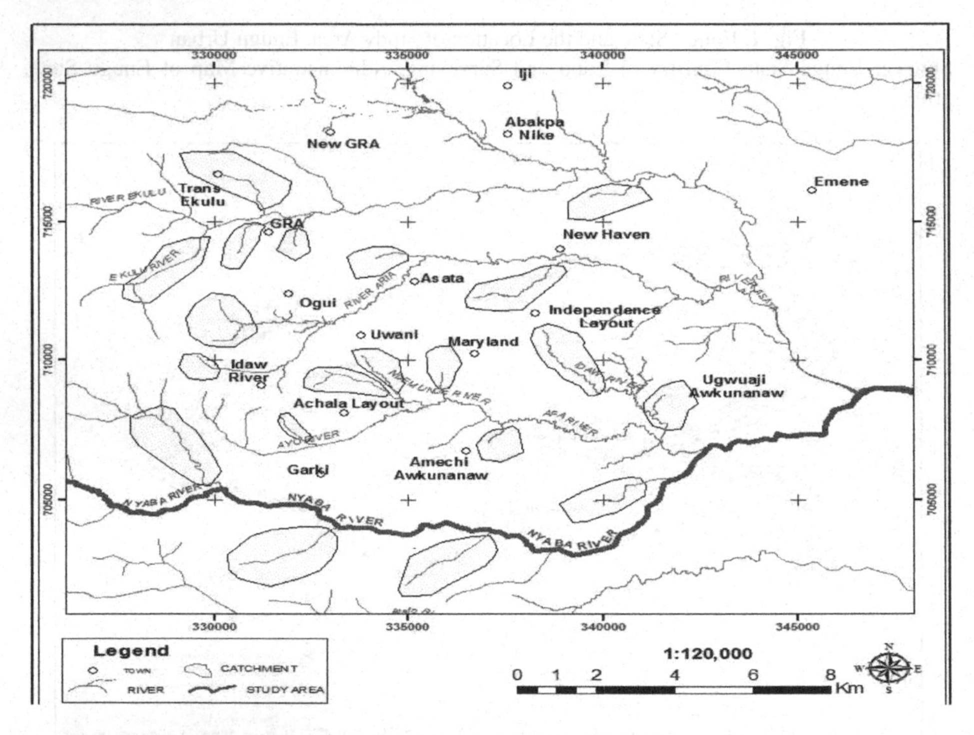

Fig. 3. Nyaba Drainage Basin with the Subcatchment Area
Source: Adapted from Chukwu 1995 with modification from Landsat ETM+ Satellite Imagery from U.S.G.S (United State Geological Survey) and Field Work 2011

3.1 Other Data Collected Were Processed Using Different Methods

3.1.1. Supervised classification using the Gaussian Maximum Likelihood technique was used to classify the LandSat image of 2011, five classes was defined (Built up area, Farm land, sand dune, Water body, Forest). GIS technology was used to produce and reclassified the land use / Landcover into two classes: other land and forest. Fig. 4 below shows the representative subcatchment and the land use /land cover type.

3.1.2. The Hydrological measurement was carried out in the divided 20 subcatchment areas within the Nyaba River Basin, some hydrological data on runoff characteristics were collected in the field by experimentation and measurement, such as stream flow discharge, drainage composition and river depth, top and bottom width of the river within the less developed area, this is because most other subcatchment in the urbanized area has been encroached by urban development. The Stream flow discharge was measured by velocity/area technique. This was based on the fact that discharge Q is directly proportioned to the product of the average stream velocity and the cross-sectional area of the channel A, at the point of measurement as represented by:

$$Q = AV \tag{1}$$

Where: Q = the stream discharge (m^3/sec)
A = the cross-sectional area of the channel (in m^3)
V = the average stream velocity (m/sec)

Each gauging site was carefully selected at such a reach where the stream bed was regular and stable without any aquatic plant to cause obstruction for a length of about 20-50m distance. Float was used to determine the mean velocity of each of the stream. The float was made up of small polythene bag filled with water. The float was dropped at one end of the channel reach to be measured. A stop watch was used to find how long it took to arrive at the other end of the chosen stream reach. The mean velocity of the stream was obtained from

$$V = s/t \tag{2}$$

Where: V = the mean stream velocity
s = the distance of the stretch travelled
t = the time taken by the float to make the travel.

Three test runs were made to obtain reliable results in each case. The cross-sectional area of the channel at the measuring site was estimated roughly by multiplying the measured channel width by depth.

The soil sample analysis was carried out in Enugu State Ministry of Transport soil laboratory to find out the particle size and the percentage of clay, silt and sand found in the catchment area. The particle size of the soil was compared with the Standard Textural Triangle (Fig. 4 below) to find the soil type obtainable in each subcatchment area.

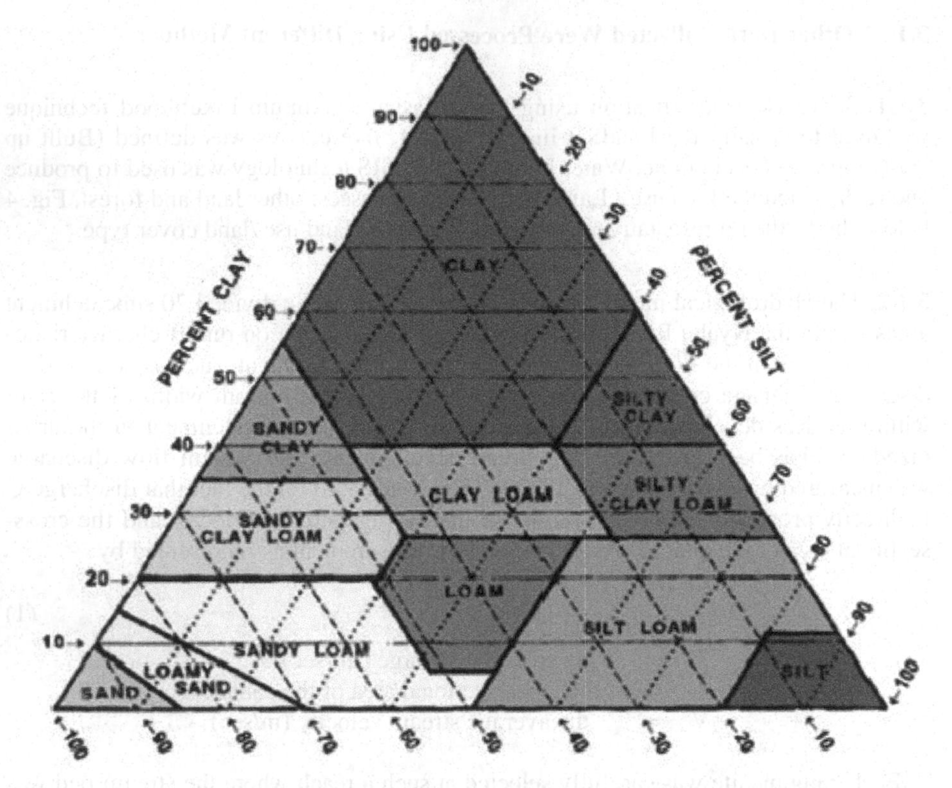

Fig. 4. Soil Textural Triangle- Based on the Triangle, a Loamy Soil has 40% Sand, 20% Clay and 4% Silt. A Sandy Loam has 60% Sand, 10% and 30% Silt
Source: David Whiting, 2011 U.D.S.A

The porosity of the soil in most basins is commonly determined by finding the ratio of the volume of the interstices from soil samples at different depths to the total volume as represented by Todd [3] in a simple relation

$$a = \frac{vi}{v} \tag{3}$$

Where a is the porosity, vi is the volume of interstices and v is the total volume. However, there are objections to the effectiveness of utilizing this formula because it does not account for the relative masses of the solid mineral. Therefore, for the present work, the following equation which gives allowance for both the relative masses and volumes of the solid materials as reviewed by Todd [3] was adopted:

$$a = \frac{P_m - P_d}{P_m} \text{ or } \frac{1-P_d}{P_m} \tag{4}$$

Where P_m is the density of mineral particles (grain density), P_d is the bulk density and a is as explained in equation 1.3. The percentage porosity in the study area was determined by multiplying the right hand side of equation of 1.4 by 100 viz:

$$P_P = [(P_m - P_d)\, P_m^{\ -1}] * 100 \tag{5}$$

Where Pp is the percentage porosity, Pm and Pd are as defined in (1.4) and are utilized in finding the percentage porosities of the various soil types at two different depths in each of the twenty subcatchment of Nyaba Draingae base.

3.2 GIS Raster Calculation

The Drainage feature of the study area was extracted from landsat imagery using polyline shapefile in Arcgis 10 and the 20 catchments area were demarcated using topographic sheet 301NE, Udi and sheet 302NW Nkalagu sourced from Survey Division ministry of Land and Transport. All the relevant information generated about the catchment area (Hydrological result, Soil type, and Land use type) was attached to the catchment attribute in GIS environment. The Hydrological map was generated by conversion of the catchment area into raster using hydrological value, soil map was generated by conversion of catchment area into raster using soil type information and also the land use land cover map was generated from Landsat ETM 2011. The Raster calculation in Arcgis 10 was used to define the criteria to generate the erosion type map. A multi-criterion was used to define the type of erosion based on the drainage size and land use of the area. The GIS logical expression (OR / AND) and arithmetic expression (=) were used in raster overlay operation,

- Criteria for area with no erosion: select true if rational method value is high, medium or low, and landuse is forest, otherwise choose false.
  (Rational method = High **OR** rational method = High **OR** rational method = High **AND** landuse = Forest)
- Criteria for area with sheet erosion: select true if rational method value is low, and landuse is other land. (rational method = low **AND** landuse = other land)
- Criteria for area with rill erosion: select true if rational method value is medium and landuse is other land. (Rational method = medium **AND** landuse = other land)
- Criteria for area with gully erosion: select true if rational method is high and landuse is other land. (rational method = high **AND** landuse = other land)

4 Results and Discussion

4.1 Land Use Land Cover

Land use pattern is the factor that contributes more to soil erosion and land degradation in the study area (Fig. 5 below), the land use classes generated from image classification were reclassified as other land while the forest cover control the erosion, thereby preventing soil erosion which leads to land degradation.

Findings further showed that most of the subcatchment has been engulfed in other land which means urbanisation has overtaken most of the river channel, especially those in more urbanized center, subcatchment with serial number 16, 11,5,16 and 10.

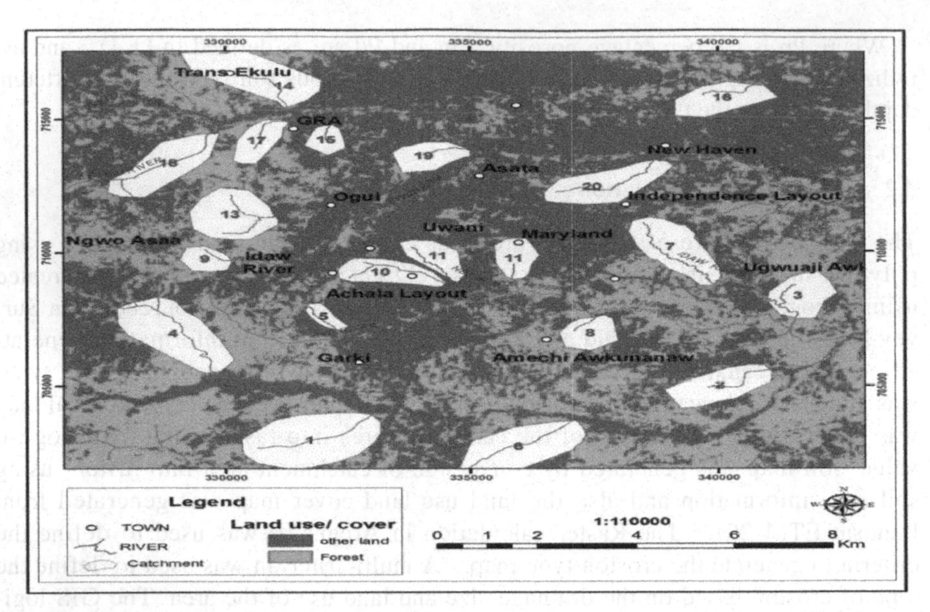

Fig. 5. Showing land use land cover 2011 of Nyaba Catchment area, in Enugu, Nigeria

4.2 Soil Types

The results of soil sample analysed using the USD Soil textural triangle (Fig. 6) and input into the GIS data base and ranked, shows that the catchment has mainly sandy clay, sandy loam, sandy clay loam, clay and clay soil types, the findings were converted into a raster format so as to digitise each catchment unit to find each catchment unit specific soil type as shown in Fig. 6 below.

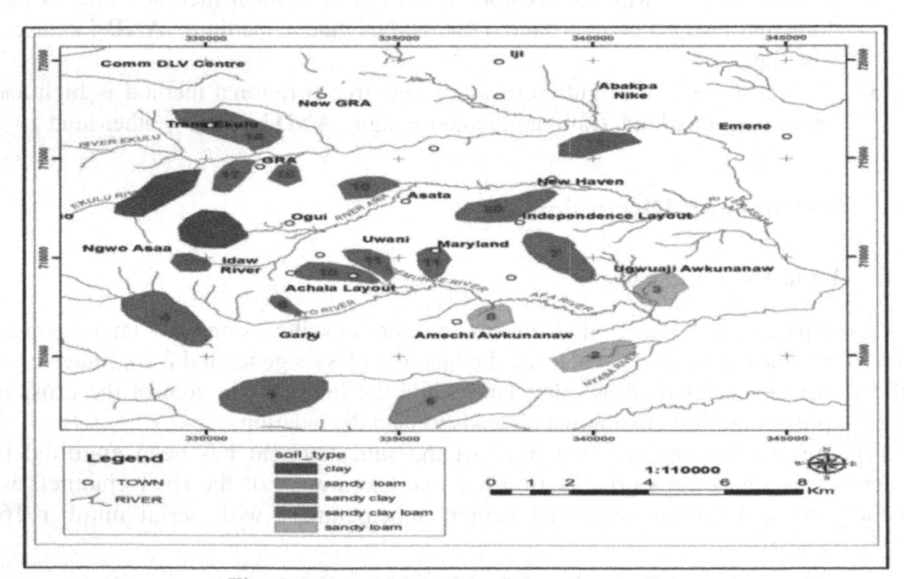

Fig. 6. Soil type Map of the Subcatchment Unit

4.3 Hydrology Data

In the 20 representative subcatchment units in Nyaba drainage basin, sixteen variable collected in the field during hydrological measurement were subjected to Principal Component analysis which brought out five factors; the overland flow, rainstorm duration, spatial variation, intensity and discharge measured by Rational Method was used in the calculation and the findings was digitized. Remote sensing technology was used to classified the LandSat image of 2011, five classes was defined (Built up area, Farm land, sand deposit, Water body, Forest), GIS technology was used to produce the rational map for the parameters Fig. 7.

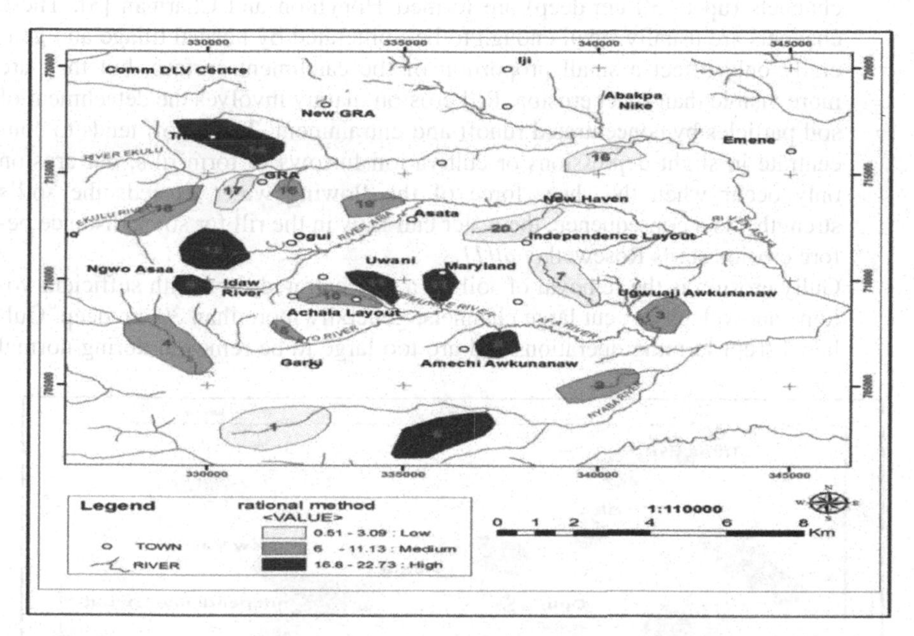

Fig. 7. Subcatchment Unit and the Values from Hydrology Data (Rational Method)

Findings showed that hydrology data are more effective at catchment units at Nyaba River at Amagu, Idaw river at Amechi, Idaw at Idaw River Layout, Idaw River at Amabia, Ogbete river at Akwata Bridge Mgbemena and Ekulu river at Trans-Ekulu Flyover, most of this subcatchment are disturbed by increase human activities of sand and stone minning in the river channel, agricultural activities at river corridors and other domestic landuse activities, which increase the sizes of the river channel ,thus increase runoff and erosion problems.

4.4 Erosion Types

The erosion type in Enugu urban was ranked into three classes from lower to higher based on soil type, the hydrological aspect, and land use/Landcover of the area. Findings shows that it confines with the general erosion view of: the erosion with lower intensity is the Rill erosion, the medium erosion type is the sheet erosion while the most advanced erosion type is the gully erosion.

The GIS Ranking analysis in this study is the process of ordering some phenomenon base on the factor that affects it; the result shows the grade of erosion from no to the higher erosion with red color intensity. (Fig. 8 below); the rank criteria are as follows:

i. Area with no erosion effect: These are the area of the study that is not affected by erosion.
ii. Sheet erosion (or inter-rill erosion): This is the removal of a fairly uniform layer of soil over a relatively wide area, with no perceptible channels where flow has been concentrated. Houghton and Charman [3].
iii. Rill erosion: This is the removal of soil by runoff, whereby numerous small channels (up to 30 cm deep) are formed Houghton and Charman [3]. These channels are usually small enough to be obliterated by normal tillage and generally only affect a small proportion of the catchment surface, but they are more visible than sheet erosion. Rill erosion mainly involves the detachment of soil particles by concentrated runoff and entrainment. The runoff tends to concentrate in slight depressions or cultivation furrows to form rills, but erosion only occur when the sheer force of the flowing water exceeds the soil's strength. As a consequence, the water can flow in the rill for some distance before erosion starts Rosewell *et al[1]*.

. Gully erosion is the removal of soil by a concentrated flow with sufficient volume and velocity to cut large channels, generally more than 30 cm deep. Gullies disrupt farming operations and are too large to be removed during normal

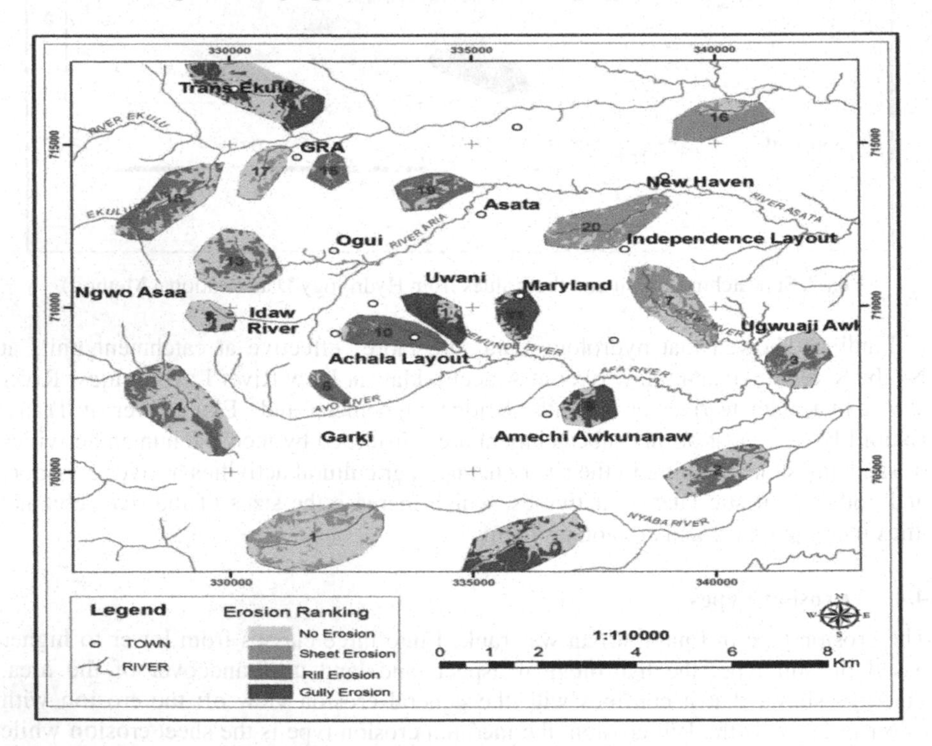

Fig. 8. Erosion Ranking for the Subcatchment Unit

cultivation. Houghton and Charman [3]. The hydrology of a catchment is altered when the land is cleared for agriculture. Gullies tend to form in natural watercourses as they adjust to the increased runoff normally generated after clearing. Soils with dispersible subsoils are particularly susceptible to gully erosion if the subsoil is exposed. Areas with gully were noticed at subcatchments with serial numbers 6,8,and11

5 Conclusion and Recommendations

In Conclusion, the results of soil erosion types identified and mapped out within the period of study clearly shows that there is a rapid increase in development of various soil erosion types (rill, sheet and gully); the gully erosion type is aided by the establishment of irregular quarries by individuals who go to the river channel to explore the sands and materials for sale. This weakens the land area and aids erosion features to develop slopes leading to increase in gully. The sandy nature of the soil and low vegetation cover around the area also contributes and makes it more prone to erosion. The gully trend relates the incessant increase of the gully erosion with the area encumbered so as to serve as a warning for possible future damage. This paper therefore serve as a guide to environmental and water resources managers involved in the mitigation of the impact of erosion to know the areas of various erosion types in the area for urgent intervention. The maps provide a strong basis for urgent establishment of erosion control measures targeted at mitigating the impact of erosion types in the Enugu Urban of Enugu State, South Eastern Nigeria

Based on findings the following are recommended for proper management and control of the erosion types at Nyaba River Catchment area of Enugu State, Southeast Nigeria: Consistent monitoring of the area to detect early stages and signs of the soil erosion development; Maintaining an adequate pasture cover by regulating grazing around the area; Restrict people who mine sands for sale; Construct waterways to stabilize the flow of the river; This appropriates specification to maintain the gullies; Avoid cultivation around those areas so as to eliminate the physical weathering caused by tillage since the soil is porous.

References

1. Akpokodje, E.G., Tse, A.C., Ekeocha, N.: Gully erosion Geohazards in Southeastern Nigeria and management Implications. Scientia Africana 9(1), 20–36 TS03B (2010)
2. Agu, A.N.: Soil Erosion Problems and Solution Strategies for Rural People. The Planet and Man 1(2), 12–19 (2005)
3. Brown, L.R., Wolf, E.C.: Soil Erosion "Quiet Crisis in the World Economy". World Watch Paper, World Watch Inst., Washington D.C., p. 49 (1984)
4. Egboka, B.C.E., Okpoko, E.I.: Gully erosion in the Agulu-Nanka region of Anambra State, Nigeria. In: Proceedings of the Harare Symposium, pp. 335–347. IAHS Publication, 144 (1984)
5. Ezezika, C.O., Adetona, O.: Resolving the gully erosion problem in Southeastern Nigeria: Innovation through public awareness and community-based approaches. Journal of Soil Science and Environmental Management 2(10), 286–291 (2011),

http://www.academicjournals.org/JSSEM ISSN 2141-2391 ©2011 Academic Journals
6. Enemuoh, C.O., Ojinnaka, O.C.: Application of Remote Sensing and GIS in Mapping and Monitoring of Gully Erosion in Anambra State, A Case Study of Nanka-oko Gully Sites. Paper presented at the 56th Annual general meeting of the 77 Nigerian Institution of Surveyors, NIS (2011)
7. Obiadi, I.I., Nwosu, C.M., Ajaegwu, N.E., Anakwuba, E.K., Onuigbo, N.E., Akpunonu, E.O., Ezim, O.E.: Gully Erosion in Anambra State, South East Nigeria:Issues and Solution. International Journal of Environmental Sciences 2(2) (2011) Integrated Publishing Association Research article ISSN 0976 – 4402 (2011)
8. Okagbue, C.O., Uma, K.O.: Performance of Gully Erosion Control Measures in Southeastern Nigeria. Forest Hydrol. Watershed Manage 167, 163–172 (1987)
9. Igbokwe, J.I., Akinyede, J.O., Dang, B., Alaga, T., Ono, M.N., Nnodu, V.C., Anike, L.O.: Mapping and Monitoring of the Impact of Gully Erosion in South-Eastern Nigeria with Satellite Remote Sensing and Geographic Information System. The International Archives of the Photogrammetry. Remote Sensing and Spatial Information Sciences XXXVII Part B8 (2008)
10. Burrough, P.A.: Principles of Geographical Information System for Land Resource Assessment, pp. 35–46. Oxford University Press, New York (1996)
11. Todd, D.K.: Groundwater Hydrology. John Wiley & Sons, New York (1980)
12. Houghton, P.D., Charrman, P.E.V.: Glossary of Terms Used in Soil Conservation. Department of Infrastructure Planning and Natural Resources, Sydney (1986)
13. Ofomata, G.E.K.: Soil Erosion in Nigeria; the Views of A Geomorphologic Inaugural Lecture Series No 7 University of Nigeria Nsukka, pp. 7–10 (1985)
14. Rosewell, C.J., Crouch, R.J., Morse, R.J., Leys, J.F., Hicks, R.W., Stanley, R.J.: Forms of Erosion. In: Charman, P.E.V., Murphy, B.W. (eds.) Soils - Their Properties and Management, South Melbourne. Oxford University Press (1991)
15. Chukwu, K.E.: Effects of Enugu Urban Environment on the Water Quality of Streams in Nyaba Catchment area of Southeastern Nigeria. Unpublished PhD. Project; University of Nigeria, Nsukka (August 2010)
16. Chukwudelunzu, V.U.: Effect of Soil Erosion on Ajalli Water Intake. Case Study of Ajali Water Pumping Station in Enugu State. Unpublished M.Sc Project, Enugu State University of Science and Technology ESUT, pp. 47–49 (2004)
17. UNESCO. The impact of global change on erosion and sediment transport by rivers: current progress and future challenges. Scientific Paper. The United Nations World Water Assessment Programme (2009), http://unesdoc.unesco.org/images/0018/001850/185078e.pdf
18. David, W.: UDSA Soil Textural Triangle (2011), http://nesoil.com/properties/texture/sld005.htm
19. Nigerian Meteorological Agency; Seasonal Rainfall Prediction 2011. An Annual Climate Review Bulletin. N0.005, pp. 4–5 (January 2012)

Emergy Analysis of Agricultural Eco-economic—A Case Study of Haidong Region of Qinghai Province

Na Wang[1], Kelong Chen[1,2,*], Jianguo Zhang[1], and Baoliang Lu[3]

[1] College of Life and Geography Science, Qinghai Normal University, Xining, 810008, China
[2] Qinghai Province Key Laboratory of Physical Geography And Environmental Process, Qinghai Normal University, Xining 810008, Tel.: + 86 18997295071, China
ck17813@163.com
[3] Key Laboratory of Salt Lake Resources and Chemistry, Qinghai Institute of Salt Lake, Chinese Academy of Sciences, Xining, 810008, China

Abstract. Emergy analysis is founded on the basis of energy system analysis by the famous American ecologist Odum. H.T, is one of the hot issues of ecological research in economics. This paper describes briefly the basic theory of the energy value, and takes agricultural ecological economic system of the eastern region of Qinghai province as an example, using the net energy output, Per capita energy quantity, important emergy indicators to analyze the problems of agricultural ecosystems in the region, in order to provide a scientific basis for agriculture(de) the sustainable and healthy development of Haidong Region.

Keywords: Emergy theory, Ecological Economic System, Sustainable Development.

1 Introduction

For the study of ecosystem and human society economy system, to analyze the real value of the resources and environment with economic activity and the relationship between them, in the 1980s, famous American ecologist Odum. H.T put forward this theory "value", and define it as: the sum of enery which the formation of the product or service consumed directly or indirectly(Odum, 1988). The theory is based on: ecological economic system is a self-organization, all of the energy in any form is from the solar, with the energy flowing in the system, part dissipated (entropy), form a higher level and new energy. Through retrospective study, any form of energy can be used to represent a form of energy(Daniele, 1998). In the process of energy conversion, odum put forward the concept of energy conversion, the energy of different types of material is different, different energy have different energy levels and quality, through the energy conversion rate can convert the energy of different types for the same energy (such as solar energy), on the basis of the input-output analysis of the study area were full scientific. The popularity of energy method in our country began in the 1990s, introduced by the famous scholar Lan Fang Sheng (Lan et al., 2001; Lan et al., 2002),

* Corresponding author.

F. Bian and Y. Xie (Eds.): GRMSE 2014, CCIS 482, pp. 593–601, 2015.

and in the country carried out on the energy value analytical and theoretical research about the region, agriculture and so on(Bai et al., 2006; Du et al., 2008;Yao et al., 2009;Fu et al., 2009). However, relatively few studies on the Loess Plateau region(Dong et al., 2004;Zhang et al., 2005;Zhang et al., 2010)and the Qinghai-Tibet Plateau region(Wei et al., 2009)as well as its agricultural ecosystems transition emergy (Zhang et al., 2011).

2 The General Situation in the Study Area

HaiDong Region is located between Qinghai province and Gansu province,and named after the east of the Qinghai Lake.Region covers: Ping'an County,Le du County, Huzhu Tu Autonomous County, Minhe Hui and Tu Autonomous County, Hua Long Hui Autonomous County, Xunhua salar autonomous county,with a total area of 1.32 square kilometers. Located in the loess plateau and Qinghai-Tibet plateau transition zone, the altitude is between 1650-4636 meters, average annual precipitation is 368 mm, annual evaporation is 1581 mm, the annual average temperature is 6.2°C. The agricultural population of the region is134 million, accounted for 84.8% of the total population and 40% of the province's agricultural population; The total land area of the region is 323.72 acres , good weather conditions, fertile land,lighter pollution, adequate light,annual sunshine hours for 2200-2800,solar radiation intensity of 580-740 MJ / square kilometers, large temperature difference between day and night. Here is the best prospects for development of Qinghai province area, advantageous geographical advantage makes it easy to accept the capital and technology from other areas of radiation, preferably natural conditions, make the district become good agricultural economic zone of farming and is the important food, vegetables and meat production base of Qinghai province. However, the development of economy, the growing of the population, the prevalence of industrial civilization in the agricultural ecological system of the district population carrying capacity, environmental carrying capacity faces a series of challenges.

This paper, based on the energy theory method, draw on the previous scholars make a in-depth comparative analysis on agricultural ecological economic system, the structure and the function of energy input and output of Qinghai region, put forward corresponding strategies to explain the movement characteristics of the agricultural ecological system, and find the key factors affecting the development of agriculture in the region. For the rational use of agricultural resources, optimizing the agricultural ecological system in this area of investment structure, and provide scientific basis for realizing the sustainable development of agriculture Haidong region in Qinghai.

3 Energy Analysis Method and Procedure

This study as following steps:

1).We gathered Environment data,geography data, a variety of socio-economic information and data about the research objects from.The date of Haidong region agricultural

ecosystem emergy analysis is mainly from the year of 2008 and 2010 Haidong Prefecture of Qinghai Statistical Yearbook,the elements and produce energy conversion coefficient mainly refer to the Agricultural Economic Technical Manua ,sorted it out, kept and processed on computer.
2).On the basis of the data of the study of Luo Shiming,ect(Luo,S.M.,2001),and the conversion rate of solar energy of the research data of Lan Sheng Fang(Lan et al., 2001; Lan et al., 2002),corresponding energy value of the conversion rate of resources, and different units of measure (J) ecological flow or economic value of the stream we converted into energy units.
3).Enumerated the main items of energy input and output,using the Excel draw the energy output and input form, we evaluated it is contribution in agriculture system.
4).According to emergy analysis table, we had been established and calculated the value of the index system about a series of ecological and economic efficiency,such as, Net Energy Yield, Percapita Energy Consumption etc.
5).Through energy indexes comparative and analysis, provide the scientific basis for the development of proper workable system management measures and economic development strategies, to guide Ecological and economic systems with a virtuous cycle and sustainable development.

4 Results and Analysis

4.1 The Emergy Input Analysis

According to all aspects dates of the agriculture ecological in 2008,2010 of Haidong region, by means of the energy conversion coefficient calculated the energy input -- output stream(Odum, 1996; Zhang et al., 2004), according to the Odum value of the conversion rate converted the energy flow data to energy data(Lan, 2002), to calculate the agriculture input energy of region in 2008, 2010(Table 1).

Table 1. 2008, 2010 Emergy agricultural eco-economic system into Haidong Region

Project	Input type	Conversion rate of solar energy (Sej·J^{-1},Sej·g$^{-1)}$	Solar energy	
			Year of 2008	Year of 2010
Renewable environmental resources	Solar radiation energy	1	1.25E+14	1.25E+14
	The wind	1.50×10^{3}	6.57E+16	6.57E+16
	The rain chemical energy	1.80×10^{4}	5.92E+17	7.89E+17

Table 1. (*Continued*)

Non-renewable environmental resources	Topsoil loss	7.04×10^4	4.71E+16	1.28E+17
Non-renewable industrial support energy	Diesel oil	5.40×10^4	5.71E+19	5.80E+19
	Nitrogen fertilizer	3.80×10^9	3.97E+20	1.85E+20
	Phosphate fertilizer	3.90×10^9	2.48E+20	6.90E+19
	Compound fertilizer	2.80×10^9	2.14E+20	8.66E+19
	Potash fertilizer	1.10×10^9	5.31E+20	1.82E+19
	Agricultural plastic film	3.80×10^8	1.04E+22	4.12E+22
	Pesticide	1.60×10^9	4.67E+22	5.04E+22
Renewable organic energy	labor	3.80×10^5	3.06E+17	3.13E+17
	Animal power	1.46×10^5	2.61E+17	2.19E+17
	Seed	6.60×10^4	3.59E+19	3.70E+19
Energy input			5.82E+22	9.23E+22

The enery input of agricultural ecological economic system in Haidong region mainly consist of Renewable environmental resources, Non-renewable environmental resources, Non-renewable industrial support energy, Renewable organic energy. Table 1 shows : Compared 2008 with 2010, energy input is increased from 5.82×10^{22}sej to 9.24×10^{22}sej. The largest value increase is renewable organic energy,the second is non-renewable industrial support energy. The largest growth rate is agricultural plastic film,from 18% to 45%, wide application of plastic sheeting, promotes the development of the greenhouse agriculture, and promotes agricultural structure changing in Haidong region,while it generates economic benefits but also inevitable produces white pollution, water pollution, soil pollution(Fu et al, 2009), makes the environment of Haidong Prefecture deteriorate(Shen et al, 2005). From 2008 to 2010 the amount of applied fertilizer decreased from 15.7% to 3.8%, affected by the natural environment, the rate of Surface soil loss is very high and the Soil fertility is weak, so it needs a lot of nitrogen fertilizer and organic fertilizer to supply the fertility. But the extensive use of fertilizers, may affect the soil structure, causing soil compaction(Shen et al, 2009).So,from 2008 to 2010 , reducing the amount of applied fertilizer means that Haidong area pays more attention on the protection of ecological environment in the process of developing of Agriculture.

Compared with 2008, the investment in agriculture of Haidong Prefecture is increased in 2010. It explains that the government of Haidong Prefecture attaches importance to the importance of agriculture.

Table 2. 2008,2010agricultural eco-economic system output in Haidong Region

Project	Output type	Conversion rate of solar energy (Sej·J^{-1}, Sej·g^{-1})	Solar energy Year of 2008	Solar energy Year of 2010
Planning enery	Wheat	6.80×10^4	1.62E+17	1.65E+17
	Corn	8.52×10^4	2.49E+16	1.49E+17
	Beans	6.90×10^5	4.29E+17	3.92E+17
	Potato	2.70×10^3	2.37E+16	2.48E+16
	Oil	6.90×10^5	2.31E+18	2.41E+18
	Vegetable	2.70×10^4	2.45E+17	2.84E+17
	Fruit	5.30×10^5	3.37E+19	4.02E+19
Forestry enery	Forest	3.49×10^4	1.53E+20	2.66E+20
Animal husbandry enery	daily	1.70×10^6	1.30E+21	1.64E+21
	Pork	1.70×10^6	1.23E+18	1.40E+18
	Mutton	2.00×10^6	3.45E+17	3.06E+18
	beef	4.00×10^6	6.60E+17	5.71E+17
	Other meat	3.17×10^6	1.75E+19	3.15E+19
	Poultry and Eggs	2.00×10^6	1.10E+17	1.21E+19
Fisheries enery	Aquatic products	2.00×10^6	1.47E+17	2.04E+17
Total energy output			1.51E+21	1.99E+21

4.2 The Emergy Output Analysis

Table 2 shows,The enery output of agricultural ecological economic system in Haidong region mainly consists of Planning enery, Forestry enery, Animal husbandry enery, Fisheries enery. Compared with 2008, the proportion of Animal husbandry in 2010 dropped to 84.4%, and the forest proportion increased to 13.4%, the proportion of farming and fishing has also increased. Among them, the proportion of Vegetable and Fruit is increased from 91.4% to 92.3%. It can be seen that,2008 and 2010, Haidong agricultural ecosystems output is still dominated by Animal husbandry, the proportion of Forestry, farming, fishing is small, but showed a rising trend. Studies have shown that China's labor-intensive agricultural products, such as beef and mutton and other animal products, fruit, vegetables and so on, has the obvious advantages of price competition in the international market(Xue et al, 2003). And these products are the dominant agricultural of Haidong region. Development of animal husbandry in eastern region, can promote the development of planting industry, the tertiary industry, and is a important way to promote agricultural internal structure rationalization and stretched industrial.The output of agricultural economic system in Haidong region increases due to the Haidong region government support and investment.In 2009, the government promoted the Plateau Lean piglets breeding,standardization of greenhouse and biological fermentation bed Breeding Application Effectiveness technology. Only the first half of 2009 the new barn insulation is 135,400 m^2, the Biological fermentation bed is more than 10000 square meters(Haidong region Statistics Bureau, 2009), Greatly improve breeding efficiency, promote the development of agricultural economy system in Haidong region.

4.3 Comparing the Main Enery Index of Haidong Prefecture with other Areas

Haidong region is located in the border region of Qinghai-Tibet plateau and the loess plateau, Ecological environment is fragile, Now research in the region is less. In order to more clearly analyze the development of agricultural eco-economic system in Haidong region,according to research fruits, The main indicators of the energy value of Haidong region comparison with other regions(table 3), and by using the dynamic indicators analysis process of agricultural development Haidong region impact on the natural environment and the dependence.

Table 3. The emergy index of Agricultural ecosystem of Haidong Region

Project	Index	Expression	Solar energy	
			Year of 2008	Year of 2010
Enery input	Renewable environmental resources	R	6.58E+17	8.55E+17
	Non-renewable environmental resources	N	4.71E+16	1.28E+17
	industrial auxiliary energy	F	5.82E+22	9.23E+22
	Renewable organic energy	R1	3.65E+19	3.75E+19
	Total investment in environmental resources	I=R+N	7.05E+17	9.83E+17
	The total auxiliary energy investment	U=F+R1	5.82E+22	9.24E+22
	The total investment in energy	T=I+U	5.82E+22	9.24E+22
Enery output	Planning	Y1	3.69E+19	4.36E+19
	Forestry	Y2	1.53E+20	2.66E+20
	Animal husbandry	Y3	1.32E+21	1.68E+21
	Fisheries	Y4	1.47E+17	2.04E+17
	The total energy output	Y=YI+Y2 +Y3+Y4	1.51E+21	1.99E+21
	Net energy yield		2.59%	2.15%
	Per capita energy consumption		1.53×10^{17} sej	2.37×10^{17} sej

Net Energy Yield. The net emergy yield ratio,one of the main indicators of agricultural economic system, is index of measuring the system output contribution on economic system. The formula is: Net energy yield Emergy output / total auxiliary energy, the higher net energy yield, the higher agricultural ecosystems born out of the energy value of the product, the higher productivity(Zhang et al, 2011).In the year of 2008, the Net energy yield of Haidong region is 2.59%, the Net energy economic system of agriculture in 2010 of Haidong region is 2.15%, is higher than Linzhou district(1.18%)(Sun et al, 2010). But the net emergy output is still low, which Indicates that the agricultural eco-economic system operation efficiency is not high, Input energy conversion rate,energy return, deep processing of products,products with added value are low, with weak competitiveness in the market(Xie et al, 2009). Haidong region developed agriculture, at the same time should pay attention to the deep processing of agricultural products to improve the added value of products, extend the industrial chain, agricultural and industrial combination, the formation of industrial management.

Percapita Energy Consumption. Per capita energy consumption values reflected the regional level of people's lives, the higher the value of the amount of per capita energy, which means that the higher standard of living in the region. The per capita energy consumption of agricultural economic system in Haidong region in the year of 2008 was 1.53×10^{17}sej,the year of 2010 was 2.37×10^{17} sej, a tread of increase in the two years.Compared With other regions (Xining amount of per capita energy value is 27.95×10^{15} sej, Urumqi value per capita energy consumption is 2.58×10^{15} sej), were higher than the other two regions, which fully showed that the current Haidong area farmer's life has been greatly improved during 2008 to 2010.

5 Conclusion

The theory of emergy analysis combines economics and ecology, unit social, economic, natural and 3 sub system of organic, analyzes the real value of nature and human society and economy, the criticism of traditional simply focus on the ecological and economic value, makes the ecological and economic value of integration. Balance will be used for agricultural ecosystem value theory, facing various problems in the process of agricultural development, making the development of economy and ecology integrated, to explore new ways of agricultural ecosystem and agriculture ecology by quantitative. Haidong region, as one of the important grain production bases in Qinghai region, the extent of the development of agriculture has a crucial significance to the whole city of Xining in Qinghai Province. By analyzing the Haidong agricultural ecosystems Emergy, drew the following conclusions:

The total investment of agricultural eco-economic system in Haidong area by 2008 of 5.82×10^{22} sej increased to by 2010 of 9.24×10^{22} sej, the structure of energy inputs is given priority to industrial auxiliary energy,but the role of the organic energy input in agricultural ecosystem is still great. Emergy output increased from 2008 of 1.51E+21 sej to 1.99E+21 sej in 2010.Among them, emergy input-output structure importance degree is the animal husbandry, forestry, farming, the proportion of fishery is very small. From 2008 to 2010 the emergy structure varied slightly , 2010 compared with 2008, the proportion of planting was increased. Further to the Haidong region Agricultural Ecosystcm emergy index analysis, showed that the production efficiency was improved in Haidong region Agricultural system, the technological content increasing, its per capita energy value was relatively high, that the farmers' living standard has been greatly improved and Haidong region level of development of the rural economy has improved significantly.

The results showed that, from 2008 to 2010, agriculture had great development, agricultural development mainly due to the benefit at agricultural policies of all levels of government: 2009, Qinghai released the development plan of "four areas with two tier", regarded Haidong Region as a "leading integrated economy of the province's economic and social development and the promotion of the development of the pilot area of the province", for the development of Haidong Region leap an important opportunity;2008 to 2010 government Haidong Region increased the investment in agricultural technology,built the modern agricultural science and Technology Park in Ledu,Huzhu district, elenium-rich agricultural demonstration park and people in Pingan district,and agriculture demonstration park in Hualong, Xunhua,and and the transformation of agricultural development as the core, the region's agricultural economic development to achieve increased quality and efficiency of agriculture and farmers continue to increase. But Haidong Region in the development of agriculture should intensify the environmental protection, in the energy value analysis system for agricultural development Haidong environmental investments account for a large proportion of the development of agriculture showed pay a great dependence on the environment. Therefore, Haidong Region in the development of the agricultural economy should strive to develop ecological agriculture, tourism agriculture, agricultural Haidong achieve healthy and harmonious development.

Acknowledgements. This work was supported by National Natural Science Foundation of China (41261020, 31260052), National Social Science Foundation of China (10CJY015), Ministry of Education Key project (2012178), Natural Science Foundation of Qinghai Province (2013-z902), Chinese Academy of Sciences, "Western Light" (Kefa taught the word [2012] No. 179) and Qinghai Normal University Science and Technology Innovation Program.

References

1. Odum, H.T.: Self—organization, Transfomity, and Information. Science 242, 1132–1139 (1988)
2. Daniele, E.C.: Emergy analysis of human carrying capacity and regional sustainability:an example using the state of Maine. Environmental Monitoring and Assessment 51, 531–569 (1998)
3. Lan, S.F., Qin, P.: Emergy Analysis of Ecosystem. Chinese Journal of Applied Ecology 12(1), 129–131 (2001) (in Chinese)
4. Lan, S.F., Qin, P.: Emergy Analysis of Ecosystem, pp. 1–4. Chemical Industry Press, Beijing (2002) (in Chinese)

5. Bai, Y., Lu, H.F., He, J.H., et al.: Based-on Emergy analysis for agricultural system of Guangdong province. Ecology and Environment 15(1), 103–108 (2006) (in Chinese)
6. Du, B.Y., Men, M.X., Xu, H., et al.: Comprehensive Evaluation of Environmental Resources and Farmland Ecosystems in Hebei Province based on Emergy Theory. Resources Science 30(8), 1236–1242 (2008) (in Chinese)
7. Yao, Z.F., Liu, X.T., Li, X.J., et al.: Analysis of agroecosystems in Jilin Province based on emergy theory. Chinese Journal of Ecology 28(10), 2076–2081 (2009) (in Chinese)
8. Fu, X., Wu, G., Shang, W.Y., et al.: Emergy analysis of agricultural eco-economic system in Chaoyang City. Liaoning Province 24(8), 902–906 (2009) (in Chinese)
9. Dong, X.B., Gao, W.S., Yan, M.C.: Emergy Analysis of Agroecosystem Productivity of Typical Valley in Loess Hilly—gully Region of the Loess Plateau: A Case Study in Zhifanggou Valley of Ansai County. Acta Geographica Sinica 59(2), 223–229 (2004) (in Chinese)
10. Zhang, X.: Emergy analysis of agricultural eco-economic systems in Jinghe river valley. Agricultural Research in the Arid Areas 23(5), 196–201 (2005) (in Chinese)
11. Zhang, F., Zhou, Z.X.: Dynamic assessment of agri-ecological system based on emergy analysis in Yan'an City. Agricultural Research in the Arid Areas 28(4), 251–257 (2010) (in Chinese)
12. Wei, F.Z., Yue, M.: Using emergy to analysis agro-ecosystem sustainability in the edge-regions of Qinghai-Tibet Plateau-A case study of Tibetan and Qiang Autonomous Perfecture of Aba,Sichuan Province. Chinese Journal of Eco-Agriculture 17(3), 580–587 (2009) (in Chinese)
13. Zhang, X.P., He, W., Fang, T.: Emergy analysis of agricultural eco-economic system in Huangshui Valley: a case of Xining City. Arid Land Geography 34(2), 344–354 (2011) (in Chinese)
14. Luo, S.M.: Agricultural ecology. China Agriculture Press (2001) (in Chinese)
15. Odum, H.T.: Environmental Accounting Emergy and Environmental Decision Making, pp. 320–370. John Wiley & Sons, New York (1996)
16. Zhang, Y.H.: Emergy analysis method of agro-ecosystem. Chinese Journal of Eco-Agriculture 12(3), 181–183 (2004) (in Chinese)
17. Fu, X., Wu, G., Shang, W.Y., et al.: Emergy analysis of agricultural eco-economic system in Chaoyang City. Liaoning Province 24(8), 902–906 (2009) (in Chinese)
18. Shen, G.X., Yang, J.J., Huang, S.F., et al.: Water pollution load of saliferous soil washed by water in plastic greenhouse. Transactions of the CSAE 21(1), 124–127 (2005) (in Chinese)
19. Xue, L.: The Regional Arrangement of the Advantageous Agricultural Product & the Agricultural International Competitiveness. Issues in Agricultural Economy (1), 34–36 (2003) (in Chinese)
20. Haidong region Statistics Bureau. The development ideas of livestock in Haidong region. Qinghai Statistics 8, 17–19 (2010) (in Chinese)
21. Sun, H.D.: Energy Analysis of Agricultural Ecosystem Eco-system in Lintao County. Gansu Agricultural Science and Technology 10, 20–22 (2010) (in Chinese)
22. Xie, Z.Y., Liu, J.B., Zhang, X.M.: Impact on ecological function caused by surface changes of Lake Ebinur. Arid Land Geography 32(2), 226–233 (2009) (in Chinese)

The Application of Spatial Information Technology Based on VR for the River Shoreline Resource Management

Longhua Gao, Weijian Li, and Long Xie

Department of Hydraulic Engineering, Pearl River Hydraulic Research Institute, Guangzhou 510642, China
liweijianshmily@126.com

Abstract. River shoreline is a precious but finite resource. With the continuous development of the social economy, all levels of local governments along the river raise the higher requirements for the utilization of shoreline. The exploitation, utilization and protection of the river shoreline resources play a significant role in maintaining the stability of the river ecosystem and promoting the sustainable development of society and economy. The spatial information technology based on VR (virtual reality) is gradually becoming an important technical support to the management of the river shoreline resource. This paper mainly introduces the characteristics and advantages of spatial information technology based on VR as well as its applications in the river shoreline resource management. In combination of the application examples of the main stream of Liujiang river (urban river), the paper discusses the role of spatial information technology based on VR in the management of river shoreline resource.

Keywords: Virtual Reality, River Shoreline, Spatial Information.

1 Introduction

River shoreline refers to a strip-shaped area with the comprehensive utilization and exploitation characteristics on both sides of the river water and land border line in a certain range as well as the natural attributes like flood discharge and water flow regulation. As well, with its resource attributes on the resource exploitation and utilization value, river shoreline becomes a kind of the finite precious resources. With the continuous development of social economy in recent years, the over-exploitation of the shoreline is increasingly apparent. How to rationally use and manage the river shoreline resources has become a major problem for the river administration departments to make a contribution to the social and economic development, protect the ecological environment of the river bank and promote the sustainable use of the resource.

Liujiang river is located in the Guangxi Zhuang Autonomous Region of which the river is a main tributary of the left bank of Xijiang River of Pearl River System and an important water traffic arteries in Guizhou and Guangxi. Its total length is 726 km and the basin area covers 57 thousand square kilometers. The river flows through the

F. Bian and Y. Xie (Eds.): GRMSE 2014, CCIS 482, pp. 602–611, 2015.

important industrial city in Guangxi, Liuzhou City. There is a certain difficulty for the daily management of the water supply administrative department on account of the long shoreline of Liujiang river and its complicated shoreline resources.

VR technology is a new computer technology which is widely used with the aid of computer hardware, software and virtual world integration technology. The vivid virtual world in the feeling form of vision, hearing, touch and others is generated by the computer. With the continuous advancement of water information in recent years, the management tools are gradually converted from the traditional introduction of electronic documents, pictures, vector map and other expressions on the water conservancy facilities to "digital watershed" and "digital river" and other water conservancy information at a higher level in the combination of the spatial information based on VR technology. In a manner of speaking, spatial information based on VR technology is constantly promoting the work efficiency and application level of the river administration departments.

1.1 The Introduction of Spatial Information Technology Based on VR

VR technology is proposed by Jaron Lanier, the founder of VPL Research (America), implying to generate a 3-D interactive environment in the computer. The users get into the environment by means of “role”. VR technology is a very active research field in recent years. A series of the advanced technology are included such as computer graphics, multimedia technology, artificial intelligence, human-computer interaction technology, sensor technology, etc. VR technology is widely applied in the fields of tourism management, resource management, city planning, military simulation and so on.

Spatial information technology based on VR refers to the vivid 3-D natural environment with the attribute of spatial information constructed by the virtual reality technology as well as a new technology upon the combination of the virtual reality technology and spatial information technology, providing the multidimensional spatial information management. The technology achieves the virtualization of the real environment in a computer system. As an important development direction of spatial information technology, VR allows users to enter a 3-D virtual world and realize the interactive operation in the virtual world.

1.2 The Characteristics of Spatial Information Technology Based on VR

As a new technology, spatial information technology based on VR integrates 3-D modeling technology, remote sensing, geographic information system technology, real-time rendering technology, database technology, software engineering and other technology into one.The basic principles of this technology include: the basic topographic data in the range of river shoreline is obtained through the measurement and remote sensing technology, establishing the three-dimensional terrain with the texture information of remote sensing satellite; the information on the shoreline resources is positioned by two-dimensional information data so as to build three-dimensional model and add the real texture properties in the model; the established 3-D data will be imported into the virtual reality simulation platform so that it is able to implement various operations on the scene. Its major characteristics are stated as follows:

Firstly, the true 3-D stereoscopic visual effect. 3-D scenario based on VR is the simulation of the real environment in the computer system in bringing the users with the experience of "immersion" feeling. With 3-D form, the model shows the users with the spatial phenomenon, expressing the planar and vertical relationships between the spatial objects. The immersion atmosphere allows the users to view the location, appearance, texture and other attributes in the simulation model as if they were in the real objective world.

Secondly, good interactivity. In the computer environment based on VR, the users are able to easily achieve the scene roaming, the fixed-point observation, quick positioning, the examination of the attributes in the model and other operations through the external device such as mouse, keyboard, joystick, etc.

Thirdly, the characteristics of spatial information.3-D data is based on the data of two-dimensional space. The generated 3-D data inherits the characteristics of the two-dimensional space attribute and manifests the attribute information in the vertical direction such as elevation information, building height information, etc. The traditional two-dimensional data with the non-intuitive and single performance characteristics is extended.

1.3 The Existing Problems of Current Management of River Shoreline

The management of river shoreline resources is related to a series of issues including flood control safety, ecological environment protection and sustainable utilization of water resources, etc. With the rapid development of our social economy, the exploitation, utilization and governance of the rivers is still in the ascendant. At the same time, the exploitation and utilization of river shoreline resources is also gradually increasing. On the one hand, because of road network, transportation and other needs for the cities along the river, it is necessary to construct all kinds of water engineering projects, such as bridges, docks, etc. In a certain extent, the projects change the attributes of river shoreline resource. On the other hand, some illegal and unreasonable use of shoreline resources is also objectively existing, for example, the illegal river sand excavation, illegal change of the current river shoreline, etc. These illegal acts have seriously damaged the shoreline resources in leading to the deterioration of river ecosystem.

At the present stage, the laws and regulations for the river shoreline exploitation and utilization management are imperfect; the consciousness of the public is weak on the watercourse protection; the illegal cases have occasionally occurred in relation of river shoreline. Based on the above reasons, the daily management works of river administration departments need to be more comprehensive and accurate to understand the shoreline resources while the efficiency and management level of the shoreline resource management shall be improved. The introduction of spatial information technology based on VR is thereof proposed in terms of the river shoreline resource management.

2 The Application of Spatial Information Technology Based on VR for the River Shoreline Management

The application of spatial information technology based on VR for the river shoreline resource management mainly refers to the advantages of spatial information technology on the geographic entity's location, shapes, relations between entities, regional spatial structures and others, presenting the river shoreline resources, realizing the digitization and virtualization of the shoreline resource management and improving the management level. To implement 3-D information management of the river shoreline resource information involves a series of 3-D visualization issues including 3-D data construction, the performance of the shoreline resources, the performance of spatial information of the shoreline, the query and analysis of the attribute information of the shoreline resources, etc.

2.1 The Construction of a Wide Range of Scenario

It needs to collect the terrain elevation data and remote sensing image data in the control research range of both sides of the river banks, use GIS data processing software to register and optimize the original data, etc. The collected data shall be converted into a unified projection coordinate system. Through the automatic biopsy, format conversion and other operations on the data, the high-precision grid 3-D is constructed.

The establishment of the scenario terrain model data cost hundreds of thousands facets even millions of facets to realize the terrain structures in undulation. With the constraints of the current computer hardware technology, there appears a certain lag and blocking scene and other issues in the rendering process of scene display. The introduction of multilevel model data accuracy is applied to alleviate the scene data loading during the system running process, which it can effectively avoid the occurrence of the phenomenon. Based on the display data accuracy of the scenarios, it is able to guarantee the smooth system running upon the loading of the scenarios.

2.2 The Performance of River Shoreline

The performance of river shoreline mainly means that 3-D river shoreline data is constructed by applying the remote sensing satellite image and two-dimensional data in terms of river shape, flow, etc. 3-D watercourse is a kind of data model superimposed on 3-D terrain data. The established watercourse model needs to refer to the terrain model in undulation, properly adjust their relative positions in a seamless way. There is no issues like dangling of the models, incorrect superposition, etc.

For the performance of the water surface of the river, the physical material of Fresnel is applied of which the kind of material can truly reflect various phenomena of water surface in the environment, properly adjust the relevant parameters of materials so that it is able to come into being flow rate, flow direction, reflection, refraction and other effects in the water surface as well as simulate the characteristics of water surface in the vivid environment and vividly display 3-D effects of the river.

2.3 The Performance of Shoreline Resources

The data shall be collected in terms of the utilization planning of the target river shoreline resources, the current utilization of shoreline resources, etc. The division of shoreline control line shall be clarified in terms of the lear water control line, the outer control line, etc. In 3-D model, the spline curve tool is used to draw the model line while the different styles and colors are adopted to distinguish 3-D display effect of the control line.

According to the division results of the functional zone of river shoreline, the Plane model is applied to establish each functional zone; The red, purple, yellow and green materials respectively represent the protection zones of shoreline, the reserve zone of shoreline, the control utilization zone of shoreline and the exploitation zone of shoreline. These four functions can be distinguished by colors. The above mentioned two performances of shoreline resources are based on a two-dimensional data of which the two-dimensional vector data is imported into 3-D modeling software. The established shoreline resource data has the relatively high accuracy.

2.4 The Performance of Evolution Trend of Watercourse

The performance of evolution trend of watercourse is the 3-D visualization analysis based on the historic riverbed terrain data, collecting the measured watercourse underwater topography of both sides of the watercourse control range at different periods. Through the comparative analysis of underwater 3-D terrain generated by the altitude Figs in the different historical periods, it is able to intuitively understand the riverbed evolution trend and provide the scientific basis for the reasonable use of shoreline resources and the sustainable development of river ecology.

The measured watercourse underwater terrain data is majorly based on the vector data in the CAD format. The vector topographic data will be converted into DEM (digital elevation model) data; The GIS software can easily realize the conversion of DEM data to 3-D TIN model data, the generated 3-D TIN data will be imported into the virtual reality platform so as to realize the establishment of the riverbed 3-D terrain.

2.5 Information Query of Spatial Attribute

Spatial attribute information mainly includes the project control point coordinates, size, name, location, etc. The attribute information on the project are stored into the introduced relational database. Through one-to-one correspondence relations on the 3-D models and relational database table, it is able to obtain the correlation of attribute information in the scenario model and database. The users can take use of attribute query tool to click the 3-D scenario model and inquire the specific attributes of the project in real time.

The river engineering project contains a variety of projects such as embankment, wharf, bridge, water inlet, etc. The corresponding attentions on the attribute information are also different. Therefore, it can appropriately classify the river engineering projects, set up the classified relational tables on the river engineering projects in terms of embankment, port, bridge, etc. The relational tables are used to store the spatial attribute information of river engineering project. In this way, the different relational tables are used to store the information of the different types of the river engineering projects.

3 The Applications of VR Simulation System in the Main Stream of Liujiang River (Urban River)

VR simulation system in the main stream of Liujiang river (Urban river) is the established virtual 3-D scenario based on a real 3-D terrain, remote sensing satellite image, shoreline utilization planning reports and other materials. It is also taken as a typical example of river shoreline resource management based on VR spatial information technology. Through VR technology, the 3-D visualization system in the main stream of Liujiang river is established, realizing the digitization management of river shoreline resource of Liujiang river and greatly improving the management level on the shoreline resource of Liujiang river. The main functions are involved: 3-D scene roaming and the fixed-point observation, two-dimensional navigation, the management and query of river shoreline resources, 3-D model observation & display, the display of the administrative licensing approval of the relevant shoreline projects, etc. The main function analysis is stated as follows.

Fig. 1. Main interface of system

3.1 3-D Model Observation and Display

Upon the construction of the 3-D scenario model, the simulation effects with light and shadow are achieved by the assignment of materials, baking and other methods. After VR system is started, the system will automatically load all the model data and scenario data browser window in bringing about the automatic display and loading of the scenario data in the program. The following Fig 2 shows the automatic loading and display results of 3-D model data upon the startup of the program. Through the display window of the scenario data, it is easy to observe the model shape, location, texture and appearance, etc.

Fig. 2. Data Display of 3-D Model Scenario

3.2 3-D Scene Roaming and Fixed-Point Observation

In the VR system, the roaming observation and the fixed-point observation modes are installed. In the roaming mode, it is to simulate the visual effect of the flight in the scene. After the start-up of the mode, the scene changes to the roaming path at the system default. The users can use the mouse and keyboard to control the flight roaming angle, speed and pitch angle, etc. In the roaming mode, users can intuitively observe the relative distributions of river engineering hydraulic projects on a global view and have a good knowledge and understanding of the project location.

The fixed-point mode is a kind of restricted camera mode. Users can use the mouse to click the name of the river engineering project in switching to the fixed-point observation model of the project. According to the relationship of the relative locations of the engineering project and watercourse, the fixed-point observation is a conFigd optimized lens in considering the actual location, engineering layout, the composition and other factors. Hereby this lens is to show the relatively complete and beautiful fixed cameras of the engineering projects. Fig 3 and Fig 4 show the scenario data in the roaming mode and the fixed-point observation mode, respectively.

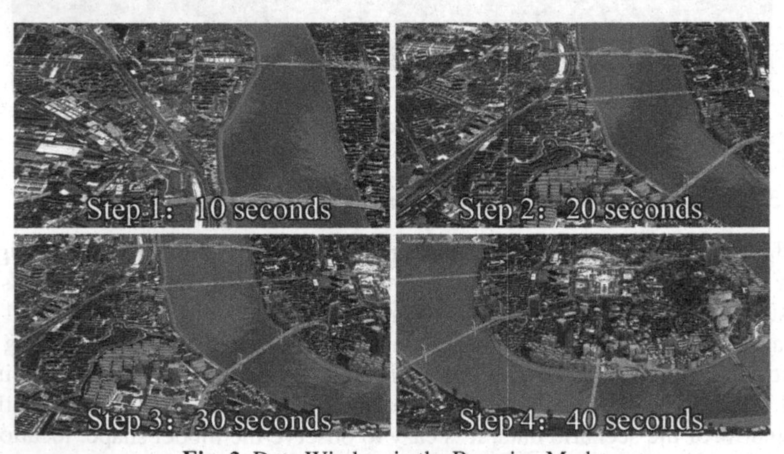

Fig. 3. Data Window in the Roaming Mode

Fig. 4. Data Window in the Fixed-Point Mode

3.3 Two-Dimensional Navigation

The interactive correlation functions of two-dimensional eagle-eye map data and 3-D scene model are introduced. The correlation is achieved through the setting of the consistency of two-dimensional and three-dimensional spatial dimensions. In three-dimensional roaming mode, the navigation function based on two-dimensional map is realized. In the eagle-eye map of two-dimensional navigation, it is able to quickly position the point of sight and view the specific position in the current angle of view of the scenario. At the same time, the layout of the two-dimensional navigation window is located in the top right-hand corner of the browser window of the scenario in achieving the "display" and "hide" function of the planar navigation window.

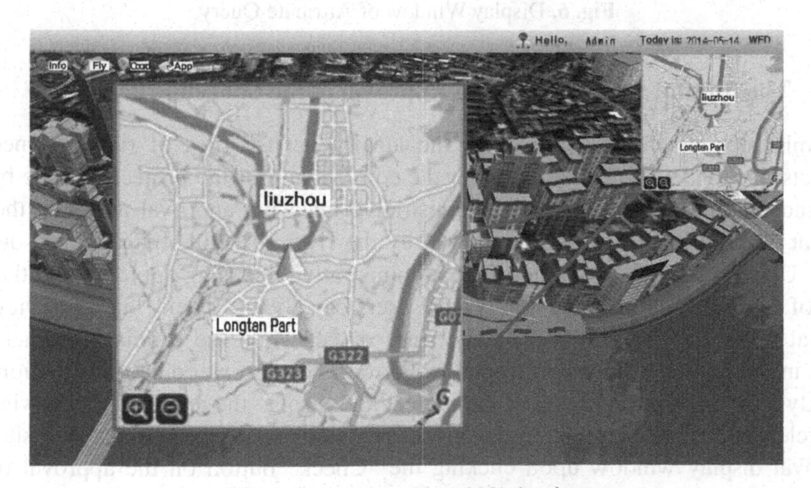

Fig. 5. Two-Dimensional Navigation

3.4 The Management and Query of Engineering Attributes

The shoreline projects contain embankment, bridge, wharf and other engineering projects. The storage ways of attribute information of the different projects are distinguished. For instance, the stored items of the embankment projects are involved in the project name, location, length, the starting point, flood control standard, protection population and zone and other attributes; the stored items of the bridge projects are the project name, location, length, year of construction, management units, water resistance and other attributes. Under such circumstances, the stored attribute information of the different project categories are distinguished. Therefore, through the establishment of the query window of the different attribute information, it is able to inquire and view the attributes of the river engineering projects in the different categories. Fig 6 respectively shows the attribute query window of the embankment project and bridge project.

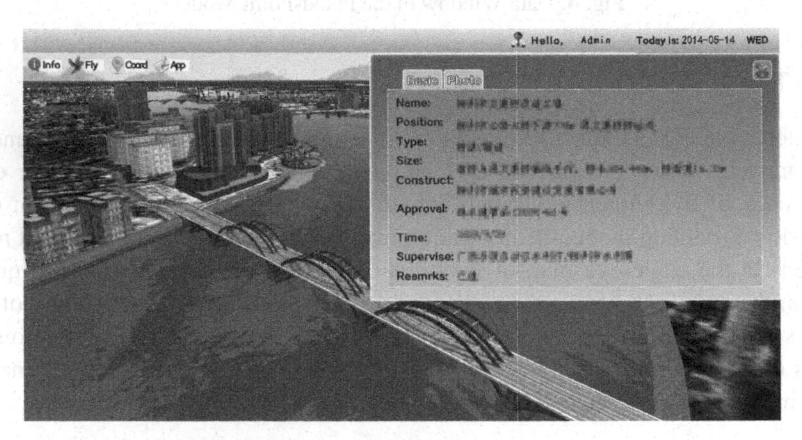

Fig. 6. Display Window of Attribute Query

3.5 The Administrative Licensing Approval of the Shoreline Projects

Administrative licensing approval is the legally license files of river engineering projects, the evidence to show whether the construction of the project is abide by the law and the rules. Therefore, the administrative licensing approval is one of the important documents that is often reviewed by the river shoreline management department. Upon the digital scanning of the administrative licensing approval and the storage of the information in the VR system, users can click "Check" button to view the official documents of the river engineering project while the official document appears in the pop-up suspension display window. The window content instruction will quickly search for the approval of the project and display the approval in the window. The relevant policies, laws and regulations are automatically reviewed. Fig 7 shows a approval display window upon clicking the "Check" button on the approval of the water engineering projects.

Fig. 7. Approval Display Window

4 Conclusions

Practice has proved that there is the extensive application prospect of VR spatial information technology on the river shoreline resource management. Based on VR technology, the river shoreline has achieved 3-D visualization effects on account of the related water conservancy engineering, river shoreline resources, river shoreline evolution and changes, etc. In this way, it is able to know the basic information of the shoreline resources in time, realize the automation management of resource information, greatly improve the management level of the shoreline resources and promote the comprehensive construction of water conservancy informatization.

References

1. Grajewski, D., Górski, F., Zawadzki, P., et al.: Application of virtual reality techniques in design of ergonomic manufacturing workplaces. Procedia Computer Science 25, 289–301 (2013)
2. Wang, J.: Research on application of virtual reality technology in competitive sports. Procedia Engineering 29, 3659–3662 (2012)
3. Seidel, T., König, C., Schäfer, M., et al.: Intuitive visualization of transient groundwater flow. Computers & Geosciences 67, 173–179 (2014)
4. Chen, G., Li, W., Kong, Q., et al.: Recent progress of marine geographic information system in China: A review for 2006-2010. Journal of Ocean University of China 11(1), 18–24 (2012)
5. Yoshida, M.: Three-dimensional visualization of numerically simulated, present-day global mantle flow. Journal of Visualization 16(2), 163–171 (2013)
6. Hou, H., Zhang, J.: Research on Real-Time Visualization of Large-scale 3D Terrain. Procedia Engineering 29, 1702–1706 (2012)

The Planning of Agricultural Space, the Use of Multivariate Geostatistics and the Schioppettino di Prepotto Native Vine Variety: Substituting Unsuitable Forestlands with Fruitful Vineyards in a Municipality of Friuli, Italy

Pier Luigi Paolillo*, Giuseppe Quattrini, Luca Festa, and Marco Picco

Politecnico di Milano, Dipartimento di Architettura e studi urbani, via Bonardi 3, 20133 Milan, Italy
pierluigi.paolillo@polimi.it

Abstract. Italian agri-environmental landscapes are often neglected – when not overlooked or exploited – by urban planning initiatives and their consequences. Often, precious agricultural land is replaced with poor quality real estate. However, with the onset of Expo Milano 2015: "*Feeding the Planet, Energy for Life*," the goal should be to protect landscapes such as these that are rich in exceptional human, alimentary and ecological resources. The following study is thus devoted to the preservation and revival of viniculture in the municipality of Prepotto (a municipality in Northeastern Italy, bordering Slovenia). These two goals can be achieved through the establishment of an urban analytical protocol for the analysis of the rural system. This protocol will enable the identification of new areas for vinicultural development that strengthen the already elevated quality of the native wine of Schioppettino. The approach presented in this paper integrates motives for the preservation of agricultural land with the economic needs of these border territories. This study offers both public officials and the producers of the unique Schioppettino wines the technical reasons to increase production, elevate vinicultural quality and guarantee environmental sustainability. The territory in question is one of tremendous beauty, exceptional landscape variety and peculiar biodiversity. Therefore, the governance of agricultural space must permeate throughout the urban plan – from the conservation of land to its environmental assets and agricultural multifunctionality. In this case, a multidisciplinary approach and a solid information base to support decision-making are indispensable. The application of Geographical Information Systems and the recourse of multivariate geostatistics have allowed the extraction of available information from regional archives. This information has been used to estimate the necessary indicators and interpret the particular complexity of these agrarian landscapes. A new role can thus be attributed to the field of urban planning: the protection of a non-renewable resource, namely land, with simultaneous attention to themes of landscape, tourism and agricultural productive quality – the only true alternative to the current industrialization crisis of these lands on the North/East border of the country.

* Corresponding author.

F. Bian and Y. Xie (Eds.): GRMSE 2014, CCIS 482, pp. 612–624, 2015.

Keywords: planning of agricultural space, vinicultural zoning, landscape appraisal, Geographical Information Systems, Multivariate Geostatistical Analysis.

1 The Foundation of the Research: Integrating a Numeric Model of Agricultural Space into the Municipal Urban Plan

It is well known that, in Italy, the dual role of planner/decision-maker contributes to the challenges in the construction of the municipal plan. The role of the planner is confined merely to that of justifier of pre-established rules. As a result, appropriate analyses that motivate decisions are not adopted, and the result is the devastation of the most important national good: Italian landscapes. Therefore, the course of action must be quickly changed. One way to do so is by measuring the efficacy of hypothesized solutions through indicators that justify the outcomes of the plan or reveal its possible pressures, posing the urban planner in a position to propose sustainable alternatives. This aspect is important because: *i)* addressing environmental complexity is no longer possible through traditional urban planning tools but requires innovative methods rooted in data analysis; and *ii)* the possibility of constructing substantive archives in Geographical Information System allows for objectivity in planning. In order to form an urban planning tool that responds to the many needs (environmental, landscape and economic) of Italian territories, it is necessary to use dynamic indicators that justify implemented decisions, monitor their effects over time and can be treated geostatistically.

The present study speculates upon an agricultural plan with a prevalence of vinicultural zoning. Thus far, the effective urban planning tool in the municipality of Prepotto (nestled between the two provinces of Udine and Gorizia) has not given the agricultural sector the importance it deserves. This fact confirms how urban planning is not at all useful in reviving agriculture. Rather, the urban planning sector proves to be indifferent to such themes (even damaging when land is appropriated for new settlements). Prepotto is a municipality of only 799 inhabitants with a territorial coverage of 33.24 kilometers. The only significant resource is agricultural production and its exceptional landscape. As a result, the accuracy of this proposed plan gives relief to the indifference toward agricultural themes. The proposed plan redirects the entire municipal space to a matrix range consisting of 330.000 cells of 10 m in width, for each of which the intensity of the 37 constructed indicators was estimated. These indicators were subsequently treated using multivariate geostatistics.

2 From Numeric Knowledge to Scenarios of Reinforcement for Multifunctional Viniculture

2.1 The Economic Crisis in Italian Municipalities Inhibits New Research: Provision for the Reuse of Available Information, Adapting It to Different Ends

An analytical protocol based on quantitative methods demands the construction of a database that allows extensive analyses that force the complex characteristics of the studied space to emerge. Inevitably, the preliminary phase of the research entails the examination of existing archives to verify their completeness, the level to which they are up-to-date, their quality and their usability in GIS domain. Upon investigation, the lack of informative material and the grave inability of the municipality of Prepotto to undertake new research ventures (given the difficult economic circumstances) emerged. The available information must thus be reutilized to its best ability. It must be transformed into an apt tool for highlighting the plurality of the economic, social and physical factors that generate territorial and directional dynamics in planning decisions.

2.2 The Components of Municipal Agriculture in a Uniquely Historical Area, the Understanding of New Possibilities for Schioppettino di Prepotto

In the First World War, the conquering of the front line at Caporetto was heroically resisted by the sanctuary of Castelmonte. Peasant women from 46 historical centers in the municipality brought munitions and rations to the trenches in the mounts. Then, at the end of the Second World War, these lands experienced Nazi oppression and tremendous suffering due to their position on the border. In 1976, three violent seismic shocks destroyed 45 municipalities of Friuli and damaged another 90, causing 1000 deaths. Today, this same municipality is renowned in the wine industry for the quality of its fine wine that has been produced since 1200. Moreover, Prepotto is the only municipality in Italy that has been commended with two Doc (controlled designation of origin) in the subzones of Schioppettino di Prepotto and Cialla. Despite a one thousand year history, in the early nineteen seventies, the native vine was nearly extinct. The native vine variety had suffered first due to the proliferation of mildew and later due to phylloxera. These circumstances obligated the producers to substitute the original vines with international ones. In those years, Schioppettino did not even appear on the list of permitted varieties. It is an indigenous vine of the Ribolle Nere family, and it has high, fixed acidity. The wine is bottled early, and it passes three years in barrel for definitive refinement. This process has made it emerge amongst quality red wines in a region, Friuli Venezia Giulia, which has always been distinguished for its high quality white wines. The areas of Colli Orientali (of which Prepotto is a part) and Collio (that borders with Prepotto along the western front) became the main Italian exporters of white wine in the world. The wine is a common asset of two macro environments: *i)* the forested hills with reduced urban episodes which constitute 2/3 of the northern area and *ii*) the densely concentrated vineyards which extend across 416 ha of the undulated southern shelf. These are joined on one side by apertures tied together by the urban framework and, on the other, by small villages surrounded by the sloping vineyards. This extremely complex situation has forced – in the absence of a common digital cadaster – the photographic interpretation of land use in the agri-environmental system. The qualitative and multifunctional development of vinicultural production in the area was hypothesized in this way.

2.3 Advancing Hypotheses on the Use of Municipal Agricultural Space, Moving Toward the Recuperation of More Land That Is Useful for Wine Production

Upon examination of the evolution of agricultural lands in the area, the proposal to amplify the wine-growing surfaces has proven to be valuable. Vineyards should be implemented in portions of the territory that have been overcome by spontaneous arboreal vegetation and stripped of areas useful for agricultural production. The change in Prepotto's land use (7 classes of pastures, woods, agrarian crops, meadows, arable lands, vineyards, and urbanized areas totaling 33,241,073 m^2 of the entire municipal territory) reveals an extensive productive evolution from 1950 onwards. In particular, the instance of vineyards grew from 2% to 13% today. This growth is a result of the radical (and intelligent) cultivation decision that favored viniculture. It is an activity that characterizes many landscapes of Friuli, even in the ever-increasing presence of forested areas (from 68% to 78%) and the accentuated spontaneous arboreal repopulation of the basin that extends from Cladrecis to Bordon.

Table 1. The Evolution of Land Uses

	1950			***1975***	
Pasture	1.181.354 m^2	3%	Pasture	842.116 (– 339.238) m^2	2%
Woods	22.727.607 m^2	68%	Woods	23.514.207 (+ 786.600) m^2	71%
Agrarian Crops	1.485.215 m^2	4%	Agrarian Crops	96.806 (– 1.388.409) m^2	2%
Meadows	40.860 m^2	1%	Meadows	75.860 (+ 35.000) m^2	1%
Arable Lands	6.871.336 m^2	21%	Arable Lands	3.951.884 (– 2.919.452) m^2	10%
Vineyards	737.750 m^2	2%	Vineyards	4.502.455 (+ 3.764.705) m^2	14%
Urbanized	196.634 m^2	1%	Urbanized	259.826 (+ 63.192) m^2	1%
	1990			***2012***	
Pasture	428.372 (+ 413.744) m^2	1%	Pasture	528.372 (+ 100.000) m^2	1%
Woods	24.812.193 (+ 1.197.986) m^2	75%	Woods	26.463.263 (+ 1.751.070) m^2	78%
Agricultural Crops	69.806 (– 27.000) m^2	0.5%	Agricul-tural Crops	69.806 m^2	0.5%
Meadows	40.860 (– 35.000) m^2	0.5%	Meadows	40.860 m^2	0.5%
Arable Lands	3.422.260 (– 729.624) m^2	10%	Arable Lands	1.479.293 (– 1.742.967) m^2	5%
Vineyards	4.107.678 (– 394.777) m^2	12%	Vineyards	4.166.174 (+ 58.496) m^2	13%
Urbanized	311.507 (+ 51.681) m^2	1%	Urbanized	485.203 (+ 173.696) m^2	2%

3 The Interdependencies between Viniculture and Agricultural Space: Estimating Them Using Multivariate Analyses

3.1 A Land between the Plain, the Hill and the Judrio Valley

The municipal territory of Prepotto presents a complex orography characterized by the inter-mountainous canal of Judrio, where the cold currents of Northern Europe collide with the hot gusts of the Adriatic Sea. This position contributes to the generation of an ideal microclimate for the production of fine wine. This can be attributed to two indicators that result in variations from 60 m to over 700 m above sea level: the strong, daily thermic excursions and the elevated and morphometric differences. The Digital Terrain model and its distribution of lands by altitude profile highlight the two macro areas of Prepotto: the southern vinicultural area, from 69 m to 250 m, and the woods that extend north from Casa Brischis to the 717 m of Bordon.

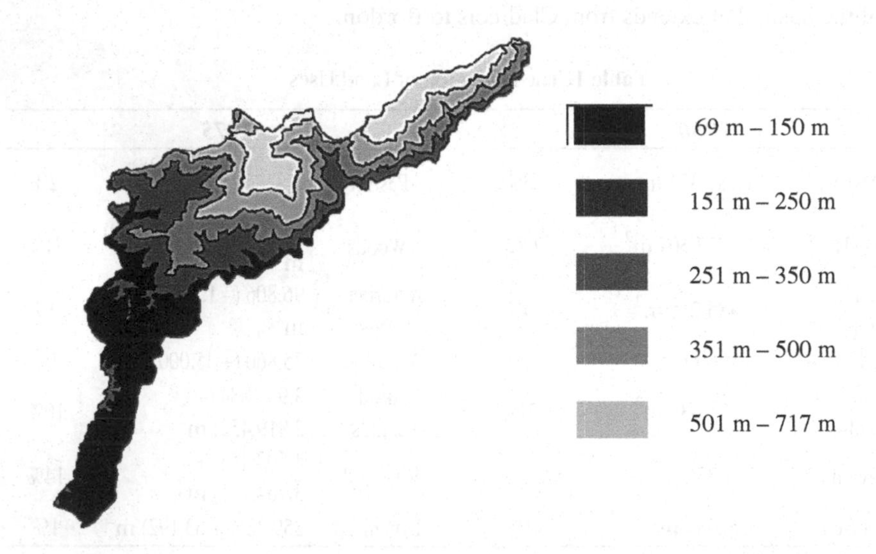

Fig. 1. The Altitude Profile of the Municipal Space of Prepotto

3.2 The Evolution of the Vine: From Wine as a Post-war Supplement to Current, Fine Quality Productions

In the estimate of the temporal permanence of agricultural uses, four land use thresholds were used: the years 1950, 1975, 1990, and 2012. The first three were obtained from the informational layers of the *Moland Land Use* project, made available by the autonomous region of Friuli Venezia Giulia. The last layer was derived from our own photographic interpretation and digitalization of the conditions in 2012. The temporal intervals were reduced to classes of land use for each cell of a matrix at the rate of 10 m, standardized based on the maximum value, classifying the cells in 5 groupings with special attention given to the level of vinicultural persistence (from low to high).

The results of the conducted analyses highlight the obvious growth in the surface area of the vineyard between 1950 and 1975. The area started out at 737,750 m^2 after World War II to eventually reach 5,502,455 m^2. This growth was a result of the elevated agricultural investments that guaranteed economic regeneration to local producers, at the expense of the existing landscape pattern and its environmental characteristics (the excavations of the intrusive interventions on the substratum of the soil profile registered in those years are, in fact, quite notable). Viniculture then remained stagnant until 1990, when it began to grow again, this time more moderately. In the last twenty years, the subsistence economy that integrated poor nutrition with wine in the 1950s evolved, giving way to the achievement of refined quality and elegance in the current production of Schioppettino di Prepotto.

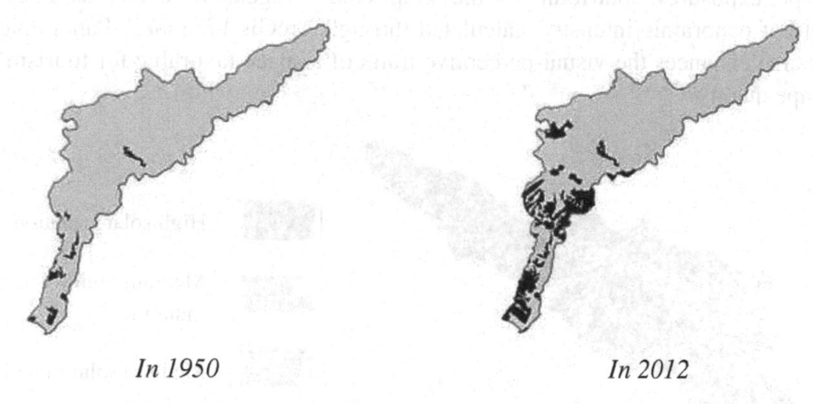

Fig. 2. The Temporal Evolution of the Wine Growing Areas

3.3 Anthropic and Environmental Pressures on Agroforestry in Prepotto

The complex characteristics of this municipal morphology were also examined with the digital terrestrial model, given: i) the direction of the exposure of the slopes (orientation of the sides with respect to the cardinal points); ii) the degrees of incline of the surfaces (expressed by % decline with respect to the plain); iii) the degree of solar radiation (Σ of direct and diffused solar radiation); and iv) the degree of panoramic intensity (expanse of visual basins from pre-established focal points).

Table 2. The Morphologic Indicators

Indicator		*Variables utilized*
Ipen = incline	$\sum_n^j (h_j/b_j)$ x 100	*h* = vertical distance between cells *b* = horizontal distance between cells
Iesp = exposure	$K* \, arctg2\left(\left[\frac{dz}{dy}\right] - \left[\frac{dz}{dx}\right]\right)$	[*dz*/*dy*] = velocity of variation in direction *y* for the cell considered [*dz*/*dx*] = velocity of variation in direction *x* for the cell considered *K* = radiant coefficient

Table 2. (*Continued*)

Irad = solar radiation	$Dir_{tot} + Diff_{tot}$	Dir_{tot} = direct solar radiation Dff_{tot} = diffused solar radiation
Ivet = panoramic intensity	$\sum_n^j (n_j/N_j)$	*n* = number of interest points visible from the single cell *N* = maximum number of interest points

In the identification of lands suitable for viniculture, the low inclines facilitate the management of the vine lowering its costs. Good levels of solar radiation and favorable slope exposures contribute to the grapevine's vegetative cycle, as does the estimate of panoramic intensity, calculated through ArcGis *Viewshed*. Panoramic intensity also enhances the visual-perceptive traits of a space favorable for tourism and landscape quality.

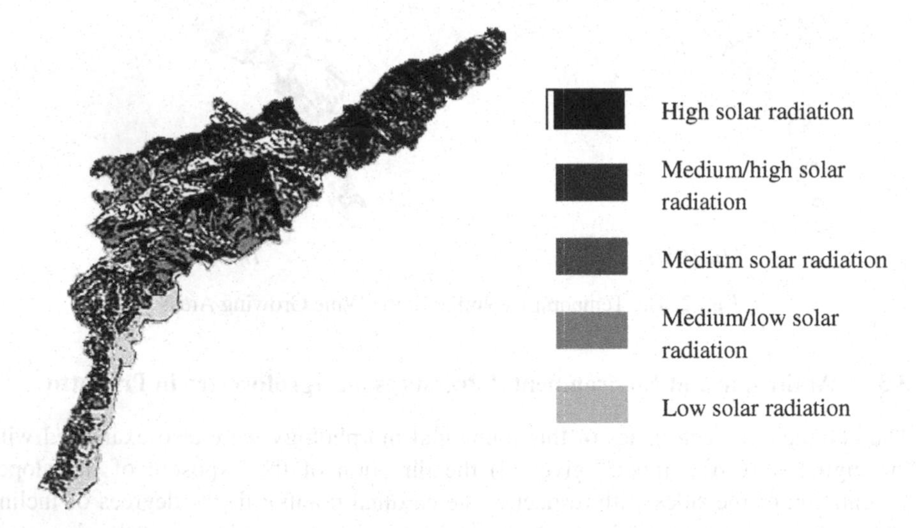

Fig. 3. Map of Solar Radiation

To limit agricultural over-exploitation, the propensity toward soil erosion is also estimated in this research through the joint examination of four basic variables: *i*) land use, *ii*) incline, *iii*) exposure and *iv*) rainfall. These variables help to determine the choice of vinicultural expansion areas where the soil erosion risk is limited.

3.4 Multivariate Applications and the Agrosilvicultural Characteristics of the Municipal Territory

In order to find new areas for the expansion of viniculture in the municipality, judgments on crop variation must stem from multidimensional estimates of the descriptive and synthetic components of the study area (done here by employing the use of Addawin software). An analytical framework that estimates the interdependencies of the

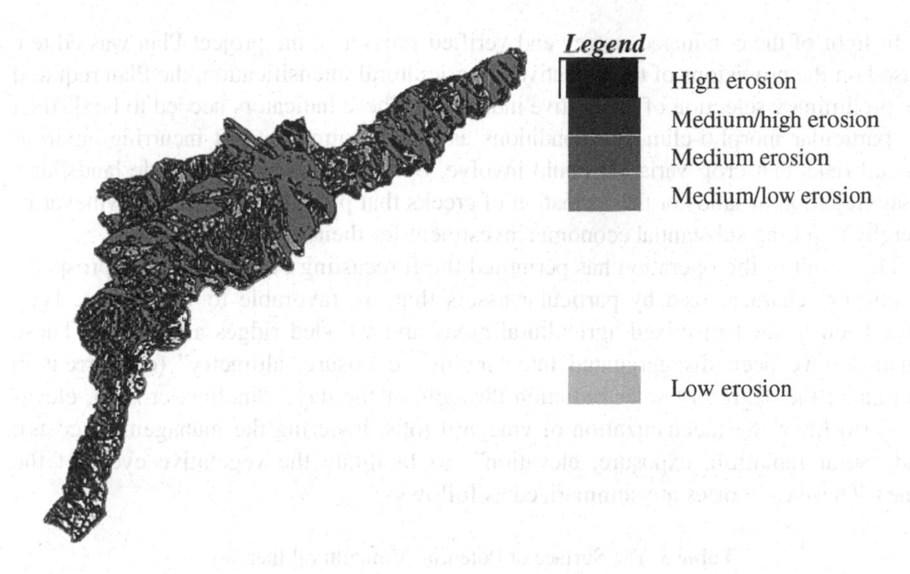

Fig. 4. Map of Soil Erosion Risk

environment and local landscapes through: *i*) state of the lands; *ii*) morphology; *iii*) orography; *iv*) structural characteristics; and *v*) accessibility is thereby advantageous.

It was first necessary to insert the matrices of the municipal areas that were in GIS format into Excel tables for multivariate treatment with *Addawin*. In *Addawin*, it was possible to then proceed to the analyses of the correlations to identify the most descriptive indicators of the present phenomena. In the subsequent phase, the analyses of the principal components reduced a more or less elevated number of variables into fewer latent variables. In such a way, the number of principal resulting components explicable through the variance quota is established, resulting in the minimum number possible of components with a minimum loss of information and a significance quota of up to 60%.

The non-hierarchical analysis required the preliminary treatment of the variables chosen so as to generate an objective function that expresses the level of inertia. Initially, *n* units were distributed into *G* clusters with an iterative procedure that would allow for the obtainment of a single, final partition. Finally, the characteristic median values of the number of classes were assigned to an output and the results were georeferenced in ArcGIS.

This step made it possible to extract 13 initial, stable classes, derived from the reading of the stable output profiles, whose differences were extremely extensive. This characteristic justified their reaggregation based on similarity of land use into 4 large distributions found across the entire municipality: *i*) mixed agricultural areas (2 reaggregated stable classes, = 9% of the municipality); *ii*) wooded ridges and slopes (7 stable reaggregated classes = 76%); *iii*) areas of peculiar vocation and specialized viticulture (3 stable reaggregated classes = 13%); and *iv*) areas of elevated anthropic and urban density (1 stable class = 2%).

In light of the conducted survey and verified pressures, the project Plan was edited. Based on the provisions of the objective of agricultural intensification, the Plan required the preliminary selection of qualitative indicators. These indicators needed to be distinct for particular morpho-climactic conditions as well as aimed at not incurring environmental risks that crop variation could involve. Some of these risks include landslides, disaggregation of lands or the formation of creeks that provoke cracks in the vineyards, thereby requiring substantial economic investment for their maintenance, etc.

The result of the operation has permitted the forecasting of new selective prospects of clusters characterized by particular assets that are favorable to viniculture. They have been located in mixed agricultural areas and wooded ridges and slopes. These clusters have been disaggregated into "incline, exposure, altimetry" (concurrent to guarantee the degree of solar radiation throughout the day); "incline, erosion, elevation" (to favor the mechanization of vineyard jobs, lowering the management costs); and "solar radiation, exposure, elevation" (to facilitate the vegetative cycle of the vine). These categories are summarized as follows:

Table 3. The Surface of Potential Vinicultural Increase

Wooded Ridges and Slopes Suitable for Conversion	*Ha*
Factors of incline, exposure, altimetry	119
Incline, erosion, altimetry	46.5
Solar radiation, exposure, altimetry	205
Total	370.5

Mixed Agricultural Plots Suitable for Conversion	*Ha*
Incline, exposure, altimetry	50
Incline, erosion, altimetry	2
Solar radiation, exposure, altimetry	13
Total	65

3.5 Phenomenal Basins and the Claiming of New Areas for Vinicultural Conversion

National and regional planning practices allow the variance in land use destination of a wooded area (in this case, a crop change) only in the absence of a burden on the landscape that protects the formation of valuable wooded areas to be preserved as essential resources. Further according to these practices, a change in land use must not cause a hydrogeological barrier; this is in order to preserve the land from forms of land use that generate loss of stability or conflict with the water currents. Stemming from this assumption, there are areas that, even while presenting peculiar and ideal morpho-climactic environments for viniculture, cannot be converted for this use. Once these were individuated in GIS domain, the alternative strategy, which was to localize the spontaneous formation of black locust forests, was adopted. These forests were chosen for conversion due to their scarce landscape quality, which usually originates from the abandonment of previously cultivated plots of land. The majority of these lots were abandoned between the nineteen seventies and nineteen eighties, a period during which the explosion of the manufacturing industry in Friuli provoked a strong decline in the rural population living in wooded areas in northern Prepotto.

Based on the findings from this extensive survey, the previous classification of predictions was further disaggregated into: *i*) areas, currently free of vines, burdened by landscape or hydrogeological barriers and, therefore, in which conversion to viticulture is not allowed (318 ha); *ii*) areas, currently free of vines, absent of such barriers and for which no administrative procedure is necessary in order to launch the vinicultural conversion (75 ha).

The final result of the research was a potential increase of the wine-growing area equal to 75 ha and corresponding to approximately 20% of the current vinicultural surface of 416 ha. This surface would allow the entirety of the vineyards to reach 491 ha, having important impacts in terms of future business growth (of businesses already well-rooted into the national and international markets). Furthermore, young agrarian entrepreneurs would also benefit and have the possibility, offered by our in-depth analysis, to establish and/or recuperate vined lots in an area that has always been suitable and recognized for its vinicultural quality in the most qualified enological panorama.

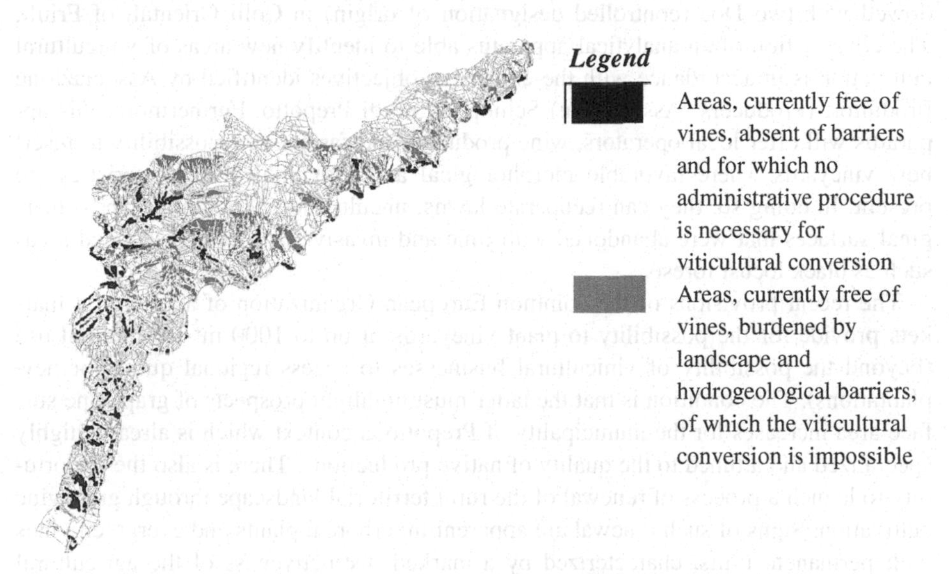

Fig. 5. Map of the Conversion from Non Qualified Woods to Viticulture

4 Conclusions: In the Agricultural Dimension of Urban Planning, the Improvement of Cultivation and Sustainable Use of Land

Expert systems allowed the potential "intervenibility" to emerge in favor of the specialized cultivation of grapevines, identifying new convertible areas. The conviction is that changes in the global economy – globalization, on one hand, and regionalism, on the other – make a different model for local development necessary. This model must avoid the appeal of growth and the past errors of fanciful, unimplemented politics. The final proposal is the pursuit of a qualitative development, where the sustainable use of physical resources is prioritized and economic ones are then reached through innovative scientific paths. In such a way, a new culture is instilled: one of valuing common goods and respecting scarce and depleted resources. Friuli-Venezia Giulia is a place battered by war, floods and earthquakes. However, it has been cared for by a strong people with a strong work ethic. Today, it has seen the enhancement of its unique agricultural products as the only way to overcome difficult economic circumstances, celebrate territorial identity, boast its local peculiarities and represent its principal product: wine. Wine has become the decisive, identifying element of this territory.

The choice of the municipality of Prepotto as the study area becomes even more meaningful after having instigated the emergence of one of the typical local products that most distinguishes it in the global wine sphere: Schioppettino. The native Italian wine was almost extinct in the nineteen seventies. Today, thanks to the hard work of the local wine producers in a municipality of less than 800 inhabitants, the wine is endowed with two Doc (controlled designation of origin) in Colli Orientali of Friuli. The construction of an analytical apparatus able to identify new areas of vinicultural cultivation is in accordance with the expansive objectives identified by Associazione Produttori (Producers' Association) Schioppettino di Prepotto. Furthermore, this apparatus will offer local operators, wine producers and farmers the possibility to insert new vineyards where favorable morphological and orographical characteristics are present. In doing so, they can recuperate lawns, uncultivated agricultural lands, marginal surfaces that were abandoned with time and invasive and foreign wooded areas such as black locust forests.

The recent provisions of the common European Organization of agricultural markets provide for the possibility to plant vineyards of up to 1000 m^2 for familial use (beyond the possibility of vinicultural businesses to access regional quota for new plantations). The condition is that the latter must highlight prospects of grapevine surface area increases for the municipality of Prepotto, a context which is already highly specialized and attuned to the quality of native productions. There is also the opportunity to launch a process of renewal of the rural territorial landscape through grapevine cultivation. Signs of such renewal are apparent in: arboreal plants and evergreen vines with permanent traits, characterized by a marked distinctiveness of the agricultural pattern.

The identification of areas with a propensity for agricultural conversion in proximity of peri-urban areas becomes an occasion to recompose the landscapes in order to diversify the visual-perceptive quality of the urban fabric. The intermediate spaces that are scarcely qualified by the fractions of Prepotto will thus be invested with agricultural uses oriented toward the expansion of quality vinicultural activities and the differentiation of products. Through diversification, the remunerative fruit and vegetable varieties can be cultivated and, simultaneously, connected to the offering of typical local products. This work, beyond providing an instrument for private vinicultural businesses, also provides informative and directional support to municipal administrators in order to govern local agricultural policies. It can also help these political actors to consciously manage the environmental resources, increasing the production on the one hand and elevating the vinicultural quality on the other hand. This research also offers useful support for directing eventual urban expansion in areas less apt for agriculture. The sacrificing of depleted and non-renewable resources such as land can be avoided, and the historical dichotomy between urbanization and agriculture can be resolved.

A protocol for agri-environmental planning based on geostatistic analytical models has here been identified. It is a versatile examination; the models consider the largest available share of data and investigate their relations. The goal here is to identify the surfaces best suited for viniculture. These identified surfaces must be enriched through detailed investigations of the stratigraphic characteristics of the soil through pedological and agronomical analyses. They must also be added to spatial plots already identified by the general plan as potentially suitable given their orographic and morpho-climactic characteristics.

References

1. Benzécri, J.P.: L'analyse des données. Analyse des correspondances. Dunod, Paris (1973)
2. Colugnati, G., Michelutti, G.: Suoli e vigneti. Ersa, Trieste (1998)
3. Costantini, E.: Metodi di valutazione dei suoli e delle terre. Cantagalli, Firenze (2006)
4. Diday, E.: New approaches in classification and data analysis. Springer, New York (1994)
5. Fabiano, N., Paolillo, P.L.: La valutazione ambientale nel piano. Maggioli, Rimini (2008)
6. Fabbris, L.: Statistica multivariata. McGraw Hill, Milano (1997)
7. Fraire, M.: Metodi di analisi multidimensionale dei dati. Cisu, Roma (1994)
8. Fregoni, M.: Viticoltura generale. Reda, Roma (1985)
9. Griguolo, S.: Addati. Un pacchetto per l'analisi esplorativa dei dati. Guida all'uso. Iuav, Venezia (2008)
10. Gruppo Chadoule: Metodi statistici nell'analisi territoriale. Clup, Milano (1983)
11. Ispra: Aree agricole ad alto valore naturale. Dall'individuazione alla gestione, Roma (2010)
12. Jones, G.V., Snead, N., Nelson, P.: Geology and Wine 8. Modeling viticultural landscapes: a Gis analysis of the terroir potential in the Umpqua Valley of Oregon. Geoscience Canada 31(4) (2004)
13. Marenghi, M.: Manuale di viticoltura. Edagricole, Bologna (2005)
14. Mazzarino, S., Pagella, M.: Agricoltura e mondo rurale tra competitività e multifunzionalità. Franco Angeli, Milano (2003)
15. Malczewski, J.: Gis-based land suitability analysis: a critical overview. Progress in Planning 62, 3–65 (2004)
16. Paolillo, P.L., Prestamburgo, M.: Il sistema agro – forestale dello spazio regionale. Indirizzi territoriali in materia di zone agricole e forestali.Collana Studi e ricerche per il piano territoriale regionale generale. Regione autonoma Friuli Venezia Giulia, Trieste (1995)
17. Paolillo, P.L.: Tipologie dell'assetto territoriale e consumo di suolo agricolo. In: Martellato, D., Sforzi, F. (a cura di) Studi di Sistemi Urbani. FrancoAngeli, Milano (1990)
18. Paolillo, P.L.: Un atlante cartografico per l'analisi multivariata delle risorse fisiche. In: Idem (a cura di) Il Programma di Diana. De Agostini, Novara (1995)

19. Paolillo, P.L.: Procedure analitico-statistiche per la valutazione dei processi strutturali agricoli e per la loro localizzazione sul territorio comunale. In: Idem (a cura di) Spazi Agricoli a Cusago. Franco Angeli, Milano (1995)
20. Paolillo, P.L.: Al confine del nord – est. Forum, Udine (1998)
21. Paolillo, P.L. (a cura di): Terre lombarde. Studi per un eco – programma in aree bergamasche e bresciane. Giuffre, Milano (2000)
22. Paolillo, P.L., Mariani, L., Rasio, R.: Climi e suoli lombardi. Il contributo dell'Ersal alla conoscenza, conservazione e uso delle risorse fisiche. Rubbettino, Catanzaro (2001)
23. Paolillo, P.L. (a cura di): Acque, suolo territorio. Esercizi di pianificazione sostenibile. Franco Angeli, Milano (2003)
24. Paolillo, P.L.: Sistemi informativi e costruzione del piano. Metodi e tecniche per il trattamento dei dati ambientali. Maggioli, Rimini (2010)
25. Paolillo, P.L.: La tecnica paesaggistica. Stimare il valore dei paesaggi nel piano. Maggioli, Rimini (2013)
26. Griguolo, S., Palermo, P.C. (a cura di): Nuovi problemi e nuovi metodi di analisi territoriale. Franco Angeli, Milano (1984)
27. Tomasi D., Marcuzzo P., Gaiotti F.: Delle terre del Piave. Uve, vini e paesaggi. Centro per la ricerca in viticoltura, Conegliano (2011)
28. Tomasi, D., Gaiotti, F., Locci, O., Goddi, E., Battista, F., Fantola, F., Tore, C.: Le terre e le vigne del Cannonau di Jerzu. Centro per la ricerca e la sperimentazione in agricoltura, Roma (2012)

Monitoring Urban Expansion and Morphology Changes of Tangshan by Using Remote Sensing

Lifeng Shi, Fang Liu, Zengxiang Zhang, and Xiaoli Zhao

P.O. Box 9718, Olympic Village Science Park W. Beichen Road, Chaoyang District
shilf@radi.ac.cn

Abstract. On the basis of multi-source satellite data on compact ratio, fractal dimension and barycentric coordinates across different periods, we analyze the spatiotemporal characteristics and morphology changes of Tangshan using geographical information system and remote sensing techniques. The speed of urban expansion slowed down from 1976 to 1996 and changed periodically from 1996 to 2013. The variation trend of the compaction index is contrary to that of the fractal dimension. The changes in the compaction index showed a downward trend, and the fluctuations were reduced after 1996. Meanwhile, the variation trend of the fractal dimension is exactly the opposite of that of the compaction index, which showed an upward trend and experienced an increase in fluctuations after 1996. The barycenter of the built-up area of Tangshan moved toward the northeast from 1976 to 2009 and turned southwest from 2009 to 2013.

Keywords: urban expansion, compact ratio, fractal dimension, barycenter, spatiotemporal characteristics.

1 Introduction

Land-use/land-cover changes (LUCCs) are related to interactions between humans and nature. Urban expansion is an effect of natural, environmental, and socio-economic factors. It is also an important part of LUCCs [1]. Since the declaration of the Reform and Open Policy in the late 1970s, China has been experiencing rapid economic development. Along with such development and rapid urbanization in the country, land use patterns have changed gradually [2]. Land use changes resulting from urban expansion affect the ecology, climate, environment, resources, economy, population, and so on. Thus, a growing number of studies focus on the relationship between urban expansion, LUCCs, and environmental resources. The use of high temporal resolution remote sensing (RS) data is insufficient to summarize the spatiotemporal characteristics and morphology changes of urban expansion, especially the changes in socio-economic circumstances resulting from the development of the urban forms of cities [14]. In the past, geographers regarded urban morphology to be disorganized until the appearance of fractal dimension [3]. In the 1990s, Batty and Longley studied urban boundary, morphology, structure, etc., from the aspect of urban morphology [4-5]. WANG Xinsheng calculated and analyzed the fractal dimension

F. Bian and Y. Xie (Eds.): GRMSE 2014, CCIS 482, pp. 625–634, 2015.

of 31 large cities based on the national resources and environmental database [6]. Changes in urban morphology not only reflect the temporal-spatial evolution of urban expansion and the movement of urban barycenter but also affect city traffic, environment, production, and so on[15-16]. Tangshan is a special city. After the Tangshan Earthquake in 1976, the city was left in ruins. While Tangshan was attempting to cope with the destruction caused by the earthquake, other cities were experiencing rapid development. As a city undergoing post-disaster reconstruction, Tangshan differs from other cities in terms of urban expansion and planning [7]. The spatiotemporal characteristics of the urban expansion in Tangshan are also different from those in cities that underwent rapid economic development during the Open Policy and Economic Reform. Therefore, the post-earthquake monitoring of urban expansion and morphological changes is significant to Tangshan. This study analyzes the spatiotemporal characteristics and morphological changes in Tangshan based on 16 RS images from 1976 to 2013.

2 Study Area and Data

2.1 Study Area

Tangshan is a regional city in Hebei Province that is located in the east and central section of the Bohai Bay region (117°31'E-119°19'E and 38°55'N-40°28'N). Tangshan serves as the bridge connecting the northeast and northern regions of China. It is also an important port city and logistics center of northeast Asia. It is one of the most important industrial cities in China, as it is home to a wide range of industries. Tangshan is the economic center of Hebei Province and is one of the core cities of the Beijing–Tianjin–Hebei city groups. It has a warm temperate semi-humid continental monsoon climate. The terrain slopes of Tangshan flatten out from the northwest to the southeast until the coast of East China Sea. The area covers 13.5 thousand km^2. It has jurisdiction over two country-level cities (Qian'an and Zhunhua), five counties (Qianxi, Yutian, Niexian, Nienan, and Laoting), seven areas, and four development zones. By 2012, the city had registered a total population of 747.40 million and a GDP of 612.121 billion yuan [8].

2.2 Data Collection and Preprocessing

Using satellite data as the main data source, we monitored the changes in the urban district area of Tangshan from 1976 to 2013, thus covering a total of 16 years and 15 time periods. The datasets used to monitor urban expansion included 2 Landsat MSS images, 22 Landsat TM images, 3 Landsat OLI images, 1 CBERS CCD image, and 2 HJ-1 CCD images.

In the Intergraph MEG platform, 1:10 million topographic maps of Tangshan were used as reference for image geometric correction using the least squares method. The nearest neighbor method or the bilinear interpolation method was used to resample the images; the geometric correction accuracy of the images was below 1 pixel. Secant conic projection, Krassowski ellipsoid of 1938, and the central meridian and double standard weft with national standards were also used. The central meridian was 105°E, and the double standard weft was set to 25°N and 47°N.

In the same platform, the man–machine interactive visual interpretation method was used to draw the boundaries of the built-up area and the dynamics of urban expansion based on the corrected images. First, the screen digital function of the Intergraph MEG was used to draw the boundary of Tangshan in 1976. Each image after 1976 was then interpreted, and the boundaries of Tangshan in different periods were obtained. These boundaries were then exported in vector format from the Intergraph MEG to the swap file of DxfArc to realize the graphics edition and the area statistics in Arc/Info. Finally, the changes in the built-up area and land use of Tangshan were obtained.

2.3 Methods

Fractal Dimension. The concept of fractal geometry was first proposed by French American mathematician B.B. Mandelbrot in 1975. Fractal geometry has been used by many scholars to study the shapes of city boundaries. The fractal dimension of urban built-up areas can be considered as an indicator of the complexity or dispersion of urban form. The value of fractal dimension is between 1 and 2. Generally, a high fractal dimension value indicates the great complexity or dispersion of a city. The area–perimeter fractal dimensions are based on the box-counting method [9-10]. The mathematical description is given below.

$$\ln N_{(r)} = C + D \ln M_{(r)}^{(1/2)} \quad (1)$$

Where $M_{(r)}$ and $N_{(r)}$ denote the number of boxes that cover the urbanized area and its boundary, respectively; D is the fractal dimension, and C is a constant.

Compact Ratio. The compactness index of urban built-up areas can be considered as an indicator of urban morphology. For a given area, a large compact ratio is equal to a highly compact shape, and vice versa. The circle is considered to be the most compact shape, given its compactness index value of 1; other shapes have compactness indexes that are less than 1. A mathematical description of the compactness index is presented below [11-12].

$$c = 2\sqrt{\pi A} / P \quad (2)$$

Where c is the compact ratio, and A and P are the area and perimeter, respectively.

Barycentric Coordinates. The concept of urban barycentric coordinates was derived from the population distribution principle in geography. These coordinates can be used to calculate the urban barycentric position at different periods. Barycentric coordinates can be calculated by the following formulae [13].

$$X_t = \sum_{i=1}^{n} (C_{ti} \times X_i) / \sum_{i=1}^{n} C_{ti} . \quad (3)$$

$$Y_t = \sum_{i=1}^{n} (C_{ti} \times Y_i) / \sum_{i=1}^{n} C_{ti} . \quad (4)$$

Where X_t and Y_t are the barycentric coordinates of the patch in year t, C_{ti} is the area of patch i in year t, X_i and Y_i are the mass centric coordinates of patch i in year t, and n is the quantity of the patch.

3 Results and Discussion

3.1 Urban Expansion in Tangshan from 1976 to 2013

Tangshan benefited from Beijing's and Tianjin's urban planning for post-disaster reconstruction and attractions from 1976 to 2013. The post-disaster reconstruction, which was based on the old urban district, was implemented along the traffic line in the southwest, northeast, and west. Covering a total area of 51.86 km^2 in 1976, the reconstruction was expanded to cover an area of 182.23 km^2 in 2013. The net increase in area reached 130.37km^2. The reconstruction expanded 2.51 times, thus yielding an urban expansion rate of 3.52km^2/year (Table.1, Fig. 1). In the span of 37 years, the expansion of the built-up area of Tangshan underwent different stages, namely, slow growth (1976-1996), initial growth (1996-2002), stable growth (2002-2009), and astable growth (2009-2013). In July 28, 1976, the industrial city of Tangshan was hit by an earthquake that caused great destruction. The Party Central Committee took over the reconstruction of the city. At first, the urbanization of Tangshan expanded rapidly. With the city reconstruction completed, the expansion of Tangshan slowed down. That is, from 1976 to 1996, Tangshan experienced a slow growth. In the span of 20 years, the built-up area of Tangshan increased to 35.91km^2. After 1996, the speed of Tangshan's urbanization continued to accelerate. The built-up area of Tangshan expanded rapidly at an annual rate of 3.68km^2 between 1996 and 2002; in six years, the built-up area increased to 22.09km^2, and was mainly distributed in Xujunzi village in Lubei District, near Songxuexinzhuang village, and along the ring road. Guided by the Tangshan city master plan (2003–2020), Tangshan expanded steadily at 3.40km^2/year between 2002 and 2009. The city's expansion experienced a short wave from 2009 to 2013; during this period, the expansion first sped up and then slowed down after 2012. The Fengnan District, originally from the built-up area of Tangshan, underwent urban expansion, which led to a spurt of built-up area from 133.67 km^2 in 2009 to 182.23 km^2 in 2013.

Table 1. Built-up area of Tangshan from 1976 to 2013 (unit: km^2)

Year	1976	1979	1987	1996	1998	1999	2000	2002
Area	51.86	61.44	80.18	87.77	91.28	93.25	95.25	109.86
Year	2004	2006	2008	2009	2010	2011	2012	2013
Area	115.25	120.52	131.18	133.67	168.49	174.17	179.68	182.23

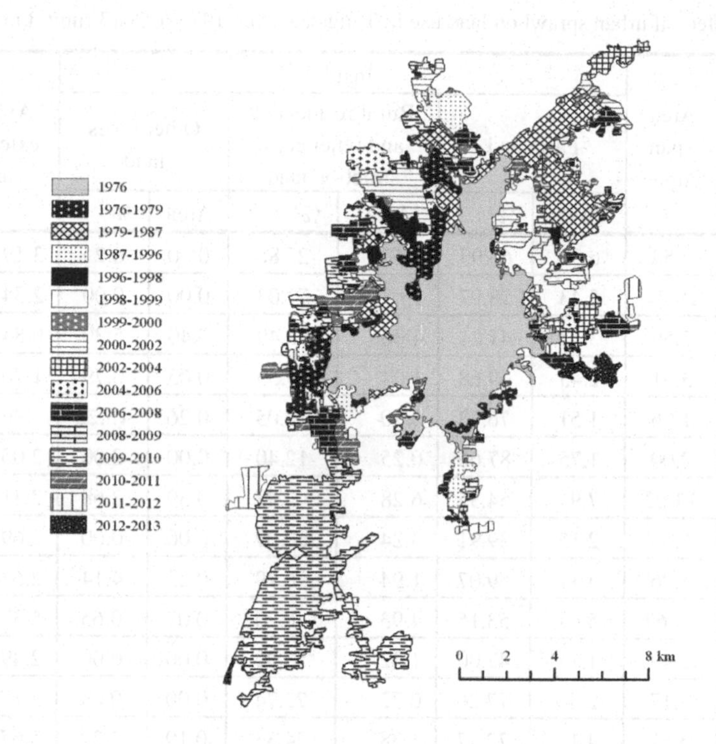

Fig. 1. Urban sprawl in Tangshan from 1976 to 2013

3.2 Land Use Changes

From 1976 to 2013, the urban expansion of Tangshan covered 98.71km^2 of land around the built-up area (Table.2). Urban expansion greatly affected the farmlands, followed by the rural residential lands and other land types for construction and then by the other land types, such as woodlands and wetlands. Urban expansion occupied 50.91km^2 of farmland, 45.64 km^2 of rural residential land and other land types for construction, and 2.17km^2 of other land types. The contribution rate of farmland, rural residential land, and other land types for construction during the urban expansion of Tangshan reached as high as 97.81%. Farmland, rural residential land and other land types for construction, and other land types were occupied by built-up areas at the following rates: 1.38, 1.23, and 0.06km^2/year, respectively. The reduction rate in farmland, rural residential land, and other land types for construction is consistent with the increase in the built-up area of Tangshan. However, the reduction rate in the other land types is rather different from that in the built-up area.

3.3 Spatiotemporal Characteristics and Morphological Changes

Along with the expansion of Tangshan from 1976 to 2013, the values of compact ratio showed a decreasing trend, whereas the values of fractal dimension demonstrated an

Table 2. Effect of urban sprawl on land use in Tangshan from 1976 to 2013 (unit: km^2)

Period	Area expansion	Inside						Average extension area
		Farmland		Rural residential and other construction land		Other types land		
		Area	%	Area	%	Area	%	
1976-1979	9.58	6.32	65.93	2.67	27.86	0.60	6.21	3.19
1979-1987	18.74	5.43	28.97	13.31	71.03	0.00	0.00	2.34
1987-1996	7.59	3.13	41.21	4.06	53.49	0.40	5.30	0.84
1996-1998	3.51	2.45	69.68	1.04	29.53	0.03	0.79	1.76
1998-1999	1.96	1.51	76.73	0.20	10.05	0.26	13.22	1.96
1999-2000	2.00	1.75	87.60	0.25	12.40	0.00	0.00	2.00
2000-2002	14.62	7.94	54.35	6.28	42.97	0.39	2.68	7.31
2002-2004	5.39	2.15	39.91	3.24	60.09	0.00	0.00	2.69
2004-2006	5.26	3.11	59.07	1.94	36.79	0.22	4.14	2.63
2006-2008	10.67	5.67	53.15	4.93	46.20	0.07	0.65	5.33
2008-2009	2.49	1.17	47.00	1.32	53.00	0.00	0.00	2.49
2009-2010	3.17	2.45	77.26	0.72	22.74	0.00	0.00	3.17
2010-2011	5.67	4.11	72.37	1.38	24.33	0.19	3.29	5.67
2011-2012	5.52	2.53	45.94	2.97	53.80	0.01	0.25	5.52
2012-2013	2.54	1.20	47.22	1.34	52.78	0.00	0.00	2.54
Total	98.71	50.91	51.58	45.64	46.23	2.17	2.19	2.67

increasing trend. Thus, the structure of Tangshan was less compact, and the city boundaries were more complex compared with the old Tangshan. The table shows a negative correlation between compact ratio and fractal dimension. The change in Tangshan's morphology can be divided into four stages. In the first stage (1976 to 2013), the values of compact ratio declined continually, but the values of fractal dimension increased. In the second stage (1996 to 2004), the third stage (2004 to 2010), and the last stage (2010 to 2013), the values showed a wave crest or trough (Fig. 2).

During periods of fast urban expansion, the values of compact ratio decrease, and those of fractal dimension increase. Thus, the urban space structure becomes less compact, and the city boundaries become more complex compared with the characteristics from previous periods. The trends were different for Tangshan. From 1976 to 1996, the urban expansion of Tangshan slowed down as the values of compact ratio decreased and as the values of fractal dimension increased continually. After 1996, the expansion of Tangshan accelerated. From 1996 to 2000, the values of compact ratio increased instead of decreasing. When the speed of the expansion of Tangshan reached its peaks in 2000-2002, 2006-2008, and 2010-2011, the values of compact ratio and fractal dimension reached their peak and valley in 2000, 2008, and 2011. A decade after the Tangshan earthquake in 1976, the city was rebuilt. Thus, urban expansion slowed down significantly. From 1979 to 1996, Tangshan showed an open expansion and annexed a vast piece of land around the built-up area. This change caused the urban space structure to become loose and the city boundaries to become extremely complex. Fig. 1 shows that in 2000-2002, 2006-2008, and 2010-2011, the expansion was mainly on the field, which led to the increase in the values of compact ratio and to the decrease in the values of fractal dimension. In the entire expansion stage of Tangshan, open expansion entailed greater effort than the field one did.

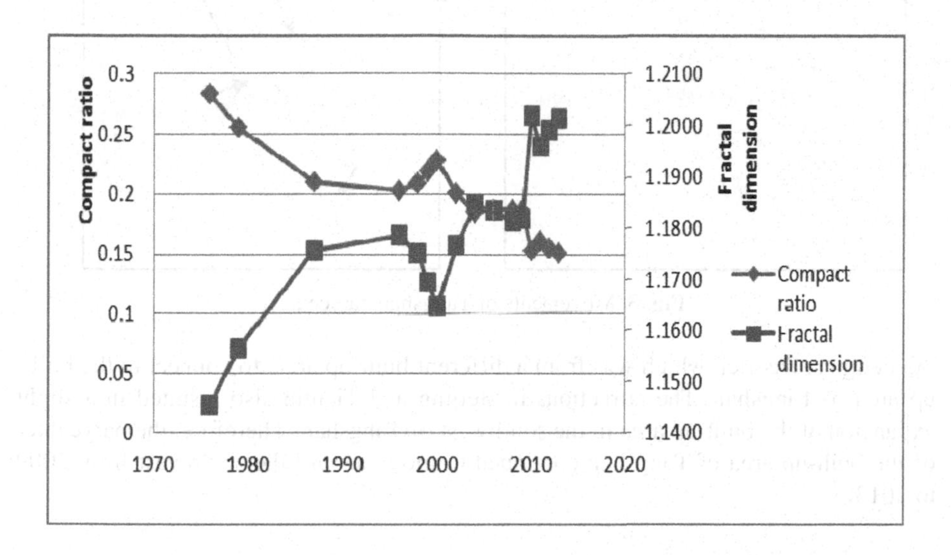

Fig. 2. Changes of compact ratio and fractal dimension from 1976 to 2013

3.4 Movements of Barycenter

From 1976 to 2009, the barycenter of the built-up area of Tangshan moved toward the northeast (Fig. 3). From 1976 to 1996, the built-up area of Tangshan annexed mainly the farmland and rural residential land in the north and northeast. In April 1992, Tangshan established the high-tech development zone. In May and July of the same year, the high-tech development zone was approved by the government of Hebei as the new provincial high-technology zone and the provincial economic and technological development zone. In April 1998, the high-tech development zone was listed by the provincial government as one of the eight key development zones in Hebei province. As the high-tech development zone is located in the northeast, the urban expansion of Tangshan from 1996 to 2009 was mainly concentrated in the northeast region of the city. As Tangshan continued to expand outward, urban expansion caused

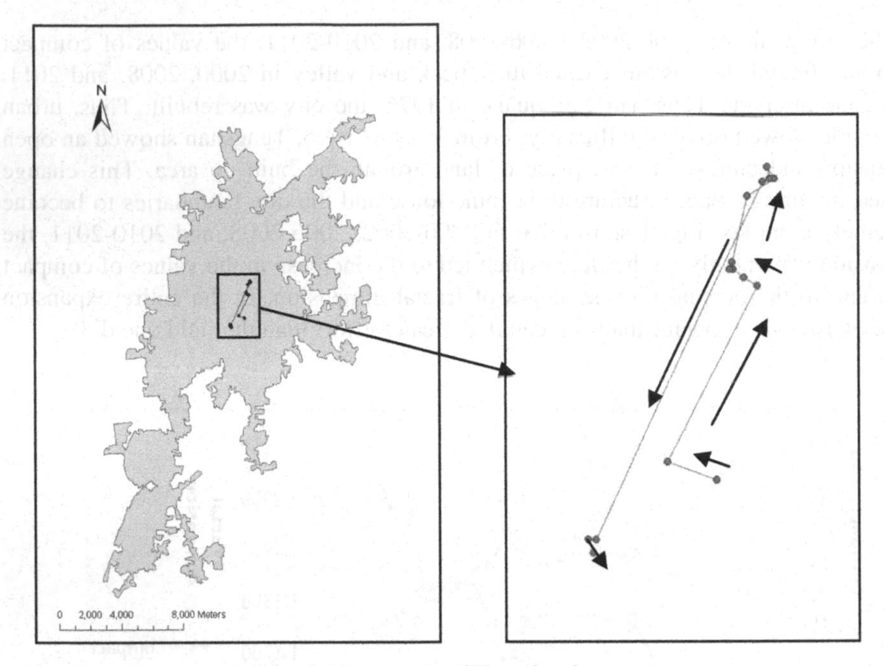

Fig. 3. Movements of Tangshan barycenter

the Fengnan District, which was from a different built-up area, to connect to the built-up area of Tangshan. The attractions in Beijing and Tianjin also resulted in a slight expansion of the built-up area in the southwest of Tangshan. Therefore, the barycenter of the built-up area of Tangshan continued to move toward the southwest from 2009 to 2013.

4 Conclusions

The urban expansion of Tangshan from 1976 to 2013 was monitored by RS. Although the urban expansion during the initial phase was slow, it was unreasonable. In the initial stage, the value of compact ratio decreased, and the morphology of Tangshan became loose. At the same time, the value of fractal dimension increased as the city boundary became complex. At the later period of the urban expansion under the guidance of the Tangshan city master plan (2003-2020), compact ratio was still low, and fractal dimension was still large. However, the change in the rates of these factors was less than that of urban expansion. In 2000, 2008, and 2011, urban expansion quickened, but the value of compact ratio tended to increase while that of fractal dimension tended to decrease. Thus, the urban expansion gradually became reasonable, urban morphology became stable, and the boundary of the built-up area became regular. In the future, Tangshan should devote itself to the establishment of a considerably compact city and to the development of relationships among its urban areas to improve socio-economic and environmental benefits. In-field development is considered as the main approach to the expansion of Tangshan, followed by the open type of development. We should protect the areas and the quality of the cultivated land surrounding the urban city of Tangshan. The barycenter of Tangshan should be moved toward the northwest. Moreover, Tangshan should exert great effort to connect with Beijing and Tianjin in the aspect of politics, economics, and culture so as to promote the city's development.

Acknowledgments. The author would like to acknowledge the support by the Natural Scientific Foundation of China (Grant No. 41101148).

References

1. Liu, J., Kuang, W., Zhang, Z., et al.: Spatiotemporal characteristics, patterns, and causes of land-use changes in China since the late 1980s. Geogr. Sci. 24(2), 195–210 (2014)
2. Ji, C.Y., Liu, Q., Sun, D., et al.: Monitoring urban expansion with remote sensing in China. International Journal of Remote Sensing, 22(8), 1441–1455
3. Chen, Q., Yin, C., Chen, G.: Spatial-temporal Evolution of Urban Morphology and Land Use Sorts in Changsha. Scientia Geographica Sinica 27(2) (2007) (in Chinese)
4. Batty, M., Longley, P.: Fractal Cities, a Geography of Form and Function. Academic Press, San Diego (1994)
5. Wang, X.-S., Liu, J.-Y., Zhuang, D.-F.: Spatial-Temporal Changes of the Shapes of Chinese Cities. Resources Science 27(3) (May 2005) (in Chinese)
6. Kuang, W.H.: Simulating dynamic urban expansion at regional scale in Beijing-Tianjin-Tangshan Metropolitan Area. Journal of Geographical Sciences 21(2) (April 2001)
7. Statistical Communique on the 2012 national economic and social development of Tangshan City (Hebei province). Hebei Provincial Bureau of Statistics (2013) (in Chinese)
8. Cai, B., Zhang, Z., Liu, B., Zhou, Q.: Spatial-temporal changes of Tianjin urban spatial morphology from 1978 to 2004. Journal of Geographical Sciences 04, 500–510 (2007)
9. Shen, G.: Fractal dimension and fractal growth of urbanized areas. International Journal of Geographical Information Science 16(5), 419–437 (2002)
10. Gert, D.R.: Environmental conflicts in compact cities: Complexity, decision making, and policy approaches. Environment and Planning B: Planning and Design 27(2), 151–162 (2000)
11. Roberto, C., Maria, C.G., Paolo, R.: Urban mobility and urban form: The social and environmental costs of different patterns of urban expansion. Ecological Economics 40(3), 199–216 (2002)
12. Li, Y., Zhua, X., Suna, X., et al.: Landscape effects of environmental impact on bay-area wetlands under rapid urban expansion and development policy: A case study of Lianyungang, China. Landscape and Urban Planning 94, 218–227 (2010)
13. Mu, F.-Y., Zhang, Z.-X., Chi, Y.-B., et al.: Dynamic Monitoring of Built-up Area in Beijing during 1973-2005 Based on Multi-original Remote Sensed Images. Journal of Remote Sensing 11(2) (2007) (in Chinese)
14. Wang, S., Ma, H., Zhao, Y.: Exploring the relationship between urbanization and the eco-environment—A case study of Beijing-Tianjin-Hebei region. Ecological Indicators 45, 171–183 (2014)

15. Hermosilla, T., Palomar-Vázquez, J., Balaguer-Beser, Á., et al.: Using street based metrics to characterize urban typologies. Computers, Environment and Urban Systems 44, 68–79 (2014)
16. Cavan, G., Lindley, S., Jalayer, F., et al.: Urban morphological determinants of temperature regulating ecosystem services in two African cities. Ecological Indicators 42, 43–57 (2014)

Evaluating Urban Expansion of Beijing during 1973-2013, by Using GIS and Remote Sensing

Fang Liu, Lifeng Shi, Zengxiang Zhang, and Xiaoli Zhao

P.O. Box 9718, Olympic Village Science Park W. Beichen Road, Chaoyang District
{Fang Liu,Lifeng Shi,Zengxiang Zhang,Xiaoli Zhao}
liufang@radi.ac.cn

Abstract. This paper presents an integrated study about the urban expansion process in Beijing from 1973 to 2013. Annual urban expansion area index, gravity center transfer model, fractal dimension index and the elastic coefficient or urban expansion-population growth are employed to explore the temporal and spatial characteristics of urban expansion. The results reveal that urban expansion of Beijing underwent seven different expansion stages, including one slow and steady expansion stage, three fast expansion stages, three shot deceleration stages, in the past 40 years. The centroid of urban lands moved to the southeast and urban boundaries became more and more complex and irregular. Urbane expansion speed was not harmonized with population growth distinctly in 1987-1998, 2000-2003 and 1998-2000. This study can provide scientific data for urban planning of Beijing and related studies of urban expansion in depth.

Keywords: Beijing, Remote Sensing, Urban Expansion, Spatial-temporal Characteristics.

1 Introduction

Urban lands refer to the space where human production, living, and civilization achievements are concentrated [1]. These spaces undergo dynamic changes constantly. Although these areas only cover 3 percent of the earth's land surface [2], their rapid growth exerted significant heavy pressure on land and other resources surrounding them, altered the natural landscape, and created enormous environmental, ecosystem, and social impact at the local, regional, and global scales [3][4][5]. These effects are particularly important in developing countries [3].

Since the implementation of the reform and opening-up policy in 1978 and particularly after the initiation of land reforms in 1987, China, as a developing country with a vast population and scarce land per capita [6][7], developed dramatically. The marked expansion in its urban land along with its rapid economic development led to a constant rise in its urbanization level, a large population migration from rural areas to urban areas, and the creation of favorable policies [8]. As the capital city of China, Beijing has dense population, good social and economic foundations, and its urban expansion process is almost consistent with that of China.

F. Bian and Y. Xie (Eds.): GRMSE 2014, CCIS 482, pp. 635–642, 2015.

GIS and Remote Sensing technologies provide efficient tools to collect and analyze the information necessary to detect changes in urban lands that conventional surveying technology cannot deliver in a timely and cost effective manner. Thus, numerous studies on urban growth were carried out using these tools [5].

The purpose of this study is to analyze the urban expansion evolution of Beijing by using remote sensing data in multiple temporal phases and long time series to determine their characteristics. Following this introduction, we describe the study area, along with the data source, and the methods used in the analysis. We then reveal spatial-temporal characteristics of urban expansion in Beijing by using some indicators and models, such as annual urban expansion area, fractal dimension index, and so on. We present our conclusions in the last section.

2 Description of the Study Area

As the capital city of China, Beijing locates in the northern part of the North China Plain, and is northwest high-lying, southeast of the low. With four distinct seasons, Beijing belongs to the warm temperate and semi-humid climate zone. Covering the total area of 16,410.54 square kilometers, Beijing is divided into fourteen municipal districts and two counties. With a resident population of 2069.30 million and the gross domestic product of 1.78794 trillion yuan, Beijing has dense population and developed economy [10].

3 Materials and Methods

3.1 Data

Multi-sources remote sensing images were ursed to extract urban lands information of Beijing from 1973 to 2013, including 3 Landsat Multi Spectral Scanner (MSS) images, 17 Landsat Thematic Mapper or Enhanced Thematic Mapper Plus (TM/ETM+) images, 1 Landsate Operational Land Imager (OLI) image, 1 China Brazil Earth Resources Satellite (CEBERS) CCD image and 1 China Environmental Satellite (HJ-1) CCD image. All remote sensing imagery were chosen based on the principles of phases within crop growing season, and a cloud cover area of less than 10% of the image [5][9]. Table 1 showed the specific year of different remote sensing data. In the past 40 years, urban lands in Beijing were extracted in 16 monitoring time and 15 periods.

Table 1. Monitoring time of different remote sensing data

Data types	Monitoring time
Landsat MSS	1973, 1975, 1978
Landsat TM/ ETM+	1984, 1987, 1992, 1996, 1998, 1999, 2000, 2001, 2002, 2003, 2004, 2005, 2006, 2007, 2009, 2010, 2011
CBERS CCD	2008
HJ-1 CCD	2012
Landsat OLI	2013

3.2 Data Processing

Firstly, accessible multi-source remote sensing images are composited to be false color composite images, near-infrared band, red band and green band of each image correspond to the red, green, and blue channels. Secondly, composited multiresource remote sensing images are geometric rectified on the MGE software platform by choosing topographic maps (scale 1:10 million) as the reference standard. During the geometric rectification process, finite element model and the nearest interpolation method are chosen as the spatial transformation model and Interpolation method respectively, and the image geometric rectification error is controlled within 1-2 pixels. Data were emendated into a unified coordinate system and equal-area cutting conic projection. The central meridian and double standard latitudes were 105°E, 25°N and 47°N respectively, and the adopted ellipsoid was KRASOVSKY. Thirdly, urban lands of Beijing were extracted by visually interpreting, according to characteristics such as the spectral characteristic and geometric characteristic, and so on. Finally, urban expansion database was constructed on the ArcInfo platform after some procedures, such as editing, data conversion, and so on [9].

3.3 Methods

Annual Urban Expansion Area Index. Annual urban growth area (AUA) index was adapted in this study to evaluate the urban growth speed of Beijing (Xiao et al., 2006; Tian et al., 2005). AUA is defined below

$$AUA=(UA_{n+i}-UA_i)/n\ . \tag{1}$$

Where UA_{n+i} and UA_i are urban areas at time i+n and i, respectively, and n is the interval of the calculating period (in years).

Gravity Center Transfer Model. In order to have a quantitative and intuitive description of the spatial change characteristics of urban growth in Beijing from 1973 to 2013, gravity center transfer model was employed in this study [11]. The equation for calculating center of gravity is given as follows.

$$X_t = \sum_{i=1}^{n}(C_{ti} \times X_i) \Big/ \sum_{i=1}^{n} C_{ti} \quad Y_t = \sum_{i=1}^{n}(C_{ti} \times Y_i) \Big/ \sum_{i=1}^{n} C_{ti}\ . \tag{2}$$

where X_t and Y_t indicate the area-weighted abscissa and ordinate of all the urban land patches in year t; C_{ti} is the area of urban land patch i; n is the total number of patches of urban land; and X_i and Y_i are the abscissa and ordinate of urban land patch i.

In order to describe the amplitude of variation of gravity center, the distance D of the shift of centroids is defined as

$$D = \sqrt{(X(t_2) - X(t_1))^2 + (Y(t_2) - Y(t_1))^2}\ . \tag{3}$$

where $X(t_1)$, $Y(t_1)$, $X(t_2)$, $Y(t_2)$ are the area-weighted abscissa and ordinate of all the urban land patches year t_1 and t_2, t_1 is the beginning of the monitoring period, and t_2 is the end of the monitoring period.

Fractal Dimension Index. To describe characteristics of the spatial morphology of urban land, fractal dimension index was used in this study to reveal the complex and non-linear changes of urban land boundaries of Beijing. Box-counting method is the most commonly used method estimating the fractal dimension [12], based on the relationship of the area and perimeter of the box covering the boundaries, and is defined below,

$$\ln N(r) = C + D \ln M(r)^{1/2}. \tag{4}$$

Where C is a constant to be determined, D is the fractal dimension index calculated. r is the length of the square grid, N(r) and M(r) are numbers of square boxes covering urban land boundaries and the whole urban land. Generally, the value of fractal dimension is between 1 and 2, the higher the value of fractal dimension, the more complex, irregular and disperse of urban land boundaries.

The Elastic Coefficient of Urban Expansion- population Growth. Population growth is an important factor affecting the urban expansion. The elastic coefficient of urban expansion- population growth was used in this study to reflect the coordination between urban expansion and population growth, the formula is,

$$R_{POP} = \left(\frac{UA_{n+i} - UA_i}{UA_i}\right) / \left(\frac{POP_{n+i} - POP_i}{POP_i}\right). \tag{5}$$

Where $R_{POP}$is the population elasticity coefficient; UA_{n+i}, UA_i, POP_{n+i} and POP_i are the urban area , population at time i+n and i, respectively, and n is the interval of the calculating period (in years).

4 Results

4.1 Temporal Characteristics of Urban Expansion in Beijing

Since the 1970s, accompanied by the accelerating process of urbanization, urban lands of Beijing expanded dramatically and had profound impacts on other land use types near the suburbs, such as arable land, forest, and so on. As the capital city of China, Beijing was chosen to be the study region, and its urban expansion reflected evolution characteristics of urban expansion in China in some extent. From 1973 to 2013, based on the old urban boundaries, urban lands in Beijing expanded rapidly in a balanced way towards different directions. The form of urban expansion in Beijing was just like an increased pie. In the past 40 years, urban lands in Beijing increased from 183.84 square kilometers in 1973 to 1478.94 square kilometers in 2013, with an annual urban expansion area of 32.38 square kilometers.

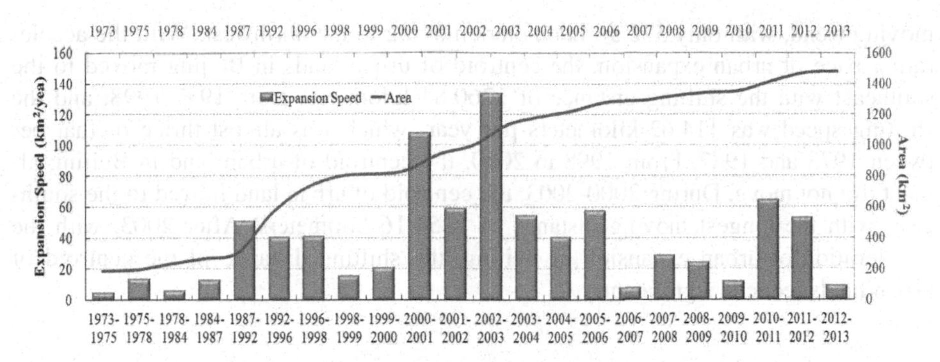

Fig. 1. The area of urban lands of Beijing and urban expansion speed during 1973-2013

Figure 1 showed different stages of urban expansion of Beijing underwent over the past 40 years. During 1973-1987, urban lands in Beijing expanded slowly and steadily, with an annual urban expansion area of 8.50 square kilometers. From 1987 to 1998, urban lands in Beijing entered the first fast expansion stage, with an annual urban expansion area of 44.77 square kilometers, which was chalked up to the rapid development of social economy, the strengthening of the production and living activities and population growth. Affected by the Asian financial crisis, urban expansion of Beijing experienced a shot deceleration stage from 1998 to 2000. Between 2000 and 2003, urban lands of Beijing entered the peak period of urban expansion, with a total expansion area of 299.58 square kilometers and an annual urban expansion area of 99.86 square kilometers, stimulated by the rapid development of economy after China joined the World Tourism Organization. During 2003-2010, urban development of Beijing focused on improving the urban traffic conditions and reducing the pollutions, and urban expansion underwent a deceleration stage caused by the successful bid for the 29th Olympic Games. After 2010, urban expansion speed of Beijing rebounded from 2010 to 2011, but reduced from 2011 to 2010, affected by the new urbanized policy and urban development strategy that encouraging the coordinative development of small, medium and large cities and priority to the development of small and medium cities. After 2013, urban expansion of Beijing is expected entering a continued deceleration stage.

4.2 Spatial Characteristics of Urban Expansion in Beijing

The area-weighted centroid method was used to represent the spatial movement of urban land from the late 1973 to 2013 in the study area. Figure 2 shows the centroid position in different monitoring time and the distance of the shift of urban land. Referring to different expansion stages of urban lands in Beijing underwent, seven time nodes (1973, 1987, 1998, 2000, 2003, 2010 and 2013) were chosen to describe and analyze the moving directions and shifting magnitudes of centroids of urban lands in Beijing during urban expansion process. Overall, the centroid of urban lands in Beijing moved to the southeast in the past 40 years. Between 1973 and 1987, centroids of urban lands in different monitoring time distributed densely, and had no obvious moving trend, with only 572.56 kilometers moving to the northwest. With the accelerating pace of urban expansion, the centroid of urban lands in Beijing moved to the southeast with the shifting distance of 1260.84 kilometers from 1987-1998, and the shifting speed was 114.62 kilometers per year, which was almost thrice of that between 1973 and 1987. From 1998 to 2000, the centroid of urban land in Beijing almost did not move. During 2000-2003, the centroid of urban land moved to the southeast, with the longest moving distance of 2254.16 kilometers. After 2003, with the deceleration of urban expansion in Beijing, the shifting distance of the centroid of urban lands reduced significantly.

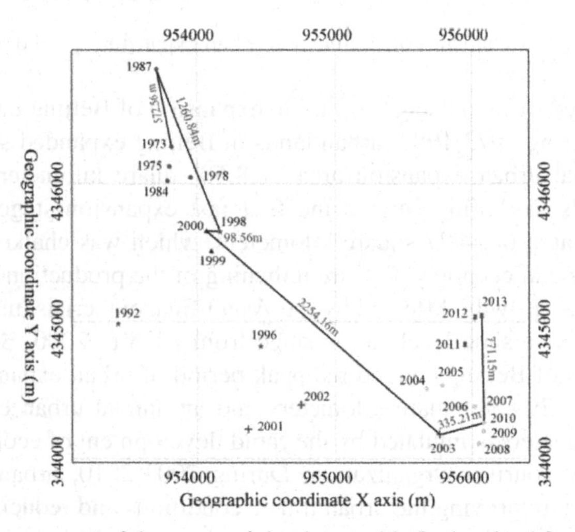

Fig. 2. Spatial movement of the area-weighted centroid of urban land from 1973 to 2013

4.3 Changes of Spatial Morphology in Beijing

Fractal dimension index can reveal the complexity of urban boundaries. The higher the value of fractal dimension index, the more complex and irregular of the corresponding urban boundaries. In this study, eight kinds of grids with the scale of 3000 meter, 2000 meter, 1000 meter, 500 meter, 250 meter, 200 meter, 150 meter and 100 meter, respectively, were employed to calculate the fractal dimension index, and values of fractal dimension index of Beijing in different monitoring time were plotted in Figure 3. The fractal dimension index of urban land in Beijing increased from 1.1469 in 1973 to 1.1742 in 2013, indicating that urban boundaries of Beijing was more complex and irregular. Though the fractal dimension index of Beijing was still fluctuating, the magnitude of changes of this index was significantly reduced after 1998. Compared to a few years before 1998, urban boundaries of Beijing were simpler after 1998, and the corresponding value of the fractal dimension index had a slow decreasing trend.

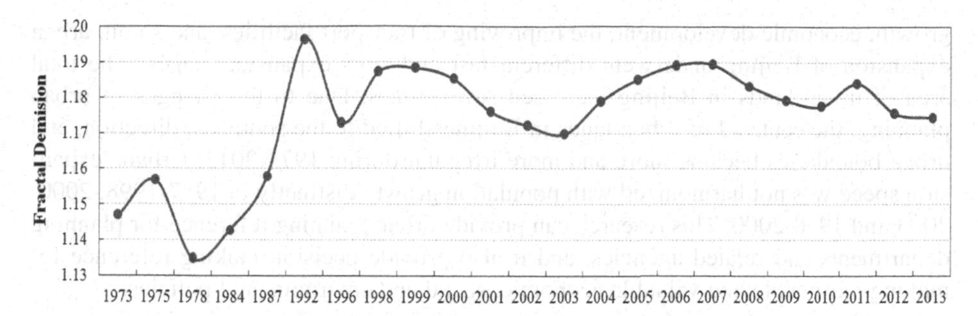

Fig. 3. Values of fractal dimension indes of Beijing from 1973 to 2013

4.4 The Relationship between Urban Expansion and Population Growth

Population is one of the important driving factors of urban expansion. Urban population growth stimulates the strongly demand of people lived in cities for urban land, and results in urban expansion constantly, so reasonable population growth closely relates to the sustainable development of cities. If the relationship of population growth and urban expansion is uncoordinated, urban development will be hindered. Table 2 revealed that the process of urban expansion in Beijing almost brought into correspondence with population growth during periods of 1978-1989 and 2003-2012, in which Beijing underwent slow and steady urban expansion stages. Urbane expansion speed was not harmonized with population growth distinctly in 1987-1998, 2000-2003 and 1998-2000. During the periods of 1987-1998 and 2000-2003, urban expansion underwent two fast expansion stages, and urban expansion speed surpassed the population rate, which resulted in the oversupply of urban lands in these two periods, had a profound impact on other land use types surround urban lands and was unfavorable for the healthy development of Beijing. Urban expansion lagged far behind the pace of population growth, and the demand for urban land exceeded the supply from 1998-2000.

Table 2. The elastic coefficient of urban expansion- population growth from the late 1970s to 2012

Peroid	1978-1987	1987-1998	1998-2000	2000-2003	2003-2010	2010-2012
P_{POP}	1.51	8.10	0.44	4.64	0.63	1.58

5 Conclusions

In this paper, Beijing, the capital city of China was selected as the study area. Urban expansion process from 1973 to 2013 was reconstructed by using remote sensing and GIS technologies. Temporal-spatial characteristics of urban expansion were fully analyzed by using many indices, such as annual urban expansion area, gravity center transfer model, fractal dimension index and the elastic coefficient of urban expansion-population growth. In the past 40 years, affected by factors, such as, population growth, economic development, the improving of transport facilities, and so on, urban expansion of Beijing underwent different fast and slow expansion stages. The total area of urban lands in Beijing increased seven times. Due to the changes of urban planning, the centroid of urban lands in Beijing shifted to the southeast direction, and urban boundaries became more and more irregular during 1973-2013. Urbane expansion speed was not harmonized with population growth distinctly in 1987-1998, 2000-2003 and 1998-2000. This research can provide urban planning reference for planning departments and related agencies, and it also provide decision making reference for making reasonable and valuable economic, social and environmental policies.

Acknowledgments. The author would like to acknowledge the support by the Natural Scientific Foundation of China (Grant No. 41101148).

References

1. Zhou, Y.: Urban Geography, pp. 20–100. Commercial Press, Beijing (1995) (in Chinese)
2. Sarvestani, M.S., Lbrahim, A.L., Kanaroglou, P.: Three decades of urban growth in the city of Shiraz, Iran: A remote sensing and geographic information systems application. Cities 28, 320–329 (2011)
3. Haregeweyn, N., Fikadu, G., Tsunekawa, A., Tsubo, M., Meshesha, D.T.: The dynamics of urban expansion and its impacts on land use/land cover change and small-scale farmers living near the urban fringe: A case study of Bahir Dar, Ethiopia. Landscape and Urban Planning 106, 149–157 (2012)
4. Berling-Wolff, S., Wu, J.: Modeling urban landscape dynamics: A case study in Phoenix, USA. Urban Ecosystems 7, 215–240 (2004)
5. Tian, G., Liu, J., Xie, Y., Yang, Z., Zhuang, D., Niu, Z.: Analysis of spatio-temporal dynamic pattern and driving forces of urban land in China in 1990s using TM images and GIS. Cities 22, 400–410 (2005)
6. Hubacek, K., Guan, D., Barua, A.: Changing lifestyles and consumption patterns in developing countries: A scenario analysis for China and India. Futures 39, 1084–1096 (2007)
7. Yang, H., Li, X.: Cultivated land and food supply in China. Land Use Policy 17, 73–88 (2000)
8. Liu, Y., Wang, Y.: Study on resource-environment response to the rapid urban expansion of China. Energy Procedia 5, 2549–2553 (2011)
9. Zhang, Z.: Remote sensing monitoring of urban expansion in China, pp. 21–50. Global Map Press, Beijing (2006) (in Chinese)
10. National Bureau of Statistics of China. China Statistical Yearbook-2013, pp. 10–25. China Statistics Press, Beijing (2013) (in Chinese)
11. Wang, S., Liu, J., Zhang, Z., Zhou, Q., Wang, C.: Spatial pattern change of land use in China in recent 10 years. Acta Geographica Sinica 57(5), 523–530 (2000)
12. Wang, X., Liu, J., Zhuang, D., Wang, L.: Spatial-temporal changes of urban spatial morphology in China. Acta Geographica Sinica 60(3), 382–400 (2005) (in Chinese)

The Yield Estimation of Rapeseed in Hubei Province by BEPS Process-Based Model and MODIS Satellite Data

Chan Ji[1], Jiahua Zhang[2], and Fengmei Yao[3,*]

[1] School of Geoscience, Yangtze University,Wuhan, 430100, China
[2] Institute of Remote Sensing Application, Chinese Academy of Sciences, Beijing, 100094, China
[3] University of Chinese Academy of Sciences, Beijing, 100049, China
zhangjh@radi.ac.cn, yaofm@ucas.ac.cn

Abstract. In this study, the BEPS (boreal ecosystem productivity simulator) model combining the Moderate Resolution Imaging Spectroradiometer (MODIS) multi-temporal satellite remote sensing data such as leaf area index (LAI) and land cover types data were used to simulate the rapeseed in Hubei province. Interpolating the 17 agricultural meteorological stations data of rapeseed in the study area, and based on the relevant data of rapeseed crop growth physiological and the process of rapeseed growth simulation technology, the estimating biomass of rapeseed was performed in a regional scale. This paper validated the result of simulated BEPS model and actual statistical data of rapeseed yield. The results showed that the model simulated values are agreed with the actual statistics yield of rapeseed, the correlation coefficient (R^2) is up to 0.877 with 0.01significant level. The simulated value with absolute mean square error (RMSE) of statistics is 416 kg/hm^2. It determined that the BEPS process-based model integrated with MODIS satellite data can estimate the rapeseed yield in a regional scale.

Keywords: BEPS model, MODIS, rapeseed, yield estimation, remote sensing, Hubei Province, China.

1 Introduction

Rapeseed planting has a long history in China, the planting area and the yield of rapeseed is the largest in the world [1]. Rapeseed is a kind of strong adaptability, wide application and high economic value oil crops, which is the important edible vegetable oil. Along with the growing demand for oil-bearing crops, however, for the rapeseed planting area has been shrinking, oil crops industry is confronted with more and more development pressure. With the increasing of population and the decreasing of the cropland, food security will urgent. The timely and accurate estimation of grain yield and annual fluctuation can help governments to plan effective strategies [2].

At present, the development and application of crop growth simulation model help to promote traditional agriculture to the digital agriculture forward. The information technology development significantly improves the comprehensive agriculture productivity.

F. Bian and Y. Xie (Eds.): GRMSE 2014, CCIS 482, pp. 643–652, 2015.

The traditional crop estimation methods mainly include sampling survey, meteorological model, based on crop growth simulation model and large area yield estimation based on remote sensing technology. The sampling survey and meteorological model yield estimation is relatively mature and stable, has been applied to estimate crop yield for many years, because there are statistical model, yield estimation results still have considerable uncertainty and limitations. As study methods have been further improved, the precision of the yield estimation method has been gradually accepted, thus it began to be used widely.

Since 1960s, the computer model has been applied to simulate all aspects of crops growth. Many scholars abroad have successfully developed many simulation models of crops and these kinds of technologies have become more mature and increasingly applied widespread [3]. The expert system, 3S technology, visualization technology and network technology have been successfully used to monitor crops growth, yield estimation, precision farming and so on [4, 5]. Crop growth models in China usually through a computer model to simulate crop growth process in the different stages of growth, such as photosynthesis and respiration, dry-matter accumulation and distribution [6, 7].

In the last years, the simulation model of rapeseed from the stage using the correlation and linear or nonlinear regression analysis and set crop shape and yield as the main research purpose to base on existing simulation model of optimization phase. Kiniry et al. (1995) developed a rapeseed simulation model (EPR95) in widely applying in the soil, climate and crop models, which in days for the time scale, considering the weather, hydrology, soil temperature, erosion, deposition, nutrient cycling, land conditions, crop management and the impact on crop growth [8]. Petersen et al. (1995) used DAISY (soil-plant system simulation model) to evaluate rapeseed growth. DAR95 (DAISY-rape) is a simulation of crop yield, soil moisture and nitrogen dynamic of soil vegetation atmosphere model [9]. The CERES-Rape developed by Gabrielle et al. (1998) [10], which simulated soil moisture balance, developmental periods, crop growth and other processes, established the relationship between leaf area index and per square flowers, pods and seeds and seeds per pod, and calculate the grain weight and yield at harvest period. In China, some rapeseed growth and yield simulated models were developed in recent years [1, 11-13].

This paper modified process-based Boreal Ecosystem Productivity Simulator (BEPS) [14] model to estimate rapeseed yield. The BEPS model was originally developed to estimate the NPP of a boreal ecosystem and has been modified to better represent canopy radiative processes [15].The related parameters in the model were used to simulate the growth process of rapeseed in order for simulating the biomass, then combining with the harvest index to compute the yield of rapeseed, at last using statistic data to validate the simulation value of rapeseed yield.

2 Materials

2.1 Study Area

The study area is in Hubei province (29°05′~33°20′N, 108°21′~116°07′E) of central China and located in the middle reach of Yangtze river. It lies in the north of Dongting

Lake. The total area of Hubei province is about 18.59 km^2, in which mountainous region, hills, downland and the plain area account for 55.5%, 24.5% and 20%, respectively. Hubei province is located in the subtropical zone, has the tropical monsoon climate. Most region except for the alpine region is a humid subtropical monsoon climate, sufficient light, frost-free period is long, abundant precipitation, rain at the same with heat and the mean annual temperature is 15~17°C.It is cold and dry in winter and has an average temperature of 2~4°Cin January(the coldest month in a year), while in summer, it is hot and rainy with an average temperature of 27~29°Cexcept for the alpine region and extreme maximum temperature can reach more than 40°C in July. The average annual precipitation is about 800~1600 mm. Sunlight is about 1100~2150 hours annually and summer is usually longer than winter.

The growth period of winter rapeseed is from October until the second year in May, due to the appropriate soil temperature, the suitable precipitation and the fertility of the vast area of Jianghan Plain. Hubei province becomes biggest production of rapeseed in China in recent ten years.

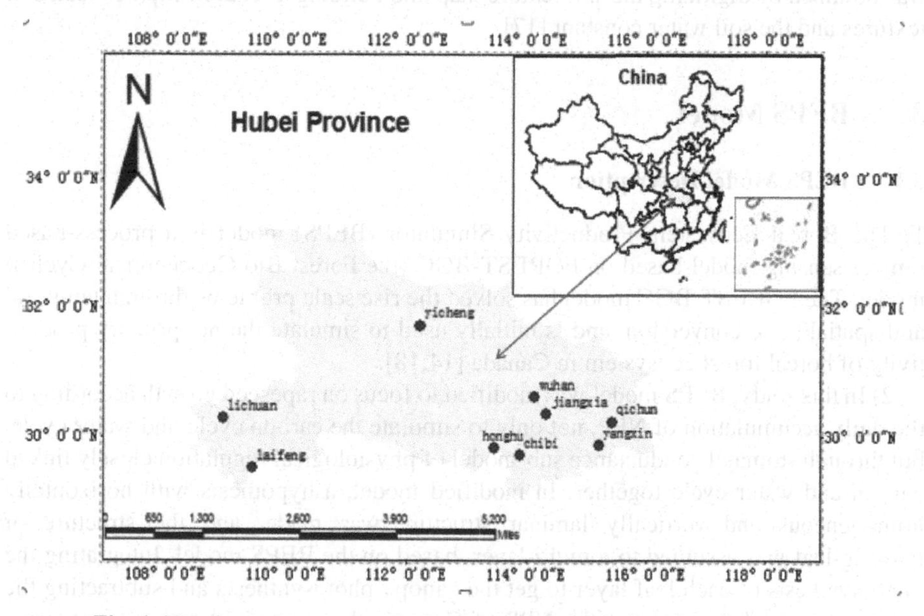

Fig. 1. Hubei province map and selected meteorological stations in the study

2.2 Data

Remote Sensing Data. The remote sensing data collected from the EOS data center in U.S. NASA including land cover and LAI (leaf area index) in the study area. Land cover data is derived from MODIS (Moderate-resolution Imaging Spectroradiometer) product (MCD12) and its spatial resolution is 500 m and time resolution is 1year. The LAI data is the MODIS level 3product at 1km spatial resolution and 8 days resolution, which range from October 2008 to May 2009. All these data were projected to the same projected coordinate system, geometric correction, atmospheric correction, sensors correction and cloud processing before being input to the model. In this study, the land cover of Hubei province in 2009 is selected.

Meteorological Data. Daily meteorological data were collected at 17 meteorological stations in Hubei province from 2008 to 2009, including maximum temperature, minimum temperature, relative humidity and precipitation which were from the National Meteorological Information Center in China. Input these meteorological data into ArcGIS by using the inverse distance weighting (IDW) method for spatial interpolation arithmetic to get the output raster dataset which is at 1km spatial resolution.

Other Data. Physical geographical data includes longitude, latitude and DEM data of the study area, which used to estimate the daily solar radiance based on the method of Windlow et al. (2001) [16]. Parameters for rapeseed in the growth period and the soil available water capacity (AWC) data are necessary in BEPS model. The soil AWC data was obtained by digitizing the soil texture map and building a relationship between soil textures and the soil water constant [17].

3 BEPS Model

3.1 BEPS Model Description

1) The Boreal Ecosystem Productivity Simulator (BEPS) model is a process-based remote sensing model based on FOREST-BGC (the Forest Bio-Geochemical Cycles) model. The FOREST-BGC model has solved the rise scale problems through temporal and spatial scale conversion, and is initially used to simulate the net primary productivity of boreal forest ecosystem in Canada [14,18].

2) In this study, BEPS model was modified to focus on rapeseed growth according to the daily accumulation of NPP, not only to simulate the carbon cycle and water cycle, but through stomatal conductance sub-model of physiological regulation closely linked carbon and water cycle together. In modified model, a hypothesis with horizontally homogeneous and vertically laminar structure was made, and the structure of two-big-leaf was assumed to a multi-layer, based on the BEPS model. Integrating the photosynthesis of each leaf layer to get the canopy photosynthesis and subtracting the autotrophic respiration to get the NPP of C_3 crop, the crop yield was estimated according the ratio of NPP [15, 19].

3.2 The Principle of Vegetation Canopy Photosynthesis

3.2.1 Photosynthetic rate of leaf-level

1) Instantaneous photosynthetic rate (A, $\mu molC m^{-2} s^{-1}$)

The model accumulates the assimilation rate of C_3 plants using the Farquhar model [20].

$$A = \min\{w_v, w_j\} - R_d \tag{1}$$

$$R_d = 0.015 V \tag{2}$$

$$w_v = V_{cmax} \tag{3a}$$

$$w_j = J \tag{3b}$$

Where, A is the net CO_2 assimilation rate, w_v is the rate limited by Rubisco, w_j is the rate limited by photoelectron transfer rate, R_d is dark respiration, $\mu molCm^{-2}s^{-1}$; V_{cmax} is the maximum rate of carboxylation, and J is the electron transfer rate.The calculation of V_{cmax} and Jcan be found in Chen *et al.* (1999) [18].

2) Daily integration of photosynthetic rate
Because C_4 photosynthesis is nearly saturated at current CO_2 concentration, the photosynthetic rate is almost free from the influence of CO_2concentration [21],and the daily total photosynthesis can be made with respect to time.

$$A_v = \frac{\alpha}{2}(V_{cmax} - R_d) \tag{4a}$$

$$A_j = \frac{\alpha}{2}(J - R_d) \tag{4b}$$

Where, $A_v$and $A_j$correspond toand w_v and w_j, respectively, after a small reduction for dark respiration. The parameter αcan be multiplied to the integral and the calculation can be expressed as [18]:

$$\alpha = \frac{1}{0.5\pi/2}\int_0^{\pi/2} \cos\theta \, d\theta = \frac{4}{\pi} \approx 1.27 \tag{5}$$

Where, θ is the solar zenith angle.

Solar Radiation. The solar radiation received by leaves is a key factor to determine their photosynthetic rate. Calculate the radiation by making leaves divided into N layers. The top of the canopy is the sunlit layer. The solar radiation received by the top sunlit leaves of the canopy includes the direct and scattered radiation from the sky. The radiation received by the lower layers is the sum of multi-scatted and reflect solar radiation from the canopy and soil surface, and the scattered radiation meet the radiation transfer equation [22].

$$S_{sun}(0) = S_{shade}(0) + S_0\mu_0 \tag{6}$$

$$-\mu\frac{dS_{in}}{dz} = -S_{in} + \frac{\omega}{2}\int_{-1}^{1} S_{in}(\mu')d\mu' + \frac{\omega S_0}{2}\exp\left(\frac{\tau}{\mu_0}\right) \tag{7}$$

Where, $S_{sun}(0)$is the photosynthesis available radiation received by the top leaves of the top of the canopy, $S_{shade}(0)$ is the scatted radiation, $S_0$is the direct radiation on the underlying surface and $\mu_0$is the cosine of the zenith angle. $S_{in}$is the scattered radiation received by the internal of the canopy, $L$is the distance between internal leaves and the top of the canopy, μis the cosine of the zenith in scatter direction, $\omega$is the single scatter albedo of leaves and $G_0$is the projection on the direction of the reflect radiation. The scattered radiation in the horizontal of L can be expressed as [19],

$$S_{shade}(L) = 2\int_0^1 S_{in}(L, \mu)\mu d\mu \tag{8}$$

Where, $S_{shade}(L)$ is the scattered radiation received by the shade leaves in L.

Photosynthetic Rate of the Canopy. The photosynthetic rate of the sunlit and shaded leaf, and the rapeseed canopy is divided into N (N>1) layers to make spatial scale expansion. Photosynthetic rate of the canopy is expressed below [23]

$$A_{canopy} = \frac{1}{N} A_{sun} LAI_{sun} + \frac{N-1}{N} A_{shade} LAI_{shade} \tag{9}$$

Where, the subscripts 'sun' and 'shade' denote the sunlit and shaded components of photosynthesis and LAI.

$$LAI_{sun} = 2\cos\theta\,(1 - \exp(-0.5\Omega LAI/co \tag{10a}$$

$$LAI_{shade} = LAI - LAI_{sun} \tag{10b}$$

Where, Ω is foliage clumping index, and Ωfor crop is 0.9 [14].

3.3 Conversion of NPP to Crop Yields

Gross primary productivity (GPP) could be calculated from photosynthetic rate A, and net primary productivity (NPP) is the difference between GPP and plant autotrophic respiration (R_a) and can be described as:

$$NPP = GPP - R_a \tag{11}$$

$$GPP = A_{canopy} \times L_{day} \times F_{GPP} \tag{12}$$

Where, $L_{day}$is the length of day, $F_{GPP}$is a conversion factor of photosynthesis to GPP; $R_a$is the autotrophic respiration, including maintenance respiration and growth respiration [18].

There is a great relationship between NPP and aboveground biomass, so the crop yield could be obtained by the NPP and harvest index (HI).

$$Yield = NPP \times \alpha \times I \tag{13}$$

Where α is the conversion ratio between carbon content and dry matter (~45%).

3.4 Key Parameters

V_{cmax}, the maximum rate of carboxylation, is one of key parameters in the model, representing the leaf of rapeseed maximum photosynthesis capacity. The value of V_{cmax} is mainly affected by the environmental factors such as temperature and fertilizer. J,the electron transfer rate, has a strong correlation with the V_{cmax}(Leuning, 1997) [24].

4 Results and Conclusion

4.1 The Distribution of Rapeseed Yield in Hubei Province

The yield of rapeseed in Hubei province in 2009 was estimated by using the modified BEPS model with MODIS remote sensing data and meteorological data (Fig. 2). The simulated results demonstrated that rapeseed yield in central and northern plains region is much higher than other places in Hubei, the western of Hubei is very small, and northeastern and southeastern is in general.

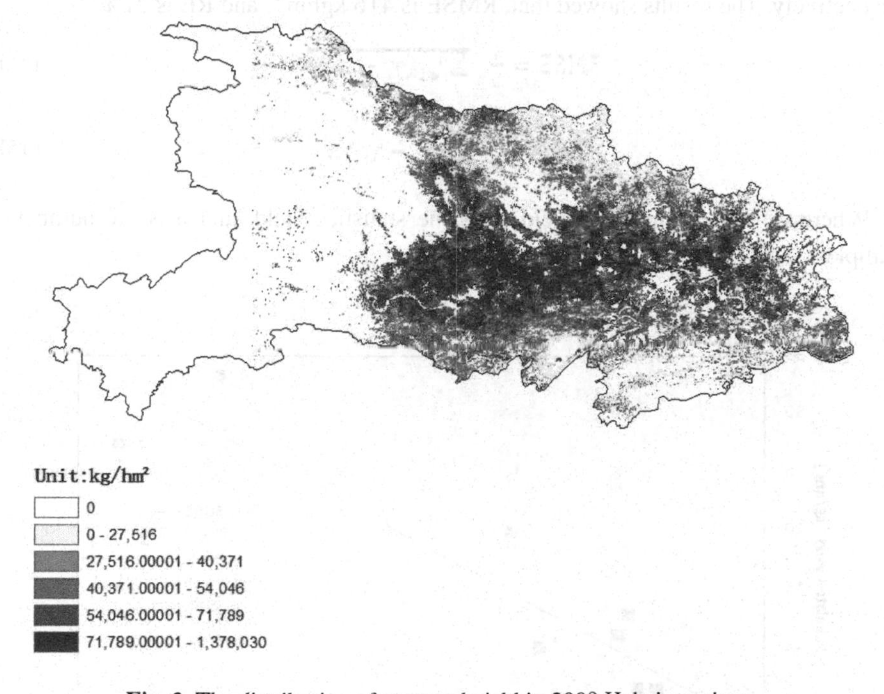

Fig. 2. The distribution of rapeseed yield in 2009 Hubei province

4.2 Validation of Results at the County Level

The county-level rapeseed yield data for model validation is from 2010Hubei Province Statistical Yearbook (Table 2).

Table 1. Comparison of statistics yield and simulated yield of rapeseed in 2009

County	Simulation yield	Actual Yield(10^4 tons)
Ezhou	3.53	4.86
Huanggang	31.813	33.05
Huangshi	6.34	6.01
Jingmen	37.71	27.27
Qianjiang	8.70	10.67
Shennongjia	0.03	0.03
Tianmen	13.32	9.67
Xiantao	8.02	12.43
Xianning	7.47	6.20
Yichang	14.89	19.63

The regression analysis between simulated yield and statistics yield in county-level was performed, and the Pearson correlation coefficient R^2 is up to 0.877(Fig.4) ($p<0.01$). The statistics yields of 11 counties/cities were used to calculate the root mean square error (RMSE) and relative error (RE) according to the Equs. (15a) and (15b), respectively. The results showed that, RMSE is 416 kghm^{-2}, and RE is 21%.

$$RMSE = \frac{1}{n}\sqrt{\sum_{i=1}^{n}(y_i - x_i)^2} \quad (15a)$$

$$RE = \frac{1}{n}\sum_{i=1}^{n}|y_i - x_i|/x_i \quad (15b)$$

Where, y_i is the simulated yield, x_i is the statistics yield, and n is the number of samples.

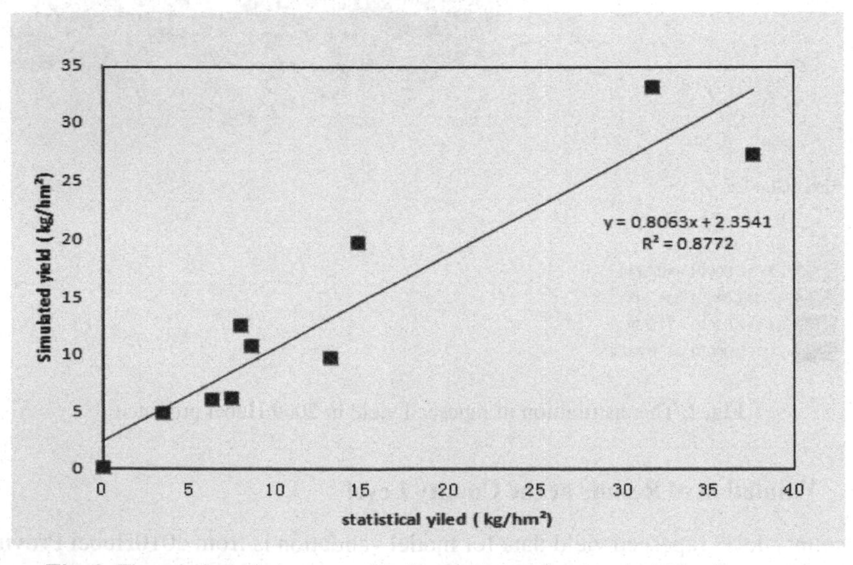

Fig. 3. The relationship between the simulated and the actual yields of rapeseed

5 Conclusion

In this study, for the difference sunlit canopy between rapeseed and forest, rapeseed canopy was assumed to have a multi-layer in modified BEPS model which combined with remote sensing data and GIS technology very well. The using of new technology and obtained data methods can contribute to transforming traditional agriculture to digital agriculture. The validation between yield simulated by the model and statistical yield at the county level showed that the developed process-based model was suitable for the estimation of rapeseed yield, with a coefficient of determination up to 0.88, the root mean square error of416 kg/hm^2. From the simulation results, the model simulation was statistically consistent with the actual data and can reflect the spatial yield distribution of rapeseed in Hubei province.

However, in the model, the developmental stages of rapeseed were set to the same that can't reflect the actual situations and only using statistical data to validate simulated result is also inadequate for regional estimation of crops yield. In future, validation can add agro-meteorological stations, and higher resolution remote sensing products and land cover of crops thematic maps are necessary to improve the estimation of crops yield.

Acknowledgements. This study is supported by the Global Change Global Research Key Project of the National Science Plan (No. 2010CB951302), 100 Talent Program of Chinese Academy of Sciences (No.Y24002101A), the 1-3-5 Innovation Project of RADI_CAS (No. Y3ZZ15101A), the Social Commonwealth Meteorological Research Project (No. GYHY201106027), and CAS-TWAS Project for Drought Monitoring and Assessment (No. Y3YI2701KB).

References

1. Liu, H., Jin, Z.Q.: Rapeseed development dynamic simulation model. J. Appl. Meteol. 14(5), 634–640 (2003)
2. Yao, F.M., Zhang, J.H.: Impact of Climate Change on Crop Yield and Its Simulation in China. China Meteo. Press, Beijing (2008)
3. De Wit, A.J.W., Van Diepen, C.: Crop model data assimilation with the Ensemble Kalman filter for improving regional crop yield forecasts. Agri. For. Meteor. 146(1), 38–56 (2007)
4. Gower, S.T., Kucharik, C.J., Norman, J.M.: Direct and Indirect Estimation of Leaf Area Index, fAPAR, and Net Primary Production of Terrestrial Ecosystems. Remote Sens. Environ. 70(1), 29–51 (1999)
5. Launay, M., Guerif, M.: Assimilating remote sensing data into a crop model to improve predictive performance for spatial applications. Agri. Ecos. & Environ. 111(1), 321–339 (2005)
6. Cao, H.X., Jin, Z.Q., Shi, C.L., Ge, D.K., Gao, L.Z.: Researches and application of crop model series in China. Agri. Net. Inform. 5, 45–50 (2006)
7. Zhang, J.H., Zhang, G.P., Wang, P.J.: Vegetation and Ecological Remote Sensing. Science Press, Beijing (2010)

8. Kiniry, J.R., Major, D.J., Izaurralde, R.C., Williams, J.R., Gassman, P.W., Morrison, M., Bergentine, R., Zentener, R.P.: EPIC model parameters forcereal, oilseed, and forage crop in the north Great Plain region. Can. J. Plant Sci. 63, 1063–1065 (1983)
9. Petersen, C.T., Svendsen, H., Hansen, S., Jensen, H.E., Nieksen, N.E.: Parameter assessment for simulation of biomass production and nitrogenuptake in winter rape. Europe. J. Agron. 4, 77–89 (1995)
10. Gabrielle, B., Denoroy, P., Gosse, G., Justes, E., Andersen, M.N.: Development and evaluation of a CERES-type model for winter oilseedrape. Field Crop Res. 57(1), 95–111 (1998)
11. Yuan, W.Z., Guan, C.Y.: The harvest index of rapeseed contributes to economical yield. J. Hunan Norm. Univ. 1, 65–67 (1998)
12. Liu, T.M., Cao, C.G., Huang, Y.: A simulation model of photosynthetic production and dry matter accumulation in rapeseed. J. Huazhong. Agric. Univ. 2(6), 126–149 (2003)
13. Wang, B.Q., Zhang, R.H., Zhang, C.L., Ma, D.L.: Rapeseed growth and development of computer simulation and application progression. Acta Agri. Bore. Sini. 13(2), 180–185 (2004)
14. Liu, J., Chen, J.M., Cihlar, J.: A process-based boreal ecosystem productivity simulator using remote sensing inputs. Remote Sensing of Environ. 62(2), 158–175 (1997)
15. Wang, P.J., Sun, R., Zhang, J.H.: Yield estimation of winter wheat in the North China Plain using the remote-sensing–photosynthesis–yield estimation for crops (RS–P–YEC) model. Int. J. Remote Sens. 32(21), 6335–6348 (2011)
16. Winslow, J.C., Hunt, E.R., Piper, S.C.: A Globally applicable model of daily solar irradiance estimated from air temperature and precipitation data. Ecol. Model. 143, 227–243 (2001)
17. Zhou, Y., Zhu, Q., Chen, J.M.: Observation and simulation of net primary productivity in Qilian Mountain, western China. J. Envir. Manag. 85(3), 574–584 (2007)
18. Chen, J., Liu, J., Cihlar, J.: Daily canopy photosynthesis model through temporal and spatial scaling for remote sensing applications. Ecol. Model. 124(2), 99–119 (1999)
19. Wang, P.J., Xie, D.H., Zhang, J.H.: Application of BEPS model in estimating winter wheat yield in North China Plain. Tran. CSAE 25(10), 148–153 (2009)
20. Farquhar, G.D., Caemmerer, S., Berry, J.A.: A biochemical model of photosynthetic CO2 assimilation in leaves of C3 species. Planta 149(1), 78–90 (1980)
21. Leakey, A.D.B., Uribelarrea, M., Ainsworth, E.A.: Photosynthesis, productivity, and yield of maize are not affected by open-air elevation of CO2 concentration in the absence of drought. Plant Phy. 140(2), 779–790 (2006)
22. Huang, H.F.: The study on Theory and Simulation of the Interactions of Soil, Vegetation and Atmosphere. Metereological Press, Beijing (1997)
23. Norman, J.M.: Simulation of microclimates. Biometeorology. In: Hatfield, J.L., Thomason, I.J. (eds.) Integrated Pest Management, pp. 65–99. Academic Press, New York (1982)
24. Leuning, R.: Scaling to a common temperature improves the correlation between the photosynthesis parameters Jmax and Vcmax. J. Expe. Botany 48(2), 345–347 (1997)

Research on the Impact of China's 'Independent Innovation Policy' on Patent Licensing's Geographical Evolution in Strategic Emerging Industries

Xue Yang, Xin Gu, and YuanDi Wang

Business School of Sichuan University, Chengdu, Sichuan 610064, China
{guxin,wangyuandi}@scu.edu.cn, cathyscu@sina.com

Abstract. China's 'National Program for Medium-to-Long-Term Scientific and Technological Development (2006-2020)' has been promulgated seven years. However, there are few works referred to it's impact on innovation capability of China's strategic emerging industries. This study investigates how China's policy affects patent licensing's geographical evolution, by empirically scrutinizing licensed patents in strategic emerging industries from 2000 to 2012. The results show that, the publication of China's independent innovation policy accelerates the evolution of patent licensing in China's strategic emerging industries though reducing the lag period between patent applying and licensing as well as decreasing the geographic distance between patent licensor and licensee. Moreover, the positive effect of policy on industries' capability of independent innovation appears distinct among different strategic emerging industries.

Keywords: Independent Innovation Policy, Patent Licensing, Geographical Evolution, Strategic Emerging Industries

1 Introduction

In year 2006, China's 'National Program for Medium-to-Long- Term Scientific and Technological Development (2006-2020)' (hereinafter refers to as the 'PROGRAM') pointed out that 'in the following 15 years, the guidelines of scientific and technological works are independent innovation, focus and breakthrough, supporting the rapid development, leading the future approach.' Among them, 'focus and breakthrough is to select key areas with certain foundation and advantages which are beneficial to people's livelihood and national security, and concentrate on them in order to reach the great leap forward development.' In September 2010, the State Council promulgated 'Decision on accelerating the fostering and development of strategic emerging industries' (hereinafter refers to as the 'DECISION') pointed out China's need to 'speed up the cultivation and development of strategic emerging industries', and confirmed that there were 7 strategic emerging industries in China, which are energy conservation and environmental protection industry, new generation of information technology industry, biotechnology industry, advanced equipment

F. Bian and Y. Xie (Eds.): GRMSE 2014, CCIS 482, pp. 653–663, 2015.

manufacturing industry, new energy sources industry, new materials industry and new energy vehicles industry. As technological patent licensing can effectively improve the independent innovation capability of enterprises in China(Wang et al., 2014 [20]), this paper using Chinese patent license's data searched by statistical standard about the classification of seven strategic emerging industries to understand the development trend of technology transfer in order to study the impact of China's 'independent innovation policy' on patent licensing's geographical evolution in China's strategic emerging industries after the publication of PROGRAM.

2 Theoretical Background

2.1 Strategic Emerging Industry

'Strategic emerging industry' is a new concept formally proposed by Chinese government in 2009. Therefore, domestic related research appeared after that with limited number. As for the content, it showed that scholars mainly did research focus on the relationship between strategic emerging industries and local regions, the selection and development models as well as the industries' development evaluation indexes(Zeng et al., 2013 [24]).

Centering on the relationship between strategic emerging industries and local regions, the study(Zhou et al., 2012 [25]) found that from the regional distribution point of view, the developing scale of strategic emerging industries in eastern area was the largest with obvious advantages. At the same time, the developing scale and degree of strategic emerging industries in central and western regions were equal, of which the western area slightly better than that in the central area while the northeast area was the smallest, and the development was relatively lagged behind. For the selection and development model of strategic emerging industries, the study (He et al., 2011 [8]) set up four indexes such as industries' overall situation, industries' guidance, industries' relevance and industries' dynamics combined the characteristics of strategic emerging industries, and then took the example of Hunan province to illustrate by making use of Weaver-Thomas evaluation model.

There are many other scholars have done researches on the evaluation indexes for the development of strategic emerging industries. The study (Wu et al., 2012 [21]) constructed the strategic emerging industrial cluster's competitiveness evaluation index system based on four one class indexes such as the promote effect of industrial science and technology, regional demand and scale growth advantages, spatial agglomeration and industrial association effect. It also took Shanxi Province as the empirical study cases for the choice of regional strategic emerging industries. The study (Huang et al., 2013 [11]) built the overall evaluation indexes of China's strategic emerging industries through three aspects including industry spread effect index, employment driven index and market potential index. It also made an empirical study by using the relevant statistical data in 2007. Other study (Yu et al., 2013 [23]) used the indexes of science talent and technology investment, technology output and economic benefit to measure the production efficiency of science and technology resources' allocation in strategic emerging industries. It also did empirical research on science and technology resources' input and output efficiency of strategic emerging industries in Jiangxi province by using the DEA method.

In foreign countries, there hasn't any concept of 'strategic emerging industry', however, this industrial classification system is similar to the new industry in developed countries. The scholars did related research about the definition and development law of these industries (Hefman et al., 2008 [9]; Brachert et al., 2013 [3]) as well as the evaluation of technological innovation in new industries (Marin et al., 2007 [16]; Guo et al., 2012 [7]).

2.2 Patent Licensing

Patent licensing is the main way of patent commercialization, it refers to the behaviour of transferring right to use patent without transferring the possession of the patent's ownership. As an important channel to obtain the technology, patent licensing has been studied for a long time. However, the existing researches mainly focus on the licensor and the license contract itself, such as the motivation of licensor's technology licensing (Gambardella et al., 2007 [5]), the licensing strategy and partner selection (Fosfuri, 2006 [4]; Kim,2009 [12]), drawing up the license contract (Anand et al., 2000 [1]), the protection of licensee's right (Laursen et al., 2010 [13]), and the relationship between technology licensor and licensee (Arora et al.,2006 [2]).

Research on patent licensing appeared to increasingly developed based on the level of industry in recent years, especially after the publication of PROGRAM. One of the quantitative analysis using macroscopic statistical data, shows that technology import has positive effect on large and medium-sized enterprises in China (Guo, 2008 [6]). A study examines the sources of technological innovation in China's industries using the 2004 economic census data and reveals that technology import has become the most important source for industrial innovation in China. In-house technological efforts are critical for developing original innovations as well as for absorbing the technologies transferred from external agencies.(Sun et al., 2010 [19]) Besides, other scholars (Li et al., 2011 [14]; Xu, 2012 [22]) point out that there are three main parts in the process of patent licensing, which are standard organization, patent licenser and patent licensee. In short term, the patent licenser will always benefit from the patent license under technology standard. The patent licensee's payoff depends on the royalties. The payoff of the whole industry will always be increased. In long term, the decrease of royalties will lower the costs of every company, which will extend the period of benefit. Therefore, communication companies should participate in the technology standards with their patents. The standard organization should set reasonable royalties.

To sum up, from the above literature review, it can be seen that the current researches on the strategic emerging industries in China or new industries overseas from home and abroad are mostly done qualitatively from the concept definition, the study of China's strategic emerging industries' geographical evolution based on patent licensing is still a gap. On the basis of existing research, this paper adds a regulated variable of policy bases on the related indexes of industrial patent licensing according to the patent license's data of China's strategic emerging industries, in order to explore the geographical characteristics of the independent innovation about technology transfer in strategic emerging industries in China.

3 Data and Methods

The patent data used in this paper are all come from Chinese patent database, which are patent license registration data offered by the State Intellectual Property Office. The State Intellectual Property Office has announced part of technology licensing data in year 1998 and 1999. Besides, it also has promulgated almost all technology licensing data from year 2000 to 2012, including the information about licensor, licensee, both name and number of the patent with authorized identification number and data, which provided rich source material for this paper.

By using the classification standard of strategic emerging industries as a criteria, the search result is industrial patent license data from January 1st 2000 to December 31th 2012 with a total number of 12910 from Chinese patent database. After excluding invalid patent data with incomplete variables, there are 10776 valid patent items remained, each item contains the information such as patent's application number, application date, authorize year, classification number, licensor, licensee, address, contract number and classification of the patent.

3.1 The Variables

Dependent Variables. For enterprises, the patents which acquired by license can effectively improve the independent innovative capability of enterprises in the process of 'technological industrialization'.(Lin, 2003 [15]) Therefore, in this paper, the dependent variable is the presence or absence of the independent innovative capability in China's strategic emerging industries as measured by industrial *patent licensing*.

Independent Variables. *Lag period*: It always exists a lag period between patent's applying and licensing, because that after certain patent has applied for and obtained authorization, the enterprise won't license the patent until it figures out the technical requirements and prepares for negotiating and signing the license contract. Therefore, it will consume some time before licensing and achieving the economic benefit through technological industrialization. In general, the lag period of licensing is shorter, the more advanced patent's technology are, and it has more obvious effect on enterprise to enhance the independent innovative capability by patent licensing. On the contrary, the effect is weaker. (Nerkar, 2003 [17])

H1: After the publication of PROGRAM, the implementation of relevant policies has significant positive impact on the rise of enterprises' independent innovative capability in China's strategic emerging industries through shortening the lag period of patent licensing.

Geographic distance: A patent license has two sides which are licensor and licensee. The geographic distance between them will influence the probability of successful licensing to some extent. Generally, the shorter the geographic distance between licensor and licensee of a patent is, the more conducive for them to communicate, negotiate and assist to learn technology, and the probability of successful patent licensing is higher. On the contrary, the effect is weaker. (Nooteboom et al., 2007 [18])

In this paper, the spherical distance between a licensor and a licensee can be calculated by equation (1):

$$d_{AB} = R \times \arccos\left[\sin(lat_A)\sin(lat_B) + \cos(lat_A)\cos(lat_B)\cos(|long_A - long_B|)\right] \quad (1)$$

In equation (1), R represents the mean radius of the Earth and uses constant of 6371km in calculations; A indexes the licensor's city, and B indexes the licensee's city; *lat* (latitude) and *long* (longitude) are respectively calculated from the mean of the above and below latitude and the mean of the left and right longitude, which are available from a city's website.

According to the latitude and longitude of Beijing and Shanghai, the calculated geographic distance between them is 1067.08km, which is similar to the linear distance (1066.55km) between the two cities measured by Google Earth. Obviously, the geographic distance alone is not sufficient to account for travel-time between Chinese cities. Travelling between two distant cities by airplane may take less time than travelling between two adjacent cities by bus. However, from a practical point of view, the flight distance between Beijing and Shanghai announced by airline company is 1084km, which is similar to the straight line distance. Therefore, taking the development of Chinese aviation industry into account (at the end of 2011, the national large-scale cities' airport coverage rate had reached to 100%, and the total number of transport airports in China's had reached to 180), we agree with that since the distribution of the coarsened distance is similar to that of log transformation, it still resolves the non-linearity problem.(Hong, 2013 [10])

H2: After the publication of PROGRAM, the implementation of relevant policies has significant positive impact on the rise of enterprises' independent innovative capability in China's strategic emerging industries through reducing the geographic distance of patent licensing.

3.2 The Basic Model

Based on previous researches, the basic model is specified by equation (2):

$$\ln Patent_{it} = \beta_1 + \beta_2 Year_i + \beta_3 Period_i + \beta_4 Distance_i + \gamma_{1i} + \gamma_{2i} Year_i + \beta_5 Policy_i + \varepsilon_i \quad (2)$$

In equation (2), ln$Patent_{it}$ represents the dependent variable which is the natural logarithm of patent license's number in different years of China's seven strategic emerging industries, including inventions, utility models and designs.

As for the independent variables, $Year_i$ represents the calendar year in which China's seven strategic emerging industries have records of patent license. $Period_i$ is a average number calculated by the difference between the year of patent licensed and the year of patent applied according to calendar years in seven strategic emerging industries. It measures the lag period of patent licensing. $Dintance_i$ is also a average number calculated by the difference between the location of licensor and licensee according to calendar years in seven strategic emerging industries. It measures the geographic distance of patent licensing. $Policy_i$ is a dummy variable about the publication of PROGRAM which assumes a value of 1 for the implementation of relevant policies and 0 otherwise.

In the basic model, β_1= the fixed intercept of the equation, β_2, β_3, β_4 and β_5= the fixed coefficient on the variables of *Year*, *Period*, *Distance* and *Policy*. γ_{1i}= the deviation of industry *i*'s intercept from the population intercept β_1. γ_{2i}= the deviation of industry *i*'s coefficient on *Policy* from the population coefficient. ε_i= the residual error term.

Table 1 shows the variables and their descriptive statistics for patent license in China's strategic emerging industries.

Table 1. Variables and descriptive statistics

Variable	Obs.	Min.	Max.	Mean	S.D.	Var.
Year	10776	2000	2012	2010.60	1.405	1.974
Policy	10776	0	1	0.99	0.091	0.008
Period	10776	0	18	3.18	2.007	4.029
Distance	10776	0	3795.16	392.445	628.708	395273.506

4 Results

The percentage of licensors and licensees in each province during the period of 2000-2006 and 2007-2012 shown in *Figure 1* and *Figure 2* illustrates the changing trend of geographical characteristics in China's strategic emerging industries.

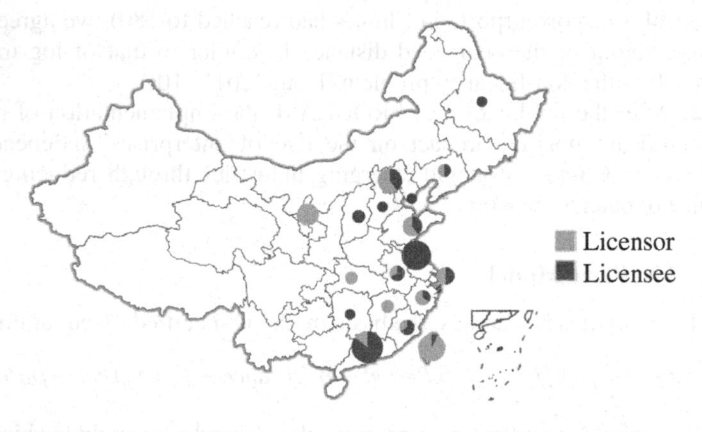

Fig. 1. Geographical characteristics of licensors and licensees in China's strategic emerging industries from 2000 to 2006

Figure 3 provides the overall situation and observations of patent licensing's average lag period in China's seven strategic emerging industries from 2000 to 2012. As we can seen, the average lag period of patent licensing in China's seven strategic emerging industries has shortened dramatically since 2008, and appears a decreasing tendency.

Figure 4 provides the overall situation and observations of patent licensing's average geographic distance in China's seven strategic emerging industries from 2000 to 2012. As we can see from figure 4, the average geographic distance of patent licensing in China's seven strategic emerging industries has also shortened dramatically since 2008, and decreased year by year.

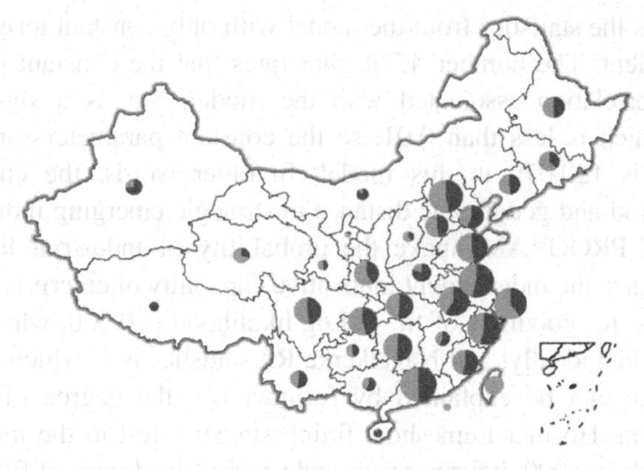

Fig. 2. Geographical characteristics of licensors and licensees in China's strategic emerging industries from 2007 to 2012

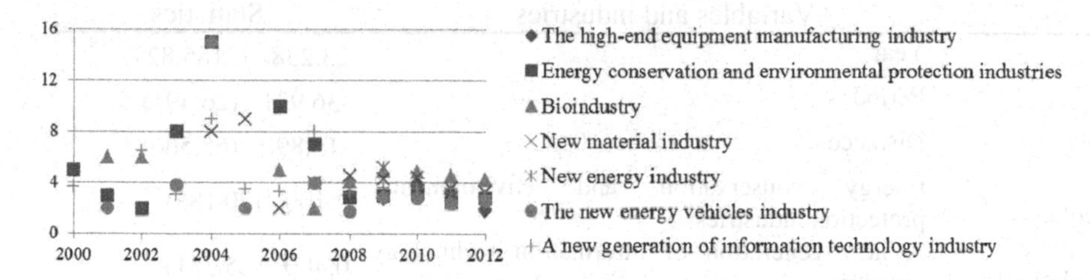

Fig.3. Average lag period of patent licensing in China's strategic emerging industries

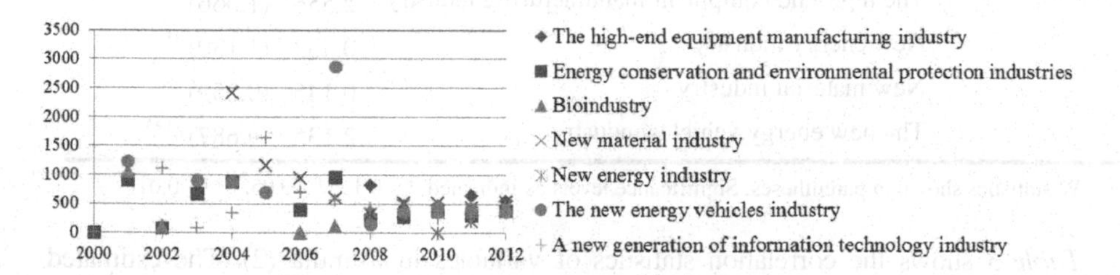

Fig. 4. Average geographic distance of patent licensing in China's strategic emerging industries

In order to further study the impact of 'independent innovation policy' on patent licensing's geographical evolution and the independent innovative capability of enterprises in China's strategic emerging industries, we use logistic regression models to do the statistical analysis for formula (2).

Table 2. The initial model statistics

	B	S.E	Wals	df	Sig.	Exp(B)
Step 0 constant	4.788	.106	2023.593	1	.000	120.079

Table 2 includes the statistics from the model with only constant term, where B is the correlation coefficient. The number 4.788 illustrates that the constant parameters has a high degree of correlation associated with the model. Sig. is a significant level of statistics Wals, which is less than 0.01, so the constant parameters are effective. The value of Exp(B) is 120.079 in this model. In other words, the changes of patent licensing's lag period and geographic distance in strategic emerging industries as well as the publication of PROGRAM, make the probability of industrial licensing become increased and enhance the independent innovative capability of enterprises.

In the model test for goodness of fit, -2 Log likelihood is 0.000, which illustrates that the model fits the data ideally; and Nagelkerke R2 statistics is 1, which shows that more regression variation can be explained by formula (2), the degree of fitting is good. Meanwhile, by using Hosmer-Lemeshow fitting statistics test to the model, we can get the Sig. value which is 0.000, it further tests and verifies the degree of fitting is good.

Table 3. The statistics of the basic model

	Variables and industries	Statistics
	Year	23.238 (2185.829)***
	Period	-36.921 (26.491)***
	Distance	-15.890 (65.566)***
policy=1 (take policy=0 as reference)	Energy conservation and environmental protection industries	0.100 (70.183)***
	A new generation of information technology industry	0.400 (25.331)***
	Bioindustry	1.387 (45.591)***
	The high-end equipment manufacturing industry	2.556 (1.066)**
	New energy industry	0.233 (1.130)**
	New material industry	0.145 (0.259)*
	The new energy vehicles industry	2.135 (9.687)***

W statistics shown in parentheses. Significance levels as indicated: * 0.1, ** 0.05, *** 0.01.

Table 3 shows the correlation statistics of variables in formula (2). The estimated coefficients (B) of each factor was significant, which means that the publication of PROGRAM has affected above factors to some extent.

The Lag Period of Patent Licensing. The coefficient of lag period in the model of China's strategic emerging industries' patent licensing is -36.921, which illustrates that the publication of PROGRAM has negative impact on this factor. In other words, the publication of PROGRAM shorten the period between the time of patent applying and patent licensing, and has a powerful influence. As for the reason, PROGRAM clearly puts forward to the target for China to get into the ranks of innovation driven countries, establishes the positon of enterprises as main part in the process of innovation, and further makes explicit classification of strategic emerging industries in China's national economic industries. All of this have greatly motivated the enterprises in strategic emerging industries and improved their enthusiasm and driving force.

The Geographic Distance Between Licensor and Licensee. In the model of China's strategic emerging industries' patent licensing, the coefficient of geographic distance is -15.890, which interprets that the publication of PROGRAM shorten the geographic distance between the location of licensor and licensee and has a powerful negative influence on this factor. As for the reason, the publication of PROGRAM and implementation of relevant industrial policies have enhanced the enterprises' driving force for technological innovation. At the same time, because of the limitation of technological development for their own, the enterprises will seek the source of technological innovation from outside. Therefore, considering the frequent negotiation and communication before and after signing technology license contract, as a licensee, the enterprise is more willing to choose a licensor with shorter geographic distance based on the consideration of efficiency, which can help to improve its independent innovation capability.

Seven Strategic Emerging Industries. Generally speaking, the publication of PROGRAM has significant positive impact on the rise of enterprises' independent innovative capability in China's strategic emerging industries. Ranking by the degree of influence, these seven industries are the high-end equipment manufacturing industry (2.556), the new energy vehicles industry (2.135), bioindustry (1.387), a new generation of information technology industry (0.400), new energy industry (0.233),new material industry (0.145), energy conservation and environmental protection industries (0.100). Moreover, the implementation of relevant policies has significant positive influence on industrial patent licensing and raises the number of licensed patent by increasing the driving force of' technological innovation. Although China's seven strategic emerging industries appear different policy results after the publication of PROGRAM, on the whole, enterprises in these industries have all improved the industries' overall independent innovative capability in the process of technology industrialization.

5 Conclusions and Recommendations

From the perspective of patent licensing, this paper discusses the impact of PROGRAM on patent licensing's geographical evolution and the capability of independent innovation in China's strategic emerging industries, by using the panel data's logistic regression method based on the information of patent licensing in seven strategic emerging industries. The results shows that, the publication and implementation of relevant policies accelerates the evolution of patent licensing in China's strategic emerging industries though reducing the lag period between the time of patent applying and licensing as well as decreasing the geographic distance between patent licensor and licensee. Moreover, the positive effect of policy on industries' capability of independent innovation appears distinct among different strategic emerging industries. Ranking by the degree of influence, these seven industries are the high-end equipment manufacturing industry, the new energy vehicles industry, bioindustry, a new generation of information technology industry, new energy industry, new material industry, energy conservation and environmental protection industries.

To certain country or a region, strategic emerging industries are the leading industries for future economic growth with the decisive function of promoting and guiding national economic development and industrial structure conversion. Because of their broad market prospects and ability to guide the progress of science and technology, strategic emerging industries are related to the national economy and industrial security. Therefore, the government should enforce industrial support according to different policy results after the implementation of relevant policies. Moreover, during the process of technology transfer by patent licensing, it also should bring market-oriented operation system into play and use the basic function of market to arouse the enterprises' initiative and actively cultivate the market, so as to shorten the lag period between the time of patent applying and licensing, and further improve the independent innovative capability of enterprises in China's strategic emerging industries.

Acknowledgments. This paper is supported by the National Natural Science Foundation of China (NO. 71302133) and Innovation Team Project of Education Department in Sichuan, China (13TD0040).

References

1. Anand, B., Khanna, T.: The structure of licensing contracts. Journal of Industrial Economics 48(1), 103–135 (2000)
2. Arora, A., Ceccagnoli, M.: Patent protection, complementary assets, and firms' incentives for technology licensing. Management Science 52(2), 293–308 (2006)
3. Brachert, M., Cantner, U., Graf, H., Gunther, J., Schwartz, M.: Which regions benefit from emerging industries? European Planning Studies 21(11), 1703–1707 (2013)
4. Fosfuri, A.: The licensing dilemma: Understanding the determinants of the rate of technology licensing. Strategic Management Journal 27(12), 1141–1158 (2006)
5. Gambardella, A., Giuri, P., Luzzi, A.: The market for patents in Europe. Research Policy 36(8), 1163–1183 (2007)
6. Guo, B.: Technology acquisition channels and industry performance: An industry-level analysis of Chinese large- and medium-size manufacturing enterprises. Research Policy 37(2), 194–209 (2008)
7. Guo, X.Y., Hui, X.F.: Research on regional strategic emerging industry selection models based on fuzzy optimization and entropy evaluation. Journal of Applied Mathematics 12(1), 1–15 (2012)
8. He, Z., Wu, Y.: Assessment index system and evaluating model of strategic emerging industries. Studies in Science of Science 29(5), 678–683, 721 (2011)
9. Hefman, P., Phal, R.: The emergence of new industries. The University of Cambridge, Cambridge (2008)
10. Hong, W., Su, Y.: The effect of institutional proximity in non-local university- industry collaborations: An analysis based on Chinese patent data. Research Policy 42(2), 454–464 (2013)
11. Huang, L., Zhang, J., Wu, F., et al.: The overall evaluation indexs and standards of strategic emerging industries. Statistics and Decisions (5), 34–37 (2013)
12. Kim, Y.: Choosing between international technology licensin partners: An empirical analysis of U.S. biotechnology firms. Journal of Engineering and Technology Management 26(1-2), 57–72 (2009)

13. Laursen, K., Leone, M., Torrisi, S.: Technological exploration through licensing: New insights from the licensee's point of view. Industrial & Corporate Change 19(3), 871–897 (2010)
14. Li, M., Li, X., Zhao, H., et al.: License model: The way to achieve a new leap forward development of pharmaceutical industry in Jilin province. Economic Vision (2), 39–40 (2011)
15. Lin, B.: Technology transfer as technological learning: A source of competitive advantage for firms with limited R&D resources. R&D Management 33(3), 327–341 (2003)
16. Marin, P., Siotis, G.: Innovation and market structure: An empirical evaluation of the 'bounds approach' in the chemical industry. Journal of Industrial Economics 55(1), 93–111 (2007)
17. Nerkar, A.: Old is gold? The value of temporal exploration in the creation of new knowledge. Management Science 49(2), 211–229 (2003)
18. Nooteboom, B., Van, W., Duysters, G., et al.: Optimal cognitive distance and absorptive capacity. Research Policy 36(7), 1016–1034 (2007)
19. Sun, Y., Du, D.: Determinants of industrial innovation in China: Evidence from its recent economic census. Technovation 30(9-10), 540–550 (2010)
20. Wang, Y., Li, Y.: When does inward technology licensing facilitate firms' NPD performance? A contingency perspective. Technovation 34(1), 44–53 (2014)
21. Wu, R.: Evaluation and selection of regional strategic emerging industry. Scientific Management Research 30(2), 42–45 (2012)
22. Xu, M.: The patent license payoffs with technology standards in communication industry. Science of Science and Management of S.&T. 33(11), 19–23 (2012)
23. Yu, D., Chen, H., Tu, G.: Sci-tech resources input-output efficiency evaluation of strategic emerging industries of Jiangxi provience. Journal of Intelligence 32(2), 178–185 (2013)
24. Zeng, F., Peng, Z., Chen, X.: The literature review and assessment for the latest policy research progress of strategic emerging industries. Science & Technology Progress and Policy 30(14), 1–7 (2013)
25. Zhou, J.: Development and distribution of strategic emerging industries in China. Statistical Research 29(9), 24–30 (2012)

Land Use/Cover Change and Its Impact on Net Primary Productivity in Huangfuchuan Watershed Temperate Grassland, China

Jiren Xu[1,2], Jihong Dong [1,3], Lixin Wu[1,3], Guoquan Shao[1,3], and Hongbing Yang[1,3]

[1] College of Environment and Spatial Informatics, China University of Mining and Technology, Xuzhou 221116, Jiangsu, China

[2] State Key Laboratory of Soil and Sustainable Agriculture, Institute of Soil Science, Chinese Academy of Sciences, Nanjing 210008, China

[3] Jiangsu Key Laboratory of Resources and Environmental Information Engineering, China University of Mining and Technology, Xuzhou 221116, China

Abstract. This study take the Huangfuchuan Watershed temperate grassland as the study area, RS and GIS techniques are used to explore the relationship between LUCC and NPP. With the combination of CASA model, the dynamic characteristics of NPP in 1987-2011 are studied. The main conclusions are as follows: (1) land use structure changes obviously in Huangfuchuan Watershed. The main trend of land use change was the gradual increase of construction land and woodland, the gradual decrease of water. Arable land, grass, shrub, bare rock and sand were fluctuant. It could be seen from land use dynamic degree. (2) Through the calculation of NPP model, the total value of NPP in 1987, 1995, 2000, 2007 and 2011 was 28.12GgC, 53.47GgC, 73.23GgC, 157.92GgC and 78.52GgC. (3) Through the analysis of land use change effects on NPP, it indicates the main reason for the increase of NPP is due to grassland transfer to shrub between 1987 and 1995. The decade of bare rock is the main reason for the increase of NPP in 1995-2000. Shrub transferring to grassland is the main reason for the increase of NPP in 2000-2007. Grassland transferring to shrub is the main reason for the reduction of NPP in 2007-2011. The results of the study is very meaningful for rational using of temperate grassland resources and improvement of the fragile ecological environment.

Keywords: LUCC, NPP, Huangfuchuan watershed, temperate grassland, CASA model.

Net Primary Productivity (NPP) means accumulated quantities of organics in green plants in unit area and time [1]. The index of NPP is used to evaluate coordination of the structure and function of eco-system and interaction with environment. Moreover, it's closely related to significant scientific issues of global changes such as carbon circulation, water circulation and food security [2, 3]. LUCC refers to quite a number of natural and human factors of earth system science, global environment change, and sustainable development and so on. It is a issue crossed closely among natural and human process which affect the structure and function of

F. Bian and Y. Xie (Eds.): GRMSE 2014, CCIS 482, pp. 664–683, 2015.

ecosystem, it has deep effects on regional climate, hydrology, vegetation, biochemical circle, biodiversity and so on [4].

Grassland ecosystem is one of the most important and widely distributed terrestrial ecosystem, in especial, temperate grassland ecosystem has a particular biochemical process with other ecosystems and it's sensible to climate and environment changes which determines its unique status and importance in the study of global carbon circulation. NPP is an important representation of carbon sequestration of temperate grassland ecosystem. It is determined by the structure (vegetation distribution and seasonal aspect) and function (earth biochemical circulation) as well as environment variables including solar radiation, concentration of atmospheric carbon dioxide, temperature, precipitation, soil fertility and so on. NPP can provide scientific basis for evaluating productivity of temperate grassland ecosystem correctly by simulating productivity of grassland ecosystem and imposing quantitative analysis on characteristics of temporal-spatial variability. So the study on the relationship between LUCC and NPP has become one of the hot pots of global environment changes issues. Many scholars have carried out research on these issues [5-11]. But study mostly focuses on estimation and evaluation of NPP in regional and global scale, less refers to the specific region of temperate grassland ecosystem. Zhao used observation data of weather stations and GIS techniques to estimate spatial distribution of NPP in northwest region [12]; Zhang studied the relationship between the change of NPP in northwest China and land cover as well as climate by constructing solar energy efficiency model based on remote sensing and ecological process [13]; Mu studied the temporal and spatial variation of NPP of grassland ecosystem in Inner Mongolia in recent ten years and its relationship with the change of climate [14]. It's concluded that the study on land cover and NPP of typical temperate grassland with long time span is deficient in China.

Most areas of temperate grassland in northern China are integrated with agriculture and livestock husbandry, predominated by livestock. It's located in ecologically fragile regions which are characterized as dry and sandy. Due to the adjustment of national policies, high-strength mining activities, climate changes and so on, water and soil lost seriously and land cover had severe changes. Huangfuchuan is a one-level branch river of the middle Yellow River. It's a typical semiarid seasonal river which is filled with coarse sediment and the problem of water and soil loss had a deep effect on ecological environment [15]. In the past twenty years, land cover of Huangfuchuan watershed changed remarkably. Forest land and farmland had an increase, grassland decreased due to expansion of urban and rural construction land. Area of water declined severely and ecosystem service value had a decreasing trend [16]. A study on land cover changes in Huangfuchuan watershed and effects on NPP helps promote proper land resource use in Huangfuchuan watershed typical as temperate grassland ecosystem, optimization of land use structure and improvement of fragile ecological environment in theory and practice.

1 Materials and Methods

1.1 General Situation of the Studied Area

Huangfuchuan is a one-level branch river of the middle Yellow River which lies in the east longitude 110.3°~111.2°, north latitude 39.2°~39.9°. Temperate semiarid continental climate in Huangfuchuan is characterized as dry, less rain and more wind, rainfall mostly comes from rain storm in summer. The widest south-north distance is 85.9km, east-west distance is 102.1km, the basin area is 3235.14 km^2 and most part of it lies in ZhunGer in Inner Mongolia; the area of 415 km^2 which accounts for 12.83% of the whole watershed area belongs to Shanxi province. Soil in watershed area gives priority to steppe chestnut soil. Climate change, massive lop in history time and agricultural reclamation since the end of the Qing dynasty caused few remaining natural forest and grassland which were replaced by artificial plantation and natural secondary grassland vegetation. The vast majority of the watershed area is grassland. Grassland occupied large percentage and it was located in ecological fragile region, besides it suffered from human disturbance, especially high-strength mining activities. So it was listed as one of the eight key harnessment areas in the whole country in 1983. The watershed area is both national harnessment area and one of the most ecologically destroyed areas. This is the reason we choose this area as the studied area in the time of fast economic development to make a study on land cover change and NPP response of temperate grassland ecosystem in China. The location and administrative division of Huangfuchuan is shown in Fig 1.

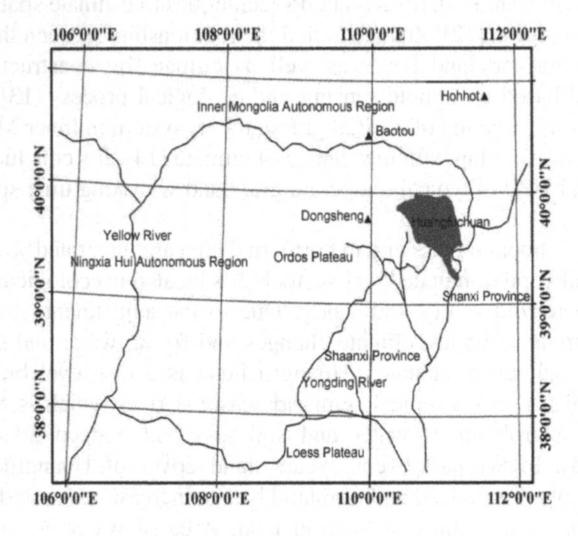

Fig. 1. Location of Huangfuchuan watershed

1.2 Data Source and Preprocessing

It rains most in July in Huangfuchan watershed generally. There are extremely significant correlations between NPP of typical grassland in Inner Mongolia and annual precipitation. The dates of the five TM images which have a time span of 25 years are 1987, 1995, 2000, 2007 and 2011. The row-column number of image track is 127/32 and 127/33. The method of system radiation correction, geometric correction based on ground control point and terrain correction is used to deal with the original images. The projected coordinate system of UTM/WGS84 were added into five images which include ten with the help of ERDAS IMAGINE9.3. The module of Raster/Mosaic was used to splice and mosaic corrected TM images and then we get five TM images covering the whole studied area. Next we convert the extracted boundary of Huangfuchuan watershed to AOI using ERDAS 9.2 in order to clip the images. In the end, the images of the studied area were obtained.

1.3 Research Method

Classification of TM Images. In order to make sure that land use classification is scientific and conform to area actual characteristics and meet the need of studying NPP, according to Land Use Dynamic RS Monitoring Regulation and Land Use Classification, LUCC of Huangfuchuan watershed were classified into eight types of farmland, forest, water, sandy land, bare rock, grass land, shrub and construction land based on the attributes of land use and the accuracy of present images in the RS survey of land resource.

CASA Model. Taking environment conditions and features of vegetation itself into consideration, CASA (Carnegie-Ames-Stanford Approach) model was drive by data of RS, temperature, precipitation, solar radiation and types of vegetation as well as soil. The principle which CASA model follows was that NPP is decided by the two variables of absorbed photosynthetically active radiation (APAR) and conversion efficiency for solar radiation (ε). The formula is as follows.

$$NPP_{(x,t)} = APAR_{(x,t)} \times \varepsilon_{(x,t)} \quad (1)$$

In Formula 1, t represents time, x represents spatial position; APAR(x,t) represents photosynthetically active radiation absorbed by pixel x in the month t (unit: MJ/m^2/month); $\varepsilon(x,t)$ represents light utilization efficiency of pixel x in the month t (unit: gC/MJ).

1) APAR

APAR is determined by total solar radiation and the absorption percentage of photosynthetically active radiation absorbed by vegetation. Formula 2 is shown below.

$$APAR_{(x,t)} = SOL_{(x,t)} \times FPAR_{(x,t)} \times 0.5 \quad (2)$$

In Formula 2, *SOL*(*x,t*) stands for total solar radiation (unit: MJ/m^2) of pixel x in the month t; *FPAR*(*x,t*) stands for the absorption percentage of incoming photosynthetically active radiation; Constant 0.5 stands for the percentage of solar active radiation used by vegetation in total solar radiation (wavelength range is 0.38~0.71um).

Within a certain range, linear relationship exists between FPAR and NDVI and it depends on the max and min NDVI value of a certain type of vegetation and corresponding FPAR value. Formula 3 is as follows.

$$FPAR(x,t) = \frac{(NDVI(x,t) - NDVI_{min}) \times (FPAR_{max} - FPAR_{min})}{(NDVI_{max} - NDVI_{min})} + FPAR_{min} \quad (3)$$

In Formula 3, NDVImax and NDVImin represents the maximum value and minimum value of a certain type of vegetation.

Further study shows that there is also notable linear relationship between FPAR and SR. it's expressed in Formula 4.

$$FPAR(x,t) = \frac{(SR(x,t) - SR_{min}) \times (FPAR_{max} - FPAR_{min})}{(SR_{max} - SR_{min})} + FPAR_{min} \quad (4)$$

In Formula 4, FPARmax and FPARmin don't matter with vegetation types, their corresponding value is 0.001 and 0.95. The value of SRmax and Srmin is 4.46 and 1.05.

Acquisition of $NDVI_{(x,t)}$:

$$NDVI = [NIR - R] / [NIR + R] \quad (5)$$

In Formula 5, NIR is near-infrared range of Landsat TM data; R is red spectral band of Landsat TM data. Considering this situation, the mean value was accepted as the estimated data of FPAR.

Comparing estimated FPAR-NDVI result with FPAR-SR result, we found that the estimated FPAR result by NDVI was higher than measured value, while the estimated FPAR result by SR was lower, but the error of the estimated data by SR was little than the error of data by NDVI.

$$FPAR(x,t) = \alpha FPAR_{NDVI} + (1 - \alpha) FPAR_{SR} \quad (6)$$

In Formula 6, FPARNDVI is the calculated result according to Formula 3, FPARSR is the calculated result according to Formula 4. α is the adjustment coefficient of the two method. The value of α is 0.5 in this study (mean value of FPARNDVI and FPARSR).

2) Conversion efficiency for solar radiation

Conversion efficiency for solar radiation means conversion efficiency for PAR to organic carbon. Generally thinking, vegetation has optimal conversion efficiency under ideal conditions, while conversion efficiency is affected by factors of temperature and moisture under factual conditions [17].

$$\varepsilon(x,t) = T_{\varepsilon 1}(x,t) \times T_{\varepsilon 2}(x,t) \times W_{\varepsilon}(x,t) \times \varepsilon^* \quad (7)$$

In Formula 7, $T_{\varepsilon 1}$ and $T_{\varepsilon 2}$ reflect the effect of temperature on conversion efficiency for solar radiation, $W_{\varepsilon 1}$ is coefficient of water stress effect which represents effects of water, ε^* means optimal conversion efficiency for solar radiation under ideal conditions which adopts the value 0.389gC/MJ of global maximum conversion efficiency for solar radiation.

a. Temperature stress factors

$T_{\varepsilon 1}$ means inherent biochemical effect of vegetation limits photosynthesis and then NPP decrease.

$$T_{\varepsilon 1}(x) = 0.8 + 0.02 Topt(x) - 0.0005 \times [Topt(x)]^2 \quad (8)$$

In Formula 8, Topt(x) is the month mean temperature when NDVI value reaches the highest in a certain area in one year. When the mean temperature in certain month is below or equal to -10°C, $T_{\varepsilon 1}$ is equal to 0. Quite a number of studies show that the value and changes of NDVI can reflect growth status of plants. Once NDVI value reaches the highest, plants grow fastest. Then temperature can be regarded as the appropriate growth temperature.

$T_{\varepsilon 2}$ means the decreasing trend of conversion efficiency for solar radiation when environment temperature changes from the most appropriate temperature Topt(x) to high and low temperature.

$$T_{\varepsilon 2}(x,t) = 1.1814/\{1+\exp[0.2\times(Topt(x)-10-T(x,t))]\}/\{1+\exp[0.3\times(-Topt(x)-10+T(x,t))]\} \tag{9}$$

In Formula 9, when the month mean temperature is 10°C higher or 13°C lower than the most appropriate temperature *Topt*(*x*).

b. Water stress factors

Water stress coefficient W^{ε} reflects the effects of effective water conditions which plants use on conversion efficiency for solar radiation. With environmental effective water increasing, W^{ε} tends to increase. The value of W^{ε} ranges from 0.5 (under extremely dry condition) to 1 (under extremely wet condition), it's calculated according to formula below:

$$W_{\varepsilon}(x,t) = 0.5 + 0.5\times E(x,t)/E_p(x,t) \tag{10}$$

In Formula 10, *E*(*x*,*t*) means regional factual evapotranspiration which can be obtained based on regional factual evapotranspiration model proposed by Zhou Guangsheng and Zhang Xinshi (Zhou and Zhang, 1995). *Ep*(*x*,*t*) means regional potential evapotranspiration which can be calculated on the basis of complementary relationship proposed by Zhang Zhiming (Zhang, 1990).

$$E(x,t) = \frac{\{P(x,t)\times R_n(x,t)\times[(P(x,t))^2+(R_n(x,t))^2+P(x,t)\times R_n(x,t)]\}}{\{[P(x,t)+R_n(x,t)]\times[(P(x,t))^2+(R_n(x,t))^2]\}} \tag{11}$$

In Formula 11, *P*(*x*,*t*) represents precipitation of pixel *x* in month *t* (mm), *Rn*(*x*,t) means surface net radiation of pixel *x* in month *t* (mm).

Common meteorological observation stations don't observe surface net radiation. Besides, quite a number of calculated meteorological factors include precipitation, field moisture capacity, wilting moisture content, percentage of soil clay and sand, depth of soil, soil volume moisture content and so on. These factors have something to do with soil types and it's hard to obtain their value. So empirical model proposed by Zhou Guangsheng and Zhang Xinshi can be put into use.

$$R_n(x,t) = [E_{p0}(x,t)\times P(x,t)]^{0.5}\times\{0.369+0.598\times[\frac{E_{p0}(x,t)}{P(x,t)}]\} \tag{12}$$

$$E_p(x,t) = [E(x,t) + E_{p0}(x,t)]/2 \quad (13)$$

In Formula 13, *Ep0(x,t)* is local potential evapotranspiration which can be calculated according to vegetation climate relationship model by Thornthwaite [18].

$$E_{p0}(x,t) = 16 \times [\frac{10 \times T(x,t)}{I(x)}]^{\alpha(x)} \quad (14)$$

$$\alpha(x) = [0.6751 \times I^3(x) - 77.1 \times I^2(x) + 17920 \times I(x) + 492390] \times 10^{-6} \quad (15)$$

$$I(x) = \sum_{i=1}^{3} [\frac{T(x,t)}{5}]^{1.514} \quad (16)$$

I(*x*) is heat quantity index, $\alpha(x)$ is a constant which varies by region and its function of *I*(*x*), this relationship come into effect just from 0°C to 26.5°C. Thornthwaite set rate of evapotranspiration as 0 when air temperature is below 0°C; When air temperature is above 26.5°C, possible rate of evapotranspiration increases with temperature and it doesn't matter with *I*(*x*). Adjusted possible rate of evapotranspiration (*APE*(*x*,*t*)) can be obtained after calculated value of *Ep0*(*x*,*t*) is adjusted on the basis of factual daily hours and monthly number of days.

$$APE(x,t) = E_{po}(x,t) \times CF(x,t) \quad (17)$$

2 Acquisition and Processing of Model Parameters

2.1 Acquisition and Processing of NDVI

After NDVI is extracted and bands go through ratio procession, the effects of irradiation condition changes relating to satellite observation angle, terrain, cloud, shadow, terrain and atmospheric conditions can be erased partly. Five NDVI images of 1987, 1995, 2000, 2007 and 2011 which went through radiometric and atmospheric correction. Figure 2 is shown below.

2.2 Acquisition and Processing of Meteorological Data

Meteorological data comes from the website http://cdc.cma.gov.cn which includes air temperature, precipitation and solar radiation data in July (1987, 1995, 2000, 2007 and 2011). The data covers Huangfuchuan watershed and surrounding meteorological stations (a number of meteorological observation stations can't get solar radiation data directly, so daily amount of solar radiation can be converted by daily total sunshine hours). After sieving these data and eliminating irreplaceable erroneous data, 13 stations of relatively complete data which distribute evenly was reserved.

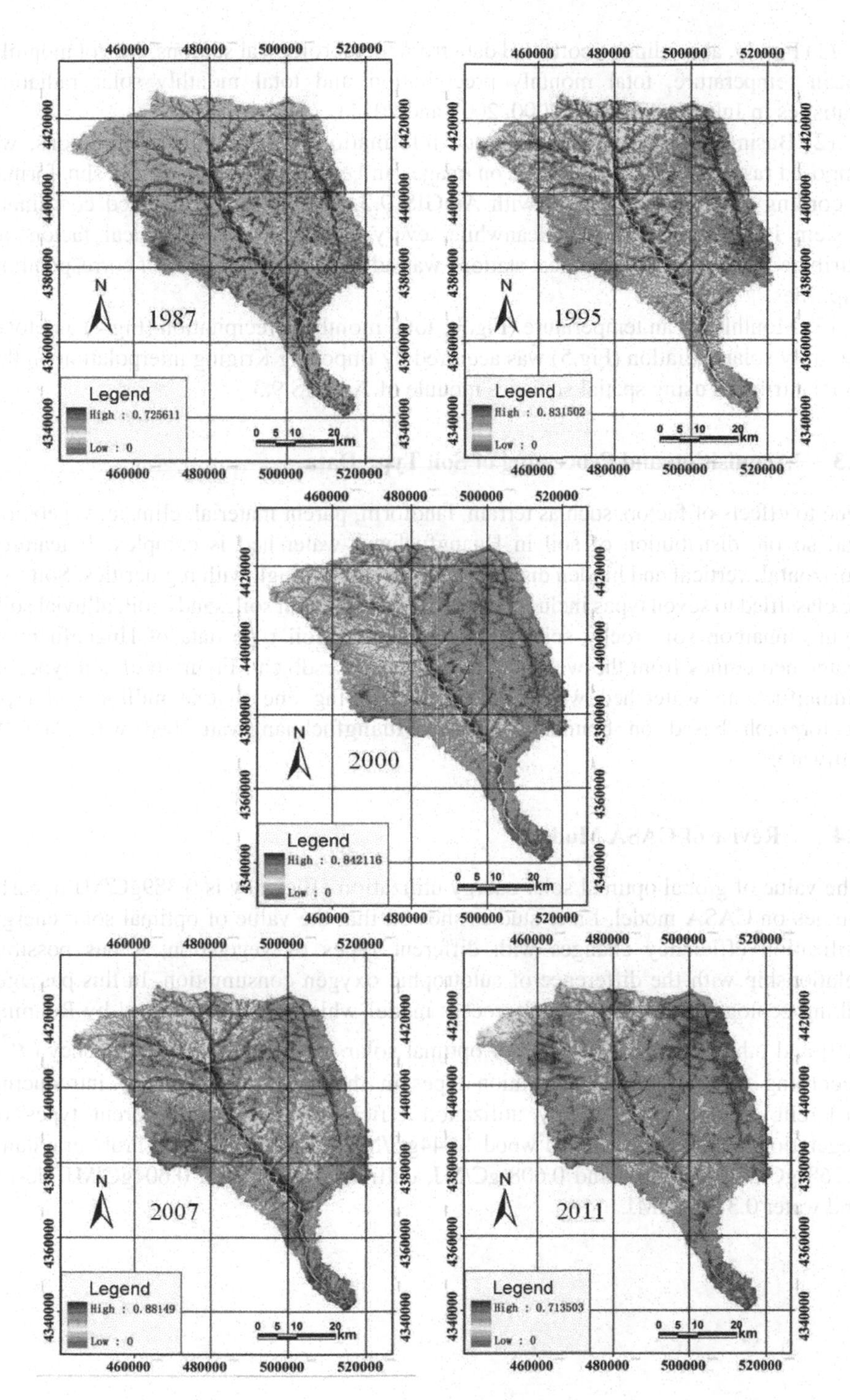

Fig. 2. NDVI during 1987-2011 in Huangfuchuan watershed

(1) Firstly, according to collected data from meteorological stations, we got monthly mean temperature, total monthly precipitation and total monthly solar radiation statistics in July (1987, 1995, 2000, 2007 and 2011).

(2) Basing on longitude and latitude information of meteorological stations, we imposed raster to vector conversion on images and exported them to data of shp. format according to point coordinates with ArcGIS 9.3. The selected projected coordinate system is UTM projection. Meanwhile, every value of meteorological factors as attribute fields of meteorological stations was added to attribute lists of corresponding shp. files.

(3) Monthly mean temperature (Fig.3), total monthly precipitation (Fig.4) and total monthly solar radiation (Fig.5) was acquired by imposing Kriging interpolation on the three attributes using spatial statistics module of ArcGIS 9.3.

2.3 Acquisition and Processing of Soil Type Data

Due to effects of factors such as terrain, landform, parent material, climate, vegetation and so on, distribution of soil in Huangfuchuan watershed is complex. It features horizontal, vertical and hidden distribution. It's shot through with regularities. Soil can be classified to seven types including calcium soil, loessial soil, sandy soil, alluvial soil, light cinnamon soil, rocky soil and skeleton soil. Soil type data of Huangfuchuan watershed comes from the website http://www.soil.csdb.cn/. Figure 6 of soil types in Huangfuchuan watershed was acquired by clipping one to one million soil type vectorgraph based on boundary line of Huangfuchuan watershed with ArcGIS software.

2.4 Revise of CASA Model

The value of global optimal solar energy utilization efficiency is 0.389gC/MJ in early studies on CASA model. Later studies showed that the value of optimal solar energy utilization efficiency changes with different types of vegetation, it has possible relationship with the difference of autotrophic oxygen consumption. In this passage, taking ecologically physiological process model which was put forward by Running [19] and other learners, the value of optimal solar energy utilization efficiency (ε^*) according to seven major vegetation types in the north of Shanxi by introducing different optimal solar energy utilization efficiency aimed at different types of vegetation is as follows: mixed wood 1.044gC/MJ, sparse forest and shrub, grassland 0.768 gC/MJ, dry grassland 0.608 gC/MJ, cultivated vegetation 0.604gC/MJ, desert and water 0.389 gC/MJ.

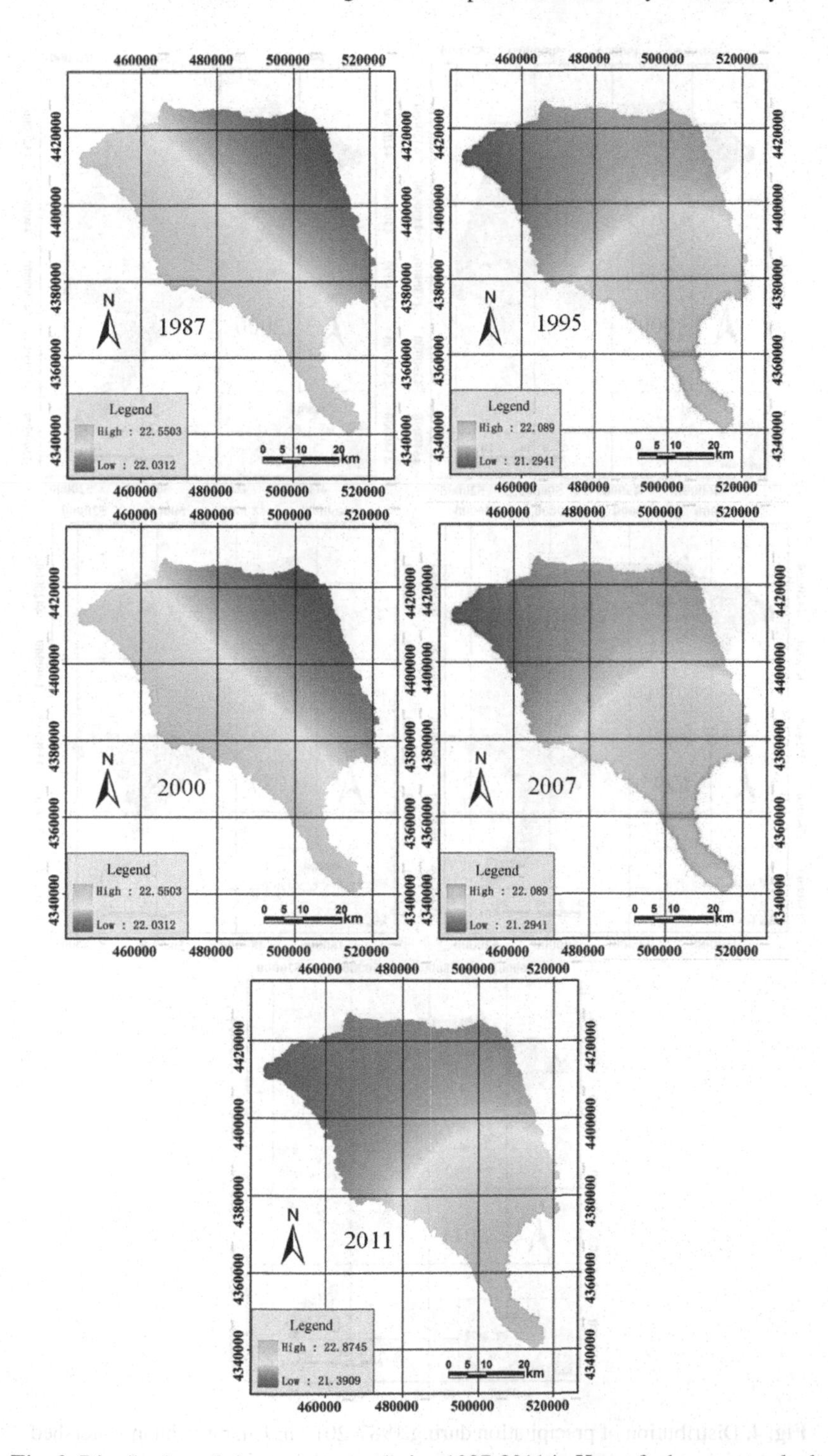

Fig. 3. Distribution of air temperature during 1987-2011 in Huangfuchuan watershed

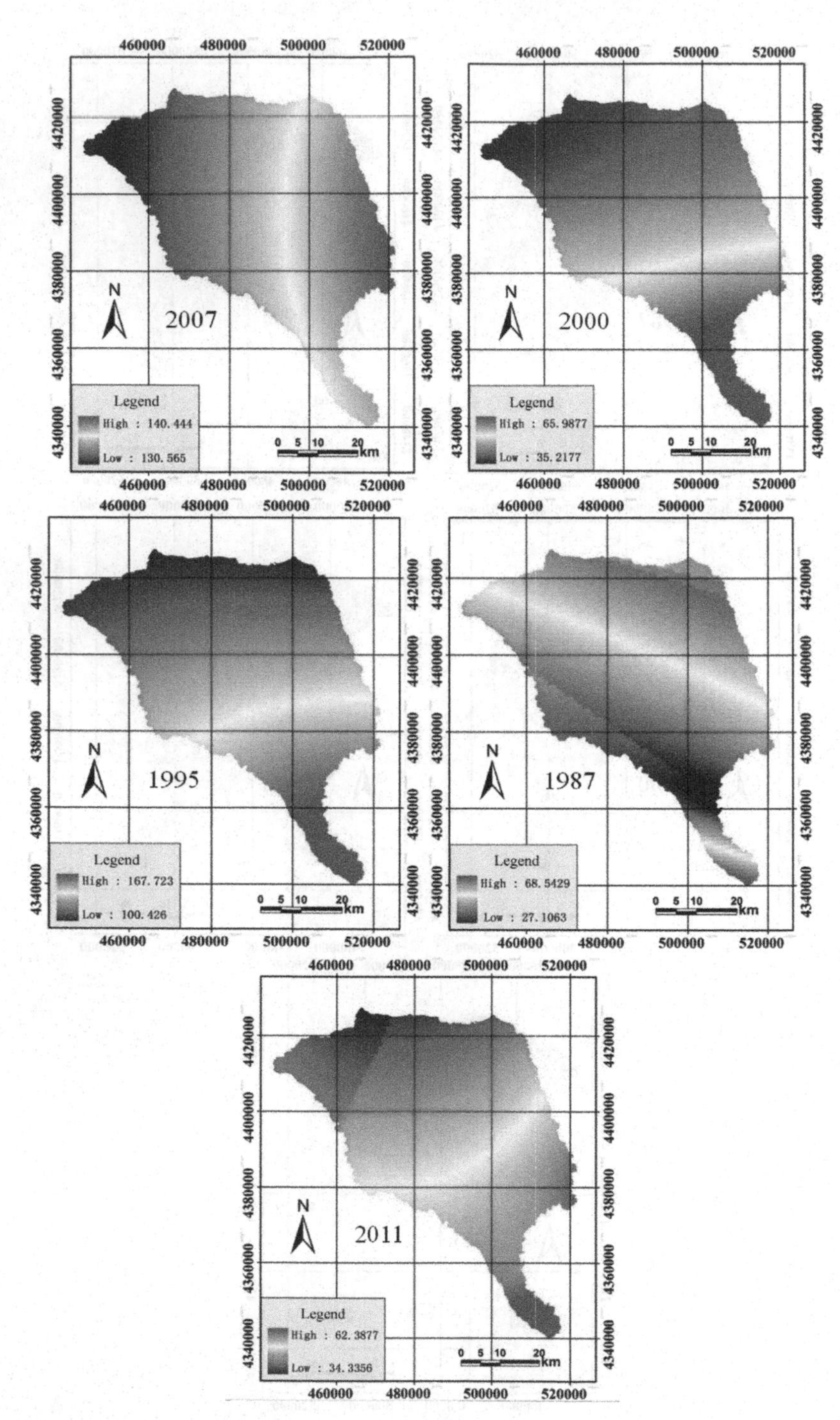

Fig. 4. Distribution of precipitation during 1987-2011 in Huangfuchuan watershed

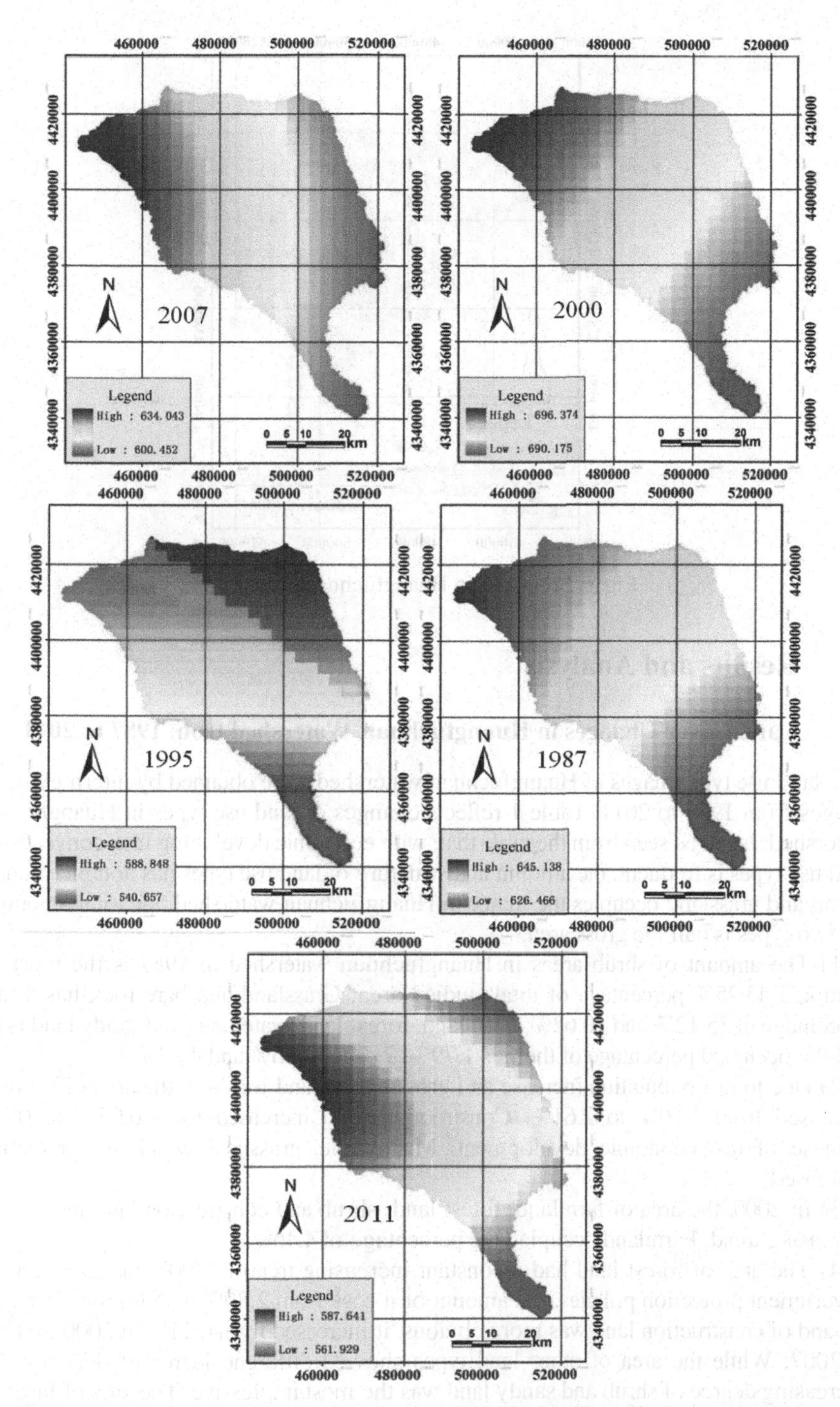

Fig. 5. Distribution of solar radiation during 1987-2011 in Huangfuchuan watershed

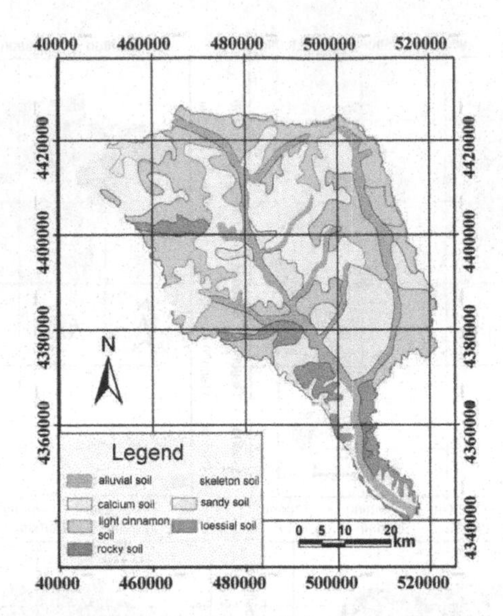

Fig. 6. Soil types in Huangfuchuan watershed

3 Results and Analysis

3.1 Land Cover Changes in Huangfuchuan Watershed from 1987 to 2011

Five land use type images of Huangfuchuan watershed were obtained by interpreting TM images from 1987 to 2011. Table 1 reflects changes of land use types in Huangfuchuan watershed. It can be seen from the table that: with economic developing fast, conversion of land use types is frequent, the amount and structure of land use types has notable changes. Shrub and grassland occupies the major in Huangfuchuan watershed, the total amount of the two types is half the gross area.

1) The amount of shrub areas in Huangfuchuan watershed in 1987 is the most and occupied 43.25% percentage of total studied area. Grassland and bare rock has a quite percentage of 35.12% and 10.62%. Farmland, forest land, water area and sandy land is less and the occupied percentage of them is 1.79%, 2.44%, 2.66% and 4.07%.

2) Due to fast population increase and growing demand for food, the area of farmland increased from 1.79% to 2.61%. Construction land increased from 0.05% to 0.08% because of fast economic development. Meanwhile, grassland which was reclaimed decreased.

3) In 2000, the area of farmland, forest land, shrub and construction land had a faster increasing trend. Farmland occupied the percentage of 4.46%.

4) The area of forest land had a constant increasing trend in 2007 due to a series of government protection policies, the amount of it rose from 2.80% in 2000 to 4.25%; The expand of construction land was more obvious, it increased from 0.11% in 2000 to 0.15% in 2007; While the area of other land types showed different degree of decrease. The decreasing degree of shrub and sandy land was the most impressive. The area of farmland in watershed increasing right along and reached the maximum value, the reason was that the increasing demand of fast growing population promoted increase of farmland.

5) The area of forest and construction land had a faster speed of increase in Huangfuchuan watershed in 2011. The percentage of the two types is 6.81% and 0.24%. Unused land (rocky land and sandy land) decreased to 5.71% and 0.99% because of growing demand for land and the increasing degree of land intensive utilization. The percentage of water area fell down to 1.61% due to aridification. The area of grassland and farmland firstly increased, then decreased to a greater degree.

Table 1. Land use change during 1987-2011 in Huangfuchuan watershed

Date		Farm land	Forest land	Wate r	Sandy land	Rocky land	Grassland	Shrub	Construction land
1987	area	58.06	78.89	85.94	131.52	343.66	1136.10	1399.47	1.50
	percentage	1.79	2.44	2.66	4.07	10.62	35.12	43.25	0.05
1995	area	84.57	84.38	79.97	139.00	392.22	989.49	1462.84	2.67
	percentage	2.61	2.61	2.47	4.30	12.12	30.59	45.22	0.08
2000	area	144.29	90.61	76.31	109.86	284.23	891.13	1635.23	3.48
	percentage	4.46	2.80	2.36	3.40	8.79	27.55	50.53	0.11
2007	area	385.4	137.6 3	68.31	36.90	339.86	896.22	1365.81	5.01
	percentage	11.91	4.25	2.11	1.14	10.51	27.71	42.22	0.15
2011	area	238.48	220.2 0	52.04	32.05	184.79	713.53	1786.18	7.87
	percentage	7.37	6.81	1.61	0.99	5.71	22.06	55.21	0.24

3.2 Estimation of NPP Based on Adjusted CASA Model

Five images of NPP distribution in July from 1987 to 2011 are shown in Fig 7.

The NPP amount of the five periods is 28.12GgC, 53.47GgC, 73.23GgC, 157.92GgC and 78.52GgC. NPP in July rose gradually from 1987 to 2007, then declined gradually. It showed a trend of firstly increase then decrease and fluctuation with time flowing.

3.3 Response of Land Cover Changes in Huangfuchuan Watershed to NPP

As a result of location of the studied area in ecologically fragile temperate grassland area, transfer of land cove types led to changes of land cover pattern and further changed the structure and function of ecological system, NPP was affected deeply. Major land use types which had a effect on NPP can be determined by studying the effects of land use conversion on NPP. Firstly, we overlaid land cover type images of two periods in studied area , then extracted changed and unchanged land cover areas; Secondly, subtracted them by corresponding NPP images and acquired distribution of NPP changes. At last, extracted corresponding NPP changes with changed and unchanged land cover area. It's show below in table 2 to 5.

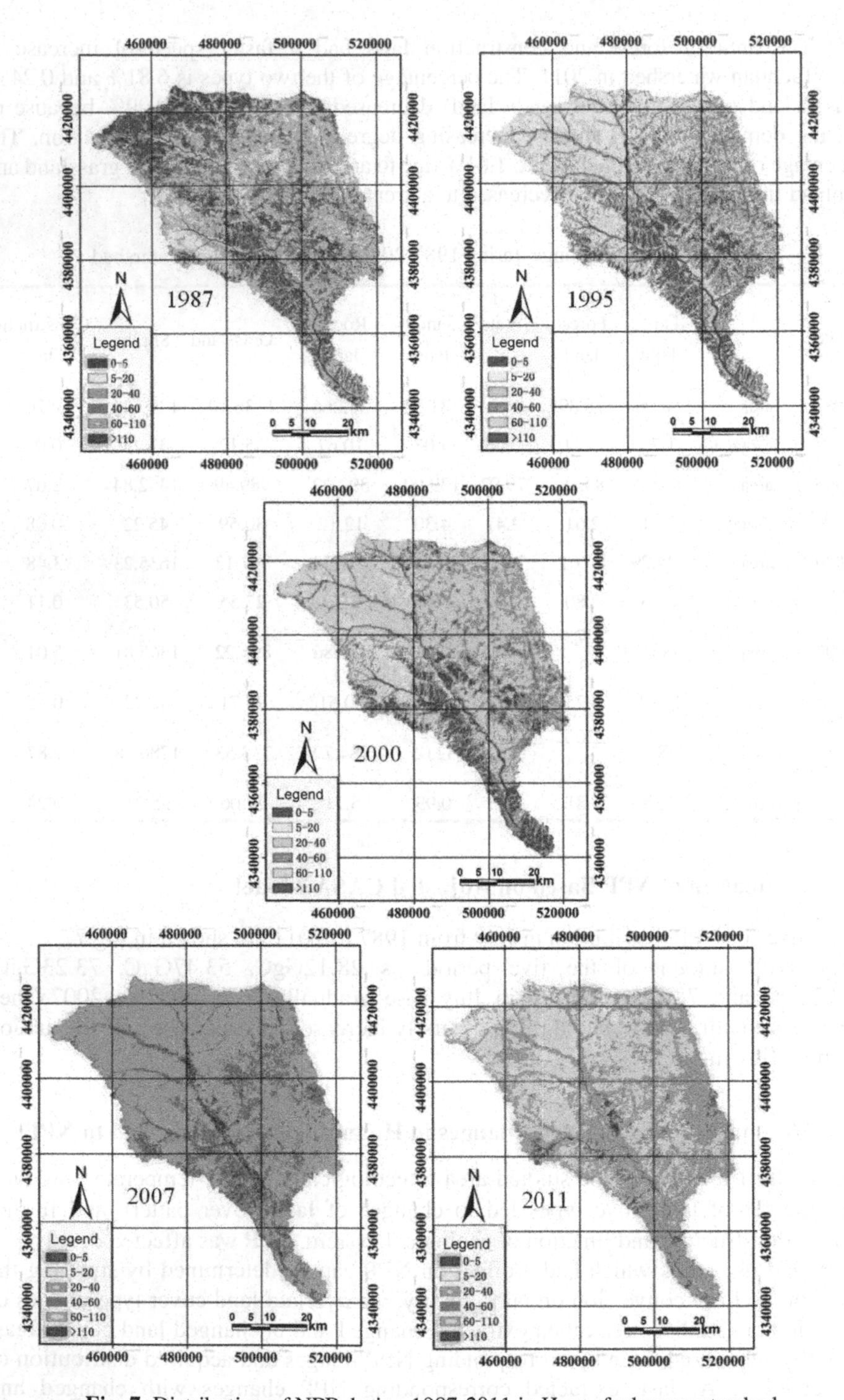

Fig. 7. Distribution of NPP during 1987-2011 in Huangfuchuan watershed

Table 2. NPP change corresponding with land cover change between 1987 and 1995 in Huangfuchuan watershed

1987 \ 1995	Farm land	Forest land	water	Sandy land	Rocky land	Grassland	Shrub	Construction land	Total
Farmland	0.53	0.04	-0.08	-	-0.05	-0.09	-0.48	-	-0.13
Forest land	0.26	0.20	-0.03	-	-0.24	0.03	-0.03	-	0.19
water	0.30	-	-	-	0.07	0.15	0.39	-	0.91
Sandy land	0.06	0.02	-	-	-	0.35	0.97	-	1.40
Rocky land	0.04	0.24	-	-	-	1.79	0.91	-	2.98
Grassland	0.31	0.52	-0.03	-0.09	-0.58	6.30	3.63	-	10.06
Shrub	1.83	1.27	-0.25	-0.54	-1.16	1.83	6.99	-0.02	9.95
Construction land	-	-	-	-	-	-	-	-	-
Total	3.33	2.29	-0.39	-0.63	-1.96	10.36	12.38	-0.02	25.36

Table 3. NPP change corresponding with land cover change between 1995 and 2000 in Huangfuchuan watershed

1995 \ 2000	Farm land	Forest land	water	Sandy land	Rocky land	Grassland	Shrub	Construction land	Total
Farmland	0.21	-	-0.23	-0.02	-0.04	-0.08	-0.67	-	-0.83
Forest land	0.16	0.20	-0.05	-0.02	-0.21	-0.05	-0.46	-	-0.43
water	0.32	0.03	-	-	-	0.13	0.39	-	0.87
Sandy land	0.16	0.04	-	0.03	-	0.30	1.67	-	2.20
Rocky land	0.20	0.46	-	-	-	2.75	2.37	-	5.78
Grassland	0.48	0.19	-0.10	-0.16	-1.23	3.74	1.08	-	4.00
Shrub	2.11	1.00	-0.28	-0.60	-0.97	1.60	5.33	-0.02	8.17
Construction land	-	-	-	-	-	-	-	-	-
Total	3.64	1.92	-0.66	-0.77	-2.45	8.39	9.71	-0.02	19.76

It can be calculated from Table 2 that from 1987 to 1995, NPP decrease by 0.66GgC, 0.01GgC because of transfer of farmland and forest land. NPP increased by 0.91GgC, 1.40 GgC, 2.98 GgC, 3.76 GgC and 2.96 GgC as a result of transfer of water, sandy land, rocky land, grassland and shrub. NPP increased by 11.34GgC totally due to transfer of land use types. Grassland transfer was the major factor leading to rise of NPP. In this stage, grassland mainly transferred to shrub and rocky land, that transfer from grassland to shrub made NPP rose by 3.63 GgC indicated transfer from grassland to shrub made the most contribution on NPP increase.

Table 4. NPP change corresponding with land cover change between 2000 and 2007 in Huangfuchuan watershed

2000 \ 2007	Farm land	Forest land	water	Sandy land	Rocky land	Grassland	Shrub	Construction land	Total
Farmland	3.32	0.42	-0.05	-0.02	-0.17	0.16	0.63	-0.03	4.26
Forest land	0.44	0.92	-0.01	-	-0.37	0.08	0.71	-0.01	1.76
water	0.89	0.09	-	-	-	0.14	0.53	-	1.65
Sandy land	0.26	0.05	-	-	-	2.23	1.45	-	3.99
Rocky land	1.15	0.89	-	-	-	2.85	4.21	-	9.10
Grassland	2.38	0.78	0.09	-0.24	-2.00	8.69	5.64	-	15.16
Shrub	13.40	5.24	-0.07	-0.08	-2.20	9.34	23.13	-0.03	48.73
Construction land	-	0.01	-	-	-	-	0.02	-	0.03
Total	21.84	8.40	-0.22	-0.34	-4.74	23.49	36.32	-0.07	84.68

Table 5. NPP change corresponding with land cover change between 2007 and 2011 in Huangfuchuan watershed

2007 \ 2011	Farm land	Forest land	water	Sandy land	Rocky land	Grass land	Shrub	Construction land	Total
Farmland	-6.59	-1.14	-0.54	-0.20	-0.52	-3.29	-9.20	-	-21.48
Forest land	-0.57	-1.85	-0.08	-0.02	-0.50	-0.48	-4.69	-	-8.19
water	0.11	0.03	-	-	-	0.12	0.22	-	0.48
Sandy land	0.01	-	-	-	-	0.29	0.07	-	0.37
Rocky land	0.27	1.13	-	-	-	1.46	2.77	-	5.63
Grassland	-0.35	-0.23	-0.08	-0.56	-2.05	-6.38	-12.50	-0.05	-22.20
Shrub	-1.63	-1.88	-0.40	-0.38	-2.54	-4.25	-22.87	-0.05	-34.00
Construction land	-	-	-	-	-	-0.01	-	-	-0.01
Total	-8.75	-3.94	-1.10	-1.16	-5.61	-12.54	-46.22	-0.10	-79.40

It was seen from Table 3 that from 1995 to 2000, NPP decrease by 1.04GgC, 0.63GgC because of transfer of farmland and forest land. NPP increased by 0.87GgC, 2.17 GgC, 5.78 GgC, 0.26 GgC and 2.84 GgC as a result of transfer of water, sandy land, rocky land, grassland and shrub. NPP increased by 10.25GgC totally due to transfer of land use types. Transfer of rocky land was the major reason for NPP increase.

It was shown in Table 4 that from 2000 to 2007, NPP increased by 0.94GgC, 0.84GgC, 1.65GgC, 3.99 GgC, 9.10 GgC, 6.47 GgC and 25.60 GgC owing to transfer of farmland ,forest land, water, sandy land, rocky land, grassland and shrub. NPP increased by 10.25GgC totally due to transfer of land use types. Transfer of shrub was the major reason for NPP increase. In this stage, shrub mainly transferred to grassland and rocky land, that transfer from shrub to grassland made NPP rose by 9.34 GgC indicated transfer from shrub to grassland made the most contribution on NPP increase.

It can be calculated from Table 5 that from 2007 to 2011, NPP decreased by 14.89 GgC, 6.34GgC because of transfer of farmland and forest land. NPP increased by 0.48GgC, 0.37 GgC, 5.63 GgC as a result of transfer of water, sandy land and rocky land. NPP decreased by 15.82 GgC and 11.13 GgC due to transfer of grassland and shrub. NPP decreased by 41.70GgC totally due to transfer of land use types. Transfer of grassland was the major reason for NPP decrease. In this stage, grassland mainly transferred to shrub and rocky land that transfer from grassland to shrub made NPP decreased by 12.50 GgC indicated transfer from shrub to grassland made the most contribution on NPP descend.

As is analyzed above, although natural and human factors gave impetus to LUCC in a long term, human activities is no doubt the most principal driving factor in short time. Land cover changes in Huangfuchuan watershed and the changing tend of NPP which firstly rose then descended and increased wholly in recent 25 years were determined by human disturbance including growing demand of population increase for food, expansion of livestock husbandry, mining and policies on improving environment in ecologically fragile area. Since 1990s, a series of ecological restoration projects such as Beijing and Tianjin sandstorm source control, grain for green, return grazing land to grassland, enclosed transfer and so on were conducted. These projects had promoted effects on environmental improvement and vegetation recovery. These ecologically restoring measures suppress degeneration and promote NPP increase of grassland. In recent ten years, as a result of high-intensity mining in Huangfuchuan watershed and surrounding areas, problems such as loss of underground water, destroyed grassland ecosystem appeared. They further led to a declining trend of NPP.

4 Conclusion

Along with increasingly intensive human activities and climate changes, quantities and structure of land use types in the Huangfuchuan watershed have had greater changes as a typical temperate grassland ecosystem in ecologically fragile areas in recent 25 years. This situation resulted in changes of NPP value.

1) Land use types majored in shrub and grassland. Land structure changed evidently. The area of construction and forest land increased gradually, the area of water was right opposite. The area of farmland, grassland, shrub, rocky land and sandy land fluctuated. Owing to aridification, water area decreased sharply and forestland area had a increase with protective policies put forward.

2) NPP value in 1987, 1995, 2000, 2007 and 2011 was 28.11GgC, 53.47GgC, 73.23GgC, 157.91GgC and 78.51GgC. It changed with a process of firstly increased then decreased. The NPP percentage of shrub was 57.84%, 52.03%, 52.33%, 44.97% and 51.87%;The NPP percentage of grassland was 27.08%, 31.66%, 29.44%, 25.74% and 22.20%. NPP of the two land types topped two among all land types in the watershed.

3) Prominent land use transfer forms were determined by analyzing NPP changes of land use type transfer. Between 1987 and 1995, transfer from grassland to shrub was the principal reason for NPP changes in Huangfuchuan watershed; Between 1995 and 2000, transfer from rocky land to grassland was the major factor; Between 2000 and 2007, transfer from shrub to grassland become the primary cause; Between 2007 and 2011, transfer from grassland to shrub was the one.

In order to support NPP in temperate grassland areas, cover area of shrub and grassland should be ensured emphatically. Land cover changed frequently in the studied area in recent 25 years. There are both positive succession and regressive succession. It indicated that temperate grassland ecosystem in the north part of China which was susceptible to outside influence and changed dynamically was quite fragile. Water resource shortage was severe additionally. A deeper study on making mining coordinate with human settlements, implementing measures of grain to green, improving water resource utilization efficiency, strengthening land management and protecting ecological environment is needed.

Acknowledgements. This research was funded by the National Science Foundation of China (No.51374208) and a project funded by the Priority Academic Program Development of Jiangsu Higher Education Institutions (SZBF2011-6-B35) and Research Fund of State Key Laboratory of Soil and Sustainable Agriculture, Nanjing Institute of Soil Science (Y412201432), Chinese Academy of Science. Also, we want to thanks the research of Yang Hongbing's degree paper. Special thanks are due to the reviewers for their important suggestions.

References

1. Leith, H., Wittaker, R.H.: Primary Productivity of the Biosphere, pp. 103–134. Springer, Heidelberg (1975)
2. Fang, J.Y.: Global Ecology: climate change and ecological response, pp. 78–89. Higher Education Press, Beijing (2000)
3. Leng, S.Y., Song, C.Q., Lu, K.J., et al.: The important scientific problems of regional environmental change research. Progress in Natural Science 11(2), 222–224 (2001)
4. Li, X.B.: A review of the international researches on land use/land cover change. Acta Geographica Sinica 51(5), 553–557 (1996)
5. Cramer, W., Bondeau, A., Woodward, F.: Global response of terrestrial ecosystem structure and function to CO2 and climate change: results from six dynamic global vegetation models. Global Change Biology 7(4), 357–373 (2001)
6. LeBauer David, S., Treseder Kathleen, K.: Nitrogen limitation of net primary productivity in terrestrial ecosystems is globally distributed. Ecology 89(2), 371–379 (2008)
7. Gilmanov, T.G., Soussana, J.E., Aires, L.: Partitioning European grassland net ecosystem CO_2 exchange into gross primary productivity and ecosystem respiration using light response function analysis. Agriculture Ecosystems & Environment 121(1), 93–120 (2007)
8. Gao, Z.Q., Liu, J.Y., Cao, M.K., et al.: The impact on the ecosystem productivity and the carbon cycle in agriculture and animal husbandry transition zone of the land use and climate change. Science in China Ser. D Earth Sciences 34(10), 946–9571 (2004)

9. Liu, Z.B., Liu, M.S., Xu, C., et al.: NPP and CO_2 Assimilation Value of Vegetation in Jiangyin, Jiangsu Province. Journal of Nanjing Forestry University (Natural Sciences Edition) 31(3), 139–1421 (2007)
10. Xu, X.B., Yang, G.S., Li, H.P.: Impacts of Land Use Change on Net Primary Productivity in the Taihu Basin, China. Resources Science 33(10), 1940–1947 (2011)
11. Wang, Y., Huang, M., Wang, X.R.: Impacts of land use and climate change on agricultural productivity in Shanghai. Acta Scientiae Circumstantiae 30(3), 641–648 (2010)
12. Zhao, C.Y., Cheng, G.D., Zhou, S.B., et al.: Spatial distribution of net primary productivity of natural vegetation in the northwest China. Journal of Lanzhou University (Natural Sciences) 45(1), 42–49, 55 (2009)
13. Zhang, J., Pan, X.L., Gao, Z.Q., et al.: Estimation of net primary productivity of the oasis-desert ecosystems in arid west China based on RS-based ecological process. Arid Land Geogra Phy. 29(2), 255–261 (2006)
14. Mu, S.J., Li, J.L., Yang, H.F., et al.: Spatio-temporal variation analysis of grassland net primary productivity and its relationship with climate over the past 10 years in Inner Mongolia. Acta Prataculturae Sinica 22(3), 6–15 (2013)
15. Zhang, Y.L., Jia, Z.B.: Effect of Different Land Use Types on Diversity and Community Structure in Huangpuchuan Watershe. Journal of Inner Mongolia University 39(3), 325–331 (2008)
16. Zhang, C., Li, X.B., Zhang, L., et al.: Impacts of land use and cover change on ecosystem services values in Huangfuchuan Watershed. Journal of Beijing Normal University (Natural Science) 45(4), 399–403 (2009)
17. Zhu, W.Q.: Estimation of net primary productivity of Chinese terrestrial vegetation based on remote sensing. Beijing Normal University, Beijing (2005)
18. Zhang, X.S.: Vegetation PE index and classification of vegetation and climate (2): Several methods and the PEP program introduced. Acta Phytoecologica and Geobotany Sinica 13(3), 197–207 (1989)
19. Running, S.W., Coughlan, J.C.: A general model of forest ecosystem Process for regional Applications. Hydrologic balance, canopy gas exchange and primary Production Process. Ecological Modelling 42, 125–154 (1988)

Applications of GIS for Evaluation Land Suitability for Development Planning of Peanut Production

Quanghien Truong[1,3], Zhiyu Ma[2], Caixue Ma[2,*], Liyuan He[1], and Thivan Luong[3]

[1] College of Resources and Environment, Huazhong Agricultural University, China
[2] College of Public Administration, Huazhong Agricultural University, China
[3] Faculty Of Geography - Land Administration, Quy Nhon University, Vietnam
atruongquanghiendhqn@gmail.com,
macaixue@mail.hzau.edu.cn

Abstract. In recent years, Vietnam is very interested in the investigation classification, soil mapping, evaluating Suitable land, which has been an important contribution for enhancing the quality of plans for agriculture land use planning. It has been also the primary goal to synthesize and build the plant restructuring orientation associated with the restructuring of land use.

Through the preliminary evaluation regarding the socio-economic conditions, the soil as well as the farming practices at Thuy Bang Commune, Huong Thuy Town, Thua Thien Hue Province, Vietnam, it shows that this is the great potential land which is to develop the peanut production, and the inhabitants who live there are also increasingly expanding the planted peanut area in recent years. However, the determination and the suitable ability partition which is the scientific basis for the implementation of land planning to develop the planted peanut is yet to be processed. The study combined with the land evaluation according to FAO and GIS technology application, specifically is ArcGIS software, which are the implements to build the thematic maps and superpose these maps constitute the land unit map, then classify appropriately for peanut planting in the studied area at present and in the future. The research results have identified that Thuy Bang Commune has 28 land map units; the current classification for peanut planting shows that the studied area appears only at low and suitable medium accounted mainly. The research has been proposed to the developed area and distributed specifically to every other area and each land unit for the peanut planting of Thuy Bang Commune. The proposed research is very important and it is the scientific basis for the reference planning of the agricultural development in the future of Thuy Bang Commune, Huong Thuy Town, Thua Thien Hue Province, Vietnam.

Keywords: Land evaluation, Land unit, Peanut, GIS, Thuy Bang.

1 Introduction

Land and land use are the strategic goals for countries around the world. By using land effectively is urgently required, especially for nation states with limited land resources such as Vietnam.

* Corresponding author.

F. Bian and Y. Xie (Eds.): GRMSE 2014, CCIS 482, pp. 684–698, 2015.

Practically, the management and use of land are still inadequate in Vietnam currently, not only at a local level. Land in general, agricultural in particular, which is the one is manageable and used mostly based on the experience of the farmers. What crops layout, on what land depends primarily on the ability to consume, product prices and the market demand. Consequently, the efficiency of agricultural production is low and often unstable. The profits from this production do not really attract the producer. Besides, the crops are not interested in maintaining and protecting soil fertile, which has made soil quality to expect serious decline.

Land use is not the main objective of soil quality which makes effective land use planning as well as feasibility of planning is not often high. This has also happened at the area of Thuy Bang Commune, Huong Thuy Town, Thua Thien Hue Province. Currently, the inhabitants opened the planted peanut area because they realize that the benefits from which this production form brings back are much higher than the others'. However, the determination and the suitable ability partition which is the scientific basis for the implementation of land planning to develop the planted peanut has not been done yet. [3]

In this study, we present the results of the combined method of FAO soil evaluation and the application of GIS technology, namely ArcGIS software for building and overlay the thematic soil map put out units land use map and use land units, coupled with the ecological requirements of peanuts for grading suitable for cultivation peanuts in Thuy Bang Commune, Huong Thuy Town, Thua Thien Hue Province, Vietnam, simultaneously put out the orientation for the development peanut production for the study area in the future.

2 Material and Method

2.1 Material

- *Source of data space:* Including soil maps ; current use of land use map; thematics maps of study area, including maps of 1:10,000 scale: soil types, slope map, soil depth map, mechanical composition map, humus map, porosity map, pH map, trafficmap. [8]

- *Sources attribute data:* Including data tables associated with spatial data and data on attributes such as data on climate and weather conditions, geographical location; The statistics of natural conditions, economic - social; Development situation agriculture and forestry of study area ; Ecological requirements of the land use type peanuts [8, 3, 4].

- **The software used:* Including Excel for data processing, ArcGis for edit, update data, overlay maps, database management and decorative map.

2.2 Method

Data Collection Methods. Collecting documents and data on natural conditions, economic, social, development of the agricultural commune in the past years, etc. from the report of the Thuy Bang Committee Commune People in order to know the orientation communal development of agriculture in the next 5 years.

Collecting documents as soil maps, topographic maps, land use maps, current status, etc. and the associated data as a basis for data processing in the future.

Investigation Interview Methods. During the study proceeded by interviewing farmers explains that besides the direct field work, in order to have a comprehensive view of the distribution of the soil groups we need the type and status land use of commune. Locations, which are to be scrutinized are the peanut-growing areas and from there comments can be made on the suitability of land for the development peanut production of these areas.

Data Processing Methods. The data processor including the two processing methods is executing the primary data and subsequently of the secondary data. The data will be processed after synthesis in the form of statistics, charts, etc. on Microsoft Excel software.

Map methods. In this study, we havc dcveloped all kinds of thematic maps such as soil type map, slopes map, soil depth map, soil texture map, etc. and overlay these maps by ArcGIS software constitute soil map units. Furthermore, we also used topographic map, the current land use map of commune serves as the base map for the proposes plant layout later.

Land Evaluation of FAO methods. We used the limiting factor combination method in the topics. This method taking the factors assessment are less most suitable than the limiting factor. Thus, suitable level overview of a land map unit for each type of land use is the lowest suitable level ranked and soil properties and based on the excellent factors and the other factors in evaluation. [5, 6],

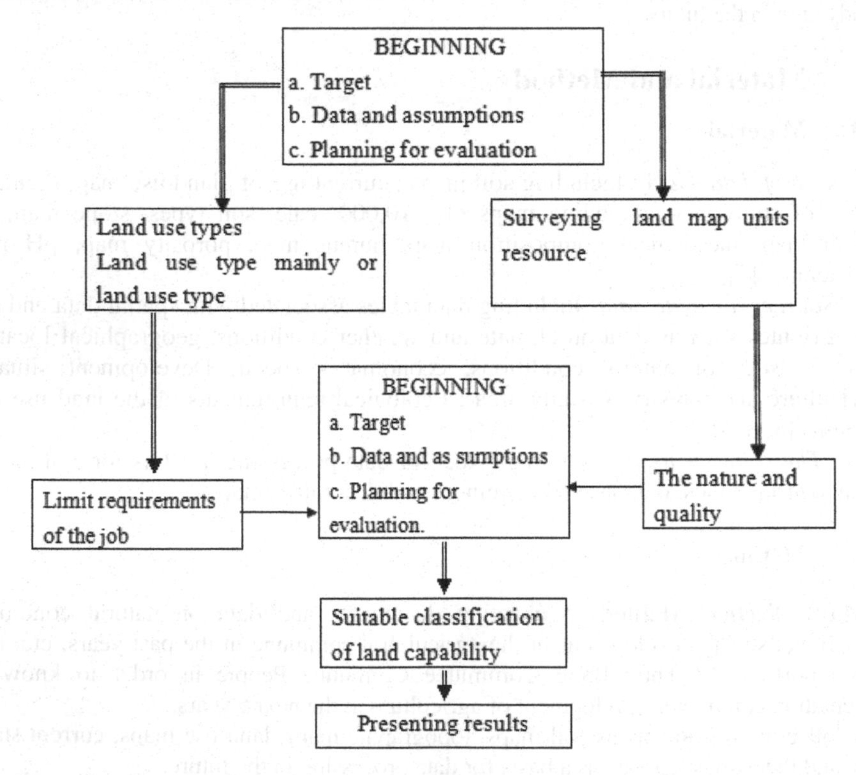

Fig. 1. Process diagram for land evaluation of FAO [7, 14]

3 Results and Discussions

3.1 Generalizing Research Area

Thuy Bang Commune is located in the western town of Huong Thuy. It is about 6 km south of Hue city center, geographic coordinates from 107 °33' to 107 °37' east longitude and from 16°20 'to 16 °26' north latitude. [3]

The natural area is 2298 ha, land used accounted for 69.72% of the total natural area, unused land accounted for 30.28% and the population is 8758 people. Thuy Bang Commune is a sparsely populated area but remains a poor commune by many causes in which are mainly due to the complex topography, population distributions not concentrated and low intellectual level [3, 4]. Reviewing generally about the study area, it shows some following emerged problems:

Advantages. + System traffic has National Highway routes, which is the favorable condition for cultural exchanges and commodity exchanges.

+ Source of social labor abundant is the good condition for economic - social development of this local in the future.

+ The structure of the trades is very diverse and consistent with the production level of the people. That is the good condition for the development of agriculture, forestry, services and other industries of this local.

+ The health and the education are invested and developed, create conditions for the people who live there will have qualifications in the future.

Constraints. + Land also lost territory because the investment is inaccurate and synchronous.

+ Roads of some areas are difficult to travel, great influence economic, social and cultural activity from people.

+ Work of agricultural and forestry remain weak, so the effective transfer of scientific and technical advances to people was being rather slowly.

3.2 Land Suitability Evaluation for Land Use Type

Developed Thematic Maps. To choose and hierarchical indicators, we had to rely on some of the following principles:

+ The natural elements, characteristics and soil characteristics as well as agro-ecological factors of the study area.

+ The assessing results of the current state land use, production efficiency analysis of the land use types in the study area.

+ Land use requirements of the land use types in the study area. [15]

Using the results of the soil analysis for land evaluation study of Vietnam Highlands Agricultural Science Institute, combined with field surveys and other materials of local to test the accuracy to developed thematic maps and accompanying attributes such as soil quality, slope, soil fertility, etc. in order to get the exact results of the spatial elements and attributes of the thematic maps.

The thematic maps which were selected to Land mapping units based on the requirements of the land use of peanut tree include: soil types map, slope map, soil texture map, soil fertility map, world maps components, traffic map, porosity maps, land use status maps. These are the materials needed to serve the Superposition the map in the further process.

About the formulation of thematic maps, we carried out homogenizing the type of spatial data and attribute of all these maps. All data has been converted to the Shapefile format can handle, edited on ArcGIS software, the map coordinating system is set up in accordance with the territory of Thua Thien Hue Province, Vietnam. Besides, we supplied and editted due to investigators, field surveys in the study area; therefore, the thematic map accuracy in service overlay mapping process to serve the appropriate classification of land for Thuy Bang Commune.

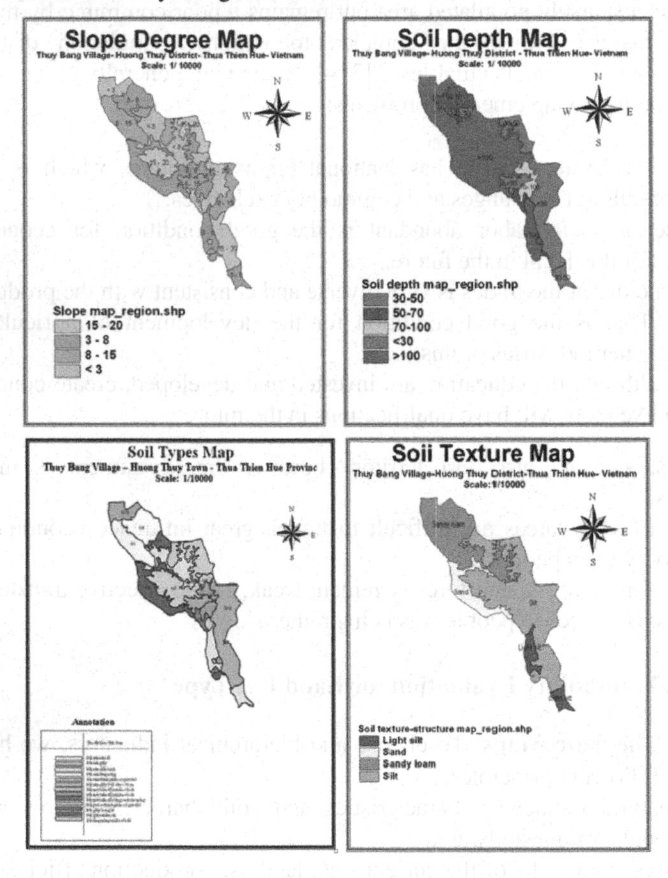

Fig. 2. Some of the illustrative thematic maps

Land Mapping Units. Thuy Bang Commune has 13 soil types belonging to 5 major groups of soil. Those are Acrisols, Cambisols, Fluvisols, Gleysols, Leptosols. Among them, Acrisols has the largest area, 776.19 hectares (33.78% of the total natural area of the commune), Gleysols has the smallest area of 151.66 hectares (6.60 % of total natural area of the commune). Thuy Bang Commune of topography is relatively complex with many hills, fertile land at an average level. [8]

To build soil map units for Thuy Bang Commune, we used ArcGIS to overlay the thematic maps which have been chosen. This is a high-performance software shown in space as well as store, update and query the data attributes. [11]

With these products that the thematic maps have been processed, uniform formats, overlay maps to give out land units. As a result, overlapping thematic map is a land unit map with 28 units of land with associated attributes, the database is stored and managed on Arc GIS software. *(Fig 3)*

Based on the data attributes of which land map units are stored and managed on ArcGIS software, the synthetic characteristics and properties of the land map units Thuy Bang Commune have been placed as a basis for grading of land suitable for crops on the study area. (*Fig 4, 5*)

Through aggregate results from attributes of land map units are stored on ArcGis software, Thuy Bang Commune has 28 land mapping units, of which Acrisols has 10 units, Cambisols has 6 units, Fluvisols has 5 units, Gleysols has 2 units and Leptosols has 5 units. The area of each land unit also has a wide disparity, the smallest land unit area is 0,09 ha (land units No. 1 and No. 5), the largest land unit area is 281,81 ha (land unit No. 8), besides each land unit shown with the accompanying factor is soil type, slope, fertility, porosity, location, etc.. This is the basis for the classification of land suitability for crop production in general and in particular peanuts in the study area.

The Results of Current Suitability Evaluation. The evaluation and classification of land suitability for crops is a combination of land mapping units and soil use types. Using the ecological requirements of the soil use types compared to units land map on the principle of FAO will give us results that match the level of the land map units for soil use types were selected. [5]

In this study, the suitability of the land map units for the production of peanuts is also estimated, so it is necessary to study the land use requirements of the type of peanuts. (Table 1)

Through the composite panel, the ecological requirements of peanuts for land whose criteria is suitable for peanuts S1 level is very little. The important factor for peanut plant is soil type, mechanical composition and porosity of the soil. These factors are dominant factors which might hardly be renovated and have directly simultaneous decision to the productivity of peanuts. For peanuts, the most suitable soil is Fluvisols, then Arenicsoil and Feralitsoil; Infertile Acrisols. Gleysols is not suitable for growing peanuts. Soil mechanical composition is soft. Sandysoils is the most suitable soil for planting peanuts. Fruiting characteristics peanuts show more difference than other crops, peanut fruits formed rays pierced through the process, so soil which has high porosity is usually more favorable for the stabs ray and more effective form.

Based on the characteristics of the land unit map of the study area, combined with the ecological requirements of the land use type peanuts, after applying appropriate soil evaluation methods of the FAO, the study obtained results of evaluation and classification of current soil suitable for the type of land use peanuts. Evaluation results were synthesized in Table 2 and editted by the map and storage, management of space and attributes on ArcGIS software (Fig 6).

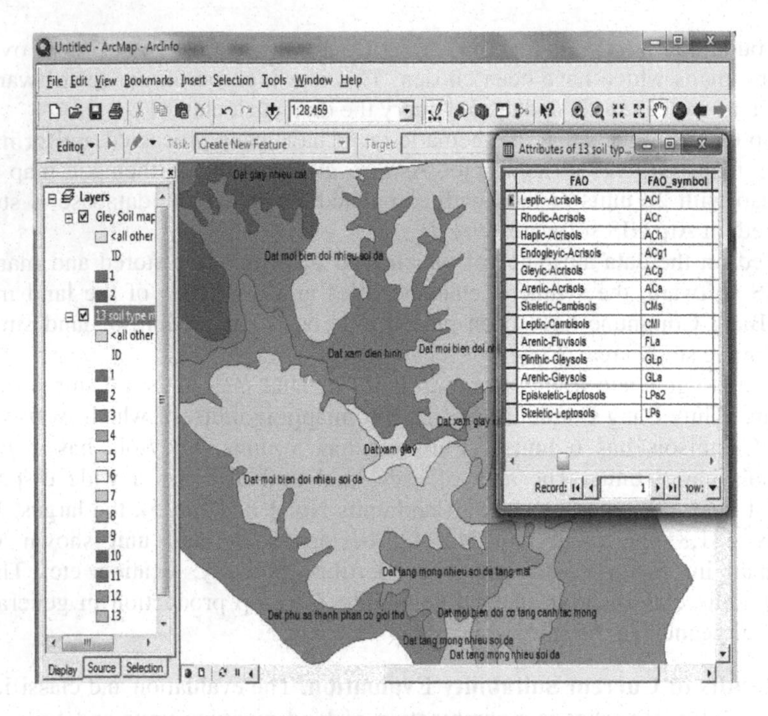

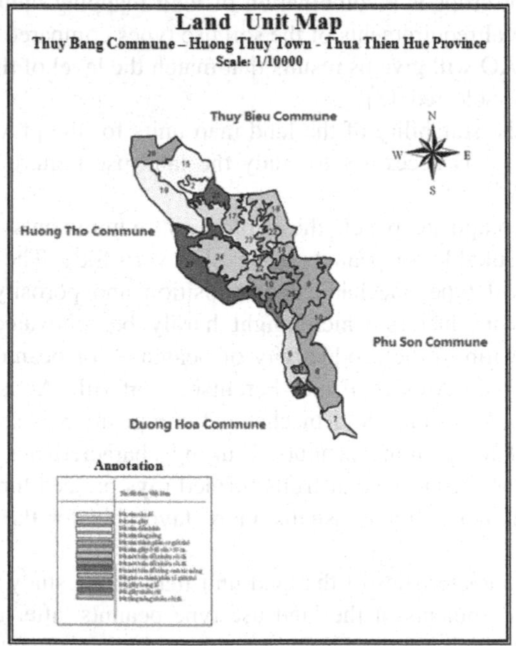

Fig. 3. The database on soil evaluation on ArcGIS software

Indicators	Decentralisation	Symbol	Area (ha)
1. Soil type	1. Leptic-Acrisols	G1	175,42
	2. Gleyic-Acrisols	G2	16,09
	3. Haplic-Acrisols	G3	27,00
	4. Arenic-Acrisols	G4	7,37
	5. Rhodic-Acrisols	G5	179,40
	6. Haplic-Acrisols	G6	370,91
	7. Leptic-Cambisols	G7	89,23
	8. Skeletic-Cambisols	G8	424,17
	9. Arenic-Fluvisols	G9	361,07
	10. Plinthic-Gleysols	G10	99,78
	11. Arenic-Gleysols	G11	51,88
	12. Episkeletic-Leptosols	G12	464,15
	13. Skeletic-Leptosols	G13	31,53
2. Texture_Soil	1. Silt	T1	1248,64
	2. Light silt	T2	279,44
	3. Sandy loam	T3	633,85
	4. Sand	T4	136,07
3. porosity	1. high	X1	361,07
	2. medium	X2	186,77
	3. low	X3	635,56
	4. very low	X4	1114,60
4. TopSoil_cm	1. >100	D1	1236,01
	2. 70 -100	D2	82,21
	3. 50 -70	D3	271,47
	4. <50	D4	708,31
5. Slope_degree	1. <3	SL1	445,53
	2. 3 -8	SL2	801,08
	3. 8 - 15	SL3	442,27
	4. 15 – 20	SL4	609,12
6. Fertility_level	1. rich	P1	0,00
	2. medium	P2	851,15
	3. poor	P3	1080,87
	4. very poor	P4	365,98
7. pH	1. 6,5-7.0	H1	0,00
	2. 5,5-6,5	H2	0,00
	3. 4,5-5,5	H3	1199,37
	4. <4,5	H4	1098,63
8. Position	1. favorable	GT1	455,73
	2. less favorable	GT2	968,49
	3. unfavorable	GT3	873,78

Fig. 4. Decentralisation Indicators land unit map of Thuy Bang Commune

Through the analysis studies and synthesis of the current level of suitable land for peanut production in Thuy Bang Commune, we drew the following comments:

+ The very suitable level of land (S1) for peanut production is not there.

+ The suitable medium level of land (S2) has 458,75 ha (accounting for 19,96% of the total natural land area of the commune). At this suitable level, mainly due to limited soil fertility, transport and pH, these factors can be overcome to enhance the level of suitable.

+ The less suitable medium level of land (S3) has 486,33 ha (accounting for 21,16% of the total natural land area of the commune). This limited suitable level is mainly due to soil fertility factors, these factors can be overcome in the course of cultivation.

+ The inappropriate level of land (N) has 1352,92 ha (accounting for 58,88% of the total natural land area of the commune). Not-suitable for peanut production shows relatively more, mainly because of the factors restrictions from the nature of land, and it is difficult to overcome in order to improve the level of suitability to the proper level for the higher.

Land unit	Characteristics and nature of the land unit								Area (ha)
	G	T	X	D	SL	P	H	GT	
1	9	2	1	1	2	3	2	3	0,09
2	1	3	3	4	3	4	3	1	28,63
3	4	1	2	1	3	4	3	3	7,37
4	13	1	4	3	1	4	3	3	31,53
5	12	2	4	1	2	2	2	2	0,09
6	6	4	3	1	2	2	3	3	79,75
7	6	4	3	4	2	2	3	3	51,91
8	12	1	4	4	4	2	2	3	281,81
9	12	4	4	3	4	2	2	3	4,41
10	12	1	4	4	3	2	2	3	177,84
11	7	1	3	3	2	3	3	3	7,02
12	9	2	1	1	2	3	3	2	50,00
13	9	1	1	3	2	3	3	3	41,72
14	6	1	3	4	4	2	2	2	60,00
15	1	3	3	3	2	4	2	1	146,79
16	3	1	4	1	1	3	2	2	27,00
17	8	1	4	1	3	3	3	1	49,03
18	8	1	4	4	4	3	3	3	108,12
19	5	3	2	1	3	3	2	1	179,40
20	2	1	4	1	1	2	2	2	16,09
21	11	1	4	1	1	4	2	1	51,88
22	8	1	4	1	2	3	3	2	112,24
23	6	3	3	1	1	2	3	2	179,25
24	8	1	4	1	4	3	3	2	154,78
25	7	1	3	2	2	3	3	3	82,21
26	9	1	1	3	1	3	2	2	40,00
27	9	2	1	1	2	3	2	2	229,26
28	10	3	4	1	1	4	2	2	99,78
Total									**2298,00**

Fig. 5. Characteristics and nature of the land unit of Thuy Bang Commune

The Results of Future Suitability Evaluation. Evaluating the suitability of land for future crops based on the current suitable assessment results of the land for crops that overcome the limited factors in order to upgrade the suitablity of land to a higher suitable level. The nature factors of land (excellent factors) such as depth, thickness floor, mechanical composition, etc. cannot be overcome, just can get over the factors such as traffic, soil fertility, pH, etc. [1, 2, 10].

From the assessment results to adapt the current classification for land use type planting peanuts, we can see: The factors limiting the ability of land suitability of land mapping unit for peanuts land use type are excellent factors. These factors are irreparable or difficult to repair in the future such as the kind of land, slope, topography level, thick floors, mechanical composition. Although the normal factors such as humus content, market orientation, technical proficiency, transport affect to the suitability level of land for this land type, they are easy to be changed or can be overcome by the impact of technical measures (humus content factor) or training activities, training (for qualified technical factors of production) due to the users' demand (market) [7]. Therefore, even though these factors can be improved to the best level in the furture, the suitable class of land map units for the land use type to plant peanuts is not be changed as current.

The conformity assessment results for future planted peanuts in the study area is shown in the following table:

Table 1. Synthesis of the required types of land use for peanut planting

Land use type	Factor rating	Suitable level			
		S1	**S2**	S3	N
Peanut Production	1. Soil type (G)	9	4, 5	1, 6, 7	2, 3, 8, 10, 11, 12, 13
	2. Texture Soil (T)	3	2	4	1
	3. Porosity (X)	1	2	3	4
	4. Slope degree (SL)	1	2	3	4
	5. pH (H)	1	2	3	4
	6. Fertility level (P)	1	2	3	4
	7. Position (GT)	1	2	3	4

[9]

The above composite panel shows that after overcoming the limited factors which may overcome such as soil fertility, traffic, the pH in the area of suitable grades unchanged but sub class were raised than before. The not-suitable land area should not be planted peanuts because the profits from that will not be high, so we can research more to give suggestions for other crops which are more suitable in order to get higher economic efficiency.

3.3 Development Orientation Peanut Production for the Study Area

According to the results of future suitable level assessment for planting peanuts on the study area, there is not any land unit reached very suitable level (S1), there are 458.75 ha at suitable average level (S2), accounting for 19.96% of the natural land area of the commune, with 4 land mapping units which are 1, 12, 27, 19. Prioritizing the development of peanut crop area on the land map units suitable at this level. However, to achieve high productivity and profit per land area unit, the following measures should be applied:

+ Researching to build the technical processes (selected varieties, fertilizer processes and care) suitable with the specific situation of natural conditions and economic – social of the locality.

+ Researching the production planning associated with the market orientation.

+ Needing to have some measures to compulsorily purchase products for farmers after the harvest. This is a positive direction to solve farmers' outputs and avoid the disadvantages for them.

Table 2. Statistical area and the current suitability of land for peanut production in Thuy Bang Commune

SUITABLE LEVEL		Quantity of land unit	Land unit	Area (ha)	Total (ha)	Percentage (%)
Class	Sub-class					
Medium suitable level (S2)	$S2_{p,g}$	1	1	0,09	458,75	19,96
	$S2_{p,h}$	1	12	50,00		
	$S2_{sl,p}$	1	19	179,40		
	$S2_{p}$	1	27	229,26		
Low suitable level (S3)	S3	3	6, 7, 23	310,91	486,33	21,16
	$S3_{p}$	2	2, 15	175,42		
Not suitable level (N)	$N_{t,p}$	1	3	7,37	1352,92	58,88
	$N_{t,x,p}$	1	4	31,53		
	$N_{g,x}$	1	5	0,09		
	$N_{g,t,x,sl}$	3	8, 18, 24	544,71		
	$N_{g,x,sl}$	1	9	4,41		
	$N_{g,t,x}$	5	10,16,17,20,22	382,20		
	N_{t}	4	11,13,25,26	170,95		
	$N_{t,sl}$	1	14	60,00		
	$N_{g,t,x,p}$	1	21	51,88		
	$N_{g,t,p}$	1	28	99,78		
Total					**2298,00**	**100,00**

Note: g, t, sl, p, x, h, gt is the the limiting factor of type of soil, the soil texture, slope, soil fertility, soil porosity, pH, location.

For the land map units that are suitable for planting peanuts at less suitable level (S3), there are 5 units. They are 2, 6, 7, 15, 23, with 486,33 ha, accounting for 21,16% of the natural land area of the commune. After assuming overcome the limiting factor, here is the map units in the less suitable form land. However, we can still plant peanuts on these areas and need to apply some the following measures:

+ Helping farmers solve their capital problems in order that they can invest in developing peanut production.

+ Stepping up the agricultural extension in this local and transfer the new scientific technique advances to farmers for applying them to production.

+ Must have the appropriate investments for building infrastructure in the district contributing to better serve agricultural production.

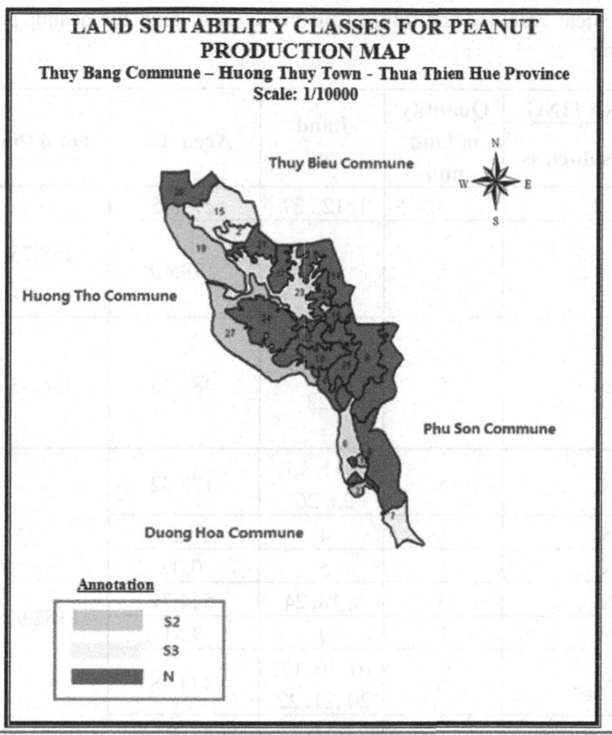

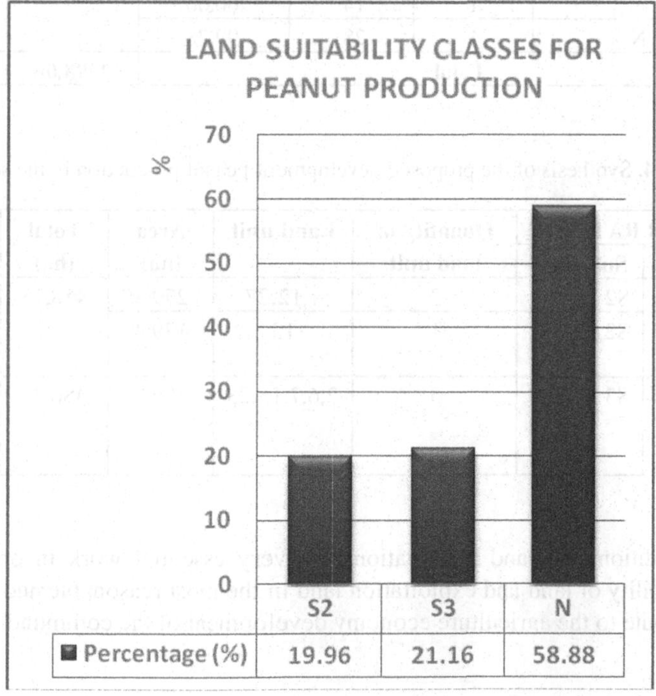

Fig. 6. Land suitability classes maps for peanut production

Table 3. Statistical area and the future suitable level of land for peanut production in Thuy Bang Commune

FACTOR RATING		Quantity of land unit	Land unit	Area (ha)	Total (ha)	Percentage (%)
Class	Sub-class					
Medium suitable level (S2)	S2	3	1, 12 , 27	279,35	458,75	19,96
	$S2_{sl}$	1	19	179,40		
Low suitable level (S3)	S3	5	2, 6, 7, 15, 23	486,33	486,33	21,16
Not suitable level (N)	N_t	5	3, 11, 13, 25, 26	178,32	1352,92	58,88
	$N_{t,x}$	1	4	31,53		
	$N_{g,x}$	1	5	0,09		
	$N_{g,t,x,sl}$	3	8, 18, 24	544,71		
	$N_{g,x,sl}$	1	9	4,41		
	$N_{g,t,x}$	6	10, 16, 17, 20 ,21, 22	434,08		
	$N_{t,sl}$	1	14	60,00		
	$N_{g,t}$	1	28	99,78		
Total					**2298,00**	**100,00**

Table 4. Synthesis of the proposed development peanut production in the study area

FACTOR RATING		Quantity of land unit	Land unit	Area (ha)	Total (ha)	Percentage (%)
Class	Sub-class					
Medium suitable level (S2)	S2	3	1,12,27	279,35	458,75	19,96
	$S2_{sl}$	1	19	179,40		
Low suitable level (S3)	S3	5	2,6,7,15,23	486,33	486,33	21,16

+ In addition, the land reclamation is a very essential work in order to utilize potential ability of land and exploitation land in the most reasonable and efficient way and contribute to the agriculture economy development of the commune.

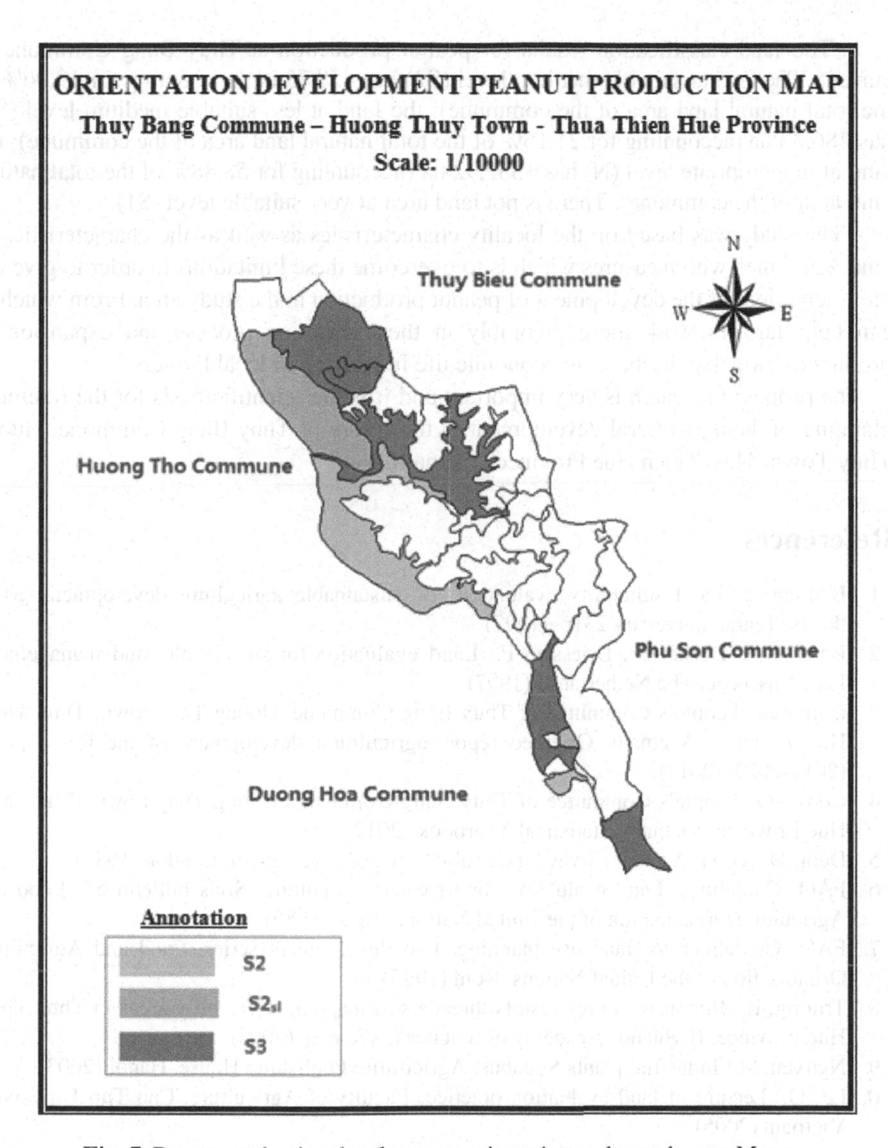

Fig. 7. Peanut production development orientation at the study area Map

4 Conclusions

Through the research results, GIS applications, namely Arc GIS software in evaluating land resources serving the peanuts development and production in Thuy Bang Commune, Huong Thuy Town, Thua Thien Hue Province, we have the following conclusion:

- The research results have identified 28 land map units. Among them, Acrisols has 10 units, Cambisols has 6 units, Fluvisols has 5 units, Gleysols has 2 units, Leptosols has 5 units.

- The land classification results for peanut production at Thuy Bang Commune in current: The land at suitable medium level (S2) has 458,75 ha (accounting for 19,96% of the total natural land area of the commune); the land at less suitable medium level (S3) has 486,33 ha (accounting for 21,16% of the total natural land area of the commune); the land at inappropriate level (N) has 1352,92 ha (accounting for 58,88% of the total natural land area of the commune). There is not land area at very suitable level (S1).
- The study was based on the locality characteristics as well as the characteristics of land, combined with measures which is to overcome these limitations in order to give out the orientation for the development of peanut production in the study area. From which, it can help farmers work more favorably in the cultivation process and expansion of production, which contributes to economic life improving of local farmers.

The proposed research is very important and it is the scientific basis for the reference planning of the agricultural development in the future of Thuy Bang Commune, Huong Thuy Town, Thua Thien Hue Province, Vietnam.

References

1. Bahmaniar. Land suitability evaluation for sustainable agriculture development. MSc. thesis, Tehran university 231p. (1997)
2. Beek, K.J., DeBie, A., Driessen, P.: Land evaluation for sustainable land management. ITC, Enschede, The Netherlands (1997)
3. Commune People's Committee of Thuy Bang Commune, Huong Thuy town, Thua Thien Hue Province, Vietnam. Oriented report agricultural development of the town, phase (2011-2020) (2010)
4. Commune People's Committee of Thuy Bang Commune, Huong Thuy town, Thua Thien Hue Province, Vietnam. Statistical Yearbook (2012)
5. Dent, D., Yong, A.: Soil survey land evaluation. Allen & Unwin, London (1981)
6. FAO. Guidelines: Land evaluation for irrigated agriculture. Soils bulletin 55. Food and Agriculture Organization of the United Nations. Rom (1985)
7. FAO. Guidelines for land use planning. Fao development series. Food and Agriculture Organization of the United Nations. Rom (1993)
8. Truong, H.: Report of survey results threads soil mapping some hilly areas in Thua Thien Hue province. Highlands Academy of Sciences, Vietnam (2003)
9. Nguyen, M.: Industrial plants Syllabus. Agriculture Publishing House, Hanoi (2003)
10. Le, Q.: Lecture of land evaluation practice, Faculty of Agriculture, Can Tho University, Vietnam (2005)
11. Nguyen, T., Tran, C.: Organization of geographic information system (GIS). Construction Publishing House, Hanoi (2000)
12. Huynh, V.: Application of GIS to criteria evaluate the most suitable land for crops at Huong Binh Commune. Scientific Journal of Hue University, Episode 16(50), 5–16 (2009)
13. Huynh, V.: Land evaluation Syllabus. Agriculture Publishing House, Hanoi (2011)
14. Hoang, V., Le, T., Tran, D.: Soil Syllabus. Agriculture Publishing House, Hanoi (1995)
15. Young, A., Golsmith, P.F.: Soil survey and land evaluation in developing countries. A case study in Malawi. The Geographical Journal 143, 407–438 (1977)

Research on System Simulation Technology for Joint Prevention and Control of Environmental Assessment Based on C4ISRE

Yunfeng Ma, Qi Wang*, Xiaofei Shi, and Zhihong Sun

Shenyang Aerospace University, College of Energy and Environment,
Shenyang, 110136, China
rcdxph @126.com

Abstract. There were many problems in Environmental Assessment (EA) application, which were inadaptable to the request of Joint Prevention and Control of EA (JPCEA),nowadays in China. Based on Environmental Assessment Automation System-C4ISRE, JPCEA integration simulation demonstration system were constructed which could product multidimensional information environment pollution situation scene which including natural geography environment, electromagnetic environment, space environment and so on, the related concepts and processes about JPCEA could ,which drove by simulation model and simulation system ,be shown in the way of all-direction, multi-angle. This method supplied one feasible technical reference for further JPCEA business research.

Keywords: Environmental Assessment, Joint Prevention and Control of Environmental Assessment, C4ISRE, Integration Simulation Demonstration System.

1 Introduction

1.1 The Problems Exist in China's Present EA

In China the EA system has been carried out for more than two decades. Due to the short period of development, there still have some problems which are inadaptable to the increasingly austere environmental position. It mainly exists in two fields: microscopic field and macroscopic field[1].

Main Problems in Microscopic Field

① The evaluation scope of EA is rather limited.

The deficiency of effective guidance from strategic environmental assessment; only attach importance to the EA for construction project, while neglect the one for residential area; the ecological environmental impact assessment hasn't been developed completely and so on.

* Corresponding author.

F. Bian and Y. Xie (Eds.): GRMSE 2014, CCIS 482, pp. 699–706, 2015.

② The technical quality and efficiency of EA should be further improved.
The inappropriate selection of evaluation standard; some assessment contents have become formalistic; the related data are inadequate; the analysis on the project is roughness; the calculation of pollution source strength is inaccurate; the unrelated contents congested in the environment evaluation reports; the environmental protection measures and requirements required by EA are of non-operating.
③ The distemperedness of EA's examination and approval institutions; the tracking assessment and supervision management are lagging.
④ Lack of public participation.
⑤ The related intergovernmental cooperation is blocked.

Main Problems in Macroscopic Field
① The environmental policies are marginalized and EA is always in a passive position.
② The ecological compensation is serious lack.
③ The relevant environmental protection system and law are distempered
④ The relevant charges of EA are short of effective criteria.
⑤ The justice and objectivity of EA can not be guaranteed effectively.
⑥ The units and the staffs' quality which engaged in EA are differently.
The solution for problems in microscopic field is the keystone of this thesis.

1.2 Information Technology Thinking for JPCEA

The conception of joint prevention and control meant that——adopting the scientific development philosophy, in order to enhance regional environmental protection, working mechanism which consisted of unified planning principle, unified monitoring principle, unified supervision principle, unified assessment principle and unified coordination principle were established. Insist on combining both environmental protection and economic development; Insist on combining both territorial management and regional linkage.

Currently, EA techniques were unable to meet the needs mentioned above of JPC because of EA's problems. The joint prevention and control integrated intelligent command and control system for EA, which constructed by modern information network technology, could meet the technical requirements of joint prevention and control.

JPCEA business involved massive systems and the relationship among systems were complicated extremely, furthermore, there existed non-reduced, antagonism, emergent-property, time-variation and so on called characteristics of complex systems in JPCEA business, so, it was difficult for the traditional modeling techniques and methods of EA which mainly constructed by Reductionism and Newton classic science to meet the need of JPCEA business.

According to JPCEA business characteristics, different relevant elements(such as Application of environmental protection force, Command and Control technology, Information network technology, and so on) should be considered synthetically, and, the related theory and method about System of Systems Engineering should be used to cope with the problems of JPCEA business. Aiming at JPCEA business features, based on C4ISR theory, C4ISRE (Command, Control, Communications, Computer, Intelligence,

Surveillance, Reconnaissance, Environmental Assessment, i.e. Environmental Assessment Automation System)[2] were constructed. This paper focused on the research of construction method for C4ISRE, striving to provide the necessary technical support for JPCEA business.

2 C4ISRE System Construction

2.1 Definition of C4ISRE

C4ISRE——Command, Control, Communications, Computer, Intelligence, Surveillance, Reconnaissance, Environmental Assessment, i.e. Environmental Assessment Automation System (ESAS).

C4ISRE were a human-computer discrete event dynamic system which synthetically utilized modern information network technology and environmental science theory who's core was C4ISR theory in order to realize the relative EA information's automatic management on collection, transference and analysis. It was C4ISRE were also that one highly-efficient complicated system that could fulfill the control, management, simulation and forecast of the related information in EA.

2.2 Typical Technical Characteristics of C4ISRE——Meta-Synthesis

C4ISRE were constructed as one automated system that managed collection and integration of data by using the data themselves to determine if more information should be collected against environmental pollutions in the database. Based on C4ISR theory[3], the meta-synthesis capabilities of C4ISRE aiming at JPCEA business characteristics were designed to integrate different hardware, software, sensors , staff, equipments, information, Datasets and so on, details were as follows : Moving Target Indicator, MTI; DB, database; SIGINT, Signals intelligence; ELINT, Electronic intelligence; IR, infrared; SAR, synthetic aperture radar; COMINT, Communications Intelligence ;IMINT, Image intelligence and so on. Reference model of C4ISRE meta-synthesis principle was shown in Fig.1.

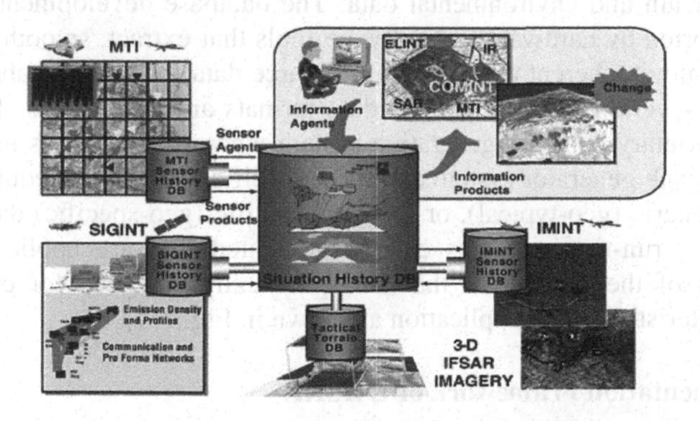

Fig. 1. Reference model of C4ISRE meta-synthesis principle

2.3 Key Concept of C4ISRE——Synthetic Natural Environment (SNE)

Synthetic Natural Environment had been defined by the Defense Modeling and Simulation Office (DMSO) as an "environment within which humans may interact through simulation(s) and/or simulators at multiple networked sites using compliant architecture, modeling, protocols, standards, and databases". An instance of a SNE coulde represent portions of the following environment Depending on the needs of the simulation: the terrain, terrain features (both natural and man-made), 3-D models of vehicles and personnel, the ocean (both on and below the surface), the ocean bottom including features on the ocean floor, the atmosphere including environmental phenomena, and near space. In addition to physical objects, the SNE included the specific attributes of these environmental objects as well as their relationships. Depending on the purpose of the simulation, the SNE may represent various aspects of the full dimension of ground, air, maritime and space operations across the entire spectrum of conflict and operations other than war (include domestic emergencies, nation assistance, disaster relief, peacekeeping, and peacemaking).Synthetic Natural Environment (SNE) was the necessary part of C4ISRE system. The complete modeling of entities within a synthetic environment was shown in Fig.2[4].

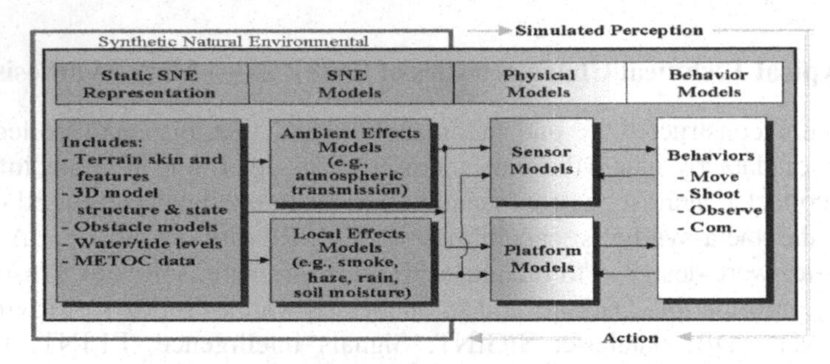

Fig. 2. Components of the Synthetic Environment

Databases containing SNE data were generated by integration and data fusion from a variety of terrain and environmental data. The database development activity was typically supported by hardware and software tools that extract, smooth, and fuse the required data into a coherent representation. Source data for these databases could be obtained from several origins in various data formats and resolutions. This data was input to a proprietary database generation system where the data was manipulated to fit a specific image generator system and meet specific application requirements. The result was a generic (geo-typical), or location specific (geo-specific) database. From these databases, run-time versions can be generated that are application specific representations of the same SNE that are specifically designed for efficiency and specific characteristics of the application as shown in Fig..3[4].

2.4 Implementation Framework of C4ISRE

Implementation framework of C4ISRE was shown in Fig. 4.

Through the system architecture(shown in Fig. 4), C4ISRE achieved the interconnection and integration for related resources including different levels of the

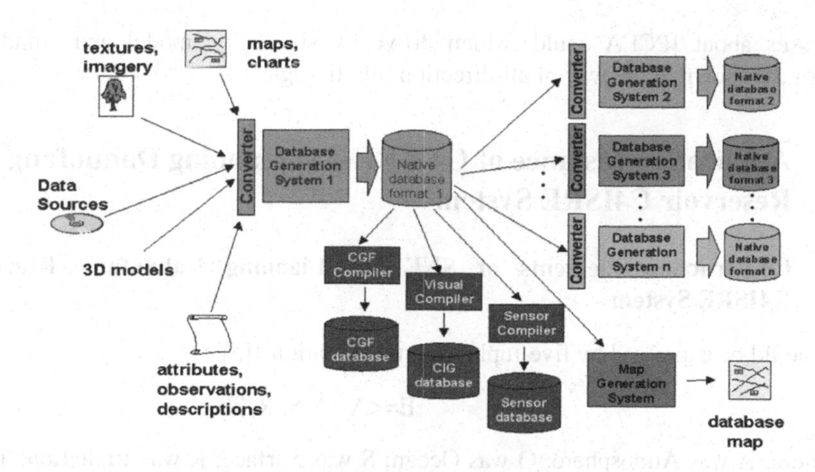

Fig. 3. Current Database Generation

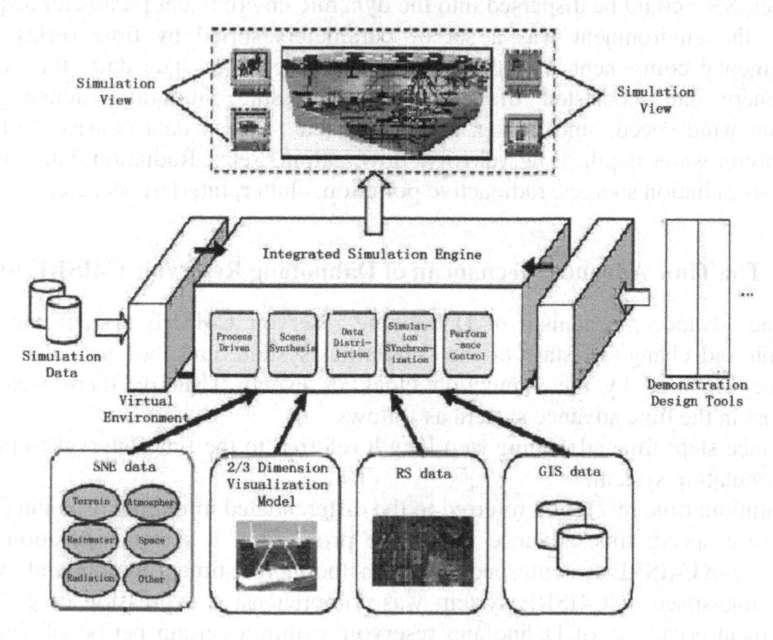

Fig. 4. Implementation Framework of C4ISRE

models, heterogeneous models, simulation Systems and distributed heterogeneous simulation resource(such as environmental model/data, environmental pollution situation scenario, visual models and resources).

Public support environment was constructed mainly by Service-Oriented Architecture (SOA)technology. Based on SOA,JPCEA integration simulation demonstration system could product multidimensional information environment pollution situation scene which including natural geography environment, electromagnetic environment, space environment and so on, the related concepts and processes about JPCEA could ,which drove by simulation model and simulation system ,be shown in the way of all-direction, multi-angle.

3 Application Instance of C4ISRE——Liaoning Dahuofang Reservoir C4ISRE System

3.1 Construction Elements of SNE for Liaoning Dahuofang Reservoir C4ISRE System

SNE could be expressed by five-tuple as in the equation 1[5].

$$E=<A, O, S, R, T> \tag{1}$$

where: A was Atmosphere; O was Ocean; S was Surface; R was Radiation; T was Time.

In fact, SNE could be dispersed into the dynamic environment parameter sequence, namely, the environment was a set of parameters sorted by time series. Every environmental component was also made up of a series of state data, for example, Atmosphere data consisted of temperature, pressure, humidity, density, wind direction, wind speed, smoke, fog, rain, snow, etc.; Ocean data consisted of water temperature, water depth, flow velocity, flow, salinity, etc.; Radiation data consisted of various radiation sources, radioactive pollution, clutter, interference, etc.

3.2 The Time Advance Mechanism of Dahuofang Reservoir C4ISRE System

The time advance mechanism of Dahuofang reservoir C4ISRE system caused the operation and change of state of the simulation system, and the propulsion mode could be triggered by the simulation clock or events. Usually, there were three indicators in the time advance system as follows:

Advance step: time advancing step length referred to the time interval to promote in the simulation system;

Resolution: time resolution referred to the differentiated minimum time interval;

Advance speed: time advance meant the proportion of system operation to the actual time in C4ISRE system, specifically ,including real-time, time-out and owed.

The time-space of C4ISRE system was comprehensive expression of ecological environment condition of Dahuofang reservoir within a certain period of time. The time-axes in the C4ISRE system Created a foundation of the state evolution of ecological environment, describing the State change process of each subsystem.

3.3 System Simulation Results of C4ISRE

Integrated simulation result of JPCEA was shown in Fig. 5 on basis of Dahuofang reservoir C4ISRE.

As shown in Fig. 5~6, the C4ISRE system generated one framework for integral comprehensive demonstration system based on multiple perspectives, multiresolution, multimodal performance technology. Comprehensive performance engines could

Fig. 5. Integrated simulation result of JPCEA

The simulation results of UAV sensor and atmospheric pollution monitoring were shown in Fig. 6 on basis of Dahuofang reservoir C4ISRE

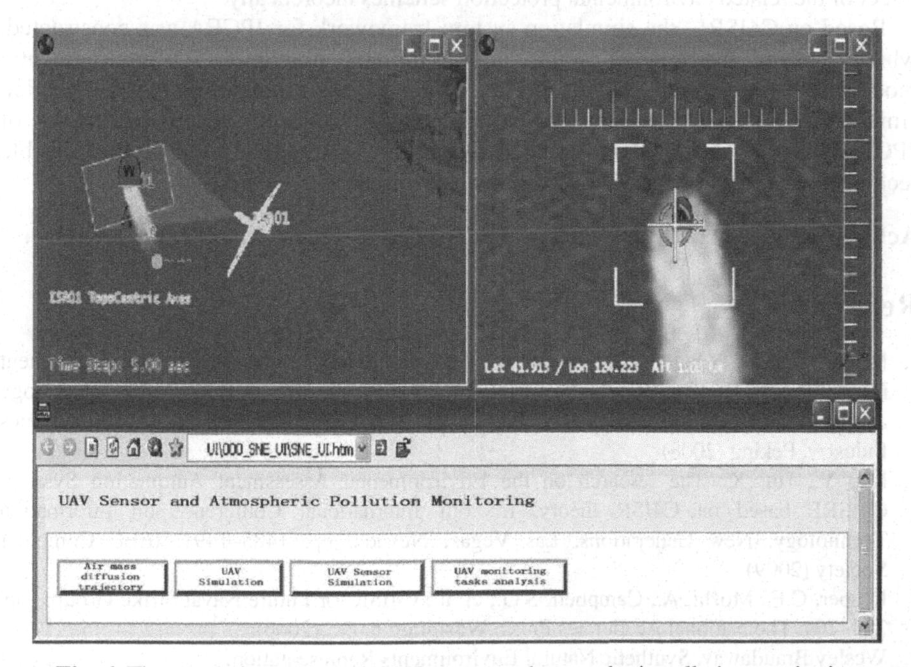

Fig. 6. The simulation results of UAV sensor and atmospheric pollution monitoring

generate the visualization from different views according to the simulation data of public view established in the study area, which consisted of different environment models/data, 2/3 Dimension visualization model, RS data, GIS data and so on. A full range of display for environmental pollution situation and JPCEA had been made from different levels and different views. Not only the overall situation of macro process, but also specific entity microcosmic action could be realized within the system.

Different models and simulation system were used to simulate the local composition of JPCEA and environmental protection tactical operations and process under the local environmental pollution situation. It was comprehensively performed for local environmental, entity , action from different simulation system in the unified SNE space, which based on synthesizing local simulation from senior regional model ,realizing the integrated demonstration of overall concept of JPCEA.

4 Conclusion

Since C4ISRE is also an open system, therefore it could be extended to other related environmental science fields, for example, environmental inspection, calamity prediction, pre-research about environmental protection devices, personnel training of EA and so on. Based on SOA, the environmental simulation system distributed in other region would be probably integrated by the large international public simulation platform. Consequently it will be able to simulate the status of the environmental pollution all over China and the effect of the related environmental protection schemes theoretically.

Based on C4ISRE, the simulation system framework for JPCEA was constructed, which realized the interconnection and combined simulation for environmental models and simulation system. In the dynamic environment pollution situation simulation, by exploratory analysis experiment, the concept, scheme and tactics of JPCEA were demonstrated and evaluated globally. This method supplied one feasible technical reference for further JPCEA business research.

Acknowledgements. This research was supported by ceeusro fund (cxy2012SH18).

References

1. Ma, Y., Hu, X., Yuan, Y.: The research on the Informatization Environmental Assessment remarked with Hypothesize and Visualize. In: 2008 Proceedings of Information Technology and Environmental System Sciences (Part 2), pp. 854–861. Publishing House of Electronics Industry, Peking (2008)
2. Ma, Y., Hu, X.: The research on the Environmental Assessment Automation System-C4ISRE based on C4ISR theory. In: 6th International Conference on Information Technology: New Generations, Las Vegas, Nevada, pp. 1485–1491. IEEE Computer Society (2009)
3. Draper, C.F., Mozhi, A., Campbell, S.G., et al.: C4ISR for Future Naval Strike Groups, pp. 202–204. The National Academies Press, Washington, DC (2006)
4. Wesley Braudaway, Synthetic Natural Environments Representation, `http://www.tmpo.nima.mil/guides/Glossary/`
5. Pang, G.: The introduction of virtual battlefield, pp. 120–135. National Defense Industry Press, Changsha (2007)

Development of the Management System of the National Sustainable Development Experimentation Areas

Huaji Zhu[1,2], Huarui Wu[1,2,*], Cheng Chen[2], and Jingqiu Gu[2]

[1] Beijing Research Center of Intelligent Equipment for Agriculture, Beijing Academy of Agriculture and Forestry Sciences, Beijing, Beijing 100097, China
[2] National Research Center of Intelligent Equipment for Agriculture, Beijing 100097, China
zhuhuaji@126.com,
{wuhr,chenc,gujq}@nercita.org.cn

Abstract. The management of the national sustainable development experimentation areas is faced with many problems, such as the complex and variable flow, affluent information and high-grade decision analysis. An online management system of the national sustainable development experimentation areas is developed based on the workflow technology and the business process management theory. With the traditional method, the system should be re-developed if the management flow of the national sustainable development experimentation areas changed. In order to overcome this repetitive work, the dynamic definition model of the business process about the management of the national sustainable development experimentation areas is put forward. Based on the new system, the traditional artificial management mode of the national sustainable development experimentation areas can be transformed to system adaptive management pattern. The on-line use of the system shows that the efficiency and normative of the experimentation area management has been greatly improved.

Keywords: workflow, sustainable development, business flow, dynamic definition model.

1 Introduction

After 20 years of construction, the national sustainable development experimentation area in China has made delightful achievements [1]. More than 150 national sustainable development experimentation areas have been established and distributed over 20 provinces, autonomous regions, municipalities directly under the central government [2, 3]. However, with the scale expanding, the management of the national sustainable development experimentation areas based on the traditional mode has been a series of problems. The management of the national sustainable development experimentation areas includes many segments, e.g. application, implementation, acceptance and track [3, 4]. Thereinto, the application segment can

* Crossponding author.

F. Bian and Y. Xie (Eds.): GRMSE 2014, CCIS 482, pp. 707–714, 2015.

be divided into application by the local government, preliminary examination by the provincial department in charge of science and technology, evaluation by experts and examine and verification by the management office of the national sustainable development experimentation areas. In every segment, many kinds of users participate the work, including, local staff of the national sustainable development experimentation area, staff in the provincial department in charge of science and technology, experts and administrator of the management office of the national sustainable development experimentation areas. Additionally, the flow of the national sustainable development experimentation areas and the format of the document have changes many times.

For a long time, the organization and operation management of the sustainable development of experimentation area always adopt the traditional manual operation mode which has many problems, e.g. long approval period, low work efficiency, not public in the approval process. Obviously, the traditional mode can't satisfy the need of examination and approval management of the national sustainable development experimentation areas under the new situation [5].

In order to better meet the needs examination and approval service of national sustainable development experimentation areas, the work proposes a national sustainable development experimentation area online management system based on workflow technology. The new system insulates the business process logic from the information support system which can overcome the redevelopment of the system resulted from the change of the workflow. The deployment and implementation of the system can promote the optimization of the management of the national sustainable development experimentation area and promote the service ability of the national sustainable development experimentation area.

2 The Management Model of the Experimentation Area

2.1 The Concept of Workflow

The workflow concept was put forward in routine work with fixed procedures which originated in the production and office automation. Through the analysis of the structured business process, the work activity can be divided into task, role, rule and process. The work activity can be executed and completed by the cooperation. By this way, the people and information, application tools can be dynamically organized and configured and the performance of the system can be optimized [6, 7]. Research and application of workflow technology has aroused general concern of many researchers, developers and users [8,9,10]. Fig.1 is a common disposal framework of the workflow system.

2.2 The Dynamical Management Model of the National Sustainable Development Experimentation Area

The management system of the national sustainable development experimentation area not only provides the functions of information report and acceptance but also provide aid decision-making function in order to improve work and management efficiency.

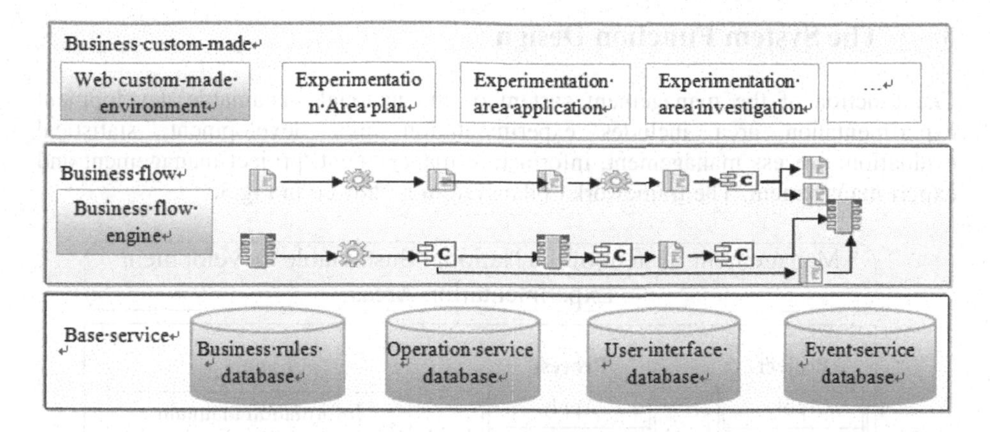

Fig. 1. A common disposal framework of the workflow system

The processing flow in the network is various. The characteristic is that the business is completed by several departments or various staff. With the workflow, every work is composed of flow relationships set in a task set [11, 12]. Based on this idea, the dynamical management model of the national sustainable development experimentation area is proposed, as showed in Fig.2. Based on the framework in Fig.1, experimentation area organization management process can be adjusted easy-going without the redevelopment of the system. If new user roles, disposal processes or businesses are added, new configurations are defined in the database and new business nodes can be assembled.

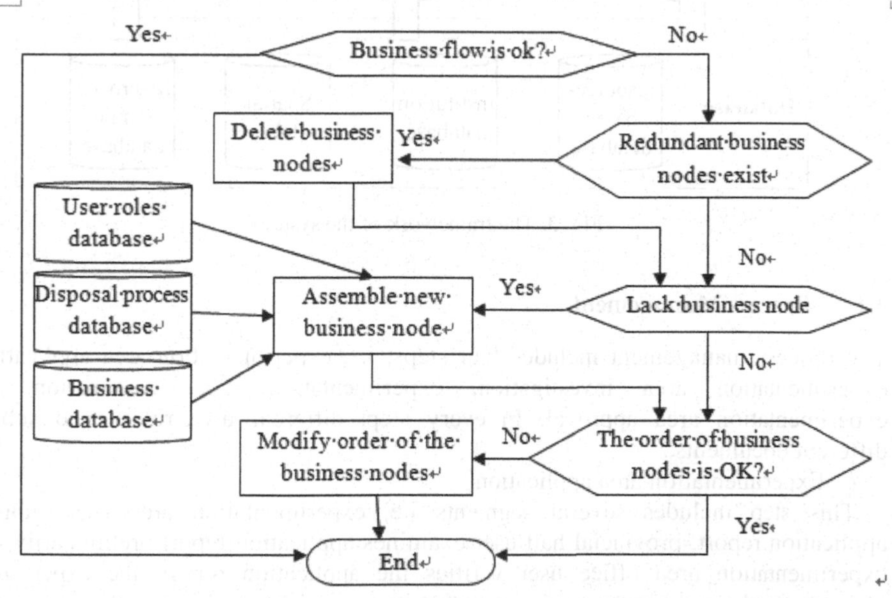

Fig. 2. The dynamical management model of the National Sustainable Development Experimentation Area

3 The System Function Design

The function of the management system of the national sustainable development experimentation area includes experimentation area development, statistical evaluation, process management, information management, project management and expert management. The framework of the system is showed in Fig.3.

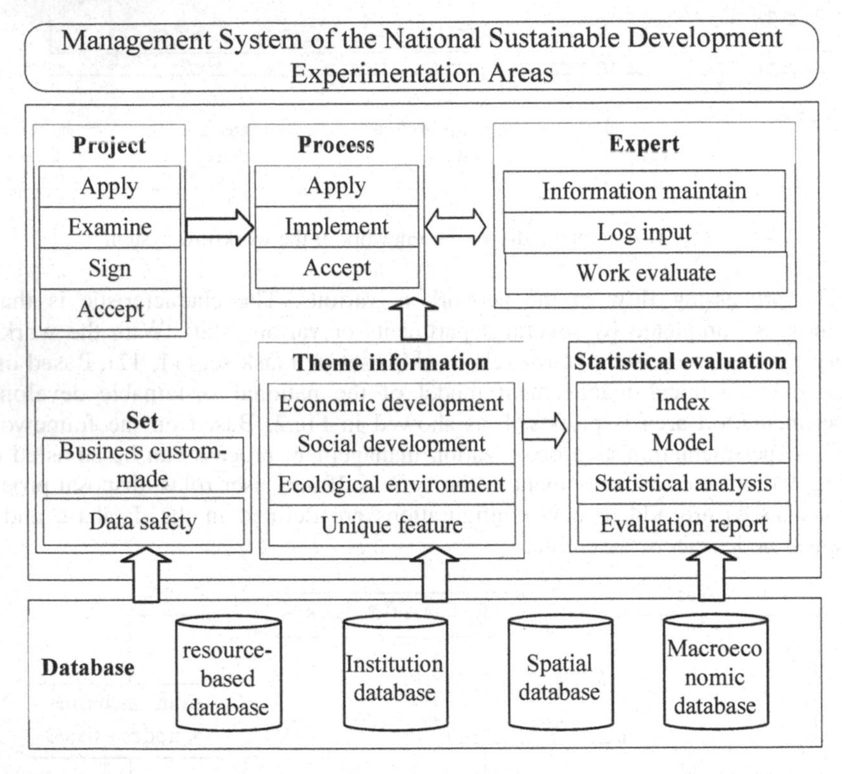

Fig. 3. The framework of the system

3.1 Process Management

The process management includes four steps, i.e. experimentation area application, experimentation area investigation, experimentation area examination and experimentation area approval. In every step, different user roles need submit different documents.

(1)Experimentation area application

This step includes several segments i.e. experimentation area user submits application report, provincial hall user examines application report preliminarily, the experimentation area office user verifies the application report, the expert user investigates the experimentation area, the expert user submits the investigation report, the expert user submits the application report and the experimentation area office user verifies the experimentation area. The detailed workflow is showed in Fig.4.

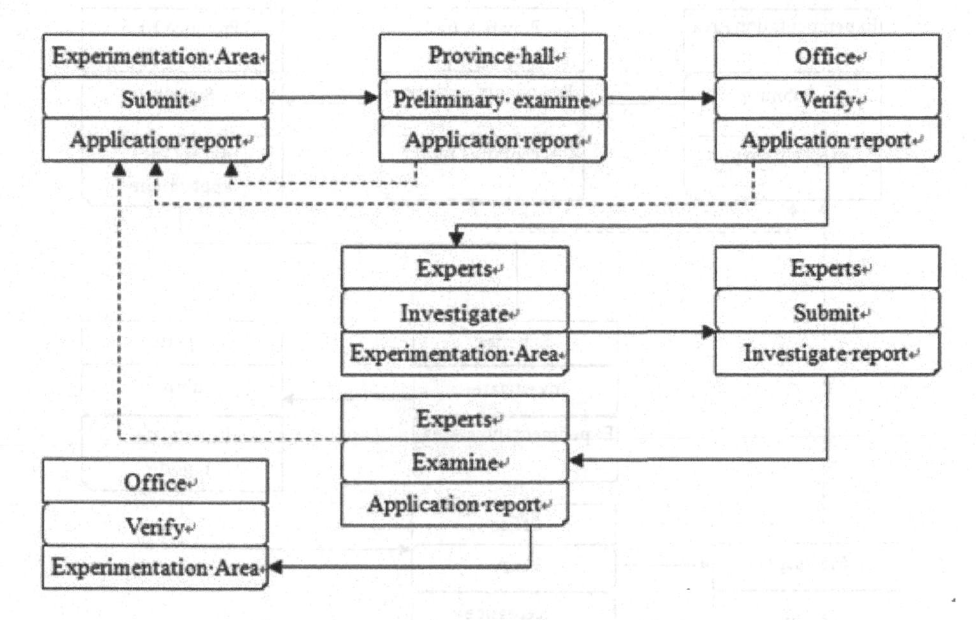

Fig. 4. Application flow of Experimentation Area

(2)Experimentation area Implementation

This step includes two segments i.e. experimentation area annual inspection and experimentation and area intermediate inspection. The experimentation area annual inspection includes the following steps, experimentation area user submits annual report, and provincial hall user examines annual report. If the annual report is qualified, the provincial hall user submits annual report to the office. Otherwise, the provincial hall user returns the annual report to the experimentation area user. The provincial hall user also submits thc annual report of the construction of the experimentation area in the province.

The intermediate inspection includes the following steps, the experimentation area user submits intermediate report, and provincial hall user examines annual report. If the intermediate report is qualified, the provincial hall user will hold review meeting and submits the review result to the office.

(3) Experimentation area approval management

The experimentation area approval management includes two segments, i.e., the experimentation area user submits the approval report; the provincial hall user examines approval report preliminarily. If the intermediate report is qualified, the provincial hall user submits the approval letter of application to the office. The office verifies the approval document. If the approval document is qualified, the office selects the expert group. The expert group investigates the experimentation area and submits the investigation report. If the investigation report is qualified, the office will hold review meeting. The detailed workflow is showed in Fig.5.

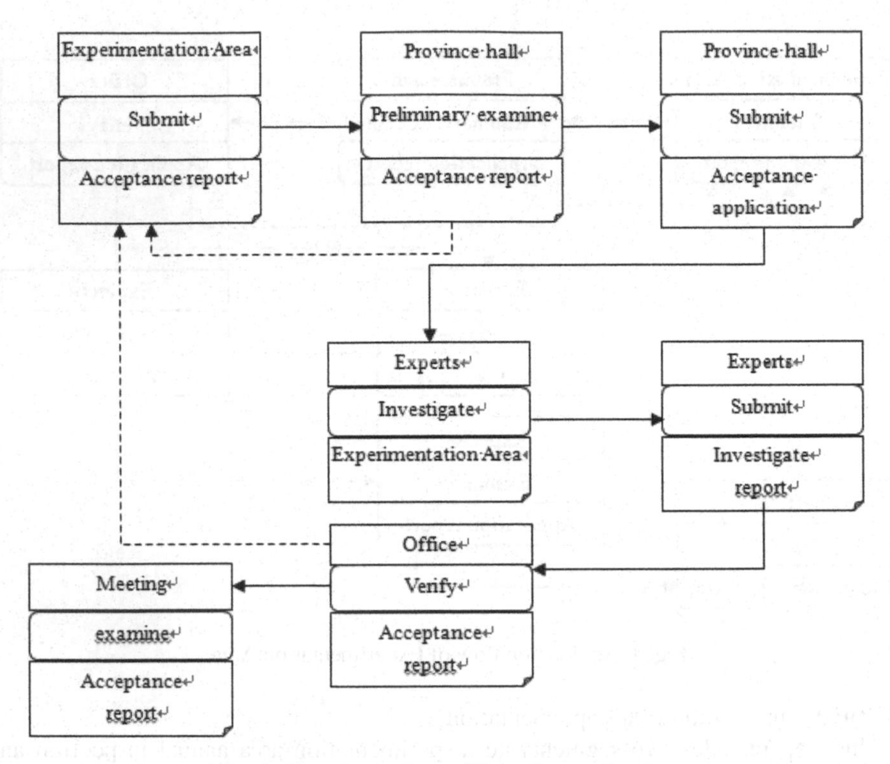

Fig. 5. Acceptance flow of Experimentation Area

3.2 Information Management and Evaluation Analysis

Project Management. The project management function is used by the experimentation area user and office user. This function provides two kinds of services, i.e. the basic information query of the project and the management of rate of process. In the application step, the basic information of the project can be input and managed. In the implement step, the rate of progress can be input and managed. The office user can view all projects and make many kinds of cartogram.

Expert Information Management. The expert information management function is used by the expert user and office user. This function provides two kinds of services, i.e. the basic information management of the expert and the management of work tasks.

Evaluation Index Management. This evaluation index management is used by the experimentation area user and office user. For every experimentation area , there are 21 confirmed indexes and 3 characteristic indexes. These indexes are input and modified in every step. The function is showed in Fig.6.

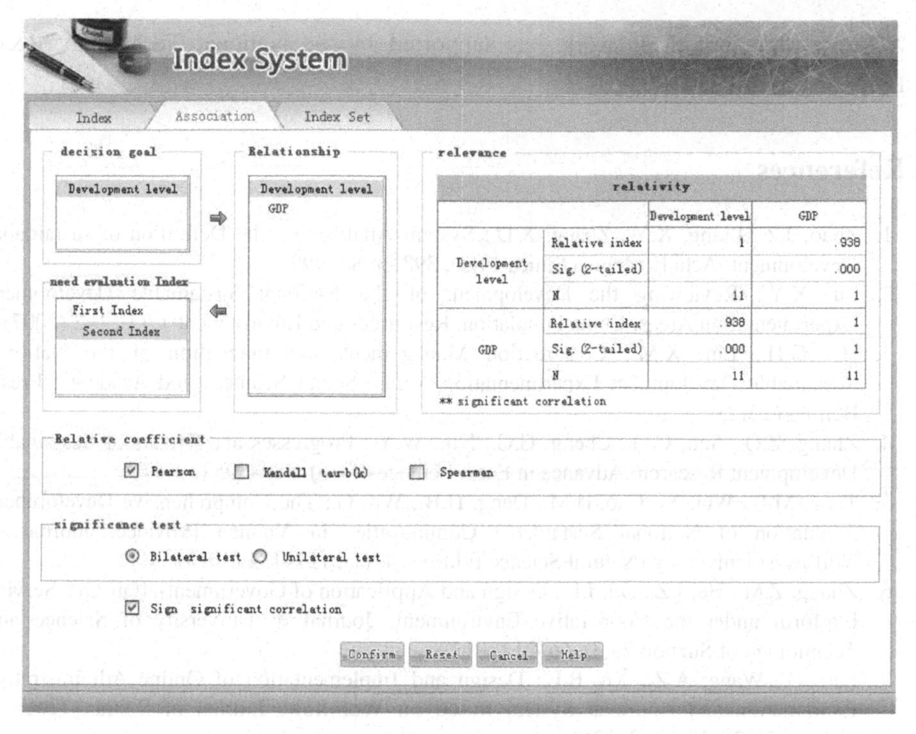

Fig. 6. Index Management

Evaluation Analysis. The evaluation analysis function is used by office user. This function can make many kinds of cartogram with the index data, generate the evaluation report dynamically. It provides many kinds of services, i.e. index management, evaluation model management; evaluation data management and evaluation report generation. The index management service can provide evaluation target management, evaluation index maintenance and definition of the attribute of evaluation index, correlation analysis between the evaluation target and evaluation index. The evaluation model management service can provide the dynamical construction of the model and analysis of the model. The evaluation data management service can provide the fusion of different data and statistical analysis. The evaluation report generation service can provide the templet custom-made, set style and generate.

4 Conclusions

After 20 years of construction, the national sustainable development experimentation area in China has made delightful achievements. However, with the scale expanding, the management of the national sustainable development experimentation areas based on the traditional mode has been a series of problems. In order to overcome these problems, a new management mode based on the workflow technology is proposed in this work.

Acknowledgement. This work was supported by the National Technology R&D Program (2013BAJ04B04).

References

1. Zhao, J.Z., Liang, X.Y., Zhang, X.D.: System Analysis on the Definition of sustainable Development. Acta Ecologica Sinica 19(3), 393–398 (1999)
2. Lu, X.Y.: Reviewing the Development of the National Sustainable Development Experimentation Areas. China Population, Resources and Environment 17(3), 1–2 (2007)
3. He, G.H., Liu, X.M.: Construction Management and Innovation of the National Sustainable Development Experimentation Areas. Social Sciences and Academic Press, Beijing (2012)
4. Zhang, Z.Q., Sun, C.Q., Cheng, G.G., Niu, W.Y.: Progresses and Trends of Sustainable Development Research. Advance in Earth Sciences 14(6), 589–595 (1999)
5. Tang, M.F., Wei, X., Cao, H.M., Deng, H.B., Wu, G.: The Comprehensive Development Evaluation of National Sustainable Communities in Yunnan Province. Journal of Northwest University (Natural Science Edition) 43(1), 127–132 (2013)
6. Zhang, Z.M., He, J.Z., Du, J.L.: Design and Application of Government-affair GIS Service Platform under the Cooperative Environment. Journal of University of Science and Technology of Suzhou 29(3), 66–71 (2012)
7. Pan, G., Wang, A.Z., Xu, B.L.: Design and Implementation of Online Administrative Examination and Approval System Based on Workflow. Journal of Daxian Teachers College 22(2), 71–73 (2012)
8. Zhang, Y.H., Zhang, L.L., Shi, G., Zhao, K., Chen, J.L., Feng, Z.W.: Applying Workflow Technology In Geological And Mineral Equipments Management System And Its Implementation. Computer Applications and Software 30(6), 13–15 (2013)
9. Yang, Y., Zhang, J.: Application Research of Dynamic Workflow in E-Government. Software Guide 9(2), 82–84 (2010)
10. Fu, Z.W., Yue, X.B.: Research On Petri Net-Based Extended Workflow Model. Computer Applications and Software 30(9), 173–175, 233 (2013)
11. Sun, C.A., Wang, R., Wang, X.W.: Design And Implementation of Modelling Tool For Flex-Based Dynamic Workflow. Computer Applications and Software 30(9), 134–136 (2013)
12. Wei, H.X., Wang, S.H., Qi, Z.H., Liu, H.: On Applying Workflow Technology in Project Contract Approval System. Computer Applications and Software 30(8), 155–157, 177 (2013)

The Application of Base State with Amendments Model in Land Survey Data Management

Yanbo Hu[1,2], Huarui Wu[2,3], and Huaji Zhu[2,3,*]

[1] Department of Surveying and Land Science, China University of Mining and Technology (Beijing), Beijing, 100083, China

[2] Beijing Research Center for Information Technology in Agriculture, Beijing Academy of Agriculture and Forestry Sciences, Beijing, 100097, China

[3] China National Engineering Research Center for Information Technology in Agriculture, Beijing, 100097, China

hyb20112012@163.com, wuhr@nercita.org.cn, zhuhuaji@126.com

Abstract. With the launch of the first and second national land surveys, it has obtained the massive historical data of land survey. However, the key condition is how to manage the mass land survey data effectively. In accordance with these problems that the current base state amendments model exists high data redundancy and low history backtracking, based on the analysis on the strengths and weaknesses of existing base state amending model, combining the characteristics of changing trivially in land survey data, this paper puts forward a new base state with amendments model which is suitable for land survey data management. The model changes the general base state as a snapshot storage mode, only stores the initial base state of the file, reducing the amount of data storage; It dynamically establishes base state and improves the retrieval efficiency. The application results show that the improved model can save, manage and backtrack land survey data with low redundancy and high efficiency.

Keywords: Temporal GIS, Base State with Amendments Model, Land Survey Data Management, History Backtracking.

1 Introduction

Land resource is the premise of human survival development, is the most important in the GIS application fields, especially in our country. With the development of economy, the use of regional land changes more frequently. From 1996, for finding out the land resources and its utilization situation, grasping the true and accurate land survey data, China began to conduct a comprehensive land survey work to meet the management requirements of the land use survey. The land survey ordinance was promulgated in 2008, which regulates: According to national economic and social development needs, it should have a national land survey every 10 years; according to the

* Corresponding author.

F. Bian and Y. Xie (Eds.): GRMSE 2014, CCIS 482, pp. 715–723, 2015.

needs of land administration, land change survey is conducted annually. Since the first national land survey start to the second national land survey end, we got three different periods of national land survey data. With the increase in the number of land survey, it will get more in different periods of the land survey data and land change data and produce a large amount of historical information that has great value for data mining and decision support government. In order to save and manage these changes data effectively, simulate and realize the historical state backtracking, track changes, predict the future and other functions, it is necessary to establish an effective management of time, space and attributes information database of space and time, its core job is to choose the appropriate spatio-temporal data model.

Since TGIS proposed so far, it has many years of history for the research of spatial-temporal data model. Scholars at home and abroad have done the corresponding theoretical research about spatio-temporal data mode. In accordance with the actual demand for the different attempts and innovation, they proposed a variety of spatial and temporal data model. Most of the existing dozens of spatio-temporal data models is put forward, which is all about how to express the space, time and attributes of the three elements and their mutual relations. The elements of historical land survey data will be layered management, In connection with the existing temporal data model, there is not a universal model which can be layered with the traditional spatial data management compatibly. The important advantage of base state with amendments model is that can be combined with traditional data structure. Based on this point, in accordance with the analysis on the advantages and disadvantages of the existing base state with amendments model, combining with the characteristics of land survey and land change survey data and for the purpose of convenient retrieval, this paper puts forward a kind of suitable for management of land investigation and historical data of spatio-temporal data model.

2 Base State with Amendments Model

Base state with amendments model was first used by Langran G in his doctoral thesis, the basic idea is to determine the initial state of the geographic phenomenon first, then according to certain time interval record the change area, the status of each change can be obtained in the end, through the superposition of the contents of each change [1]. Base state with amendments model is the model that improves the sequence snapshot model, avoiding the duplicate records of the nature of the unchanged part. Fig.1 shows the initial base state correction model schematics.

Base state with amendments model complete information for each object is stored only once, each change, just a small change in the amount of records, a better solution to the problem of data redundancy, but at the same time it has brought difficulties to time and space object index, express the relationship between time and space and space-time analysis [2]. Temporal data redundancy and complexity of the operation is a contradiction about the spatio-temporal data model, many scholars have been committed to addressing the balance between the two. As a breakthrough point, any domestic and foreign scholars have made a series of improvements on base state with amendments model, they proposed a variety of base state with amendments models. Fig.2 shows the comparison of various base state with amendments model.

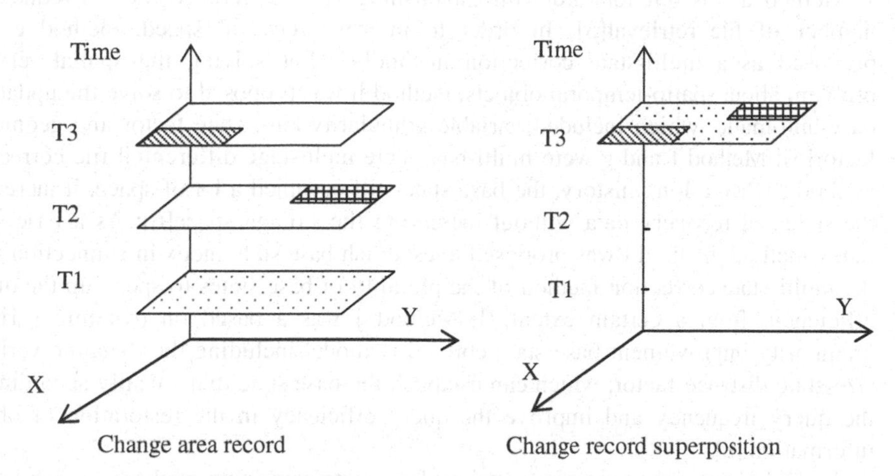

Fig. 1. The initial base state with amendments model

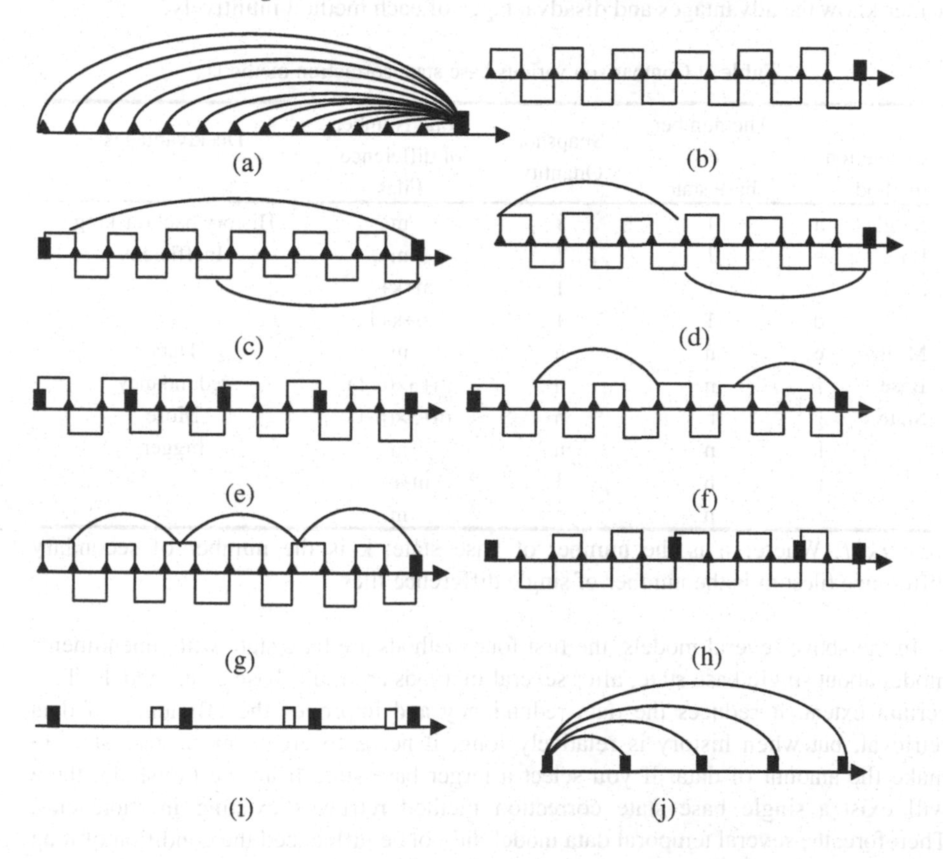

Fig. 2. Comparison of various base state with amendments model

Method a was put forward with establishing more difference files to reduce the number of file retrieval[3]; In order to improve retrieval speed, method e was proposed as a multi-state correction method[4]; For solving the spatial relation problem about spatio-temporal objects, method h was proposed to solve the update of base automatic ,which included variable granularity base state factor and geometric factor[5]; Method f and g were multi-base state multistage differential file correction method , when a long history, the base state still occupied a lot of space. it increased the speed of recovery data without increasing the storage space[6]; As a base state index method ,method i was proposed to establish base state index in connection with the multi-state correction method of the plurality of base states to speed up the query efficiency, from a certain extent[7]; Method j was a based on dynamic variable granularity improvement base state correction model including the dynamic variable size-state distance factor, which can establish the base state dynamically according to the query frequency and improve the query efficiency in the restoration of object information [8].

Table 1 shows the comparison various base state correction methods, which let the reader know the advantages and disadvantages of each method intuitively.

Table 1. Comparison various base state correction methods

Correction method		The number of base state	Snapshot Quantity	Total Number of difference Files	Disadvantages
Single	a	1	1	m	History backtracking
Base	b	1	1	m	Inefficient
State	c	1	1	m+k+1	
	d	1	1	m+k+1	
Multi-	e	n	n	m	Data
Base	f	n	n	m+3×(n-1)	redundancy
State	g	n	n	m+3×(n-1)	More
	h	n	n	m	bigger
	i	n	1	m+n-1	
	j	n	n	m	

Remark 1. Where: n is the number of base state; k is the number of secondary difference files; m is the number of single difference files.

In the above several models, the first four methods are base state with amendments model about single base state, after several methods are multi-base state models. To a certain extent, it reduces the data redundancy and improves the efficiency of data retrieval, but when history is relatively long, it needs to create more base state to make the amount of data. If you select a larger base state from the threshold, there will exist a single base state correction method retrieves existing inefficiencies. Therefore, the several temporal data model did not be influenced the condition of long history. Be aimed at massive land survey data for the history of the problem, it is necessary to find a suitable base state correction model to solve the above problems.

3 Land Survey Data Characteristics

The Land survey is the investigation that we can get the land of natural, social, economic, legal and many other conditions and its dynamic change. According to the time limit can be divided into the national land survey and change survey, its purpose is a comprehensive identification and utilization of land resources, it should grasp the real underlying data for scientific planning, rational use, effective protection of land resources, implement the strictest arable land protection system, strengthen and improve macroeconomic regulation and control to provide the basis and promote comprehensive, balanced and sustainable economic and social development .

Geographic Characteristics. Land with geospatial features, therefore, land survey data is a description of a geographic feature GIS data which can generally be divided by type of spatial data and non-spatial data.

The spatial data is also known as graphic data; non-spatial data, also known as the attribute data.

The Temporal Characteristics. The occurrence of anything or evolution has its characteristics. In the land information management, time and space is inextricably linked together. Land information has significant temporal features, which reflects in the land information changes frequently, complexity and land information management requirements of the land and the impact of the change process or not.

Massive Feature. The land survey data includes a variety of elements, is a large amount of data for each of the national land survey, including spatial data and attribute data.

Frequent Changes in Resistance. On the basis of national land survey, in order to provide accurate and reliable information, such as land based data, maps and other data for the government at all levels and relevant departments. It should find out the annual land use status and change situation comprehensively to meet the needs of the daily management of land and resources, data update relatively frequent.

Trivial Change, Diversity. Land survey database update includes and use types and traffic flow, changes in land ownership and boundaries of land properties and changes in content and graphics. A section of the change is related to changes in the final outcome data, including polygons, linear features, and sporadic feature of basic farmland parcels and other layers of data. For each recording layer data, it must change behavior which includes adding, loss, property changes.

Topological Relations Complexity. Topology elements include points (nodes), lines (chain, arcs, edges), plane (polygon).Topological relationships include association, adjacency relationships, including relationships, geometry, hierarchy. Land survey data contains a variety of graphic data, which involves a lot of point, line, surface elements, and particularly complex topological relations.

4 Improved Base State with Amendments Model

4.1 Thinking of Improved Model

Land survey data is relatively strong current. The data query frequency in close distance now is on the high side, so this paper let a recent land survey time as the initial base state and a snapshot storage. The improved model is improved only in the storage method, when retrieved it still choose to establish the dynamic state search. Combined with the state storage of method i, in addition to the initial base state, the other base states are both in the form of file storage, which sets the time of each land survey database construction to other base state and stores the moment base state relative to the base before the difference Number of Files. Through a simple superposition calculation, we can get the base state of complete information. Land change survey will be conducted annually, the land change survey data in the other historical moments are stored. The advantage of this model: with the development of history, land survey frequency increases, it will produce a lot of the base state relatively. At the time of the retrieve the model won't be affected by the number of the base state. we can build any base according to the specific needs; in addition to the initial base state is stored in the sequence mode, other base state are in difference file storage, which significantly reduces the data redundancy. Fig.3 shows the effect diagram of improved base state with amendments model .

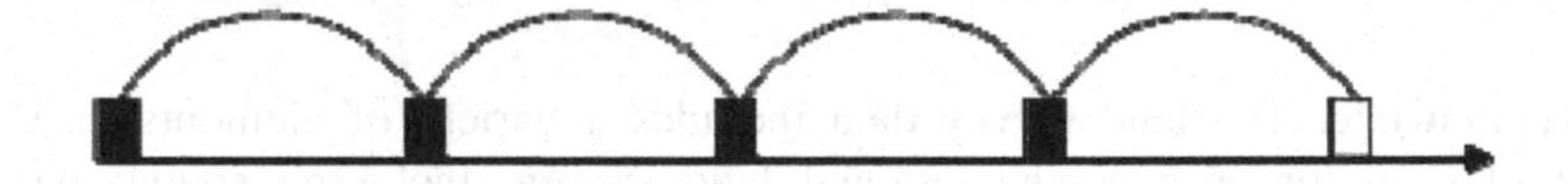

Fig. 3. Improved base state with amendments model

4.2 Retrieval of the Improved Model

According to the above method, if you want to set up n base states, the required storage space is only a snapshot of the N-1 base state and a differential file space occupied, amount of data is much smaller than the method of 5, 6, 7, n a snapshot of the data quantity.

Database construction is completed, you want to retrieve a particular moment in history of the state, when the user enters search time (*tj*),the first judgment is to find the difference files for the least amount of base state (*base state n*), then the user can get the desired result. superimposed on the base based on differential file. Fig.4 shows the retrieval process of temporal data visually in graphical form.

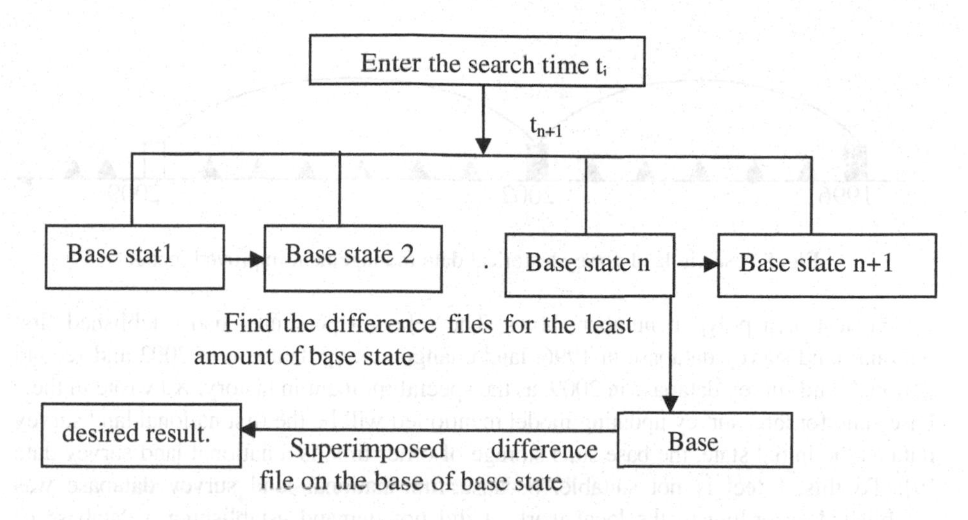

Fig. 4. Temporal data retrieval implementation process

5 Application

Since 1996, Wuhan has carried out land use change survey annually. The relevant personnel carried out 1996,2002 urban domain 1:10 000 land use database first in the country and established the land use management information system; In 2003,on the basis of 1:10 000 land use database The relevant personnel made full use of satellite remote sensing images, fully carried out the update of land survey work and established updated land survey database in 2002; In 2007,in accordance with the unified deployment of national and provincial the second national land survey department, The relevant personnel made use of DMC digital aerial photographic images to establish the second land survey database after preliminary data preparation, base map production, reproduction field, data processing, database construction in the industry, annotation nearly three years.

In conjunction with the test area with the data situation, the selection of land types polygon data volume of data to verify the improved base state with amendments model in the land survey data has a strong advantage. Table 2 is the list of experimental data and Fig.5 is the improved model used in land survey historical data.

Table 2. A region of Wuhan in recent years types polygon number

Year	The number of polygons	Year	The number of polygons
1996	1907	2004	1497
1997	1854	2005	1511
1998	1864	2006	1517
1999	1777	2007	1703
2000	1812	2008	1693
2001	1828	2009	2305
2002	1629	2010	2305
2003	1640	2011	2305

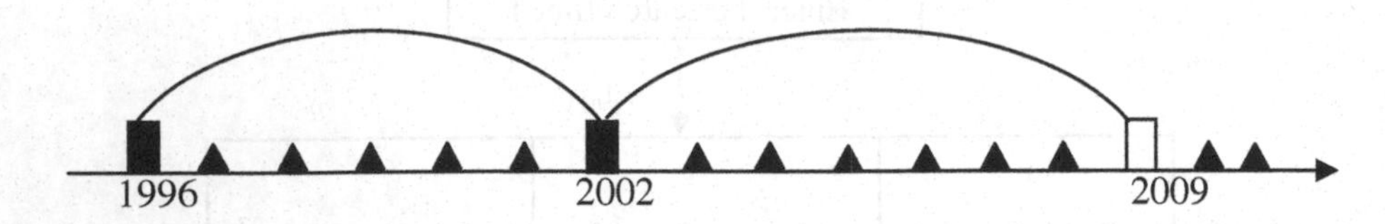

Fig. 5. Used in land survey historical data management improved model

The test area polygon number shows, The relevant personnel had established first national land survey database in 1996, land change survey database in 2002 and second national land survey database in 2009, as the special moment in history. Xu wrote in the base state for land survey updating model mentioned will be the first national land survey data as the initial state, the base state storage of other time is a national land survey data [9]. To this, I feel is not suitable, because first national land survey database was established according to the local work, it did not demand establishing a database of national unity, database logical relation was not tight; when the second national land survey database was being embellished, the country had established industry standards (TD/T1016-2007),with unity, rigor; various historical periods and frequency of queries is different, the farther away from the current time period of history, frequent queries the smaller degree; the nearer the current time period of history, frequent queries the greater degree. Table 3 is the compare results of improved method in this paper and other methods.

Table 3. Experimental data under various base state correction method comparison

Correction method		The number of base state	Snapshot quantity	Total number of difference Files	Disadvantage
Single	a	1	1	15	History
Base	b	1	1	15	backtracking
State	c	1	1	18	Inefficient
	d	1	1	18	
Multi-	e	3	3	15	Data redundancy
Base	f	3	3	21	More bigger
State	g	3	3	21	
	h	3	3	15	
	i	3	1	17	
	j	3	3	15	
The improved method		3	1	17	No data redundancy, high search efficiency

6 Conclusions

Based on the advantages and disadvantages of various base state with amendments models, this paper proposes an improved base state with amendments model which is suitable for the management of land survey historical data, combining the characteristics of the land survey data. The model takes the land survey database moment for the base state, overcomes the base state change caused by a large amount of data to modify the shortcomings, introduces the dynamic establishment of base state index, and storages the non-initial base state and the former state differential file. It greatly reduces the data redundancy, the improved model is more suitable for the investigation of land history data management. In view of short-term land survey data, the model has strong advantages. As time goes by, more and more of the land survey data, for the same area data, it can achieve low redundancy and efficient management like theory or not, which needs further research and testing.

References

1. Langran, G.: A framework for temporal geographic information. Cartographica 25, 1–14 (1988)
2. Qi, Q.C.: Study on spatio-temporal data model in TGIS. M.S. thesis, Dept. Cartography and GIS, Central South Univ., Changsha, China (2008)
3. Zhang, Z.X., Huang, M.Z.: The discussion on TGIS data structure. Bulletin of Surveying and Mapping 1, 19–22 (1996)
4. Cao, Z.Y., Liu, Y.: An object oriented spatio temporal data model. Acta Geodaetica et Cartographica Sinica 31, 87–92 (2002)
5. Zhang, B.G.: Application of spatio-temporal data model in city surveying and mapping in the database. Ph.D. dissertation, Dept. Cartography and Geographic Information Engi., Geosciences Univ., Beijing, China (2005)
6. Yu, Z.W., Zhang, L.T., Lin, Y.H.: Further expansion of the base state spatio-temporal data model. Journal of Acta Scientiarum Naturalium Universitatis Sunyatseni 42, 100–103 (2004)
7. Qi, Q.C., Zeng, Y.N., Wu, G.P.: An improved method of multi state spatio-temporal data model. Science of Surveying and Mapping 33, 178–180 (2008)
8. Yao, M.H., Zhou, W.T.: Improved base dynamic variable size correction model of, based on Silicon Valley 17, 43–45 (2009)
9. Xu, C.R., Yu, Z.W., Shen, J.H.: A base for land survey spatio-temporal data model. Journal of Anhui Agricultural Sciences 37, 10097–10099 (2009)

Simulation of Urban Land Development Based on Multi-agent System and GIS Technology

XinYan Li and JinFu Chen

School of Architecture and Urban Planning, Huazhong University of Science and Technology, 1037 Luoyu Road, Wuhan, China
lixinyanemail@263.net, jfchen_01@126.com

Abstract. MAS(Multi-agent System) is a new modeling method developed in recent years. This article aims to use MAS technology to simulate the development process of urban land. Five factors, which may have impact on the choices of agents, are defined. Spatial analysis and multi-factor evaluation technique is employed to determine the weights and utility value for each factors. On the basis of an MAS platform named RePast, combining with other open source GIS component, the model proposed is implemented through second development using Java language. Finally, the status quo of Wuhan City in 2003 as initial value, future land-use changes are simulated. The result of primary experiment seems promising, and the outcome of model simulation is interpretable and reasonable. So we can conclude that MAS can be used to simulate urban land development, and the simulation results will provide decision basis for urban managers.

Keywords: urban land development, multi-agent system, GIS.

1 Introduction

Multi-agent system (MAS) is a modeling and simulation technology developed in the recent years on the basis of artificial intelligence and the theory of complex adaptive system (CAS). The CAS is a theory proposed by the computer scientist John H. Holland in 1994. Inspired by biological science, its fundamental concept is the agent. An agent is proactive, adaptive, and in mutual interaction with its environment, which differentiates it from the elements or subsystems in traditional models. The theory immediately attracted attention and responses from a wide range of researchers, and was soon employed by economists in the observation and study of economic systems. The most prominent example of such an application is the Aspen system developed by Sandia National Laboratories, an agent-based macroeconomic model of the United States [1]. The system modeled the relations and interactions of tens of thousands of agents encompassing the companies, the banks and the state, and simulated the development path of the entire American economy. The fact that realistic phenomena, such as the cycles and fluctuations of economic growth, have been observed in its simulation, have garnered worldwide interest. To researchers of geography and urban

F. Bian and Y. Xie (Eds.): GRMSE 2014, CCIS 482, pp. 724–731, 2015.

planning, the breakthrough in both theory and modeling methodology brought on by MAS has ensured its place as an important approach to urban simulation after the cellular automata method [2-8].

This paper is aimed at the dynamic simulation of urban land development based on MAS and GIS technology.

2 Theoretical Model

2.1 Defining the Agents

The dynamic agents that influence China's urban land development have been categorized as the following:

①.the government; ②.owners of collective-owned land; ③. real estate developers; ④. industrial entrepreneurs; ⑤. commercial entrepreneurs; ⑥.urban residents; ⑦. urban planners.

2.2 Environment of the Model

The environment of the agents is defined by three elements: point, line and area. The point refers to the nodal points of the city network, which are generally large centers of commerce or major transportation hub in the city. Since the connections between the nodes form the main development axis of a city, the relative location of a given land to the nodes is an important factor in its development. The line refers to the roads and other traffic routes of the city. Obviously, the main traffic routes have a significant effect on the development of surrounding areas, and the development potential of a land depends heavily on its accessibility. The area refers to the attributes of the lands neighboring a land in particular, and to the conditions and attributes of the location of the land in general.

In the model, the nodal points, the traffic routes, and the land blocks are represented using the entity types of point, line and area respectively. An agent-based model (MAS) system was utilized for the construction and running of the agents. The GIS and the MAS are integrated through a loose coupling, where data are shared using files.

2.3 Socioeconomic Analysis of the Model

From the perspective of urban spatial structure and urban economic system analysis, the evolution of the model should be guided by two principles, namely the competition and symbiosis of agents [9][10]. As the spatial structure of the city changes due to the effects of both competition and symbiosis, the usage of urban land would also change.

The principle of competition represents the natural selection in the urban economic system, which results in the replacement and change the function of urban land. In the process of urban development, with the rising land value at city center and improving living standard of its residents, continuing industrial development in city center becomes unfeasible in terms of both economic and environmental benefits, meaning the manufacturing industries would gradually move out of the center. This is why many Chinese cities have employed a development policy of "exiting the secondary, entering the tertiary". That is to say, the industries of the secondary sector are moved out from within the city limits, replaced by the services of the tertiary sector, with the effects of reducing pollution, protecting the environment, increasing the efficiency of the businesses, and improving urban land use. The economic basis for the removal and replacement of industries in city center is the difference in profit per unit area between industrial and commercial land. Running a commercial business in city center would gain location benefits: first, it reduces the consumers' travel distance and traffic costs; second, it allows the largest coverage area for commerce, meaning the highest turnover; additionally, the overlap of commerce and other tertiary sector businesses is more attractive than a single type of business, which would also increase the revenue; lastly, commerce and other tertiary sector businesses make better use of the vertical space, which would both increase the effective area of business and decrease the land cost.

In addition to competition, symbiosis, or harmony or cooperation is another important type of relation between agents. As Gu Chaolin observed [9], aggregation and dispersion can serve as a basic framework of system dynamics for the distribution of population within in the urban spatial structure. Aggregation creates an economy of scale, as the proximity of various businesses in the city both reduces the costs of investment and strengthens the social communication between businesses, allowing the creation of many institutions of commerce, finance, research and education which provide a good environment for businesses, and as a result create a great boost to social productivity.

In addition to the self-governed "adaptive" changes, human intervention and regulation are indispensable for urban land development. In the model, such regulation and control are represented collectively as urban administrators. Each group of the urban system seeks to maximize its own profit, but as the total amount of resources is limited, the urban system can only stay in a virtuous circle by maintaining some sort of dynamic balance. The administrators act as levers that balance the interests of all groups to maximize the total economic, social and environmental benefits of the city. In the model, the function of the land occupied by urban administrators is fixed and unchanging.

3 Implementation of the Model

The model was built in Java language using the Repast modeling toolkit [11], in combination with open-source GIS components, including functional components from Geotools and Java Topology Suite. The model consisted of three main modules: Agent, Main and GIS. Two main functions were implemented, on one hand the input, output, and basic management of spatial data, on the other hand the construction of agents, and simulation of the agents' dynamic decision making processes. The Agent module is responsible for the simulation of the agents; the Main module is responsible for creating the model framework, and is also the external interface of the model; the GIS module manages the spatial information. Data is inputted in Esri's "shapefile" format, and outputted in MapInfo's MIF/MID file format.

3.1 Parameter Settings

(1) Designation of target set (estimated future function of the land):

$A=\{A_k\}$, k=1, 2, 3, 4, 5, 6, 7, 8

A_1 is residential, A_1=R;

A_2 is commercial and public services, A_2=C2;

A_3 is other public works, A_3=C;

A_4 is industrial, A_4=M;

A_5 is warehouse, A_5=W;

A_6 is green space, A_6=G;

A_7 is for other urban constructions, A_7=Z;

A_8 is land to be developed, A_8=E.

(2) The following parameters are defined:

I: Node, representing the relation between a land and urban nodal points, or the attraction from nodal points, with $0 \le Node \le 1$. The attraction is determined using IDW interpolation or buffers based on the spatial distribution of the tiered nodal points.

II: Traffic, representing the relation between a land and traffic routes, or the attraction from traffic routes, with $0 \le Traffic \le 1$. This is determined using buffers.

III: Grade, representing the grade of urban land development. There are 8 grades, represented by the numbers 1 to 8.

IV: Radius, representing the radius of a given land's neighbors.

V: XRate, representing the ratio of X-type land in the area centered around the given land, with the Radius as its radius, where X represents the function of land development. $0 \le XRate \le 1$.

The first 3 parameters are attributes of the land development, and inputted into the model as initial values. The 4th and 5th parameters are computed through running.

(3) The behavior rules of agents

I: The closer an agent is to a nodal point or a main traffic route, the more likely the land will be developed to commercial, with residential as the second highest possibility.

Nodal points and traffic routes exert a powerful draw upon the surrounding land development. They bring in the flows of population, materials and information to create the core of the city with the highest density and the most energy. These core zones would generally be composed of commerce, services and administration sectors, and provide various services for residents, businesses and institutions. The growth of manufacturing industries located in city center tends to be limited by their heavy pollution, obsolete manufacturing processes, or lack of space for further development. Therefore it is likely for them to move away from nodal points, with their land utilized for commerce, services or residence.

II: The land development tends to be the same function as surrounding lands.

The principles mentioned above mean that urban land development tend to form into homogenous zones. It can be seen through studying the spatial structure of cities over time that while the function and degrees of homogeneity differed for zones, and the degrees of homogeneity also differed for the same type of zones in different eras or stages of urban development, the zones have always tend to maintain their homogeneity and reject heterogeneity within themselves, creating continuous spatial strips that serve clearly different purposes from their neighbors.

III: The irreversibility of urban land development means different behavior rules for different types of agents, as shown in Table 1:

Table 1. Changes to agents

Agent Type	Possible States in the Next Moment
E	R, C2, C, M, W, Z, G
M	R, C2, C, M, Z, G
W	R, C2, C, W, Z, G
Z	R, C2, C, Z, G
R	R, C2, C, G
C	R, C2, C, G
C2	C2, G

A linearly weighted evaluation equation can be used:

$$p_i = \sum_{j-1}^{n} w_j I_{ij} (i = 1, 2, ..., m)$$

Where w_j is the weight for n parameters, and I_{ij} is the effect of the ith unit for the jth parameter.

The probabilities for an agent to switch types can be determined using the equation. Two methods can be used to determine its final state, either using the result with the greatest probability, or normalizing the probabilities for random determination. Through multiple tests, the former proved better than the latter method.

Fig. 1. Initial state of the model (t=0)

3.2 Simulation Experiment

The city of Wuhan was chosen as the experiment area. The land usage data of Wuhan in 2003 were used to simulate its future urban land development. Taking into account the general urban planning of Wuhan, the ratio of each type of land was controlled at:

Fig. 2. State of the model at moment t

residential 30%, commerce and services 5%, other public works 9%, manufacturing industry 18%, storage 3%, greenery 31%, other urban constructions 4%. The results from the experiment using these values include an overly large greenery area, which is not realistic. Hence the ratio of green space was adjusted, resulting in the following ratios: residence 32%, commercial and public services 5%, other public works 11%, industrial 20%, warehouse 3%, green space 25%, other urban constructions 4%.

The future land development of Wuhan was simulated on the basis of these parameters:

3.3 Result Analysis

A land balance sheet can be obtained from the simulation results in Fig. 2:

Table 2. Land balance sheet

Type Code	Type Name	Present Area (m2)	Ratio
R	Residential	109,183,272	31.80%
C2	Commercial and Public services	17,273,953	5.03%
C	Other public works	37,861,824	11.03%
M	Industrial	69,115,838	20.13%
W	Warehouse	10,397,910	3.03%
Z	Other urban constructions	13,683,632	3.99%
G	Green space	85,789,264	24.99%
Total		343,305,693	100.00%

Result analysis:

It can be seen from the map that the manufacturing industries have mostly moved out of inner city, now concentrated at fridge areas, and the replacement of functions for land at city center has taken place. Other public works also tend to be concentrated at city fridge, which may be explained as certain administrative or education institutions moving out of city center. However, it is difficult to tell whether the over-concentration of residence land in inner city, and continuous commerce zones at fridge shown in the results would occur in the future. In overall, the results of the model are explainable, and prove to be of use for reference.

4 Conclusion

In this paper, a city model was constructed based on MAS technology and GIS technology, the model is used to simulate the evolution of urban land development. The following factors that affecting choice of agents is defined: the relation with nodal point, the relation with main traffic route, urban land level, the relation with its neighbors, etc. Spatial analysis and comprehensive evaluation method are used to quantitatively determine the weight and utility values of factors, which can determine behavior rules of agents. The model was built in Java language using the Repast modeling toolkit, in combination with open-source GIS components, including functional components from Geotools and Java Topology Suite. The city of Wuhan was chosen as the experiment area, the land usage data of Wuhan in 2003 were used to simulate its future urban land development. By preliminary tests, the results of the model are explainable, and the model proved to be of use for reference

Acknowledgments. This paper is based in part on work supported by the National Natural Science Foundation of China (50808089), National Social Science Fund (09BZZ045), National Natural Science Foundation of China (51278211).

References

1. Chen, Y.: New Progress in Complexity Research – Agent Based Modeling Method and its enlightenment. Journal of Systematic Dialectics 11(1), 43–50 (2003) (in Chinese)
2. http://www.casa.ucl.ac.uk/agent.htm
3. Benenson, I., Omer, I.: Agent-Based Modeling of Residential Distribution, http://www.demogr.mpg.de/Papers/workshops/010221_paper01.pdf
4. Jiang, B.: Agent-based approach to modeling environmental and urban systems within GIS, http://www.hig.se/~bjg/JiangSDH.pdf
5. Dijkstra, J.: A Multi-Agent Model for Network Decision Analysis, http://www.ds.arch.tue.nl/General/Staff/jan/
6. Ettema, D.: A multi-agent model of urban processes: Modelling relocation processes and price setting in housing markets. Computers, Environment and Urban Systems 35, 1–11 (2011)

7. Li, S., Li, X., Liu, X., Wu, Z., Ai, B., Wang, F.: Simulation of spatial population dynamics based on labor economics and multi-agent systems_a case study on a rapidly developing manufacturing metropolis. International Journal of Geographical Information Science 27(12), 2410–2435 (2013)
8. Zou, Y., Torrens, P.M., Ghanem, R.G., Kevrekidis, I.G.: Accelerating agent-based computation of complex urban systems. International Journal of Geographical Information Science 26(10), 1917–1937 (2012)
9. Gu, C.: Centralization and Diffusion: New Theory of Urban Spatial Structure. Southeast University Press (January 2000)
10. Yao, S., Shuai, J.: Urban Land Use and Urban Development——A Case Study of South East Coastal City Developments. Press of Chinese University of Science and Technology (December 1995)
11. http://repast.sourceforge.net

Spatial Distribution of Heavy Metals in Roadside Soils Based on Voronoi Diagram: A Case Study of Wuhan City

Ruili Shen[1], Jianping Li[2], Mingzheng Yang[2], Mingzhong Zeng[1], and Min Zhou[2]

[1] Hubei Geological Survey Institute, Gutian 5th Road, 430034, Wuhan, China
[2] School of Resource and Environment Science, Wuhan University, 430072, Wuhan, China
{Shen Ruil,shenruili}_sab@163.com

Abstract. The central part of Wuhan City was chosen as soil sampling region to investigate the spatial distribution of heavy metals in roadside soils and the correlation between city road net and the spatial distribution of heavy metals. The total number of samples collected is 224 and the concentrations of As, Cr, Cu, Ni, Pb and Zn in soils were detected. The spatial distribution of heavy metals in roadside soils was characterized by applying Universal Kriging interpolation model. The correlation between road net and the spatial distribution of heavy metals was analyzed based on Voronoi diagram. The results show that there is high correlation between the road net and the distribution of heavy metals in the range of 400 m on both sides of the main road of the study area. There is no strong relationship between road net and heavy metals in the soil of the whole study area which indicates that the road net of Wuhan is not the main source of heavy metal contaminate in soils.

Keywords: soil contamination, heavy metal, spatial analysis, Voronoi diagram.

1 Introduction

Gas pollutants and particle pollutants produced in the process of car running on the road is one of the main resource of urban pollution[1]. As for the particle pollutants, they contain heavy metals which mainly result from gasoline combustion and abrasion of auto parts. Zhang[2] analyzed the characteristics of the heavy metal contamination in the soil surrounding the city road of Tokyo. Zhi[3] assess the soil surrounding four parts of roads of Erdos. But these methods just considerate the soil near the road without the effect of the city road net on the heavy metal contamination of the whole city. The effect on whole city based on the Voronoi diagram is analyzed.

Sampling points are laid using the method of uniformed grid[4]. The samples of soil are collected for the laboratory tests and the GPS points are collected at the same time. Based on the data, this paper will interpolate using the method of Krigin based on the soil sampling of the study area and analysis the distribution of heavy metal contamination pattern in soil surrounding urban net. This paper will establish Voronoi

F. Bian and Y. Xie (Eds.): GRMSE 2014, CCIS 482, pp. 732–739, 2015.

diagram based on the soil sampling of the study area and calculate the road density of each part, then analysis the correlation between the distribution of heavy metal contamination in the soils and the road net.

2 Materials and Methods

2.1 Sample Collection and Analysis

The research area is in the central part of Wuhan (as shown in Fig.1). In 2013, the 224 soil samples were collected in this area using the method of uniform grid.

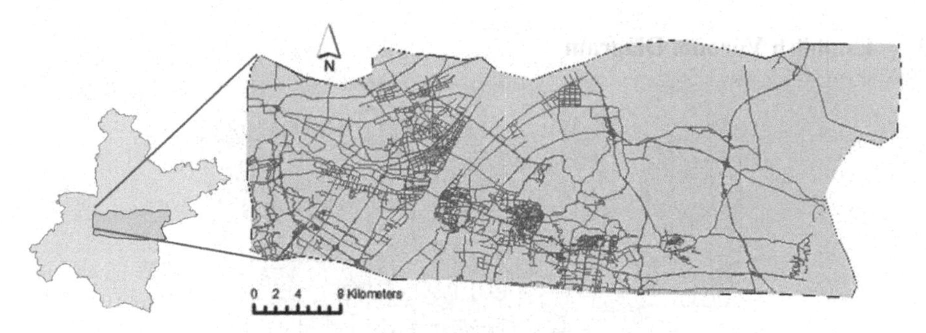

Fig. 1. Research Area and the Road Net of Wuhan City

The sampling density is 1 point/km2 and the sampling depth from earth surface is from 0cm to 20cm. The content of As, Cr, Cu, Ni, Pb and Zn in the samples were measured, these 6 kinds of heavy metal are supposed to cause main heavy metal contamination in Wuhan[5].

2.2 Spatial Analysis Method

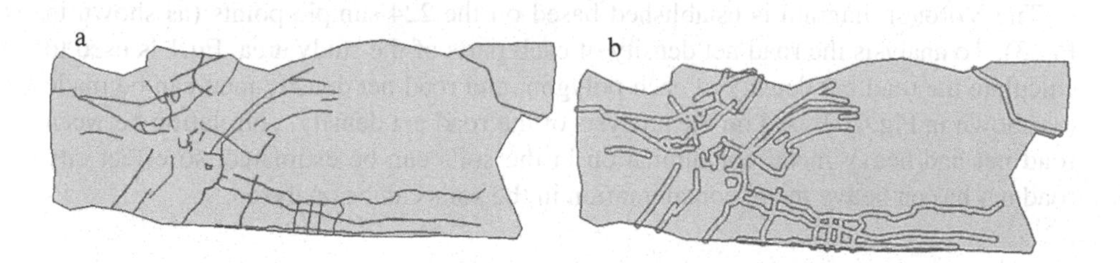

Fig. 2. The City Main Road (a) and the Buffer of 350m-400m between Both Sides of the Main Roads (b)

The distribution models of different kinds of heavy metals are fitted using the ArcGIS Geostatistical Analyst. Based on the results of cross validation[6] (according to decision criteria that standard average prediction error close to zero and standard root mean square prediction error is close to 1), the optimal spatial distribution model (Exponential model, Gaussion model and Sphere model) for each metal is selected. According to the selected model, the distribution map of each metal is made in ArcGIS.

Based on the distribution maps, buffers range from 0m to 400m with 50m interval (0-50m, 50-100m, 100-150m, 150-200m, 200-250m, 250-300m, 300-350m, 350-400m) of the main roads are created (as shown in Fig.2), we calculate the average values of heavy metal concentration in each interval and distribution characteristics can be analyzed.

2.3 Establish Voronoi Diagram

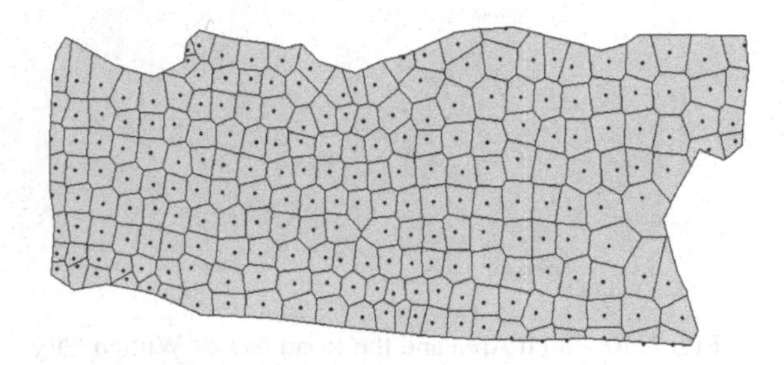

Fig. 3. Voronoi diagram of Research Area

Voronoi diagram, setting geographical spatial entities as growth targets, divides the continuous space into several Voronoi regions, according to the principle of being most nearby the objects, and each Voronoi region contains only one growth target[7]. On the basis of the definition and formation, the Voronoi model, integrated with vector model and grid model, has many features such as influence region, lateral adjacency and local dynamization[8].

The Voronoi diagram is established based on the 224 sample points (as shown in Fig.3). To analysis the road net density of each parts of the study area, Eq.1 is used to calculate the road net density of each polygon, and road net density map can be made (as shown in Fig.4). Based on the analysis of the road net density, correlation between road net and heavy metal contamination in the soils can be estimated, so effect city road net has on heavy metal contamination in the soils can be analyzed.

$$D = \frac{L}{S} \tag{1}$$

D means road net density. L means sum of road length of a polygon. S means area of a polygon.

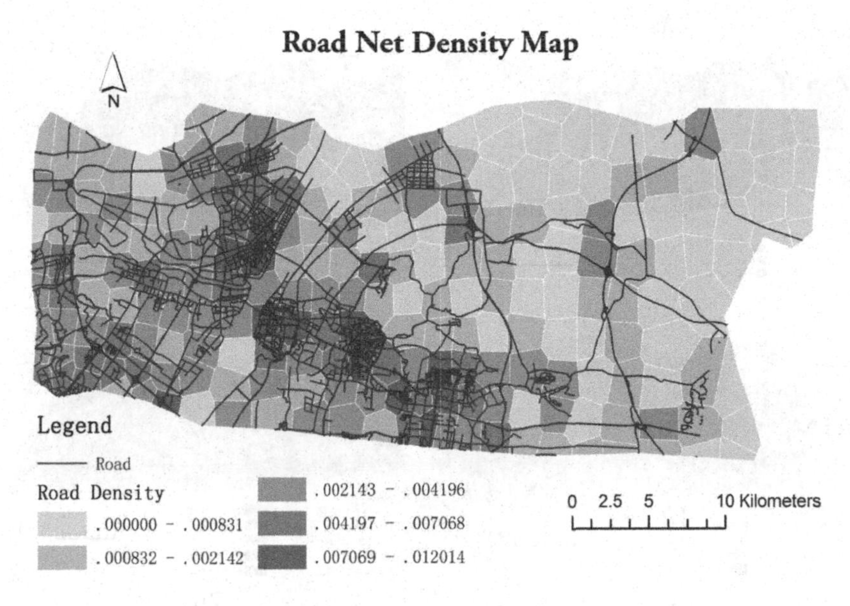

Fig. 4. Road Net Density Map of Study Area

3 Results and Analysis

3.1 Distribution Pattern Analysis

The best semi variogram model of each kind of heavy metal is analyzed in ArcGIS (as shown in Table.1). According to the selected models, the map of the spatial distribution of heavy metals in soils of study area was made in ArcGIS (as shown in Fig.5).

Table 1. Semi-variation models for heavy metals in soils of study area

Element	Model	Nugget	DrillBase	Nugget/%	Standard Average	Forecast Error
As	Exponential	0.046	0.076	61	-0.003462	0.9915
Cr	Exponential	158.41	195	81	0.00149	1.074
Cu	Gaussion	0.157	0.171	92	-0.04105	1.327
Ni	Gaussion	57.796	59.417	97	0.005362	1.001
Pb	Exponential	0.037	0.158	25	-0.03786	1.229
Zn	Exponential	0.031	0.197	15	0.95029	1.186

Based on the spatial distribution maps, the average metal content of each external on the both sides of the main roads is analyzed (as shown in Fig.6).

The source of heavy metals in the soils near road is mainly made up of two parts, which are nature source and human source[9]. The heavy metals of nature source,

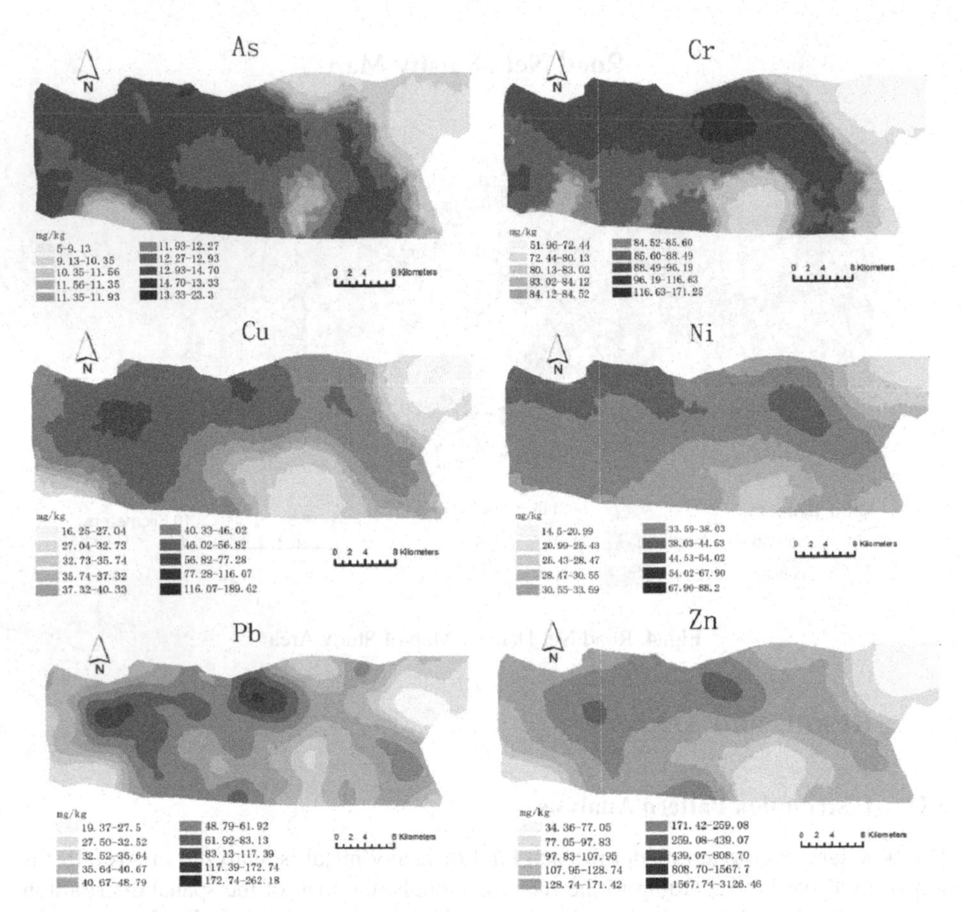

Fig. 5. The Spatial Distribution of Heavy Metals in Soils of Study Area

including Ni, Cr are mainly related to the soil parent material. The study area is narrow and the soil parent material is similar, so the nature source has little impact on the spatial variation of soil heavy metal content. Human source consists of two parts[10], the traffic source controlled by the road net and the industry source along the road. And there is little industry within 400m of road on both sides. So the change of the heavy metal content in the study area is mainly controlled by traffic source.

From the analysis on content, the concentrations of Pb and Zn are much higher than the CNS and the concentration of Zn is twice more than local background value, which indicates that Pb and Zn in soils near the main road is likely to come from the vehicles. For example, leaded fuel releases Pb while burning and tires release Zn while wearing[11].

It is obvious in the graphs that the distributions of metal in the soils are related to the distance from the main roads. The concentration of As increase from the road, after reaching the maximum value at 50-100m, drop to the average. This distribution may be caused by the border trees which are planted on the both sides of the roads and containing the migration of the main particle pollutants of As. As settles down at

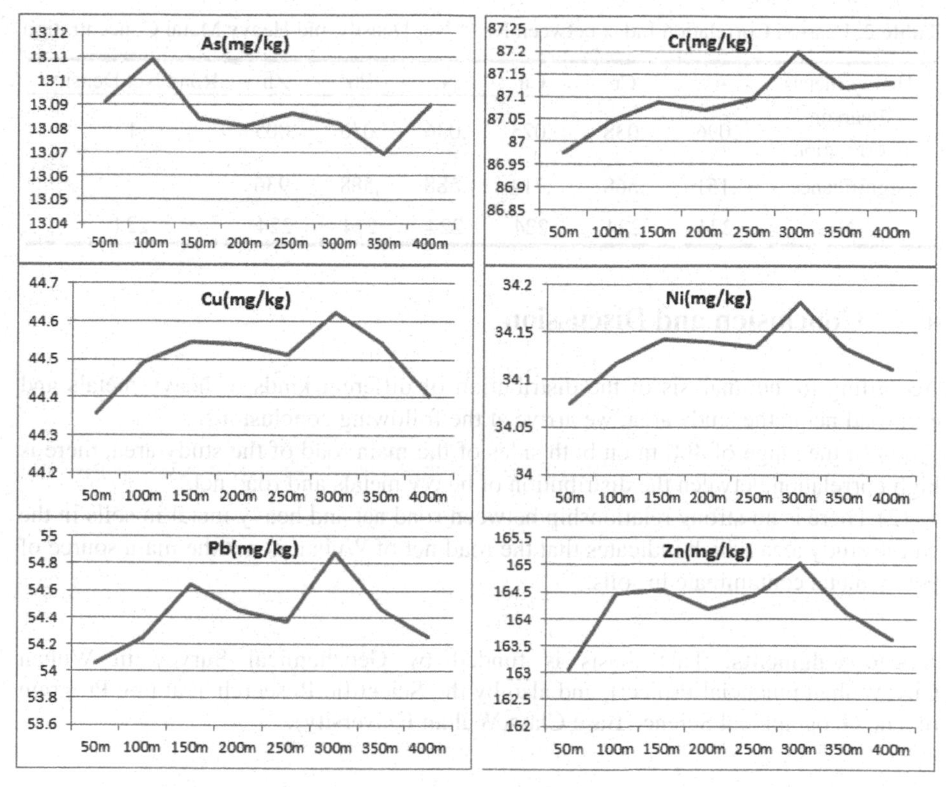

Fig. 6. The Distribution Pattern of Heavy Metal Contamination near Main City Roads

50-100m leading to the As content of this scope is the largest. The concentrations of Cu, Ni, Pb, Zn increase from the road and peak at 250-300m then gradually decline, which presents the skewness distribution, because the pollution particle mobility of Cu, Ni, Pb and Zn is greater than the pollution particle mobility of As, and the pollution particle can migrate across the border trees. The migration mobility of Cr particles is strongest so its migration distance is outside the range of research area.

3.2 Spatial Correlation Analysis

The effect different kinds of heavy metals have on the soils of the study area is analyzed based on the Pearson correlation index between road net density and heavy metals concentration (as shown in Table.2). The correlation index of each metal is lower than 0.1, which indicates that there are low correlations between the distributions of heavy metal and the road net. So roads of city is not the main source of heavy metal contaminate in soils.

Through on-the-spot investigation, we find that there are many heavy industries, such as large power equipment factory, mechanical and electrical factory and metal manufacturing company, located in the study area, so the distribution of the heavy metal in the soils of the whole study area may have certain relations with the companies[12] and the influence of city road net is relatively weak.

Table 2. Pearson Correlation Index between Road Net Density and Heavy Metal Concentration

Heavy metals	As	Cr	Cu	Ni	Pb	Zn	Road Net Density
Pearson Correlation	.096	.038	.025	.036	.036	.005	1
significance	.151	.568	.714	.588	.588	.936	
N	224	224	224	224	224	224	224

4 Conclusion and Discussion

According to the analysis of the distribution of different kinds of heavy metals and city road net in the study area, we arrive at the following conclusions:

(1) In the range of 400 m on both sides of the main road of the study area, there is high correlation between the distribution of heavy metals and road net.

(2) There is no strong relationship between road net and heavy metal in soils in the whole study area which indicates that the road net of Wuhan is not the main source of heavy metal contaminate in soils.

Acknowledgments. This thesis is funded by Geochemical Survey in Wuhan City(Wuhan Financial Project), and also by the Scientific Research Training Program of The Geographical Science Base Class(Wuhan University).

References

1. Ren, Y., Wang, X., Ouyang, Z., Wang, Q., Hou, P.: The pollution characteristics of Beijing urban road sediments. Acta Ecologica Sinica 08, 2365–2371 (2013)
2. Zhang, J.-Q., Shiraishi, S., Watanabe, I.: Heavy Metal Pollution of Dust, Topsoil and Roadside Tree Nearby Main City Roadways. Journal of Southwest Jiaotong University 01, 68–73 (2006)
3. Zhi, Y.-B., Wang, Z.-L., Ma, Z., Wang, Z.-S., Deng, Z.-F., Li, H.-L.: The speciation and bioavailability of heavy metals pollutions in soil along highway in Erdos. Acta Ecologica Sinica 05, 2030–2039 (2007)
4. Guo, J.-P., Zhang, Y.-X.: Spatial Sampling Methods and Their Applications in landscape pattern analysis for landscape ecological research. Scientia Geographica Sinica 05, 74–79 (2005)
5. Huang, M., Yang, H.-Z., Yu, C., Li, J.-J.: Accumulation Characteristics and Pollution Evaluation of Heavy Metals in Soils of Wuhan City. Journal of Soil and Water Conservation 24(4), 135–139 (2010)
6. Jaber, S.M., Ibrahim, K.M., Al-Muhtaseb, M.: Comparative evaluation of the most common kriging techniques for measuring mineral resources using Geographic Information Systems. GIScience & Remote Sensing 50(1), 93–111 (2013)
7. Deza, M.M., Deza, E.: Voronoi Diagram Distances. In: Encyclopedia of Distances, pp. 339–347. Springer, Heidelberg (2013)
8. Gold, C.M.: The Meaning of “Neighbour”. In: Frank, A.U., Formentini, U., Campari, I. (eds.) GIS 1992. LNCS, vol. 639, pp. 220–235. Springer, Heidelberg (1992)

9. Luo, Y.: Trends in soil environmental pollution and the prevention-controlling-remediation strategies in China. Environmental Pollution & Control 12, 27–31 (2009)
10. Ding, X.W., Shen, Z.Y., Liu, R.M., et al.: Effects of ecological factors and human activities on nonpoint source pollution in the upper reach of the Yangtze River and its management strategies. Hydrology and Earth System Sciences Discussions 11(1), 691–721 (2014)
11. Carrero, J.A., Goienage, N., Barrutia, O., Artetxe, U.: Diagnosing the Impact of Traffic on Roadside Soils Through Chemometric Analysis on the Concentrations of More Than 60 Metals Measured by ICP/MS. Springer Science Business Media B.V. (2010)
12. Sun, X.-B., Li, Y.-C.: The Spatial Distribution of Soil Heavy Metals and Variation Characteristics of Datong Abandoned Coal Mine Area in Huainan City. Scientia Geographica Sinica 10, 1238–1244 (2013)

Assessment of Heavy Metal Pollution in Surface Soils of Hankou Region in Wuhan, China

Min Zhou[1], Yi Lv[1], Ruili Shen[2], Zhehao Zhou[1], Jing Zhou[1], Shaoxiang Hu[2], and Xiaojuan Zhou[2]

[1] School of Resource and Environment Science, Wuhan University, 430072, Wuhan, China
[2] Hubei Geological Survey Institute, Gutian 5th Road, 430034, Wuhan, China
{Min Zhou,zhoumin}@whu.edu.cn

Abstract. The purpose of this study is to investigate the current status of heavy metal pollution in surface soils from Hankou Region, the old industrial district in the city of Wuhan, China. The contents of eight heavy metals As, Cd, Cr, Cu, Hg, Ni, Pb and Zn were tested for each 132 soil samples. The results show that the average value of Cd and Zn in soils of Hankou Region were up to 0.548 and 220.7 ug/g, respectively, which were much higher than their natural background values. Their spatial patterns were analyzed by the geographical information system (GIS) technology. The spatial patterns of these elements revealed that Cd, Zn and Cu might originate from the familiar traffic and industrial, and the elements of As, Cr and Ni could originate from the natural sources. The values of pollution index (PI) indicated that metal pollution level was Cd>Zn>Cu>Hg>Pb>Ni>Cr>As. Potential ecological risk indexes (RI) further indicated that Wuhan was suffering from serious metal contamination.

Keywords: urban surface soil, heavy metals pollution, GIS, Risk assessment.

1 Introduction

Urban soil, as an important part of urban environment, is the most important reservoir or sink of heavy metals and other pollutants in urban areas. However, the rapid urbanization and industrialization that has occurred in China in recent years has been accompanied by unprecedented environmental changes. Heavy metal pollution of urban soils is one of the fastest growing types of environmental pollution due to automobile emissions, industrial discharges and other activities in China. Therefore, excessive inputs of heavy metals into urban soils can impose a long-term burden on the biogeochemical cycle in the urban ecosystem by causing effects such as soil function deterioration, changes in soil properties and other environmental problems. Past studies have revealed that although urban soils are rarely used for food production, heavy metals in urban soils are non-degradable and can accumulate in the human body, where they cause damage to the nervous system and internal organs. Moreover, the long-term

F. Bian and Y. Xie (Eds.): GRMSE 2014, CCIS 482, pp. 740–750, 2015.

input of metals could result in decreased buffering capacity of soil and ground- water contamination. Thus, trace metal contamination of the urban environment can have long-term and far-reaching environmental and health implications [1].

For human and ecological risk assessment, a growing body of evidence has shown the necessity of determining the spatial distribution of pollutants. Such spatial data help scientists in defining the areas where risks are high and then help decision-makers in identifying locations where remediation efforts should be focused [2]. However, one of the characters of the spatial distribution of contaminants lies in its often high spatial heterogeneity, paralleling variations in soil parameters. To describe these spatial structures, mapping based on geographical information systems (GIS) is a useful approach. GIS is a system for managing, manipulating, analyzing and presenting geographically related information. It is a new approach to refining and confirming geochemical interpretations of statistical out-put. The application of GIS to the evaluation of soil environmental quality enables the geo-statistics information to be visualized and provides a reliable means of monitoring environmental conditions and identifying problem areas.

2 Materials and Methods

2.1 Sample Collection

A total of 132 soil samples were collected in the Hankou Region. The sampling points were randomly distributed in the study area based on a regular grid of 1×1km, and each grid had at least one sampling point (Fig. 1). The surface soil (0-15cm) was collected. All of the sample sites were recorded using a hand-held global positioning system (GPS). Related information, such as land use history, vegetation, and soil type were also recorded in detail.

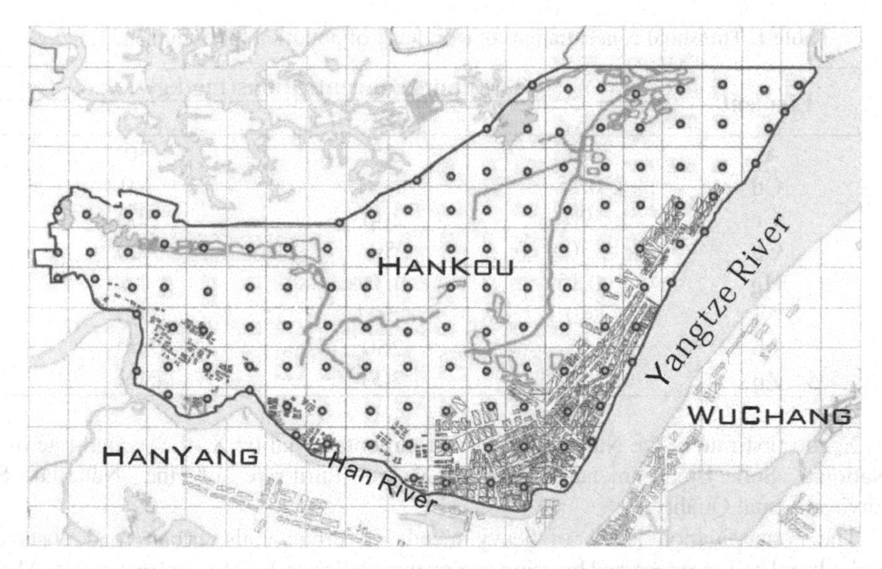

Fig. 1. Sampling locations of surface soil on Hankou Region

2.2 Laboratory Analysis

All analyses were completed in Hubei Geological Research Laboratory. The concentrations of As and Hg were analyzed by atomic fluorescence spectrometry (AFS), whereas Cd was measured by flameless Atomic Absorption Spectrometry (AAS), and the total concentration of Zn, Cu, Pb, Ni and Cr were analyzed by Inductively Coupled Plasma Atomic Emission Spectrometry. Limits of detection were 0.03ug/g for Cd and 1.0, 5.0, 1.0, 0.0005, 2.0, 2.0 and 2.0 ug/g for As, Cr, Cu, Hg, Ni, Pb and Zn, respectively.

2.3 Methods

Geostatistics provides a set of statistical tools for incorporating the spatial and temporal coordinates of observations in data processing, and its increasing use in environmental applications testifies to its utility and success. A semivariogram is a basic tool of geostatistics and is the mathematical expression of the square of regional variables $z(x_i)$ and $z(x+h_i)$, namely the variance of regional variables [3].

Many indices and calculation methods, such as geo-accumulation index (Igeo), pollution index (PI), Nemerow pollution index and Potential ecological risk index have been proposed for quantifying the degree of metal enrichment or pollution in soils, sediments.

The PI of each element was defined as the ratio of the heavy metal concentration in the study to the background concentration of the corresponding metal of the city. Table 1 shows the threshold concentrations of these metals in China. The PI of each metal was classified as: non-pollution ($PI \leq 1$), which indicates that the level of metals was below the threshold concentration but does not necessarily mean there was no pollution from anthropogenic sources or other enrichment over the background, low level pollution ($1<PI \leq 2$), moderate level of pollution ($2<PI \leq 3$) and high level of pollution ($PI>3$) [4][5].

Table 1. Threshold concentrations of each level of pollution for each metal (mg/kg)

Element	The limits concentrations(mg/kg)		
	X_a	X_c	X_p
As	15	25	30
Cd	0.15	0.30	1.0
Cr	90	200	300
Cu	30	50	400
Hg	0.15	0.30	1.5
Ni	40	50	200
Pb	35	250	500
Zn	85	200	500

X_a-the first-rate of the National Soil Environmental Quality; X_c-the second-rate of the National Soil Environmental Quality; X_p-the third-rate of the National Soil Environmental Quality.

The contamination levels of heavy metals in urban soils, urban road dusts and agricultural soils are assessed by using geo-accumulation index (I_{geo}) introduced by Muller (1969). The method has been widely employed in European trace metal studies since the late 1960s. The I_{geo} is used to assess heavy metal contamination in urban soils by comparing current and pre-industrial concentrations, although it is not always easy to reach pre-industrial sediment layers. It is also employed in pollution assessment of heavy metals in urban road dust. Here the focus is between the concentration obtained and the concentration of elements in the Earth's crust, because soil is a part of the layer of Earth's crust and its chemical composition if related to the one of the crust [6]. The According to Muller (1969), the I_{geo} for each metal is calculated and classified as: uncontaminated ($I_{geo} \leq 0$); uncontaminated to moderately contaminated ($0 < I_{geo} \leq 1$); moderately contaminated ($1 < I_{geo} \leq 2$); moderately to heavily contaminated ($2 < I_{geo} \leq 3$); heavily contaminated ($3 < I_{geo} \leq 4$); heavily to extremely contaminated ($4 < I_{geo} \leq 5$); extremely contaminated ($I_{geo} \geq 5$) [7].

Potential ecological risk index (RI) was introduced to assess the degree of heavy metal pollution in soil, which was originally introduced by Hankinson, according to the toxicity of heavy metals and response of the environment:

$$E_r^i = C_f^i \times T_r^i$$

$$RI = \sum_{i=1}^{n} E_r^i = \sum_{i=1}^{n} T_r^i \cdot C_f^i = \sum_{i=1}^{n} T_r^i \cdot \frac{C^i}{C_n^i}$$

where RI is calculated as the sum of all four risk factors for heavy metals in soils, E_r^i is the monomial potential ecological risk factor, T_r^i is the metal toxic factor. Based on the standardized heavy metal toxic factor developed by Hakanson, the order of the level of heavy metal toxicity is Hg> Cd > As> Pb = Cu= Ni> Cr> Zn. The toxic factor for Zn is 1, 2 for Cr, 5 for Pb, Ni and Cu, 10 for As, 30 for Cd and 40 for Hg [8]. C_f^i is the metal pollution factor, C^i is the concentration of metals in soil, and C_n^i is a reference value for metals. In this study, the adjustment of factor standards was made according to Zhu et al. The ecological risk (E_r^i) degree of a given substance was divided into 5 grades and the potential ecological risk (RI) degree of a given region was divided into 4 grades, as shown in Table 2.

Table 2. The relationship of Eri, RI and pollution degree

E_r^i	Pollution Degree	RI	Pollution Degree
$E_r^i < 40$	Low potential ecological risk	$RI < 150$	Low ecological risk
$40 \leq E_r^i < 80$	Moderate potential ecological risk	$150 \leq RI < 300$	Moderate ecological risk
$80 \leq E_r^i < 160$	Considerable ecological risk	$300 \leq RI < 600$	Considerable ecological risk
$160 \leq E_r^i < 320$	High potential ecological risk	$RI \geq 600$	Very high ecological risk
$E_r^i \geq 320$	Very high ecological risk		

3 Results and Discussions

The characteristic of the contamination of heavy metals were presented in Table 3. The concentration of each element varied greatly, the range of As and Zn were observed to be 6.7 to 37.8 ug/g and 70.0 to 3227.0 ug/g. The mean value of Cd and Hg were 7.8 and 9.6 times of their background levels and twice as the national first level standard in Table 1. Except for that, the other four elements exceeded their background values more or less. Except As, the concentration of other seven elements were much higher than the national first level standards. This indicated that the surface soil of Hankou Region, Wuhan was contaminated in different degrees. The contamination of Cd, Hg, Pb and Zn were the most extreme. CV reflected the average variable of each sampling point in the overall samples. The CV of Hg is 218%, the highest in this district. The following elements were Pb and Zn, which were 136% and 140.3%. The lowest elements was Cr, which was 16.33%. This demonstrated that the concentration of Hg, Pb and Zn were quite different sampling points while the distribution of Hg, Pb and Zn were relatively concentrated.

Table 3. Heavy metal concentrations and background values (ug/g) of urban surface soil in the Hankou Region, Wuhan

Heavy metal	As	Cd	Cr	Cu	Hg	Ni	Pb	Zn
Range	31.1	2.12	88.9	321.4	5.967	36.6	971.0	3227.0
Minimum	6.7	0.161	55.5	21.2	0.004	23.6	18.1	70.0
Maximum	37.8	2.277	144.4	342.6	5.971	60.2	989.1	3297.0
Mean	13.72	0.548	92.3	56.2	0.288	38.68	68.62	220.7
Median	12.70	0.447	91.7	47.8	0.14	37.85	48.35	152.4
SD	4.928	0.324	15.07	34.92	0.628	8.629	93.35	309.8
CV%	35.92	59.12	16.33	62.14	218.0	22.31	136.0	140.3
Skewness	2.556	2.428	0.491	5.340	7.072	0.405	7.775	7.944
Kurtosis	9.800	8.148	1.341	38.14	57.50	-0.269	73.41	75.54
Background value	12.73	0.07	88.24	28.57	0.03	37.98	23.44	68.33

The spatial distribution of metal concentrations is a useful aid to assess the possible sources of the enrichment and to identify hot-spot, where the heavy metal element has a high concentration. Distribution patterns of the studied elements in the study area are represented in Fig. 2.

Similar patterns of spatial distribution were observed for Cu, Hg and Cd. For these elements, they had relatively high spatial variability, suggested that they were from the same sources.

In urban areas, many isolated sites with an industrial history indicate extremely high values or high traffic density; these are termed hotspots. There were obvious hotspots for As, Cr, Cd, Cu, Hg, Ni, Pb and Zn in the middle of the study area. The highest concentration of As, Cr, Cu, Ni, Pb and Zn all appeared at this location, up to 37.8, 144.4, 342.6, 60.2, 989.1 and 3287 mg/kg, respectively. There is a railway station in the hotspot. The metal line 2 and the main road of Hankou cross here. Moreover, there were

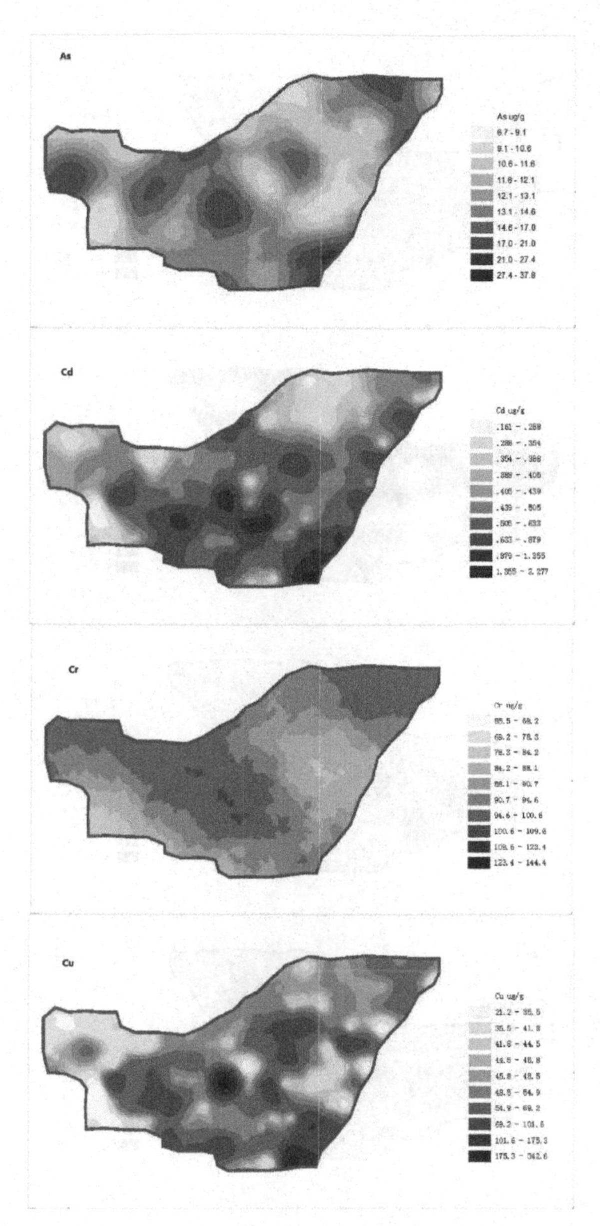

Fig. 2. Estimated Ordinary Kriging concentration maps for As, Cd, Cr, Cu, Hg, Ni, Pb and Zn (ug/g)

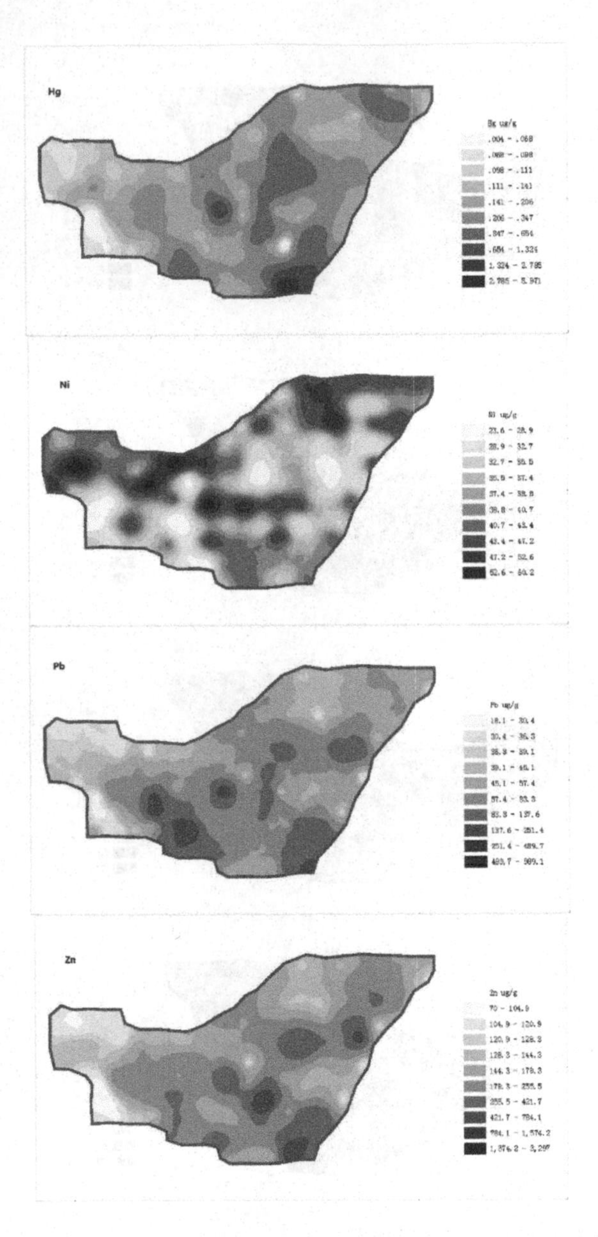

Fig. 2. (*continued*)

many manufacturing industries in the hotspot about ten years ago. Therefore, the concentrations of these heavy metals were a results of high traffic density and industries. There was another pollution hotspot at the south of the study area, where is the hotspot of Cd, Cu and Hg. The highest concentration of Cd and Hg appeared at this location, up to 2.277 and 60.2. The location is along the Yangtze River, and the high concentration of Cd and Hg appeared along the Yangtze River are the highest concentration of Cd, which has something to do with the migration and accumulation of Cd. Therefore, the high concentration of Cd and Hg were caused by the industrial history of the district along the Yangtze River.

4 Metal Pollution Assessment

The minimum, maximum and mean values of PI for each elements are shown in Table 4.1. Except As, the PI values of elements were more than 1. The mean values of PI for Cr, Cu, Hg, Ni, Pb and Zn was 1.012, 1.769, 1.192, 1.018, 1.121 and 1.810, respectively, and for Cd was 2.317, indicating that urban surface soil in the study area was low contaminated. The PI values varied greatly among metals. The PI values for As, Cd, Hg, Pb and Zn in urban surface soil ranged from 0.447 to 4.56, from 1.073 to 4.824, from 1.073 to 4.824, from 0.027 to 6.726, from 0.517 to 4.956 and from 0.823 to 12.322, respectively, indicating that urban surface soil in study area was uncontaminated to highly contamination. Indeed, the PI values of As, Cr, Pb were moderately contaminated, some of the samples had non-pollution level PI values for these elements, and the other samples had low-pollution level PI values. Most samples had low-pollution level or moderate-pollution level PI values for Cu, Ni, Cd, Hg and Zn, and some samples had high level of pollution for As, Cd, Hg and Zn.

In total, there were no high level contaminated elements in study are, the PI values of Cd was highest, which had moderately pollution level, followed Zn and Cu. The values of Cr, Hg, Ni and Pb had low-pollution level, the PI value of As had no-pollution.

Table 4. Statistical results of pollution index (PI) of metals in urban soils of the Hankou Region, Wuhan

	Max	Min	Mean	Polluted Degree
As	4.560	0.447	0.983	Non-polluted
Cd	4.824	1.073	2.317	Moderate polluted
Cr	1.495	0.617	1.012	Low polluted
Cu	2.836	0.707	1.769	Low polluted
Hg	6.726	0.027	1.192	Low polluted
Ni	2.068	0.590	1.018	Low polluted
Pb	4.956	0.517	1.121	Low polluted
Zn	12.323	0.823	1.810	Low polluted

The result of the Nemerow pollution index was shown in the Table 4. The $P_{Nemerow}$ of all samples varied from 0.76 to 28.77 with an average of 3.02. Assessment of the data shows that there were no sample with an $P_{Nemerow}<0.7$, only 4% of all samples had a $P_{Nemerow}$ value between 0.7 and 1, while approximately 45% of all samples had a $P_{Nemerow}$ value between 1 and 2, about 25% of all samples had a $P_{Nemerow}$ value between 2 and 3 and about 26% of all samples had an $P_{Nemerow}>3$.

The calculated results of I_{geo} of heavy metals in Hankou Region are presented in Fig. 3. The I_{geo} ranges from -1.51 to 0.98 with a mean value of -0.54 for As, 0.68 to 4.5 with a mean value of 2.27 for Cd, -1.25 to 0.12 with a mean value of -0.53 for Cr, -1.02 to 3.00 with a mean value of 0.26 for Cu, -3.61 to 6.9 with a mean value of 1.68 for Hg, -1.27 to

0.08 with a mean value of -0.59 for Ni, -0.96 to 4.81 with a mean value of 0.61 for Pb and -0.55 to 5.00 with a mean value of 0.76 for Zn. The mean value of I_{geo} decrease in the order of Cd> Hg> Zn> Pb> Cu> Cr> As> Ni. The mean I_{geo} and over 90% I_{geo} of As, Cr and Ni falling into calss 1 indicates uncontaminated, while less than 10% I_{geo} falling into class 2 reveals uncontaminated to moderately contaminated, suggested the contamination of As, Cr and Ni are controlled by le natural factor. The mean value of I_{geo} for Cd and Hg point to strong polluted. Percentage I_{geo} values of Cd and Hg falling class 2 is 2% and 27%, class 3 is 36% and 37%, class 4 is 48% and 16%, class 5 is 12% and 11%, class 6 is 2% and 2% and class 7 is 0% and 2%, respectively. The percentage I_{geo} of Cd and Hg show that over half of the surface soil of Hankou Region were polluted by Cd and Hg. The percentage of Hg, Pb and Zn is the most special, the I_{geo} values of Hg, Pb and Zn are wide fluctuating. Most of the I_{geo} values of Pb and Zn fall in class 2, indicated the I_{geo} value of Pb and Zn are controlled by two or more factors. The wide distribution of Pb and Zn were controlled by the natural factor, and the focus contamination were controlled by the anthropogenic factor.

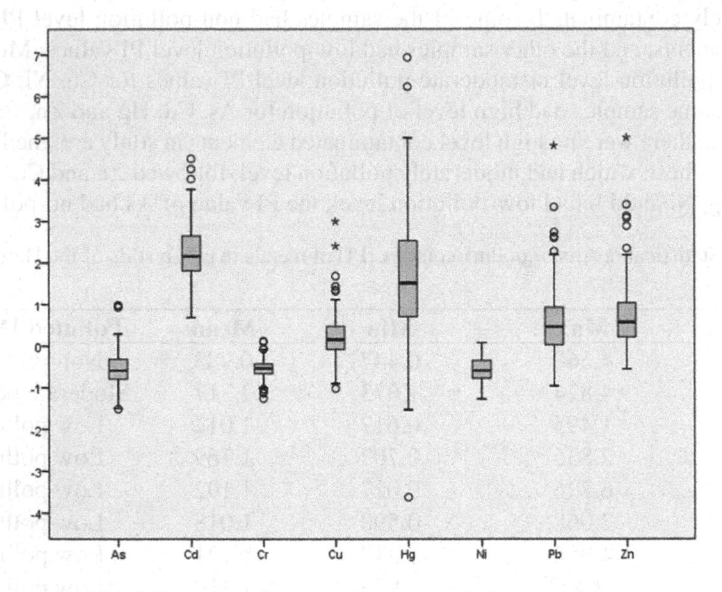

Fig. 3. Box-plot of Igeo for heavy metals in street dust

The calculated result of the heavy metal ecological risk index and the potential ecologic risk are shown

in Table 5 and Table 6. The E_r^i values of all samples for As, Cr, Ni and Zn are less than 40, which suggested that the four elements are low ecological risk. The E_r^i values of most samples for Cu and Pb are less than 40, and less than 2% samples are in moderate ecological risk. Copper and mercury pollution was the most serious. Only 8% samples for Cd are in low ecological risk degree, 58% in moderate degree, 26% in considerable degree, 7% in high degree and 1% in very high degree. The heavy metal ecological risk degree for Hg are 58% of samples in low, 19% in moderate, 14% in considerable, 6% in high and 3% in very high. The result shows that the copper and mercury are the main pollution metal in Hankou Region. The two elements are the most harmful heavy metal to human.

RI represents the sensitivity of various biological communities to toxic substances and illustrates the potential ecological risk caused by heavy metal.

Table 5. The heavy metal ecological risk index

E_r^i	As	Cd	Cr	Cu	Hg	Ni	Pb	Zn	Degree
<40	100%	8%	100%	99%	58%	100%	98%	100%	Low
40~80	0	58%	0	1%	19%	0	2%	0	Moderate
80~160	0	26%	0	0	14%	0	0	0	Considerable
160~320	0	7%	0	0	6%	0	0	0	High
≥320	0	1%	0	0	3%	0	0	0	Very high

Table 6. The potential ecological risk

RI	Percentage(%)	Degree
RI<150	51	Low
150≤RI<300	38	Moderate
300≤RI<600	8	Considerable
RI≥600	3	Very high

The potential ecological risk degree of 51% samples are low, 38% are moderate, 8% are considerable and 3% are very high. The result suggested that about a half of the study area are under the threat of heavy metal pollution, which was harmful to the health.

5 Summary

This study investigated heavy metal soil concentrations of the Hankou Region, Wuhan. The results showed high values of Cd, Hg, Pb and Zn and revealed the impact of industrial and anthropogenic activities on heavy metal accumulation in the soil of the study area. The distributions of the Cd, Cu and Hg concentration showed a very similar spatial pattern. As, Cr and Ni soil contamination levels are relatively low, resulting from the local natural background.

The calculated results of PI showed that the polluted level of Cd was moderate polluted, As was non-polluted and other metals are low polluted. The mean value of I_{geo} decrease in the order of Cd> Hg> Zn> Pb> Cu> Cr> As> Ni. The Nemerow pollution index of metals showed that the pollution of the study area was serious and the potential ecological risk (RI) revealed the about a half surface soil of the Hankou Region were harmful to the health.

Acknowledgments. The research was supported by the Program for the Undergraduate Training Programs for Innovation and Entrepreneurship (Wuhan University) (No.S2013368) and the Scientific Research Training Program of The Geographical Science Base Class (Wuhan University).

References

1. Wong, C.S.C., Li, X., Thornton, I.: Urban environmental geochemistry of trace metals. Environmental Pollution 142(1), 1–16 (2006)
2. Maas, S., Scheifler, R., Benslama, M., et al.: Spatial distribution of heavy metal concentrations in urban, suburban and agricultural soils in a Mediterranean city of Algeria. Environmental Pollution 158(6), 2294–2301 (2010)
3. Li, X., Liu, L., Wang, Y., et al.: Heavy metal contamination of urban soil in an old industrial city (Shenyang) in Northeast China. Geoderma 192, 50–58 (2013)
4. Chen, T.B., Zheng, Y.M., Lei, M., et al.: Assessment of heavy metal pollution in surface soils of urban parks in Beijing, China. Chemosphere 60(4), 542–551 (2005)
5. Yang, Z., Lu, W., Long, Y., et al.: Assessment of heavy metals contamination in urban surface soil from Changchun City, China. Journal of Geochemical Exploration 108(1), 27–38 (2011)
6. Srinivasa Gowd, S., Ramakrishna Reddy, M., Govil, P.K.: Assessment of heavy metal contamination in soils at Jajmau (Kanpur) and Unnao industrial areas of the Ganga Plain, Uttar Pradesh, India. Journal of Hazardous Materials 174(1), 113–121 (2010)
7. Muller, G.: Index of geoaccumulation in sediments of the Rhine River. Geojournal 2(3), 108–118 (1969)
8. Xu, Z.Q., Ni, S.J., Tuo, X.G., et al.: Calculation of heavy metals' toxicity coefficient in the evaluation of potential ecological risk index. Environmental Science and Technology 31(2), 112–114 (2008)

Spatial Distribution Analysis of Chinese Soccer Players' Birthplaces

Yang Zhang[1], Cheng Hu[2], and Jingwen Liu[2]

[1] Eastern Michigan University, Ypsilanti, MI, USA, 48197
[2] International School of Software, Wuhan University, Wuhan, Hubei, China, 430079
yzhang19@emich.edu, chu@whu.edu.cn, endy_liu@qq.com

Abstract. Success in any commercial events of sport in today's world is built and refined largely based on abundant and exceedingly diversified information. Player as the principal element of any sport has long been measured by many attributes and further examined by multiple disciplines and resulted with abundant findings and inspiration. The study tries to utilize information regarding birthplace at prefecture city level of a player in those records to identify some distributional characteristics of those places and examine the certain factors that leading to the findings. The results of spatial Distribution Analysis of Chinese Soccer Players' birthplaces show a very unbalanced distribution, and are promoting that differentiated background of soccer developments, geographic seating with advantageous means of outward communication as well as conventional social-economic measures interact closely with a place in promoting its chance of breeding more soccer players of top level in nation.

Keywords: Chinese Super League, spatial autocorrelation, Regional Analysis, Multiple Linear Regression Model.

1 Introduction

Globally, soccer is the world's most prevalent sport event today with international games covering five continents and countless divisional and sub-divisional games participated by billions of people (McCarroll&Meaney&Sieber, 1984; Backous&Friedl&Smith&Parr&Carpine, 1988; Høy&Lindblad& Terkelsen&Helleland& Terkelsen, 1992; Inklaar,1994; Tucker, 1997). Therefore, multiple disciplines are long involved in study of soccer as amateur sport or as commercial events including perspectives as biological, sociological and administrative, etc. Particularly, in a geographic perspective, an extensive number of research regarding of human character' involvement and impact into the development of soccer. Similar rationale could be found in piles of programs conducted in EU documenting a series of issues of the wide-spread European soccer games and national leagues in recent years. Also, traditional thoughts and ideas that long exist in professional-soccer-related social-economic environment of China. As a matter of fact, ever since the professionalization in 1994, youth training and scouting of young soccer players in

F. Bian and Y. Xie (Eds.): GRMSE 2014, CCIS 482, pp. 751–761, 2015.

according sports systems of China, both private sector and state sector, it seems inevitably an heavy consideration of geographical origin and distributional diversity has worked its way through out the process of the growth of a youth player into professional arena at adult stage. Moreover, the geographical distributional element even interacts with policy-making level of this sport in China, all the way from the Planned Economy Era to late 1990s supported by the controversial difference existed between the north and the south part of China in fundamentally guiding ideology and operational methods, one of the results seems to be two seemingly completely different structure and hierarchy of Chinese Football Association and its affiliated local-level organizations. However, behind the two ways of youth training focused on very different aspects and measure of youth players divided by from the north of China and the south of China, the supporting evidence and knowledgeable understanding of this issue has no yet been cleared and thoroughly presented to the general public.

2 Datasets

1. 94 Chinese Super League(CSL) squad entry forms with 30 participating professional soccer clubs from 2009 to 2014,recording basic information of a total of 989 Chinese players that has ever entered CSL level.

2. The columns includes name, date of birth (a requirements of 'at least 16 years old' applied since 2004), place of birth (the current place of national household registration' is not strictly excluded by regulation of CSL's Commission of Playing Certificate) and a number limits of 38 imposed since 2005.

3. Forms collected from inside the dataset of China Football Association (F.A.) that provides authorized access to participating clubs in each season, and according to data source, the forms stored are of clear official signature and dated verification of the association, proving the validity of those forms.

4. Detailed to each player in those forms of a 989 total, 863 of the records have been through a process of sorting, modifying and verifying regarding 'place of birth' in its literal meaning with information collected by per player from the official website of the club he belongs to and regional organization of the F.A. and finally confirm with a place of origin at prefectural city level (87.26 %), 79 of them at provincial level (7.99%) and 47 of them no other confirming source than their record living place in the national household registration system (4.75%).

3 Results

3.1 Overall Distribution and Focus

Figure-1 indicates the majority of the birthplaces are areas in the east of China and stretching northeast-southwest following the shape of the eastern coastlines in general, along with two areas as distant exceptions to the west end of China in Xinjiang Autonomous Province.

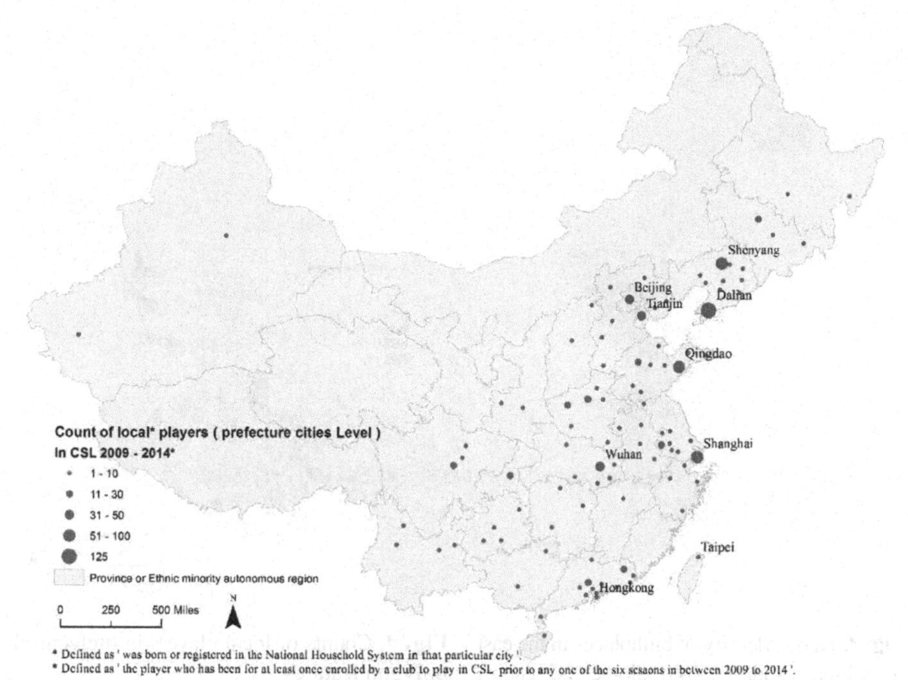

Fig. 1. Birthplaces of Chinese players in Chinese Super League (CSL) 2009 - 2014

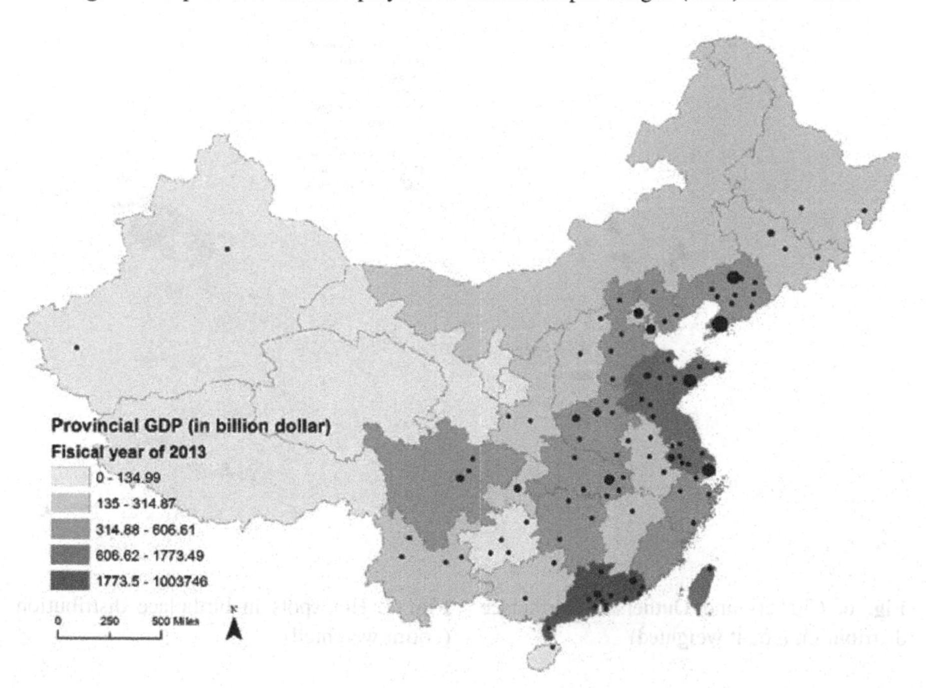

Fig. 2. Birthplaces of Chinese players & 2013 provincial GDP

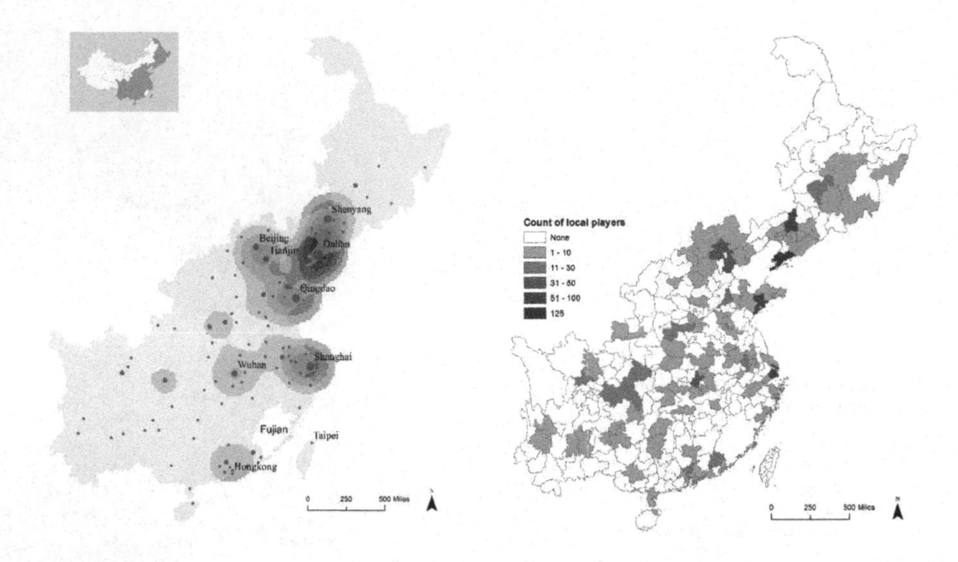

Fig. 3. Kernel density of birthplaces in the east of China (Radius: 400 kilometers, counts weighted)

Fig. 4. Counts of local players in prefectural polygon features

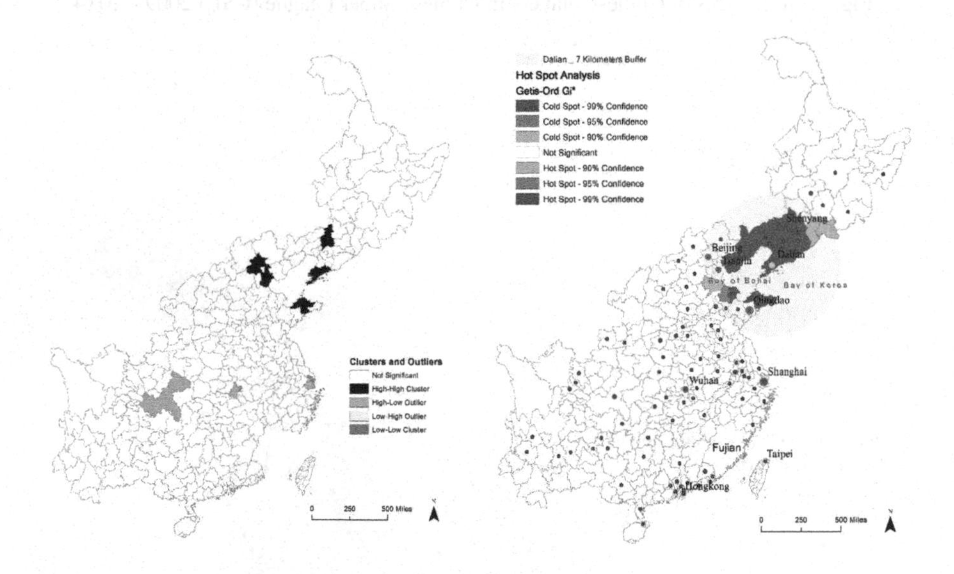

Fig. 6. Clusters and Outliers of birthplace distribution (count weighted)

Fig. 7. Hot spots in birthplace distribution (count weighted)

Figure-3 illustrates the unequivocally main area of distribution observed in Figure-1 which consists of 26 provinces and 99 contributing prefecture cities (Fujian Province excluded since there is no CSL player in recent 6 years). A 400 kilometers radius is generated by comparing and selecting around the mean distance between each and any pair of the 99 cities. This whole area reaches Jilin Province to the north and Hainan Province to the south including Taiwan, and according to counts it covers the origins of 99% (855 out of 863) players in recent six years of CSL. It seems obvious a larger density of major prefecture cities roughly in the middle and northern part of the area, along or adjacent to the coast of Yellow sea and Bo Sea have bred a relatively high volume of CSL players, particularly including Liaodong Peninsula (where the city of Dalian is at), Jiaodong Peninsula (where the city of Qingdao is at) and closely Beijing and Tianjin, in comparison with the southern region by the coast of South China Sea (also known as Pearl Delta region, including city of Guangzhou and Hong Kong), which also bears considerable density level. Further to the west from the coastlines, there are high density observed surrounding the city of Wuhan, Chongqing and Luoyang.

Noticeably, there are 8 players from Xinjiang Autonomous Province, being less than 1% of the total, it is in fact advantageous compared with its neighbors and provinces in the conventional 'Western China ' including Inner Mongolia, Tibet, Qinghai and several other provinces. A study on development and cultivate pattern of youth soccer players in Xinjiang reveals that although being extremely competitive through dozens of years' youth games at national level, players from Xinjiang used to found it difficult entering adult level stage due to gaps and disadvantages in local adult training system and lack of participation of professional franchises, sponsorship and talent scouting in local soccer market (Li&Wu&Zhang&Shu, 2011). Cases alike widely exist in neighboring provinces in west of China, namely regions contain a large population of certain minority ethnics or populate a splendid diversity in terms of ethnic groups involving unique but prevalent traditional sport events, might be subject to according impacts on popular sports like soccer, which is indicated in a study on status quo and countermeasures about sports in the rural areas of minority region in northwest of China (Yang&Zheng&Liu, 2010).

Moreover, although no sufficient evidence and research yet directly support any study in how minority ethnic belonging lowers the chance for them going pro in China, Figure-1 inspires knowledgeable links to the long existed geographic gaps and pattern involving social-economic factors in China's fast growing era that triggered years ago in late 1980s, which help to explain the phenomena of this east-west unbalance in professional soccer development in terms of discovery, cultivation and financing on a extensive term of local players, supported by Figure-2 as an example.

3.2 Test of Spatial Autocorrelation

Figure-4 puts the subjects of this study into a prefecture cities level. Patrick Alfred Pierce Moran develop a spatial autocorrelation measurement called Moran's I, in the field of statistics (Li&Calder&Cressie, 2007). The result of using ArcGIS to examine whether there is an overall spatial autocorrelation exists in this area by means of Global Morans'I, a pair of Z-Value and P-Value of 1.087905 and 0.276637 and the results from ArcGIS Semivariogram tool show that the subjects might just be randomly distributed, there is no evident impact or dependence imposed or be subject to by any neighboring prefectural area to its surrounding areas in a sense of causation, which means the chance for going pro and reaching the national top stage of the 863 players in this study might not be, in a nation level, clearly promoted or declined whether when they were born or growing up in areas that have relatively more or less other players born or shall be born close to them and how close to them.

3.3 Regional Analysis

However, certain patterns discovered when the density of birthplaces as point features being mapped are further confirmed when exploring clustering and hotspots in this distribution (Figure-6, Figure-7). There are 8 major cities that bred players in recent years of CSL, 5 of them are seen concentrated mainly surrounding the bay of Bohai, while three other dispersed but sharing an observable common feature of lying alongside Yangtze River.

A. *Surrounding Areas of Bohai Bay*

Except for the mentioned coastal location should be considered, a prominent tribute for the prevail of both the city of Qingdao and Beijing, along with their adjacent regions including Tianjin and Yantai, all as being highlighted in Figure-6 and Figure-7 in producing high volume of CSL players in recent years, might be paid to the consistency and coherence of local organizations carrying out youth training programs in extensive time span. Noticeably, consists mostly of 'self-made' local and neighboring players, Beijing Guoan Club and Shandong Luneng Club are the only two soccer clubs that have never lost their place in the early-called League-A or later-called CSL games ever since the very beginning of 1994 and both made their ways to the throne several times. Records indicate that not only the adult level games draw the lasting attention of them, youth teams of under-15 and under-13 level coming from these regions also well performed in recent years.

However, there are other factors should be included to reveal this clustering. As the nation's capital and its satellite coastal city, the generalized Beijing and Tianjin region possess not only a radically high population density of locally registered household (Statistical Yearbook of China, 2007):

Beijing: 741 / km²
Tianjin: 683 / km²
National Average: 142 / km²
Jiangsu Province (2nd highest in mainland China): 695 / km²
Xizang Autonomous Province: < 8 / km²

This region also bears a population of a higher ratio on settled-first-generation-migrant than any other city in this country, providing them more than sufficient resource and subject when conducting scouting and training, what adds to this point is the fact that the local population is also made extremely diversified. On the other side, behind Shandong Lunneng club's considerable contributions of players by their training facilities and regulation, the major sponsor and also owner of this club is, the only of its kind and scale in their arena, a state-own giant in energy and electric transmission business, Luneng Group, which for years guarantees the continuing development of this club against well-known commercial insecurities lie behind a youth training system of this size (Luneng group operates more than 56 local and non-local training campus and programs all over the country, taking up more than a fourth of the national sum).

Fig. 8. Yangtze River and three major birthplaces of players

B. Along Middle and Lower Yangtze River

Besides, there are 75, 43 and 29 players in recent seasons of CSL are from Shanghai, Wuhan and Chongqing respectively, ranking a 3rd, 6th and 8th in this study. Although not being labeled as 'Hotspot' due to a relatively lower profile in their surrounding areas, these 3 cities are identified as High-Low outliers by measuring Anselin Local Morans' I, which refers to the exact common distributional character of the three subjects as standing alone with exceptionally large counts high above all of their neighboring areas, even more so to the case of Chongqing, judging from Figure-3. Moreover, alongside other common social-economic characters shared between the three including high and fast-growing population, large economic scales and be of giant size in terms of area compared with other major cities in this distribution, another potential candidates referred to by Figure-8 as a contributing factor is that of having convenient access to waterborne transportation. That relates even to more of the observed high ranking regions in this study when the concept further expanded as places of having overall better means of outward communication.

3.4 Identifying Major Subjects of Change from 2009 to 2014

A key index in population study is PGR (Pop-Growth-Rate), conventionally defined as the rate at which the number of individuals in a population increases in a given time period as a fraction of the initial population. In this study, average change rate of each polygon's attribute of count is calculated fox six seasons referring to five individual growth rates in all, however, some birthplaces are newly raised meaning players of some origins have not reached the level of CSL until lately and some disappeared newly meaning that players of that particular origin might have retired or transferred to clubs participated in games other than CSL (that could only be lower levels in China) and in a sufficient number to leave the count zero, anyway, the averages are carefully calculated based on an according starting point and ending point and breaking point considered also, representing actuality. The result is represented by Figure-9.

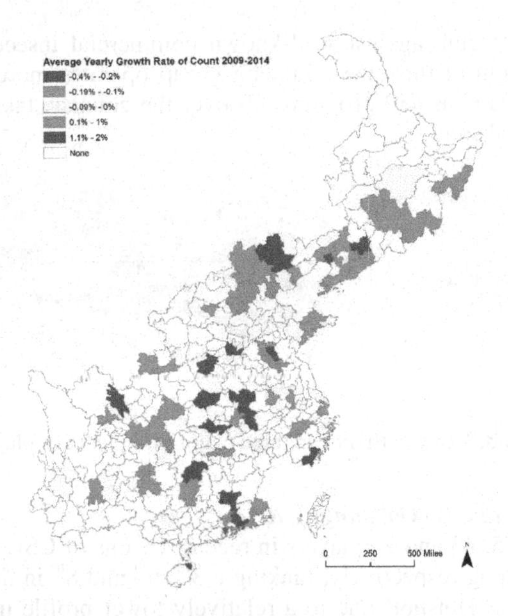

Fig. 9. Pop-Growth-Rate (calculated as yearly average since 2009 to 2014)

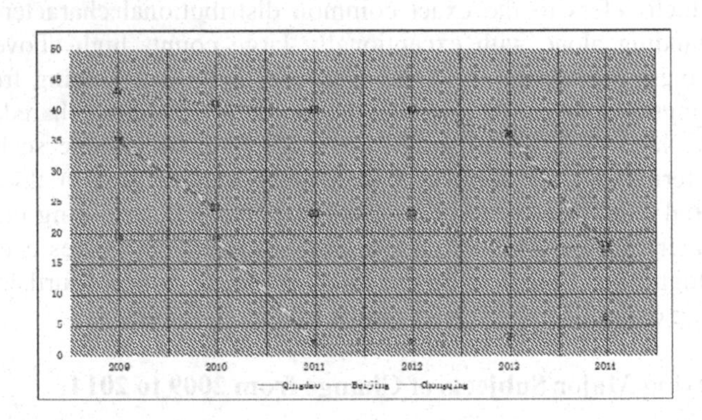

Fig. 10. Top 3 cities with fastest declining count by average yearly change of count

No evident pattern is identified by looking at a scale of overall distribution of PGRs in the focused area. Nonetheless, a further look combining analyzing original records provides more hints and information.

Comparing PGR eliminates the impact of difference between subjects' initial size when measuring and comparing their growth in a period of time. Judging by PGR, by taking a further look, what might be concerning in this study are the fastest declining numbers of Chengde; Chengdu; Changsha; Huizhou and Shaoguan in Guangdong Province; Jingzhou, Xiang Fan and Huanggang surrounding Wuhan; Wenzhou; Fushun and Panjin in Liaoning Province, highlighted by dark blue, as well as a less radical declining one in Beijing. Also the fastest growing number of Luoyang is set to be further examined.

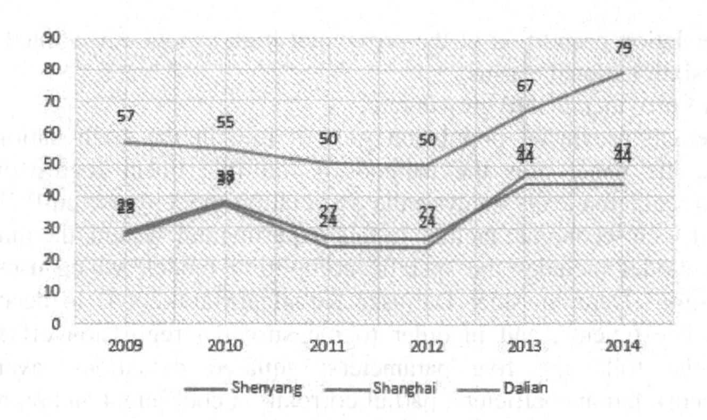

Fig. 11. Top 3 cities with fastest growing count by average yearly change of count

It is not unexpected that when examining each polygon's attribute of count by average yearly change of count, top 3 cities in either declining or growing sorts appear to be the cities with obviously larger initial size of counts as we mentioned hereinbefore (Figure-10, Figure-11). Noticeably, there are observably radical downslope for Chongqing from 2010 to 2011, for Qingdao from 2013 to 2014, as well as radical upslope for Shanghai from 2012 to 2013 and continuous two for Dalian from 2012 to 2013 and 2013 to 2014, all filtered through a change of at least 20 count between two continuous seasons. It turns out that those swelling and plunging of count are caused by, as conventional policy and phenomena in many commercial sport event, promotion and relegation.

Therefore, what remain after filtering out the dramatic impact on count by promotion, relegation and the rest of the changing counts of unmentioned birthplaces resulted in data of this study which are considered minor and also regular given characteristics of commercial sports. Subjects that were with high probability verified as whose contributing factors being in considerable variation and in the other hand statistically prominent and more noteworthy are:

For declining of count:
Beijing and its northeast neighbor city, Chengde
Shaoguan and Huizhou in Guangdong Province

For growing of count:
Yanbian Korea Autonomous City
Meizhou in Guangdong Province

3.5 Multiple Linear Regression Model of the Change from 2009 to 2014

In experimental studies, often need to look at the relationship between an independent variable and a set of independent variables. Multiple linear regression analysis is one of the very flexible study methods. In this part of the study were precisely determine the relationship between PGR of soccer players calculated in section 3, and the set of independent variables includes:

1. The population proportion of the provincial high school educational attainment rate in the sixth national census;
2. The 2013 provincial GDP growth rate;
3. The average provincial population growth rate in the sixth national census. Therefore, this paper uses the method of multiple linear regression analysis, attempts to mathematically analysis the association model between PGR of soccer players and socio-economic factors. In this experimental system, the multiplelinear regression model includes the specific computing model, which uses Cholesky decomposition (Dereniowski& Dariusz&Kubale&Marek2004) to decompose the regression coefficients, and in order to measure the regression effect, we also calculate the following five parameters: squared deviations, average stand deviation, correlation coefficient, partial correlation coefficient and regression sum of squares. The regression equation is:

$$Y = -0.1333 * X0 + (-2.7372) * X1 + (-1.5752) * X2 + 33.3484 \quad (1)$$

In the formula, Y represents PGR value of 2009-2014 Years of soccer population, X0 stands for the population proportion of the provincial high school educational attainment rate in the sixth national census, X1 represents the 2013 provincial GDP growth rate, X2 represents the average provincial population growth rate in the sixth national census. In addition, the model generates some other descriptive variables for this formula: deviation squared q = 4.6225, average standard deviation s = 0.9615, multiple correlation coefficient r = 0.9022, multiple correlation coefficient close to 1, showing that the relative error q / t is close to zero, which indicating good linear regression results, the partial correlation coefficients of each variables, respectively, v (0) = 0.9721, v (1) = 0.9993, v (2) = 0.9879. Which means the three variables have the similar influence to the PGR of soccer population. In addition, it get the regression sum of squares u = 20.2254.

4 Conclusion

This research further develops the data identifying certain birthplaces, with some changing counts of locally bred players within this 6-year time span, are statistically prominent and noteworthy regarding their potential contributing factors probably being in considerable variation and according further investigations also put forward by this research.

References

1. McCarroll, J., Meaney, C., Sieber, J.: Profile of Youth Soccer Injuries. Phys. Sportsmed 12, 113–117 (1984)
2. Backous, D.D., Friedl, K.E., Smith, N.J., Parr, T.J., Carpine Jr., W.D.: Soccer Injuries and Their Relation to Physical Maturity. Am. J. Dis. Child. 142(8), 839–842 (1988)
3. Høy, K., Lindblad, B.E., Terkelsen, C.J., Helleland, H.E., Terkelsen, C..: European soccer injuries. A prospective epidemiologic and socioeconomic study. Am. J. Sports Med. 20(3), 318–322 (1992)

4. Inklaar, H.: Soccer injuries. I: Incidence and severity. Sports Med. 18(1), 55–73 (1994)
5. Tucker, A.: Common soccer injuries. Diagnosis, treatment and rehabilitation. Sports Med. 23, 21–32 (1997)
6. Li, X.-T., Wu, J.-H., Zhang, X., Shu, Y.-H.: Study on Development and Cultivate Pattern of Young Football Players in Xin Jiang. Science & Technology Information (29), 565–566 (2011)
7. Yang, J.-S., Zheng, W.-H., Liu, R.-S.: Research on the status quo and countermeasures about sports in the rural areas of minority region in Northwest. Journal of Xi'an Institute of Physical Education 23(2), 48–55 (2006)
8. Li, H., Calder, C.A., Cressie, N.: Beyond Moran's I: Testing for Spatial Dependence Based on the Spatial Autoregressive Model. Geographical Analysis 39(4), 357–375
9. Dereniowski, D., Kubale, M.: Cholesky factorization of matrices in parallel and ranking of graphs. In: Wyrzykowski, R., Dongarra, J., Paprzycki, M., Waśniewski, J. (eds.) PPAM 2004. LNCS, vol. 3019, pp. 985–992. Springer, Heidelberg (2004)

Author Index